THEORETICAL PHYSICS

SECOND EDITION

A. S. KOMPANEYETS

TRANSLATED FROM THE RUSSIAN

EDITED BY GEORGE YANKOVSKY

DOVER PUBLICATIONS, INC.
Mineola, New York

This translation has been read and approved
by the author, Professor A. S. Kompaneyets

Bibliographical Note

This Dover edition, first published in 1962 and reissued in 2012, is an unabridged and corrected republication of the English translation first published by the Foreign Languages Publishing House, Moscow, in 1961.

Library of Congress Catalog Card Number: 63-005516
International Standard Book Number
ISBN-13: 978-0-486-60972-0
ISBN-10: 0-486-60972-3

Manufactured in the United States by Courier Corporation
60972342
www.doverpublications.com

CONTENTS

FROM THE PREFACE TO THE FIRST EDITION

This book is intended for readers who are acquainted with the course of general physics and analysis of nonspecializing institutions of higher education. It is meant chiefly for engineer-physicists, though it may also be useful to specialists working in fields associated with physics—chemists, physical chemists, biophysicists, geophysicists, and astronomers.

Like the natural sciences in general, physics is based primarily on experiment, and, what is more, on quantitative experiment. However, no series of experiments can constitute a theory until a rigorous logical relationship is established between them. Theory not only allows us to systematize the available experimental material, but also makes it possible to predict new facts which can be experimentally verified.

All physical laws are expressed in the form of quantitative relationships. In order to interrelate quantitative laws, theoretical physics appeals to mathematics. The methods of theoretical physics, which are based on mathematics, can be fully mastered only by those who have acquired a very considerable volume of mathematical knowledge. Nevertheless, the basic ideas and results of theoretical physics are readily comprehensible to any reader who has an understanding of differential and integral calculus, and is acquainted with vector algebra. This is the minimum of mathematical knowledge required for an understanding of the text that follows.

At the same time, the aim of this book is not only to give the reader an idea about what theoretical physics is, but also to furnish him with a working knowledge of the basic methods of theoretical physics. For this reason it has been necessary to adhere, as far as possible, to a rigorous exposition. The reader will more readily agree with the conclusions reached if their inevitability has been made obvious to him. In order to activize the work of the student, some of the applications of the theory have been shifted into the exercises, in which the line of reasoning is not so detailed as in the basic text.

In compiling such a relatively small book as this one it has been necessary to cut down on the space devoted to certain important sections of theoretical physics, and omit other branches entirely. For instance, the mechanics of solid media is not included at all since to set out this branch, even in the same detail as the rest of the text, would mean doubling the size of the book. A few results from the mechanics of continuous media are included in the exercises as illustrations in thermodynamics. At the same time, the mechanics and electrodynamics of solid media are less related to the fundamental, gnosiological problems of physics than microscopic electrodynamics, quantum theory, and statistical physics. For this reason, very little space is devoted to macroscopic electrodynamics: the material has been selected in such a way as to show the reader how the transition is made from microscopic electrodynamics to the theory of quasi-stationary fields and the laws of the propagation of light in media. It is assumed that the reader is familiar with these problems from courses of physics and electricity.

On the whole, the book is mainly intended for the reader who is interested in the physics of elementary processes. These considerations have also dictated the choice of material; as in all nonencyclopaedic manuals, this choice is inevitably somewhat subjective.

In compiling this book, I have made considerable use of the excellent course of theoretical physics of L. D. Landau and E. M. Lifshits. This comprehensive course can be recommended to all those who wish to obtain a profound understanding of theoretical physics.

I should like to express my deep gratitude to my friends who have made important observations: Ya. B. Zeldovich, V. G. Levich, E. L. Feinberg, V. I. Kogan and V. I. Goldansky.

A. Kompaneyets

PREFACE TO THE SECOND EDITION

In this second edition I have attempted to make the presentation more systematic and rigorous without adding any difficulties. In order to do this it has been especially necessary to revise Part III, to which I have added a special section (Sec. 30) setting out the general principles of quantum mechanics; radiation is now considered only with the aid of the quantum theory of the electromagnetic field, since the results obtained from the correspondence principle do not appear sufficiently justified.

Gibbs' statistics are included in this edition, which has made it necessary to divide Part IV into something in the nature of two cycles: Sec. 39-44, where only the results of combinatorial analysis are set out, and Sec. 45-52, an introduction to the Gibbs' method, which is used as background material for a discussion of thermodynamics. A phenomenological approach to thermodynamics would nowadays appear an anachronism in a course of theoretical physics.

In order not to increase the size of the book overmuch, it has been necessary to omit the theory of beta decay, the variational properties of eigenvalues, and certain other problems included in the first edition.

I am greatly indebted to A. F. Nikiforov and V. B. Uvarov for pointing out several inaccuracies in the first edition of the book.

A. Kompaneyets

PART I

MECHANICS

Sec. 1. Generalized Coordinates

Frames of reference. In order to describe the motion of a mechanical system, it is necessary to specify its position in space as a function of time. Obviously, it is only meaningful to speak of the relative position of any point. For instance, the position of a flying aircraft is given relative to some coordinate system fixed with respect to the earth; the motion of a charged particle in an accelerator is given relative to the accelerator, etc. The system, relative to which the motion is described, is called a frame of reference.

Specification of time. As will be shown later (Sec. 20), specification of time in the general case is also connected with defining the frame of reference in which it is given. The intuitive conception of a universal, unique time, to which we are accustomed in everyday life, is, to a certain extent, an approximation that is only true when the relative speeds of all material particles are small in comparison with the velocity of light. The mechanics of such slow movements is termed Newtonian, since Isaac Newton was the first to formulate its laws.

Newton's laws permit a determination of the position of a mechanical system at an arbitrary instant of time, if the positions and velocities of all points of the system are known at some initial instant, and also if the forces acting in the system are known.

Degrees of freedom of a mechanical system. The number of independent parameters defining the position of a mechanical system in space is termed the number of its degrees of freedom.

The position of a particle in space relative to other bodies is defined with the aid of three independent parameters, for example, its Cartesian coordinates. The position of a system consisting of N particles is determined, in general, by $3N$ independent parameters.

However, if the distribution of points is fixed in any way, then the number of degrees of freedom may be less than $3N$. For example, if two points are constrained by some form of rigid nondeformable coupling, then, upon the six Cartesian coordinates of these points, $x_1, y_1, z_1, x_2, y_2, z_2$, is imposed the condition

$$(x_2 - x_1)^2 + (y_2 - y_1)^2 + (z_2 - z_1)^2 = R_{12}^2, \tag{1.1}$$

where R_{12} is the given distance between the points. It follows that the Cartesian coordinates are no longer independent parameters: a relationship exists between them. Only five of the six values $x_1, \ldots, z_2$ are now independent. In other words, a system of two particles, separated by a fixed distance, has five degrees of freedom. If we consider three particles which are rigidly fixed in a triangle, then the coordinates of the third particle must satisfy the two equations:

$$(x_3 - x_1)^2 + (y_3 - y_1)^2 + (z_3 - z_1)^2 = R_{31}^2, \tag{1.2}$$

$$(x_3 - x_2)^2 + (y_3 - y_2)^2 + (z_3 - z_2)^2 = R_{32}^2. \tag{1.3}$$

Thus, the nine coordinates of the vertices of the rigid triangle are defined by the three equations (1.1), (1.2) and (1.3), and hence only six of the nine quantities are independent. The triangle has six degrees of freedom.

The position of a rigid body in space is defined by three points which do not lie on the same straight line. These three points, as we have just seen, have six degrees of freedom. It follows that any rigid body has six degrees of freedom. It should be noted that only such motions of the rigid body are considered as, for example, the rotation of a top, where no noticeable deformation occurs that can affect its motion.

Generalized coordinates. It is not always convenient to describe the position of a system in Cartesian coordinates. As we have already seen, when rigid constraints exist, Cartesian coordinates must satisfy supplementary equations. In addition, the choice of coordinate system is arbitrary and should be determined primarily on the basis of expediency. For instance, if the forces depend only on the distances between particles, it is reasonable to introduce these distances into dynamical equations explicitly and not by means of Cartesian coordinates.

In other words, a mechanical system can be described by coordinates whose number is equal to the number of degrees of freedom of the system. These coordinates may sometimes coincide with the Cartesian coordinates of some of the particles. For example, in a system of two rigidly connected points, these coordinates can be chosen in the following way: the position of one of the points is given in Cartesian coordinates, after which the other point will always be situated on a sphere whose centre is the first point. The position of the second point on the sphere may be given by its longitude and latitude.

Together with the three Cartesian coordinates of the first point, the latitude and longitude of the second point completely define the position of such a system in space.

For three rigidly bound points, it is necessary, in accordance with the method just described, to specify the position of one side of the triangle and the angle of rotation of the third vertex about that side.

The independent parameters which define the position of a mechanical system in space are called its generalized coordinates. We will represent them by the symbols q_α, where the subscript α signifies the number of the degree of freedom.

As in the case of Cartesian coordinates, the choice of generalized coordinates is to a considerable extent arbitrary. It must be chosen so that the dynamical laws of motion of the system can be formulated as conveniently as possible.

Sec. 2. Lagrange's Equation

In this section, equations of motion will be obtained in terms of arbitrary generalized coordinates. In such form they are especially convenient in theoretical physics.

Newton's Second Law. Motion in mechanics consists in changes in the mutual configuration of bodies in time. In other words, it is described in terms of the mutual distances, or lengths, and intervals of time. As was shown in the preceding section, all motion is relative; it can be specified only in relation to some definite frame of reference.

In accordance with the level of knowledge of his time, Newton regarded the concepts of length and time interval as absolute, which is to say that these quantities are the same in all frames of reference. As will be shown later, Newton's assumption was an approximation (see Sec. 20). It holds when the relative speeds of all the particles are small compared with the velocity of light; here Newtonian mechanics is based on a vast quantity of experimental facts.

In formulating the laws of motion a very convenient concept is the material particle, that is, a body whose position is completely defined by three Cartesian coordinates. Strictly speaking, this idealization is not applicable to any body. Nevertheless, it is in every way reasonable when the motion of a body is sufficiently well defined by the displacement in space of any of its particles (for example, the centre of gravity of the body) and is independent of rotations or deformations of the body.

If we start with the concept of a particle as the fundamental entity of mechanics, then the law of motion (Newton's Second Law) is formulated thus:

$$m\frac{d^2\mathbf{r}}{dt^2} = \mathbf{F}. \tag{2.1}$$

Here, **F** is the resultant of all the forces applied to the particle (the vector sum of the forces) $\frac{d^2\mathbf{r}}{dt^2}$ is the vector acceleration, the Cartesian components of which are

$$\frac{d^2x}{dt^2}, \quad \frac{d^2y}{dt^2}, \quad \frac{d^2z}{dt^2}.$$

The quantity m involved in equation (2.1) characterizes the particle and is called its mass.

Force and mass. Equality (2.1) is the definition of force. However, it should not be regarded as a simple identity or designation, because (2.1) establishes the form of the interaction between bodies in mechanics and thereby actually describes a certain law of nature. The interaction is expressed in the form of a differential equation that includes only the second derivatives of the coordinates with respect to time (and not derivatives, say, of the fourth order).

In addition, certain limiting assumptions are usually made in relation to the force. In Newtonian mechanics it is assumed that forces depend only on the mutual arrangement of the bodies at the instant to which the equality refers and do not depend on the configuration of the bodies at previous times. As we shall see later (see Part II), this supposition about the character of interaction forces is valid only when the speeds of the bodies are small compared with the velocity of light.

The quantity m in equality (2.1) is a characteristic of the body, its mass. Mass may be determined by comparing the accelerations which the same force imparts to different bodies; the greater the acceleration, the less the mass. In order to measure mass, some body must be regarded as a standard. The choice of a standard body is completely independent of the choice of standards of length and time. This is what makes the dimension (or unit of measurement) of mass a special dimension, not related to the dimensions of length and time.

The properties of mass are established experimentally. Firstly, it can be shown that the mass of two equal quantities of the same substance is equal to twice the mass of each quantity. For example, one can take two identical scale weights and note that a stretched spring gives them equal accelerations. If we join two such weights and subject them to the action of the same spring, which has been stretched by the same amount as for each weight separately, the acceleration will be found to be one half what it was. It follows that the overall mass of the weights is twice as great, since the force depends only on the tension of the spring and could not have changed.

Thus, mass is an *additive* quantity, that is, one in which the whole is equal to the sum of the quantities of each part taken separately. Experiment shows that the principle of additivity of mass also applies to bodies consisting of different substances.

In addition, in Newtonian mechanics, the mass of a body is a constant quantity which does not change with motion.

It must not be forgotten that the additivity and constancy of masses are properties that follow only from experimental facts which relate to very specific forms of motion. For example, a very important law, that of the conservation of mass in chemical transformations involving rearrangement of the molecules and atoms of a body, was established by M. V. Lomonosov experimentally.

Like all laws deduced from experiment, the principle of additivity of mass has a definite degree of precision. For such strong interactions as take place in the atomic nucleus, the breakdown of the additivity of mass is apparent (for more detail see Sec. 21).

We may note that if instead of subjecting a body to the force of a stretched spring it were subjected to the action of gravity, then the acceleration of a body of double mass would be equal to the acceleration of each body separately. From this we conclude that the force of gravity is itself proportional to the mass of a body. Hence, in a vacuum, in the absence of air resistance, all bodies fall with the same acceleration.

Inertial frames of reference. In equation (2.1) we have to do with the acceleration of a particle. There is no sense in talking about acceleration without stating to which frame of reference it is referred. For this reason there arises a difficulty in stating the cause of the acceleration. This cause may be either interaction between bodies or it may be due to some distinctive properties of the reference frame itself. For example, the jolt which a passenger experiences when a carriage suddenly stops is evidence that the carriage is in nonuniform motion relative to the earth.

Let us consider a set of bodies not affected by any other bodies, that is, one that is sufficiently far away from them. We can suppose that a frame of reference exists such that all accelerations of the set of bodies considered arise only as a result of the interaction between the bodies. This can be verified if the forces satisfy Newton's Third Law, i.e., if they are equal and opposite in sign for any pair of particles (it is assumed that the forces occur instantaneously, and this is true only when the speeds of the particles are small compared with the speed of transmission of the interaction).

A frame of reference for which the acceleration of a certain set of particles depends only on the interaction between these particles is called an *inertial* frame (or inertial coordinate system). A free particle, not subject to the action of any other body, moves, relative to such a reference frame, uniformly in a straight line or, in everyday language, by its own momentum. If in a given frame of reference Newton's Third Law is not satisfied we can conclude that this is not an inertial system.

Thus, a stone thrown directly downwards from a tall tower is deflected towards the east from the direction of the force of gravity. This direction can be independently established with the aid of a suspended weight. It follows that the stone has a component of acceleration which is not caused by the force of the earth's attraction. From this we conclude that the frame of reference fixed in the earth is noninertial. The noninertiality is, in this case, due to the diurnal rotation of the earth.

On the forces of friction. In everyday life we constantly observe the action of forces that arise from direct contacts between bodies. The sliding and rolling of rigid bodies give rise to forces of friction. The action of these forces causes a transition of the macroscopic motion of the body as a whole into the microscopic motion of the constituent atoms and molecules. This is perceived as the generation of heat. Actually, when a body slides an extraordinarily complex process of interaction occurs between the atoms in the surface layer. A description of this interaction in the simple terms of frictional forces is a very convenient idealization for the mechanics of macroscopic motion, but, naturally, does not give us a full picture of the process. The concept of frictional force arises as a result of a certain averaging of all the elementary interactions which occur between bodies in contact.

In this part, which is concerned only with elementary laws, we shall not consider averaged interactions where motion is transferred to the internal, microscopic, degrees of freedom of atoms and molecules. Here, we will study only those interactions which can be completely expressed with the aid of elementary laws of mechanics and which do not require an appeal to any statistical concepts connected with internal, thermal, motion.

Ideal rigid constraints. Bodies in contact also give rise to forces of interaction which can be reduced to the kinematic properties of rigid constraints. If rigid constraints act in a system they force the particles to move on definite surfaces. Thus, in Sec. 1 we considered the motion of a single particle on a sphere, at the centre of which was another particle.

This kind of interaction between particles does not cause a transition of the motion to the internal, microscopic, degrees of freedom of bodies. In other words, motion which is limited by rigid constraints is completely described by its own macroscopic generalized coordinates q_α.

If the limitations imposed by the constraints distort the motion, they thereby cause accelerations (curvilinear motion is always accelerated motion since velocity is a vector quantity). This acceleration can be formally attributed to forces which are called reaction forces of rigid constraints.

Reaction forces change only the direction of velocity of a particle but not its magnitude. If they were to alter the magnitude of the velocity, this would produce a change also in the kinetic energy of the particle. According to the law of conservation of energy, heat would then be generated. But this was excluded from consideration from the very start.

To summarize, the reaction forces of ideally rigid constraints do not change the kinetic energy of a system. In other words, they do not perform any work on it, since work performed on a system is equivalent to changing its kinetic energy (if heat is not generated).

In order that a force should not perform work, it must be perpendicular to the displacement. For this reason the reaction forces of constraints are perpendicular to the direction of particle velocity at each given instant of time.

However, in problems of mechanics, the reaction forces are not initially given, as are the functions of particle position. They are determined by integrating equations (2.1), with account taken of constraint conditions. Therefore, it is best to formulate the equations of mechanics so as to exclude constraint reactions entirely. It turns out that if we go over to generalized coordinates, the number of which is equal to the number of degrees of freedom of the system, then the constraint reactions disappear from the equations. In this section we shall make such a transition and will obtain the equations of mechanics in terms of the generalized coordinates of the system.

The transformation from rectangular to generalized coordinates. We take a system with a total of $3N \equiv n$ Cartesian coordinates of which ν are independent. We will always denote Cartesian coordinates by the same letter x_i, understanding by this symbol all the coordinates x, y, z; this means that i varies from 1 to $3N$, that is, from 1 to n. The generalized coordinates we denote by q_α $(1 \leqslant \alpha \leqslant \nu)$. Since the generalized coordinates completely specify the position of their system, x_i are their unique functions:

$$x_i = x_i(q_1, q_2, \ldots q_\alpha, \ldots, q_\nu) . \tag{2.2}$$

From this it is easy to obtain an expression for the Cartesian components of velocity. Differentiating the function of many variables $x_i(\ldots q_\alpha)$ with respect to time, we have

$$\frac{dx_i}{dt} = \sum_{\alpha=1}^{\nu} \frac{\partial x_i}{\partial q_\alpha} \frac{dq_\alpha}{dt} .$$

In the subsequent derivation we shall often have to perform summations with respect to all the generalized coordinates q_α, and double and triple sums will be encountered. In order to save space we introduce the following *summation convention.*

If a Greek symbol is met twice on one side of an equation, it will be understood as denoting a summation from unity to ν, *that is, over all the generalized coordinates.* (It is not convenient to use this convention for the Latin characters which denote the Cartesian coordinates.) Then the velocity $\frac{dx_i}{dt}$ can be rewritten thus:

$$\frac{dx_i}{dt} = \frac{\partial x_i}{\partial q_\alpha}\frac{dq_\alpha}{dt}. \tag{2.3}$$

Here the summation sign is omitted.

The total derivative with respect to time is usually denoted by a dot over the corresponding variable:

$$\frac{dx_i}{dt} \equiv \dot{x}_i; \qquad \frac{dq_\alpha}{dt} \equiv \dot{q}_\alpha. \tag{2.4}$$

In this notation, (2.3) is written in an even more abbreviated form:

$$\dot{x}_i = \frac{\partial x_i}{\partial q_\alpha}\dot{q}_\alpha. \tag{2.5}$$

Differentiating (2.5) with respect to time again, we obtain an expression for the Cartesian components of acceleration:

$$\ddot{x}_i = \frac{d}{dt}\left(\frac{\partial x_i}{\partial q_\alpha}\right)\dot{q}_\alpha + \frac{\partial x_i}{\partial q_\alpha}\ddot{q}_\alpha.$$

The total derivative in the first term is written as usual:

$$\frac{d}{dt}\left(\frac{\partial x_i}{\partial q_\alpha}\right) = \frac{\partial^2 x_i}{\partial q_\beta \partial q_\alpha}\dot{q}_\beta.$$

The Greek symbol over which the summation is performed is denoted by the letter β to avoid confusion with the symbol α, which denotes the summation in the expression for velocity (2.5). Thus, we obtain the desired expression for $\ddot{x}_i$:

$$\ddot{x}_i = \frac{\partial^2 x_i}{\partial q_\beta \partial q_\alpha}\dot{q}_\beta\dot{q}_\alpha + \frac{\partial x_i}{\partial q_\alpha}\ddot{q}_\alpha. \tag{2.6}$$

The first term on the right-hand side contains a double summation with respect to α and β.

Potential of a force. We now consider components of force. In many cases, the three components of the vector of a force acting on a particle can be expressed in terms of *one* scalar function U according to the formula:

$$F_i = -\frac{\partial U}{\partial x_i}. \tag{2.7}$$

Such a function can always be chosen for the force of Newtonian attraction, and for electrostatic and elastic forces. The function U is called the *potential* of the force.

It is clear that by far not every system of forces can, in the general case, be represented by a set of partial derivatives (2.7), since if

$$F_i = -\frac{\partial U}{\partial x_i}, \quad F_k = -\frac{\partial U}{\partial x_k},$$

then we must have the equality

$$\frac{\partial F_i}{\partial x_k} = \frac{\partial F_k}{\partial x_i} = -\frac{\partial^2 U}{\partial x_i \partial x_k}$$

for all i, k, which is not, in advance, obvious for the arbitrary functions F_i, F_k. The definite form of the potential, in those cases where it exists, will be given below for various forces.

Expression (2.7) defines the potential function U to the accuracy of an arbitrary constant term. U is also called the *potential energy* of the system. For example, the gravitational force $F = -mg$, while the potential energy of an elevated body is equal to mgz, where $g \sim 980$ cm/sec^2 is the acceleration of a freely falling body and z is the height to which it has been raised. It can be calculated from any level, which in the given case corresponds to a determination of U to the accuracy of a constant term. A more precise expression for the force of gravity than $F = -mg$ (with allowance made for its dependence on height also admits of a potential, which we shall derive a little later [see (3.4)].

We denote the component reaction forces of rigid constraints by F_i. We now note that

$$\sum_{i=1}^{n} F'_i \, dx_i = 0, \tag{2.8}$$

if the displacements are compatible with the constraints. Indeed, (2.8) expresses precisely the work performed by the reaction forces for a certain possible displacement of the system; but this work has been shown to be equal to zero.

Lagrange's equations.* We will now write down the equations of motion with the aid of (2.7) and (2.8) as

$$m_i \left(\frac{\partial^2 x_i}{\partial q_\beta \, \partial q_\alpha} \dot{q}_\beta \, \dot{q}_\alpha + \frac{\partial x_i}{\partial q_\alpha} \ddot{q}_\alpha \right) = -\frac{\partial U}{\partial x_i} + F'_i. \tag{2.9}$$

Here, of course, $m_1 = m_2 = m_3$ is equal to the mass of the first particle, $m_4 = m_5 = m_6$ equals the mass of the second particle, etc.

---

* In the first reading, the subsequent derivation up to equation (2.18) need not be studied in detail.

Let us multiply both sides of this equation by $\frac{\partial x_i}{\partial q_\nu}$ and sum from 1 to n over i.

Let us first consider the right-hand side. We obviously have

$$\sum_i \frac{\partial U}{\partial x_i}\frac{\partial x_i}{\partial q_\gamma} = \frac{\partial U}{\partial q_\gamma} \tag{2.10}$$

in accordance with the law for differentiating composite functions. For the forces of reaction we obtain

$$\sum_i F_i \frac{\partial x_i}{\partial q_\gamma} = 0, \tag{2.11}$$

since this equality is a special case of (2.8), in which the displacements dx_i are taken for all constants q except for q_γ; that is why we retain the designation of the partial derivative $\frac{\partial x_i}{\partial q_\gamma}$. It is clear that in such a special displacement the work done by the reaction forces of the constraints is equal to zero, as in the case of a general displacement.

After multiplication by $\frac{\partial x_i}{\partial q_\gamma}$ and summation, the left-hand side of equality (2.9) can be written in a more compact form, without resorting to explicit Cartesian coordinates. It is precisely the purpose of this section to give such an improved notation. To do this we express the kinetic energy in terms of generalized coordinates:

$$T = \frac{1}{2}\sum_{i=1}^{n} m_i \left(\frac{dx_i}{dt}\right)^2 = \frac{1}{2}\sum_{i=1}^{n} m_i \dot{x}_i^2 . \tag{2.12}$$

Substituting the generalized velocities by using (2.5), we obtain

$$T = \frac{1}{2}\sum_{i=1}^{n} m_i \frac{\partial x_i}{\partial q_\alpha}\dot{q}_\alpha \frac{\partial x_i}{\partial q_\beta}\dot{q}_\beta .$$

The summation indices for q must, of course, be denoted by different letters, since they independently take all values from 1 to ν inclusive. Changing the order of summation for Cartesian and generalized coordinates we have

$$T = \frac{1}{2}\dot{q}_\alpha \dot{q}_\beta \sum_{i=1}^{n} m_i \frac{\partial x_i}{\partial q_\alpha}\frac{\partial x_i}{\partial q_\beta} . \tag{2.13}$$

Henceforth, T will have to be differentiated both with respect to generalized coordinates q_α and generalized velocities $\dot{q}_\alpha$. The coordinate q_α and its corresponding generalized velocity $\dot{q}_\alpha$ are independent of each other since, in the given position in which the coordinate has a given value q_α, it is possible to impart to the system an arbitrary velocity $\dot{q}_\alpha$ permitted by the constraints. Naturally, q_α and $\dot{q}_{\beta \neq \alpha}$ are also independent. It follows that in calculating $\frac{\partial T}{\partial q_\alpha}$ all the remaining velocities $\dot{q}_{\beta \neq \alpha}$ and all the coordinates, including q_α, should be regarded as constant.

Let us calculate the derivative $\frac{\partial T}{\partial \dot{q}_\gamma}$. In the double summation (2.13), the quantity γ can be taken as the index α and also the index β, so that we obtain

$$\frac{\partial T}{\partial \dot{q}_\gamma} = \frac{1}{2} \dot{q}_\beta \sum_{i=1}^{n} m_i \frac{\partial x_i}{\partial q_\gamma} \frac{\partial x_i}{\partial q_\beta} + \frac{1}{2} \dot{q}_\alpha \sum_{i=1}^{n} m_i \frac{\partial x_i}{\partial q_\alpha} \frac{\partial x_i}{\partial q_\gamma} .$$

Both these sums are the same except that in the first the index is denoted by β and in the second by α. They can be combined, replacing β by α in the first summation; naturally the value of the sum does not change due to renaming of the summation sign. Then we obtain

$$\frac{\partial T}{\partial \dot{q}_\gamma} = \dot{q}_\alpha \sum_{i=1}^{n} m_i \frac{\partial x_i}{\partial q_\alpha} \frac{\partial x_i}{\partial q_\gamma} . \tag{2.14}$$

Let us calculate the total derivative of this quantity with respect to time:

$$\frac{d}{dt} \frac{\partial T}{\partial \dot{q}_\gamma} = \ddot{q}_\alpha \sum_{i=1}^{n} m_i \frac{\partial x_i}{\partial q_\alpha} \frac{\partial x_i}{\partial q_\gamma} + \dot{q}_\alpha \sum_{i=1}^{n} m_i \frac{\partial^2 x_i}{\partial q_\alpha \partial q_\beta} \dot{q}_\beta \frac{\partial x_i}{\partial q_\gamma} +$$

$$+ \dot{q}_\alpha \sum_{i=1}^{n} m_i \frac{\partial x_i}{\partial q_\alpha} \frac{\partial^2 x_i}{\partial q_\gamma \partial q_\beta} \dot{q}_\beta . \tag{2.15}$$

Here we have had to write down the derivatives of each of the three factors of all the terms in the summation (2.14) separately.

Now let us calculate the partial derivative $\frac{\partial T}{\partial q_\gamma}$. As has been shown, $\dot{q}_\alpha$, $\dot{q}_\beta$ are regarded as constants. Like $\frac{\partial T}{\partial \dot{q}_\gamma}$, the derivative $\frac{\partial T}{\partial q_\gamma}$ consists of two terms which may be amalgamated into one. Differentiating (2.13), we obtain

$$\frac{\partial T}{\partial q_\gamma} = \dot{q}_\alpha \dot{q}_\beta \sum_{i} m_i \frac{\partial^2 x_i}{\partial q_\gamma \partial q_\beta} \frac{\partial x_i}{\partial q_\alpha} . \tag{2.16}$$

Subtracting (2.16) from (2.15), we see that (2.16) and the last term of (2.15) cancel. As a result we obtain

$$\frac{d}{dt}\frac{\partial T}{\partial \dot{q}_\gamma} - \frac{\partial T}{\partial q_\gamma} = \ddot{q}_\alpha \sum_{i=1}^{n} m_i \frac{\partial x_i}{\partial q_\alpha}\frac{\partial x_i}{\partial q_\gamma} + \dot{q}_\beta \dot{q}_\alpha \sum_{i=1}^{n} m_i \frac{\partial^2 x_i}{\partial q_\alpha \partial q_\beta}\cdot\frac{\partial x_i}{\partial q_\gamma}. \quad (2.17)$$

However, the expression on the right-hand side of (2.17) can also be obtained from (2.9) if we multiply its left-hand side by $\frac{\partial x_i}{\partial q_\gamma}$ and sum over i. For this reason, (2.17), in accordance with (2.10), is equal to $-\frac{\partial U}{\partial q_\gamma}$. Thus we find

$$\frac{d}{dt}\frac{\partial T}{\partial \dot{q}_\gamma} - \frac{\partial T}{\partial q_\gamma} = -\frac{\partial U}{\partial q_\gamma}. \quad (2.18)$$

In mechanics it is usual to consider interaction forces that are independent of particle velocities. In this case U does not involve $\dot{q}_\alpha$, so that (2.18) may be rewritten in the following form:

$$\frac{d}{dt}\frac{\partial}{\partial \dot{q}_\gamma}(T-U) - \frac{\partial}{\partial q_\gamma}(T-U) = 0. \quad (2.19)$$

The difference between the kinetic and potential energy is called the *Lagrangian function* (or, simply, Lagrangian) and is denoted by the letter L:

$$L \equiv T - U. \quad (2.20)$$

Thus we have arrived at a system of ν equations with ν independent quantities q_α, the number of which is equal to the number of degrees of freedom of the system:

$$\frac{d}{dt}\frac{\partial L}{\partial \dot{q}_\alpha} - \frac{\partial L}{\partial q_\alpha} = 0, \qquad 1 \leqslant \alpha \leqslant \nu. \quad (2.21)$$

These equations are called *Lagrange's equations.* Naturally, in (2.21) L is considered to be expressed solely in terms of q_α and $\dot{q}_\alpha$, the Cartesian coordinates being excluded. It turns out that this type of equation holds also in cases when the forces depend on the velocities (see Sec. 21).*

The rules for forming Lagrange's equations. Since the derivation of equations (2.21) from Newton's Second Law is not readily evident we will give the order of operations which, for this given system, lead to the Lagrange equations.

---

* In this case, the Lagrangian function does not have the form of (2.20), where U is a function of generalized coordinates only. However, the form of equations (2.21) is still valid.

1) The Cartesian coordinates are expressed in terms of generalized coordinates:

$$x_i = x_i\,(q_1, \ldots, q_\alpha, \ldots, q_\nu).$$

2) The Cartesian velocity components are expressed in terms of generalized velocities:

$$\dot{x}_i = \frac{\partial x_i}{\partial q_\alpha}\,\dot{q}_\alpha\,.$$

3) The coordinates are substituted in the expression for potential energy so that it is defined in relation to generalized coordinates:

$$U = U\,(q_1, \ldots, q_\alpha, \ldots, q_\nu).$$

4) The velocities are substituted in the expression for kinetic energy

$$T = \frac{1}{2}\sum_{i=1}^{n} m_i \dot{x}_i^2\,,$$

which is now a function of q_α and $\dot{q}_\alpha$. It is essential that in generalized coordinates, T is a function both of q_α and $\dot{q}_\alpha$.

5) The partial derivatives $\frac{\partial L}{\partial \dot{q}_\alpha}$ and $\frac{\partial L}{\partial q_\alpha}$ are found.

6) Lagrange's equations (2.21) are formed according to the number of degrees of freedom.

In the next section we will consider some examples in forming Lagrange's equations.

Exercises

1) Write down Lagrange's equation, where the Lagrangian function has the form:

$$L = -\sqrt{1 - \dot{q}^2} + q\dot{q}\,.$$

2) A point moves in a vertical plane along a given curve in a gravitational field. The equation of the curve in parametric form is $x = x\,(s)$, $z = z\,(s)$. Write down Lagrange's equations.

The velocities are

$$\dot{x} = \frac{dx}{ds}\,\dot{s} = x'\,\dot{s}\,, \qquad \dot{z} = \frac{dz}{ds}\,\dot{s} = z'\,\dot{s}\,.$$

The Lagrangian has the form:

$$L = \frac{m}{2}\,(x'^2 + z'^2)\,\dot{s}^2 - mgz\,(s)\,.$$

Lagrange's equation is

$$\frac{d}{dt}\,m\,[(x'^2 + z'^2)\,\dot{s}] - m\,\dot{s}^2\,(x'x'' + z'z'') + mgz' = 0\,.$$

Sec. 3. Examples of Lagrange's Equations

Central forces. *Central* forces is the name given to those whose directions are along the lines joining the particles and which depend only on the distances between them. Corresponding to such forces, there is always a potential energy, U, dependent on these distances. As an example, we consider the motion of a particle relative to a fixed centre and attracting it according to Newton's law. We shall show how to find the potential energy in this case by proceeding from the expression for a gravitational force.

Gravitational force is known to be inversely proportional to the square of the distance between the particles and is directed along the line joining them:

$$\mathbf{F} = -a \frac{1}{r^2} \frac{\mathbf{r}}{r} . \tag{3.1}$$

Here a is the factor of proportionality which we will not define more precisely at this point, r is the distance between the particles, and $\frac{\mathbf{r}}{r}$ is a unit vector. The minus sign signifies that the particles attract each other, so that the force is in the opposite direction to the radius vector $\mathbf{r}$. According to (3.1), the attractive-force component along x is equal to

$$F_x = -a \frac{x}{r^3} , \tag{3.2}$$

since x is a component of $\mathbf{r}$. But $r = \sqrt{x^2 + y^2 + z^2}$, so that

$$F_x = -\frac{\partial}{\partial x}\left(-\frac{a}{r}\right) \tag{3.3}$$

and similarly for the two other component forces. Comparing (3.3) and (2.7), we see that in the given case

$$U = -\frac{a}{r} . \tag{3.4}$$

We note that the potential energy U is chosen here in such a way that $U(\infty) = 0$ when the particles are separated by an infinite distance. The choice of the arbitrary constant in the potential energy is called its gauge. In this case it is convenient to choose this constant so that the potential energy tends to zero at infinity.

It is obvious that an expression similar to (3.4) is obtained for two electrically charged particles interacting in accordance with Coulomb's law.

Spherical coordinates. Formula (3.4) suggests that in this instance it is best to choose precisely r as a generalized coordinate. In other words, we must transform from Cartesian to spherical coordinates. The relationship between Cartesian and spherical coordinates is shown in Fig. 1. The z-axis is called the polar axis of the spherical coordinate system. The angle ϑ between the radius vector and the polar axis is called the polar angle; it is complementary (to 90°) to the "latitude." Finally, the angle φ is analogous to the "longitude" and is called the azimuth. It measures the dihedral angle between the plane zOx and the plane passing through the polar axis and the given point.

Let us find the formulae for the transformation from Cartesian to spherical coordinates. From Fig. 1 it is clear that

$$z = r \cos \vartheta. \tag{3.5}$$

The projection ρ of the radius vector onto the plane xOy is

$$\rho = r \sin \vartheta. \tag{3.6}$$

Whence,

$$x = \rho \cos \varphi = r \sin \vartheta \cos \varphi, \tag{3.7}$$

$$y = \rho \sin \varphi = r \sin \vartheta \sin \varphi. \tag{3.8}$$

We will now find an expression for the kinetic energy in spherical coordinates. This can be done either by a simple geometrical construction or by calculation according to the method of Sec. 2.

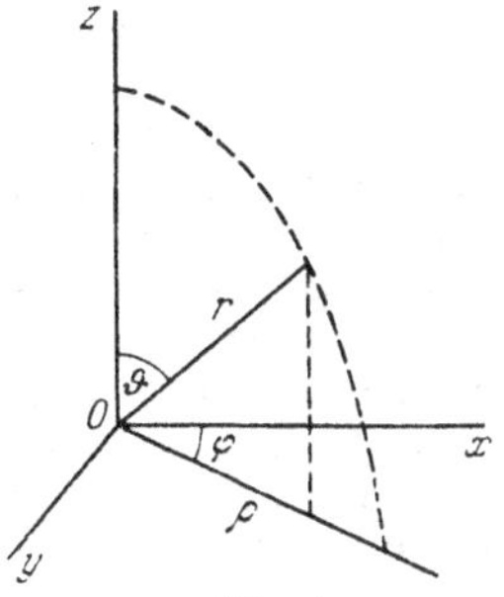

Fig. 1

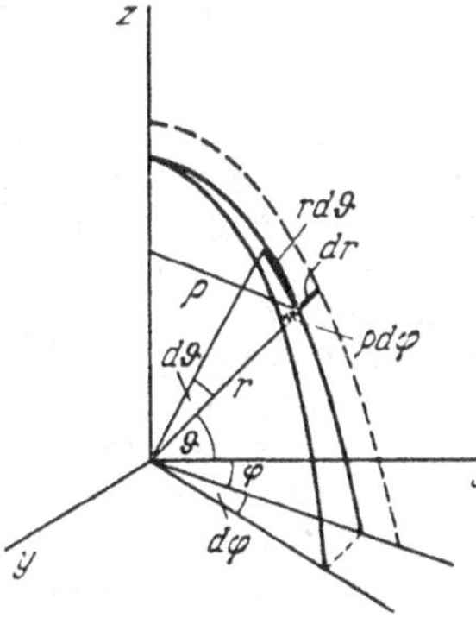

Fig. 2

Although the construction is simpler, let us first follow the computation procedure in order to illustrate the general method. We have:

$$\dot{z} = \dot{r} \cos \vartheta - r \sin \vartheta \, \dot{\vartheta},$$
$$\dot{x} = \dot{r} \sin \vartheta \cos \varphi + r \cos \vartheta \cos \varphi \, \dot{\vartheta} - r \sin \vartheta \sin \varphi \dot{\varphi},$$
$$\dot{y} = \dot{r} \sin \vartheta \sin \varphi + r \cos \vartheta \sin \varphi \, \dot{\vartheta} + r \sin \vartheta \cos \varphi \dot{\varphi}.$$

Squaring these equations and adding, we obtain, after very simple manipulations, the following:

$$T=\frac{1}{2}m(\dot{x}^2+\dot{y}^2+\dot{z}^2)=\frac{m}{2}(\dot{r}^2+r^2\dot{\vartheta}^2+r^2\sin^2\vartheta\dot{\varphi}^2). \qquad (3.9)$$

The same is clear from the construction shown in Fig. 2. An arbitrary displacement of the point can be resolved into three mutually perpendicular displacements: dr, $rd\vartheta$ and $\rho d\varphi = r\sin\vartheta d\varphi$. Whence

$$dl^2 = dr^2 + r^2 d\vartheta^2 + r^2\sin^2\vartheta\, d\varphi^2. \qquad (3.10)$$

Since the square of the velocity $v^2=\left(\frac{dl}{dt}\right)^2$, (3.9) is obtained from (3.10) simply by dividing by $(dt)^2$ and multiplying by $\frac{m}{2}$.

Hence, in spherical coordinates, the Lagrangian function is expressed as

$$L=\frac{m}{2}(\dot{r}^2+r^2\sin^2\vartheta\dot{\varphi}^2+r^2\dot{\vartheta}^2)-U(r). \qquad (3.11)$$

Now in order to write down Lagrange's equations it is sufficient to calculate the partial derivatives. We have:

$$\frac{\partial L}{\partial \dot{r}}=m\dot{r}, \quad \frac{\partial L}{\partial \dot{\vartheta}}=mr^2\dot{\vartheta}, \quad \frac{\partial L}{\partial \dot{\varphi}}=mr^2\sin^2\vartheta\dot{\varphi};$$

$$\frac{\partial L}{\partial r}=mr\sin^2\vartheta\dot{\varphi}^2+mr\dot{\vartheta}^2-\frac{\partial U}{\partial r}, \quad \frac{\partial L}{\partial \vartheta}=mr^2\sin\vartheta\cos\vartheta\dot{\varphi}^2, \quad \frac{\partial L}{\partial \varphi}=0.$$

These derivatives must be substituted into (2.21), which, however, we will not now do since the motion we are considering actually reduces to the plane case (see beginning of Sec. 5).

Two-particle system. So far we have considered the centre of attraction as stationary, which corresponds to the assumption of an infinitely large mass. In the motion of the earth around the sun, or of an electron in a nuclear field, the mass of the centre of attraction is indeed large compared with the mass of the attracted particle. But it may happen that both masses are similar or equal to each other (a binary star, a neutron-proton system, and the like). We shall show that the problem of the motion of two masses interacting only with one another can always be easily reduced to a problem of the motion of a single mass.

Let the mass of the first particle be m_1 and of the second m_2. We call the radius vectors of these particles, drawn from an arbitrary origin, $\mathbf{r}_1$ and $\mathbf{r}_2$, respectively. The components of $\mathbf{r}_1$ are x_1, y_1, z_1; the components of $\mathbf{r}_2$ are x_2, y_2, z_2. We now define the radius vector of the centre of mass of these particles $\mathbf{R}$ by the following formula:

$$\mathbf{R}=\frac{m_1\mathbf{r}_1+m_2\mathbf{r}_2}{m_1+m_2}. \qquad (3.12)$$

Synonymous terms for the "centre of mass" are the "centre of gravity" and the "centre of inertia."

In addition, let us introduce the radius vector of the relative position of the particles

$$\mathbf{r} = \mathbf{r}_1 - \mathbf{r}_2 \tag{3.13}$$

Let us now express the kinetic energy in terms of $\dot{\mathbf{R}}$ and $\dot{\mathbf{r}}$. From (3.12) and (3.13) we have

$$\mathbf{r}_1 = \mathbf{R} + \frac{m_2 \mathbf{r}}{m_1 + m_2}, \tag{3.14}$$

$$\mathbf{r}_2 = \mathbf{R} - \frac{m_1 \mathbf{r}}{m_1 + m_2}. \tag{3.15}$$

The kinetic energy is equal to

$$T = \frac{m_1}{2} \dot{\mathbf{r}}_1^2 + \frac{m_2}{2} \dot{\mathbf{r}}_2^2. \tag{3.16}$$

Differentiating (3.14) and (3.15) with respect to time and substituting in (3.16), we obtain, after a simple rearrangement,

$$T = \frac{m_1 + m_2}{2} \dot{\mathbf{R}}^2 + \frac{m_1 m_2}{2(m_1 + m_2)} \dot{\mathbf{r}}^2. \tag{3.17}$$

If we introduce the Cartesian components of the vectors $R\,(X, Y, Z)$ and $r\,(x, y, z)$, then we obtain an expression for the kinetic energy in terms of Cartesian components of velocity.

Since no external forces act on the particles, the potential energy can be a function only of their relative positions: $U = U(x, y, z)$. Thus, the Lagrangian is

$$L = \frac{m_1 + m_2}{2} (\dot{X}^2 + \dot{Y}^2 + \dot{Z}^2) + \frac{m_1 m_2}{2(m_1 + m_2)} (\dot{x}^2 + \dot{y}^2 + \dot{z}^2) - U(x, y, z).$$

Transition to the centre-of-mass system. Let us write down Lagrange's equations for the coordinates of the centre of mass. We have

$$\frac{\partial L}{\partial \dot{X}} = (m_1 + m_2)\dot{X}, \quad \frac{\partial L}{\partial \dot{Y}} = (m_1 + m_2)\dot{Y}, \quad \frac{\partial L}{\partial \dot{Z}} = (m_1 + m_2)\dot{Z};$$

$$\frac{\partial L}{\partial X} = 0, \quad \frac{\partial L}{\partial Y} = 0, \quad \frac{\partial L}{\partial Z} = 0.$$

Hence, in accordance with (2.21)

$$\ddot{X} = \ddot{Y} = \ddot{Z} = 0.$$

These equations can be easily integrated:

$$X = \dot{X}_0 t + X_0, \quad Y = \dot{Y}_0 t + Y_0, \quad Z = \dot{Z}_0 t + Z_0, \tag{3.18}$$

where the letters with the index 0 signify the corresponding values at the initial time.

Combining the coordinate equations into one vector equation, we obtain

$$\mathbf{R} = \dot{\mathbf{R}}_0 t + \mathbf{R}_0 \, .$$

Thus, the centre of mass moves uniformly in a straight line quite independently of the relative motion of the particles.

Reduced mass. If we now write down Lagrange's equations for relative motion in accordance with (2.21) the coordinates of the centre of mass do not appear. It follows that the relative motion occurs as if it were in accordance with the Lagrangian

$$L_{\text{rel}} = \frac{m_1 m_2}{2\,(m_1 + m_2)} \, \dot{\mathbf{r}}^2 - U(r) \tag{3.19}$$

(where $\dot{\mathbf{r}}^2 = \dot{x}^2 + \dot{y}^2 + \dot{z}^2$), formed in exactly the same way as the Lagrangian for a single mass m equal to

$$m = \frac{m_1 m_2}{m_1 + m_2} \, . \tag{3.20}$$

This mass is called the *reduced* mass.

The motion of the centre of mass does not affect the relative motion of the masses. In particular, we can consider, simply, that the centre of mass lies at the coordinate origin $\mathbf{R} = 0$.

In the case of central forces (for example, Newtonian forces of attraction) acting between the particles, the potential energy is simply equal to $U(r)$ [this is taken into consideration in (3.19)], where $r = \sqrt{x^2 + y^2 + z^2}$. Then, if we describe the relative motion in spherical coordinates, the equations of motion will have the same form as for a single particle moving relative to a fixed centre of attraction.

The centre of mass can now be considered as fixed, assuming $\mathbf{R} = 0$. From this, in accordance with (3.14) and (3.15), we obtain the distance of both masses from the centre of mass:

$$r_1 = \frac{m_2 r}{m_1 + m_2} \, ; \quad r_2 = \frac{m_1 r}{m_1 + m_2} \, .$$

We see that if one mass is much smaller than the other, $m_2 \ll m_1$, then $r_1 \ll r_2$, i.e., the centre of mass is close to the larger mass. This is the case for a sun-planet system. At the same time the reduced mass can also be written thus:

$$m = \frac{m_2}{1 + \dfrac{m_2}{m_1}} \, . \tag{3.21}$$

From here it can be seen that it is close to the smaller mass. That is why the motion of the earth around the sun can be approximately described as if the sun were stationary and the earth revolved about it with its own value of mass, independent of the mass of the sun.

Simple and compound pendulums. In concluding this section we shall derive the Lagrangian for simple and compound pendulums.

The simple plane pendulum is a mass suspended on a flat hinge at a certain point of a weightless rod of length l. The hinge restricts the swing of the pendulum. Let us assume that swinging occurs in the plane of the paper (Fig. 3). It is clear that such a pendulum has only one degree of freedom. We can take the angle of deflection of the pendulum from the vertical φ as a generalized coordinate. Obviously the velocity of particle m is equal to $l\dot{\varphi}$, so that the kinetic energy is

$$T = \frac{m}{2} l^2 \dot{\varphi}^2 .$$

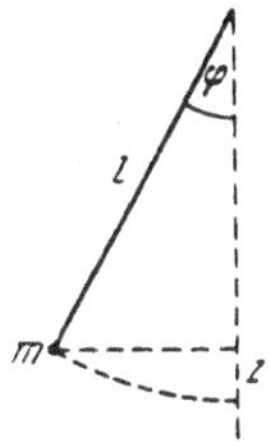

Fig. 3

The potential energy is determined by the height of the mass above the mean position $z = l\,(1 - \cos\,\varphi)$. Whence, the Lagrangian for the pendulum is

$$L = \frac{m}{2} l^2 \dot{\varphi}^2 - mgl\,(1 - \cos\varphi)\,. \tag{3.22}$$

A double pendulum can be described in the following way: in mass m there is another hinge from which another pendulum, which is forced to oscillate in the same plane (Fig. 4), is suspended. Let the mass and length of the second pendulum be m_1 and l_1, respectively, and its angle of deflection from the vertical, ψ. The coordinates of the second particle are

$$x_1 = l \sin\varphi + l_1 \sin\psi\,,$$

$$z_1 = l\,(1 - \cos\varphi) + l_1\,(1 - \cos\psi)\,.$$

Whence we obtain its velocity components:

$$\dot{x}_1 = l \cos\varphi\,\dot{\varphi} + l_1 \cos\psi\,\dot{\psi}\,,$$

$$\dot{z}_1 = l \sin\varphi\,\dot{\varphi} + l_1 \sin\psi\,\dot{\psi}\,.$$

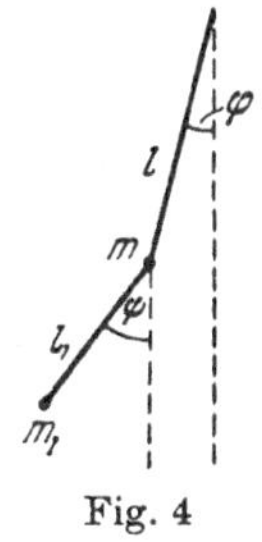

Fig. 4

Squaring and adding them we express the kinetic energy of the second particle in terms of the generalized coordinates φ, ψ and the generalized velocities $\dot{\varphi}$, $\dot{\psi}$:

$$T_1 = \frac{m_1}{2} [l^2 \dot{\varphi}^2 + l_1^2 \dot{\psi}^2 + 2\,l l_1 \cos(\varphi - \psi)\,\dot{\varphi}\,\dot{\psi}]\,.$$

The potential energy of the second particle is determined in terms of z_1. Finally, we get an expression of the Lagrangian for a double pendulum in the following form:

$$L = \frac{m + m_1}{2} l^2 \dot{\varphi}^2 + \frac{m_1}{2} l_1^2 \dot{\psi}^2 + m_1 l l_1 \cos(\varphi - \psi)\,\dot{\varphi}\,\dot{\psi} -$$
$$- (m + m_1)\,gl\,(1 - \cos\varphi) - m_1 g l_1\,(1 - \cos\psi)\,. \tag{3.23}$$

All the formulae for the Lagrangian functions (3.11), (3.22), and (3.23) will be required in the sections that follow.

Exercises

1) Write down Lagrange's equation for an elastically suspended pendulum. For such a pendulum, the potential energy of an elastic force is $U = \frac{k}{2}(l - l_0)^2$, where l_0 is the equivalent length of the unstretched rod and k is a constant, characteristic of its elasticity.

2) Write down the kinetic energy for a system of three particles with masses m_1, m_2, and m_3 in the form of a sum of the kinetic energy of the centre of mass and the kinetic energy of relative motion, using the following relative coordinates:

$$\rho_1 = \mathbf{r}_1 - \mathbf{r}_2, \quad \rho_2 = \mathbf{r}_1 - \mathbf{r}_3.$$

Sec. 4. Conservation Laws

The problem of mechanics. If a mechanical system has ν degrees of freedom, then its motion is described by ν Lagrangian equations. Each of these equations is of the second order with respect to the time derivatives $\ddot{q}$ [see (2.17)]. From general theorems of analysis we conclude that after integration of this system we obtain 2ν arbitrary constants. The solution can be represented in the following form:

$$\left.\begin{array}{l} q_1 = q_1(t; C_1, \ldots, C_{2\nu}), \\ q_2 = q_2(t; C_1, \ldots, C_{2\nu}), \\ \cdots\cdots\cdots\cdots \\ q_\alpha = q_\alpha(t; C_1, \ldots, C_{2\nu}), \\ \cdots\cdots\cdots\cdots \end{array}\right\} \tag{4.1}$$

Differentiating these equations with respect to time, we obtain expressions for the velocities:

$$\left.\begin{array}{l} \dot{q}_1 = \dot{q}_1(t; C_1, \ldots, C_{2\nu}), \\ \dot{q}_2 = \dot{q}_2(t; C_1, \ldots, C_{2\nu}), \\ \cdots\cdots\cdots\cdots \\ \dot{q}_\alpha = \dot{q}_\alpha(t; C_1, \ldots, C_{2\nu}), \\ \cdots\cdots\cdots\cdots \end{array}\right\} \tag{4.2}$$

Let us assume that equations (4.1) and (4.2) are solved with respect to the constants $C_1, \ldots, C_{2\nu}$, so that these values are expressed in terms of t and $q_1, \ldots, q_\nu, \dot{q}_1, \ldots, \dot{q}_\nu$.

Then

$$\left.\begin{array}{l} C_1 = C_1(t; q_1, q_2, \ldots, q_\nu; \dot{q}_1, \dot{q}_2, \ldots, \dot{q}_\nu), \\ C_2 = C_2(t; q_1, q_2, \ldots, q_\nu; \dot{q}_1, \dot{q}_2, \ldots, \dot{q}_\nu), \\ \cdots\cdots\cdots\cdots\cdots\cdots\cdots\cdots \\ C_{2\nu} = C_{2\nu}(t; q_1, q_2, \ldots, q_\nu; \dot{q}_1, \dot{q}_2, \ldots, \dot{q}_\nu). \end{array}\right\} \tag{4.3}$$

From the equations (4.3) we see that in any mechanical system described by 2ν second-order equations there must be 2ν functions of generalized coordinates, velocities, and time, which remain constant in the motion. These functions are called *integrals of motion.*

It is the main aim of mechanics to determine the integrals of motion.

If the form of the function (4.3) is known for a given mechanical system, then its numerical value can be determined from the initial conditions, that is, according to the given values of generalized coordinates and velocities at the initial instant.

In the preceding section we obtained the so-called centre-of-mass integrals $\mathbf{R}_0$ and $\dot{\mathbf{R}}_0$ (3.18).

Naturally, Lagrange's equations cannot be integrated in general form for an arbitrary mechanical system. Therefore the problem of determining the integrals of motion is usually very complicated. But there are certain important integrals of motion which are given directly by the form of the Lagrangian. We shall consider these integrals in the present section.

Energy. The quantity

$$\mathscr{E} = \dot{q}_\alpha \frac{\partial L}{\partial \dot{q}_\alpha} - L^{*}) \tag{4.4}$$

is called the total energy of a system. Let us calculate its total derivative with respect to time.

We have

$$\frac{d\mathscr{E}}{dt} = \ddot{q}_\alpha \frac{\partial L}{\partial \dot{q}_\alpha} + \dot{q}_\alpha \frac{d}{dt}\frac{\partial L}{\partial \dot{q}_\alpha} - \frac{\partial L}{\partial q_\alpha}\dot{q}_\alpha - \frac{\partial L}{\partial \dot{q}_\alpha}\ddot{q}_\alpha - \frac{\partial L}{\partial t}.$$

The last three terms on the right-hand side are the derivatives of the Lagrangian L, which, in the general form, depend on q, $\dot{q}$ and t. In determining $\mathscr{E}$ and its derivative we have made use of the summation convention. The quantity $\frac{d}{dt}\frac{\partial L}{\partial \dot{q}_\alpha}$ in Lagrange's equations can be replaced by $\frac{\partial L}{\partial q_\alpha}$. The result is, therefore,

$$\frac{d\mathscr{E}}{dt} = -\frac{\partial L}{\partial t}. \tag{4.5}$$

Consequently, if the Lagrangian does not depend explicitly on time $\left(\frac{\partial L}{\partial t} = 0\right)$, the energy is an integral of motion. Let us find the conditions for which time does not appear explicitly in the Lagrangian.

If the formulae expressing the generalized coordinates q in terms of Cartesian coordinates x do not contain time explicitly (which corre-

---

* α is summed from 1 to ν (see Sec. 2).

sponds to constant, time-independent constraints) then the transformation from x to q cannot introduce time into the Lagrangian.

Besides, in order that $\frac{\partial L}{\partial t} = 0$, the external forces must also be independent of time. When these two conditions—constant constraints and constant external forces—are fulfilled, the energy is an integral of motion. To take a particular case, when no external forces act on the system its energy is conserved. Such a system is called *closed.*

When frictional forces act inside a closed system, the energy of macroscopic motion is transformed into the energy of molecular microscopic motion. The total energy is conserved in this case, too, though the Lagrangian, which involves only the generalized coordinates of macroscopic motion of the system, no longer gives a complete description of the motion of the system. The *mechanical* energy of only macroscopic motion, determined by means of such a Lagrangian, is not an integral. We will not consider such a system in this section.

Let us now consider the case when our definition of energy (4.4) coincides with another definition, $\mathscr{E} = T + U$. Let the kinetic energy be a homogeneous quadratic function of generalized velocities, as expressed in equation (2.13). For this it is necessary that the constraints should not involve time explicitly, otherwise equation (2.5) would have the form

$$\dot{x}_i = \frac{\partial x_i}{\partial q_\alpha} \dot{q}_\alpha + \frac{\partial x_i}{\partial t},$$

where the partial derivative of the function (2.2) with respect to time is taken for all constants q_α. But then terms containing $\dot{q}_\alpha$ in the first degree would appear in the expression for T.

Since we assume that the potential energy U does not depend on velocity [see (2.18) and (2.19)], then

$$\frac{\partial L}{\partial \dot{q}_\alpha} = \frac{\partial T}{\partial \dot{q}_\alpha},$$

and the energy is

$$\mathscr{E} = \dot{q}_\alpha \frac{\partial T}{\partial \dot{q}_\alpha} - L. \tag{4.6}$$

But according to Euler's theorem, the sum of partial derivatives of a homogeneous quadratic function, multiplied by the corresponding variables, is equal to twice the value of the function (this can easily be verified from the function of two variables $ax^2 + 2\,bxy + cy^2$). Thus,

$$\mathscr{E} = 2\,T - T + U = T + U, \tag{4.7}$$

that is, the total energy is equal to the sum of the potential energy and the kinetic energy, in agreement with the elementary definition.

We note that the definition (4.4) is more useful and general also in the case when the Lagrangian is not represented as the difference

$L=T-U$. Thus, in electrodynamics (Sec. 15) L contains a linear term in velocity. For the energy integral to exist, only one condition is necessary and sufficient: $\frac{\partial L}{\partial t}=0$ (if, of course, there are no frictional forces).

The application of the energy integral to systems with one degree of freedom. The energy integral allows us, straightway, to reduce problems of the motion of systems with one degree of freedom to those of quadrature. Thus, in the pendulum problem considered in the previous section we can, with the aid of (4.7), write down the energy integral directly:

$$\mathscr{E}=\frac{m}{2}\cdot l^2\dot{\varphi}^2+mgl(1-\cos\varphi). \tag{4.8}$$

The value $\mathscr{E}$ is determinable from the initial conditions. For example, let the pendulum initially be deflected at an angle φ_0 and released without any initial speed. It follows that $\dot{\varphi}_0=0$. Whence

$$\mathscr{E}=mgl(1-\cos\varphi_0). \tag{4.9}$$

Substituting this in (4.8), we have

$$mgl(\cos\varphi-\cos\varphi_0)=\frac{m}{2}l^2\dot{\varphi}^2. \tag{4.10}$$

From this, the relationship between the deflection angle and time is determined by the quadrature

$$t=-\sqrt{\frac{l}{2g}}\int_{\varphi_0}^{\varphi}\frac{d\varphi}{\sqrt{\cos\varphi-\cos\varphi_0}}. \tag{4.11}$$

It is essential that the law governing the oscillation of a pendulum depend only on the value of the ratio l/g and is independent of the mass. The integral in (4.11) cannot be evaluated in terms of elementary functions.

A system in which mechanical energy is conserved is sometimes called a *conservative* system. Thus, the energy integral permits reducing to quadrature the problem of the motion of a conservative system with one degree of freedom.

In a system with several degrees of freedom the energy integral allows us to reduce the order of the system of differential equations and, in this way, to simplify the problem of integration.

Generalized momentum. We shall now consider other integrals of motion which can be found directly with the aid of the Lagrangian. To do this we shall take advantage of the following, quite obvious, consequence of Lagrange's equations. If some coordinate q_α does not appear explicitly in the Lagrangian $\left(\frac{\partial L}{\partial q_\alpha}=0\right)$, then in accordance with Lagrange's equations

$$\frac{d}{dt}\frac{\partial L}{\partial \dot{q}_\alpha} = 0. \tag{4.12}$$

But then

$$p_\alpha \equiv \frac{\partial L}{\partial \dot{q}_\alpha} = \text{const}, \tag{4.13}$$

i.e., it is an integral of motion. The quantity p_α is called the *generalized momentum* corresponding to a generalized coordinate with index α. This definition includes the momentum in Cartesian coordinates: $p_x = mv_x = \frac{\partial L}{\partial v_x}$. Summarizing, if a certain generalized coordinate does not appear explicitly in the Lagrangian, the generalized momentum corresponding to it is an integral of motion, i.e., it remains constant for the motion.

In the preceding section we saw that the coordinates X, Y, Z of the centre of gravity of a system of two particles, not subject to the action of external forces, do not appear in the Lagrangian. From this it is evidentthat

$$(m_1+m_2)\dot{X} = P_X, \quad (m_1+m_2)\dot{Y} = P_Y, \quad (m_1+m_2)\dot{Z} = P_Z \tag{4.14}$$

are constants of motion.

The momentum of a system of particles. The same thing is readily shown also for a system of N particles. Indeed, for N particles we can introduce the concept of the centre of mass and the velocity of the centre of mass by means of the equations:

$$\mathbf{R} = \frac{\sum_i m_i \mathbf{r}_i}{\sum m_i}, \tag{4.15a}$$

$$\dot{\mathbf{R}} = \frac{\sum_i m_i \dot{\mathbf{r}}_i}{\sum m_i}. \tag{4.15b}$$

The velocity of the ith mass relative to the centre of mass is

$$\dot{\mathbf{r}}_i' = \dot{\mathbf{r}}_i - \dot{\mathbf{R}} \tag{4.16}$$

(by the theorem of the addition of velocities). The kinetic energy of the system of particles is

$$T = \frac{1}{2}\sum_{i=1}^{N} m_i \dot{\mathbf{r}}_i^2 = \frac{1}{2}\sum_{i=1}^{N} m_i (\dot{\mathbf{r}}_i' + \dot{\mathbf{R}})^2 =$$

$$= \frac{1}{2}\sum_{i=1}^{N} m_i \dot{\mathbf{r}}_i'^2 + \dot{\mathbf{R}}\sum_{i=1}^{N} m_i \dot{\mathbf{r}}_i' + \frac{1}{2}\sum_{i=1}^{N} m_i \dot{\mathbf{R}}^2. \tag{4.17}$$

However, from (4.15b) and (4.16) it can immediately be seen that $\sum_{i=1}^{N} m_i \dot{\mathbf{r}}_i' = 0$, by the definition of $\mathbf{r}_i'$ and $\mathbf{R}$. Therefore, the kinetic energy of a system of particles can be divided into a sum of two terms: the kinetic energy of motion of the centre of mass

$$T_{\text{c. m.}} = \frac{1}{2}\left(\sum_{i=1}^{N} m_i\right)\dot{\mathbf{R}}^2$$

and the kinetic energy of motion of the mass relative to the centre of mass

$$T_{\text{rel}} = \frac{1}{2}\sum_{i=1}^{N} m_i \dot{\mathbf{r}}_i'^2.$$

The vectors $\dot{\mathbf{r}}_i'$ are not independent; as has been shown, they are governed by one vector equation $\sum_{i=1}^{N} m_i \mathbf{r}_i' = 0$. Consequently, they can be expressed in terms of an N-1 independent quantity by determining the relative positions of the ith and first masses. For this reason the kinetic energy of N particles relative to the centre of mass is, in general, the kinetic energy of their relative motion, and is expressed only in terms of the relative velocities $\dot{\mathbf{r}}_1 - \dot{\mathbf{r}}_i$. By definition, no external forces can act on the masses in a closed system, and the interaction forces inside the system can be determined only by the relative positions $\mathbf{r}_1 - \mathbf{r}_i$.

Thus, only $\dot{\mathbf{R}}$ appears in the Lagrangian, and $\mathbf{R}$ does not. Therefore, the overall momentum is conserved:

$$\mathbf{P} = \frac{\partial L}{\partial \dot{\mathbf{R}}} = \left(\sum_{i=1}^{N} m_i\right)\dot{\mathbf{R}} = \text{const}. \tag{4.18}$$

Equality (4.18), which contains a derivative with respect to a vector, should be understood as an abbreviated form of three equations:

$$P_X = \frac{\partial L}{\partial \dot{X}}, \quad P_Y = \frac{\partial L}{\partial \dot{Y}}, \quad P_Z = \frac{\partial L}{\partial \dot{Z}}.$$

For more detail about vector derivatives see Sec. 11.

We have seen that the overall momentum of a mechanical system not subject to any external force is an integral of motion. It is important that it is what is known as an *additive* integral of motion, i.e., it is obtained by adding the momenta of separate particles.

It may be noted that the momentum integral exists for any system in which only internal forces are operative, even though they may be frictional forces causing a conversion of mechanical energy into heat.

If we integrate (4.18) with respect to time once again, the result will be the centre-of-mass integral similar to (3.18). This will be the so-called second integral (for it contains two constants); it contains only coordinates but not velocities. (3.18) and (4.11) are also second integrals.

Properties of the vector product. The angular momentum of a particle is defined as

$$\mathbf{M}=[\mathbf{r}\,\mathbf{p}]\,. \tag{4.19}$$

Here the brackets denote the vector product of the radius vector of the particle and its momentum. We know that (4.19) takes the place of three equations,

$$M_x=yp_z-zp_y, \quad M_y=zp_x-xp_z, \quad M_z=xp_y-yp_x,$$

for the Cartesian components of the vector **M**.

Recall the geometrical definition for a vector product. We construct a parallelogram on the vectors **r** and **p**. Then [**rp**] denotes a vector numerically equal to the area of the parallelogram with direction perpendicular to its plane. In order to specify the direction of [**rp**] uniquely, we must agree on the way of tracing the parallelogram contour. We shall agree always to traverse the contour beginning with the first factor (in this case beginning with **r**). Then that side of the plane will be considered *positive* for which the direction is anticlockwise. The vector [**rp**] is along the normal to the positive side of the plane. In still another way, if we rotate a corkscrew from **r** to **p**, then it will be displaced in the direction of [**rp**]. The direction of traverse changes if we interchange the positions of **r** and **p**. Therefore, unlike a conventional product, the sign of the vector product changes if we interchange the factors. This can also be seen from the definition of Cartesian components of angular momentum.

The area of the parallelogram is $rp \sin \alpha$, where α is the angle between **r** and **p**. The product $r \sin \alpha$ is the length of a perpendicular drawn from the origin of the coordinate system to the tangent to the trajectory whose direction is the same as **p**. This length is sometimes called the "arm" of the moment.

The vector product possesses a distributive property, i.e.,

$$[\mathbf{a}, \mathbf{b}+\mathbf{c}]=[\mathbf{ab}]+[\mathbf{ac}]\,.$$

Hence, a binomial product is calculated in the usual way, but the order of the factors is taken into account.

$$[\mathbf{a}+\mathbf{d}, \mathbf{b}+\mathbf{c}]=[\mathbf{ab}]+[\mathbf{ac}]+[\mathbf{db}]+[\mathbf{dc}]\,.$$

The angular momentum of a particle system. The angular momentum of a system of particles is defined as the sum of the angular momenta of all the particles taken separately. In doing so we must, of course, take the radius vectors related to a coordinate origin common to all the particles:

$$\mathbf{M} = \sum_{i=1}^{N} [\mathbf{r}_i \, \mathbf{p}_i] . \tag{4.20}$$

We shall show that the angular momentum of a system can be separated into the angular momentum relative to the motion of the particles and the angular momentum of the system as a whole, similar to the way that it was done for the kinetic energy. To do this we must represent the radius vector of each particle as the sum of the radius vector of its position relative to the centre of mass and the radius vector of the centre of mass; we must expand the expression for the particle velocities in the same way. Then, the angular momentum can be written in the form

$$\mathbf{M} = \sum_{i=1}^{N} [\mathbf{R} + \mathbf{r}_i', m_i \dot{\mathbf{R}} + \mathbf{p}_i'] =$$

$$= \sum_{i=1}^{N} m_i [\mathbf{R}\dot{\mathbf{R}}] + \sum_{i=1}^{N} [m_i \mathbf{r}_i', \dot{\mathbf{R}}] + \sum_{i=1}^{N} [\mathbf{R}\mathbf{p}_i'] + \sum_{i=1}^{N} \mathbf{r}_i' \, \mathbf{p}_i'] .$$

In the second and third sums, we can make use of the distributive property of a vector product and introduce the summation sign inside the product sign. However, both these sums are equal to zero, by definition of the centre of mass. This was used in (4.17) for velocities. Thus, the angular momentum is indeed equal to the sum of the angular momenta of the centre of mass ($\mathbf{M}_0$) and the relative motion ($\mathbf{M}'$):

$$\mathbf{M} = [\mathbf{R}\mathbf{P}] + \sum_{i} [\mathbf{r}_i' \, \mathbf{p}_i'] \equiv \mathbf{M}_0 + \mathbf{M}'. \tag{4.21}$$

Let us perform these transformations for the special case of a system of two particles. We substitute $\mathbf{r}_1$ and $\mathbf{r}_2$ expressed [from (3.14) and (3.15)] in terms of $\mathbf{r}$ and $\mathbf{R}$. This gives

$$\mathbf{M} = [\mathbf{r}_1 \, \mathbf{p}_1] + [\mathbf{r}_2 \, \mathbf{p}_2] = [\mathbf{R}, \mathbf{p}_1 + \mathbf{p}_2] + \frac{1}{m_1 + m_2} (m_2 [\mathbf{r}\mathbf{p}_1] - m_1 [\mathbf{r}\mathbf{p}_2]) .$$

Further, we replace $\mathbf{p}_1$ by $m_1 \dot{\mathbf{r}}_1$, $\mathbf{p}_2$ by $m_2 \dot{\mathbf{r}}_2$ and $\mathbf{p}_1 + \mathbf{p}_2$ by $\mathbf{P}$, after which the angular momentum reduces to the required form:

$$\mathbf{M} = [\mathbf{R}\mathbf{P}] + \frac{m_1 m_2}{m_1 + m_2}[\mathbf{r}\dot{\mathbf{r}}] = [\mathbf{R}\mathbf{P}] + [\mathbf{r}\mathbf{p}]. \tag{4.22}$$

Here, $\frac{m_1 m_2}{m_1 + m_2}\dot{\mathbf{r}} = m\dot{\mathbf{r}} = \mathbf{p}$ is the momentum of relative motion.

We shall now show that the determination of angular momentum of relative motion does not depend on the choice of the origin. Indeed, if we displace the origin, then all the quantities $\mathbf{r}_i'$ change by the same amount $\mathbf{r}_i' = \mathbf{r}_i'' + \mathbf{a}$.

Accordingly, angular momentum for relative motion will be

$$\mathbf{M}' = \sum_{i=1}^{N}[\mathbf{r}_i'\,\mathbf{p}_i'] = \sum_{i=1}^{N}[\mathbf{r}_i''\,\mathbf{p}_i'] + \sum_{i=1}^{N}[\mathbf{a}\,\mathbf{p}_i'] =$$

$$= \sum_{i=1}^{N}[\mathbf{r}_i''\,\mathbf{p}_i''] + \left[\mathbf{a}\sum_{i=1}^{N}\mathbf{p}_i'\right] = \mathbf{M}'',$$

because

$$\sum_{i=1}^{N}\mathbf{p}_i' = \sum_{i}^{N} m_i\dot{\mathbf{r}}_i' = 0.$$

Thus, the determination of angular momentum for relative motion does not depend on the choice of the origin of the coordinate system.

Conservation of angular momentum. We shall now show that the angular momentum of a system of particles not acted upon by any external forces is an integral of motion.

Let us begin with the angular momentum of the system as a whole. Its time derivative is equal to zero:

$$\frac{d\mathbf{M}_0}{dt} = [\dot{\mathbf{R}}\mathbf{P}] + [\mathbf{R}\dot{\mathbf{P}}] = 0,$$

because $\dot{\mathbf{P}} = 0$ for any system not acted upon by external forces, and $\dot{\mathbf{R}}$ is in the direction of $\mathbf{P}$, so that the vector product $[\dot{\mathbf{R}}\mathbf{P}] = 0$.

We shall now prove that angular momentum is conserved for relative motion. The total derivative with respect to time is

$$\frac{d\mathbf{M}'}{dt} = \sum_{i=1}^{N}[\dot{\mathbf{r}}_i'\mathbf{p}_i'] + \sum_{i=1}^{N}[\mathbf{r}_i'\,\dot{\mathbf{p}}_i']. \tag{4.23}$$

Here, the first term on the right-hand side is equal to zero since $\dot{\mathbf{r}}_i'$ is in the direction of $\mathbf{p}_i'$. We consider the second term. Let us choose the origin to be coincident with any particle, for example, the first. As a result, $\mathbf{M}'$, as we have already seen, does not change. The potential energy can only depend on the differences $\mathbf{r}_1$—$\mathbf{r}_2$, $\mathbf{r}_1$—$\mathbf{r}_3$, ..., $\mathbf{r}_1$—$\mathbf{r}_k$, ...

The other differences are expressed in terms of these, for example, $\mathbf{r}_k - \mathbf{r}_l = (\mathbf{r}_1 - \mathbf{r}_l) - (\mathbf{r}_1 - \mathbf{r}_k)$. We introduce the abbreviations.

$$\begin{aligned} \boldsymbol{\rho}_1 &= \mathbf{r}_1 - \mathbf{r}_2, \\ \boldsymbol{\rho}_2 &= \mathbf{r}_1 - \mathbf{r}_3, \\ &\cdots\cdots \\ \boldsymbol{\rho}_{k-1} &= \mathbf{r}_1 - \mathbf{r}_k, \\ &\cdots\cdots\cdots \end{aligned}$$

Then the derivatives $\frac{\partial U}{\partial \mathbf{r}_1}, \frac{\partial U}{\partial \mathbf{r}_2}, \ldots, \frac{\partial U}{\partial \mathbf{r}_k} \ldots$ will be expressed in terms of the variables $\boldsymbol{\rho}_1, \ldots, \boldsymbol{\rho}_{k-1}, \ldots$ as follows:

$$\frac{\partial U}{\partial \mathbf{r}_1} = \frac{\partial}{\partial \mathbf{r}_1} U(\mathbf{r}_1 - \mathbf{r}_2 \ldots \mathbf{r}_1 - \mathbf{r}_k \ldots) = \sum_{k=1}^{N-1} \frac{\partial U}{\partial \boldsymbol{\rho}_k};$$

$$\frac{\partial U}{\partial \mathbf{r}_2} = \frac{\partial}{\partial \mathbf{r}_2} U(\mathbf{r}_1 - \mathbf{r}_2 \ldots \mathbf{r}_1 - \mathbf{r}_k \ldots) = -\frac{\partial U}{\partial \boldsymbol{\rho}_1}; \quad \frac{\partial U}{\partial \mathbf{r}_k} = -\frac{\partial U}{\partial \boldsymbol{\rho}_{k-1}}.$$

Substituting this in (4.23) we obtain

$$\frac{d\mathbf{M}'}{dt} = \sum_{k=1}^{N} [\mathbf{r}_k' \dot{\mathbf{p}}_k'] = -\sum_{k=1}^{N} \left[\mathbf{r}_k' \frac{\partial U}{\partial \mathbf{r}'_k}\right] =$$

$$= -\left[\mathbf{r}_1' \sum_{k=1}^{N-1} \frac{\partial U}{\partial \boldsymbol{\rho}_k}\right] + \sum_{k=1}^{N-1} \left[\mathbf{r}_k' \frac{\partial U}{\partial \boldsymbol{\rho}'_k}\right] = -\sum_{k=1}^{N-1} \left[\boldsymbol{\rho}_k \frac{\partial U}{\partial \boldsymbol{\rho}_k}\right]. \quad (4.24)$$

In this expression, only the relative coordinates $\boldsymbol{\rho}_1, \ldots, \boldsymbol{\rho}_{N-1}$ remain. We shall now show that, for a closed system, the right-hand side is identically equal to zero. The potential energy is a scalar function of coordinates. Hence, it can depend only on the scalar arguments ρ_k^2, ρ_l^2, $(\boldsymbol{\rho}_l \boldsymbol{\rho}_k)$, totally irrespective of whether the initial expression was a function only of the absolute values $|\mathbf{r}_i - \mathbf{r}_k|$, or whether it also involved scalar products of the form $(\mathbf{r}_i - \mathbf{r}_k, \mathbf{r}_l - \mathbf{r}_n)$. An essential point is that the system is closed (in accordance with the definition, see page 32), and the forces in it are completely defined by the relative positions of the points and by nothing else. Therefore, the potential energy depends only on the quantities $\mathbf{r}_i - \mathbf{r}_k$, and only in scalar combinations $(\mathbf{r}_i - \mathbf{r}_k, \mathbf{r}_l - \mathbf{r}_n)$ (in particular, the subscripts $i, l; k, n$ can even be the same; then the scalar product becomes the square of the distance between the particles i, k).

To summarize, the potential energy U depends on the following arguments:

$$U = U[\rho_1^2, \rho_2^2, \ldots, \rho_k^2, \ldots, \rho_{N-1}^2; (\boldsymbol{\rho}_1 \boldsymbol{\rho}_2), \ldots, (\boldsymbol{\rho}_k \boldsymbol{\rho}_l)].$$

In order to save space we will, in future, perform the operation for two vectors, though this operation can be directly generalized to any arbitrary number. We obtain

$$\frac{\partial U}{\partial \boldsymbol{\rho}_1} = \frac{\partial U}{\partial(\rho_1^2)} \frac{\partial(\rho_1^2)}{\partial \boldsymbol{\rho}_1} + \frac{\partial U}{\partial(\boldsymbol{\rho}_1 \boldsymbol{\rho}_2)} \frac{\partial(\boldsymbol{\rho}_1 \boldsymbol{\rho}_2)}{\partial \boldsymbol{\rho}_1},$$
$$\frac{\partial U}{\partial \boldsymbol{\rho}_2} = \frac{\partial U}{\partial(\rho_2^2)} \frac{\partial(\rho_2^2)}{\partial \boldsymbol{\rho}_2} + \frac{\partial U}{\partial(\rho_1 \rho_2)} \frac{\partial(\boldsymbol{\rho}_1 \boldsymbol{\rho}_2)}{\partial \boldsymbol{\rho}_2}.$$

The partial derivatives of the scalar quantities ρ_1^2, $(\rho_1 \rho_2)$ with respect to the vector arguments are in the given case easily evaluated. Thus,

$$\frac{\partial \rho_1^2}{\partial \boldsymbol{\rho}_1} = 2\boldsymbol{\rho}_1; \quad \frac{\partial(\boldsymbol{\rho}_1 \boldsymbol{\rho}_2)}{\partial \boldsymbol{\rho}_1} = \boldsymbol{\rho}_2.$$

Each of these equations is a shortened form of three equations referring to the components (the components of ρ_i are ξ_i, η_i, ζ_i):

$$\frac{\partial}{\partial \xi_1}(\boldsymbol{\rho}_1 \boldsymbol{\rho}_2) = \frac{\partial}{\partial \xi_1}(\xi_1 \xi_2 + \eta_1 \eta_2 + \zeta_1 \zeta_2) = \xi_2;$$
$$\frac{\partial}{\partial \eta_1}(\boldsymbol{\rho}_1 \boldsymbol{\rho}_2) = \eta_2; \quad \frac{\partial}{\partial \zeta_1}(\boldsymbol{\rho}_1 \boldsymbol{\rho}_2) = \zeta_2.$$

Hence,

$$\frac{\partial U}{\partial \boldsymbol{\rho}_1} = 2\boldsymbol{\rho}_1 \frac{\partial U}{\partial(\rho_1^2)} + \boldsymbol{\rho}_2 \frac{\partial U}{\partial(\boldsymbol{\rho}_1 \boldsymbol{\rho}_2)}; \quad \frac{\partial U}{\partial \boldsymbol{\rho}_2} = 2\boldsymbol{\rho}_2 \frac{\partial U}{\partial(\rho_2^2)} + \boldsymbol{\rho}_1 \frac{\partial U}{\partial(\boldsymbol{\rho}_1 \boldsymbol{\rho}_2)}. \quad (4.25)$$

Substituting (4.25) into (4.24) for the case of the two variables, we obtain

$$\frac{d\mathbf{M}'}{dt} = -2[\boldsymbol{\rho}_1 \boldsymbol{\rho}_1] \frac{\partial U}{\partial(\rho_1^2)} - 2[\boldsymbol{\rho}_2 \boldsymbol{\rho}_2] \cdot \frac{\partial U}{\partial(\rho_2^2)} - ([\boldsymbol{\rho}_1 \boldsymbol{\rho}_2] + [\boldsymbol{\rho}_2 \boldsymbol{\rho}_1]) \frac{\partial U}{\partial(\boldsymbol{\rho}_1 \boldsymbol{\rho}_2)}.$$

But the sign of a vector product depends on the order of the factors $[\boldsymbol{\rho}_1 \boldsymbol{\rho}_2] = -[\boldsymbol{\rho}_2 \boldsymbol{\rho}_1]$. Hence it can also be seen that $[\boldsymbol{\rho}_1 \boldsymbol{\rho}_1] = -[\boldsymbol{\rho}_1 \boldsymbol{\rho}_1] \equiv 0$ and $[\boldsymbol{\rho}_2 \boldsymbol{\rho}_2] \equiv 0$. Therefore, $\frac{d\mathbf{M}'}{dt} = 0$, as stated.

The integral, like the angular momentum, can also be formed when the forces are determined not only by the relative position of the particles but also by their relative velocities. This is the case, for example, in a system of elementary currents interacting in accordance with the Biot-Savart law.

Additive integrals of motion for closed systems. We have thus shown that a closed system has the following first integrals of motion: energy, three components of the momentum vector and three components of the angular-momentum vector. Momentum and angular momentum are always additive, while energy is additive only for the noninteracting parts of a system.

All the other integrals of motion are found in a much more complicated fashion and depend on the specific form of the system (in the sense that one cannot give a general rule for their definition).

Exercises

Describe the motion of a point moving along a cycloid in a gravitational field.

The equation of the cycloid in parametric form is

$$z = -R \cos s, \quad x = Rs + R \sin s.$$

The kinetic energy of the point is

$$T = \frac{m}{2}(\dot{x}^2 + \dot{z}^2) = 2mR^2 \cos^2 \frac{s}{2} \cdot \dot{s}^2.$$

The potential energy is $U = -mgR \cos s$.

The total-energy integral is

$$\mathscr{E} = 2m\,R^2 \cos^2 \frac{s}{2} \cdot \dot{s}^2 - mg\,R \cos s = \text{const}.$$

The value $\mathscr{E}$ can be determined on the condition that the velocity $\dot{s}$ is equal to zero when the deflection is maximum $s = s_0$; the particle moves along the cycloid from that position. Hence,

$$\mathscr{E} = -mg\,R \cos s_0.$$

After separating the variables and integrating, we obtain

$$t = R\sqrt{2} \int \frac{\cos \frac{s}{2}\, ds}{\sqrt{\frac{\mathscr{E}}{m} + gR \cos s}} = \sqrt{\frac{2R}{g}} \int \frac{\cos \frac{s}{2}\, ds}{\sqrt{\cos s - \cos s_0}}.$$

Calling $\sin \frac{s}{2} = u$, we rewrite the integral in the form:

$$t = \sqrt{\frac{R}{g}} \int \frac{2\,du}{\sqrt{u_0^2 - u^2}} = 2\sqrt{\frac{R}{g}} \arcsin \frac{u}{u_0}.$$

In order to find the period of the motion, we must take the integral between the limits $-u_0$ and $+u_0$ and double the result. This corresponds to the oscillation of the particle within the limits $s = -s_0$ and $s = s_0$.

Thus the total period of oscillation is equal to $4\pi\sqrt{\frac{R}{g}}$

Hence, as long as the particle moves on the cycloid, the period of its oscillation does not depend on the oscillation amplitude (Huygens' cycloidal pendulum). The period of oscillation of an ordinary pendulum, describing an arc of a circle, is known, in the general case, to depend on the amplitude [see (4.11)].

Sec. 5. Motion in a Central Field

The angular-momentum integral. We shall now consider the motion of two bodies in a frame of reference fixed in the centre of mass. If the origin coincides with the centre of mass, then $\mathbf{R} = 0$. As was shown in the preceding section, the angular momentum of relative motion is conserved in any closed system; specifically, it is also conserved in a system of two particles. If the radius vector of the relative position of the particles $\mathbf{r}=\mathbf{r}_1-\mathbf{r}_2$, and the momentum of relative motion is

$$\mathbf{p}=\frac{m_1 m_2}{m_1+m_2}\mathbf{v}=m\mathbf{v}\,, \tag{5.1}$$

then the angular-momentum integral is reduced to the simple form:

$$\mathbf{M}=[\mathbf{rp}]=\text{const.} \tag{5.2}$$

It follows that the velocity vector and the relative position vector all the time remain perpendicular to the constant vector **M**; in other words, the motion takes place in a plane perpendicular to **M** (Fig. 5).

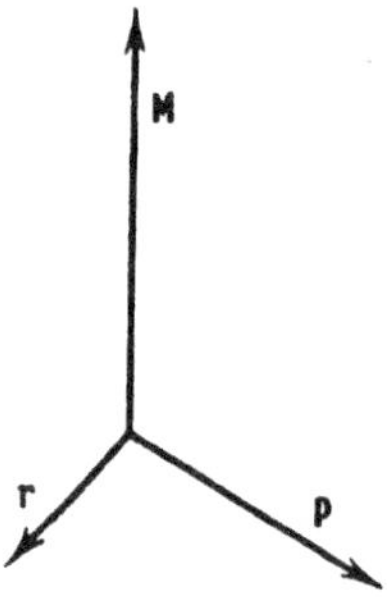

Fig. 5

When transforming to a spherical coordinate system, it is advisable to choose the polar axis along **M**. Then the motion will take place in the plane xy or $\vartheta=\frac{\pi}{2}$, $\sin\vartheta=1$, $\dot{\vartheta}=0$.

The potential energy can depend only on the absolute value r, because this is the only scalar quantity which can be derived solely from the vector **r**. In accordance with (3.9), the Lagrangian for plane motion, with $\dot{\vartheta}=0$, $\sin\vartheta=1$, is

$$L=\frac{m}{2}(\dot{r}^2+r^2\dot{\varphi}^2)-U(r)\,, \tag{5.3}$$

where m is the reduced mass.

Angular momentum as a generalized momentum. We shall now show that $M_z=M$ is nothing other than $\frac{\partial L}{\partial\dot{\varphi}}$, i.e., the component of angular momentum along the polar axis is a generalized momentum, provided the angle of rotation φ around that axis is a generalized coordinate. Indeed, in accordance with (5.2), the angular momentum M is

$$M=M_z=xp_y-yp_x=mr\cos\varphi\,(\dot{r}\sin\varphi+r\cos\varphi\dot{\varphi})-$$
$$-mr\sin\varphi\,(\dot{r}\cos\varphi-r\sin\varphi\dot{\varphi})=mr^2\dot{\varphi}\,(\cos^2\varphi+\sin^2\varphi)=mr^2\dot{\varphi}\,.$$

On the other hand, differentiating L with respect to $\dot{\varphi}$ we see that

$$p_\varphi=\frac{\partial L}{\partial\dot{\varphi}}=mr^2\dot{\varphi}=M_z\,. \tag{5.4}$$

The expression for angular momentum in polar coordinates can also be derived geometrically (Fig. 6). In unit time, the radius vector **r** moves to the position shown in Fig. 6 by the dashed line. Twice the area of the sector OAB, multiplied by the mass m, is by definition equal to the angular momentum [cf. (5.2)]. But, to a first approximation the area of the sector is equal to the product of the modulus r and $\frac{h}{2}$. The height h is proportional to the angle of rotation in unit time and to the radius itself so that the area of the sector is $1/2\, r^2 \dot{\varphi}$. Thus, a doubled area multiplied by the mass m is indeed equal to the angular momentum.

The quantity $1/2\, r^2 \dot{\varphi}$ is the so-called areal velocity, or the area described by the radius vector in unit time. The law of conservation of angular momentum, if interpreted geometrically, expresses constancy of areal velocity (Kepler's Second Law).

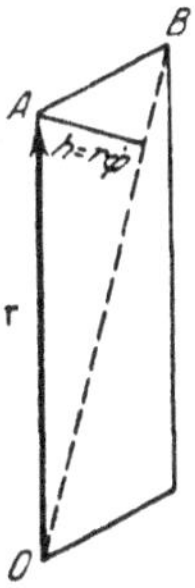

Fig. 6

The central field. If one of two masses is very much greater than the other, the centre of mass coincides with the larger mass (see Sec. 3). In this case, the particle with the smaller mass moves in the *given central field* of the heavy particle. The potential energy depends only on the distance between the particles and does not depend on the angle φ. Then, in accordance with (4.12), $p_\varphi = M_z$ is the integral of motion. However, since one particle is considered at rest, the origin should be chosen coincident with that particle and not with some arbitrary point, as in the case of the relative motion of two particles. In the case of motion in a central field, angular momentum is conserved only relative to the centre.

Elimination of the azimuthal velocity component. The angular-momentum integral permits us to reduce the problem of two-particle motion, or the problem of motion of a single particle in a central field, to quadrature. To do this we must express $\dot{\varphi}$ in terms of angular momentum and thus get rid of the superfluous variable, in as much the angle φ itself does not appear in the Lagrangian. Such variables, which do not appear in L, are termed *cyclic*.

In accordance with (4.7), we first of all have the energy integral

$$\mathscr{E} = \frac{m}{2}\left(\dot{r}^2 + r^2\dot{\varphi}^2\right) + U(r)\,. \tag{5.5}$$

Eliminating $\dot{\varphi}$ with the aid of (5.4) we obtain

$$\mathscr{E} = \frac{m\dot{r}^2}{2} + \frac{M^2}{2mr^2} + U(r)\,. \tag{5.6}$$

This first-order differential equation (in r) is later on reduced to quadrature. Before writing down the quadrature, let us examine it graphically.

The dependence of the form of the path on the sign of the energy. For such an examination, we must make certain assumptions about the variation of potential energy.

From (2.7), force is connected with potential energy by the relation

$$F = -\frac{\partial U}{\partial r}, \quad U = \int F\, dr\,.$$

The upper limit in the integral can be chosen arbitrarily. If $F(r)$ tends to zero at infinity faster than $\frac{1}{r}$, then the integral $\int_r^\infty F\,dr$ is convergent. Then we can put $U(r)=\int_r^\infty F\,dr$, or $U(\infty)=0$. In other words, the potential energy is considered zero at infinity. The choice of an arbitrary constant in the expression for potential energy is called its *gauge*.

In addition we shall consider that at $r=0$, $U(r)$ does not tend to infinity more rapidly than $\frac{1}{r}$, as, for example, for Newtonian attraction $U=-\int_r^\infty \frac{a}{r^2}\,dr=-\frac{a}{r}$.

Let us now write (5.6) as

$$\frac{m\dot{r}^2}{2}=\mathscr{E}-\frac{M^2}{2mr^2}-U(r). \tag{5.7}$$

The left-hand side of this equation is essentially positive. For $r=\infty$ the last two terms in (5.7) tend to zero. Thus, for the particles to be able to recede from each other an infinite distance, the total energy must be positive when the gauge of the potential energy satisfies $U(\infty)=0$.

Given a definite form of U, we can now plot the curve of the function

$$U_M(r)\equiv\frac{M^2}{2mr^2}+U(r). \tag{5.8}$$

The index M in U denotes that the potential energy includes the "centrifugal" energy $\frac{M^2}{2mr^2}$. The derivative of this quantity with respect to r, taken with the opposite sign, is equal to $\frac{M^2}{mr^3}$. If we put $M=mr^2\dot{\varphi}$, the result will be the usual expression for "centrifugal force." However, henceforth, we will call a mechanical quantity of different origin the "centrifugal force" (see Sec. 8). Let $U<0$ and monotonic. Since $U(\infty)=0$, we see that $U(r)$ is an increasing function of r. It follows that the force has a negative sign $\left(\text{since } \mathbf{F}=-\frac{\partial U}{\partial \mathbf{r}}\right)$, i.e., it is an attractive force. Let us assume, in addition, that at infinity $|U(r)|>\frac{M^2}{2mr^2}$. Let us summarize the assumptions that we have made concerning $U_M(r)$:

1) $U_M(r)$ is positively infinite at zero, where the centrifugal term is predominant.

2) at infinity, where $U(r)$ predominates, $U_M(r)$ tends to zero from a negative direction.

Consequently, the curve $U_M(r)$ has the form shown in Fig. 7, since we must go through a minimum in order to pass from a decrease for small values of r to an increase at large values of r.*

In this figure we can also plot the total energy $\mathscr{E}$. But since the total energy is conserved, the curve of $\mathscr{E}$ must have the form of a horizontal straight line lying above or below the abscissa, depending on the sign of $\mathscr{E}$.

For positive values of energy, the line $\mathscr{E}=$const lies above the curve $U_M(r)$ everywhere to the right of point A. Hence the difference $\mathscr{E}-U_M(r)$ to the right of A is positive. The particles can approach each other from infinity and recede from each other to infinity. Such motion is termed *infinite*. As we will see later in this section, in the case of Newtonian attraction, we obtain hyperbolic orbits.

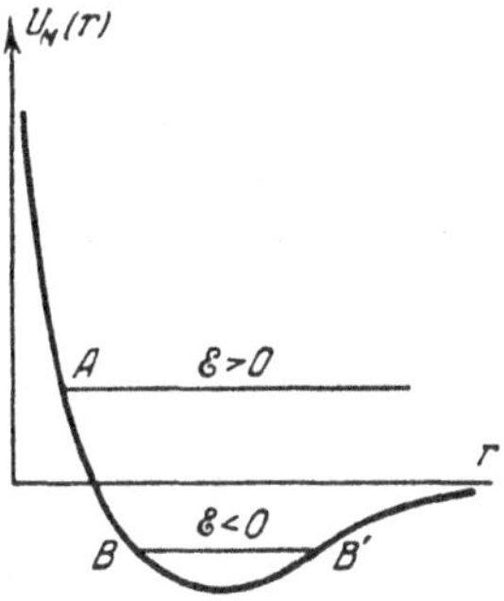

Fig. 7

For $\mathscr{E}<0$, but higher than the minimum of the curve $U_M(r)$, the difference $\mathscr{E}-U_M(r)$, i.e., $\frac{m\dot{r}^2}{2}$, remains positive only between the points B and B' (*finite* motion). Thus, between these values of the radius there is included a physically possible region of motion, to which there correspond elliptical orbits in the case of Newtonian attraction. In the case of planetary motion around the sun, point B is called the *perihelion* and point B' the *aphelion*.

For $\mathscr{E}=0$ the motion is infinite (parabolic motion).

If $U(r)>0$, which corresponds to repulsion, the curve $U_M(r)$ does not possess a minimum. Then finite motion is clearly impossible.

Falling towards the centre. For Newtonian attraction, $U(0)$ tends to infinity like $-1/r$. If we suppose that $U(0)$ tends to $-\infty$ more rapidly than $-1/r^2$, then the curve $U_M(r)$ is negative for all r close to zero. Then, from (5.7), $\dot{r}^2$ is positive for infinitesimal values of r and tends to infinity when r tends to zero. If $\dot{r}<0$ initially, then $\dot{r}$ does not change sign and the particles now begin to move towards collision. In Newtonian attraction this is possible only when the particles are directed towards each other; then "the arm" of the angular momentum is equal to zero and, hence, the angular momentum itself is obviously equal to zero, too, so that $U_M(r)=U(r)$. If an initial "arm" exists within the distance of minimum approach, then

* If $|U(r)|<\frac{M^2}{2mr^2}$ at infinity, then the curve approaches zero on the positive side, and there can be a further small maximum after the minimum. This form of $U_M(r)$ applies to the atoms of elements with medium and large atomic weights.

the angular momentum $M = mv\rho \neq 0$ (ρ is the "arm") and the motion can in no way become radial.

In the case of Newtonian or Coulomb attraction for a particle with angular momentum not equal to zero, there always exists a distance r_0 for which $\frac{M^2}{2mr_0^2}$ becomes greater than $\mathscr{E} - U(r_0)$. This distance determines the perihelion for the approaching particles.

However, if $U(r)$ tends to infinity more rapidly than $-\frac{1}{r^2}$ then, as $r \to 0$, there will be no point at which $U_M(r)$ becomes zero. In place of a hyperbolic orbit, as in the case of Newtonian attraction, a spiral curve leading to one particle falling on the other results. The turns of the spiral diminish, but the speed of rotation increases so that the angular momentum is conserved, as it should be in any central field. But the "centrifugal" repulsive force turns out to be less than the forces of attraction, and the particles approach each other indefinitely.

Of course, the result is the same if the energy is negative (for example, part of the energy is transferred to some third particle, which then recedes). In the case of attractive forces increasing more rapidly than $1/r^3$, no counterpart to elliptical orbits exists.

If three bodies in motion are subject to Newtonian attraction, two of them may collide even if, initially, the motion of these particles was not purely radial. Indeed, in the case of three bodies, only the total angular momentum is conserved, and this does not exclude the collision of two particles.

Reducing to quadrature. Let us now find the equation of the trajectory in general form. To do this we must, in (5.6), change from differentiation with respect to time to differentiation with respect to φ. Using (5.4) we have

$$dt = \frac{mr^2}{M} d\varphi . \tag{5.9}$$

Separating the variables and passing to φ in (5.6) gives

$$\varphi = \int_{r_0}^{r} \frac{M}{mr^2} \frac{dr}{\sqrt{\frac{2}{m}\left(\mathscr{E} - \frac{M^2}{2mr^2} - U(r)\right)}} . \tag{5.10}$$

Here, the lower limit of the integral corresponds to $\varphi = 0$. If we calculate φ with respect to the perihelion, then the corresponding value $r = r_0$ can be easily found by noting that the radial component of velocity $\dot{r}$ changes sign at perihelion (r has a minimum, and so $dr = 0$). From this we find the equation for the particle distance at perihelion:

$$\mathscr{E} = \frac{M^2}{2mr_0^2} + U(r_0) . \tag{5.11}$$

Kepler's problem. Thus, the problem of motion in a central field is reduced to quadrature. The fact that the integral sometimes cannot be solved in terms of elementary functions is no longer so essential. Indeed, the solution of the problem in terms of definite integrals contains all the initial data explicitly; if these data are known, the integration can be performed in some way or other.

But, naturally, if the integral is expressed in the form of a well-known function, the solution can be more easily examined in the general form. In this sense an explicit solution is of particular interest.

Such a solution can be found in only a few cases. One of these is the case of a central force diminishing inversely as the square of distance. The forces of Newtonian attraction between point masses (or bodies possessing spherical symmetry) are subject to this law.

It will be recalled that the laws of motion in this case were found empirically by Kepler before Newton deduced them from the equations of mechanics and the law of gravitation. It was the agreement of Newton's results with Kepler's laws that was the first verification of the truth of Newtonian mechanics. The problem of the motion of a particle in a field of force diminishing inversely with the square of the distance from some fixed point, is called Kepler's problem. The problem of the motion of two bodies with arbitrary masses always reduces to the problem of a single body when passing to a frame of reference fixed in the centre of mass.

The expression "Kepler's problem" can also be applied to Coulomb forces acting between point charges. These can either be forces of attraction or repulsion. In all cases we shall write $U = \frac{a}{r}$, where $a < 0$ for attractive forces and $a > 0$ for repulsive forces.

If we replace $\frac{M}{mr}$ in (5.10) by a new variable x, the integral in the Kepler problem is reduced to the form

$$\varphi = -\int \frac{dx}{\sqrt{-x^2 - \frac{2a}{M}x + \frac{2\mathscr{E}}{m}}} = \text{arc cos} \left. \frac{x + \frac{a}{M}}{\sqrt{\frac{a^2}{M^2} + \frac{2\mathscr{E}}{m}}} \right|_{x = \frac{M}{mr_0}}^{x = \frac{M}{mr}} .$$

At the lower limit, the expression inside the arc cos sign is equal to unity [as will be seen from (5.11)], since the lower limit was chosen on the condition that $dr = 0$. But arc cos $1 = 0$. Rearranging the result of integration and reverting to r, we obtain, after simple manipulations,

$$r = \frac{\frac{M^2}{am}}{-1 + \frac{M}{a}\sqrt{\frac{a^2}{M^2} + \frac{2\mathscr{E}}{m}} \cos \varphi} . \tag{5.12}$$

(5.12) represents the standard equation for a conic section, the eccentricity being equal to $\sqrt{1+\frac{2\mathscr{E}M^2}{ma^2}}$. As long as this expression is less than unity, the denominator in (5.12) cannot become zero, because $\cos\varphi \leqslant 1$. But this is true for $-\frac{ma^2}{2M^2} < \mathscr{E} < 0$. Thus, when $\mathscr{E} < 0$, the result is elliptical orbits. For this it is necessary that $a < 0$, i.e., that there be an attraction, otherwise (5.12) would lead to $r < 0$, which is senseless.

For $\mathscr{E} > 0$ the eccentricity is greater than unity and the denominator in (5.12) becomes zero for a certain $\varphi = \varphi\infty$. Thus, the orbit goes to infinity (a hyperbola). The direction of the asymptote is obtained by putting $r = \infty$ in (5.12). This requires that $\cos\varphi_\infty = \frac{a}{M}\frac{1}{\sqrt{\frac{a^2}{M^2}+\frac{2\mathscr{E}}{m}}}$.

The angle between the asymptotes is equal to $2\,\varphi\infty$, when the particles repel each other, and to $2\,(\pi - \varphi_\infty)$, when the particles attract. An example of a trajectory, when the forces are repulsive, is shown in Fig. 8, Sec. 6.

Exercise

Obtain the equation of the trajectory when $U = \frac{ar^2}{2}\ \mathscr{E} > 0$.

Sec. 6. Collision of Particles

The significance of collision problems. In order to determine the forces acting between particles, it is necessary to study the motion of particles caused by these forces. Thus, Newton's gravitational law was established with the aid of Kepler's laws. Here, the forces were determined from finite motion. However, infinite motion can also be used if one particle can, in some way, be accelerated to a definite velocity and then made to pass close to another particle. Such a process is termed "collision" of particles. It is not at all assumed, however, that the particles actually come into contact in the sense of "collision" in everyday life.

And neither is it necessary that the incident particle should be artificially accelerated in a machine: it may be obtained in ejection from a radioactive nucleus, or as the result of a nuclear reaction, or it may be a fast particle in cosmic radiation.

Two approaches are possible to problems on particle collisions. Firstly, it may be only the velocities of the particles long before the collision (before they begin to interact) that are given, and the problem is to determine only their velocities (magnitude and direction) after they have ceased to interact. In other words, only the result of the collision is obtained without a detailed examination of the process. In this case, some knowledge of the final state must be available (or specified) beforehand: it is not possible to determine, from the initial velocities alone, all the integrals of motion which characterize the collision, and, hence, it is likewise impossible to predict the final state. With this approach to collision problems, only the momentum and energy integrals are known.

However, another approach is possible: it is required to precalculate the final state where the precise initial state is given.

Let us first consider collisions by the first method. It is clear that if only the initial velocities of the particles are known, the collision is not completely determined: it is not known at what distance the particles were when they passed each other. This is why some quantity relating to the final state of the system must be given. Usually the problem is stated as follows: the initial velocities of the colliding particles and also the direction of velocity of one of them after the collision are specified. It is required to determine all the remaining quantities after the collision. In such a form the problem is solved uniquely. Six quantities are unknown, namely the six momentum components of both particles after the collision. The conservation laws provide four equalities: conservation of the scalar quantity (energy) and conservation of the three components of the vector quantity (momentum). Therefore, with six unknowns, it is necessary to specify two quantities which refer to the final state. They are contained in the determination of the unit vector which specifies the direction of the velocity of one of the particles; an arbitrary vector is defined by three quantities, but a unit vector, obviously, only by two. Actually, only the angle of deflection of the particles after the collision need be given, i.e., the angle which the velocity of the particle makes with the initial direction of the incident particle. The orientation (in space) of the plane passing through both velocity vectors is immaterial.

Elastic and inelastic collisions. A collision is termed *elastic*, if the initial kinetic energy is conserved when the particles separate after the collision at infinity, and *inelastic*, if, as a result of the collision, the kinetic energy changes at infinity. In nuclear physics, studies are very often made of collisions of a more general character, in which the nature of the colliding particles changes. These collisions are also inelastic. They are called nuclear reactions.

The laboratory and centre-of-mass frames of reference. When collisions are studied in the laboratory, one of the particles is usually at rest prior to collision. The frame of reference fixed in this particle (and in the laboratory) is termed the laboratory frame. However, it is more convenient to perform calculations in a frame of reference, relative to which the centre of mass of both particles is at rest. In accordance with the law of conservation of the centre of mass (3.18), it will also be at rest in its own frame after the collision. The velocity of the centre of mass, relative to the laboratory frame of reference, is

$$\mathbf{V}=\frac{m_1\mathbf{v}_0}{m_1+m_2}. \tag{6.1}$$

Here $\mathbf{v}_0$ is the velocity of the first particle (of mass m_1) relative to the second (with mass m_2). In so far as the second particle is at rest in the laboratory system, $\mathbf{v}_0$ is also the velocity of the first particle relative to this system.

The general case of an inelastic collision. The velocity of the first particle relative to the centre of mass is, according to the law of addition of velocities, equal to

$$\mathbf{v}_{10}=\mathbf{v}_0-\mathbf{V}=\frac{m_2\mathbf{v}_0}{m_1+m_2}, \tag{6.2}$$

and in the same system, the velocity of the second particle is

$$\mathbf{v}_{20}=-\mathbf{V}=-\frac{m_1\mathbf{v}_0}{m_1+m_2}. \tag{6.3}$$

Thus, $m_1\mathbf{v}_{10}+m_2\mathbf{v}_{20}=0$, as it should be in the centre-of-mass system.

In accordance with (3.17), the energy in the centre-of-mass system is

$$\mathscr{E}_0=\frac{m_1m_2}{2(m_1+m_2)}v_0^2\equiv\frac{m_0v_0^2}{2}. \tag{6.4}$$

Here, the reduced mass is indicated by a zero subscript, since in nuclear reactions it may change.

Let the masses of the particles obtained as a result of the reaction be m_3 and m_4, and the energy absorbed or emitted Q (the so-called "heat" of the reaction). If Q is the energy released in radiation, then, strictly speaking, one should take into account the radiated momentum (see Sec. 13). But it is negligibly small in comparison with the momenta of nuclear particles.

Thus, the law of conservation of energy must be written in the following form:

$$\frac{m_0v_0^2}{2}+Q=\frac{mv^2}{2}. \tag{6.5}$$

Here, $m=\dfrac{m_3m_4}{m_3+m_4}$ is the reduced mass of the particles produced in the nuclear reaction, and v is their relative velocity.

In order to specify the collision completely, we will consider that the direction of $\mathbf{v}$ is known, since the value of v is determined from (6.5). Then the velocity of each particle separately will be

$$\mathbf{v}_{30}=\frac{m_4\mathbf{v}}{m_3+m_4} \qquad \mathbf{v}_{40}=-\frac{m_3\mathbf{v}}{m_3+m_4}. \tag{6.6}$$

They satisfy the requirement $m_3\mathbf{v}_{30}+m_4\mathbf{v}_{40}=0$, i. e., the law of conservation of momentum in the centre-of-mass system, and give the necessary value for the kinetic energy

$$\frac{m_3 v_{30}^2}{2} + \frac{m_4 v_{40}^2}{2} = \frac{m v^2}{2}.$$

Now, it is not difficult to revert to the laboratory frame of reference. The velocities of the particles in this system will be

$$\left.\begin{aligned} \mathbf{v}_3 &= \mathbf{v}_{30} + \mathbf{V} = \frac{m_4 \mathbf{v}}{m_3 + m_4} + \frac{m_1 \mathbf{v}_0}{m_1 + m_2}, \\ \mathbf{v}_4 &= \mathbf{v}_{40} + \mathbf{V} = -\frac{m_3 \mathbf{v}}{m_3 + m_4} + \frac{m_1 \mathbf{v}_0}{m_1 + m_2}. \end{aligned}\right\} \qquad (6.7)$$

Equations (6.7) give a complete solution to the problem provided the direction of $\mathbf{v}$ is given.

Elastic collisions. The computations are simplified if the collision is elastic, for then $m_3 = m_1$, $m_4 = m_2$, $Q = 0$. It follows from (6.5) that the relative velocity changes only in direction and not in magnitude. Let us suppose that its angle of deflection χ is given. We take the axis Ox along $\mathbf{v}_0$, and let the axis Oy lie in the plane of the vectors $\mathbf{v}$ and $\mathbf{v}_0$ (which are equal in magnitude in the case of elastic collisions). Then

$$v_x = v_0 \cos \chi, \quad v_y = v_0 \sin \chi .$$

From (6.6), the components of particle velocity in the centre-of-mass system will, after collision, be correspondingly equal to

$$v_{10x} = \frac{m_2 v_0 \cos \chi}{m_1 + m_2}, \qquad v_{10y} = \frac{m_2 v_0 \sin \chi}{m_1 + m_2},$$

$$v_{20x} = -\frac{m_1 v_0 \cos \chi}{m_1 + m_2}, \quad v_{20y} = -\frac{m_1 v_0 \sin \chi}{m_1 + m_2}.$$

Since the velocity of the centre of mass is in the direction of the axis Ox, we obtain, from (6.7), the equations for the velocities in the laboratory frame of reference:

$$v_{1x} = \frac{(m_1 + m_2 \cos \chi)\, v_0}{m_1 + m_2}, \qquad v_{1y} = v_{10y} = \frac{m_2 v_0 \sin \chi}{m_1 + m_2},$$

$$v_{2x} = \frac{m_1 (1 - \cos \chi)\, v_0}{m_1 + m_2}, \qquad v_{2y} = v_{20y} = -\frac{m_1 v_0 \sin \chi}{m_1 + m_2}.$$

By means of these equations, the deflection angle θ of the first particle in the case of collision in the laboratory system can be related to the angle χ, (i. e., its deflection angle in a centre-of-mass system):

$$\operatorname{tg} \theta = \frac{v_{1y}}{v_{1x}} = \frac{m_2 \sin \chi}{m_1 + m_2 \cos \chi}. \qquad (6.8)$$

The "recoil" angle of the second particle θ' is defined as

$$\begin{aligned} \tan \theta' &= -\frac{v_{2y}}{v_{2x}} = \frac{\sin \chi}{1 - \cos \chi} = \cot \frac{\chi}{2}, \\ \theta' &= \frac{\pi}{2} - \frac{\chi}{2}. \end{aligned} \qquad (6.9)$$

The minus sign in the definition of tg θ' is chosen because the signs of v_{1y} and v_{2y} are opposite.

The case of equal masses. Equation (6.8) becomes still simpler if the masses of the colliding particles are equal. This is approximately true in the case of a collision between a neutron and proton. Then, from (6.8),

$$\tan\theta = \tan\frac{\chi}{2}, \qquad \theta = \frac{\chi}{2},$$

$$\theta' = \frac{\pi}{2} - \frac{\chi}{2}, \qquad \theta + \theta' = \frac{\pi}{2},$$

i.e., the particles fly off at right angles and the deflection angle of the neutron in the laboratory system is equal to half the deflection angle in the centre-of-mass system. Since the latter varies from 0 to 180°, θ cannot exceed 90°. And, in addition, the velocity of the incident particle is plotted as the "resultant" velocity of the diverging particles. The collision of billiard balls resembles the collision considered here of particles of equal mass, provided that the rotation of the balls about their axes is neglected.

The energy transferred in an elastic collision. The energy received by the second particle in a collision is

$$\mathscr{E}_2 = \frac{m_2 m_1^2 (1 - \cos\chi) v_0^2}{(m_1 + m_2)^2}.$$

Its portion, relative to the initial energy of the first particle, is

$$\frac{\mathscr{E}_2}{\mathscr{E}_0} = \frac{2 m_1 m_2 (1 - \cos\chi)}{(m_1 + m_2)^2}. \qquad (6.10)$$

From this we obtain $\frac{\mathscr{E}_2}{\mathscr{E}_0} = \sin^2\frac{\chi}{2} = \sin^2\theta$ for particles of equal mass. Accordingly, the portion of the energy retained by the first particle is $\frac{\mathscr{E}_1}{\mathscr{E}_0} = \cos^2\theta$. In a "head-on" collision $\chi = 180°$, $\theta = 90°$. The first particle comes to rest and the second continues to move forward with the same velocity. This can easily be seen when billiard balls collide.

The problem of scattering. Let us now examine the problem of collision in more detail. We shall confine ourselves to the case of elastic collision and perform the calculations in the centre-of-mass system. The transformation to the laboratory system by equations (6.7) is elementary.

It is obvious that for a complete solution of the collision problem, one must know the potential energy of interaction between the particles $U(r)$ and specify the initial conditions, so that all the integrals of motion may be determined. The angular-momentum integral is found in the following way. Fig. 8, which refers to repulsive forces, represents the motion of the first particle relative to the second. The path at an infinite distance is linear, because no forces act between the particles at infinity. Since the path is linear at infinity, it possesses asymptotes. The asymptote for the part of the trajectory over which the particles are approaching is represented by the straight line AF, and FB is the asymptote for the part where they recede.

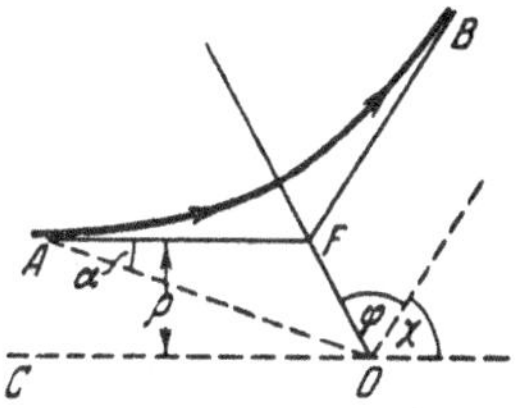

Fig. 8

The collision parameter. The distance of an asymptote from the straight line OC, drawn through the second point and parallel to the relative velocity of the particles at infinity, is called the *collision parameter* ("aiming distance"). It has been denoted by ρ, since, as can be seen from Fig. 8, ρ is also the "arm" of the angular momentum. If there were no interaction between the particles, they would pass each other at a distance ρ; this is why ρ is called the collision parameter. But we know that the angular momentum is very simply expressed in terms of ρ. In the preceding section it was shown that it is equal to $mv\rho$. Let us draw the radius vector OA to some very distant point A. Then the angular momentum is

$$M = mvr \sin \alpha$$

(the angle α is shown on the diagram). But $r \sin \alpha = \rho$, so that

$$M = mv\rho \, . \tag{6.11}$$

Recall that here m is the reduced mass of the particles and v is their relative velocity at infinity.

The energy integral is expressed in terms of the velocity at infinity thus:

$$\mathscr{E} = \frac{mv^2}{2}, \tag{6.12}$$

since $U(\infty) = 0$.

The deflection angle. The deflection angle χ is equal to $|\pi - 2\varphi_\infty|$, where φ_∞ is half the angle between the asymptotes. The angle φ_∞ corresponds to a rotation of the radius vector from the position OA, where it is infinite, to a position OF, where it is a minimum. Hence, from equation (5.10) the angle φ_∞ is expressed as

$$\varphi_\infty = \int_{r_0}^{\infty} \frac{M}{mr^2} \frac{dr}{\sqrt{\frac{2}{m}\left(\mathscr{E} - \frac{M^2}{2mr^2} - U(r)\right)}} \, ; \tag{6.13}$$

r_0 is determined from (5.11). In place of M and $\mathscr{E}$ we must substitute into (6.13) the expressions (6.11) and (6.12).

The differential effective scattering cross-section. Let us suppose that the integral (6.13) has been calculated. Then φ_∞, and therefore χ, are known as functions of the collision parameter ρ. Let this relationship be inverted, i.e., ρ is determined as a function of the deflection angle:

$$\rho = \rho(\chi). \tag{6.14}$$

In collision experiments, the collision parameter is never defined in practice; a parallel beam of scattering particles is directed with identical velocity at some kind of substance, the atoms or nuclei of which are scatterers. The distribution of particles as to deflection angles χ (or, more exactly, as to angles θ in the laboratory system) is observed. Thus a scattering experiment is, as it were, performed very many times one after the other with the widest range of aiming distances.

Let one particle pass through a square centimetre of surface of the scattering substance. Then, in an annulus contained between ρ and $\rho + d\rho$, there pass $2\pi\rho\, d\rho$ particles. We classify the collisions according to the aiming distances, similar to the way that it is done on a shooting target with the aid of a concentric system of rings. If ρ is known in relation to χ, then it may be stated that $d\sigma \equiv 2\pi\rho\, d\rho = 2\pi\rho \frac{d\rho}{d\chi}\, d\chi$ particles will be deflected at the angle between χ and $\chi + d\chi$.

Let us suppose that the scattered particles are in some way detected at a large distance from the scattering medium. Then the whole scatterer can be considered as a point and we can say that after scattering the particles move in straight lines from a common centre. Let us consider those particles which occupy the space between two cones that have the same apex and a common axis; the half-angle of the inner cone is equal to χ, and the external cone $\chi + d\chi$. The space between the two cones is called a solid angle, similar to the way that the plane contained between two straight lines is called a plane angle. The measure of a plane angle is the arc of a circle of unit radius drawn about the vertex of the angle, while the measure of a solid angle is the area of a sphere of unit radius drawn about the centre of the cone. An elementary solid angle is shown in Fig. 9 as that part of the surface of a sphere covered by an element of arc $d\chi$ when it is rotated about the radius OC. Since $OC = 1$, the radius of rotation of the element $d\chi$ is equal to $\sin\chi$. Therefore, the surface of the sphere which it covers is equal to $2\pi \sin\chi d\chi$. Thus, the elementary solid angle is

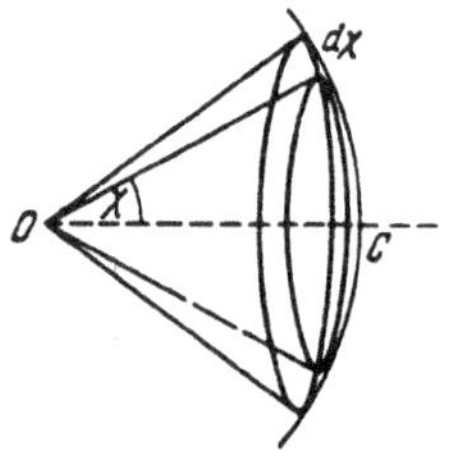

Fig. 9

$$d\Omega = 2\pi \sin \chi \, d\chi . \tag{6.15}$$

The number of particles scattered in the element of solid angle is, thus,

$$d\sigma = \rho \frac{d\rho}{d\chi} \frac{d\Omega}{\sin \chi} . \tag{6.16}$$

The quantity $d\sigma$ has the dimensions of area. It is the area in which a particle must fall in order to be scattered within the solid-angle element $d\Omega$. It is called the *effective differential scattering cross-section* in the element of solid angle $d\Omega$.

Experimentally we determine just this value, because it is the angular distribution of the scattered particles that is dealt with [in (6.16) we consider that ρ is given in relation to χ]. If there are n scatterers in unit volume of the scattering substance, then the attenuation of the primary beam J in passing through unit thickness of the substance, due to scattering in an elementary solid angle $d\Omega$, is

$$dJ_\Omega = -Jn\, d\sigma = -Jn\rho \frac{d\rho}{d\chi} \frac{d\Omega}{\sin \chi} \text{ particles/cm.}$$

If we examine $d\sigma$ as a function of χ, we find a relationship between the collision parameter and the deflection angle. And this allows us to draw certain conclusions about the nature of the forces acting between the particle and the scattering centre.

Rutherford's formula. A marvellous example of the determination of forces from the scattering law is given by the classical experiments of Rutherford with alpha particles. As was pointed out in Sec. 3, the Coulomb potential acting on particles decreases with distance according to a $\frac{1}{r}$ law, in the same way as the Newtonian potential. Consequently, the deflection angle can be calculated from the equations of Sec. 5. Let us first of all find the angle φ_∞. It can be determined from equation (5.12) by putting $r = \infty$, $a > 0$ (the charges on the nucleus and alpha particle are like charges). Hence,

$$\cos \varphi_\infty = \frac{1}{\sqrt{1 + \frac{2M^2\mathscr{E}}{ma^2}}}, \quad \tan \varphi_\infty = \frac{M}{a}\sqrt{\frac{2\mathscr{E}}{m}} . \tag{6.17}$$

The integrals of motion $\mathscr{E}$ and M are determined with the aid of (6.11) and (6.12). We therefore have

$$\rho = \frac{a}{mv^2} \operatorname{tg} \varphi = \frac{a}{mv^2} \cot \frac{\chi}{2} \tag{6.18}$$

$\left(\text{since } \varphi = \frac{\pi}{2} - \frac{\chi}{2}\right)$. We now form the equation for the effective differential scattering cross-section in the centre-of-mass system with the aid of (6.16):

$$d\sigma = \frac{a^2}{Am^2 v^4} \frac{d\Omega}{\sin^4 \frac{\chi}{2}} . \tag{6.19}$$

If the scattering nucleus is not too light, this equation, to a good approximation, also holds in the laboratory system.

Thus, the number of particles scattered in the elementary solid angle $d\Omega = 2\pi \sin \chi \, d\chi$, is inversely proportional to the fourth power of the sine of the deflection half-angle. This law is uniquely related to the Coulomb nature of the forces between the particles.

Studying the scattering of alpha particles by atoms, Rutherford showed that the law (6.19) is true for angles up to χ, corresponding to collision parameters less than 10^{-12} cm. It was thus experimentally proved that the whole mass of the atom is concentrated in an exceedingly small region (recall that the size of an atom is $\sim 10^{-8}$ cm.). Thus, experiments on the scattering of alpha particles led to the discovery of the atomic nucleus and to an estimation of the order of magnitude of its dimensions.

Isotropic scattering. As may be seen from equation (6.19), the scattering has a pronounced maximum for small deflection angles. This maximum relates to large aiming distances since particles passing each other at these distances are weakly deflected, while large distances predominate since they define a larger area. Thus, if the interaction force between the particles does not *identically* convert to zero at a finite distance, then, for small deflection angles, the expression for $d\sigma$ will always have a maximum. This maximum is the more pronounced, the more rapidly the interaction force decreases with distance, for in the case of a rapidly diminishing force, large aiming distances correspond to very small deflection angles.

However, particles that are very little deflected can in no way be detected experimentally as deflected particles. Indeed, the initial beam cannot be made ideally parallel. For this reason, when investigating a scattered beam, one must always neglect those angles which are comparable with the angular deviation of the particles in the initial beam from ideal parallelism.

For a sufficiently rapid attenuation of force with distance, the region of the maximum of $d\sigma$ in relation to the angle χ can refer to such small angles that the particles travelling within these angles will not be distinguished as being scattered because of their small deflection angles. On the other hand, the remaining particles will be the more uniformly distributed as to scattering angle, the more rapidly the forces fall off with distance.

This can be seen in the example of particles scattered by an impermeable sphere (Exercise 1). Such a sphere may be regarded as the limiting case of a force centre repulsing particles according to the law $U(r) = U_0 \left(\frac{r_0}{r}\right)^n$; when n tends to infinity: if $r < r_0$, then $U(r) \to \infty$, and if $r > r_0$, then $U(r) \to 0$. When $n = \infty$ the scattering is completely isotropic. If n is large, the angular distribution of the particle is almost isotropic, and only for very small deflection angles has the distribution a sharp maximum. Hence, a scattering law that is almost isotropic indicates a rapid diminution of force with distance.

The scattering of neutrons by protons in the centre-of-mass system is isotropic up to energy values greater than 10 Mev (1 Mev equals 1.6×10^{-6} erg). An analysis of the effective cross-section shows that nuclear forces are short-range forces; they are very great at close distances and rapidly diminish to zero at distances larger than 2×10^{-13} cm. It must be mentioned, however, that a correct investigation of this case is only possible on the basis of the quantum theory of scattering (Sec. 37).

Exercises

1) Find the differential effective scattering cross-section for particles by an impermeable sphere of radius r_0.

The impermeable sphere can be represented by giving the potential energy in the form $U(r) = 0$ for $r > r_0$ (outside the sphere) and $U(r) = \infty$ for $r < r_0$ (inside the sphere). Then, whatever the kinetic energy of the particle, penetration into the region $r < r_0$ is impossible.

In reflection from the sphere, the tangential component of momentum is conserved and the normal component changes sign. The absolute value of the momentum is conserved since the scattering is elastic. A simple construction shows us that the collision parameter is related to the deflection angle by

$$\rho = r_0 \cos \frac{\chi}{2},$$

if $\rho < r_0$. Hence the general equation gives

$$d\sigma = \frac{r_0^2}{4} d\Omega,$$

so that the scattering occurs isotropically for all angles. The total effective scattering cross-section σ is equal, in this case, to πr_0^2, as expected. Note that if the interaction converted to zero not at a finite distance but at infinity, the total scattering cross-section would tend to infinity since to any arbitrarily large approach distance ρ there would correspond a certain deflection angle, and the integral $\int 2\pi\rho\, d\rho$ diverges. In the quantum theory of scattering, σ is also finite when the forces diminish fast enough with distance.

2) The collision of particles with masses m_1 and m_2 is considered (the mass of the incident particle is m_1). As a result of the collision, particles with the same masses are obtained whose paths make certain angles φ and ψ with the initial flight direction of the particle of mass m_1. Determine the energy Q which is absorbed or emitted in the collision.

Sec. 7. Small Oscillations

In applications of mechanics, we very often meet a special form of motion known as small oscillations. We devote a separate section to the theory of small oscillations.

The definition of small oscillations of a pendulum. In the problem of pendulum oscillations in Sec. 4 it was shown that the equation relating the deflection angle φ to time led, in the general case, to a nonelementary (elliptical) integral (4.11). A simple graphical investigation shows that the function $\varphi(t)$ is periodic. Fig. 10 shows the curve $U(\varphi) = mgl(1 - \cos\varphi)$, which gives the relationship between potential energy and deflection angle. The horizontal straight line corresponds to a certain constant value of $\mathcal{E}$. If $\mathcal{E} < 2\,mgl$, the motion occurs periodically with time between the points $-\varphi_0$ and φ_0.

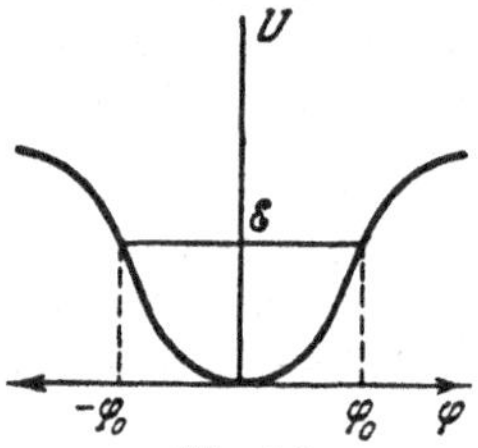

Fig. 10

The problem is greatly simplified if $\varphi_0 \ll 1$, i.e., the angle φ_0 is small in comparison with a radian. Then $\cos\varphi_0$ can be replaced by the expansion $1 - \frac{\varphi_0^2}{2}$. Since $|\varphi| < \varphi_0$, $\cos\varphi$ can also be replaced by $1 - \frac{\varphi^2}{2}$. After this the integral (4.11) can be easily evaluated:

$$t = -\sqrt{\frac{l}{g}} \int_{\varphi_0}^{\varphi} \frac{d\varphi}{\sqrt{\varphi_0^2 - \varphi^2}} = \sqrt{\frac{l}{g}} \arccos\frac{\varphi}{\varphi_0}. \tag{7.1}$$

Inverting relation (7.1), we get the angle as a function of time

$$\varphi = \varphi_0 \cos\sqrt{\frac{g}{l}}\, t. \tag{7.2}$$

The result is a periodic function. As can be seen from (7.2), the period φ is equal to $2\pi\sqrt{\frac{l}{g}}$. The quantity $\sqrt{\frac{g}{l}}$ is called the frequency of oscillation

$$\omega \equiv \sqrt{\frac{g}{l}}. \tag{7.3}$$

This quantity gives the number of radians by which the argument of the cosine in (7.2) changes in one second. Sometimes the term frequency denotes a quantity that is 2π times less and equal to the number of oscillations performed by the pendulum in one second. The inverse value $\frac{2\pi}{\omega}$, is the period of small oscillations of the pendulum. An important point is that the period and the frequency of small oscillations do not depend on the amplitude of oscillation φ_0.

The general problem of small oscillations with one degree of freedom. In order to solve the small-oscillation problem, we need not, initially, reduce to quadrature the problem of arbitrary oscillations; we can first perform an appropriate simplification of the Lagrangian.

First of all, we note that any oscillations, both large and small, always occur about a position of equilibrium. Thus, a pendulum oscillates about a vertical. On deflection from the position of stable equilibrium, a restoring force acts on the system in the opposite direction to the deflection. In the equilibrium position, this force obviously becomes zero (by definition of the "equilibrium" concept).

Force is equal to the derivative of potential energy with respect to the coordinate taken with opposite sign. The equilibrium condition written in terms of this derivative is

$$\frac{\partial U}{\partial q} = 0 . \tag{7.4}$$

Let us denote the solution of this equation by $q = q_0$. We assume, initially, that the system has only one degree of freedom and expand U in a Taylor series in the vicinity of the point q_0:

$$U(q) = U(q_0) + \left(\frac{\partial U}{\partial q}\right)_0 (q - q_0) + \frac{1}{2}\left(\frac{\partial^2 U}{\partial q^2}\right)_0 (q - q_0)^2 + \ldots \tag{7.5}$$

The linear term relative to $q - q_0$ vanishes in accordance with (7.4). We denote $\left(\frac{\partial^2 U}{\partial q^2}\right)_0$ by the letter β. Then, confining ourselves to these terms of the series, we obtain

$$U(q) = U(q_0) + \frac{\beta}{2}(q - q_0)^2 . \tag{7.6}$$

The force near the equilibrium position is

$$F(q) = -\frac{\partial U}{\partial q} = -\beta (q - q_0) . \tag{7.7}$$

For this force to be a restoring force (i.e., for it to act in the direction opposite to the deflection), the following inequality must hold:

$$\beta = \left(\frac{\partial^2 U}{\partial q^2}\right)_0 > 0 . \tag{7.8}$$

This is the stability condition for the equilibrium; the function $U(q)$ must increase on both sides of the point $q = q_0$, It follows that the potential energy at that point must be a minimum. This is shown in Fig. 10 at $\varphi = 0$.

Let us now examine the expression for kinetic energy. If, in the general formula for the kinetic energy of a particle,

$$T = \frac{m}{2}(\dot{x}^2 + \dot{y}^2 + \dot{z}^2)$$

we put $x = x(q)$, $y = y(q)$, $z = z(q)$, then T reduces to the form

$$T = \frac{m}{2}\left[\left(\frac{dx}{dq}\right)^2 + \left(\frac{dy}{dq}\right)^2 + \left(\frac{dz}{dq}\right)^2\right]\dot{q}^2 .$$

The quantity in the brackets depends only on q; and so the kinetic energy of a particle can be represented in the form

$$T=\frac{1}{2}\alpha(q)\dot{q}^2. \tag{7.9}$$

Let us now expand the coefficient $\alpha(q)$ in a series, in terms of $q=q_0$, in the vicinity of the equilibrium position:

$$T=\frac{1}{2}\alpha(q_0)\dot{q}^2+\frac{1}{2}\left(\frac{\partial\alpha}{\partial q}\right)_{q=q_0}(q-q_0)\dot{q}^2+\ldots$$

In order that the particle should not move far from the equilibrium position, its velocity must be small. In other words, the zero member of the kinetic energy expansion $\frac{1}{2}\alpha(q_0)\dot{q}^2$ is already of the same order of smallness for small oscillations as the second term in the expansion of $U(q)$, i.e., $\frac{\beta}{2}(q-q_0)^2$. When $q=q_0$ all the energy of oscillation is kinetic, while for maximum deflection all the energy is potential. Therefore $\frac{1}{2}\alpha(q_0)\dot{q}^2$ and $\frac{\beta}{2}(q-q_0)^2$ are of the same order of magnitude, and the remaining terms in the series [including those containing $(q-q_0)\dot{q}^2$] can be neglected. We shall show that the mean values of both the quantities $\frac{1}{2}\alpha(q_0)\dot{q}^2$ and $\frac{\beta}{2}(q-q_0)^2$ are the same after we determine q as an explicit function of time.

In future, the coordinate q will be measured from the equilibrium position, i.e., we shall put $q_0=0$. Then [omitting $U(0)$, which does not affect the equations of motion] the Lagrangian can be written in the following form:

$$L=\frac{1}{2}\alpha(0)\dot{q}^2-\frac{1}{2}\beta q^2. \tag{7.10}$$

Thus, Lagrange's equation will be written as

$$\alpha(0)\ddot{q}+\beta q=0. \tag{7.11}$$

Denoting

$$\omega^2\equiv\frac{\beta}{\alpha(0)}=\frac{\left(\frac{\partial^2 U}{\partial q^2}\right)_0}{\alpha(0)}, \tag{7.12}$$

we reduce (7.12) to the general form for the oscillation equation:

$$\ddot{q}+\omega^2 q=0. \tag{7.13}$$

Various forms for the solutions of small-oscillation problems. The general solution of this equation, which contains two arbitrary constants, may be written in one of three forms:

$$q = C_1 \cos \omega t + C_2 \sin \omega t, \tag{7.14a}$$

$$q = C \cos (\omega t + \gamma), \tag{7.14b}$$

$$q = \mathrm{Re} \{C' e^{i\omega t}\}. \tag{7.14c}$$

The symbol Re{} signifies the real part of the expression inside the braces. The constant C' inside the braces is complex: $C' = C_1 - iC_2$. The constants are chosen in accordance with the initial conditions. The constant γ is called the initial phase, and C is the amplitude.

If we are only interested in the frequency of small oscillations, and not the phase or amplitude, it is sufficient to use equation (7.12), verifying that the second derivative $\left(\frac{\partial^2 U}{\partial q^2}\right)_0$ is positive.

A system which is described by equation (7.13) is called a *linear harmonic oscillator.*

It can be seen from equations (7.10), (7.12), and (7.14b) that the averages of the potential energy and kinetic energy of the oscillator during one period are the same because the averages of the squares of a sine or cosine are equal to one half:

$$\overline{\sin^2 (\omega t + \gamma)} = \frac{\omega}{2\pi} \int_0^{\frac{2\pi}{\omega}} \sin^2 (\omega t + \gamma)\, dt = \frac{1}{2}; \quad \overline{\cos^2 (\omega t + \gamma)} = \frac{1}{2};$$

$$\overline{T} = \overline{U} = \frac{1}{4} \alpha (0)\, \omega^2 C^2 = \frac{1}{4} \beta C^2 .$$

Small oscillations with two degrees of freedom. We shall now consider oscillations with several degrees of freedom. As an example, let us first take the double pendulum of Sec. 3. If we confine ourselves to small oscillations, we must consider that the deflections φ and ψ are close to zero, (i.e., the pendulum is close to a vertical position). Then, by substituting the equilibrium values of the coordinates φ and ψ, $\cos (\varphi - \psi)$ in the kinetic energy must be replaced by $\cos 0 = 1$ as in the problem of oscillations with one degree of freedom where $\cos \varphi$ and $\cos \psi$, in the expression for potential energy, must be replaced by $1 - \frac{\varphi^2}{2}$ and $1 - \frac{\psi^2}{2}$. Then the Lagrangian will have the form

$$L = \frac{m + m_1}{2} l^2 \dot{\varphi}^2 + \frac{m_1}{2} l_1^2 \dot{\psi}^2 + m_1 l l_1 \dot{\varphi} \dot{\psi} - \frac{m + m_1}{2} l g \varphi^2 - \frac{m_1 l_1 g}{2} \psi^2. \tag{7.15}$$

Let us examine this in a somewhat more general form:

$$L = \frac{1}{2} (\alpha_{11} \dot{q}_1^2 + 2\alpha_{12} \dot{q}_1 \dot{q}_2 + \alpha_{22} \dot{q}_2^2) - U(0) - \frac{1}{2} (\beta_{11} q_1^2 + 2\beta_{12} q_1 q_2 + \beta_{22} q_2^2). \tag{7.16}$$

Here, the coefficients α_{11}, α_{12}, and α_{22} are assumed to be constant numbers expressed in terms of the equilibrium values q_1 and q_2. Comparing (7.15) and (7.16), we find that in the problem of the double pendulum

$$\alpha_{11}=(m+m_1)\,l^2, \quad \alpha_{12}=m_1\,ll_1, \quad \alpha_{22}=m_1\,l_1^2;$$

$$\beta_{11}=(m+m_1)\,lg, \quad \beta_{12}=0, \quad \beta_{22}=m_1 l_1 g.$$

In the general case, the coefficients β_{11}, β_{12} and β_{22} are expressed by the equations

$$\beta_{11}=\left(\frac{\partial^2 U}{\partial q_1^2}\right)_0, \quad \beta_{12}=\left(\frac{\partial^2 U}{\partial q_1 \partial q_2}\right)_0, \quad \beta_{22}=\left(\frac{\partial^2 U}{\partial q_2^2}\right)_0,$$

where the derivatives are also taken in the equilibrium positions. For the equilibrium to be stable, we must demand that the following inequality be satisfied:

$$U(q)-U(0)=\frac{1}{2}(\beta_{11}q_1^2+2\beta_{12}q_1q_2+\beta_{22}q_2^2)>0. \tag{7.17}$$

Under this condition, U has a minimum at the point $q_1=0$, $q_2=0$.

Let us rewrite the left-hand side of (7.17) in identical form

$$\frac{1}{2}(\beta_{11}q_1^2+2\beta_{12}q_1q_2+\beta_{22}q_2^2)=\frac{\beta_{11}}{2}\left(q_1+\frac{\beta_{12}q_2}{\beta_{11}}\right)^2+\frac{\beta_{22}\beta_{11}-\beta_{12}^2}{\beta_{11}}\,q_2^2.$$

This expression remains positive for all values of q_1 and q_2, provided the coefficients of both quadratics in q are greater than zero:

$$\beta_{11}>0, \tag{7.18}$$

$$\beta_{11}\beta_{22}-\beta_{12}^2>0. \tag{7.19}$$

In future, we shall consider that the conditions (7.18) and (7.19), together with analogous conditions for α_{11}, α_{12}, and α_{22}, are satisfied.

We shall now write down Lagrange's equations. We have

$$\frac{\partial L}{\partial \dot q_1}=\alpha_{11}\dot q_1+\alpha_{12}\dot q_2, \quad \frac{\partial L}{\partial \dot q_2}=\alpha_{12}\dot q+\alpha_{22}\dot q_2;$$

$$-\frac{\partial L}{\partial q_1}=\beta_{11}q_1+\beta_{12}q_2, \quad -\frac{\partial L}{\partial q_2}=\beta_{12}q_1+\beta_{22}q_2.$$

Whence

$$\left.\begin{aligned}\alpha_{11}\ddot q_1+\alpha_{12}\ddot q_2+\beta_{11}q_1+\beta_{12}q_2=0,\\ \alpha_{12}\ddot q_1+\alpha_{22}\ddot q_2+\beta_{12}q_1+\beta_{22}q_2=0.\end{aligned}\right\} \tag{7.20}$$

In order to satisfy these equations, we shall look for a solution in the form

$$q_1=A_1e^{i\omega t}, \quad q_2=A_2e^{i\omega t}. \tag{7.21}$$

As in (7.14c), the real part of the solution (7.21) must be taken.

The equation for frequency. Substituting (7.21) in (7.20), we obtain equations relating A_1 and A_2:

$$\left.\begin{aligned}(\beta_{11}-\alpha_{11}\omega^2)A_1+(\beta_{12}-\alpha_{12}\omega^2)A_2=0\,,\\(\beta_{12}-\alpha_{12}\omega^2)A_1+(\beta_{22}-\alpha_{22}\omega^2)A_2=0\,.\end{aligned}\right\}\qquad(7.22)$$

Transferring terms in A_2 to the right-hand side of the equation and dividing one equation by the other, we eliminate A_1 and A_2:

$$\frac{\beta_{11}-\alpha_{11}\omega^2}{\beta_{12}-\alpha_{12}\omega^2}=\frac{\beta_{12}-\alpha_{12}\omega^2}{\beta_{22}-\alpha_{22}\omega^2}\,.\qquad(7.23)$$

Reducing (7.23) to a common denominator, we arrive at the biquadratic equation

$$(\alpha_{11}\alpha_{22}-\alpha_{12}^2)\,\omega^4-(\beta_{11}\alpha_{22}+\beta_{22}\alpha_{11}-2\alpha_{12}\beta_{12})\,\omega^2+\\+\beta_{11}\beta_{22}-\beta_{12}^2=0.\qquad(7.24)$$

Substituting here the expressions for α_{ik}, β_{ik} from (7.15), we obtain an equation for the frequencies of a double pendulum

$$mm_1l^2l_1^2\omega^4-(m+m_1)\,m_1ll_1\,(l_1+l)\,g\omega^2+(m+m_1)\,m_1ll_1g^2=0.$$

If we introduce still another contraction in notation (for the given problem) $\frac{l_1}{l}=\lambda$, $\frac{m_1}{m}=\mu$, the expression for frequencies will be of the following form:

$$\omega^2=\frac{g}{2l\lambda}\left[(1+\mu)(1+\lambda)\pm\sqrt{(1+\mu)^2(1+\lambda)^2-4\lambda(1+\mu)}\,\right].$$

It is easy to see that this expression yields only the real values of the frequencies. However, we shall show this in more general form for equation (7.24). Let us assume that the following function is given:

$$F(\omega^2)\equiv(\alpha_{11}\alpha_{22}-\alpha_{12}^2)\,\omega^4-(\beta_{11}\alpha_{22}+\beta_{22}\alpha_{11}-2\beta_{12}\alpha_{12})\,\omega^2+\beta_{11}\beta_{22}-\beta_{12}^2,$$

which passes through zero for all values of ω that satisfy equation (7.24). $F(\omega^2)$ is positive for $\omega^2=0$ and for $\omega^2=\infty$, since $\beta_{11}\beta_{22}-\beta_{12}^2>0$, $\alpha_{11}\alpha_{22}-\alpha_{12}^2>0$. Let us now substitute into this function the positive number $\omega^2=\frac{\beta_{11}}{\beta_{22}}$. After a simple rearrangement we obtain

$$\alpha_{11}^2\,F\left(\frac{\beta_{11}}{\alpha_{11}}\right)=-(\alpha_{12}\beta_{22}-\beta_{12}\alpha_{22})^2\leqslant 0\,.$$

Thus, as ω^2 varies from 0 to ∞, $F(\omega^2)$ is first positive, then negative, and then again positive. Hence, it changes sign twice, so that equation (7.24) has two positive roots ω_1^2, ω_2^2 and, as was asserted, all the values for frequency are real.

The quantity ω has four values, both pairs of which are equal in absolute value. If we represent the solution in the form (7.21), it is sufficient to take only positive ω.

Normal coordinates. Let us put these roots in (7.22). To each of them there will correspond a definite ratio of the coefficients $\frac{A_2}{A_1}$. For $i=1,\ 2$ we have

$$\zeta_i = \frac{A_2^{(i)}}{A_1^{(i)}} = -\frac{\beta_{11}-\alpha_{11}\omega_i^2}{\beta_{12}-\alpha_{12}\omega_i^2} \qquad (i=1,\ 2)\,. \tag{7.25}$$

According to (7.23), the same ratio is also obtained from the second equation of (7.22). For example, for the double pendulum $\zeta_i = \frac{\omega_i^2}{g/l-\lambda\omega_i^2}$; i is equal to 1 or 2 depending on the sign in front of the root in the solution for ω^2.

Each frequency ω_i defines one partial solution of the system (7.20). Since the system is linear, the general solution is the sum of these particular solutions. Let us write this as

$$\begin{aligned} q_1 &= A_1^{(1)}e^{i\omega_1 t} + A_1^{(2)}e^{i\omega_2 t}\,, \\ q_2 &= \zeta_1 A_1^{(1)}e^{i\omega_1 t} + \zeta_2 A_1^{(2)}e^{i\omega_2 t}\,. \end{aligned} \tag{7.26}$$

We must, of course, take only the real parts of the expressions on the right.

We now introduce the following notation:

$$A_1^{(1)}e^{i\omega_1 t} \equiv Q_1\,, \quad A_1^{(2)}e^{i\omega_2 t} \equiv Q_2\,. \tag{7.27}$$

According to (7.27), the quantities Q_1 and Q_2 satisfy the differential equations

$$\ddot{Q}_1 + \omega_1^2 Q_1 = 0; \ \ddot{Q}_2 + \omega_2^2 Q_2 = 0. \tag{7.28}$$

Each of these equations can be obtained from the Lagrangian

$$L_i = \frac{1}{2}\dot{Q}_i^2 - \frac{1}{2}\omega_i^2 Q_i^2\,, \tag{7.29}$$

which describes oscillation with one degree of freedom.

Thus, in terms of the variables Q_i, the problem of two related oscillations with two degrees of freedom q_1, q_2 has been reduced to the problem of two independent harmonic oscillations with one degree of freedom Q_1 and Q_2. The coordinates Q_1 and Q_2 are termed *normal*.

In equations (7.20), we cannot arbitrarily put $q_1=0$ or $q_2=0$: if the quantity q_1 oscillates, then it must cause q_2 to oscillate. In contrast, the oscillations of the quantities Q_1 and Q_2 are in no way related [as long as we limit ourselves to the expansion (7.16) for L].

From equations (7.26), we can express Q_1 and Q_2 in terms of q_1 and q_2:

$$Q_1 = \frac{\zeta_2 q_1 - q_2}{\zeta_2 - \zeta_1}, \qquad Q_2 = \frac{\zeta_1 q_1 - q_2}{\zeta_1 - \zeta_2}. \tag{7.30}$$

If, for example, we choose the initial values of q and $\dot{q}$ so that at this instant $Q_1 = 0$ and $\dot{Q}_1 = 0$, then the oscillation with frequency ω_1 will not occur at all. For this it is sufficient, at $t = 0$, to take the coordinates and velocities in a relationship such that $\zeta_2 q_1 - q_2 = 0$ and $\zeta_2 \dot{q}_1 - \dot{q}_2 = 0$. In other words, only the frequency ω_2 will occur, and the oscillations will be strictly periodic. When both frequencies ω_1 and ω_2 are excited they are generally speaking incommensurable, i.e., their ratio cannot be expressed as a rational fraction), the oscillation q is no longer periodic, since the sum of two periodic functions with incommensurable periods is not periodic.

Expressing energy in normal coordinates. From the form of the Lagrangian (7.29) it can be immediately concluded that the expression for energy in normal coordinates reduces to the form

$$\mathscr{E} = \frac{1}{2} \sum_i (\dot{Q}_i^2 + \omega_i^2 Q_i^2), \tag{7.31}$$

since $L = T - U$ and $\mathscr{E} = T + U$. This result is true for small oscillations with any number of degrees of freedom.

We must note that if the normal coordinates are expressed directly by equations (7.30), then the separate energy terms $\frac{1}{2}\,(\dot{Q}_i^2 + \omega_i^2 Q_i^2)$, will also be multiplied by certain numbers a_i. However, if we replace Q_i by $Q_i \sqrt{a_i}$, then these numbers are eliminated from the expression for energy, which is then reduced to the form (7.31). An example of this procedure is given in the exercises.

Thus, the energy of any system performing small oscillations is reduced to the sum of the energies of separate, independent linear harmonic oscillators. As a result of this, consideration of oscillation problems is greatly simplified since the linear harmonic oscillator is, in many respects, one of the most simple mechanical systems.

The reduction to normal coordinates turns out to be a very fruitful method in studies of the oscillations of polyatomic molecules, in the theory of crystals, and in electrodynamics. In addition, normal coordinates are useful in technical applications of oscillation theory.

The case of equal frequencies. If the roots of equation (7.24) coincide, the general solution must not be written in the form (7.26), but somewhat differently, namely,

$$\left.\begin{aligned} q_1 &= A \cos \omega t + B \sin \omega t, \\ q_2 &= A' \cos \omega t + B' \sin \omega t. \end{aligned}\right\} \tag{7.32}$$

Four arbitrary constants appear in this solution, and this is as it should be in a system with two degrees of freedom.

An example of such a system is a pendulum suspended by a string instead of a hinge. In the approximation (7.32), it turns out that the pendulum describes an ellipse centred about the equilibrium position. Account taken of the subsequent terms in the expansion of the potential energy in powers of deflection shows that the axes of the ellipse do not remain stationary, but rotate.

Exercise

Find the natural frequencies and normal oscillations of a double pendulum, taking the ratios of load masses $\mu = 3/4$ and the rod lengths $\lambda = 5/7$.

From the equation for the oscillation frequencies of a double pendulum, we obtain $\omega_1^2 = \frac{7}{2}\frac{g}{l}$, $\omega_2^2 = \frac{7}{10}\frac{g}{l}$. Further, $\zeta_1 = -7/3$, $\zeta_2 = 7/5$.

Let us now write down the expression for kinetic energy. For simplicity, we write $l = g = m = 1$ so that only ratios of λ and μ will appear in all the equations. This gives $\alpha_{11} = 1 + \mu = 7/4$, $\alpha_{12} = \mu\lambda = 15/28$, $\alpha_{22} = \mu\lambda^2 = 75/196$; $\beta_{11} = 1 + \mu = 7/4$, $\beta_{12} = 0$, $\beta_{22} = \mu\lambda = 15/28$. Let us determine the coefficients α_i. To do this we must calculate the kinetic energy

$$2T = \frac{7}{4}(\dot{Q}_1 + \dot{Q}_2)^2 + \frac{15}{14}(\dot{Q}_1 + \dot{Q}_2)\left(-\frac{7}{3}\dot{Q}_1 + \frac{7}{5}\dot{Q}_2\right) + \\ + \frac{75}{196}\left(-\frac{7}{3}\dot{Q}_1 + \frac{7}{5}\dot{Q}_2\right)^2 = \frac{4}{3}\dot{Q}_1^2 + 4\dot{Q}_2^2.$$

Consequently, we must put $a_1 = \frac{\sqrt{3}}{2}$, $a_2 = 1/2$.

Denoting $\frac{Q_1\sqrt{3}}{2}$ and $\frac{Q_2}{2}$ again by the letters Q_1 and Q_2, we have the expression for potential energy

$$2U = \frac{7}{4}\left(\frac{\sqrt{3}}{2}Q_1 + \frac{1}{2}Q_2\right)^2 + \frac{15}{28}\left(-\frac{7}{2\sqrt{3}}Q_1 + \frac{7}{10}Q_2\right)^2 = \frac{7}{2}Q_1^2 + \frac{7}{10}Q_2^2,$$

as it should be according to (7.10). The generalized coordinates are related to the normal coordinates by

$$Q_1 = \frac{5\sqrt{3}}{28}\left(\frac{7}{5}\varphi - \psi\right), \quad Q_2 = \frac{5\sqrt{3}}{28}\left(\frac{7}{\sqrt{3}}\varphi + \sqrt{3}\,\psi\right).$$

Thus, if $7\varphi = -3\psi$, and $7\dot{\varphi} = -3\dot{\psi}$ initially, then we have $Q_2 = 0$ for all time, so that both pendulums oscillate with one frequency ω_1, with the constant relationship between the deflection angles $7\varphi = -3\psi$ holding all the time. Both pendulums are deflected to opposite sides of the vertical. The other normal oscillation, with frequency ω_2, occurs for a constant angular relationship $7\varphi = 5\psi$.

Sec. 8. Rotating Coordinate Systems. Inertial Forces

The equivalence of inertial coordinate systems. The particular significance of inertial coordinate systems in mechanics was pointed out in Sec. 2. In such systems, all accelerations are produced by interaction between bodies. It is impossible to find a strictly inertial system in nature (any system is noninertial if the motions of bodies in it are observed over a sufficiently long period of time).

In the exercise at the end of this section we shall consider the Foucault pendulum, whose plane of oscillation rotates with a speed depending only on the geographical latitude of its location. This rotation cannot be explained by an interaction with the earth, because the gravitational force cannot make the pendulum rotate from east to west instead of from west to east.* However, if we consider several oscillations, then the rotation of the plane is still insignificant and can be ignored. Then it is sufficient to consider that gravity alone is acting on the pendulum and that the coordinate system fixed in the earth is approximately inertial over a period of several oscillations.

The concept of an inertial system is meaningful as an approximation and is a very convenient idealization in mechanics. In such a coordinate system, the interaction forces are measured by the accelerations of the bodies.

Let a coordinate system be defined for which it is known that, to the required degree of accuracy, it can be regarded as inertial. Then another coordinate system, moving uniformly relative to it, is also inertial within the same degree of accuracy. Indeed, if all the accelerations in the first system are due to interaction forces between bodies, then no additional accelerations can appear in the second system either. Therefore, both systems are inertial. Either of them may be considered at rest and the other moving, since motion is always relative.

The principle of relativity. One of the basic principles of mechanics is that all laws of motion have an identical form in all inertial coordinate systems, since these systems are, physically, completely equivalent. This principle of the equivalence of all inertial systems is known as the relativity principle, for it is connected with the relativity of motion.

It should be noted that this in no way signifies that inertial and noninertial coordinate systems are equivalent: in the latter, not all the accelerations can be reduced to interaction forces, so that there is no physical equivalence between two such systems.

Mathematically, the principle of relativity is expressed by the fact that equations of motion for one inertial system preserve their form after the variables have been transformed to another inertial system.

The equations for the transformation from one inertial system to another can be obtained only on the basis of certain physical

* In this case the plane of oscillation must pass through the vertical, since, otherwise, the pendulum would have an initial angular momentum relative to the vertical and would describe an ellipse whose semiaxes rotate (see end of Sec. 7).

assumptions. In Newtonian mechanics it is always taken that the interaction forces between bodies, in particular, gravitational forces, are transmitted instantaneously over any distance. Thus, the displacement of any body immediately transmits a certain momentum to any other body, no matter where it is located. As a result, a clock located in a certain inertial system can be instantaneously synchronized with a clock moving in another inertial system. Thus, in Newtonian mechanics, time is considered universal. In transforming from one inertial system to another (the latter with a velocity V relative to the former) it is taken that the time t is the same in both systems. Later on we shall see that this assumption is approximate and holds only when the relative velocity of the systems is considerably less than the velocity of light.

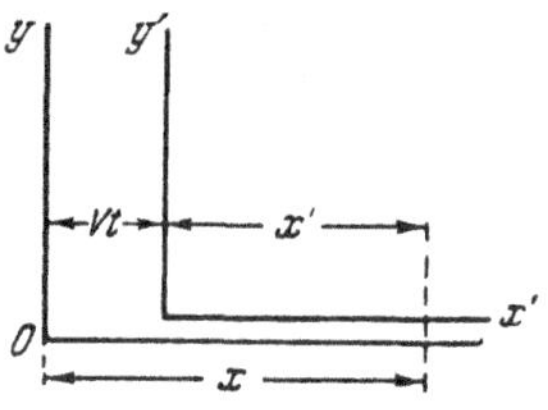

Fig. 11

The Galilean transformation. Let us construct coordinate systems in two inertial frames of reference such that their abscissae are in the direction of the relative velocity V and the other coordinate axes are also mutually parallel. Then from Fig. 11 it will be immediately seen that the abscissa of point x in the system which we shall call stationary is related to the abscissa in the moving system by the simple relation

$$x = x' + Vt, \tag{8.1}$$

provided that the origins coincided at the instant $t = 0$. The coordinate construction does not impose any limitations on the generality of the transformation equations. The remaining transformations lead simply to the identities

$$y = y', \quad z = z'. \tag{8.2}$$

The relationship $t = t'$ is a hypothesis which is correct only for values of V considerably less than the velocity of light (Sec. 20).

Condition (8.1) is absolutely symmetrical with respect to both inertial systems: if we consider that the one in which the variables are primed is stationary and the other in motion, (8.1) retains the same form; one should, of course, replace V by $-V$. In the given case, symmetry exists because $t' = t$. If $t \neq t'$, the transformation equations $x = x' + Vt$ and $x' = x - Vt'$ would contradict each other. But it would seem that equation (8.1) is obtained, quite obviously, from Fig. 11. Thus, if we do not consider time as identical for all inertial systems, the mathematical formulation of the relativity principle should be more complicated than that obtained on the basis of equation (8.1); and, we must definitely give up this "obviousness," which is so rooted in our everyday experience with velocities that are small compared with the velocity of light.

The equations of Newtonian mechanics involve, on the right-hand side, the forces of interaction between particles. These forces depend on the relative coordinates of the particles and, for this reason, they do not change with transformation (8.1), since Vt is cancelled in the formation of differences between the coordinates of any pair of particles. The left-hand sides of the equations contain accelerations, i.e., the second derivatives of the coordinates with respect to time. But since time enters linearly in (8.1) and is the same in both systems, $\ddot{x}=\ddot{x}'$. Thus, the equations of mechanics are of identical form in any inertial frame of reference.

To summarize, the equations of mechanics do not change their form when the variables undergo transformations (8.1). In other words, it is common to say that the equations of mechanics are *invariant* to these transformations, which are usually called Galilean transformations.

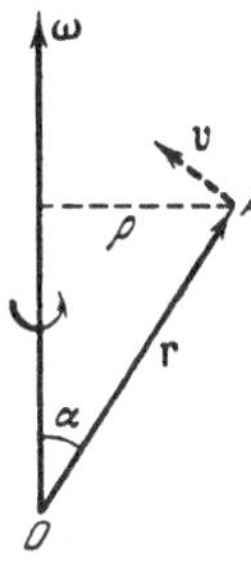

Fig. 12

The constancy (invariance) of mechanical laws under Galilean transformations is the essence of the relativity principle of Newtonian mechanics.

Here we must bear in mind that the relativity principle, which expresses the equivalence of all inertial coordinate systems, expresses a far more general law of nature than the approximate equations of transformation (8.1), (8.2). The extension of the relativity principle to electromagnetic phenomena involves the replacement of these equations by more general ones, which reduce to the former equations only when all velocities are much smaller than that of light.

Rotating coordinate systems. Several new terms appear in the equations of mechanics when transforming to rotating coordinate systems. Let us first obtain the equations for this transformation.

In Fig. 12 the axis of rotation is represented by a vertical line. The origin 0 is on the axis of rotation. Let $\mathbf{r}$ be the radius vector of a point A rotating around the axis. Then, for a rotation angular velocity ω (radians per second) the linear speed of the point will be

$$v=\omega \cdot r \sin \alpha\,, \tag{8.3}$$

since the radius of rotation is $\rho=r \sin \alpha$ (see Fig. 12). Let the rotation be anticlockwise. If point A lies in the plane of the paper, then the velocity $\mathbf{v}$ is perpendicular to the plane of the paper and directed towards the back of the paper. This permits us to obtain a relationship between the linear and angular velocities in vector form. We represent the angular velocity by a vector directed along the axis of rotation and associated with the direction of rotation by the corkscrew rule.

Then, if the rotation occurs in an anticlockwise direction, the vector $\boldsymbol{\omega}$ is directed upwards from the paper. From this it follows that

$$\mathbf{v} = [\boldsymbol{\omega} \mathbf{r}] . \tag{8.4}$$

This expression ensures a correct magnitude and direction for the linear velocity of the point.

Let us assume that point A, in addition to rotation, is somehow displaced relative to the origin 0 with velocity $\mathbf{v}' = \dot{\mathbf{r}}$. The resultant velocity of the point relative to a nonrotating system will be represented as the sum $\mathbf{v}' + \mathbf{v}$. The kinetic energy of the point relative to the nonrotating system is $\frac{m}{2}(\mathbf{v} + \mathbf{v}')^2$, and the Lagrangian is

$$L = \frac{m}{2}(\mathbf{v} + \mathbf{v}')^2 - U(\mathbf{r}) = \frac{m}{2}(\mathbf{v}' + [\boldsymbol{\omega} \mathbf{r}])^2 - U(\mathbf{r}) . \tag{8.5}$$

Let us now write down Lagrange's equations for motion relative to a rotating system, i.e., considering $\mathbf{r}$ a generalized coordinate. In order to do this we must calculate the derivatives $\frac{\partial L}{\partial \dot{\mathbf{r}}}$ and $\frac{\partial L}{\partial \mathbf{r}}$; let it be noted that differentiation with respect to a vector denotes a shortened way of writing down the differentiation with respect to all of its three components. The general rules for such differentiations will be given in Sec. 11; here we shall calculate the derivatives for each component separately.

Let ω be along the direction of the z-axis. Then, in vector components, L will be of the form

$$L = \frac{m}{2}[(\dot{x} - \omega y)^2 + (\dot{y} + \omega x)^2 + \dot{z}^2] - U(x, y, z) . \tag{8.6}$$

Whence we obtain

$$\frac{\partial L}{\partial \dot{x}} = m(\dot{x} - \omega y) , \quad \frac{\partial L}{\partial \dot{y}} = m(\dot{y} + \omega x) , \quad \frac{\partial L}{\partial \dot{z}} = m\dot{z} ;$$

$$\frac{\partial L}{\partial x} = m\omega(\dot{y} + \omega x) - \frac{\partial U}{\partial x} , \quad \frac{\partial L}{\partial y} = -m\omega(\dot{x} - \omega y) - \frac{\partial U}{\partial y} ,$$

$$\frac{\partial L}{\partial z} = -\frac{\partial U}{\partial z} .$$

Lagrange's equations in component form appear thus:

$$m(\ddot{x} - \omega \dot{y}) - m\omega(\dot{y} + \omega x) - m\dot{\omega} y + \frac{\partial U}{\partial x} = 0 ,$$

$$m(\ddot{y} + \omega \dot{x}) + m\omega(\dot{x} - \omega y) + m\dot{\omega} x + \frac{\partial U}{\partial y} = 0 ,$$

$$m\ddot{z} + \frac{\partial U}{\partial z} = 0 .$$

Let us leave on the left only the second derivatives and rewrite the last three equations as a single vector equation:

$$m\ddot{\mathbf{r}} = m\,[\mathbf{r}\,\dot{\boldsymbol{\omega}}] + 2m\,[\dot{\mathbf{r}}\boldsymbol{\omega}] + m\,[\boldsymbol{\omega}\,[\mathbf{r}\boldsymbol{\omega}]] - \frac{\partial U}{\partial \mathbf{r}}. \tag{8.7}$$

Expanding the double vector product on the right by means of the equation $[\mathbf{A}\,[\mathbf{BC}]] = \mathbf{B}\cdot(\mathbf{AC}) - \mathbf{C}\,(\mathbf{AB})$, and transforming to components, we can see that (8.7) is equivalent to the preceding system of three equations. A direct differentiation with respect to the vectors $\mathbf{r}$ and $\dot{\mathbf{r}}$ would have led to (8.7), without the expression in terms of components.

Inertial forces. The first three terms on the right in (8.7) essentially distinguish the equations of motion, written relative to a rotating coordinate system, from the equations written relative to a non-rotating system.

The use of a noninertial system is determined by the nature of the problem. For example, if the motion of terrestrial bodies is being studied, it is natural to choose the earth as the coordinate system, and not some other system related to the Galaxy (the aggregate of stars in the Milky Way). If we consider the reaction of a passenger to a train that suddenly stops, we must take the train as frame of reference and not the station platform. When the train is braked sharply, the passenger continues to move forwards "inertially" or, as we have agreed to say, he continues to move uniformly relative to an inertial system attached to the earth. Thus, relative to the carriage, it is the familiar jerk forward. At the same time it is obvious that the noninertial system is the train and not the earth, since no one experiencies any jerk on the platform.

The additional terms on the right of equation (8.7) have the same origin as the jerk when the train stopped; they are produced by noninertiality (in the given case, rotation) of the coordinate system. Naturally, the acceleration of a point caused by noninertiality of the system is absolutely real, relative to that system, in spite of the fact that there are other, inertial, systems relative to which this acceleration does not exist. In equation (8.7) this acceleration is written as if it were due to some additional forces. These forces are usually called *inertial forces*. In so far as the acceleration associated with them is in every way real, the discussion (which sometimes arises) about the reality of inertial forces themselves must be considered as aimless. It is only possible to talk about the difference between the forces of inertia and the forces of interaction between bodies.

But if we consider the force of Newtonian attraction, we cannot ignore the striking fact that, like the forces of inertia, it is proportional to the mass of the body. As a result of this, the equations of mechanics can be formulated in such a way that the difference between gravitational forces and inertial forces does not at all appear in the equations; all these forces turn out to be physically equivalent. However, this formulation is, of course, connected with a re-evaluation and a substantial revision of the basis of mechanics. It is the subject of Einstein's general theory of relativity, which is discussed in somewhat more detail at the end of Sec. 20.

Coriolis force. Let us now consider in more detail the inertial forces appearing in (8.7), which are due to a rotating coordinate system.

The first term in (8.7) occurs as a result of nonconstancy of angular velocity. It will not interest us. The second term is called the Coriolis force. For a Coriolis force to appear, the velocity of a point relative to a rotating coordinate system must have a projection, other than zero, on a plane perpendicular to the axis of rotation. This velocity projection can, in turn, be separated into two components: one, perpendicular to the radius drawn from the axis of rotation to the moving point, and the other, directed along the radius. The most interesting, as to its action, is the component of the Coriolis force due to the radial component of velocity. It is perpendicular both to the radius and to the axis of rotation. If a body moves perpendicularly to a radius, then its Coriolis acceleration is radial, and therefore analogous in its action to the centripetal acceleration which will be considered a little further on.

We note that the Coriolis force cannot be related, even formally, to the gradient of a potential function U.

There are many examples of the deflecting action of the Coriolis force in nature. The water of rivers in the Northern Hemisphere which flow in the direction of the meridian, i.e., from north to south, or from south to north, experience a deflection towards the right-hand bank (if we are looking in the direction of flow). This is why the right-hand bank of such rivers is steeper than the left. It is easy to form the corresponding component of the Coriolis force. The angular-velocity vector of the earth's rotation is directed along the earth's axis, "upwards" from the north pole. The waters of a river, which flows southwards at the mean latitudes of the Northern Hemisphere, have a velocity component perpendicular to the earth's axis and directed away from the axis. This means that the Coriolis acceleration of the water, relative to the earth, is in a westerly direction or, relative to a river flowing southwards, to the right. If the river flows in a northerly direction, the deflection will be towards the east, i.e., again to the right. In the southern hemisphere the deflection occurs leftwards.

The warm Gulf Stream which flows northwards is deflected towards the east, which is of tremendous importance for the climate of Europe. In general, the Coriolis force considerably affects the motion of air and water masses on the earth, though when compared in magnitude with the gravitational force it is very insignificant. Indeed, the angular velocity of the earth, as it completes one rotation about its axis in 24 hours, is a little less than 10^{-4} rad/sec, while the velocity of a particle of water or air can be taken as having an order of magnitude of 10^2 cm/sec. From this the Coriolis acceleration has an order of magnitude of 10^{-2} cm/sec^2, which is one hundred thousand times less than the acceleration caused by the force of gravity.

The Coriolis force also causes the rotation of the plane of oscillation of a Foucault pendulum. With the aid of the Foucault pendulum, we can prove the rotation of the earth about its axis without astronomical observations. In a nonrotating system, the plane of oscillation must be invariable in accordance with the law of conservation of angular momentum.

Centrifugal force. The third vector term in equation (8.7) is the usual centrifugal force. Indeed, it is perpendicular to the axis of rotation and, in absolute value, is equal to

$$|m[\boldsymbol{\omega}[\boldsymbol{\omega}\mathbf{r}]]| = m\boldsymbol{\omega}\,|[\boldsymbol{\omega}\mathbf{r}]| = m\omega(\omega r \sin\alpha) = m\omega^2 r \sin\alpha. \quad (8.8)$$

Here, the first equality takes account of the fact that the vectors $\boldsymbol{\omega}$ and $[\boldsymbol{\omega}\mathbf{r}]$ are perpendicular to each other, so that the absolute value of the vector product is equal to the product of their absolute values.

But $r\sin\alpha$ is equal to the distance from the axis of rotation, so that this force satisfies the usual definition of a centrifugal force.

Exercise

Let us consider the rotation of the plane of oscillation of a Foucault pendulum under the action of the earth's rotation about its axis.

The axis Ox at a given point on the earth is drawn in a northerly direction and the axis Oy in an easterly direction. Then, if $\omega_B = \omega \sin\theta$, where θ is the latitude of the locality, we have the equation of motion

$$\ddot{x} = -\omega_0^2 x - 2\dot{y}\omega_B, \quad \ddot{y} = -\omega_0^2 y + 2\dot{x}\omega_B, \quad \omega_0^2 = \frac{g}{l}.$$

Multiplying the first equation by y and the second by x and then subtracting, we get

$$\frac{d}{dt}(y\dot{x} - x\dot{y}) = -\frac{d}{dt}(y^2 + x^2)\omega_B.$$

Integrating and transforming to polar coordinates ($x = r\cos\varphi$, $y = r\sin\varphi$):

$$r^2\dot{\varphi} = r^2\omega_B.$$

Whence, after cancelling the r^2's, we have

$$\dot{\varphi} = \omega_B = \omega\sin\theta,$$

which gives the angular velocity of rotation of the plane of oscillation.

Sec. 9. The Dynamics of a Rigid Body

The dynamics of a rigid body is a large independent chapter of mechanics and is very rich in technical applications. Our aim is to give only a brief account of the basic concepts of this branch of mechanics inasmuch as it contains instructive examples of general laws. In addition, certain mechanical quantities that characterize a rigid body are necessary for an understanding of molecular spectra.

The kinetic energy of a rigid body. As was shown in Sec. 1, a rigid body has six degrees of freedom. Three of them relate to the translational motion of the centre of mass of a body in space. The remaining three degrees of freedom correspond to rotation (relative to this centre of mass).

In Sec. 4, it was shown that the kinetic energy of a system consists of the kinetic energy of the motion of the whole mass of the body concentrated at the centre of mass, and the kinetic energy of the relative motion of the separate particles of the system. In the case of a rigid body, relative motion reduces to rotation with the value of angular velocity $\boldsymbol{\omega}$ the same for all particles. Naturally, both the magnitude and the direction of $\boldsymbol{\omega}$ may vary with time.

Let us calculate the kinetic energy of rotation of a rigid body. In the general case, the density ρ of the body may not be uniform over the whole volume of the body, and may depend on the coordinates: $\rho = \rho(x, y, z) = \rho(\mathbf{r})$. The mass of an element of volume dV is equal to $dm = \rho\, dV$. The velocity of rotation $\mathbf{v}$ is, from (8.4), $[\boldsymbol{\omega}\mathbf{r}]$. Therefore, the kinetic energy of the volume element is equal to $\frac{1}{2}\rho[\boldsymbol{\omega}\mathbf{r}]^2 dV$. The kinetic energy of the whole body is represented by the integral of this quantity with respect to the volume

$$T = \frac{1}{2}\int \rho [\boldsymbol{\omega}\mathbf{r}]^2 dV. \tag{9.1}$$

Expressing the square of the vector product in terms of the components $\boldsymbol{\omega}$, we have

$$[\boldsymbol{\omega}\mathbf{r}]^2 = \omega^2 r^2 \sin^2\alpha = \omega^2 r^2 - \omega^2 r^2 \cos^2\alpha = \omega^2 r^2 - (\boldsymbol{\omega}\mathbf{r})^2.$$

Here α is the angle between $\boldsymbol{\omega}$ and $\mathbf{r}$. But

$$\omega^2 = \omega_x^2 + \omega_y^2 + \omega_z^2,$$

$$(\omega r)^2 = (\omega_x x + \omega_y y + \omega_z z)^2 =$$
$$= \omega_x^2 x^2 + \omega_y^2 y^2 + \omega_z^2 z^2 + 2\omega_x\omega_y xy + 2\omega_x\omega_z xz + 2\omega_y\omega_z yz.$$

Since the body is rigid, the components ω_x, ω_y, ω_z can be taken out of the volume integral. Combining terms which are similar in the components $\boldsymbol{\omega}$, we obtain for T:

$$T = \frac{1}{2}\omega_x^2 \int \rho(y^2 + z^2)\, dV + \frac{1}{2}\omega_y^2 \int \rho(x^2 + z^2)\, dV +$$
$$+ \frac{1}{2}\omega_z^2 \int \rho(x^2 + y^2)\, dV - \omega_x\omega_y \int \rho xy\, dV - \omega_x\omega_z \int \rho xz\, dV -$$
$$- \omega_y\omega_z \int \rho yz\, dV. \tag{9.2}$$

Moments of inertia. All the integrals appearing in (9.2) depend only on the shape of the body and its density distribution, and do not depend on the motion of the body (in a coordinate system fixed in the body). We denote them as follows:

$$\left.\begin{aligned} J_{xx} &= \int \rho\,(y^2+z^2)\,dV, & J_{xy} &= -\int \rho\,xy\,dV, \\ J_{yy} &= \int \rho\,(x^2+z^2)\,dV, & J_{xz} &= -\int \rho\,xz\,dV, \\ J_{zz} &= \int \rho\,(x^2+y^2)\,dV, & J_{yz} &= -\int \rho\,yz\,dV \end{aligned}\right\}. \tag{9.3}$$

The quantities with the same indexes are called *moments of inertia*, while those with different indexes are called *products of inertia.*

In the notation of (9.3), the kinetic energy has the form

$$T = \frac{1}{2}\left(J_{xx}\omega_x^2 + J_{yy}\omega_y^2 + J_{zz}\omega_z^2 + 2J_{xy}\omega_x\omega_y + 2J_{xz}\omega_x\omega_z + 2J_{yz}\omega_y\omega_z\right). \tag{9.4}$$

With the aid of the summation convention used in Sec. 2, when evaluating Lagrange's equations the kinetic energy can be written in the following concise form:

$$T = \frac{1}{2} J_{\alpha\beta}\,\omega_\alpha\,\omega_\beta\,.$$

Principal axes of inertia. Let us suppose that $Oxyz$ is a coordinate system fixed in a body. In this system all the quantities $J_{xx}, \ldots, J_{yz}$ are constant. Let us take another coordinate system $Ox'y'z'$ which is also fixed in the body. The old coordinates of any point are expressed in terms of its new coordinates by the well-known formulae of analytical geometry:

$$\begin{aligned} x &= x' \cos \angle (x', x) + y' \cos \angle (y', x) + z' \cos \angle (z', x), \\ y &= x' \cos \angle (x', y) + y' \cos \angle (y', y) + z' \cos \angle (z', y), \\ z &= x' \cos \angle (x', z) + y' \cos \angle (y', z) + z' \cos \angle (z', z), \end{aligned}$$

or, if we denote $\cos < (x_\alpha' x_\beta)$ by the symbol $A_{\alpha\beta}$*, then, with the aid of the summation convention

$$x_\beta = x_\alpha' A_{\alpha\beta}\,.$$

The same formulae are used to express also the components of any vector, and in particular ω_β, relative to the old axes, in terms of the components ω_α' relative to the new axes.

Let us substitute these expressions into the kinetic energy (9.4) and collect the terms containing the products $\omega_x'\omega_y'$, $\omega_x'\omega_z'$, $\omega_y'\omega_z'$ and

* $x_1 \equiv x$, $x_2 \equiv y$, $x_3 \equiv z$.

the squares $\omega_x'^2$, $\omega_y'^2$, $\omega_z'^2$. We shall now show that we can always rotate the coordinate axes so that the coefficients of the new products $\omega_x'\omega_y'$, $\omega_x'\omega_z'$, $\omega_y'\omega_z'$ become zero. Indeed, any rotation of the coordinate system can be described with the aid of three independent parameters, for a coordinate system is like an imaginary rigid body and its position in space is defined by the three angles of rotation (see Sec. 1). These three angles can be chosen so that the sums of the products of the cosines of the angles between the axes, for $\omega_x'\omega_y'$, $\omega_x'\omega_z'$ and $\omega_y'\omega_z'$ become zero. The remaining expressions for $\omega_x'^2$, $\omega_y'^2$, and $\omega_z'^2$ will be called J_1, J_2, J_3, so that

$$J_1 = A_{1\alpha} A_{1\beta} J_{\alpha\beta}, \quad J_2 = A_{2\alpha} A_{2\beta} J_{\alpha\beta}, \quad J_3 = A_{3\alpha} A_{3\beta} J_{\alpha\beta}.$$

The kinetic energy is written in the following form in the new coordinate axes:

$$T = \frac{1}{2}(J_1\omega_1^2 + J_2\omega_2^2 + J_3\omega_3^2). \tag{9.5}$$

These axes are called the *principal axes of inertia of the body*; they can be defined relative to any point connected with the body. By definition, the products of inertia convert to zero in the principal axes of inertia. The moments of inertia in the principal axes are called principal moments of inertia. They are denoted by J_1, J_2, J_3.

The angular momentum of a rigid body. Let us now calculate a projection of the angular momentum of a rigid body. From the definition of angular momentum we obtain

$$\begin{aligned} M_x &= \int \rho\,[\mathbf{r}\mathbf{v}]_x\,dV = \int \rho\,[\mathbf{r}\,[\boldsymbol{\omega}\mathbf{r}]]_x\,dV = \int \rho\,(\omega_x r^2 - x\,(\boldsymbol{\omega}\mathbf{r}))\,dV = \\ &= \omega_x \int \rho\,(y^2+z^2)\,dV - \omega_y \int \rho\,xy\,dV - \omega_z \int \rho\,xz\,dV = \\ &= J_{xx}\,\omega_x + + J_{xy}\,\omega_y + J_{xz}\,\omega_z \end{aligned} \tag{9.6}$$

or, in shortened form,

$$M_\alpha = J_{\alpha\beta}\,\omega_\beta.$$

Comparing (9.6) and (9.4), we see that

$$M_x = \frac{\partial T}{\partial \omega_x}. \tag{9.7}$$

M_y and M_z appear analogous. In vector form, we may write

$$\mathbf{M} = \frac{\partial T}{\partial \omega}. \tag{9.8}$$

Equations (9.7) and (9.8) again express the fact that the angular momentum is a generalized momentum related to rotation. In this sense, (9.7) corresponds to (5.4). The only difference is that the components $\boldsymbol{\omega}$ are not total time derivatives of some quantities. This will be shown a little later in the present section. In that sense, $\boldsymbol{\omega}_x$, in (9.7), is not altogether similar to $\dot{\varphi}$ in (5.4).

If the coordinate axes coincide with the principal axes of inertia, then the expression for angular momentum is even simpler than (9.6):

$$M_1 = \frac{\partial T}{\partial \omega_1} = J_1 \omega_1 \tag{9.9}$$

and similarly for the other components.

Moment of forces. Let us now find equations which describe the variation of angular momentum with time. The derivative of angular momentum of a particle is

$$\frac{d\mathbf{M}}{dt} = \frac{d}{dt}[\mathbf{r}\mathbf{p}] = [\dot{\mathbf{r}}\mathbf{p}] + [\mathbf{r}\dot{\mathbf{p}}] = [\mathbf{r}\mathbf{F}],$$

where the first term becomes zero since $\dot{\mathbf{r}}$ and $\mathbf{p}$ are parallel. Integrating this equation over the volume of the rigid body and taking advantage of the additive property of angular momentum, we have

$$\dot{\mathbf{M}} = \int [\mathbf{r}\mathbf{F}]\, dV = \mathbf{K}. \tag{9.10}$$

The right-hand side of (9.10), which we denote by $\mathbf{K}$, is called the resultant moment of the forces applied to the body. If $\mathbf{F}$ is the gravitational force (which occurs in the majority of cases) then $\mathbf{K}$ can also be written as

$$\mathbf{K} = -\int \rho g [\mathbf{r}\mathbf{z}_0]\, dV,$$

where $\mathbf{z}_0$ is the unit vector in a vertical direction. But since the vector z_0 is a constant, it should be put outside the integration sign:

$$\mathbf{K} = \left[\mathbf{z}_0, \int \rho g \mathbf{r}\, dV\right].$$

If the body is supported at its centre of mass, then, by the definition of centre of mass, the integral for all three projections $\rho\mathbf{r}$ will be zero. Then $\mathbf{K} = 0$ and the total angular momentum will be conserved. This occurs in the case of a gyroscope.

For the conservation of angular momentum of a rigid body it is sufficient that $\mathbf{K} = 0$; but for any arbitrary mechanical system, angular momentum is conserved only when there are no external forces.

Euler's equations. Equation (9.6) gives a relationship between $\mathbf{M}$ and $\boldsymbol{\omega}$. The quantities $J_{xx}, \ldots, J_{yz}$ are constant only in a coordinate system fixed in the rigid body itself. If we write equation (9.10) for a stationary coordinate system, then, differentiating $\mathbf{M}$ with respect to time, we must also find the derivatives of $J_{xx}, \ldots, J_{yz}$ with respect to time, which is very inconvenient. Therefore, it is preferable to transform the equation to a coordinate system fixed in the body, taking into account the accelerated motion of that system. The variation of the vector **M** relative to the moving axes consists of two components: one is due to the variation of the vector itself, while the other is due to the motion of the axes onto which it is projected. For the vector **M** this variation is equal to $[\boldsymbol{\omega}\mathbf{M}]$, similar to the way that it was equal to $[\boldsymbol{\omega}\mathbf{r}]$ for the radius vector **r** in Sec. 8. When the coordinate system is rotated, any vector varies like a radius vector.

Let the coordinate axes be taken in the direction of the principal axes of inertia. Obviously, the moments of inertia relative to these coordinates are constant. For this reason, the time derivative of $M_1 = J_1 \omega_1$ is

$$\dot{M}_1 = J_1 \dot{\omega}_1 + [\boldsymbol{\omega}\mathbf{M}]_1 = J_1 \dot{\omega}_1 + \omega_2 M_3 - \omega_3 M_2 = J_1 \dot{\omega}_1 + (J_3 - J_2)\,\omega_3 \omega_2 .$$

Equating this expression to the magnitude of the projection of the moment of force on the first axis of inertia, and doing the same for the other axes, we obtain the required system of equations

$$\left.\begin{aligned} J_1 \dot{\omega}_1 + (J_3 - J_2)\,\omega_2 \omega_3 &= K_1 , \\ J_2 \dot{\omega}_2 + (J_1 - J_3)\,\omega_3 \omega_1 &= K_2 , \\ J_3 \dot{\omega}_3 + (J_2 - J_1)\,\omega_1 \omega_2 &= K_3 . \end{aligned}\right\} \tag{9.11}$$

These equations were obtained by L. Euler and are named after him. They can be reduced to quadrature for any arbitrary values of integrals of motion in the following cases:

1) $K_1 = K_2 = K_3 = 0$ (point of support at the centre of mass) for arbitrary values of the moments of inertia;

2) $J_2 = J_3 \neq J_1$ and the point of support lies on the axis of symmetry, relative to which two moments of inertia are equal. This is the so-called symmetrical top.

For more than a hundred years, no other case of a solution of system (9.11) by quadratures was known. Only in 1887 did S. V. Kovalevskaya find another example (see G. K. Suslov, *Theoretical Mechanics*, Gostekhizdat, 1944). Kovalevskaya showed that the three listed cases exhaust all the possibilities of integrating the system (9.11) by quadratures for arbitrary constants (integrals) of motion.

A free symmetrical top. All three cases, and in particular the Kovalevskaya case, are very complicated to integrate. Therefore, we shall only consider the simplified first case, when $J_2 = J_3$ (a free symmetrical top).

From the first equation of (9.11), it immediately follows that $\boldsymbol{\omega}_1 = \text{const}$. For brevity, we write the value

$$\omega_1 \left(\frac{J_1}{J_2} - 1 \right) \equiv \Omega . \tag{9.12}$$

The second two equations of (9.11) are written thus:

$$\dot{\omega}_2 + \Omega\omega_3 = 0\,, \quad \dot{\omega}_3 - \Omega\omega_2 = 0\,. \tag{9.13}$$

Equations (9.13) are easily integrated if we represent the components ω_2 and ω_3 in the following form:

$$\omega_2 = \omega_\perp \cos \Omega t\,, \quad \omega_3 = \omega_\perp \sin \Omega t\,. \tag{9.14}$$

Here, $\omega_2^2 + \omega_3^2 = \omega_\perp^2$ is a constant quantity. Thus, the angular-momentum projection on the axis of symmetry and the sum of the squares of the angular-momentum projections on the other two axes are conserved. This means that the angular-momentum vector rotates about the axis of symmetry, i. e., the first axis of inertia, with angular velocity Ω; the vector makes with it a constant angle, the tangent of which is $\frac{\omega_1}{\omega_\perp}$. This is the situation in a system of moving axes.

Of course, in a system of stationary axes, the total angular momentum is conserved in magnitude and direction, since the resultant moment of force is equal to zero. In this system, the axis of symmetry of the top rotates about the angular-momentum direction making a constant angle with it. Such motion is called precession. Precessional motion is only stable for relatively small external perturbations. The stabilizing action of gyroscopes is based on this principle.

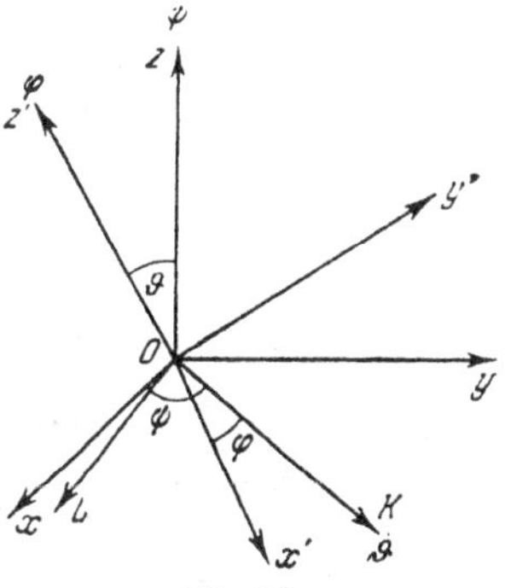

Fig. 13

Eulerian angles. We shall now show how to describe the rotation of a rigid body with the aid of parameters which specify its position. Such parameters are the Eulerian angles shown in Fig. 13. The figure depicts two coordinate systems: a fixed system $Oxyz$ and a system $Ox'y'z'$ fixed in the rigid body. It is most convenient to take x', y', z' along the principal axes of inertia through the point of support. Then the Eulerian angles are:

ϑ is the angle between the axes z and z',

φ is the angle between the line OK of intersection between the planes xOy and $x'Oy'$ and the x'-axis,

ψ is the angle between the line OK and the x-axis.

If the angle ψ varies, then the angular-velocity vector $\dot{\psi}$ is directed along the axis Oz since that vector is perpendicular to the plane of angle of rotation ψ. Thus $\dot{\varphi}$ must be taken along the axis Oz' and $\dot{\vartheta}$ along the line OK.

Let us now express the angular-velocity projections (i. e., ω_1, ω_2, ω_3,) onto the principal axes of inertia in terms of the generalized velocities $\dot{\psi}$, $\dot{\varphi}$, $\dot{\vartheta}$.

ω_3 is the projection of the angular velocity on the axis Oz' (z' is the third axis). As was shown, $\dot{\varphi}$ is projected exclusively on this axis and the projection of $\dot{\psi}$ is equal to $\dot{\psi}\cos\vartheta$, since ϑ is the angle between the axes Oz and Oz'. Hence,

$$\omega_3 = \dot{\varphi} + \dot{\psi}\cos\vartheta . \tag{9.15}$$

In order to find the projections of the angular velocity on the other two axes, we draw a line OL which lies in the plane $x'Oy'$ and is perpendicular to OK.

From Fig. 13 it can be seen that

$$\angle LOx' = \frac{\pi}{2} - \varphi \text{ and } \angle zOL = \frac{\pi}{2} + \vartheta ,$$

since the straight line OL lies in the plane zz', as do all lines perpendicular to OK. The projection of $\dot{\psi}$ on OL is equal to $-\dot{\psi}\sin\vartheta$, and the projection on Ox' is equal to $-\dot{\psi}\sin\vartheta\cos\left(\frac{\pi}{2} - \varphi\right) = -\dot{\psi}\sin\vartheta$ $\sin\varphi$. The projection of $\dot{\psi}$ on Oy' is $\dot{\psi}\sin\vartheta\cos\varphi$. The projection of $\dot{\vartheta}$ on Ox' and Oy' can be directly found by means of the diagram; they are $\dot{\vartheta}\cos\varphi$ and $\dot{\vartheta}\sin\varphi$. The result is therefore

$$\omega_1 = \dot{\vartheta}\cos\varphi - \dot{\psi}\sin\vartheta\sin\varphi , \tag{9.16}$$

$$\omega_2 = \dot{\vartheta}\sin\varphi + \dot{\psi}\sin\vartheta\cos\varphi . \tag{9.17}$$

From equations (9.15), (9.16), and (9.17) it will be seen that ω_1, ω_2 and ω_3 are not total time derivatives of any quantities and, in that sense, do not exactly agree with the usual notion of generalized velocities (as do $\dot{\varphi}$, $\dot{\psi}$, $\dot{\vartheta}$).

If we substitute into (9.5) the expressions for ω_1, ω_2, ω_3 in terms of the Eulerian angles, we obtain the kinetic energy of a rigid body as a function of the generalized coordinates φ, ψ, ϑ.

The symmetrical top in a gravitational field. We shall find the Lagrangian for a symmetrical top whose point of support lies on the axis of symmetry at a distance l below the centre of mass. Then the height of the centre of mass above the point of support is $z = l\cos\vartheta$. Hence, the potential energy of the top is

$$U = mgz = mgl\cos\vartheta . \tag{9.18}$$

The kinetic energy of the top, expressed in terms of the Eulerian angles, is

$$\begin{aligned} T &= \frac{1}{2}J_1(\omega_1^2 + \omega_2^2) + \frac{1}{2}J_3\omega_3^2 = \\ &= \frac{1}{2}J_1(\dot{\vartheta}^2 + \dot{\psi}^2\sin^2\vartheta) + \frac{1}{2}J_3(\dot{\varphi} + \dot{\psi}\cos\vartheta)^2 . \end{aligned} \tag{9.19}$$

The difference between the quantities (9.19) and (9.18) gives the Lagrangian for a symmetrical top. The sum gives the total energy $\mathscr{E}$.

Since L does not contain time explicitly, the energy is an integral of motion:

$$\mathscr{E} = T + U = \text{const}\,. \tag{9.20}$$

We can find two more integrals of motion, noting that the angles φ and ψ do not appear explicitly in L (φ is eliminated only in the case of a symmetrical top). These integrals of motion are

$$p_\varphi = \frac{\partial L}{\partial \dot{\varphi}} = J_3 (\dot{\varphi} + \dot{\psi} \cos \vartheta) = \text{const}\,, \tag{9.21}$$

$$p_\psi = \frac{\partial L}{\partial \dot{\psi}} = J_1 \sin^2 \vartheta \dot{\psi} + J_3 \cos \vartheta\, (\dot{\varphi} + \dot{\psi} \cos \vartheta) = \text{const}\,. \tag{9.22}$$

If we eliminate $\dot{\varphi}$ and $\dot{\psi}$ from equations (9.21) and (9.22) and substitute them into the energy integral, the latter will contain only the variable ϑ, which allows us to reduce the problem to quadrature.

Substituting (9.21) in (9.22), we obtain

$$p_\psi = J_1 \sin^2 \vartheta\, \dot{\psi} + p_\varphi \cos \vartheta\,,$$

whence

$$\dot{\psi} = \frac{1}{J_1 \sin^2 \vartheta} (p_\psi - p_\varphi \cos \vartheta)\,.$$

The energy integral, after substituting p_φ and p_ψ is

$$\mathscr{E} = \frac{1}{2} J_1 \dot{\vartheta}^2 + \frac{1}{2 \sin^2 \vartheta} \frac{(p_\psi - p_\varphi \cos \vartheta)^2}{J_1} + \frac{p^2_\varphi}{2J_3} + mgl \cos \vartheta\,. \tag{9.23}$$

Thus, the problem is reduced to motion with one degree of freedom ϑ, as it were. The corresponding "kinetic energy" is $\frac{1}{2} J_1 \dot{\vartheta}^2$, and the "potential energy" is represented by those energy terms which depend on ϑ. This potential energy becomes infinite for $\vartheta = 0$, and $\vartheta = \pi$. Hence, for $0 < \vartheta < \pi$ it has at least one minimum. If this minimum corresponds to $\vartheta > \frac{\pi}{2}$, then the rotation of the top, whose centre of mass is above the point of support, is stable. Small oscillations are possible near the potential energy minimum. These oscillations are superimposed on the precessional motion of the top which we have already noted. They are called *nutations*.

Sec. 10. General Principles of Mechanics

In this part of the book, mechanics is explained mainly through the use of Newton's equations (2.1). Going over to generalized coordinates, we obtain from them Lagrange's equations and a series of further deductions. In this section it will be shown that the system of Lagrange's equations can be obtained not only from Newton's Second Law, but also from a very simple assertion about the value of the integral of the Lagrangian taken with respect to time. The basic laws of mechanics thus formulated are usually called *integral principles.*

The particular importance of these principles is that they allow us to understand, in a unified manner, the laws relating to various areas of theoretical physics (mechanics and electrodynamics), thus opening up a field for broad generalizations.

Action. For a certain mechanical system, let it be possible to define the Lagrangian

$$L = L(q, \dot{q}, t), \tag{10.1}$$

as dependent on the generalized coordinates q, velocities $\dot{q}$, and the time t. We shall consider that all the coordinates and all the velocities are independent. Let us choose some continuous, but otherwise arbitrary, dependence of the coordinates upon the time $q(t)$. The functions $q(t)$ can be in complete disagreement with the actual law of motion. The only requirement imposed on $q(t)$ is that the functions $q(t)$ should be smooth, i. e., that they should provide for differentiation and should correspond to the rigid constraints present in the system.

The time integral of the Lagrangian is called the *action* of the system:

$$S = \int_{t_0}^{t_1} L(q, \dot{q}, t)\, dt. \tag{10.2}$$

The magnitude of this integral depends upon the law chosen for $q(t)$, and is, in that sense, arbitrary. In order to examine the relationship between the action and the function $q(t)$, it is convenient to calculate the change of S for a transition from some arbitrary law $q(t)$ to another, infinitely close but also arbitrary, law $q'(t)$.

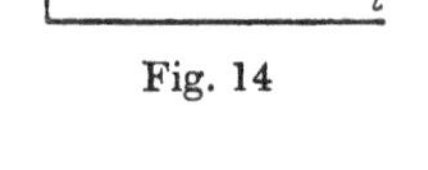

Fig. 14

Variation. Fig. 14 shows two such conceivable paths. Time is taken along the abscissa, and one of the generalized coordinates q, representing the totality of generalized coordinates, is plotted on the ordinate axis.

For the specification of future operations, we shall consider that both paths pass through the same points, q_0 and q_1, at the initial and final instants of time.

The vertical arrow shows the difference between two conceivable, infinitely close paths at some instant of time other than initial or final. This difference is usually called the *variation* of q and is denoted by δq. The symbol δ should emphasize the difference between variation and the differential d; the differential is taken for the same path at various instants of time, while the variation is taken for the same instant of time between different paths.

Since the neighbouring paths in Fig. 14 have different forms, the speed of motion along them will also differ. Together with the variation of the coordinate δq between paths, we can also find the variation in velocity $\delta \dot{q}$. We shall show that $\delta \dot{q} = \frac{d}{dt}\delta q$. Indeed, $\delta q = q'(t) - q(t)$, where q' and q are values of the coordinates for neighbouring paths. But the derivative of the difference $\frac{d}{dt}\delta q$ is equal to the difference of the derivatives $\dot{q}'(t) - \dot{q}(t) = \delta \dot{q}$.

Let us now find the variation of the Lagrangian, i. e., the difference of the function for two adjacent paths. Since $L = L(q, \dot{q}, t)$ and the variation is taken at the same instant of time, i. e., $\delta t = 0$, we obtain

$$\delta L = \frac{\partial L}{\partial q}\delta q + \frac{\partial L}{\partial \dot{q}}\delta \dot{q}. \tag{10.3}$$

Let us rearrange the second term. Taking advantage of the fact that $\delta \dot{q} = \frac{d}{dt}\delta q$, we can write it thus:

$$\frac{\partial L}{\partial \dot{q}}\delta \dot{q} = \frac{\partial L}{\partial \dot{q}}\frac{d}{dt}\delta q = \frac{d}{dt}\left(\frac{\partial L}{\partial \dot{q}}\delta q\right) - \delta q \frac{d}{dt}\frac{\partial L}{\partial \dot{q}}.$$

The last equation simply expresses a transformation by parts. Substituting it into (10.3), we find

$$\delta L = \frac{d}{dt}\left(\frac{\partial L}{\partial \dot{q}}\delta q\right) + \delta q\left(\frac{\partial L}{\partial q} - \frac{d}{dt}\frac{\partial L}{\partial \dot{q}}\right). \tag{10.4}$$

The integral of the variation of L is equal to the variation of action δS, since the difference between integrals taken between the same limits is equal to the difference between the integrands.

The first term in (10.4) can be integrated with respect to time, because it is a total derivative. The variation of action is then reduced to the form

$$\delta S = \frac{\partial L}{\partial \dot{q}}\delta q \Big|_{t_0}^{t_1} + \int_{t_0}^{t_1} \delta q\left(\frac{\partial L}{\partial q} - \frac{d}{dt}\frac{\partial L}{\partial \dot{q}}\right)dt. \tag{10.5}$$

We have agreed to consider only those paths which pass through the same points, q_0 and q_1, at the initial and final instants of time. Hence, at these instants the variation δq becomes zero by convention, and the integrated term disappears. The expression δS is reduced to the following integral:

$$\delta S = \int_{t_0}^{t_1} \delta q\left(\frac{\partial L}{\partial q} - \frac{d}{dt}\frac{\partial L}{\partial \dot{q}}\right)dt. \tag{10.6}$$

The extremal property of action. If the chosen path coincides with the actual path of motion, the coordinates satisfy Lagrange's equation:

$$\frac{d}{dt}\frac{\partial L}{\partial \dot{q}} - \frac{\partial L}{\partial q} = 0\,. \tag{10.7}$$

Substituting this in (10.6), we see that the variation of action tends to zero close to the actual path. The change in magnitude is equal to zero either close to its extreme, or close to the "stationary point" (for example, the function $y = x^3$ has such a point at $x = 0$, where $y' = 0$, $y'' = 0$). Three cases can, in general, be realized: a minimum, a maximum and a stationary point.

For example, let a point, not subject to the action of any forces other than constraint reactions, move freely on a sphere. Then its path will be an arc of a great circle. But through any two points on the sphere there pass *two* arcs of a great circle representing the largest and smallest sections of the circumference. One corresponds to a maximum, and the other, to a minimum, S. If the beginning and end of the path are diametrically opposite, the result is a stationary point.

The principle of least action. We have proven, on the basis of Lagrange's equations, that $\delta S = 0$. We can proceed in a different way: by asserting that close to the actual path passing between the given initial and final positions of the system the increment of action is equal to zero, we can derive Lagrange's equations. Ordinarily, the action on an actual path is minimal, and therefore the assertion we have made is called the *principle of least action.* Action was written in the form (10.2) by Hamilton. Much earlier, the principle of least action was mathematically formulated by Euler for the special case of paths corresponding to constant energy.

For us, it is not essential that the action should be a minimum, but that it should be steady, $\delta S = 0$.

Lagrange's equations are derived from the principle of least action by means of proving the opposite. We assume the right-hand side of equation (10.6) to be zero, $\delta S = 0$, and the variation δq to be *arbitrary*. Then, if the expression inside the parentheses is not equal to zero, the sign of the variation δq can always be chosen to be the same as for the quantity $\frac{\partial L}{\partial q} - \frac{d}{dt}\frac{\partial L}{\partial \dot{q}}$, because the variation is arbitrary. If, for example, the sign of the quantity $\frac{\partial L}{\partial q} - \frac{d}{dt}\frac{\partial L}{\partial \dot{q}}$ changes several times along the path of integration, then the sign of δq must also be changed accordingly at those points so that the integrand of (10.6) should everywhere be non-negative. But the integral of a non-negative function cannot equal zero unless the function is equal to zero everywhere. Therefore, $\delta S = 0$ only when $\frac{\partial L}{\partial q} - \frac{d}{dt}\frac{\partial L}{\partial \dot{q}}$ becomes zero along the whole path of integration, for otherwise the variation δq can be so chosen that $\delta S > 0$. We have shown that if we proceed from the principle of least action as a requirement for the motion along an actual path, then that path must satisfy Lagrange's equations.

The advantages of using action. The principle of least action may at first sight appear artificial or, in any case, less obvious than Newton's laws, to whose form we are accustomed. For this reason we shall try to explain where its advantages lie.

First of all, let it be noted that Lagrange's or Newton's equations are always associated with some coordinates whose choise is, to a significant extent, arbitrary. In addition, the choice of coordinate system, relative to which the motion is described, is also arbitrary. Yet the motion of particles along actual paths in a mechanical system expresses a certain set of facts which cannot depend on the arbitrary manner of their description. For example, if the motion leads to a collision of particles, that fact must always be represented in any description of the system.

But it is precisely the integral principle that is especially useful in a formulation of laws of motion not related to any definite choice of coordinates, the value of the integral between the given limits being independent of the choice of integration variables. The extremal property of an integral cannot be changed by the way in which it is calculated.

The integral principle $\delta S = 0$ is equivalent, purely mathematically, to Lagrange's equations (2.21). But in order to apply it to any actual system, we must have an explicitly expressed Lagrangian. It may be found from those physical requirements which should be imposed on an invariant law of motion that is independent of the choice of coordinate axes and the frame of reference.

As a result of the invariance of the principle of least action, we can consider the laws of mechanics in a very general form, and this, therefore, opens the way for further generalizations.

The determinacy of the Lagrangian. Before finding an explicit form for the Lagrangian, we must put the question: Is the determined function we are looking for single-valued? We shall show that if we add the total time derivative of any function of coordinates and time, $\frac{d}{dt} f(q, t)$, then Lagrange's equations remain unchanged. This can be verified either by simple substitution into (10.7), or directly from the integral principle. Writing

$$L = L' + \frac{d}{dt} f(q, t), \tag{10.8}$$

we see that

$$S = \int_{t_0}^{t_1} L dt = \int_{t_0}^{t_1} \left(L' + \frac{df}{dt}\right) dt = \int_{t_0}^{t_1} L' di + f \Big|_{t_0}^{t_1}. \tag{10.9}$$

The variations of f appear in the variation of S only at the limits of integration. But since we have arranged that f depends on the coordinates and time, but not on the velocities, the variation of f is expressed linearly in terms of the variations of the coordinates, and is zero at the limits of integration. Therefore,

$$\delta\int_{t_0}^{t_1} Ldt=\delta\int_{t_0}^{t_1} L'dt\,. \qquad (10.10)$$

Hence, the Lagrangian is determined only to the accuracy of the total time derivative of the function of coordinates and time.

Defining forms of the Lagrangian. We shall now formulate in more detail those requirements which the integral principle expressing laws of mechanics must satisfy.

First of all we note that the form of this principle must be the same for different inertial systems, since all such systems are equivalent. This statement follows from the relatively principle (see Sec. 8). The essence of the relativity principle consists in the fact that the choice of an inertial coordinate system is arbitrary, while the physical consequences of the equations of motion cannot be arbitrary.

Similarly arbitrary is the choice of the origin and the initial instant of time and also orientation of the coordinate axes in space.

It must, of course, be borne in mind that the form of action is by no means determined by speculation; this form represents no less a generalization of physical experience than the laws of Newton. However, the principle of least action, best expresses the invariance of physical laws to the method of their formulation. Quite naturally, the form of the invariance (in relation to rotations, translations, reflections, etc.) is itself a certain, very broad, generalization of experience, and must by no means be considered as *a priori*.

Considering now the problem of finding the form of the Lagrangian, let us first of all determine the action of a free particle in an inertial coordinate system.* In such a system, the particle moves uniformly in a straight line, i. e., with constant velocity. (This statement is based on the experimental fact that inertial systems exist in nature). Thus, the Lagrangian for a free particle in an inertial system cannot contain any coordinate derivatives other than velocity.

By definition, a free particle is very far away from any other bodies with which it could interact. Therefore, its Lagrangian must not change its form upon displacement of the origin to any arbitrary point fixed in the given inertial system. In other words, the Lagrangian of such a particle does not depend explicitly on the coordinates.

In this way, one can conclude that the Lagrangian does not depend explicitly on time.

---

* See L. D. Landau and E. M. Lifshits, *Mechanics*, Fizmatgiz, 1958.

The orientation of the coordinate axes is arbitrary as well as the choice of the origin. For the Lagrangian to be independent of the orientation of coordinate axes, it must be scalar quantity.

To summarize, then, the Lagrangian is a scalar that depends only on the velocity of the free particle relative to the given inertial system. The only scalar quantity which can be formed from a vector is the absolute value of the vector. Therefore,

$$L = L\,(v^2).$$

The form of this function can be found from the relativity principle, in accordance with which the Lagrangian must not change with the transformation from one inertial system to another. In Newtonian mechanics, this transformation is effected with the aid of equations (8.1), (8.2), i. e., Galilean transformations. The Galilean transformations led to the law of addition of velocities:

$$\mathbf{v} = \mathbf{v}' + \mathbf{V},$$

where $\mathbf{V}$ is the relative velocity of the inertial systems. Therefore, the Lagrangian must remain invariant with respect to Galilean transformations.

Since the Lagrangian is determined to a total derivative, it is sufficient (for its invariance) for the following equality to be satisfied:

$$L = L\,(v^2) = L\,[(\mathbf{v}' + \mathbf{V})^2] = L\,(v'^2) + \frac{df}{dt}, \tag{10.11}$$

where the functions $L\,(v^2)$ and $L\,(v'^2)$ have the same form in accordance with the principle of relativity.

Any transformation (8.1), (8.2), in which the relative velocity $\mathbf{V}$ is finite, can be obtained by a set of infinitely small transformations applied successively. It is, therefore, sufficient to consider a transformation in which the relative velocity of the inertial systems $\mathbf{V}$ is very much smaller than the particle velocity $\mathbf{v}$. Then, to a very good approximation, the quantity $(\mathbf{v}' + \mathbf{V}^2)$ is equal to

$$(\mathbf{v}' + \mathbf{V})^2 = v'^2 + 2\,\mathbf{v}'\mathbf{V},$$

where the term of the second order of smallness is discarded.

Expanding $L\,[(\mathbf{v}' + \mathbf{V})^2)]$ in a series, we obtain, to the same approximation,

$$L\,[(\mathbf{v} + \mathbf{V})^2] = L\,(v'^2) + \frac{\partial L}{\partial (v'^2)}\,2\,\mathbf{v}'\mathbf{V}.$$

Comparing this with (10.11), we find:

$$\frac{\partial L}{\partial (v'^2)}\,2\,\mathbf{v}'\mathbf{V} = \frac{\partial L}{\partial (v'^2)}\,2\,\mathbf{V}\frac{d\mathbf{r}'}{dt} = \frac{df(\mathbf{r}')}{dt}.$$

However, the expression on the left-hand side of the equation can be a total derivative of the function of coordinates only if $\frac{\partial L}{\partial(v'^2)}$ is independent of velocity. Introducing the notation

$$\frac{\partial L}{\partial(v'^2)}=\frac{m}{2}=\text{const},$$

we obtain

$$\frac{df(\mathbf{r}')}{dt}=\frac{d}{dt}m\mathbf{V}\mathbf{r}',$$

for, otherwise, $\frac{\partial L}{\partial(v'^2)}$ could not be put inside the derivative sign.

In this way we have shown that the Lagrangian for a free particle is equal to

$$L(v^2)=\frac{m}{2}v^2. \tag{10.12}$$

The Lagrangian for a system of noninteracting particles is equal to the sum of the Lagrangians of these particles taken independently, since it is the only sum of quadratic expressions of the type (10.12) that changes by a total derivative when $\mathbf{v}_i=\mathbf{v}_i'+\mathbf{V}$ (where i is the particle number) is substituted.

In order to write down L for a system of interacting particles, we must, of course, make certain physical assumptions about the nature of the interaction.

1) The interaction does not depend on the particle velocities. This assumption is justified for gravitational and electrostatic forces, and is not justified for electromagnetic forces. It should, however, be noted that electromagnetic interactions involve ratios of particle velocities and the velocity of light c, and therefore, to the approximation of Newtonian mechanics, they must be considered as negligibly small. The Lagrangian of Newtonian mechanics is not universal and is applicable only to a limited group of phenomena, when all $v_i \ll c$.

2) The interaction does not change the masses of the particles.

3) The interaction is invariant with respect to Galilean transformations.

From these conditions it can be seen that the interaction appears in the Lagrangian in the form of a scalar function determined only by the relative distribution of the particles:

$$L=\sum_i\frac{m_iv_i^2}{2}-U(\ldots\mathbf{r}_i-\mathbf{r}_k\ldots). \tag{10.13}$$

From this expression, we can find the conservation laws for energy, linear momentum, and angular momentum (see Sec. 4).

The Hamiltonian function. We shall now use the principle of least action in order to transform a system of equations of motion to other variables. Namely, in place of coordinates and velocities we shall employ coordinates and momenta. Let us assume that velocities are eliminated from the relations

$$p = \frac{\partial L}{\partial \dot{q}} . \tag{10.14}$$

Since the Lagrangian depends quadratically on the velocities, equations (10.14) are linear in the velocities and can always be solved. We shall obtain for coordinates and momenta a more symmetrical system of equations than Lagrange's equations.

The passing from velocities to momenta was performed to some extent when we substituted the integrals of motion in the expression for energy, for example, in (5.4), (9.21), (9.22).

Now, in place of the velocities we shall introduce into the energy the momenta for all the degrees of freedom, (and not only for the cyclic ones, i. e., those, whose coordinates do not appear explicitly in L). Energy expressed in terms of coordinates and momenta only is called the Hamiltonian function of the system or, for short, the *Hamiltonian*:

$$\mathscr{E}\,[q, \dot{q}\,(p)] \equiv \mathscr{H}\,(q, p) = \dot{q}\,p - L . \tag{10.15}$$

Thus, for example, if we replace $\dot{\vartheta}$ by $\frac{p_\vartheta}{J_1}$ in (9.23), we obtain the Hamiltonian for a symmetrical top:

$$\mathscr{H} = \frac{p_\vartheta^2}{2J_1} + \frac{(p_\psi - p_\varphi \cos\vartheta)^2}{2J_1 \sin^2\vartheta} + \frac{p_\varphi^2}{2J_3} + mgl\cos\vartheta . \tag{10.16}$$

Hamilton's equations. In order to derive the required system of equations, we write the expression for the principle of least action, expressing L in terms of $\mathscr{H}$:

$$\delta S = \delta \int_{t_0}^{t_1} (p\dot{q} - \mathscr{H})\, dt = 0 . \tag{10.17}$$

Here it is assumed that $\dot{q}$ is expressed in terms of p and q.

Let us calculate the variation δS:

$$\delta S = \int_{t_0}^{t_1} \left(\delta p \dot{q} + p \delta \dot{q} - \frac{\partial \mathscr{H}}{\partial p} \delta p - \frac{\partial \mathscr{H}}{\partial q} \delta q\right) dt = 0 .$$

The second term inside the parentheses can be integrated by parts, similar to the way that it was done in (10.5). This gives

$$\delta S = p\delta q \Big|_{t_0}^{t_1} + \int_{t_0}^{t_1} \left[\delta p \left(\dot{q} - \frac{\partial \mathscr{H}}{\partial p}\right) - \delta q \left(\dot{p} + \frac{\partial \mathscr{H}}{\partial q}\right)\right] dt .$$

The integrated part becomes zero when limits of integration have been substituted. The independent variables are now p and q. The variation of p, as well as the variation of q, is completely arbitrary in sign. For δS to be equal to zero, the following equations must be satisfied:

$$\dot{p}+\frac{\partial \mathscr{H}}{\partial q}=0\,, \quad \dot{q}-\frac{\partial \mathscr{H}}{\partial p}=0\,. \tag{10.18}$$

This system of equations is more symmetrical than Lagrange's equations. Instead of ν second-order Lagrangian equations, we have $2\,\nu$ first-order equations (10.18). They are called Hamilton's equations.

Reducing the order with the aid of the energy integral. If $\mathscr{H}$ does not depend on time, we can exclude time completely from the equations by dividing all the equations (10.18), except one, by the said equation. Then we have

$$\frac{dp}{dq}=-\frac{\dfrac{\partial \mathscr{H}}{\partial q}}{\dfrac{\partial \mathscr{H}}{\partial p}}\,. \tag{10.19}$$

Here, for simplicity, this operation has been performed for a system with one degree of freedom. The integration of (10.19) yields one constant. The second constant will be determined by quadrature from the equation

$$\frac{dt}{dq}=\frac{1}{\partial \mathscr{H}/\partial p}\,, \tag{10.20}$$

where $\dfrac{\partial H}{\partial p}$ is a certain function q which can be obtained by integrating (10.19). The constant of integration in (10.20) is the initial instant t_0.

The connection between momentum and action. We shall now show that if action is calculated for the actual paths of a system, then momentum can be very simply expressed in terms of this action. For this we shall consider the change in action when the ends of the integration interval are displaced along the actual paths. From (10.7), the expression under the integral sign in (10.5) is equal to zero on such paths. But the integrated part does not become zero; only the variations in it must be replaced by differentials, since we are considering the displacement of the ends of the integration interval along given paths. Therefore,

$$dS=\frac{\partial L}{\partial \dot{q}}\,dq-\frac{\partial L}{\partial q_0}\,dq_0=p\,dq-p_0\,dq_0 \tag{10.21}$$

in agreement with the definition of momentum (4.13).

But action calculated along an actual path is uniquely determined by its initial and final points $S=S\,(q_0,\,q)$. So

$$dS=\frac{\partial S}{\partial q_0}\,dq_0+\frac{\partial S}{\partial q}\,dq\,. \tag{10.22}$$

Comparing (10.21) and (10.22), we obtain the very important relationship between momentum and action

$$p = \frac{\partial S}{\partial q}, \qquad p_0 = -\frac{\partial S}{\partial q_0}, \tag{10.23}$$

which is very essential for the formulation of quantum mechanics.

Exercise

Write down the Hamiltonian and Hamilton's equations for a particle in a central field.

PART II

ELECTRODYNAMICS

Sec. 11. Vector Analysis

The equations of electrodynamics gain considerably in conciseness and vividness if they are written in vector notation. In vector notation, the arbitrariness associated with the choice of one or another coordinate system disappears, and the physical content of the equations becomes more apparent.

We have assumed that the reader is acquainted with the elements of vector algebra, such as the definition of a vector and the various forms of vector products. However, in electrodynamics, vector differential operations are also used. This section is devoted to a definition of vector differential operations and to proofs of their fundamental properties, which will be needed later.

The vector of an area. We first of all give a definition of the vector of an elementary area $d\mathbf{s}$. This is a vector in the direction of the normal to the area, numerically equal to its surface and related to the direction of traverse of the contour around the area by the corkscrew rule (Fig. 15).

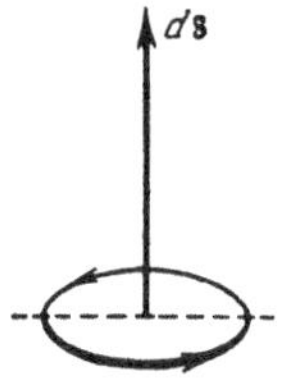

Fig. 15

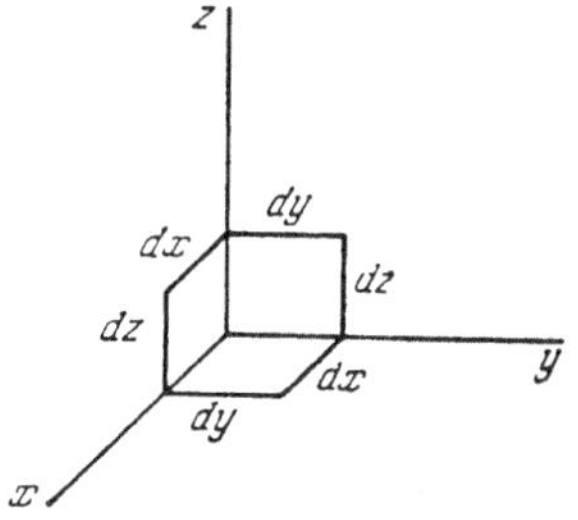

Fig. 16

We shall make use of a right-handed coordinate system x, y, z, in which, if we look from the direction of the z-axis, the x-axis is rotated towards the y-axis in an anticlockwise sense (Fig. 16). In this system, the vector area can be resolved into components which are expressed thus:

$$ds_x = dy\,dz,\quad ds_y = dz\,dx,\quad ds_z = dx\,dy.$$

Vector flux. Now suppose that a liquid of density 1 ("water") flows through the area, the flow velocity being represented by the vector **v**. We shall call the angle between **v** and *d*s, α. Fig. 17 shows the flow lines of the liquid passing through *ds*. They are parallel to the velocity **v**. Let us calculate the amount of liquid that passes through the area *ds* every second. Obviously, it is equal to the amount that passes through the area *ds'*, placed perpendicular to the flux and intersected by the same flow lines as pass through *ds*. This quantity is simply equal to $v\,ds'$, because every second a liquid cylinder of base ds' and height v passes through the area ds'. But $ds' = ds \cos \alpha$, whence the quantity of liquid we are concerned with is

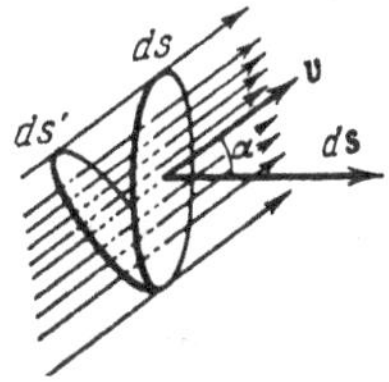

Fig. 17

$$dJ = v\,ds' = v\,ds \cos \alpha = \mathbf{v}\,d\mathbf{s}. \tag{11.1}$$

By analogy, the scalar product of any vector **A** (taken at the point of infinitesimal area) on *d***s** is called the flux of the vector **A** across the area *ds*. Similar to the way that the flow of liquid across a finite area *s* is equal to the integral of *dJ* with respect to the surface,

$$J = \int \mathbf{v}\,d\mathbf{s}, \tag{11.2}$$

the integral

$$J = \int \mathbf{A}\,d\mathbf{s} \tag{11.3}$$

is called the flow (flux) of the vector **A** across any area.

The area vector is introduced so that we can make use of the noncoordinate and convenient notation of (11.3). The integrals appearing in (11.3) are double. In terms of the projections of (11.3) we can write

$$J = \int \mathbf{A}\,d\mathbf{s} \equiv \iint A_x\,dy\,dz + \iint A_y\,dz\,dx + \iint A_z\,dy\,dx,$$

where the limits of the double integrals are determined from the corresponding projections, onto the coordinate planes, of the contour bounding the surface.

The Gauss-Ostrogradsky theorem. Let us now calculate the vector flux through a closed surface. For this we shall consider, first of all, the infinitesimal closed surface of a parallelepiped (Fig. 18). We shall make the convention that the normal to the closed surface will always be taken outwards from the volume.

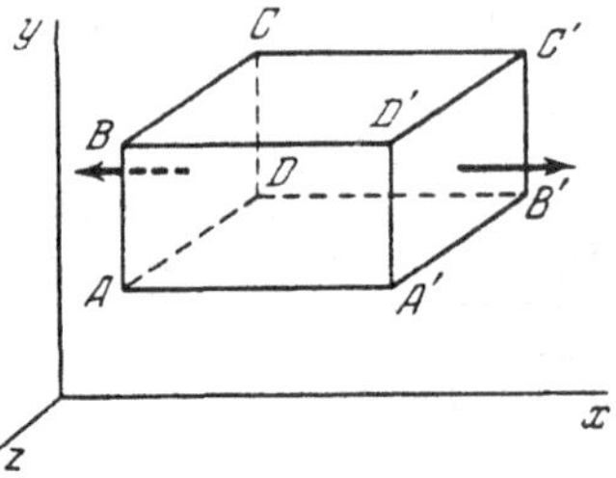

Fig. 18

Let us calculate the flux of the vector **A** across the area $ABCD$ (the direction of traverse being in agreement with the direction of the normal). Since the flux is equal to the scalar product of A by the vector area $ABCD$, in the negative x-direction (and hence equal to $dydz$), we obtain for this infinitely small area

$$dJ_{ABCD} = -A_x(x)\,dy\,dz.$$

We get a similar expression for the area $A'B'C'D'$, only in this case the projection ds_x is equal to $dy\,dz$, and A_x is taken at the point $x+dx$ instead of x. And so

$$dJ_{A'B'C'D'} = A_x(x+dx)\,dy\,dz.$$

Thus the resultant flux through both areas, perpendicular to the x-axis, is

$$dJ_{A'B'C'D'} + dJ_{ABCD} = [A_x(x+dx) - A_x(x)]\,dy\,dz = \frac{\partial A_x}{\partial x}\,dx\,dy\,dz. \tag{11.4}$$

We have utilized the fact that dx is an infinitely small quantity, and we have expanded $A_x(x+dx)$ in a series. The resultant fluxes across the boundaries perpendicular to the y and z axes are formed similarly. The resultant flux across the whole parallelepiped is

$$dJ = \left(\frac{\partial A_x}{\partial x} + \frac{\partial A_y}{\partial y} + \frac{\partial A_z}{\partial z}\right)dx\,dy\,dz. \tag{11.5}$$

A finite closed volume can be divided into small parallelepipeds, and the relationship (11.5) applied to each one of them separately. If we sum all the fluxes, the adjacent boundaries do not give any contribution, since the flux emerging from one parallelepiped enters the neighbouring one. Only the fluxes through the outer surface of the selected volume remain, since they are not cancelled by others. But the right-hand sides of (11.5) will be additive for all the elementary volumes $dV = dx\,dy\,dz$, yielding the very important integral theorem:

$$\int \mathbf{A}\,d\mathbf{s} = \int\left(\frac{\partial A_x}{\partial x} + \frac{\partial A_y}{\partial y} + \frac{\partial A_z}{\partial z}\right)dV. \tag{11.6}$$

It is called the Gauss-Ostrogradsky theorem.

The divergence of a vector. The expression appearing on the right-hand side under the integral sign can be written down in a much shorter form. We first of all notice that it is a scalar expression, since there is a scalar on the left-hand side in (11.6) and dV is also a scalar. This expression is called the *divergence* of the vector $\mathbf{A}$ and is written thus:

$$\operatorname{div} \mathbf{A} \equiv \frac{\partial A_x}{\partial x} + \frac{\partial A_y}{\partial y} + \frac{\partial A_z}{\partial z}. \tag{11.7}$$

The divergence can be defined independently of any coordinate system, if (11.5) is used. Indeed, from (11.5) the definition for divergence follows as

$$\operatorname{div} \mathbf{A} \equiv \lim_{V \to 0} \frac{\int \mathbf{A}\, d\mathbf{s}}{V}. \tag{11.8}$$

The divergence of a vector at a given point is equal to the limit of the ratio of the vector flux through the surface surrounding the point to the volume enveloped by the surface, when the surface is contracted into the point.

Let us suppose that the vector $\mathbf{A}$ denotes the velocity field of some fluid. Then, from the definition (11.8), it can be seen that the divergence of the vector $\mathbf{A}$ is a measure of the density of the sources of the fluid, for it is obvious that the more sources there are in unit volume, the more fluid will flow out of the closed volume. If div $\mathbf{A}$ is negative, we can speak of the density of vents. But it is more convenient to define the source density with arbitrary sign. We note that from (11.7) there follows the quantity

$$\operatorname{div} \mathbf{r} = \frac{\partial x}{\partial x} + \frac{\partial y}{\partial y} + \frac{\partial z}{\partial z} = 1 + 1 + 1 = 3, \tag{11.9}$$

since $\mathbf{r}$ has components x, y, z.

Contour integrals. We shall now consider the vector integral of a closed contour having the following form:

$$C = \int \mathbf{A}\, d\mathbf{l} = \int (A_x\, dx + A_y\, dy + A_z\, dz). \tag{11.10}$$

This single integral is called the *circulation* of the vector over the given contour. For example, if $\mathbf{A}$ is the force acting on any particle, then $\mathbf{A}\, d\mathbf{l} = A\, dl \cos \alpha$ is the work done by the force on the contour element $d\mathbf{l}$ and C is the work performed in covering the whole contour.

Stokes' theorem. We shall now prove that the circulation of the vector $\mathbf{A}$ around the contour can be replaced by the surface integral "pulled over" the contour.

Let us consider the projection of an infinitely small rectangular contour onto the plane yz. Let this projection also have the form of a rectangle shown in Fig. 19. We shall calculate the circulation of $\mathbf{A}$ around

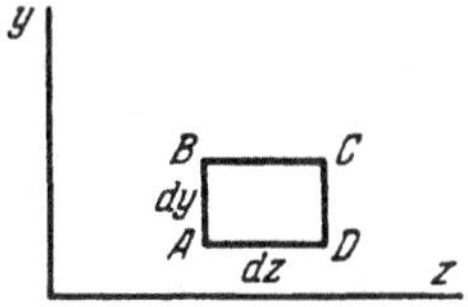

Fig. 19

this rectangle. The side AB contributes a component $A_y(z)\,dy$ and side CD the component $-A_y(z+dz)\,dy$, where the minus sign must be written because the direction of the vector CD is opposite to that of the vector AB. We obtain, for the sum due to the sides AB and CD,

$$-A_y(z+dz)\,dy+A_y(z)\,dy=-\frac{\partial A_y}{\partial z}\,dy\,dz$$

(we have expanded $A_y(z+dz)$ in a series for dz), while for the sides BC and DA,

$$A_z(y+dy)\,dz-A_z(y)\,dz=\frac{\partial A_z}{\partial y}\,dy\,dz\,.$$

The resultant value for circulation in the yz-plane is

$$dC=\left(\frac{\partial A_z}{\partial y}-\frac{\partial A_y}{\partial z}\right)dy\,dz=\left(\frac{\partial A_z}{\partial y}-\frac{\partial A_y}{\partial z}\right)ds_x\equiv B_x\,ds_x\,. \qquad (11.11)$$

The notation B_x is clear from the equation. Let us now find out what meaning this expression has. From the definition of (11.10), circulation is a scalar quantity and, hence, on the right side of equation (11.11) there must also be a scalar quantity. If the contour lies in the plane yz, this quantity is of the form $dC=B_x\,ds_x$; consequently, for an arbitrary orientation of the contour, the relationship (11.11) must have the form of the scalar product

$$dC=B_x\,ds_x+B_y\,ds_y+B_z\,ds_z=\mathbf{B}\,d\mathbf{s}, \qquad (11.12)$$

where B_x, B_y, B_z must necessarily be the components of a vector, since, otherwise, dC could not be a scalar. From (11.11),

$$B_x=\frac{\partial A_z}{\partial y}-\frac{\partial A_y}{\partial z}\,. \qquad (11.13)$$

In order to find the circulation for infinitely small contours in the xz, yz planes, it is sufficient to perform a cyclic permutation of the indices x, y, z. This permutation yields the components B_y, B_z:

$$B_y=\frac{\partial A_x}{\partial z}-\frac{\partial A_z}{\partial x}\,, \qquad (11.14)$$

$$B_z=\frac{\partial A_y}{\partial x}-\frac{\partial A_x}{\partial y}\,. \qquad (11.15)$$

The vector $\mathbf{B}$ has a special name: it is called the *rotation* or *curl* of the vector $\mathbf{A}$ and is denoted thus:

$$\mathbf{B}=\operatorname{rot}\mathbf{A}.$$

rot $\mathbf{A}$ is expressed in terms of unit vectors $\mathbf{i}$, $\mathbf{j}$, $\mathbf{k}$, directed along the coordinate axes:

$$\mathbf{B} = \operatorname{rot} \mathbf{A} = \mathbf{i}\left(\frac{\partial A_z}{\partial y} - \frac{\partial A_y}{\partial z}\right) + \mathbf{j}\left(\frac{\partial A_x}{\partial z} - \frac{\partial A_z}{\partial x}\right) + \mathbf{k}\left(\frac{\partial A_y}{\partial x} - \frac{\partial A_x}{\partial y}\right). \quad (11.16)$$

Changing to the notation (11.16), we see that the component of rot **A** normal to the area appears in equation (11.11):

$$\int \mathbf{A}\, d\mathbf{l} = \operatorname{rot}_n \mathbf{A}\, ds\,, \quad (11.17)$$

where the subscript n of rot **A** indicates that we must take the projection of rot **A** normal to the area, i.e., coinciding with the vector $d\mathbf{s}$. (11.17) permits us to define rot **A** in a noncoordinate manner, similar to the way that we defined div **A** in (11.8), namely:

$$\operatorname{rot}_n \mathbf{A} = \lim_{s \to 0} \frac{\int \mathbf{A}\, d\mathbf{l}}{s}\,, \quad (11.18)$$

or the projection of rot **A**, normal to the area at the given point, is the limit of the ratio of the circulation of **A**, over the contour of the area, to its value when the contour is contracted into the point.

So that the integral $\int \mathbf{A}\, d\mathbf{l}$ should not become zero, we must have closed vector lines, to some extent following the integration contour, which lines are similar to the closed lines of flow in a liquid during vortex motion. Hence the term curl, or rotation.

If the circulation is calculated from a finite contour then the contour can be broken up into infinitely small cells to form a grid. For the sides of adjacent cells, the circulations mutually cancel since each side is traversed twice in opposite directions; only the circulation along the external contour itself remains. The integral on the right-hand side of equation (11.17) gives the flux of rot **A** across the surface "pulled over" the contour. Thus, we obtain the desired integral theorem

$$\int \mathbf{A}\, d\mathbf{l} = \int \operatorname{rot} \mathbf{A}\, d\mathbf{s}\,, \quad (11.19)$$

which is called Stokes' theorem.

Differentiation along a radius vector. The divergence and rotation of a vector are its derivatives with respect to the vector argument. They can be reduced to a unified notation by means of the following. We introduce the vector symbol ∇ (nabla*) with components

$$\nabla_x \equiv \frac{\partial}{\partial x}, \quad \nabla_y \equiv \frac{\partial}{\partial y}, \quad \nabla_z \equiv \frac{\partial}{\partial z}. \quad (11.20)$$

Then, from (11.7), we obtain for the divergence of A:

---

* Nabla is an ancient musical instrument of triangular shape. This symbol is also called del.

$$\operatorname{div} \mathbf{A} \equiv \nabla_x A_x + \nabla_y A_y + \nabla_z A_z = (\nabla A), \tag{11.21}$$

i.e., a scalar product of nabla and **A**.

From (11.16), we have for the rotation

$$\operatorname{rot} \mathbf{A} \equiv \mathbf{i}\,(\nabla_y A_z - \nabla_z A_y) + \mathbf{j}\,(\nabla_z A_x - \nabla_x A_z) + \mathbf{k}\,(\nabla_x A_y - \nabla_y A_x) \equiv \equiv [\nabla\,\mathbf{A}]. \tag{11.22}$$

We use the identity symbol $\equiv$ here in order to emphasize the fact that we are simply dealing with a new system of notation. We shall see, however, that this system is very convenient in vector analysis. We note, with reference to algebraic operations, that nabla is in all cases similar to a conventional vector. We shall use the expression "multiplication by nabla" if, when nabla operates on any expression, that expression is differentiated. Sometimes, nabla is multiplied by a vector without operating on it as a derivative. In that case it is applied to another vector [see (11.30), (11.32)].

Gradient. If we operate with ∇ on a scalar φ, we obtain a vector which is called the gradient of the scalar φ:

$$\operatorname{grad} \varphi \equiv \nabla\varphi = \mathbf{i}\frac{\partial\varphi}{\partial x} + \mathbf{j}\,\frac{\partial\varphi}{\partial y} + \mathbf{k}\frac{\partial\varphi}{\partial z}. \tag{11.23}$$

Its components are:

$$\nabla_x \varphi = \frac{\partial\varphi}{\partial x}, \quad \nabla_y \varphi = \frac{\partial\varphi}{\partial y}, \quad \nabla_z \varphi = \frac{\partial\varphi}{\partial z}. \tag{11.24}$$

From equations (11.24), it can be seen that the vector $\nabla\varphi$ is perpendicular to the surface $\varphi = \text{const.}$ Indeed, if we take a vector $d\mathbf{l}$ lying on this surface, then, in a displacement $d\mathbf{l}$, φ does not change. This is written as

$$d\varphi = \frac{\partial\varphi}{\partial x}\,dl_x + \frac{\partial\varphi}{\partial y}\,dl_y + \frac{\partial\varphi}{\partial z}\,dl_z = (\nabla\varphi dl) = 0, \tag{11.25}$$

i.e., $\nabla\varphi$ is perpendicular to any vector which lies in the plane tangential to the surface $\varphi = \text{const.}$ at the given point, which accords with our assertion.

Differentiation of products. We now give the rules governing differential operations with ∇.

First of all, the gradient of the product of two scalars is calculated as the derivative of a product:

$$\nabla\varphi\psi = \varphi\nabla\psi + \psi\nabla\varphi. \tag{11.26}$$

The divergence of a product of a scalar with a vector is calculated thus:

$$\operatorname{div} \varphi\,\mathbf{A} = (\nabla_\varphi, \varphi\,\mathbf{A}) + (\nabla_A, \varphi\,\mathbf{A}) = (\mathbf{A}\nabla\varphi) + \varphi\,(\nabla\mathbf{A}) = = \mathbf{A} \operatorname{grad} \varphi + \varphi \operatorname{div} \mathbf{A}. \tag{11.27}$$

Here the indices φ and **A** attached to ∇ show what ∇ is applied to.

We find the rotation of $\varphi\,\mathbf{A}$ in a similar manner:

$$\operatorname{rot}\varphi\,\mathbf{A}=[\nabla_{\varphi},\varphi\,\mathbf{A}]+[\nabla_{A},\varphi\,\mathbf{A}]=[\operatorname{grad}\varphi,\mathbf{A}]+\varphi\operatorname{rot}\mathbf{A}\,. \quad (11.28)$$

Now we shall operate with ∇ on the product of two vectors:

$$\operatorname{div}[\mathbf{AB}]=(\nabla\,[\mathbf{AB}])=(\nabla_{A}\,[\mathbf{AB}])+(\nabla_{B}\,[\mathbf{AB}]).$$

We perform a cyclic permutation in both terms, since ∇ can be treated in the same way as an ordinary vector. In addition, we put $\mathbf{B}$ after ∇_B in the second term, and here, as usual, we must change the sign of the vector product. The result is

$$\operatorname{div}[\mathbf{AB}]=(\mathbf{B}\,[\nabla_{\mathbf{A}}\,\mathbf{A}])-(\mathbf{A}\,[\nabla_{\mathbf{B}}\,\mathbf{B}])=\mathbf{B}\operatorname{rot}\mathbf{A}-\mathbf{A}\operatorname{rot}\mathbf{B}. \quad (11.29)$$

Let us find the rotation of a vector product. Here we must use the relationship $[\mathbf{A}[\mathbf{BC}]]=\mathbf{B}\,(\mathbf{AC})-\mathbf{C}\,(\mathbf{AB})$:

$$\operatorname{rot}[\mathbf{AB}]=[\nabla_{\mathbf{A}}\,[\mathbf{AB}]]+[\nabla_{\mathbf{B}}\,[\mathbf{AB}]]=(\nabla_{\mathbf{A}}\,\mathbf{B})\,\mathbf{A}-(\nabla_{\mathbf{A}}\,\mathbf{A})\,\mathbf{B}+(\nabla_{\mathbf{B}}\,\mathbf{B})\,\mathbf{A}-$$
$$-(\mathbf{A}\nabla_{\mathbf{B}})\,\mathbf{B}=(\mathbf{B}\nabla)\,\mathbf{A}-\mathbf{B}\operatorname{div}\mathbf{A}+\mathbf{A}\operatorname{div}\mathbf{B}-(\mathbf{A}\nabla)\,\mathbf{B}. \quad (11.30)$$

Here we note the new symbols $(\mathbf{B}\nabla)$ and $(\mathbf{A}\nabla)$ operating on the vectors $\mathbf{A}$ and $\mathbf{B}$. Obviously, $(\mathbf{A}\nabla)$ and $(\mathbf{B}\nabla)$ are symbolic scalars, equal, by definition of ∇, to

$$(\mathbf{A}\nabla)=A_x\nabla_x+A_y\nabla_y+A_z\nabla_z\equiv A_x\frac{\partial}{\partial x}+A_y\frac{\partial}{\partial y}+A_z\frac{\partial}{\partial z}\,, \quad (11.31)$$

and similarly for $(\mathbf{B}\nabla)$. Then, $(\mathbf{A}\nabla)\,\mathbf{B}$ is a vector which is obtained by application of the operation (11.31) to all the components of $\mathbf{B}$.

Of the operations of this kind, we have yet to calculate grad $\mathbf{AB}$:

$$\operatorname{grad}(\mathbf{AB})=\nabla_{\mathbf{A}}\,(\mathbf{AB})+\nabla_{\mathbf{B}}\,(\mathbf{AB}).$$

We use the same transformation as in the preceding case:

$$\operatorname{grad}(\mathbf{AB})=(\mathbf{B}\nabla_{\mathbf{A}})\,\mathbf{A}+[\mathbf{B}\,[\nabla_{\mathbf{A}}\mathbf{A}]]+(\mathbf{A}\nabla_{\mathbf{B}})\,\mathbf{B}+[\mathbf{A}\,[\nabla_{\mathbf{B}}\mathbf{B}]]=$$
$$=(\mathbf{B}\nabla)\,\mathbf{A}+(\mathbf{A}\nabla)\,\mathbf{B}+[\mathbf{B}\operatorname{rot}\mathbf{A}]+[\mathbf{A}\operatorname{rot}\mathbf{B}]. \quad (11.32)$$

Certain special formulae. We note certain essential cases of operations involving ∇.

From the definition of divergence (11.7), we obtain from (11.27) and (11.9)

$$\operatorname{div}\frac{\mathbf{r}}{r^3}=\frac{1}{r^3}\operatorname{div}\mathbf{r}+\mathbf{r}\operatorname{grad}\frac{1}{r^3}=\frac{3}{r^3}-\frac{3\,(\mathbf{rr})}{r^5}=0\,. \quad (11.33)$$

Further,

$$\operatorname{rot}_x\mathbf{r}=\frac{\partial z}{\partial y}-\frac{\partial y}{\partial z}=0$$

and in general

$$\operatorname{rot}\mathbf{r}=0. \quad (11.34)$$

We now take

$$(\mathbf{A}\nabla)\, x = A_x \frac{\partial x}{\partial x} + A_y \frac{\partial x}{\partial y} + A_z \frac{\partial x}{\partial z} = A_x ,$$

and for all components of **r** at once

$$(\mathbf{A}\nabla)\, \mathbf{r} = \mathbf{A}. \qquad (11.35)$$

In addition, we apply ∇ to a vector depending only on the absolute value of the radius vector. We note first of all that

$$\frac{\partial r}{\partial x} = \frac{\partial}{\partial x} \sqrt{x^2 + y^2 + z^2} = \frac{x}{\sqrt{x^2 + y^2 + z^2}} = \frac{x}{r}$$

[Cf. (3.3.), where $1/r$ is differentiated], so that

$$\nabla r = \frac{\mathbf{r}}{r}. \qquad (11.36)$$

Using the rule for differentiating a function of a function, we have

$$\operatorname{div} \mathbf{A}(r) = \left(\frac{d\mathbf{A}}{dr} \nabla r\right) = \frac{\dot{\mathbf{A}}\mathbf{r}}{r}. \qquad (11.37)$$

Here $\dot{\mathbf{A}}$ is a total derivative of $\mathbf{A}(r)$ with respect to the argument r, i.e., a vector whose components are the derivatives of the three components of $\mathbf{A}(r)$ with respect to r: $\dot{A}_x$, $\dot{A}_y$, $\dot{A}_z$.

Further,

$$\operatorname{rot} \mathbf{A}(r) = \left[\Delta r \frac{d\mathbf{A}}{dr}\right] = \frac{[\mathbf{r}\dot{\mathbf{A}}]}{r}. \qquad (11.38)$$

Repeated differentiation. Let us investigate certain results concerning repeated operations with ∇.

The rotation of the gradient of any scalar is equal to zero:

$$\operatorname{rot} \operatorname{grad} \varphi = [\nabla, \nabla\varphi] = [\nabla\nabla]\, \varphi = 0 , \qquad (11.39)$$

since the vector product of any vector (including ∇) by itself is equal to zero. This can also be seen by expanding rot grad φ in terms of its components. The divergence of a rotation is also equal to zero:

$$\operatorname{div} \operatorname{rot} \mathbf{A} = (\nabla\, [\nabla\mathbf{A}]) = ([\nabla\nabla]\, \mathbf{A}) = 0 . \qquad (11.40)$$

Let us write down the divergence of the gradient of a scalar φ in component form. From equations (11.7) and (11.24) we have

$$\operatorname{div} \operatorname{grad} \varphi = (\nabla\nabla)\, \varphi = \frac{\partial^2 \varphi}{\partial x^2} + \frac{\partial^2 \varphi}{\partial y^2} + \frac{\partial^2 \varphi}{\partial z^2} \equiv \Delta \varphi . \qquad (11.41)$$

Here Δ (delta) is the so-called Laplacian operator, or Laplacian:

$$\Delta = \frac{\partial^2}{\partial x^2} + \frac{\partial^2}{\partial y^2} + \frac{\partial^2}{\partial z^2} .$$

Finally, the rotation of a rotation can be expanded as a double vector product:

$$\text{rot rot } \mathbf{A} = [\nabla [\nabla \mathbf{A}]] = \nabla (\nabla \mathbf{A}) - (\nabla \nabla) \mathbf{A} = \text{grad div } \mathbf{A} - \Delta \mathbf{A} . \quad (11.42)$$

The last equation can be regarded as a definition of $\Delta \mathbf{A}$. In curvilinear coordinates, $\Delta \varphi$ and $\Delta \mathbf{A}$ are expressed differently.

Curvilinear coordinates. We shall further show how the gradient, divergence, and rotation, as well as Δ of a scalar appear in curvilinear coordinates.

Curvilinear coordinates q_1, q_2, q_3 are termed orthogonal if only the quadratic terms dq_1^2, dq_2^2, dq_3^2 appear in the expression for the element of length dl^2, and not the products $dq_1 \, dq_2$, $dq_1 \, dq_3$, $dq_2 \, dq_3$, similar to the way that $dl^2 = dx^2 + dy^2 + dz^2$ in rectangular coordinates. In orthogonal coordinates

$$dl^2 = h_1^2 \, dq_1^2 + h_2^2 \, dq_2^2 + h_3^2 \, dq_3^2 . \quad (11.43)$$

For example, in spherical coordinates $q_1 = r$, $q_2 = \vartheta$, $q_3 = \varphi$. The element of length is

$$dl^2 = dr^2 + r^2 \, d\vartheta^2 + r^2 \sin^2 \vartheta \, d\varphi^2 ,$$

so that

$$h_1 = 1, \; h_2 = r, \; h_3 = r \sin \vartheta .$$

Let us construct an elementary parallelepiped (Fig. 20). Then the components of the gradient will be

$$\left. \begin{aligned} \text{grad}_1 \psi &= \frac{1}{h_1} \frac{\partial \psi}{\partial q_1} , \\ \text{grad}_2 \psi &= \frac{1}{h_2} \frac{\partial \psi}{\partial q_2} , \\ \text{grad}_3 \psi &= \frac{1}{h_3} \frac{\partial \psi}{\partial q_3} . \end{aligned} \right\} \quad (11.44)$$

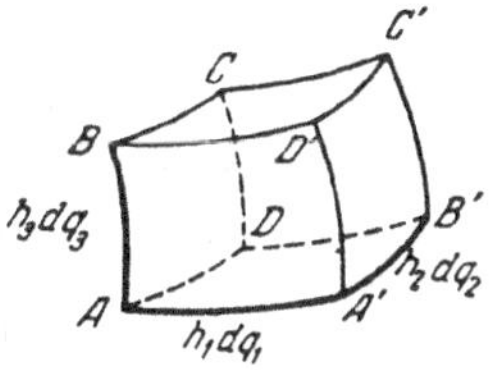

Fig. 20

In order to find the divergence we repeat the proof of the Gauss-Ostrogradsky theorem for Fig. 20. The area $ADCB$ is equal to $h_2 \, h_3 \, dq_2 \, dq_3$. The flux of vector $\mathbf{A}$ through it is

$$A_1 (q_1) \, h_2 \, h_3 \, dq_2 \, dq_3 .$$

Here, h_2 and h_3 are also taken for a definite value of q_1. The sum of the fluxes through the areas $ADCB$ and $A'B'C'D'$ is

$$\frac{\partial}{\partial q_1} (h_2 \, h_3 \, A_1) \, dq_1 \, dq_2 \, dq_3 ,$$

where we have used the expansion of the quantity $h_2 \, h_3 \, A_1$ at the point $q_1 + dq_1$ in terms of dq_1, in a way similar to (11.4). The total flux across all the boundaries is

$$dJ=\left[\frac{\partial}{\partial q_1}(h_2 h_3 A_1)+\frac{\partial}{\partial q_2}(h_3 h_1 A_2)+\frac{\partial}{\partial q_3}(h_1 h_2 A_3)\right]dq_1\,dq_2\,dq_3\,.$$

Let us now take advantage of the definition of divergence (11.8):

$$dJ=\operatorname{div}\mathbf{A}\cdot h_1 h_2 h_3\,dq_1\,dq_2\,dq_3=\operatorname{div}\mathbf{A}\,dV\,.$$

Hence,

$$\operatorname{div}\mathbf{A}=\frac{1}{h_1 h_2 h_3}\left[\frac{\partial}{\partial q_1}(h_2 h_3 A_1)+\frac{\partial}{\partial q_2}(h_3 h_1 A_2)+\frac{\partial}{\partial q_3}(h_1 h_2 A_3)\right]. \tag{11.45}$$

If, instead of A_1, A_2, A_3, we substitute the expressions (11.44), the result will be the Laplacian of a scalar in orthogonal curvilinear coordinates. Thus, in spherical coordinates it is

$$\Delta\psi=\frac{1}{r^2}\frac{\partial}{\partial r}r^2\frac{\partial\psi}{\partial r}+\frac{1}{r^2\sin\vartheta}\frac{\partial}{\partial\vartheta}\sin\vartheta\frac{\partial\psi}{\partial\vartheta}+\frac{1}{r^2}\frac{\partial^2\psi}{\partial\varphi^2}\,. \tag{11.46}$$

With the aid of Stokes' theorem, we can also calculate the rotation in curvilinear coordinates. We shall give it for reference without proof:

$$\left.\begin{aligned}\operatorname{rot}_1\mathbf{A}&=\frac{1}{h_2h_3}\left(\frac{\partial}{\partial q_2}A_3h_3-\frac{\partial}{\partial q_3}A_2h_2\right),\\ \operatorname{rot}_2\mathbf{A}&=\frac{1}{h_3h_1}\left(\frac{\partial}{\partial q_3}A_1h_1-\frac{\partial}{\partial q_1}A_3h_3\right),\\ \operatorname{rot}_3\mathbf{A}&=\frac{1}{h_1h_2}\left(\frac{\partial}{\partial q_1}A_2h_2-\frac{\partial}{\partial q_2}A_1h_1\right).\end{aligned}\right\} \tag{11.47}$$

Exercises

Where (from the requirements of the problem) expressing in terms of coordinates is not demanded, it is recommended that only the vector equations of the present section (11.26)-(11.42) be used.

1) Calculate the expressions: Answers:

a) $\Delta\frac{1}{r}$ $(r\neq 0)$.	$\Delta\frac{1}{r}=\operatorname{div}\operatorname{grad}\frac{1}{r}=-\operatorname{div}\frac{\mathbf{r}}{r^3}=0$.
b) $\operatorname{div}\varphi(r)\,\mathbf{r}$, $\operatorname{rot}\varphi(r)\,\mathbf{r}$.	$3\varphi+r\dot{\varphi}$; 0.
c) $\nabla(\mathbf{Ar})$, где $\mathbf{A}=\text{const}$.	$\mathbf{A}$.
d) $\nabla(\mathbf{A}(r)\,\mathbf{r})$.	$\mathbf{A}+\frac{\mathbf{r}}{r}(\mathbf{r}\dot{\mathbf{A}})$.
e) $\operatorname{div}\varphi(r)\,\mathbf{A}(r)$, $\operatorname{rot}\varphi(r)\,\mathbf{A}(r)$.	$\frac{\dot{\varphi}}{r}(\mathbf{rA})+\frac{\varphi}{r}(\mathbf{r}\dot{\mathbf{A}})$; $\frac{\dot{\varphi}}{r}[\mathbf{rA}]+\frac{\varphi}{r}[\mathbf{r}\dot{\mathbf{A}}]$.
f) $\operatorname{div}[\mathbf{r}[\mathbf{Ar}]]$, $\mathbf{A}=\text{const}$.	$-2(\mathbf{Ar})$.
g) $\operatorname{rot}[\mathbf{r}[\mathbf{Ar}]]$, $\mathbf{A}=\text{const}$.	$3[\mathbf{rA}]$.
h) $\Delta\mathbf{A}(r)$ [см. (11,42)].	$\ddot{\mathbf{A}}+\frac{2}{r}\dot{\mathbf{A}}$.
i) $\nabla(\mathbf{A}(r)\,\mathbf{B}(r))$.	$\frac{\mathbf{r}}{r}(\mathbf{A}\dot{\mathbf{B}})+\frac{\mathbf{r}}{r}(\dot{\mathbf{A}}\mathbf{B})$.
j) $\operatorname{rot}[\mathbf{Ar}]$, $\mathbf{A}=\text{const}$.	$2\mathbf{A}$.
k) $\operatorname{div}[\mathbf{Ar}]$, $\mathbf{A}=\text{const}$.	0.
l) $\Delta\frac{\mathbf{r}}{r}$.	$-\frac{2\mathbf{r}}{r^3}$.

2) Write down $\Delta\psi$ in cylindrical coordinates.

3) Write down the three components of $\Delta\mathbf{A}$ in spherical coordinates.

4) Two closed contours are given. The radius vector of points of the first contour is $\mathbf{r}_1$, of the second contour, $\mathbf{r}_2$. The elements of length along each contour are $d\mathbf{l}_1$ and $d\mathbf{l}_2$, respectively. Prove that the integral

$$\alpha = \int \left(d\mathbf{l}_2 \int \left[d\mathbf{l}_1 \nabla_1 \frac{1}{|\mathbf{r}_1 - \mathbf{r}_2|} \right] \right)$$

is equal to zero, 4π, 8π, and $n \cdot 4\pi$, depending upon how many times the first contour is wound round the second, linking up with the latter. ∇_1 denotes differentiation with respect to $\mathbf{r}_1$ (Ampere's theorem).

Changing the order of integration and performing a cyclic permutation of the factors, we have

$$\alpha = \int \int \left(d\mathbf{l}_1 \left[\nabla_1 \frac{1}{|\mathbf{r}_1 - \mathbf{r}_2|}, \; d\mathbf{l}_2 \right] \right).$$

We apply Stokes' theorem (11.19) to the integral in $d\mathbf{l}_1$:

$$\alpha = \int \int \left(\operatorname{rot}_1 \left[\nabla_1 \frac{1}{|\mathbf{r}_1 - \mathbf{r}_2|}, \; d\mathbf{l}_2 \right] d\mathbf{s}_1 \right).$$

We use equation (11.30); rot_1 denotes differentiation with respect to the components $\mathbf{r}_1$; and $d\mathbf{l}_2$ in such a differentiation may be regarded as a constant vector:

$$\operatorname{rot}_1 \left[\nabla_1 \frac{1}{|\mathbf{r}_1 - \mathbf{r}_2|}, \; d\mathbf{l}_2 \right] = (d\mathbf{l}_2 \nabla_1) \nabla_1 \frac{1}{|\mathbf{r}_1 - \mathbf{r}_2|} - d\mathbf{l}_2 \operatorname{div}_1 \nabla_1 \frac{1}{|\mathbf{r}_1 - \mathbf{r}_2|}.$$

In accordance with exercise 1a, the last term containing $\Delta_1 \dfrac{1}{|\mathbf{r}_1 - \mathbf{r}_2|}$ is equal to zero. There remains, therefore,

$$\operatorname{rot}_1 \left[\nabla_1 \frac{1}{|\mathbf{r}_1 - \mathbf{r}_2|}, \; d\mathbf{l}_2 \right] = (d\mathbf{l}_2 \nabla_1) \nabla_1 \frac{1}{|\mathbf{r}_1 - \mathbf{r}_2|} = -(d\mathbf{l}_2 \nabla_2) \nabla_1 \frac{1}{|\mathbf{r}_1 - \mathbf{r}_2|},$$

since a function of the difference $\mathbf{r}_1 - \mathbf{r}_2$ is differentiated.

For short, we write $\mathbf{r} \equiv \mathbf{r}_1 - \mathbf{r}_2$. Then the required integral will be

$$\alpha = -\int (d\mathbf{l}_2 \nabla_2) \int d\mathbf{s}_1 \nabla_1 \frac{1}{|\mathbf{r}_1 - \mathbf{r}_2|} = -\int (d\mathbf{l}_2 \nabla_2) \int d\mathbf{s}_1 \nabla \frac{1}{r}.$$

We shall now explain the geometrical sense of the second integrand, i.e., $d\mathbf{s}_1 \nabla \dfrac{1}{r} = -\dfrac{(d\mathbf{s}_1 \mathbf{r})}{r} \cdot \dfrac{1}{r^2}$. The scalar product $\dfrac{(d\mathbf{s}_1 \mathbf{r})}{r}$ is the projection of an element of the surface ds_1 pulled over the first contour, on the radius vector $\mathbf{r}$ drawn from a point on the second contour. In other words, $\dfrac{(d\mathbf{s}_1 \mathbf{r})}{r}$ is equal to the projection of the area ds_1 on a plane perpendicular to $\mathbf{r}$. This projection, divided by r^2, is equal to the solid angle $d\Omega$ at a point $\mathbf{r}_2$ on the second contour subtended by the area ds_1. The integral $\displaystyle\int \frac{(d\mathbf{s}_1 \mathbf{r})}{r^3}$ is, therefore, that solid angle Ω which is obtained if a cone is drawn with vertex at the point $\mathbf{r}_2$, so that the generating line of the cone formed the contour l_1.

The differential $(d\mathbf{l}_2 \nabla_2)\,\Omega$ is the increment of solid angle Ω obtained in shifting along the contour l_2 a distance $d\mathbf{l}_2$. Thus,

$$\alpha = \int (d\mathbf{l}_2 \nabla_2)\,\Omega = \int d\Omega = \Omega_1 - \Omega_2.$$

The integral of this quantity around a closed contour is equal to the total change in solid angle in traversing the contour l_2. Let the initial point of circumvention lie on the surface s_1. Then the solid angle subtended by the surface at the origin is -2π. If the contours are linked, then the solid angle will be 2π after the circumvention, since the area is observed from a terminal point on the other side. If the contours are not linked, then the solid angle is once again its initial value, -2π, and the integral is equal to zero. Thus, when the contours are linked n times, the integral in $d\mathbf{l}_2$ is equal to $4\pi n$.

Sec. 12. The Electromagnetic Field. Maxwell's Equations

Interaction in mechanics and in electrodynamics. The interaction of charged bodies in electrodynamics is principally an interaction of charges with an electromagnetic field. However, the physical concept of the field in electrodynamics differs essentially from the field concept in Newtonian mechanics.

We know that the space in which gravitational forces act is called a gravitational field. The values of these forces at any point of the field is determined, in Newtonian mechanics, by the instantaneous positions of the gravitating bodies, no matter how far they are from the given point. In electrodynamics, such a field representation is not satisfactory: during the time that it takes an electromagnetic disturbance to move from one charge to another, the latter can move a very great distance. Elementary charges (electrons, protons, mesons) very often have velocities close to the velocity of propagation of electromagnetic disturbances.

Modern gravitational theory (the general theory of relativity, see Sec. 20) shows that gravitational interaction, too, propagates with a finite velocity. But since macroscopic bodies move considerably slower, within the scale of the solar system, the finite velocity of propagation of gravitational forces introduces only an insignificant correction to the laws of motion of Newtonian mechanics.

In the electrodynamics of elementary charges, the finite velocity of propagation of electromagnetic disturbances is of fundamental significance. When speaking of point charges, the action of a field on the charge is always determined only by the field at the point where the charge is located, and only at the instant when the charge is at this point. As opposed to the "action at a distance" of Newtonian mechanics, such interactions are termed "short-range."

If the energy or momentum of a charged particle is changed under the action of a field, they can be imparted directly only to the electromagnetic field, since a finite interval of time is necessary for the energy and momentum of other particles to be changed. But this means that the electromagnetic field itself possesses energy and momentum, whereas in Newtonian mechanics it was sufficient to assume that only the interacting particles possessed energy and momentum. It follows from this that the electromagnetic field is itself a real physical entity to exactly the same extent as the charged particles. The equations of electrodynamics must describe directly the propagation of electromagnetic disturbances in space and the interaction of charges with the field.

Interaction between charges is effected through the electromagnetic field. Such laws as the Coulomb or Biot-Savart laws (in which only the instantaneous positions and the instantaneous velocities of the charges appear) are of an approximate nature and are valid only when the relative velocities of the charges are small compared with the propagation velocity of electromagnetic disturbances.

It will be shown later that this velocity is a fundamental constant which appears in the equations of electrodynamics. It is equal to the velocity of light *in vacuo* and, to a high degree of precision, is 3×10^{10} cm/sec.

A field in the absence of charges. The independent reality of the electromagnetic field is particularly evident from the fact that electrodynamic equations admit of a solution in the absence of charges. These solutions describe electromagnetic waves, in particular light waves, in free space. Thus, electrodynamics has shown that light is electromagnetic in nature.

In the course of two centuries, the protagonists of the wave theory of light considered that light waves were propagated by a special elastic medium permeating all space, the so-called "ether." In order to represent the spread of oscillations it was, naturally, necessary to have something oscillating. This "something" was called the *ether*. Proceeding from an analogy with the propagation of sound waves in a continuous medium, the ether was endowed with the properties of a fluid, physical phenomena being explained simply by reducing them to definite mechanical displacements of bodies. In particular, light phenomena were regarded as displacements of particles of the special medium, the ether.

In this, a peculiar "abhorrence of a vacuum" was apparent or, more exactly, a purely speculative representation of empty space where "nothing exists" and, hence, where nothing can occur. Physicists did not at once come to realize that the electromagnetic field itself was just as real as the more tangible "ponderable matter." Electrodynamic laws are those elementary concepts, from which the interaction of atoms should be deduced, which interaction accounts for the properties of real fluids that are incomparably more complicated than the properties of a field in "empty space," i.e., in the absence of charges. There is no sense in reducing a field to an imaginary fluid merely in order to avoid the idea of "empty space." Physical space is the carrier of the electromagnetic field and is, therefore, inseparable from the state and motion of real objects. As regards the term "ether," which still persists in the field of radio, it expresses nothing other than the electromagnetic field.

The electromagnetic field. Let us now establish the basic equations of electrodynamics. We shall proceed from certain elementary laws, which we assume the reader knows from a general course of physics or electricity. These laws will first be used in the absence of matter consisting of atoms or, as is usually said in electrodynamics, in the absence of a "material medium." By this term we must not understand any encroachment on the material nature of the electromagnetic field itself. From the electrodynamic equations for free space we shall, later on, derive the equations for an electromagnetic field in a medium (a conductor or dielectric).

As is known, the electromagnetic field in a medium is described by four vector quantities: the electric field, the electric induction, the magnetic field, and the magnetic induction. The force acting on unit electric charge at a given point in space is called the electric field intensity. In future, instead of the field intensity, we shall simply speak of the *field* at a given point in space. The magnetic field intensity or, for short, the magnetic field is defined analogously. Separate magnetic charges, unlike electric charges, do not exist in nature; however, if we make a long permanent magnet in the form of a needle, then the magnetic force acting at its ends will be the same as if there existed point charges at the ends.

A rigorous definition of the electric and magnetic induction vectors will be given in Sec. 16, where the field equations in a medium will be derived from the equations for point charges in free space. It need only be recalled that in free space there is no need to use four vectors for a description of the electromagnetic field, only two vectors being sufficient: the electric and magnetic fields.

System of units. We shall consider that all electromagnetic quantities are expressed in the CGSE system, i.e., in the absolute electrostatic system of units. In this system the dimensions of electric charge are $\text{gm}^{1/2} \cdot \text{cm}^{3/2}/\text{sec}$ and the dimensions of field intensity, both electric and magnetic, are $\text{gm}^{1/2}/\text{cm}^{1/2} \cdot \text{sec}$. If we substitute charge, expressed in this system of units, into the equation for Coulomb's law, then the interaction force between charges is expressed in dynes ($\text{gm} \cdot \text{cm}/\text{sec}^2$).

Electromotive force. Let us recall the definition for electromotive force in a circuit: this is the *work* performed by the forces of the electric field when unit charge is taken along the given *closed* circuit. And it is absolutely immaterial what the given circuit represents: whether it is filled with a conductor or whether it is merely a closed line drawn in space. Let us write down the expression for electromotive force (abbreviated as e. m. f.) in the notation of Sec. 11. The force acting on unit charge at a given point is the electric field $\mathbf{E}$. The work done by this force on an element of path $d\mathbf{l}$ is the scalar product $\mathbf{E}d\mathbf{l}$. Then, the work done on the whole closed circuit, or the e. m. f., is equal to the integral

$$\text{e.m.f.} = \int \mathbf{E}\, d\mathbf{l}\,. \tag{12.1}$$

Magnetic-field flux across a surface. Let us suppose that some surface is bounded by the given circuit. We shall denote the magnetic field by the letter $\mathbf{H}$. The magnetic-field flux through an element of the chosen surface is, by the definition given in Sec. 11, $d\,\mathbf{\Phi} = \mathbf{H} d\mathbf{S}$. The magnetic-field flux through the whole surface, bounded by the circuit, is

$$\Phi = \int \mathbf{H} d\mathbf{s}\,. \tag{12.2}$$

It can be conveniently represented thus. Let us consider a section of the surface through which unit flux $\Delta\,\mathbf{\Phi} = 1$ (in the CGSE system) passes. We draw through this section of the surface a line tangential to the direction of the field at some point on the surface. A line which is tangential to the direction of the field at its points is called a *magnetic line of force*. For this reason, the total flux $\mathbf{\Phi}$ is equal, by definition, to the number of magnetic lines of force crossing the surface.

Magnetic lines of force are either closed or extended to infinity. Indeed, a magnetic line of force may begin or end only at a single charge, but separate magnetic charges do not exist in nature. In a permanent magnet the lines of force are completed inside the magnet.

From this it follows that a magnetic flux through any surface, bounded by a circuit, is the same at a given instant. Otherwise, a number of the magnetic lines of force would have to begin or end in the space between the surfaces through which different fluxes pass. Consequently, at a given instant, a constant number of magnetic lines of force, i.e., a constant magnetic field flux passes across any surface bounded by the circuit. Therefore, the flux can be ascribed to the circuit itself, irrespective of the surface for which it is calculated.

Faraday's induction law. Faraday's induction law is written in the form of the following equation:

$$\text{e.m.f.} = -\frac{1}{c}\frac{\partial \mathbf{\Phi}}{\partial t}\,. \tag{12.3}$$

If all the quantities are expressed in the CGSE system, then the constant of proportionality c is a universal constant with the dimensions of velocity equal to 3×10^{10} cm/sec.

Usually, Faraday's law is applied to circuits of conductors; however, e. m. f. is simply the quantity of work performed by unit charge in moving along the circuit, and, for a given field value through the circuit, cannot depend upon the form of the circuit. The e. m. f. is simply equal to the integral $\int \mathbf{E} d\mathbf{l}$. In a conducting circuit, this work can be dissipated in the generation of Joule heat ("an ohmic load"). However, it is completely justifiable to consider the circuit in a vacuum also. In this case, the work performed on the charge is spent in increasing the kinetic energy of the charged particle, as, for instance in the case in an induction accelerator, the betatron.

Maxwell's equation for rot E. Thus, equation (12.3) refers to any arbitrary closed circuit. We substitute the definitions (12.1) and (12.2) into this equation:

$$\int \mathbf{E}\, d\mathbf{l} = -\frac{1}{c}\frac{\partial}{\partial t}\int \mathbf{H}\, d\mathbf{s}. \tag{12.4}$$

The left-hand side of the equation can be transformed by the Stokes theorem (11.19) and, on the right-hand side, the order of the time differentiation and surface integration can be interchanged, since they are performed for independent variables. In addition, taking this integral over to the left-hand side, we obtain

$$\int \left(\operatorname{rot} \mathbf{E} + \frac{1}{c}\frac{\partial \mathbf{H}}{\partial t}\right) d\mathbf{s} = 0. \tag{12.5}$$

But, the initial circuit is completely arbitrary, i.e., it can have arbitrary magnitude and shape. Let us assume that the integrand, in parentheses, of (12.5) is not equal to zero. Then we can choose the surface and the circuit that bounds it so that the integral (12.5) does not become zero. Thus, in all cases, the following equation must be satisfied:

$$\operatorname{rot} \mathbf{E} + \frac{1}{c}\frac{\partial \mathbf{H}}{\partial t} = 0. \tag{12.6}$$

In comparison with (12.3), this equation does not contain anything new physically; it is the same induction law, but rewritten in differential form for an infinitely small circuit (contour). In many applications the differential form is more convenient than the integral form.

We shall see later that the constant c is equal to the velocity of light in free space.

The equation for div H. As we have already said, magnetic lines of force are either closed or go off to infinity. Hence, in any closed surface, the same number of magnetic-field lines enter as leave. The magnetic-field flux in free space, across any closed surface, is equal to zero:

$$\int \mathbf{H}\, d\mathbf{s} = 0. \tag{12.7}$$

Transforming this integral to a volume integral according to the Gauss-Ostrogradsky theorem (11.6), we obtain

$$\int \operatorname{div} \mathbf{H}\, dV = 0. \tag{12.8}$$

Due to the fact that the surface bounding the volume is completely arbitrary, we can always choose this volume to be so small that the integral is taken over the region in which div **H** has constant sign if it is not equal to zero. But then, in spite of (12.7) and (12.8), $\int \operatorname{div} \mathbf{H}\, dV$ will not be equal to zero. For this reason, the divergence of **H** must become zero:

$$\operatorname{div} \mathbf{H} = 0 \,. \tag{12.9}$$

(12.9) is the differential form of (12.7) for an infinitely small volume. In Sec. 11 it was shown that the divergence of a vector is the density of sources of a vector field. The sources of the field are free charges from which the vector (force) magnetic-field lines originate. Thus, (12.9) indicates the absence of free magnetic charges.

Equations (12.6) and (12.9) are together called the *first pair of Maxwell's equations.*

Let us now introduce the second pair.

The equation for div E. The electric-field flux through a closed surface is not equal to zero, but to the total electric charge e inside the surface multiplied by 4π (Gauss' theorem):

$$\int \mathbf{E} \, d\mathbf{s} = 4\pi e \,. \tag{12.10}$$

This theorem is derived from Coulomb's law for point charges. The field due to a point charge e is expressed by the following equation:

$$\mathbf{E} = \frac{e}{r^2} \frac{\mathbf{r}}{r} \,.$$

Here, $\mathbf{r}$ is a radius vector drawn from the point situated at the charge to the point where the field is defined. The field is inversely proportional to r^2 and is directed along the radius vector.

Let us surround the charge by a spherical surface centred on the charge. The element of surface for the sphere $d\mathbf{s}$ is $r^2 \, d\Omega \, \frac{\mathbf{r}}{r}$, where $d\Omega$ is an elementary solid angle and $\frac{\mathbf{r}}{r}$ indicates the direction of the normal to the surface. The flux of the field across the surface element is

$$\mathbf{E} \, d\mathbf{s} = \frac{e}{r^2} \cdot \frac{\mathbf{r}}{r} \cdot r^2 \, d\Omega \frac{\mathbf{r}}{r} = e \, d\Omega \,.$$

The flux across the whole surface of the sphere is $\int e \, d\Omega = e \int d\Omega = 4\pi e$. But since lines of force begin only at a charge, the flux will be the same through the sphere as through any closed surface around the charge. Therefore, if there is an arbitrary charge distribution e inside a closed surface, then equation (12.10) holds.

In order to rewrite this equation in differential form, we introduce the concept of charge density. The charge density ρ is the charge contained in unit volume, so that the total charge in the volume is related to the density by the following equation:

$$e = \int \rho \, dV \,. \tag{12.11}$$

Hence, $\rho = \lim\limits_{\Delta V \to 0} \frac{\Delta e}{\Delta V}$. Introducing the charge density in (12.10), we obtain

$$\int (\operatorname{div} \mathbf{E} - 4\pi\rho)\, dV = 0\,. \tag{12.12}$$

Repeating the same argument for this integral as we used for (12.8), we have

$$\operatorname{div} \mathbf{E} = 4\,\pi\,\rho\,. \tag{12.13}$$

According to (11.8) we can say that the density of sources of an electric field is equal to the electric charge density multiplied by $4\,\pi$.

The density function for point charges. The density function for point charges is obtained by a limiting process. Let us initially assume that a finite quantity of charge is distributed in a small, but finite, volume ΔV. Then ρ must be regarded as the ratio $\frac{e}{\Delta V}$.

If we let the volume ΔV tend to zero, then the density will have a very peculiar form: it will turn out to be equal to zero everywhere except at the place where the charge is situated, and at that point it will convert to infinity, since the numerator of the fraction $\frac{e}{\Delta V}$ is finite and the denominator is infinitely small. However, the integral

$$\int \rho\, dV = \frac{e}{\Delta V} \int dV = \frac{e}{\Delta V} \cdot \Delta V = e$$

remains equal to the charge e itself. Thus, the concept of charge density can also be used in the case of a point charge. In this case, ρ is understood to be a function which is equal to zero everywhere except at the point of the charge. The volume integral of this function is either equal to the charge e itself, if the charge is situated inside the integration region, or zero, if the charge is outside the region of integration.

The charge conservation law. One of the most important laws of electrodynamics is the law of conservation of charge. The total charge of any system remains constant if no external charges are brought into it. In all charge transformations occurring in nature, the law of conservation of charge is satisfied with extreme precision (while the law of conservation of mass is approximate!).

In order to formulate the charge-conservation law in differential form, we must introduce the concept of *current density*. This vector quantity is defined as

$$\mathbf{j} = \rho\mathbf{v}\,, \tag{12.14}$$

where $\mathbf{v}$ is the charge velocity at the point where the density ρ is defined. The dimensions of charge density are charge/cm^3, and of current density, charge/cm$^2 \cdot$ sec (i.e., the dimensions of charge passing in unit time through unit area). For a point charge, $\mathbf{v}$ denotes its velocity and ρ the density function defined above.

The total current emerging from an area is

$$I = \int \mathbf{j}\, d\mathbf{s} = \int \rho\,(\mathbf{v}\, d\mathbf{s})\,. \tag{12.15}$$

It must be equal to the reduction of charge inside the surface in unit time, i.e.,

$$I = -\frac{\partial e}{\partial t} \tag{12.16}$$

(the charge-conservation law in integral form). Substituting e from (12.11) and transforming I by the Gauss-Ostrogradsky theorem, we obtain

$$\int \left(\frac{\partial \rho}{\partial t} + \operatorname{div} \mathbf{j}\right) dV = 0 . \tag{12.17}$$

Since the volume, over which the integration is performed is arbitrary, the conservation law for charge in differential form follows from (12.17):

$$\frac{\partial \rho}{\partial t} + \operatorname{div} \mathbf{j} = \frac{\partial \rho}{\partial t} + \operatorname{div} \rho \mathbf{v} = 0 . \tag{12.18}$$

Displacement current. From direct-current theory, it is known that current lines are always closed. Indeed, open circuited lines indicate that there is either an accumulation or deficiency of charge at their ends. But we can also define vector lines such that they will always be closed (or will have to go off to infinity) in the case of alternating currents. For this we substitute the derivative $\frac{\partial \rho}{\partial t}$, according to (12.13), into the equation of the charge-conservation law (12.18). This derivative is equal to $-\frac{1}{4\pi} \operatorname{div} \frac{\partial \mathbf{E}}{\partial t}$. Hence, we always have the relation

$$\operatorname{div}\left(\mathbf{j} + \frac{1}{4\pi} \frac{\partial \mathbf{E}}{\partial t}\right) = 0 . \tag{12.19}$$

Comparing (12.19) and (12.9), we see that the vector lines

$$\mathbf{j} + \frac{1}{4\pi} \frac{\partial \mathbf{E}}{\partial t}$$

are always closed. The vector $\frac{1}{4\pi} \frac{\partial \mathbf{E}}{\partial t}$ is called the *displacement* current density. Together with the charge-transport current $\mathbf{j}$, the displacement current forms a closed system of current lines.

The displacement current can be more vividly demonstrated in the following way. Fig. 21 shows a capacitor whose plates are joined by a conductor. The current I flowing in the conductor is equal to the change of charge on the plates in unit time:

$$I = \frac{\partial e}{\partial t} .$$

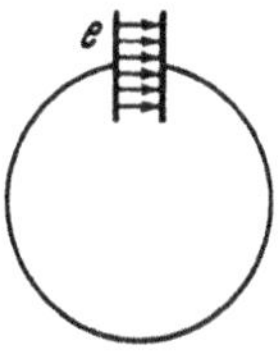

Fig. 21

But the charge on the plates is related to the field in the capacitor by the relationship

$$E = \frac{4\pi e}{f},$$

where f is the area of the plates. Whence

$$I = \frac{f}{4\pi}\frac{\partial E}{\partial t}.$$

Consequently, the quantity $\frac{1}{4\pi}\frac{\partial E}{\partial t}$ can be interpreted as the density of some current which completes the conduction current, of density $j = \frac{I}{f}$. This corresponds to the more general equation (12.19).

The conception that the magnetic action of a displacement current does not differ from the magnetic action of ordinary current is basic to Maxwellian electrodynamics.

Magnetomotive force. By analogy with electromotive force $\int \mathbf{E}\,d\mathbf{l}$, we define the *magnetomotive force* $\int \mathbf{H}\,d\mathbf{l}$, where the integration is performed over a closed circuit. Using the Biot-Savart law for direct currents, it may be shown that the magnetomotive force in a closed circuit is equal to the electric current, I, crossing a surface bounded by the circuit, multiplied by $\frac{4\pi}{c}$. In other words,

$$\int \mathbf{H}\,d\mathbf{l} = \frac{4\pi I}{c}. \tag{12.20}$$

This relationship can be shown most simply by assuming that a direct current I is flowing through an infinitely long straight-line circuit. We shall calculate the magnetomotive force in a circuit of circular form, the current line passing through the centre of the circle perpendicular to its plane. The magnetic field is tangential to the circle and equal to $H = \frac{2I}{cr}$ in accordance with the Biot-Savart law, where r is the radius of the circle. Thus, for the circuit we have chosen, the absolute value of H is constant and the magnetomotive force is equal to $2\pi r H = \frac{4\pi I}{c}$.

For a circuit of arbitrary form, we should use the Biot-Savart law in differential form:

$$d\mathbf{H} = \frac{I\,[d\mathbf{l}_1\,\mathbf{r}]}{c\,r^3}.$$

Then the magnetomotive force is represented by the integral

$$\int \mathbf{H}\,d\mathbf{l} = \frac{I}{c}\int\left(d\mathbf{l}\cdot\int\frac{[d\mathbf{l}_1\,\mathbf{r}]}{r^3}\right).$$

However, in accordance with Ampere's theorem (exercise 4, Sec. 11), this integral is equal to $\frac{4\pi I}{c}$ if the circuit in which the magnetomotive force is calculated is linked with the current-carrying circuit.

The equation for rot H. Following Maxwell, we shall assume that equation (12.20) is also true for displacement current if the field is variable. The current lines will then always be closed, and in calculating magnetomotive force we can use all the arguments that we have used for continuous current. In the case of varying fields and currents, I, in the formula for magnetomotive force, denotes the total current passing through the circuit, i.e., the sum $\int \mathbf{j}\, ds + \frac{1}{4\pi} \int \frac{\partial \mathbf{E}}{\partial t} ds$. Naturally, this assumption is not obvious beforehand and is justified by the fact that Maxwell's equations provide for the explanation or prediction of the entire assemblage of phenomena relating to rapidly changing electromagnetic fields (the displacement current does not usually exist for slowly varying fields).

Let us assume that the total current is formed by the current produced by charge transport (of density equal to $\mathbf{j}$) combined with a displacement current with density $\frac{1}{4\pi} \frac{\partial E}{\partial t}$. In accordance with our assumption,

$$\int \mathbf{H}\, d\mathbf{l} = \frac{4\pi}{c} \int \left(\mathbf{j} + \frac{1}{4\pi} \frac{\partial \mathbf{E}}{\partial t}\right) d\mathbf{s} \equiv \frac{4\pi}{c} I. \qquad (12.21)$$

Transforming the left-hand side of (12.20) in accordance with Stockes' theorem (11.19) and combining with I, we obtain

$$\int \left(\operatorname{rot} \mathbf{H} - \frac{1}{c} \frac{\partial \mathbf{E}}{\partial t} - \frac{4\pi \mathbf{j}}{c}\right) d\mathbf{s} = 0. \qquad (12.22)$$

Applying the same argument to equation (12.22) as applied to (12.5), we arrive at the differential equation

$$\operatorname{rot} \mathbf{H} - \frac{1}{c} \frac{\partial \mathbf{E}}{\partial t} - \frac{4\pi \mathbf{j}}{c} = 0. \qquad (12.23)$$

It is easy to see that this equation agrees with the law of charge conservation. Indeed, we operate on it by div. According to (11.40), div rot $\mathbf{H} = 0$, so that we are left with

$$\frac{1}{c} \frac{\partial}{\partial t} \operatorname{div} \mathbf{E} + \frac{4\pi}{c} \operatorname{div} \mathbf{j} = 0.$$

Substituting div $\mathbf{E}$ from (12.13), we arrive once again at (12.18), i.e., the law of conservation of charge.

Equation (12.23) is not merely an expression for the Biot-Savart law in differential form. In (12.23), we have introduced the displacement current, which is not involved in the theory of continuous currents.

The Maxwell system of equations. Let us once again write down the system of Maxwell's equations for free space.

The first pair:

$$\operatorname{rot} \mathbf{E} = -\frac{1}{c}\frac{\partial \mathbf{H}}{\partial t}, \tag{12.24}$$

$$\operatorname{div} \mathbf{H} = 0. \tag{12.25}$$

The second pair:

$$\operatorname{rot} \mathbf{H} = \frac{1}{c}\frac{\partial \mathbf{E}}{\partial t} + \frac{4\pi \mathbf{j}}{c}, \tag{12.26}$$

$$\operatorname{div} \mathbf{E} = 4\pi\rho. \tag{12.27}$$

In these equations we consider ρ and $\mathbf{j}$, i.e., the charge and current distributions in space, to be known. The unknowns, to be determined, are the fields **E** and **H**. Each of them has three components.

In spite of the fact that both pairs form, together, eight equations, only six of them are independent, according to the number of field components. Indeed the three components of each rotation are constrained by div rot = 0 and, hence, are not independent of one another.

Electromagnetic potentials. We can introduce new unknown quantities such that each equation will contain only one unknown. In this way the overall number of equations is reduced. These new quantities are called electromagnetic potentials.

We choose the potentials so that the first pair of Maxwell's equations are identically satisfied. In order to satisfy equation (12.25), it is sufficient to put

$$\mathbf{H} = \operatorname{rot} \mathbf{A}, \tag{12.28}$$

where **A** is a vector called the *vector potential*. Then, according to (11.40), the divergence of **H** will be equal to zero identically. We shall look for the electric field in the form:

$$\mathbf{E} = -\frac{1}{c}\frac{\partial \mathbf{A}}{\partial t} - \nabla\varphi, \tag{12.29}$$

where φ is a quantity called the *scalar potential*.

From (11.39) rot $\nabla\varphi = 0$. Substituting (12.28) and (12.29) in (12.24), we obtain the identity.

The determinacy of potentials. The electromagnetic fields **E** and **H** are physically determinate quantities since, through them, the forces acting on charges and currents can be expressed. The fields are expressed in terms of potential derivatives. Therefore, potentials are determined only to the accuracy of the expressions that cancel in differentiation. These expressions should be chosen so that the potentials satisfy equations of the simplest form. We shall now find the most general potential transformation which does not change the fields.

From equation (12.28) it can be seen that if we add the gradient of any arbitrary function to the vector potential, the magnetic field will not change, since the rotation of a gradient is identically equal to zero. Putting

$$\mathbf{A} = \mathbf{A}' + \nabla f(x, y, z, t), \tag{12.30}$$

we see that the magnetic field, expressed in terms of such a modified potential, remains unchanged:

$$\mathbf{H} = \operatorname{rot} \mathbf{A} = \operatorname{rot} \mathbf{A}'.$$

In order that the addition of ∇f should not affect the electric field, we must also change the scalar potential:

$$\varphi = \varphi' - \frac{1}{c}\frac{\partial f}{\partial t}, \tag{12.31}$$

where f is the same function as in (12.30). Then, for the electric field, we obtain

$$\mathbf{E} = -\frac{1}{c}\frac{\partial \mathbf{A}}{\partial t} - \nabla\varphi = -\frac{1}{c}\frac{\partial \mathbf{A}'}{\partial t} - \frac{1}{c}\frac{\partial}{\partial t}\nabla f - \nabla\varphi' + \nabla\frac{1}{c}\frac{\partial f}{\partial t} = \\ = -\frac{1}{c}\frac{\partial \mathbf{A}'}{\partial t} - \nabla\varphi'.$$

Consequently, the electric field does not change either. Thus, the potentials are determined to the accuracy of the transformations (12.30), (12.31), which are called *gauge* transformations.

The Lorentz condition. Let us now choose an arbitrary function f such that the second pair of Maxwell's equations leads to equations for the potentials of the simplest possible form. Substituting (12.28) and (12.29) in (12.26) gives

$$\operatorname{rot}\operatorname{rot}\mathbf{A} = -\frac{1}{c^2}\frac{\partial^2 \mathbf{A}}{\partial t^2} - \frac{1}{c}\frac{\partial}{\partial t}\nabla\varphi + \frac{4\pi\mathbf{j}}{c}. \tag{12.32}$$

We express rot rot $\mathbf{A}$ with the aid of (11.42). Then (12.32) is reduced to the following form:

$$-\Delta\mathbf{A} + \frac{1}{c^2}\frac{\partial^2 \mathbf{A}}{\partial t^2} + \nabla\left(\operatorname{div}\mathbf{A} + \frac{1}{c}\frac{\partial\varphi}{\partial t}\right) = \frac{4\pi\mathbf{j}}{c}. \tag{12.33}$$

We shall now try to eliminate the quantity inside the brackets. We denote it, for brevity, by the letter a, and we perform the transformations (12.30) and (12.31) on the potentials. Then the quantity a is reduced to the form

$$a = \operatorname{div}\mathbf{A} + \frac{1}{c}\frac{\partial\varphi}{\partial t} = \operatorname{div}\mathbf{A}' + \frac{1}{c}\frac{\partial\varphi'}{\partial t} + \Delta f - \frac{1}{c^2}\frac{\partial^2 f}{\partial t^2}. \tag{12.34}$$

The function f has, so far, remained arbitrary. Let us now assume that it has been chosen so as to satisfy the equation

$$\Delta f - \frac{1}{c^2}\frac{\partial^2 f}{\partial t^2} = a. \tag{12.35}$$

Then, from (12.34), it is obvious that the potentials will be subject to the condition

$$\operatorname{div} \mathbf{A}' + \frac{1}{c}\frac{\partial \varphi'}{\partial t} = 0 . \tag{12.36}$$

This is called the *Lorentz condition.*

As was shown, the expression of fields in terms of potentials is not changed by a gauge transformation. For this reason we shall always consider, in future, that this transformation is performed so that the Lorentz condition is satisfied; the primes in the potentials can then be omitted.

The equations for potentials. From the Lorentz condition and (12.33), we obtain the equation for a vector potential:

$$\Delta \mathbf{A} - \frac{1}{c^2}\frac{\partial^2 \mathbf{A}}{\partial t^2} = -\frac{4\pi \mathbf{j}}{c} . \tag{12.37}$$

It is now also easy to obtain the equation for a scalar potential. From (12.27) we have

$$\operatorname{div} \mathbf{E} = -\frac{1}{c}\frac{\partial}{\partial t}\operatorname{div} \mathbf{A} - \Delta\varphi = 4\pi\rho .$$

Substituting div **A** from the Lorentz condition (12.36), we obtain

$$\Delta\varphi - \frac{1}{c^2}\frac{\partial^2 \varphi}{\partial t^2} = -4\pi\rho . \tag{12.38}$$

Equations (12.37) and (12.38) each contain only one unknown. Therefore, each equation for potential does not depend on the rest and can be solved separately.

The equations for potential are second order with respect to coordinate and time derivatives. For a solution, it is necessary to give not only the initial values of the potentials, but also the initial values of their time derivatives.

Gauge invariance. As we shall see later, especially in the following section and in Sec. 21, which is devoted to the motion of charges in an electromagnetic field, it is necessary, in many cases, to use equations involving potentials. But, since potentials are ambiguous, we must take care that the form of any equation involving potentials does not change under gauge transformations (12.30) and (12.31),* since such transformations involve a completely arbitrary function f which can be chosen to be of any form. It is clear that no physical result can depend on the choice of this function, i.e., on an arbitrary gauge transformation. In other words, equations involving potentials must be *gauge invariant.*

---

* This does not refer to equations (12.37) and (12.38), from which the potentials are determined in accordance with the condition (12.36).

Sec. 13. The Action Principle for the Electromagnetic Field

The variational principle for the electromagnetic field. In the first part of this book it was shown that the equations of mechanics, obtained from Newton's laws (Sec. 2), lead to the principle of least action (Sec. 10). We obtained the equations of electrodynamics in the preceding section by proceeding from certain simple physical laws and the assumption about the magnetic effect due to displacement current. In this section, Maxwell's equations will be reduced to the variational principle, which is the principle of least action for the electromagnetic field.

Electrodynamics is not equivalent to the mechanics of particle systems or to the mechanics of liquids, which are based on Newton's laws. All the same, to a very considerable extent, electrodynamical laws are analogous to the laws of mechanics. This analogy can best be seen from the principle of least action for the electromagnetic field.

The variational formulation best of all allows us to derive the conservation laws for the electromagnetic field. The corresponding integrals of motion for a field coincide with the well-known mechanical integrals—energy, linear momentum, and angular momentum. In a closed system consisting of charged particles and a field, the total energy, total linear momentum, and total angular momentum of the charges and field are conserved.

In this sense, electrodynamics is indeed "a dynamics" of the electromagnetic field, though this by no means signifies that the laws of electrodynamics can be obtained from Newton's laws. Both are equivalent to certain integral variational principles, but the action functions are, of course, of entirely different form.

It is a noteworthy fact that Maxwell at first tried to construct mechanical models of the ether, but in his later work he rejected them and obtained the general equations of electrodynamics by means of a generalization of known elementary laws of electromagnetism.

The Lagrangian function for a field. In order to formulate the principle of least action it is necessary to have an expression for the Lagrangian. The choice of Lagrangian in mechanics is determined by considerations based on the relativity principle of Newtonian mechanics, which is formulated with the aid of Galilean transformations (Sec. 8). As will be explained in detail in Secs. 20 and 21, Galilean transformations are not valid in electrodynamics and are replaced by the more general Lorentz transformations, based on the Einstein relativity principle. These transformations allow the Lagrangian for the electromagnetic field to be uniquely found; this will be done in Sec. 21. In this section, the choice of Lagrangian is justified by the fact that the already familiar Maxwell equations are obtained from it. Similarly, in Part I, the principle of least action was formulated after Lagrange's equations had been obtained on the basis of Newton's laws. This confirmed the truth of the integral principle.

In finding the Lagrangian for a system of free particles, a summation is performed over the coordinates of the particles. The electromagnetic field, if we use the terminology of mechanics, is a system with an infinite number of degrees of freedom because, for a complete description of the field, we must know all its components at all points of space, where they differ from zero. But the points of space form a nondenumerable set, i.e., they cannot be numbered in any order. For this reason, for the electromagnetic field the summation in the Lagrangian is replaced by an integration with respect to continuously varying parameters, i.e., coordinates of points in which the field is given. The point coordinates are analogous to the indices which label the degrees of freedom of a mechanical system.

The equations of mechanics are second order in time with respect to generalized coordinates q_k. The equations for potentials (12.37) and (12.38) are also second order in time. Therefore, potential quantities should be chosen as the generalized coordinates.

In other words, $\mathbf{A}(\mathbf{r}, t)$, $\varphi(\mathbf{r}, t)$ correspond to $q_k(t)$, where $\mathbf{A}$ and φ are potentials which are generalized coordinates of an electromagnetic field. The value of the radius vector $\mathbf{r}$ for the point at which the potential is taken corresponds to the number of the generalized coordinate k.

In order to write down the complete Lagrangian function, we must first of all define it in an element of volume dV and integrate over the volume occupied by the field. It has already been mentioned that in this section we will proceed immediately from a Lagrangian that leads to correct Maxwell equations; the choice of this Lagrangian as based on considerations related to the relativity principle will be left to Sec. 21. The Lagrangian is of the following form:

$$L = \int \left(\frac{\mathbf{E}^2 - \mathbf{H}^2}{8\pi} + \frac{\mathbf{A}\mathbf{j}}{c} - \rho\varphi \right) dV. \tag{13.1}$$

Since the potentials are liable to be generalized coordinates of the field, expression (13.1) should be rewritten thus:

$$L = \int \left[\frac{1}{8\pi} \left(\frac{1}{c} \frac{\partial \mathbf{A}}{\partial t} + \nabla\varphi \right)^2 - \frac{1}{8\pi} (\mathrm{rot}\, \mathbf{A})^2 + \frac{\mathbf{A}\mathbf{j}}{c} - \rho\varphi \right] dV. \tag{13.2}$$

Here, in place of a summation over the degrees of freedom, an integration over the volume has been performed.

The extremal property of action in electrodynamics. We shall now show that action, i.e., $S = \int L\, dt$, possesses the same variational property in electrodynamics as it does in mechanics: its variation becomes zero if the field satisfies the correct equations of motion (in this case, the Maxwell equations).

We shall begin with variation with respect to the scalar potential φ:

$$\delta_\varphi L = \int \left[\frac{1}{4\pi} \left(\frac{1}{c} \frac{\partial \mathbf{A}}{\partial t} + \nabla \varphi \right)^2 \delta \nabla \varphi - \rho \delta \varphi \right] dV. \tag{13.3}$$

As was shown in Sec. 10, variation and differentiation are commutative so that $\delta \nabla \varphi = \nabla \delta \varphi$. According to (12.29) we replace the term $\left(\frac{1}{c} \frac{\partial \mathbf{A}}{\partial t} + \nabla \varphi \right)$ by $-\mathbf{E}$. Therefore,

$$\delta_\varphi L = \int \left(- \frac{\mathbf{E} \nabla \delta \varphi}{4\pi} - \rho \delta \varphi \right) dV. \tag{13.4}$$

We shall now make use of equation (11.27), in accordance with which

$$\mathbf{E} \nabla \delta \varphi = \operatorname{div} (\mathbf{E} \delta \varphi) - \delta \varphi \operatorname{div} \mathbf{E}. \tag{13.5}$$

We then obtain

$$\delta_\varphi L = \int \left[\frac{-\operatorname{div} (\delta \varphi \mathbf{E})}{4\pi} + \delta \varphi \left(\frac{\operatorname{div} \mathbf{E}}{4\pi} - \rho \right) \right] dV. \tag{13.6}$$

The first term in (13.6) can be transformed into a surface integral, so that $\delta_\varphi L$ will have the form

$$\delta_\varphi L = -\int \delta \varphi \mathbf{E} \, d\mathbf{s} + \int \delta \varphi \left(\frac{\operatorname{div} \mathbf{E}}{4\pi} - \rho \right) dV. \tag{13.7}$$

We shall consider that the first integral is taken over a surface on which $\delta \varphi$ becomes zero, similar to the way that, in Sec. 10, δq was equal to zero at the limits of integration (a surface is the limit for a volume integral).

Therefore,

$$\delta_\varphi L = \int \delta \varphi \left(\frac{\operatorname{div} \mathbf{E}}{4\pi} - \rho \right) dV. \tag{13.8}$$

However, since

$$\operatorname{div} \mathbf{E} = 4\pi\rho \tag{13.9}$$

[see (12.27)], $\delta_\varphi L$ and hence $\delta_\varphi S$, becomes zero as expected.

Let us now vary L with respect to $\mathbf{A}$. This variation has the form

$$\delta_{\mathbf{A}} L = \int \left[\frac{1}{4\pi} \left(\frac{1}{c} \frac{\partial \mathbf{A}}{\partial t} + \nabla \varphi \right) \frac{1}{c} \delta \frac{\partial \mathbf{A}}{\partial t} - \frac{1}{4\pi} \operatorname{rot} \mathbf{A} \, \delta \operatorname{rot} \mathbf{A} + \frac{\mathbf{j} \delta \mathbf{A}}{c} \right] dV. \tag{13.10}$$

Once again we interchange the differentiation and variation signs and, where possible, we replace the potentials by fields after variation. We obtain $\delta_A L$ in the following form:

$$\delta_{\mathbf{A}} L = \int \left[-\frac{1}{4\pi c} \mathbf{E} \frac{\partial}{\partial t} \delta \mathbf{A} - \frac{1}{4\pi} \mathbf{H} \operatorname{rot} \delta \mathbf{A} + \frac{\mathbf{j} \delta \mathbf{A}}{c} \right] dV. \tag{13.11}$$

Let us write down the transformation by parts:

$$\mathbf{E}\frac{\partial}{\partial t}\delta\mathbf{A}=\frac{\partial}{\partial t}\mathbf{E}\,\delta\mathbf{A}-\delta\mathbf{A}\frac{\partial\mathbf{E}}{\partial t}, \tag{13.12}$$

$$\mathbf{H}\,\mathrm{rot}\,\delta\mathbf{A}=-\mathrm{div}\,[\mathbf{H}\,\delta\mathbf{A}]+\delta\mathbf{A}\,\mathrm{rot}\,\mathbf{H}. \tag{13.13}$$

The last equation follows from (11.29). To take advantage of (13.12), we must write down the variation of the action S instead of the variation of L. Then the first term of (13.12) can be directly integrated with respect to time, and $\delta_A S$ will be

$$\delta_{\mathbf{A}}S=\int_{t_0}^{t_1}\delta_{\mathbf{A}}L\,dt=-\frac{1}{4\pi c}\int\mathbf{E}\delta\,\mathbf{A}\,dV\Bigg|_{t_0}^{t_1}+\frac{1}{4\pi}\int_{t_0}^{t_1}dt\int[\mathbf{H}\,\delta\mathbf{A}]\,ds+$$
$$+\int_{t_0}^{t_1}dt\int dV\left(-\frac{\mathrm{rot}\,\mathbf{H}}{4\pi}+\frac{1}{4\pi c}\frac{\partial\mathbf{E}}{\partial t}+\frac{\mathbf{j}}{c}\right)\delta\mathbf{A}. \tag{13.14}$$

The variation $\delta\,\mathbf{A}$ is equal to zero at the initial and final instants of time t_0 and t_1, as well as over the surface bounding the field. Therefore,

$$\delta_{\mathbf{A}}S=\int_{t_0}^{t_1}dt\int dV\left(-\frac{\mathrm{rot}\,\mathbf{H}}{4\pi}+\frac{1}{4\pi c}\frac{\partial\mathbf{E}}{\partial t}+\frac{\mathbf{j}}{c}\right)\delta\mathbf{A}, \tag{13.15}$$

and since the field satisfied the equation

$$\mathrm{rot}\,\mathbf{H}=\frac{1}{c}\frac{\partial\mathbf{E}}{\partial t}+\frac{4\pi\mathbf{j}}{c} \tag{13.16}$$

[see (12.26)], $\delta_A S$ is equal to zero.

The first pair of Maxwell's equations is, of course, satisfied identically if the fields are expressed in terms of potentials in accordance with (12.28) and (12.29).

Thus, Maxwell's equations can be interpreted as equations of the mechanics of an electromagnetic field. They could be obtained from the variational method, starting from the Lagrangian (13.1) and the requirement that the variation of action should be equal to zero for any arbitrary variations of the scalar and vector potentials. For this it is sufficient to repeat the arguments set out in Sec. 10, as applied to integrals (13.8) and (13.16).

The invariance of action with respect to a potential gauge transformation. We shall now show that action is invariant under gauge transformations (12.30) and (12.31), despite the fact that it involves not only fields, but also potentials contained in the last two terms of equation (13.1). We shall call the corresponding part of the action S_1:

$$S_1=\int dt\int dV\left(\frac{\mathbf{A}\mathbf{j}}{c}-\rho\varphi\right). \tag{13.17}$$

Let us now apply gauge transformations (12.30) and (12.31) to $\mathbf{A}$ and φ. This gives

$$S_1 = \int dt \int dV \left(\frac{\mathbf{A}'\mathbf{j}}{c} - \rho \varphi' + \frac{\mathbf{j} \nabla f}{c} + \rho \frac{1}{c} \frac{\partial f}{\partial t} \right). \tag{13.18}$$

We transform by parts terms containing f:

$$\mathbf{j} \nabla f = \operatorname{div} (f \mathbf{j}) - f \operatorname{div} \mathbf{j}, \quad \rho \frac{\partial f}{\partial t} = \frac{\partial}{\partial t} (\rho f) - f \frac{\partial \rho}{\partial t}.$$

Substituting this in the integral S_1 and performing the integration, as in (13.14), we have

$$S_1 = \frac{1}{c} \int dt \int ds f \mathbf{j} - \frac{1}{c} \int dV f \rho \Big|_{t_0}^{t_1} + \\ + \int dt \int dV \left[\frac{\mathbf{A}'\mathbf{j}}{c} - \rho \varphi' - f \left(\operatorname{div} \mathbf{j} + \frac{\partial \rho}{\partial t} \right) \right]. \tag{13.19}$$

However, the integrated terms do not affect the Maxwell equations since, when performing a variation of S_1, both $\delta_\varphi L$ and $\delta_A L$ are equal to zero at the boundaries of the integration region. We have already encountered this in (10.9). The term, proportional to f, under the integral sign, is multiplied by the quantity $\operatorname{div} \mathbf{j} + \frac{\partial \rho}{\partial t}$, which is identically equal to zero according to the charge conservation law (12.18). Thus, S_1 retains the form (13.17).

The energy of a field. Maxwell's equations also apply to a free electromagnetic field not containing charges or currents. For this it is sufficient to omit from them the terms $\frac{4\pi \mathbf{j}}{c}$ and $4\pi\rho$. In accordance with (13.1) and (13.2), the Lagrangian for a free field is

$$L_0 = \int \frac{\mathbf{E}^2 - \mathbf{H}^2}{8\pi} dV = \int \frac{1}{8\pi} \left[\left(\frac{1}{c} \frac{\partial \mathbf{A}}{\partial t} + \nabla \varphi \right)^2 - (\operatorname{rot} \mathbf{A})^2 \right] dV. \tag{13.20}$$

We shall now determine the energy of an electromagnetic field by proceeding from the general equation (4.4). First, let it be recalled that the values of potentials at all points of space are generalized coordinates. But then the derivatives $\frac{\partial \mathbf{A}}{\partial t}$ are generalized velocities. Consequently, the expression

$$\mathscr{E} = \sum_k \dot{q}_k \frac{\partial L}{\partial \dot{q}_k} - L$$

reduces to the form

$$\mathscr{E} = \int dV \left[\frac{1}{4\pi c} \left(\frac{1}{c} \frac{\partial \mathbf{A}}{\partial t} + \nabla \varphi \right) \frac{\partial \mathbf{A}}{\partial t} - \left(\frac{\mathbf{E}^2 - \mathbf{H}^2}{8\pi} \right) \right]$$

by means of the comparison

$$\dot{q}_k \sim \frac{\partial \mathbf{A}}{\partial t}, \quad \frac{\partial L}{\partial \dot{q}_k} \sim \frac{1}{4\pi c} \left(\frac{1}{c} \frac{\partial \mathbf{A}}{\partial t} + \nabla \varphi \right), \quad \sum_k \sim \int dV.$$

We shall now show that energy is expressed only in terms of the field, and not in terms of potentials. Using (12.29) we write down the energy thus:

$$\mathscr{E} = \int dV \left[\frac{1}{4\pi} \mathbf{E} (\mathbf{E} + \nabla\varphi) - \left(\frac{\mathbf{E}^2 - \mathbf{H}^2}{8\pi}\right)\right].$$

This expression is not invariant with respect to a gauge transformation and must be transformed.

Transforming the term $\mathbf{E}\nabla\varphi$ by parts, we have, from (11.27),

$$\mathbf{E}\nabla\varphi = \operatorname{div}(\mathbf{E}\varphi) - \varphi \operatorname{div}\mathbf{E} = \operatorname{div}\varphi\mathbf{E},$$

since $\operatorname{div}\mathbf{E} = 0$ for a field free of charges. The volume integral of $\operatorname{div}\varphi\mathbf{E}$ is transformed into a surface integral. However, according to the meaning attached to L_0 and $\mathscr{E}$, the integration should be performed over the whole region occupied by the field (this is analogous to summation over all the degrees of freedom of the system). At the boundary of this region, the field is equal to zero by definition, so that the surface integral in the expression for energy also becomes zero. From this we obtain the required expression for the energy of an electromagnetic field in the absence of charges:

$$\mathscr{E} = \frac{1}{8\pi} \int (\mathbf{E}^2 + \mathbf{H}^2)\, dV. \tag{13.21}$$

Hence, the quantity

$$w = \frac{\mathbf{E}^2 + \mathbf{H}^2}{8\pi} \tag{13.22}$$

may be interpreted as the energy density of the electromagnetic field. It is invariant with respect to a gauge transformation of a potential.

Conservation of the total energy of field and charges. We shall now show that the energy $\mathscr{E}$ (13.21), together with the energy of the charges contained in the field, is conserved, i.e., $\mathscr{E}$ is the energy in the usual, mechanical, sense of the word, and not some quantity which is formally analogous to it only as regards its derivation from the variational principle.

To do this we multiply equation (12.26) scalarly by $\mathbf{E}$ and (12.24) scarlarly by $\mathbf{H}$, and subtract the second from the first. This gives the following relationship:

$$\frac{1}{c}\left(\mathbf{E}\frac{\partial \mathbf{E}}{\partial t} + \mathbf{H}\frac{\partial \mathbf{H}}{\partial t}\right) = \mathbf{E}\operatorname{rot}\mathbf{H} - \mathbf{H}\operatorname{rot}\mathbf{E} - \frac{4\pi \mathbf{j}\mathbf{E}}{c}.$$

Now taking advantage of equation (11.29), we reduce the equation obtained to the form

$$\frac{\partial}{\partial t}\left(\frac{\mathbf{E}^2 + \mathbf{H}^2}{8\pi}\right) = -\operatorname{div}\frac{c}{4\pi}[\mathbf{E}\mathbf{H}] - \rho\mathbf{v}\mathbf{E}. \tag{13.23}$$

Here we have put $\mathbf{j} = \rho \mathbf{v}$ by definition. We now integrate (13.23) over some volume, though not necessarily the whole volume occupied by the field, and transform the integral of div to a surface integral:

$$\frac{\partial}{\partial t}\int \frac{\mathbf{E}^2 + \mathbf{H}^2}{8\pi} dV = -\int \frac{c}{4\pi} [\mathbf{E}\mathbf{H}]\, d\mathbf{s} - \int \rho \mathbf{v} \mathbf{E}\, dV. \qquad (13.24)$$

Let us first consider the second integral on the right. By definition, the quantity $\rho\, dV$ is the charge element de. The product $\mathbf{E}\, de$ is the electric force acting on this charge element. The scalar product $de\, \mathbf{E}\, \mathbf{v} = de \cdot \mathbf{E} \cdot \frac{d\mathbf{r}}{dt}$ is equal to the work done on the element of charge in unit time or—put in another way—to the change in kinetic energy T of the charge in unit time. Later, we shall show that the magnetic field does not perform work on charges (Sec. 21). To summarize, equation (13.24) can also be written as follows [the last integral in (13.24) will be represented in the form $\frac{dT}{dt}$, i.e., the work done in unit time]:

$$\frac{d}{dt}(\mathscr{E} + T) = -\int \frac{c}{4\pi} [\mathbf{E}\mathbf{H}]\, dV. \qquad (13.25)$$

The Poynting vector. Thus, the decrease in energy, in unit time, of an electromagnetic field and of the charged particles contained therein is equal to the vector flux $\frac{c}{4\pi}[\mathbf{E}\,\mathbf{H}]$ across the surface bounding the field. If this surface is infinitely distant and the field on it is equal to zero, what we have is simply the energy conservation law for an electromagnetic field and for the charges within it. Otherwise, if the volume is finite, the right-hand side of equation (13.25) indicates what energy passes in unit time through the surface bounding the volume. Hence, the quantity

$$\mathbf{U} = \frac{c}{4\pi} [\mathbf{E}\mathbf{H}] \qquad (13.26)$$

represents the energy crossing unit area in unit time or, more simply, the energy density flux vector (the Poynting vector).

Field momentum. Similar computations, which we shall not give, show that an electromagnetic field possesses momentum. The momentum of a field is given by the following integral:

$$\mathbf{p} = \int \frac{1}{4\pi c} [\mathbf{E}\mathbf{H}]\, dV. \qquad (13.27)$$

If the electromagnetic field interacts with some obstacle, for example, the walls of the enclosure in which it is contained, or with a screen, then the momentum of the field is transmitted to the obstacle. The momentum transmitted normally to unit area in unit time is nothing other than the pressure (since momentum transmitted in unit time is force). For this reason, electrodynamics predicts that electromagnetic fields (and, as a particular case, light waves) are capable of exerting a pressure on matter.

Angular momentum of a field. According to (13.27), the field-momentum density is

$$\frac{1}{4\pi c}[\mathbf{E}\mathbf{H}].$$

From this it follows that the angular-momentum density is

$$\frac{1}{4\pi c}[\mathbf{r}[\mathbf{E}\mathbf{H}]],$$

and the total angular momentum of the field is

$$\mathbf{M} = \frac{1}{4\pi c}\int [\mathbf{r}[\mathbf{E}\mathbf{H}]]\,dV. \tag{13.28}$$

Linear momentum and angular momentum of a field satisfy the conservation laws together with similar quantities for the charge contained in the field. The value of the angular momentum of a field is very essential in the quantum theory of radiation.

Sec. 14. The Electrostatics of Point Charges. Slowly Varying Fields

An important class of approximate solutions of electro-dynamical equations comprises slowly varying fields, for which the terms $\frac{1}{c}\frac{\partial \mathbf{E}}{\partial t}$ and $\frac{1}{c}\frac{\partial \mathbf{H}}{\partial t}$ in Maxwell's equations can be neglected. The remaining terms form two sets of equations, which are entirely independent of each other:

$$\operatorname{div}\mathbf{E} = 4\pi\rho, \tag{14.1}$$

$$\operatorname{rot}\mathbf{E} = 0 \tag{14.2}$$

and

$$\operatorname{div}\mathbf{H} = 0, \tag{14.3}$$

$$\operatorname{rot}\mathbf{H} = \frac{4\pi\mathbf{j}}{c}. \tag{14.4}$$

The first two equations contain only the electric field and the density of the charge producing the field; the second two equations involve only the magnetic field and current density, the right-hand sides of the equations being regarded as known functions of coordinates and time. Since there are no time derivatives in (14.1)-(14.4), the time dependence of the electric field is the same as the electric-charge densities, and the time dependence of the magnetic field is the same as the current densities. Hence, to the approximation of (14.1)-(14.4), the field is, as it were, established instantaneously, in correspondence with the charge and current distribution that generated it.

The fact is that any change in the field is transmitted in space with the velocity of light c. If we consider the field at a distance R from a charge, the electromagnetic disturbance will reach it in a time $\frac{R}{c}$. The charge, of velocity v, will be displaced, during that time, through a distance $v\frac{R}{c}$. The approximation (14.1)-(14.4) can be applied only when the displacement $v\frac{R}{c}$ does not lead to any essential redistribution of the charge. For example, let a system consist of two equal charges of opposite sign, which succeed in changing places in a time $\frac{R}{c}$. Then, the electric field at a distance R, at the instant $t=\frac{R}{c}$, will have a direction opposite to the one it had during the instantaneous propagation at the instant $t=0$.

Hence, if the dimensions of the system of charges are r and their velocities v (r and v determine the orders of magnitude), then equations (14.1)-(14.4) can be used at the distance R from the system, for which the inequality $\frac{r}{v} \gg \frac{R}{c}$ or $R \ll r\frac{c}{v}$ is satisfied.

We shall consider the limiting case, when $v \ll c$. Then the region of applicability of our approximation will be very large.

Equations (14.1), (14.2) are called the equations of electrostatics, and (14.3) and (14.4), the equations of magnetostatics.

Scalar potential in electrostatics. In order to satisfy equation (14.2), we put

$$\mathbf{E} = -\nabla\varphi \, . \tag{14.5}$$

According to (14.29), φ is the scalar potential. The equation for the scalar potential is obtaining from (14.1)

$$\operatorname{div}\operatorname{grad}\varphi = \Delta\varphi = -4\pi\rho \, , \tag{14.6}$$

which also follows from (12.38) if we equate to zero the nonstatic term $\frac{1}{c^2}\frac{\partial^2\varphi}{\partial t^2}$.

Let us find the solution to equation (14.6) for a point charge, i.e., we put ρ equal to zero everywhere except at one point of space. Let us put the origin at this point. Then φ can depend only on the distance from the origin r.

In Sec. 11 an expression for the Laplacian Δ was obtained in spherical coordinates (11.46). In the special case, when the required function depends only on r, we obtain from (11.46)

$$\frac{1}{r^2}\frac{d}{dr}r^2\frac{d\varphi}{dr} = -4\pi\rho\,. \tag{14.7}$$

Let us integrate this equation between r_1 and r_2, first multiplying it by r^2. Since the region of integration does not contain the origin, where the point charge is situated, the integral of the right-hand side becomes zero. Hence,

$$r_2^2\frac{d\varphi}{dr_2} = r_1^2\frac{d\varphi}{dr_1}\,; \quad r^2\frac{d\varphi}{dr} = A = \text{const}\,.$$

Therefore the potential is

$$\varphi = -\frac{A}{r} + B\,.$$

The constant B is equal to zero if we take the potential to be equal to zero at an infinite distance away from the charge. Let us now determine the constant A. For this, we integrate equation (14.6) over a certain sphere surrounding the origin. Since the Laplacian $\Delta\,\varphi$ is div grad φ, the volume integral can be transformed into an integral over the surface of the sphere. This integral is

$$\int \text{grad}\,\varphi\, d\mathbf{s} = \int \frac{A}{r^2}\, r^2 d\,\Omega = 4\,\pi A\,.$$

On the right-hand side we have

$$-\int 4\pi\rho\, dV = -4\pi e\,,$$

since the integration region includes the point where the charge is situated. Thus $A = -e$.

The potential of the point charge is

$$\varphi = \frac{e}{r}\,. \tag{14.8}$$

We obtain the same thing for a spherically symmetrical volume-charge distribution, if the potential is calculated outside the volume occupied by the charge. In other words, the potential of a charged sphere at all external points is the same as the potential of an equal point charge situated at the centre of the sphere. A similar result is, of course, obtained for the gravitational potential. This fact is used in most astronomical problems, where celestial bodies are considered as gravitating points.

If the origin does not coincide with the charge, and the charge coordinates are x, y, z (radius vector $\mathbf{r}$) then the potential at point X, Y, Z (radius vector $\mathbf{R}$) is

$$\varphi = \frac{e}{|\mathbf{R} - \mathbf{r}|} = \frac{e}{\sqrt{(X-x)^2 + (Y-y)^2 + (Z-z)^2}}\,. \tag{14.9}$$

The potential of a system of charges. The potential due to several charges $e_1, e_2, e_3, \ldots, e_i, \ldots$, whose positions are given by the radius vectors $\mathbf{r}^1, \mathbf{r}^2, \ldots, \mathbf{r}^i$, at the point $\mathbf{R}$, is

$$\varphi = \sum_i \frac{e_i}{|\mathbf{R} - \mathbf{r}^i|} = \sum_i \frac{e_i}{\sqrt{(X - x^i)^2 + (Y - y^i)^2 + (Z - z^i)^2}}.$$

Using the summation convention, any radicand in this formula can be rewritten in the form

$$(X_\lambda - x^i_\lambda)\ (X_\lambda - x^i_\lambda).$$

But, in order to save space, we shall use the notation $(X_\lambda - x^i_\lambda)^2$ instead of $(X_\lambda - x^i_\lambda)\ (X_\lambda - x^i_\lambda)$. Then the potential due to a system of point charges is written as

$$\varphi = \sum_i e_i\, [(X_\lambda - x^i_\lambda)^2]^{-1/2} \tag{14.10}$$

But we must remember that inside the brackets is a summation for λ from 1 to 3.

Note also that the potentials due to separate charges at the point $\mathbf{R}$ are additive, since equation (14.6) is linear in φ. And so the full solution, due to all the charges, is equal to the sum of all the partial solutions for each charge separately.

The potential due to a charge system at a large distance. Let us now assume that the origin is situated somewhere inside the region occupied by the charges (for example, at the centre of the smallest sphere embracing all charges), and that all the radius vectors $\mathbf{r}^i$ satisfy the inequalities

$$|\mathbf{R}| \gg |\mathbf{r}^i|. \tag{14.11}$$

In other words, we shall look for the potential of a system of charges at a great distance from it. Then the function (14.10) can be expanded in a Taylor's series in terms of x^i, y^i, z^i. We shall perform the expansion up to the quadratic term inclusively, but we shall first write it only for one term of the sum over all the charges, omitting the index i. The expansion is of the form:

$$[(X - x)^2 + (Y - y)^2 + (Z - z)^2]^{-1/2} = [(X_\lambda - x_\lambda)^2]^{-1/2} =$$
$$= [X_\lambda^2]^{-1/2} - x_\mu \frac{\partial}{\partial X_\mu} [X_\lambda^2]^{-1/2} + \frac{1}{2} x_\mu x_\nu \frac{\partial^2}{\partial X_\mu \partial X_\nu} [X_\lambda^2]^{-1/2}. \tag{14.12}$$

The summation convention permits writing in concise form the Taylor series for a function of several variables. Since $X_\lambda^2 = R^2$, we obtain the expression for the first derivative

$$\frac{\partial}{\partial X_\mu} [X_\lambda^2]^{-1/2} = \frac{\partial}{\partial X_\mu} \frac{1}{R} = \frac{\partial R}{\partial X_\mu} \frac{\partial}{\partial R} \frac{1}{R} = -\frac{X_\mu}{R^3}, \tag{14.13}$$

where we have used equation (11.36) which, in the notation of this section, is of the form $\frac{\partial R}{\partial X_\mu} = \frac{X_\mu}{R}$.

Thus, the term in the sum (14.12), which is linear in x_μ, is equal to

$$\frac{x_\mu X_\mu}{R^3} = \frac{xX + yY + zZ}{R^3} = \frac{\mathbf{rR}}{R^3}. \tag{14.14}$$

It is somewhat more difficult to calculate the term which is quadratic in x_μ, x_ν. We first write down the second derivative:

$$\frac{\partial^2}{\partial X_\mu \partial X_\nu}\frac{1}{R} = -\frac{\partial}{\partial X_\nu}\frac{X_\mu}{R^3} = -\frac{1}{R^3}\frac{\partial X_\mu}{\partial X_\nu} - X_\mu \frac{\partial}{\partial X_\nu}\frac{1}{R^3}.$$

The partial derivative $\frac{\partial X_\mu}{\partial X_\nu}$ is equal to zero for $\mu \neq \nu$ and to 1 for $\mu = \nu$. Further,

$$\frac{\partial}{\partial X_\nu}\frac{1}{R^3} = \frac{\partial R}{\partial X_\nu}\frac{\partial}{\partial R}\frac{1}{R^3} = -\frac{X_\nu}{R}\cdot\frac{3}{R^4} = -\frac{3X_\nu}{R^5}$$

by the rule for differentiation of involved functions.

Thus, we obtain

$$\frac{\partial^2}{\partial X_\mu \partial X_\nu}\frac{1}{R} = -\frac{1}{R^3}\frac{\partial X_\mu}{\partial X_\nu} + \frac{3X_\mu X_\nu}{R^5}.$$

Hence, the required expansion $|\mathbf{R} - \mathbf{r}|^{-1}$ is of the form

$$\frac{1}{|\mathbf{R}-\mathbf{r}|} = \frac{1}{R} + \frac{\mathbf{rR}}{R^3} + \frac{1}{2}x_\mu x_\nu\left(\frac{3X_\mu X_\nu}{R^5} - \frac{1}{R^3}\frac{\partial X_\mu}{\partial X_\nu}\right) \tag{14.15}$$

We shall now subtract from the quadratic term a quantity equal to zero:

$$\frac{r^2}{2}\cdot\frac{1}{3}\frac{\partial^2}{\partial X_\nu^2}\frac{1}{R} \equiv \frac{x^2+y^2+z^2}{6}\Delta\frac{1}{R}$$

[$\Delta\frac{1}{\mathrm{R}} = 0$ from (14.8)]. Then the last term in (14.15), written in terms of components, is (taking advantage of the fact that $\frac{\partial X}{\partial X} = \frac{\partial Y}{\partial Y} = \frac{\partial Z}{\partial Z} = 1$, $\frac{\partial X}{\partial Y} = \frac{\partial X}{\partial Z} = \frac{\partial Y}{\partial Z} = 0$):

$$\frac{1}{2}\left[x^2\left(\frac{3X^2}{R^5} - \frac{1}{R^3}\right) + y^2\left(\frac{3Y^2}{R^5} - \frac{1}{R^3}\right) + z^2\left(\frac{3Z^2}{R^5} - \frac{1}{R^3}\right) + 2xy\left(\frac{3XY}{R^5}\right) + \right.$$
$$\left. + 2xz\left(\frac{3XZ}{R_5}\right) + 2yz\left(\frac{3YZ}{R^5}\right) - \left(\frac{x^2+y^2+z^2}{3}\right)\left(\frac{3(X^2+Y^2+Z^2)}{R^5} - \frac{3}{R^3}\right)\right].$$

Here, it is quite obvious that a term equal to zero has been subtracted, for $X^2 + Y^2 + Z^2 = R^2$. Rearranging the terms, we have

$$\frac{1}{2}\left[\left(x^2 - \frac{r^2}{3}\right)\left(\frac{3X^2}{R^5} - \frac{1}{R^3}\right) + \left(y^2 - \frac{r^2}{3}\right)\left(\frac{3Y^2}{R^5} - \frac{1}{R^3}\right) + \left(z^2 - \frac{r^2}{3}\right)\times\right.$$
$$\left.\times\left(\frac{3Z^2}{R^5} - \frac{1}{R^3}\right) + 2xy\frac{3XY}{R^5} + 2xz\frac{3XZ}{R^5} + 2yz\frac{3YZ}{R^5}\right].$$

The expansion (14.15) must be substituted in the equation for potential (14.10) and summed over all the charges. We introduce the following abbreviated notation:

$$\mathbf{d} = \sum_i e_i \mathbf{r}^i ; \tag{14.16}$$

$$\left.\begin{aligned} q_{xx} &= \sum_i e_i \left(x^{i2} - \frac{r^{i2}}{3}\right), \\ q_{yy} &= \sum_i e_i \left(y^{i2} - \frac{r^{i2}}{3}\right), \\ q_{zz} &= \sum_i e_i \left(z^{i2} - \frac{r^{i2}}{3}\right); \end{aligned}\right\} \tag{14.17}$$

$$\left.\begin{aligned} q_{xy} &= \sum_i e_i x^i y^i , \\ q_{xz} &= \sum_i e_i x^i z^i , \\ q_{yz} &= \sum_i e_i y^i z^i . \end{aligned}\right\} \tag{14.18}$$

The vector $\mathbf{d}$ (i.e., the three quantities d_x, d_y, d_z) and the six quantities q_{xx}, q_{yy}, q_{zz}, q_{xy}, q_{xz}, q_{yz}, depend only on the charge distribution in the system, and not on the place at which the potential is determined. In the notation of (14.16)-(14.18), the potential at large distances away from the system is of the form

$$\varphi = \frac{\Sigma e_i}{R} + \frac{(\mathbf{d}\mathbf{R})}{R^3} + \frac{1}{2} q_{\mu\nu} \left(\frac{3 X_\mu X_\nu}{R^5} - \frac{1}{R^3} \frac{\partial X_\mu}{\partial X_\nu}\right), \tag{14.19}$$

with the terms of different indices of the type q_{xy} actually appearing twice in the summation (for example, q_{12} and the equal term q_{21}).

The vector $\mathbf{d}$ is called the *dipole moment* of the charge system. The six quantities q are called the *quadrupole moment* components.

The dipole moment. We shall now examine the expression obtained for potential. The zero term $\dfrac{\sum_i e_i}{R}$ corresponds to the approximation according to which all the charge is considered to be concentrated at the origin. In other words, it corresponds to a substitution of the entire charge system by a single point charge. This approximation is clearly insufficient when the system is neutral, i.e., if $\sum_i e_i = 0$. This case is very usual, since atoms and molecules are neutral (their electronic charge balances the charge on the nuclei).

Let us assume that the total charge is equal to zero and then consider the first term of the expansion, involving the dipole moment. This term decreases like $\frac{1}{R^2}$, i.e., more rapidly than the potential of the charged system. Besides, it is proportional to the cosine of the angle between **d** and **R**. The simplest thing is to produce a neutral system by taking two equal and opposite charges. Such a system is called a dipole. Its moment is

$$\mathbf{d} = \sum e_i \mathbf{r}^i = e(\mathbf{r}^1 - \mathbf{r}^2) \tag{14.20}$$

in accordance with the definition (used in general courses of physics) that the dipole moment is the product of charge by the vector joining the positive and negative charges.

It can be seen from equation (14.20) that the definition of dipole moment does not depend on the choice of coordinate origin, since it involves only the relative position of charges. We shall show that the dipole moment always possesses this property.

Indeed, if we displace the origin through some distance **a**, then the radius vectors of all the charges change thus:

$$\mathbf{r}^i = \mathbf{r}'^i + \mathbf{a}$$

Substituting this in the expression for the dipole moment, we obtain

$$\mathbf{d} = \sum_i e_i \mathbf{r}^i = \sum_i e_i \mathbf{r}'^i + \mathbf{a} \sum_i e_i = \sum_i e_i \mathbf{r}'^i, \tag{14.21}$$

because $\sum e_i = 0$.

But if the system is not neutral, then we choose **a** in the following manner:

$$\mathbf{a} = \frac{\sum_i e_i \mathbf{r}^i}{\sum_i e_i} \tag{14.22}$$

This choice is analogous to the choice of centre of mass for a system of masses. Thus, we can say that the vector **a** determines the electrical centre of a system of charges. For a neutral system it is impossible to determine **a**, since the denominator of (14.22) is equal to zero. If for a charged system we choose **a** according to (14.22), then $\sum_i e_i \mathbf{r}'^i = 0$, i.e., the dipole moment of a charged system, relative to its electric centre, is equal to zero.

We thus have the following alternatives: either the system is neutral, and then the expansion (14.29) begins with a dipole term independent of the choice of coordinate origin, or the system is charged, and then the dipole term in the expansion is equal to zero for a corresponding choice of origin.

Quadrupole moment. In the expansion (14.19), we now consider the second term containing the quadrupole moment. A quadrupole is a system of two dipoles of moment **d**, which are equal in magnitude and opposite in direction. It is clear that a potential expansion for such a system will have neither a zero nor first term, so that equation (14.19) contains only a second term on the right-hand side. The simplest quadrupole can be formed by placing four charges at the vertices of a parallelogram, where the charges are of equal magnitude but with pairs of charges having opposite signs. The charges alternate when we traverse the vertices of the parallelogram. Such a system is neutral. However, a charged system, too, can have a quadrupole moment. It indicates to what extent the charge distribution in the system differs from spherical symmetry. Indeed, in this section it was shown that the potential due to a spherically symmetrical charge system decreases in strict accordance with a $\frac{1}{R}$ law, and the potential due to a quadrupole follows a $\frac{1}{R^3}$ law. For this reason, the quadrupole term in the potential expansion can arise only in the case of a nonspherical charge distribution.

The principal axes of a quadrupole moment. Let us now determine in what sense the quadrupole moment characterizes a nonspherical distribution. In equation (14.22), an analogy was established between the centre of inertia of a mass system and the electric centre of a system of charges. In a similar way, equations (14.17), (14.18) allow us to establish a certain correspondence between the components of quadrupole moment and moments of inertia of a system of masses $J_{xx}, \ldots, J_{yz}$ (see Sec. 9), defined in equations (9.3).

We can disregard the fact that a summation appears in equations (14.17) and (14.18), while (9.3) involves integration. This difference will not exist if we take a continuous charge distribution or a discrete mass distribution (as for nuclei in a molecule). We shall take the latter. In addition, we shall forget for a moment that the components of the moment of inertia involve masses and not charges. Then the relationship between the quadrupole moment and the moment of inertia is of the form:

$$q_{xx} \cong -J_{xx} + \frac{1}{3}(J_{xx} + J_{yy} + J_{zz}), \quad q_{xy} \cong -J_{xy},$$

$$q_{yy} \cong -J_{yy} + \frac{1}{3}(J_{xx} + J_{yy} + J_{zz}), \quad q_{xz} \cong -J_{xz},$$

$$q_{zz} \cong -J_{zz} + \frac{1}{3}(J_{xx} + J_{yy} + J_{zz}), \quad q_{yz} \cong -J_{yz}.$$

The sign $\sim$ above the equality symbol indicates correspondence between terms. Indeed, according to (9.3) the first line gives

$$-\sum_i e_i (y^{i^2} + z^{i^2}) + \frac{2}{3}\sum_i e_i (x^{i^2} + y^{i^2} + z^{i^2}) =$$

$$= \sum_i \frac{e_i}{3} (2x^{i^2} - y^{i^2} - z^{i^2}) = \frac{1}{3}\sum_i e_i (3\,x^{i^2} - r^{i^2}) = q_{xx}\,.$$

The relations in the second column are obvious.

In Sec. 9 it was shown that moments of inertia can be reduced to principal axes, i.e., a coordinate system can be found for which the products of inertia are zero. But since the relations between q and J are true for any coordinate system, the components of the quadrupole moment of different signs also become zero in these same principal axes. In the principal axes, the quadrupole moment is expressed, in terms of moments of inertia, as

$$\left.\begin{aligned} q_1 &\cong \frac{1}{3}(J_2 + J_3 - 2J_1)\,, \\ q_2 &\cong \frac{1}{3}(J_1 + J_3 - 2J_2)\,, \\ q_3 &\cong \frac{1}{3}(J_1 + J_2 - 2J_3)\,. \end{aligned}\right\} \qquad (14.23)$$

If the system possesses spherical symmetry, then $J_1 = J_2 = J_3$, so that $q_1 = q_2 = q_3 = 0$. Therefore, the presence of a quadrupole moment in a system of charges indicates that the charge distribution is not spherically symmetrical. However, a reverse assertion would not be true: if the quadrupole moment is equal to zero, the system of charges may not be spherically symmetrical. It will then be necessary, in expansion (14.19), to take into account terms of higher order.

It will be noted that from (14.23) there follows directly the identity $q_1 + q_2 + q_3 = 0$, i.e., only two of the three principal components of a quadrupole moment are independent.

The relations (14.23) should be regarded literally if we are talking about a gravitational potential. We know that the earth is not strictly spherical, but is flattened at the poles. Therefore, the earth's gravitational force contains terms which are not governed by an inverse square law. This affects the motion of the moon, and all the more so that of artificial satellites moving closer to the earth.

Quadrupole moment when axes of symmetry exist. Equations (14.23) become simpler if two moments of inertia of a body are equal, for example, $J_1 = J_2$. Then

$$\begin{aligned} q_1 &\cong \frac{1}{3}(J_3 - J_1) \equiv -\frac{q}{2}\,, \\ q_2 &\cong \frac{1}{3}(J_3 - J_1) \equiv -\frac{q}{2}\,, \\ q_3 &\cong \frac{2}{3}(J_1 - J_3) \equiv q\,. \end{aligned}$$

In this case, the quadrupole moment has only one independent component q. Its sign is called the sign of the quadrupole moment. The quantity $q = \sum_i e_i \left(z^{i^2} - \frac{r^{i^2}}{3}\right)$.

If the charges were distributed with spherical symmetry, we would have the equality $\sum_i e_i r^{i^2} = 3 \sum_i e_i z^{i^2}$, for then $\sum_i e_i x^{i^2} = \sum_i e_i y^{i^2} = \sum_i e_i z^{i^2}$ and q, too, would be equal to zero.

The positive sign of q shows that $\sum_i e_i z^{i^2} > \frac{1}{3} \sum_i e_i r^{i^2}$, i.e., it indicates a charge distribution extending along the z-axis. From (14.19), the potential due to such a quadrupole with one component q is

$$\varphi_q = -\frac{1}{4} q \left(\frac{3X^2}{R^5} - \frac{1}{R^3}\right) - \frac{1}{4} q \left(\frac{3Y^2}{R^5} - \frac{1}{R^3}\right) + \frac{q}{2}\left(\frac{3Z^2}{R^5} - \frac{1}{R^3}\right) =$$

$$= -\frac{3}{4} q \left(\frac{X^2 + Y^2 - 2Z^2}{R^5}\right) = -\frac{3}{4} q \left(\frac{R^2 - 3Z^2}{R^5}\right) =$$

$$= -\frac{3}{4}\frac{q}{R^3}(1 - 3\cos^2\vartheta). \tag{14.24}$$

Thus, the potential of a quadrupole depends on the angle ϑ according to the law $1 - 3\cos^2\vartheta$, where ϑ is the angle between the axis of symmetry of the quadrupole and the radius vector of the point at which the potential is determined.

Similar deviations from spherical symmetry have been found in the electrostatic potential of many nuclei. The quadrupole moments of nuclei give us an insight into their structure.

The energy of a system of charges in an electrostatic field. We shall now calculate the energy of a system of charges in an external electric field. The potential energy of a charge in a field is equal to $U = e\varphi$, because the force acting on the charge is equal to $\mathbf{F} = -\nabla U = -e\nabla\varphi = e\mathbf{E}$. The energy of a system of charges is thus

$$U = \sum_i e_i \varphi(\mathbf{r}^i), \tag{14.25}$$

where $\mathbf{r}^i$ is the radius vector for the ith charge.

Let us suppose that the field does not change much over the space occupied by the charges, so that the potential at the site of the ith charge can be expanded in a Taylor's series:

$$\varphi(\mathbf{r}) = \varphi(0) + x_\mu \left(\frac{\partial\varphi}{\partial x_\mu}\right)_0 + \frac{1}{2} x_\mu x_\nu \left(\frac{\partial^2\varphi}{\partial x_\mu \partial x_\nu}\right)_0. \tag{14.26}$$

We shall transform the last term in the same way as in the expansion (14.15); taking advantage of the fact that φ is the potential of the *external* field (and not the field produced by the given charges), so that $\Delta\varphi = 0$, we subtract from φ the quantity $\frac{r^2}{3}\Delta\varphi$ equal to zero. Then, after summation over i, we obtain

$$U = \varphi(0)\sum_i e_i - (\mathbf{d}\mathbf{E}_0) + \frac{1}{2}\left(q_{xx}\frac{\partial^2\varphi}{\partial x^2} + q_{yy}\frac{\partial^2\varphi}{\partial y^2} + q_{zz}\frac{\partial^2\varphi}{\partial z^2} + \right.$$

$$\left. + 2\,q_{xy}\frac{\partial^2\varphi}{\partial x\,\partial y} + 2\,q_{xz}\frac{\partial^2\varphi}{\partial x\,\partial z} + 2\,q_{yz}\frac{\partial^2\varphi}{\partial y\,\partial z}\right) =$$

$$= \varphi(0)\sum_i e_i - (\mathbf{d}\mathbf{E}_0) + \frac{1}{2}q_{\mu\nu}\frac{\partial^2\varphi}{\partial x_\mu\,\partial x_\nu}. \qquad (14.27)$$

Here, the value of the field $(\nabla\varphi)_0 = \mathbf{E}_0$ at the origin has been substituted into the term involving the dipole moment. Relating equation (14.27) to the principal axes of the quadrupole moment, we get

$$U = \varphi(0)\sum_i e_i - (\mathbf{d}\mathbf{E}_0) - \frac{1}{2}\left(q_1\frac{\partial E_x}{\partial x} + q_2\frac{\partial E_y}{\partial y} + q_3\frac{\partial E_z}{\partial z}\right). \qquad (14.28)$$

In the case of a neutral system, the term involving dipole moment is especially important. The quadrupole term accounts for the extension of the system, since it involves field derivatives. It is interesting to note that if the system is spherically symmetrical, i.e., if it has a quadrupole moment equal to zero, there is no correction to finite dimensions. Higher order corrections are also absent, so that the potential energy will always depend only on the value of the potential at the centre. This is why spherical bodies not only attract, but are also attracted, as points. Of course, these assertions are mutually related by Newton's Third Law which holds for electrostatics, since the field is determined by the instantaneous configuration of charges.

Exercises

1) Show that the mean value of the potential over a spherical surface is equal to its value at the centre of the sphere, if the equation $\Delta\varphi = 0$ is satisfied over the whole volume of the sphere. Relate this to the result obtained for the potential energy of a spherically symmetrical system of charges in an external field.

The potential should be expanded in a series involving the radius powers of the sphere. In integration over the surface, all the terms containing x, y, and z an odd number of times become zero. The terms containing x, y, and z an even number of times can be rearranged so that they are proportional to $\Delta\varphi$, $\Delta\Delta\varphi$, and so on. There remains only the zero term of the expansion, which proves the theorem.

2) Calculate the electric field of a dipole.

Sec. 15. The Magnetostatics of Point Charges

The equations of magnetostatics. In the previous section it was shown that if the velocities of the charges are small in comparison with the velocity of light, then the magnetic field satisfies the following system of equations:

$$\operatorname{div} \mathbf{H} = 0\,, \tag{15.1}$$

$$\operatorname{rot} \mathbf{H} = \frac{4\pi}{c} \mathbf{j}\,. \tag{15.2}$$

They are called the equations of magnetostatics. From equation (15.2) it follows that

$$\operatorname{div} \operatorname{rot} \mathbf{H} = \frac{4\pi}{c} \operatorname{div} \mathbf{j} = 0\,. \tag{15.3}$$

Thus, for (15.2) to make strict sense, the currents must be subjected to the condition div $\mathbf{j} = 0$. But this condition is not directly satisfied for point charges, and only the charge conservation equation (12.18) holds.

Mean values. The condition div $\mathbf{j} = 0$ for moving point charges can only be satisfied for an average over some interval of time t_0. We shall define the mean value of a certain function of coordinates and velocities of charges $f(\mathbf{r}, \mathbf{v})$ in the following way:

$$\overline{f} = \frac{1}{t_0} \int_0^{t_0} f(\mathbf{r}, \mathbf{v})\, dt\,. \tag{15.4}$$

This averaging operation is commutative with differentiation of the function with respect to coordinates, since it is performed with respect to a fixed frame of reference and not to the coordinates of the charges.

Let us now average equation (12.28):

$$\frac{1}{t_0} \int_0^{t_0} \operatorname{div} \mathbf{j}\, dt = \operatorname{div} \overline{\mathbf{j}} = -\frac{1}{t_0} \int_0^{t_0} \frac{\partial \rho}{\partial t}\, dt\,.$$

Integration of the right-hand side part can be performed directly:

$$-\frac{1}{t_0} \int_0^{t_0} \frac{\partial \rho}{\partial t}\, dt = -\frac{\rho(t_0) - \rho(0)}{t_0}\,.$$

Let us now assume that the difference $\rho(t_0) - \rho(0)$ increases more slowly than the time interval t_0 itself. Then, if we choose t_0 sufficiently large, the ratio $\frac{\rho(t_0) - \rho(0)}{t_0}$ can be as small as required. Because of this, the mean value of the current may indeed satisfy the equation

$$\operatorname{div} \bar{\mathbf{j}} = 0 . \tag{15.5}$$

Consequently, equation (15.2) and all subsequent equations in this section must be regarded as mean with respect to time; this will be denoted by a bar over each quantity relating to the motion of charges (no bar will be put over **H**).

The definition of steady motion. Let us assume that the condition $\lim_{t_0 \to \infty} \frac{f(t_0) - f(0)}{t_0} = 0$ is satisfied not only for charge density, but also for any function relating to the motion of charges. Such motion is termed *stationary* or *steady*.

A special case of stationary motion is periodic motion, for example, cyclic motion. However, for a stationary state it is sufficient that the charges remain all the time in a limited region of space, for the difference $f(t_0) - f(0)$ then remains finite.

The equations of this section will relate to the stationary motion of point charges.

The equations for vector potential. In order to satisfy equation (15.1), we put, as in Sec. 12 [see (12.8)],

$$\mathbf{H} = \operatorname{rot} \mathbf{A} , \tag{15.6}$$

where **A** is the vector potential. Equation (15.6) does not fully determine **A** since, if we add to **A** the gradient of any arbitrary function f, as in (12.30), the expression for **H** will not change. Thus, an additional condition must be imposed on **A**. The Lorentz condition suggests that we must have

$$\operatorname{div} \mathbf{A} = 0 . \tag{15.7}$$

Then, substituting (15.6) in (15.2), we obtain

$$\operatorname{rot} \mathbf{H} = \operatorname{rot} \operatorname{rot} \mathbf{A} = \frac{4\pi \bar{\mathbf{j}}}{c} . \tag{15.8}$$

But according to (11.42)

$$\operatorname{rot} \operatorname{rot} \mathbf{A} = \operatorname{grad} \operatorname{div} \mathbf{A} - \Delta \mathbf{A} = -\Delta \mathbf{A} , \tag{15.9}$$

and we have used condition (15.7). Therefore, **A** satisfies the equation

$$\Delta \mathbf{A} = -\frac{4\pi}{c} \bar{\mathbf{j}} = -\frac{4\pi}{c} \overline{\rho \mathbf{v}} , \tag{15.10}$$

which is entirely analogous to equation (14.6) for the scalar potential.

Equation (15.10) can be obtained from (12.37) directly if we discard the term $\frac{1}{c^2} \frac{\partial^2 \mathbf{A}}{\partial t^2}$ which is superfluous in magnetostatics.

The vector potential for a point charge. The solution of (15.10) appears exactly the same as the solution of (14.6) given by equation (14.8) for a separate point charge: each component of **A** satisfies equation (14.6), the only difference being that on the right-hand side there appear the functions $\frac{4\pi}{c}\overline{\rho v_x}, \frac{4\pi}{c}\overline{\rho v_y}, \frac{4\pi}{c}\overline{\rho v_z}$. It follows that the vector potential for a point charge is

$$\mathbf{A} = \overline{\frac{e\mathbf{v}}{c\,|\mathbf{R}-\mathbf{r}|}}\,. \tag{15.11}$$

We shall now show that $\mathbf{A}$ satisfies condition (15.7). The divergence must be taken with respect to the radius vector $\mathbf{R}$ at the point at which $\mathbf{A}$ is determined.

But $\nabla_{\mathbf{R}}|\mathbf{R}-\mathbf{r}| = -\nabla_{\mathbf{r}}|\mathbf{R}-\mathbf{r}|$, so that

$$\operatorname{div}\mathbf{A} = \frac{e}{c}\,\overline{\mathbf{v}\nabla_{\mathbf{R}}\frac{1}{|\mathbf{R}-\mathbf{r}|}} = -\frac{e}{c}\,\overline{\mathbf{v}\nabla_{\mathbf{r}}\frac{1}{|\mathbf{R}-\mathbf{r}|}} = -\frac{e}{c}\,\overline{\frac{d}{dt}\frac{1}{|\mathbf{R}-\mathbf{r}|}}\,. \tag{15.12}$$

The expression on the right-hand side of this equation is the total time derivative of the quantity $\frac{1}{|\mathbf{R}-\mathbf{r}(t)|}$. From the steady-state condition, it is equal to zero.

The Biot-Savart law. We shall now calculate the magnetic field of a point charge. Using equation (11.28), we obtain

$$\mathbf{H} = \operatorname{rot}\mathbf{A} = \frac{e}{c}\,\overline{\left[\nabla_{\mathbf{R}}\frac{1}{|\mathbf{R}-\mathbf{r}|}\,\mathbf{v}\right]} = \frac{e}{c}\,\overline{\frac{[\mathbf{v},\mathbf{R}-\mathbf{r}]}{|\mathbf{R}-\mathbf{r}|^3}}\,.$$

This equation refers, of course, only to steady motion. In particular it is applied to constant current.

Vector potential at large distances from a system of stationary currents. The vector potential for a system of point charges is equal to the sum of vector potentials for each charge separately:

$$\mathbf{A} = \sum_i \overline{\frac{e_i\,\mathbf{v}^i}{c\,|\mathbf{R}-\mathbf{r}^i|}}\,. \tag{15.13}$$

We shall now obtain approximate formulae which are valid at large distances from the system, similar to those obtained in electrostatics. For this, we substitute an expansion of inverse distance in (15.13) [see (14.14)].

$$\frac{1}{|\mathbf{R}-\mathbf{r}^i|} = \frac{1}{R} - \frac{\mathbf{r}^i\mathbf{R}}{R^3}\,. \tag{15.14}$$

The quadratic term is not taken into account this time. The expression for vector potential, to the approximation of (15.14), is of the form

$$\mathbf{A} = \frac{1}{Rc}\overline{\sum_i e_i \mathbf{v}^i} + \frac{1}{cR^3}\sum_i \overline{e_i (\mathbf{R}\mathbf{r}^i)\,\mathbf{v}^i} =$$

$$= \frac{1}{Rc}\frac{d}{dt}\overline{\sum_i e_i \mathbf{r}^i} + \frac{1}{cR^3}\sum_i \overline{e_i (\mathbf{R}\mathbf{r}^i)\,\mathbf{v}^i}, \tag{15.15}$$

since $\dot{\mathbf{r}}^i = \mathbf{v}^i$.

The zero term of the expansion is a total derivative and, by the steady-state condition, is equal to zero. We now transform the first term of the expansion, using the identity

$$0 = \overline{\frac{d}{dt}\sum_i e_i (\mathbf{R}\mathbf{r}^i)\,\mathbf{r}^i} = \sum_i e_i \overline{(\mathbf{R}\mathbf{r}^i)\,\mathbf{v}^i} + \sum_i e_i \overline{(\mathbf{R}\mathbf{v}^i)\,\mathbf{r}^i}\,. \tag{15.16}$$

From this identity it follows that in (15.15) we can substitute half the difference of the expressions on the right-hand side of (15.16). Then the vector potential will be

$$\mathbf{A} = \frac{1}{2R^3c}\sum_i \overline{(\mathbf{v}^i(\mathbf{r}^i\mathbf{R})} - \overline{\mathbf{r}^i(\mathbf{v}^i\mathbf{R}))} = -\frac{1}{2R^3c}\sum_i e_i[\mathbf{R}\,\overline{[\mathbf{r}^i\,\mathbf{v}^i]}]\,. \tag{15.17}$$

We now interchange the signs of the summation and vector product:

$$\mathbf{A} = -\left[\frac{\mathbf{R}}{R^3}\overline{\sum_i \frac{e_i[\mathbf{r}^i\,\mathbf{v}^i]}{2c}}\right]. \tag{15.18}$$

Magnetic moment. The sum appearing inside the brackets in (15.18) is called the magnetic moment of the system of charges (or system of currents). The mean magnetic moment is written thus:

$$\overline{\boldsymbol{\mu}} = \overline{\sum_i e_i \frac{[\mathbf{r}^i\,\mathbf{v}^i]}{2c}}. \tag{15.19}$$

The equation for the vector potential (15.18) can be written, by means of the magnetic moment, in the following form:

$$\mathbf{A} = -\left[\frac{\mathbf{R}}{R^3}\,\overline{\boldsymbol{\mu}}\right] = \left[\nabla\frac{1}{R},\,\overline{\boldsymbol{\mu}}\right]. \tag{15.20}$$

The field of a magnetic dipole. Let us now calculate the magnetic field. By definition

$$\mathbf{H} = \operatorname{rot}\mathbf{A} = \operatorname{rot}\left[\nabla\frac{1}{R},\,\overline{\boldsymbol{\mu}}\right].$$

Since $\overline{\boldsymbol{\mu}}$ is a constant vector, equation (11.30) gives

$$\mathbf{H} = (\overline{\boldsymbol{\mu}}\nabla)\,\nabla\frac{1}{R} - \overline{\boldsymbol{\mu}}\Delta\frac{1}{R} = -(\boldsymbol{\mu}\nabla)\frac{R}{R^3},$$

because $\Delta \frac{1}{R} = 0$. Further, $(\bar{\mu}\,\nabla)\,\mathbf{R} = \bar{\mu}$ [see (11.35)]. In order to calculate $(\bar{\mu}\,\nabla)\frac{1}{\mathbf{R}3}$, we use equation (11.36). This yields

$$(\bar{\mu}\,\nabla)\,\frac{1}{R^3} = \left(\bar{\mu}\,\nabla\,\frac{1}{R^3}\right) = -\frac{3}{R^4}\,(\bar{\mu}\,\nabla\,R) = -\frac{3\,(\bar{\mu}\mathbf{R})}{R^5}\,.$$

Finally, collecting both terms, we arrive at an equation for $\mathbf{H}$:

$$\mathbf{H} = \frac{3\,\mathbf{R}\,(\mathbf{R}\,\bar{\mu}) - R^2\,\bar{\mu}}{R^5} \tag{15.21}$$

For comparison, we deduce the expression for the electric field of a dipole:

$$\mathbf{E} = -\nabla\varphi = -\nabla\,\frac{(\mathbf{R}\,\mathbf{d})}{R^3} = \frac{3\,\mathbf{R}\,(\mathbf{R}\,\mathbf{d}) - R^2\,\mathbf{d}}{R^5}\,. \tag{15.22}$$

Thus, both expressions for the field (both electric and magnetic) are of entirely analogous form. The only difference is that, instead of the electric moment, the equation for magnetic field involves the magnetic moment. This explains its name.

In the case of a charge moving in a flat closed orbit, the definition of magnetic moment (15.19) coincides with the elementary definition of moment in terms of "magnetic shell." As was shown in Sec. 5 [see (5.2), (5.4)], the product [rv] is twice the area swept out by the radius vector of the charge in unit time, or $[\mathbf{r}\mathbf{v}] = 2\,\frac{d\mathbf{s}}{dt}$. By definition of the mean value (15.4)

$$\bar{\mu} = \frac{1}{t_0}\int\limits_0^{t_0} \frac{e}{c}\,\frac{d\mathbf{s}}{dt}\,dt = \frac{e}{c\,t_0}\,\mathbf{s}\,. \tag{15.23}$$

Here, t_0 is the time of orbital revolution of the charge. In this time, the charge passes every point on the orbit once; hence the mean current is equal to $I = \frac{e}{t_0}$. This yields the definition of magnetic moment familiar from general physics:

$$\bar{\mu} = \frac{I\mathbf{s}}{c}\,. \tag{15.24}$$

The similarity of equations (15.21) and (15.22) shows the equivalence of a closed current (i.e., a magnetic shell) and a fictitious dipole with the same moment μ. The field at large distances from a system of currents is produced, as it were, by the effective dipole.

The relationship between magnetic and mechanical moments. An especially interesting case is that when all the charges of a system are of the same kind (for example, when they are all electrons).

Then the magnetic moment is proportional to the mechanical moment. Indeed, for a system of charges with identical ratios $\frac{e}{m}$, we obtain

$$\boldsymbol{\mu} = \frac{e}{2c}\sum_i [\mathbf{r}^i \mathbf{v}^i] = \frac{e}{2mc}\sum_i m\,[\mathbf{r}^i \mathbf{v}^i] = \frac{e}{2mc}\sum_i [\mathbf{r}^i \mathbf{p}^i] = \frac{e}{2mc}\mathbf{M}\,. \tag{15.25}$$

Equation (15.25) has very important applications.

A system of point charges in an external magnetic field. We now consider the question of the interaction of a system of currents with an external magnetic field. For this we must have an equation describing the interaction of a point charge with the field. We obtained equation (13.17) for the general spatial charge distribution. In this equation, the transition to point charges is obtained by changing the integral to a summation over the charges. The term obtained for action is of the form

$$S_1 = \int dt \sum \left(\frac{e_i\,(\mathbf{v}^i \mathbf{A}_i)}{c} - e_i\,\varphi_i \right), \tag{15.26}$$

where the indices i in $\mathbf{A}$ and φ denote that the potentials are taken at the same point as the ith charge.

In magnetostatics, only slowly moving charges for which $v \ll c$ are studied. Newtonian mechanics can then be applied to their motion. In the absence of a field, the action function of the particles is of the form

$$S_0 = \int dt \sum_i \frac{m_i v^{i2}}{2}\,. \tag{15.27}$$

It will be shown in Sec. 21 that this expression holds only when $v \ll c$. In magnetostatics, where this condition is satisfied, the action function of a system of charges in an external field is obtained by adding (15.27) and (15.26):

$$S = S_0 + S_1 = \int dt \left(\sum_i \frac{m_i v^{i2}}{2} + e_i\,\frac{\mathbf{A}_i \mathbf{v}^i}{c} - e_i\,\varphi_i \right). \tag{15.28}$$

The field due to the charges themselves does not appear in this equation. The expression for the integrand is the Lagrangian of the system. It involves velocity linearly as well as quadratically (in the expression for kinetic energy), and, for this reason, does not have the form that we used in Part I, $L = T - U$.

However, the general relationships still hold. Therefore, from the Lagrangian the expression for momentum is obtained in terms of velocity

$$\mathbf{p}_i = \frac{\partial L}{\partial \mathbf{v}^i} = m_i \mathbf{v}^i + \frac{e}{c}\mathbf{A}_i\,. \tag{15.29}$$

Let us determine the energy in terms of momentum using the basic equation (4.4):

$$\mathscr{E} = \sum \mathbf{v}^i \mathbf{p}_i - L = \sum_i \left(m_i v^{i^2} + \frac{e_i}{c} \mathbf{A}_i \mathbf{v}^i - \frac{m_i v^{i^2}}{2} - \frac{e_i}{c} \mathbf{A}_i \mathbf{v}^i + e_i \varphi_i \right) =$$
$$= \sum_i \frac{m_i v^{i^2}}{2} + e_i \varphi_i \Big), \qquad (15.30)$$

so that the term which is linear with respect to velocity is eliminated from the expression for energy in terms of velocity.

The Hamiltonian function for a system of charges in an external magnetic field. The linear term in velocity of the Lagrangian affects the form of the Hamiltonian. Let us write down $\mathscr{H}$ from its definition (10.15). To do this it is necessary to substitute into the energy expression, by means of equation (15.29), momenta instead of velocities:

$$\mathbf{v}^i = \frac{1}{m_i} \left(\mathbf{p}_i - \frac{e}{c} \mathbf{A}_i \right). \qquad (15.31)$$

The Hamiltonian is

$$\mathscr{H} = \sum_i \left[\frac{1}{2m_i} \left(\mathbf{p}_i - \frac{e_i}{c} \mathbf{A}_i \right)^2 + e_i \varphi_i \right]. \qquad (15.32)$$

Let us assume that the magnetic field, in which the system is situated, is weak and uniform (at least within the limits of the system). The vector potential for a homogeneous field will be represented as

$$\mathbf{A} = \frac{1}{2} [\mathbf{H}\mathbf{r}]. \qquad (15.33)$$

Indeed, then rot $\mathbf{A} = \mathbf{H}$ [from (11.30), (11.35) and (11.33)]. And also div $\mathbf{A} = 0$ from (11.29) and (11.34).

Since the magnetic field is weak we can neglect in (15.32) the term involving A_i^2. Then, substituting (15.33) in (15.32), we find an expression for $\mathscr{H}$:

$$\mathscr{H} = \sum_i \left(\frac{p_i^2}{2m_i} + e_i \varphi_i - \frac{e_i}{2m_i c} (\mathbf{p}_i [\mathbf{H}\mathbf{r}^i]) \right). \qquad (15.34)$$

The last term in (15.34) gives the required addition to the Hamiltonian. Since this term is proportional to $\mathbf{H}$, we can replace $\mathbf{p}_i$ by $m_i \mathbf{v}^i$ in it to the same accuracy, i.e., neglecting terms of order H^2. Performing a cyclic permutation of the factors in (15.34) and putting the sum inside a vector product sign, we obtain an expression for the addition to the Hamiltonian:

$$\mathscr{H}' = -\left(\mathbf{H} \sum_i \frac{e_i}{2c} [\mathbf{r}^i \mathbf{v}^i] \right) = -(\mathbf{H} \boldsymbol{\mu}). \qquad (15.35)$$

This expression is very similar to the energy of a system of charges in a homogeneous electric field which involves only the electric dipole moment of the system of charges. Note that this term is of the form

—(dE) [see (14.28)]. This indicates a further similarity between electric and magnetic moments.

Larmor's theorem. Let us now compare the expression for the momentum of a charge placed in a constant homogeneous magnetic field with that for the momentum of a particle relative to a rotating coordinate system. From (15.29) and (15.33), the former is

$$\mathbf{p} = m\mathbf{v} + \frac{e\mathbf{A}}{c} = m\mathbf{v} + \frac{e}{2c}[\mathbf{H}\mathbf{r}], \tag{15.36}$$

the latter can be easily found from (8.5):

$$\mathbf{p} = m\mathbf{v}' + m[\boldsymbol{\omega}\mathbf{r}]. \tag{15.37}$$

We now consider a steadily moving system of identical charges (for example, an atom or molecule); the nuclei, being heavier, are regarded as fixed. Let us assume that in the absence of any external magnetic field, the motion in the system is known. Then, comparing equations (15.36) and (15.37), it is easy to see that if we consider the motion of these charges relative to axes rotating with angular velocity

$$\boldsymbol{\omega} = -\frac{e\mathbf{H}}{2mc}, \tag{15.38}$$

it will not differ from motion relative to fixed axes in the absence of a magnetic field. The equations of motion relative to rotating axes will have their usual form $\dot{\mathbf{p}}_i = \mathbf{F}_i$, where $\mathbf{F}_i$ is the force acting on the ith charge in the absence of any external magnetic field, because the correction to the momentum due to the angular velocity $\boldsymbol{\omega}$ [defined in accordance with (15.38)] will cancel with the correction due to the magnetic field. The magnetic field must be sufficiently weak so that the change in magnetic force with rotation can be neglected.

We can say, therefore, that, with the application of a constant and uniform weak external magnetic field, a system of charges with identical $\frac{e}{m}$ ratios begins to rotate with a constant angular velocity $|\omega_L| = \frac{eH}{2mc}$. This statement is called Larmor's theorem, and ω_L is the Larmor frequency.

Precession of the magnetic moment. If a system possesses a magnetic moment $\boldsymbol{\mu}$ for motion which is undisturbed by a magnetic field, then, when a magnetic field is superimposed, this moment will move around the direction of the magnetic field, similar to the free top in equations (9.14) (note that $\boldsymbol{\omega}$ is in the direction of the axis of rotation, i.e., in the direction of the magnetic field). The precession of magnetic moment about the field is called the Larmor precession.

Magnetic moment in an inhomogeneous field. Let us suppose that the magnetic field possesses a small inhomogeneity. Then, in the equations of motion, the term

$$\mathbf{F} = -\nabla \mathscr{H}' = \nabla (\mathbf{H}\boldsymbol{\mu}) \tag{15.39}$$

denotes the force acting on the moment and tending to move it as a whole. Expanding (15.39) by (11.32), we obtain

$$\mathbf{F} = (\boldsymbol{\mu}\nabla)\,\mathbf{H} + [\boldsymbol{\mu}\,\mathrm{rot}\,\mathbf{H}]\,.$$

But for an external field rot **H** is equal to zero so that the force is

$$\mathbf{F} = (\boldsymbol{\mu}\nabla)\,\mathbf{H}\,. \tag{15.40}$$

This is the well-known force of attraction to a magnet. It is maximum near the poles of the magnet where the inhomogeneity of the field is greatest.

Exercise

Study the magnetic moment μ moving in a magnetic field given by the components $H_z = -H_0$; $H_x = H_1 \cos \omega t$; $H_y = H_1 \sin \omega t$. Consider the cases $\omega = \omega_0 = \dfrac{eH_0}{2mc}$, $\omega \to 0$.

The equation describing a vector rotating with angular velocity ω_L is, according to Secs. 8 and 9,

$$\frac{d\mu}{dt} = [\omega_L \mu]\,.$$

Whence we obtain an equation for the precession of magnetic moment in a magnetic field

$$\frac{d\mu}{dt} = \frac{e}{2mc}[\mu H]\,.$$

By multiplying both sides of this equation by μ, we convince ourselves that the absolute value of magnetic moment μ is conserved. It is, therefore, sufficient to write the equations only for the components μ_x and μ_y, replacing μ_z by $\sqrt{\mu^2 - \mu_x^2 - \mu_y^2}$.

Using the abbreviated notation $\omega_0 = \dfrac{eH_0}{2mc}$, $\omega_1 = \dfrac{eH_1}{2mc}$ [Cf. (15.38)], we multiply the equation for μ_y by $\pm i$ and combine with the equation for μ_x to obtain

$$\frac{d}{dt}(\mu_x \pm i\mu_y) = \pm i\omega_0(\mu_x \pm i\mu_y) \pm i\omega_1 e^{i\omega}\sqrt{\mu^2 - \mu_x^2 - \mu_y^2}\,.$$

We seek the solution in the form $\mu_x \pm i\mu_y = A_\pm e^{\pm i\omega t}$, and get the following equation for amplitudes $A_\pm$:

$$(\omega - \omega_0) A_\pm = \omega_1 \sqrt{\mu^2 - A_+ A_-}\,.$$

Multiplying these equations, we find

$$A_+ A_- = \frac{\mu^2\omega_1^2}{(\omega - \omega_0)^2 + \omega_1^2}\,; \quad A_\pm = \frac{\mu\omega_1}{\sqrt{(\omega - \omega_0)^2 + \omega_1^2}}$$

When $\omega = \omega_0$ ("paramagnetic resonance"), the moment rotates in the plane xy with frequency ω_0. When $\omega \to 0$, i.e., in the case of an infinitely slow rotation of the field, the moment strictly follows the field, its direction all the time being the same as that of the field.

Sec. 16. Electrodynamics of Material Media

Field in a medium. We know that material media consist of nuclei and electrons, i.e., of very small charges in very rapid motion. Therefore, in a small region of a body—a region having atomic dimensions—all electromagnetic quantities (field, charge density, and current) change very rapidly with time. In two neighbouring small regions these quantities may, at the very same instant, have completely different values. Therefore, if we examine the field in a medium full of charges in detail, then we will observe only a rapidly and irregularly varying function of coordinates and time.

Mean values. The inhomogeneities of a field are of atomic dimensions. However, such a detailed picture of the field is not usually of any interest. As usual, in any description of macroscopic bodies, it is essential to know mean values for a large number of atoms. For example, in mechanics, mean density values are used. For this mean to have any significance, we must isolate a volume of the body containing a large number of atoms, determine its mass and divide by the volume.

This volume must be so large that the microscopic atomic structure of the substance cannot affect the mean value of the density. At the same time, the mean macroscopic value must be constant over that volume. This will be readily seen from the following. Let the volume be arbitrarily divided into two equal parts. Then the mean for each part should not differ from the mean for the whole volume.

Such a volume is termed physically infinitesimal. We shall call it V_0. If we take all its dimensions to be large compared with atomic dimensions, then the mean value should not depend on the shape of the surface bounding the volume; the latter may be spherical, cubical, etc.

Besides averaging over volume, it is also necessary to perform averaging over time. The interval of time over which the average is to be taken must be large compared with the times of atomic motions, though still sufficiently small so that the mean values over two semi-intervals do not differ from one another.

Let the volume have the form of a cube of side a. We shall denote the coordinates of its centre by x, y, z. The time interval, over which the averaging is performed, will be called t_0, and the instant corresponding to the centre of the interval will be denoted by t. The coordinates of any point inside the cube, relative to the centre, will be called ξ, η, ζ, and instants of time measured from t, will be denoted by ϑ. Thus, the limits of variation of the quantities are given by the following inequalities:

$$-\frac{a}{2} \leqslant \xi \leqslant \frac{a}{2}, \quad -\frac{a}{2} \leqslant \eta \leqslant \frac{a}{2}, \quad -\frac{a}{2} \leqslant \zeta \leqslant \frac{a}{2}, \quad -\frac{t_0}{2} \leqslant \vartheta \leqslant \frac{t_0}{2}.$$

The actual value of any quantity at a definite instant of time is $f(x+\xi, y+\eta, z+\zeta, t+\vartheta)$. It is related to a mathematically infinitely small volume $dV = d\xi\, d\eta\, d\zeta$ and an interval of time $d\vartheta$. The average value, over a physically infinitesimal volume V_0 and interval of time t_0, is obtained if f is integrated over $dV\, dt$ and the integral divided by $V_0 t_0$, in accordance with the usual definition of an average:

$$\bar{f}(x, y, z, t) = \\ = \frac{1}{V_0 t_0} \int\limits_{-a/2}^{a/2} d\xi \int\limits_{-a/2}^{a/2} d\eta \int\limits_{-a/2}^{a/2} \vartheta\zeta \int\limits_{-t_0/2}^{t_0/2} d\vartheta\, f(x+\xi,\ y+\eta,\ z+\zeta,\ t+\vartheta)\,. \tag{16.1}$$

This average value $\bar{f}(x, y, z, t)$ refers to the point x, y, z, and time t. The electrodynamics of such mean values is termed macroscopic, as opposed to microscopic, which has to do with a field due to separate point charges and a field in free space.

The mean thus determined is differentiable with respect to time and coordinates. As parameters it involves the coordinates of the centre of a physically infinitely small volume x, y, z, and the time t. Obviously, we can differentiate with respect to these values:

$$\frac{\partial}{\partial x}\bar{f}(x, y, z, t) = \\ = \frac{1}{V_0 t_0} \int\limits_{-a/2}^{a/2} d\xi \int\limits_{-a/2}^{a/2} d\eta \int\limits_{-a/2}^{a/2} d\zeta \int\limits_{-t_0/2}^{t_0/2} d\vartheta\, \frac{\partial}{\partial x} f(x+\xi,\ y+\eta,\ z+\zeta,\ t+\vartheta) = \\ = \overline{\frac{\partial f}{\partial x}}\,. \tag{16.2}$$

In other words, the mean value of the derivative of a quantity is equal to the derivative of its mean value.

Density of charge and current in a medium. Under the action of the electric and magnetic field, there occurs a redistribution of the charges and currents in any substance. When Maxwell's equations are averaged, the mean density of the redistributed charge is $\bar{\rho}$ and that of the current is $\bar{\mathbf{j}}$. We shall express $\bar{\rho}$ and $\bar{\mathbf{j}}$ in terms of other values which will later make it possible to give the averaged Maxwell equations a very symmetrical form.

We define the dipole-moment density in a substance by the following formula:

$$\mathbf{d} = \int \mathbf{P}\, dV\,. \tag{16.3}$$

The dipole moment $\mathbf{P}$ in unit volume is called the *electric polarization* of the medium. If the substance is completely neutral, its dipole moment is uniquely determined as $\sum\limits_i e_i \mathbf{r}^i$ [see (14.21)]. Going over to a continuous charge distribution, we write

$$\mathbf{d} = \int \rho \, \mathbf{r} \, dV. \tag{16.4}$$

Integral (16.3) can be identically written in the form:

$$\int \mathbf{P} \, dV = -\int \mathbf{r} \operatorname{div} \mathbf{P} \, dV. \tag{16.5}$$

This relationship can most simply be proved by writing in terms of components, for example,

$$\int\int\int x \left(\frac{\partial P_x}{\partial x} + \frac{\partial P_y}{\partial y} + \frac{\partial P_z}{\partial z}\right) dx\, dy\, dz = \int\int (xP_x) \Big|_{x_1}^{x_2} dy\, dz + $$
$$+ \int\int x (P_y) \Big|_{y_1}^{y_2} dx\, dz + \int\int x (P_z) \Big|_{z_1}^{z_2} dx\, dy - \int\int\int P_x dx\, dy\, dz.$$

The limits of integration are at the external boundaries of the medium, where the values P_x, P_y, P_z are zero. This proves (16.5). Comparing (16.4) and (16.5) we obtain

$$\int \mathbf{r} (\operatorname{div} \mathbf{P} + \bar{\rho}) \, dV = 0. \tag{16.6}$$

However, since the shape and dimensions of the body are arbitrary, the quantity

$$\operatorname{div} \mathbf{P} + \bar{\rho} = 0 \tag{16.7}$$

must be zero.

Thus, the mean density of a charge "induced" by the field is equal to the divergence of the electric polarization vector taken with opposite sign.

In a similar way, we can express the mean density of induced current. To do this we define the *magnetic polarization* vector, equal to the magnetic-moment density, as

$$\boldsymbol{\mu} = \int \mathbf{M} \, dV. \tag{16.8}$$

But the magnetic moment, by definition, is expressed as $\boldsymbol{\mu} = \sum_i \frac{e_i [\mathbf{r}^i \mathbf{v}^i]}{2c}$. Applied to the current distribution, this gives

$$\boldsymbol{\mu} = \int \frac{[\mathbf{r}\bar{\mathbf{j}}] \, dV}{2c}. \tag{16.9}$$

We shall now prove an identity analogues to (16.5):

$$\int \mathbf{M} \, dV = \frac{1}{2} \int [\mathbf{r} \operatorname{rot} \mathbf{M}] \, dV. \tag{16.10}$$

For this, it is simplest to go over to components:

$$\int [\mathbf{r}\,\mathrm{rot}\,\mathbf{M}]_x dV = \int (y\,\mathrm{rot}_z\,\mathbf{M} - z\,\mathrm{rot}_y\,\mathbf{M})\,dV =$$
$$= \iiint \left\{ y\left(\frac{\partial M_y}{\partial x} - \frac{\partial M_x}{\partial y}\right) - z\left(\frac{\partial M_x}{\partial z} - \frac{\partial M_z}{\partial x}\right)\right\} dx\,dy\,dz\,.$$

The terms $-\,y\dfrac{\partial M_x}{\partial y}$ and $-\,z\dfrac{\partial M_x}{\partial z}$ are integrated by parts. All the integrated quantities become zero when the limits are inserted, so that, in agreement with (16.10), only $2\,M_x$ remains. Now, comparing (16.9) and (16.10) we obtain

$$\boldsymbol{\mu} = \int \frac{[\mathbf{r}\bar{\mathbf{j}}]}{2c}\,dV = \int \frac{[\mathbf{r}\,\mathrm{rot}\,\mathbf{M}]}{2}\,dV\,. \tag{16.11}$$

In order to determine $\bar{\mathbf{j}}$ fully, we calculate its divergence and apply the charge conservation law (12.18) written for mean values [see (16.2)]:

$$\mathrm{div}\,\bar{\mathbf{j}} = -\,\frac{\partial \bar{\rho}}{\partial t} = \mathrm{div}\,\frac{\partial \mathbf{P}}{\partial t}\,. \tag{16.12}$$

From (16.11) and (16.12), $\bar{\mathbf{j}}$ is uniquely determined as

$$\bar{\mathbf{j}} = \frac{\partial \mathbf{P}}{\partial t} + c\,\mathrm{rot}\,\mathbf{M}\,. \tag{16.13}$$

Indeed, this expression satisfies both equations. Finding the divergence of both parts of (16.13), we arrive at (16.12), since div rot $\mathbf{M} = 0$. Further, substituting the quantity $\dfrac{\partial \mathbf{P}}{\partial t}$ in the left-hand side of (16.10), we get

$$\frac{1}{2c}\int \left[\mathbf{r}\,\frac{\partial \mathbf{P}}{\partial t}\right] dV = \frac{1}{2c}\,\frac{\partial}{\partial t}\int [\mathbf{r}\mathbf{P}]\,dV\,. \tag{16.14}$$

According to (16.3) and (16.4), $\mathbf{P}$ is replaced by $\rho\mathbf{r}$. But, $[\mathbf{r}, \rho\mathbf{r}] = 0$, so that the term $\dfrac{\partial \mathbf{P}}{\partial t}$ does not contribute to equation (16.11). The identity (16.13) is thus proved.

Averaging Maxwell's equations. We shall now consider the averaging of Maxwell's equations. From (16.2), differentiation and averaging are commutative, so that a bar can simply be put over the first pair in order to denote that they have been averaged:

$$\mathrm{rot}\,\bar{\mathbf{E}} = -\,\frac{1}{c}\,\frac{\partial \bar{\mathbf{H}}}{\partial t} \tag{16.15}$$

$$\mathrm{div}\,\bar{\mathbf{H}} = -\,0\,. \tag{16.16}$$

The mean value of an electric field $\bar{\mathbf{E}}$ is called the electric field in a medium. We shall hereafter write it without the bar, which denotes that it has been averaged, taking it for granted that only mean values will always be taken in a medium. The mean value of the magnetic field is called the *magnetic induction* and is denoted by the letter **B**. It is all the more unnecessary to write a bar over it because the concept of induction, which is not equal to field, makes sense only for a medium. The asymmetry in the terminology for electric and magnetic fields will be explained later.

In this notation, the first pair of Maxwell's equations takes the following form:

$$\operatorname{rot} \mathbf{E} = -\frac{1}{c}\frac{\partial \mathbf{B}}{\partial t}, \tag{16.17}$$

$$\operatorname{div} \mathbf{B} = 0. \tag{16.18}$$

Now let us average the second pair of Maxwell's equations:

$$\operatorname{rot} \overline{\mathbf{H}} = \frac{1}{c}\frac{\partial \overline{\mathbf{E}}}{\partial t} + \frac{4\pi}{c}\overline{\mathbf{j}}, \tag{16.19}$$

$$\operatorname{div} \overline{\mathbf{E}} = 4\pi\overline{\rho}. \tag{16.20}$$

We substitute $\overline{\rho}$ and $\overline{\mathbf{j}}$ from (16.7) and (16.13) and rearrange the terms somewhat, obtaining the two following equations (the bars are again omitted):

$$\operatorname{rot}(\mathbf{B} - 4\pi\mathbf{M}) = \frac{1}{c}\frac{\partial}{\partial t}(\mathbf{E} + 4\pi\mathbf{P}), \tag{16.21}$$

$$\operatorname{div}(\mathbf{E} + 4\pi\mathbf{P}) = 0. \tag{16.22}$$

We introduce the following new designations:

$$\mathbf{E} + 4\pi\mathbf{P} \equiv \mathbf{D}. \tag{16.23}$$

D is called the *electric induction*.

Further,

$$\mathbf{B} - 4\pi\mathbf{M} \equiv \mathbf{H}. \tag{16.24}$$

H is called the *magnetic field* in a medium, which, therefore, does not equal the mean value of the magnetic field in a vacuum.

In the notations (16.23) and (16.24), the second pair is written similar to the first pair:

$$\operatorname{rot} \mathbf{H} = \frac{1}{c}\frac{\partial \mathbf{D}}{\partial t}, \tag{16.25}$$

$$\operatorname{div} \mathbf{D} = 0. \tag{16.26}$$

The similarity between (16.26) and (16.18) explains why it was convenient to call the mean value of the magnetic field the magnetic induction: here, both electric and magnetic induction vectors have no sources in the medium. The similarity between (16.25) and (16.17) justifies the term magnetic field given to the vector **H** (16.24).

The incompleteness of the system of equations in a medium. Thus, due to a suitable system of notation, the first and second pairs of

Maxwell's equations in a medium have, as it were, become more symmetrical than those in a vacuum. But we must not forget that this system has now ceased to be complete: as before, there are eight equations (of which only six are independent) and twelve unknowns **B**, **E**, **D**, **H** (with three components for each vector). Consequently, the system (16.17), (16.18), (16.25), and (16.26) cannot be solved until a relationship is found between inductions and fields. This relationship cannot be obtained without knowing the specific structure of the material medium.

Dielectrics and conductors. We shall consider, first of all, how charges behave in a medium in the presence of a constant electric field. The field will displace the positive charges in one direction, and the negative ones in another. As a result, polarization **P** will arise. Two essentially different cases can occur here.

1) Under the action of the field inside the body, a certain finite polarization **P**, dependent on the field, is established. This polarization may be represented vividly (though in very simplified fashion!) as a displacement of charges from the equilibrium positions which they occupied in the absence of the field to new equilibrium positions—much like the way a load suspended on a spring is displaced in a gravitational field. If a finite polarization (dependent on the field inside the body) is established, that body is called a nonconductor or *dielectric*.

2) In a constant electric field acting inside the body, the charges do not arrive at equilibrium, and a definite rate of polarization increase, $\frac{\partial \mathbf{P}}{\partial t}$, is established. In this case, through every section of the conductor perpendicular to the vector $\frac{\partial \mathbf{P}}{\partial t}$, there pass electric charges or, what amounts to the same thing, a flow of current. From equation (16.13), the derivative $\frac{\partial \mathbf{P}}{\partial t}$ may indeed be interpreted as a current component. As regards the second current component, c rot **M**, it relates to the instantaneous value of quantities and cannot characterize the change of anything with time. For this reason, the classification of bodies into conductors and dielectrics is obtained from the behaviour of the quantity $\frac{\partial \mathbf{P}}{\partial t}$.

The displacement of charges under the action of a field can roughly be likened to a load falling in a viscous medium with friction, when, as is known, a definite speed of fall is established.

A medium, in which a constant electric field produces a constant electric current, is called a *conductor*.

If a constant electric field is produced in free space and a conducting body of finite dimensions is introduced into it (for example, a conducting sphere or ellipsoid), the charges in the body will be displaced so that a field equal to zero will be established inside the conductor. For this, the mean charge density inside the body must also equal zero, because the lines of force of the field originate and terminate at charges. Under the action of such a field, the charges inside the conductor would be displaced. This means that an equilibrium will be established inside the conductor only when all the induced charges emerge to the surface. They will be distributed on the surface of the body so that the mean field inside the conductor is zero, and the lines of force outside the conductor will arrive normal to every point of the surface.

Continuous current in a conductor. A continuous current can flow in a conductor only along a closed conducting circuit. And the electric field must always have a component along the direction of the circuit. Then charges of given sign will always move in one direction, thereby producing a closed current. The work performed by unit charge in moving around the circuit is called the e.m.f. acting in the circuit:

$$\text{e. m. f.} = \int \mathbf{E}\, d\mathbf{l}\,. \qquad (16.27)$$

This formula differs from (12.1) in that **E** denotes the field acting inside the conductor.

External sources of e.m.f. In a conductor, a constant e.m.f. can only exist at the expense of some external source of energy, for example, a primary cell. When a current passes in the circuit of the cell, ions are neutralized on the electrodes, thus yielding the source of energy that maintains the e.m.f.

If, as usual, we put $\mathbf{E} = -\nabla\varphi$, then the expression for e.m.f. will be

$$\text{e. m. f.} = -\int \nabla\varphi\, d\mathbf{l} = -\int \left(\frac{\partial\varphi}{\partial x}dx + \frac{\partial\varphi}{\partial y}dy + \frac{\partial\varphi}{\partial z}dz\right) =$$

$$= -\int d\varphi = \varphi_1 - \varphi_2\,. \qquad (16.28)$$

Therefore, the e.m.f. may be defined as the change in potential in going round a closed path. Thus, the potential is not a unique function of a point: for each traverse, it changes by the value of the e.m.f. in the circuit.

The magnetic properties of bodies. We shall now consider the magnetic properties of bodies. In a constant magnetic field, a definite equilibrium state will always be established in the medium. Here we must distinguish between the following two cases.

1) In the absence of a field, the atoms or molecules of a substance possess certain characteristic magnetic moments that differ from zero.

As was shown in the previous section, the energy of every separate elementary magnet in a magnetic field is $-\mu\mathbf{H}$. Hence, the energy of elementary magnets that have a positive moment projection on the field is less than the energy of elementary magnets with a negative moment projection. Atoms and molecules are in random thermal motion. As a result of this motion and of the action of a magnetic field, an advantageous energy state is established in which positive moment projections on the field predominate. For more detail about this equilibrium see Part IV.

It will be noted that the projection of an isolated magnetic moment on the magnetic field is constant—it merely performs a Larmor precession around the field. But the interaction between molecules disturbs the motion of separate moments, and results in the establishment of a state with a mean magnetic polarization other than zero.

2) The atoms or molecules of a body do not possess their own magnetic moments in the absence of a field.

As was shown in the previous section, when an external magnetic field is applied, the motion of charges in atoms or molecules changes due to the Larmor precession. Indeed, a precession of angular velocity $\omega = \frac{eH}{2mc}$ is superimposed on motion undisturbed by a magnetic field. In exercise 4 of this section it will be shown that this precession leads to the appearance of a magnetic moment in a system of charges. We shall only note here that the direction of the magnetic moment induced by the field must be in opposition to the direction of the magnetic field; this follows from Lenz's induction law. Indeed, an induced current produces a magnetic field in a direction opposite to that of the inducing field.

A substance in which an external magnetic field produces a resultant moment in the same direction, is called *paramagnetic.* If the magnetization is in the opposite direction to the field, the substance is *diamagnetic.*

Ferromagnetism. There are crystalline bodies in which the magnetic moments are aligned spontaneously, i.e., in the absence of any external magnetic field. Such bodies are called *ferromagnetic.* The magnetic polarization of the body itself is related to the directions of the crystalline axes. For example, in iron, whose crystals have cubic symmetry, the intrinsic magnetization coincides with one of the sides of the cube. This direction is called the direction of ready magnetization. In order to deflect the magnetic polarization from the direction of ready magnetization, work must be performed.

A single crystal of a ferromagnetic substance will be magnetized so that the resultant energy is a minimum—equilibrium always corresponds to minimum energy. However, it does not necessarily follow from this that all of the single crystal is magnetized in one direction: in this case it will possess an external magnetic field whose energy is $\frac{1}{8\pi}\int H^2\,dV$. This quantity is always positive and increases the total energy. But if the single crystal is divided into regions or layers whose magnetization alternates in direction, then the external field can be eliminated since neighbouring layers (or, as they are called, *domains*) produce fields of opposite sign. In the transition region between domains, the polarization gradually turns from the direction of ready magnetization in one domain to the reverse direction in the other domain. Clearly, if a certain direction is the direction of ready magnetization, then the directly opposite direction also possesses this property. The structure of the transition region has been studied theoretically by L. D. Landau and E. M. Lifshits.

The domain structure of crystals was later demonstrated experimentally. If a very thin emulsion of particles of a ferromagnetic substance is spread over the smooth surface of a ferromagnetic single crystal, the particles will be distributed along lines where the interfaces between domains intersect the surface of the crystal.

Since, between domains, the polarization is deflected from the direction of ready magnetization, it is necessary to perform work to establish the transition region. Summarizing, if the whole single crystal consists of one domain, its energy increases at the expense of work done in creating an external field, equal to $\frac{1}{8\pi}\int H^2 dV$; if, however, the crystal consists of many domains, the energy increases at the expense of the additional energy of the transition regions. The equilibrium state will be that state in which the energy is least. The energy of a field increases with the volume it occupies, that is, as the cube of the linear dimensions of the crystal. The energy of the transition regions increases in proportion to their total area. In a crystal of sufficiently small dimensions, there can exist only one transition region whose area is proportional to the square of the linear dimensions of the single crystal. Therefore, in such a small crystal, the volume energy changes according to a cubic law with respect to dimensions, while the surface energy varies according to a square law. In a sufficiently small crystal, the volume energy becomes less than the surface energy; such a crystal is not separated into domains but is magnetized as a whole. This has been experimentally established in crystals with dimensions of 10^{-4}-10^{-6} cm. The thickness of domains in large crystals of appropriate ferromagnetics is of the same order.

An example of the shape of a domain as proposed by Landau and Lifshits is shown in Fig. 22. The arrows denote the direction of the polarization in each domain. The serrations at the boundary almost completely destroy the external magnetic field; the lines of magnetic induction inside the crystal are closed through them, and do not emerge.

Fig. 22

The magnetization of a ferromagnetic in an external field. If a magnetic field is applied to a ferromagnetic crystal in the direction of ready magnetization, then those regions, for which the polarization is in opposition to the field, are contracted with displacement of the interfaces and may disappear completely in a comparatively small field. Then the crystal is magnetized to saturation. In order to magnetize the crystal to saturation in a direction that is not coincident with the initial direction of polarization in the domains, considerably larger fields are required.

In a polycrystalline body, such as ordinary steel, the separate single crystals are oriented more or less at random relative to one another. In any case, the directions of ready magnetization are not the same for the separate crystals. When an external magnetic field is applied, the different crystals are magnetized differently and the magnetization curve is not as steep as is possible in the case of a separate crystal. The magnetic interaction between separate crystals results in a definite magnetic polarization remaining after steel has been magnetized and the field subsequently removed. This is what is known as hysteresis.

Magnetic interaction of atoms. Let it be noted, in addition, that the magnetic interaction between separate elementary (atomic) magnets is not at all adequate in explaining the cause of ferromagnetism. The energy of interaction between two elementary magnetic moments is of the order 10^{-16} erg, while the energy of thermal motion at room temperature is about 10^{-14} erg (see Part IV). This is why random thermal motion should destroy the orderly magnetization already at a temperature of about 1° above absolute zero. Actually, ferromagnetism of steel disappears in the neighbourhood of 1,000° above absolute zero, thus corresponding to an interaction energy between elementary magnets in the order of 10^{-13} erg.

Ferromagnetism is of quantum origin and cannot be explained with the aid of classicial analogues.

The relationship between fields and inductions. A substance is always in equilibrium in a constant external magnetic field. To this equilibrium there corresponds a very definite induction and polarization. In a weak field, the relationship between the quantities is linear. For this reason, the magnetic induction is expressed linearly in terms of the magnetic field in a medium:

$$\mathbf{B} = \chi \mathbf{H} . \qquad (16.29a)$$

In a dielectric, where a static equilibrium polarization corresponds to a definite electric field, there is a similar relationship for weak fields.

$$\mathbf{D} = \varepsilon \mathbf{E} . \qquad (16.29b)$$

The quantity χ is called the permeability and ε is the dielectric constant.

It should be noted that in ferromagnetics the region for which a linear law is applicable has an upper limit of not very large fields

(10^2-10^4 CGSE), since saturation sets in; in diamagnetic and paramagnetic substances at room temperatures a linear law applies for all actually attainable fields.

The vector nature of electric and magnetic fields. The question may arise: Why is magnetic induction expressed linearly solely in terms of magnetic field, while electric induction is expressed solely in terms of electric field?

In order to answer this question we must examine the vector properties of electromagnetic quantities in more detail.

Two separate systems of rectangular coordinates exist in space: a right-hand system and a left-hand system. They are related to each other like left and right hands, if the thumbs are in the direction of the x-axes, the forefingers along the y-axes, and the middle fingers along the z-axes. It is obvious that no rotation in space can make these two systems coincide. However, one system transforms to the other if the signs of the coordinates in it are reversed. Of course, both coordinate systems are completely equivalent physically. The choice of any one of them is completely arbitrary. Therefore, the form of any equation expressing a law in electrodynamics should not change under a transformation from a right-hand to a left-hand system.

Let us now take Maxwell's equation (12.24). In order to perform a transformation to another coordinate system, it is sufficient to change the signs of the coordinates. This changes the sign of the vector operation rot, because this operation denotes a differentiation with respect to coordinates. What happens, then, to the electric field components? Since only one of two vectors is differentiated with respect to the coordinates, namely **E**, the sign of one of them must change in order to retain the form of the equation. It is easy to see that vector **E** will change sign. Indeed, the right-hand side of equation $\operatorname{div} \mathbf{E} = 4\pi\rho$ is a scalar and does not change in sign. On the left-hand side, the sign of the div operation changes and, hence, the signs of all the components of **E** must also change. Therefore, the components of the magnetic field do not change sign in a transformation from a left-hand to a right-hand system.

In a rotation of a coordinate system, the projections of any vector are transformed by the same equations of analytical geometry as the coordinates. As was shown, the change of sign for all three coordinates is not equivalent to any rotation. It turns out that some vectors, such as **E**, behave quite similarly to a radius vector **r**; when the signs of the components of the radius vector **r** are changed, the signs of all the components of **E** also change. Other vectors, such as **H**, behave like a radius vector under coordinate rotations, and not like a radius vector in the transformation from a right-hand to a left-hand system.

Vectors that behave like **E** are called true or polar vectors, while those behaving like **H** are called pseudovectors or axial vectors.

Velocity, force, acceleration, current density, and vector potential are, in addition to the electric field, real vectors while magnetic moment, angular momentum and angular velocity are pseudovectors.

The fact that angular momentum is a pseudovector can easily be seen from its definition: $\mathbf{M} = [\mathbf{rp}]$. Both factors of the vector product, $\mathbf{r}$ and $\mathbf{p}$, change signs, so that $\mathbf{M}$ does not change in sign.

A pseudovector cannot be linearly related to a real vector in electrodynamics because the sign in any such equality would depend on the choice of coordinate system, which contradicts physical facts. For this reason the vectors $\mathbf{B}$ and $\mathbf{H}$, $\mathbf{D}$ and $\mathbf{E}$ appear separately in the linear laws (16.29a) and (16.29b).

The equations for conductors in a constant field. We shall now consider the equations of electrodynamics of constant fields for conductors. As has already been indicated, it is not a constant value of polarization that is established in a conductor in a constant field, but a constant rate of increase of polarization $\frac{\partial \mathbf{P}}{\partial t}$. This quantity has the meaning of current density $\mathbf{j}'$. The derivative $\frac{\partial \mathbf{D}}{\partial t}$, appearing on the right-hand side of Maxwell's equations, may be replaced thus:

$$\frac{\partial \mathbf{D}}{\partial t} = 4\pi \frac{\partial \mathbf{P}}{\partial t} = 4\pi \mathbf{j}', \tag{16.30}$$

because the field is constant. Here, the current $\mathbf{j}'$ is also continuous.

The magnetic field is a pseudovector and cannot be linearly related to the current density. We note that for metals the linear relationship between field and current (Ohm's law) does not break down, no matter how strong the field.

The quantity σ is called the specific conductance, or conductivity. In the CGSE system its dimensions are inverse seconds. For metals σ is of the order 10^{17} sec^{-1}.

Slowly varying fields. So far, an electromagnetic field in a medium has been regarded as strictly constant with time. But if the field varies sufficiently slowly with time, it may also be considered as constant. Let us give a general criterion whereby we can say what field may be regarded as slowly varying.

We assume that a constant field is switched on at some initial instant of time $t = 0$. A stationary state is not established in the medium at once but only after a certain interval of time θ has elapsed. If, for example, the medium is a dielectric, then, in that time, a definite polarization is established corresponding to the given field; in a metal, θ characterizes the time taken for a constant current to be established. θ is called the relaxation time. If, during the relaxation time, the field changes by only a small fraction of its value it can be regarded as constant within the accuracy of that small fraction. In other words, the criterion of slowness of variation of a field is this: within the relaxation time a stationary state, corresponding to the given value of field, has time to establish itself in the medium. Such fields are termed slowly varying. For them, the same values of permeability, dielectric constant, and conductivity can be substituted into Maxwell's equations, as for constant fields.

Let us write down Maxwell's equations for a slowly varying field in a conductor. In the expression

$$\frac{\partial \mathbf{D}}{\partial t} = \frac{\partial \mathbf{E}}{\partial t} + 4\pi \frac{\partial \mathbf{P}}{\partial t} = \frac{\partial \mathbf{E}}{\partial t} + 4\pi\sigma \mathbf{E} \tag{16.31}$$

the first term can be neglected in the majority of cases because it in no way exceeds $\frac{E}{\theta}$; if, as occurs in metals, σ is of the order of 10^{17}, then $\sigma E \gg E/\theta$. Whence Maxwell's equations are obtained for a slowly varying field in a conductor:

$$\operatorname{rot} \mathbf{H} = \frac{4\pi \mathbf{j}'}{c} = \frac{4\pi\sigma \mathbf{E}}{c}, \tag{16.32}$$

$$\operatorname{rot} \mathbf{E} = -\frac{\chi}{c}\frac{\partial \mathbf{H}}{\partial t}, \tag{16.33}$$

$$\operatorname{div} \chi \mathbf{H} = 0. \tag{16.34}$$

This system is complete and, together with the boundary conditions (see exercises 1 and 5 in this section), is sufficient for the determination of slowly varying fields in a conductor.

Rapidly varying fields. Let us now consider the case of rapidly varying fields, i.e., fields which change more rapidly than the relaxation process or the establishment of a definite stationary state in the medium. Then the state of the medium depends not only on the instantaneous value of the field, but also on its values at previous instants of time; in other words, it depends on the way in which the field changes with time. Such a relationship is very complicated in the general case. It is simplified if the field is weak; then, at any rate, we may expect the relationship to be linear.

Expansion in harmonic components. Let us examine the general form of the linear relationship. To do so we represent the field as follows:

$$\mathbf{E}(t) = \sum_k \mathbf{E}_k \cos(\omega_k t + \varphi_k), \tag{16.35}$$

i.e., we expand it in harmonic components. The more values of the amplitudes $\mathbf{E}_k$, frequencies ω_k, and phases φ_k, we take, the better the approximation for the variation of $\mathbf{E}$. However, if the relationship between induction and field is linear, then the induction is also of the form of a sum of harmonic components: $\mathbf{D}_k \cos(\omega_k t + \varphi_k)$, where (and this is most important) each term in the sum for induction is determined by the term of the same frequency in (16.35).

This does not contradict the general statement that the induction is determined by the entire time dependence of a rapidly varying field; for a harmonic relationship between field and time, it is fully given by its amplitude, phase, and frequency. And a component of the field with a certain frequency can in no way give rise to induction components with another frequency if the relationship between field and induction is linear, for no linear relationship exists between trigonometric functions of different arguments. Therefore, if we write the functions even with the same frequency but with different phases φ_k and ψ_k, we do not directly obtain a linear relationship either. However, if we use a complex form and express the field and induction in terms of exponentials by means of the equations

$$\left.\begin{aligned} \mathbf{E}(t) &= \sum_k (\mathbf{E}_k e^{-i\omega_k t} + \mathbf{E}_k^* e^{i\omega_k t}), \\ \mathbf{D}(t) &= \sum_k (\mathbf{D}_k e^{-i\omega_k t} + \mathbf{D}_k^* e^{i\omega_k t}) \end{aligned}\right\} \tag{16.36}$$

(the star denotes a complex conjugate quantity), then a linear relationship between field and induction can be written in the following form:

$$\mathbf{D}_k = \varepsilon_k \mathbf{E}_k \text{ or } \mathbf{D}(\omega) = \varepsilon(\omega)\mathbf{E}(\omega). \tag{16.37}$$

The quantities $\mathbf{D}_k$ and $\mathbf{E}_k$ are complex. From a comparison of (16.35) and (16.36) it can be seen that they differ from the real field and induction amplitudes by the complex factors $\frac{1}{2}e^{i\varphi_k}$ and $\frac{1}{2}e^{i\psi_k}$. Thus, the dielectric constant $\varepsilon_k \equiv \varepsilon(\omega)$ must also be a complex quantity. A complex permeability $\chi(\omega)$ is similarly introduced.

Maxwell's equations in complex form. Let us now write down Maxwell's equations for complex field components. It must be noted that they are a function of time according to an $e^{-i\omega t}$ law. We shall divide the equations by these factors and get the following system of Maxwell's equations for rapidly varying harmonic fields:

$$\operatorname{rot}\mathbf{H} = -i\frac{\omega}{c}\varepsilon(\omega)\mathbf{E}, \tag{16.38}$$

$$\operatorname{rot}\mathbf{E} = i\frac{\omega}{c}\chi(\omega)\mathbf{H}, \tag{16.39}$$

$$\operatorname{div}\varepsilon\mathbf{E} = 0, \tag{16.40}$$

$$\operatorname{div}\chi\mathbf{H} = 0. \tag{16.41}$$

The imaginary parts of the dielectric constant and permeability lead to the energy of a rapidly varying field being spent on the generation of heat in the substance (see exercise 18).

We note, in addition, that for rapidly varying fields the division of bodies into conductors and dielectrics is conditional and is determined by the relationship between the imaginary and real parts of the dielectric constant. A substance retains the character of a conductor up to such frequencies of a rapidly varying field as satisfy the inequality $\frac{1}{\omega} \gg \theta$.

Exercises

1) Show that on the boundary between two media the tangential components of the fields and the normal components of induction are continuous.

Integrate the equations for the inductions over a small flat cylinder, and the equations for fields over a narrow quadrilateral bounding the interface (Fig. 23a and b).

2) Show that a magnetic field varying sinusoidally with time is damped with depth in the conductor ($\chi = 1$).

From equations (16.32)-(16.34) we have

$$\frac{dH_z}{dx} = -\frac{4\pi\sigma}{c}E_y, \quad \frac{dE_y}{dx} = -\frac{i\omega}{c}H_z, \quad \frac{d^2H_z}{dx^2} = \frac{4\pi\sigma i\omega}{c^2}H_z,$$

whence

$$H_z = H_z^0 e^{-\sqrt{2\pi\sigma\omega}\,(1+i)\,x/c},$$

where x is a coordinate, normal to the surface of a conductor.

3) Show that equations (16.32)-(16.34) are formally applicable to the case of rapidly varying fields, if the relationship between field and time is considered harmonic. In this case the conductivity σ is proportional to the imaginary part of ε, and the real part of ε is equal to zero.

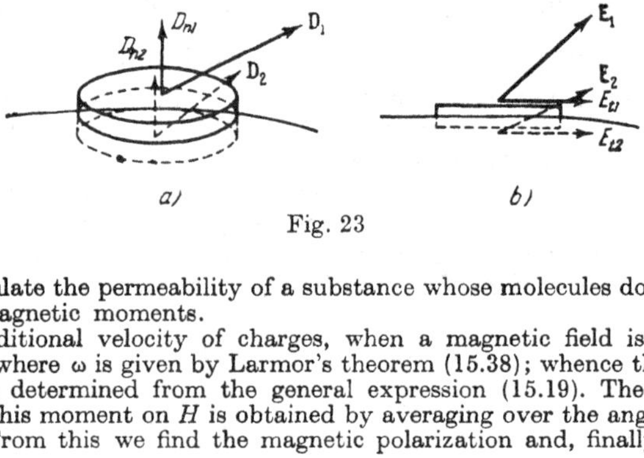

Fig. 23

4) Calculate the permeability of a substance whose molecules do not possess intrinsic magnetic moments.

The additional velocity of charges, when a magnetic field is applied, is $\mathbf{v} = [\omega\, \mathbf{r}]$, where ω is given by Larmor's theorem (15.38); whence the magnetic moment is determined from the general expression (15.19). The mean projection of this moment on H is obtained by averaging over the angles between $\mathbf{H}$ and $\mathbf{r}$. From this we find the magnetic polarization and, finally, χ:

$$\chi = 1 - 4\pi N \sum_i \frac{e_i^2\,\overline{(r^i)^2}}{6\, m_i c^2},$$

where N is the number of molecules in unit volume and $\overline{(r^i)^2}$ is the mean square of the radius of rotation of the ith charge.

5) Show that an electric field near the surface of a charged conductor is equal to $4\pi\gamma$, where γ is the surface density of static charge on the conductor.

We use the same method as in exercise 1 of integrating equation (12.27) over a small flat cylinder bounding the conductor on both sides, with account taken of the fact that the field inside the conductor is equal to zero. For quasi-stationary fields $\frac{\partial \gamma}{\partial t} = j'_n$, where j'_n is the projection of $\mathbf{j}'$ on the external normal.

6) Calculate the energy of a system of charged conductors in a vacuum.

We substitute $E = -\nabla \varphi$ in the definition of energy $\mathscr{E}_{\text{electr}} = \int \frac{E^2}{8\pi} dV$ and integrate by parts, taking advantage of the fact that in a vacuum $\Delta \varphi = 0$. Since, from exercise 5 in this section, the field close to a conductor is equal to $4\pi\gamma$, and the surface of a conductor is equipotential, we reduce $\mathscr{E}_{\text{electr}}$ to the form

$$\mathscr{E}_{\text{electr}} = \frac{1}{2} \sum_i e_i \varphi_i ,$$

where e_i is the charge on the ith conductor and φ_i is its potential.

7) Determine how a constant uniform electric field changes if a conducting sphere is introduced into it.

The field potential must be sought in the form $\varphi = -\mathbf{E}_0 \mathbf{r} + \frac{\mathbf{r}\mathbf{d}}{r^3}$, where the vector $\mathbf{d}$ is in the same direction as the initial uniform field $\mathbf{E}_0$. $\mathbf{d}$ is determined from the condition that the tangential component $-\mathbf{E} = \nabla \varphi$ on the sphere is equal to zero.

8) Determine in what way a constant uniform electric field $\mathbf{E}_0$ varies if a dielectric sphere of dielectric constant ε is introduced into it.

The field potential outside the sphere must be sought in the same form as in problem 7, but, inside the sphere, it must be sought in the form $-\mathbf{E}'\mathbf{r}$, where $\mathbf{E}' = \text{const}$. Determine the vectors $\mathbf{d}$ and $\mathbf{E}'$ from the boundary conditions derived in exercise 1.

9) Find the electric field which arises in space when a point charge e is brought to a distance a from an infinite flat conducting surface.

We drop a perpendicular from the point at which the charge is situated to the surface of the conductor and, at a distance a inside the conductor, we place an equal and opposite (fictitious) charge e. Then, the field component tangential to the surface becomes zero. The field outside the conductor is everywhere equal to the vector sum of the fields due to the real and fictitious charges.

10) Find the electric field when a point charge e is brought to a distance a from an infinite flat surface of a dielectric with constant ε. The dielectric is infinitely deep.

We make the same construction as in exercise 9. We look for the field inside the dielectric of the form $\frac{e' \mathbf{r}_1}{r_1^3}$, where $\mathbf{r}_1$ is a radius vector from the point at which the real charge is situated; the field outside the dielectric is of the form $\frac{e\,\mathbf{r}_1}{r_1^3} + \frac{e''\mathbf{r}_2}{r_2^3}$ where $\mathbf{r}_2$ is a radius vector drawn from the "image" of the charge. The constants e' and e'' are determined from the boundary conditions of exercise 1.

11) Assuming that $\chi = 1$, determine the magnetic energy of a system of conductors carrying a steady current.

Starting from the equation $\mathscr{E}_{\text{mag}} = \frac{1}{8\pi} \int H^2 \, dV$, we substitute $\mathbf{H} = \text{rot}\,\mathbf{A}$ and integrate by parts using (11.29). The surface integral at infinity is equal to zero. Then we use (16.32) and reduce $\mathscr{E}_{\text{mag}}$ to the form

$$\mathscr{E}_{\text{mag}} = \frac{1}{2c}\int \mathbf{A}\mathbf{j}\,dV.$$

12) Express the magnetic energy for a system of currents in terms of a double integral over the volumes of the conductors.

We replace the summation over the charges, in equation (15.13), by a volume integration. This gives

$$\mathscr{E}_{\text{mag}} = \frac{1}{2c^2}\iint \frac{\mathbf{j}(\mathbf{r})\,\mathbf{j}(\mathbf{r}')}{|\mathbf{r}-\mathbf{r}'|}\,dV\,dV'.$$

For line conductors we can substitute $I\,d\mathbf{l}$ instead of $\mathbf{j}\,dV$, provided dV and dV' are volume elements of different conductors. Then the mutual magnetic energy for two line conductors i, k is

$$\mathscr{E}^{ik}{}_{\text{mag}} = \frac{I_i I_k}{c^2}\iint \frac{d\mathbf{l}_i\,d\mathbf{l}_k}{|\mathbf{r}_i-\mathbf{r}_k|} = M_{ik} I_i I_k,$$

where $|\mathbf{r}_i - \mathbf{r}_k|$ is the distance between the elements of contours $d\mathbf{l}_i$ and $d\mathbf{l}_k$. When $i = k$ we must regard the conductor as thin, though not infinitely thin, otherwise the integral diverges. The intrinsic magnetic energy for one conductor is

$$\mathscr{E}^{ii}{}_{\text{mag}} = \frac{1}{2} M_{ii} I_i^2.$$

M_{ik} is called the mutual induction coefficient, and M_{ii} is the self-induction coefficient.

13) Write down the Lagrangian for a system of currents, assuming that there is capacity coupling between the conductors by virtue of capacitors connected in the circuit.

Because of the linearity of electrodynamical equations, the potential of the ith conductor is expressed linearly in terms of the charges on all the conductors:

$$\varphi_i = \sum_k C_{ik} e_k.$$

From exercise 6, we obtain the electric energy.

$$\mathscr{E}_{\text{electr}} = \frac{1}{2}\sum_{i,k} C_{ik} e_i e_k.$$

The charge on the plates of a capacitor is related to the incoming current by $\dot{e}_k = I_k$. With the aid of exercise 12 we obtain the magnetic energy

$$\mathscr{E}_{\text{mag}} = \frac{1}{2}\sum_{i,k} M_{ik} I_i I_k = \frac{1}{2}\sum_{i,k} M_{ik}\dot{e}_i\dot{e}_k.$$

From Sec. 13 the Lagrangian (neglecting sign) is $\frac{1}{8\pi}\int (H^2 - E^2)\,dV$, whence

$$L = \mathscr{E}_{\text{mag}} - \mathscr{E}_{\text{electr}} = \frac{1}{2}\sum_{i,k}(M_{ik}\dot{e}_i\dot{e}_k - C_{ik} e_i e_k)$$

[cf. (17.16)].

14) Determine the work performed in unit time by a varying electromagnetic field in a medium.

We write equation (13.23) for the external space not occupied by the substance, where $\rho = 0$. From the boundary conditions of exercise 1, it follows that the normal component of the Poynting vector **U** is continuous at the boundary of the body. From this, applying the same transformations to equations (16.17) and (16.25), as lead to (13.23), we find

$$\frac{dA}{dt} = \frac{1}{4\pi}\int\left(\mathbf{E}\frac{\partial \mathbf{D}}{\partial t} + \mathbf{H}\frac{\partial \mathbf{B}}{\partial t}\right)dV,$$

where $\frac{dA}{dt}$ may be expressed in terms of the change in energy of the field in the external space.

15) Calculate the energy transformed into heat in unit time in a conductor situated in a constant field. Assume **H** to be a single-valued function of **B**.

For such a body, where **H** and **B** are related uniquely, $\mathbf{H}\frac{\partial \mathbf{B}}{\partial t} = \frac{\partial}{\partial t} f(\mathbf{B})$, where $f(\mathbf{B})$ is some function of **B**. For example, for $\mathbf{B} = \chi\mathbf{H}$, $f = \frac{B^2}{2\chi}$. The result of the preceding exercise gives

$$\frac{dA}{dt} - \frac{d}{dt}\int \frac{f(\mathbf{B})}{4\pi}dV = \frac{1}{4\pi}\int \mathbf{E}\frac{\partial \mathbf{D}}{\partial t}dV = \int \sigma E^2\, dV,$$

see (16.30) and (16.31). For a constant field, there is zero on the left-hand side of the equation, while the right-hand side is an essentially positive quantity. This energy must therefore be converted into heat according to the energy conservation law.

16) Write down Lagrange's equation for a system of currents taking into account the conversion of energy into heat.

From exercise 14, the heat generated in unit time may be written as $\sum_i r_i I_i^2$, where r_i is the resistance of the ith conductor. We search for Lagrange's equation with the right-hand side in the form

$$\frac{d}{dt}\frac{\partial L}{\partial \dot{e}_i} - \frac{\partial L}{\partial e_i} = \nu_i .$$

From the definition (4.4) we find $\frac{d\mathscr{E}}{dt}$ as

$$\frac{d\mathscr{E}}{dt} = \sum_i \dot{e}_i\left[\frac{d}{dt}\frac{\partial L}{\partial \dot{e}_i} - \frac{\partial L}{\partial e_i}\right] = -\sum r_i \dot{e}_i^2 .$$

Whence $\nu_i = -r_i I_i = -r_i \dot{e}_i$.

17) Reasoning in the same way as for exercise 15, show that if **H** is a double-valued function of **B**, having one value for $\frac{\partial \mathbf{B}}{\partial t} < 0$ and the other for $\frac{\partial \mathbf{B}}{\partial t} > 0$, then, for a periodic variation of **B**, the heat generated in one period is equal to $\int \frac{\mathbf{H}\, d\mathbf{B}}{4\pi}$, where the integral is taken over one period.

18) Show that if $\varepsilon(\omega)$ and $\chi(\omega)$ possess imaginary parts, heat is generated in a rapidly varying field.

The density of heat generation may be represented as the divergence from the Poynting vector $\mathbf{U} = \frac{c}{4\pi}[\mathbf{E}\,\mathbf{H}]$. In forming quadratic complex quantities, we must take into account their time dependence. For example, if we take $\mathbf{E}$ and $\mathbf{H}$ with a factor $e^{-i\omega t}$, their product will be proportional to $e^{-2i\omega t}$. After time averaging, this factor will yield zero. Therefore we must take only products of the form $[\mathbf{E}\mathbf{H}^*] + [\mathbf{E}^*\mathbf{H}]$. Now, using equations (16.38) and (16.39), we obtain

$$\frac{c}{4\pi}\operatorname{div}([\mathbf{E}^*\mathbf{H}] + [\mathbf{E}\,\mathbf{H}^*]) = \frac{1}{4\pi} i\omega(\varepsilon - \varepsilon^*)\,\mathbf{E}\mathbf{E}^* + \frac{1}{4\pi} i\omega(\chi - \chi^*)\,\mathbf{H}\mathbf{H}^*.$$

Here, both parts of the equation are real, and if $\varepsilon = \varepsilon_1 + i\varepsilon_2$ and $\chi = \chi_1 + i\chi_2$, then on the right-hand side there is the expression $-\frac{1}{2\pi}\omega(\varepsilon_2\mathbf{E}\mathbf{E}^* + \chi_2\mathbf{H}\mathbf{H}^*)$. From this it can be seen that $\varepsilon_2 > 0$ and $\chi_2 > 0$, since the energy of the field is absorbed by the medium.

19) Calculate the dielectric constant of a medium, considering that all the charges in it are connected by elastic forces with the equilibrium positions. The characteristic oscillation frequency of the charges is ω_0 and the frequency of the field is ω.

The radius vector of a charge satisfies the differential equation

$$m(\ddot{\mathbf{r}} + \omega_0^2\mathbf{r}) = e\,\mathbf{E}^0 e^{-i\omega t} = e\,\mathbf{E}.$$

Its solution has the frequency of the external field and may be written as

$$\mathbf{r} = \frac{e\,\mathbf{E}}{m(\omega_0^2 - \omega^2)}.$$

The polarization can be obtained from this by multiplying by the number of charges in unit volume N and by e. Since the induction D is equal to $\varepsilon\,\mathbf{E}$ or $\mathbf{E} + 4\pi\mathbf{P}$, we find that $\varepsilon = 1 + \frac{4\pi N e^2}{m(\omega_0^2 - \omega^2)}$. For $\omega \to 0$, we obtain the static dielectric constant $\varepsilon_0 = 1 + \frac{4\pi N e^2}{m\omega_0^2}$. For $\omega \to \infty$, $\varepsilon(\omega)$ is obtained for very large frequencies or for free charges $\varepsilon = 1 - \frac{4\pi N e^2}{m\omega^2}$.

Sec. 17. Plane Electromagnetic Waves

General equations. In this section we shall first consider the solutions of Maxwell's equations for free space, i.e., in the absence of charges. These solutions, as we shall see, are of the form of travelling waves. Analogous solutions also exist for a nonabsorbing material medium. These solutions will also be found in the present section.

In the absence of charges or currents, the equations for scalar and vector potentials are written thus:

$$\Delta\mathbf{A} - \frac{1}{c^2}\frac{\partial^2\mathbf{A}}{\partial t^2} = 0, \tag{17.1}$$

$$\Delta\varphi - \frac{1}{c^2}\frac{\partial^2\varphi}{\partial t^2} = 0, \tag{17.2}$$

with the additional condition (12.36)

$$\operatorname{div} \mathbf{A} + \frac{1}{c}\frac{\partial \varphi}{\partial t} = 0. \tag{17.3}$$

Equations of the form (17.1) or (17.2) are called wave equations.

The solution of a wave equation. We shall look for particular solutions of equations (17.1) and (17.2) which depend only on one coordinate (for example, x) and on time. Then the wave equations can be rewritten in the following manner:

$$\frac{\partial^2 \mathbf{A}}{\partial x^2} - \frac{1}{c^2}\frac{\partial^2 \mathbf{A}}{\partial t^2} = 0, \tag{17.4}$$

$$\frac{\partial^2 \varphi}{\partial x^2} - \frac{1}{c^2}\frac{\partial^2 \varphi}{\partial t^2} = 0, \tag{17.5}$$

and the supplementary condition takes the form

$$\frac{\partial A_x}{\partial x} + \frac{1}{c}\frac{\partial \varphi}{\partial t} = 0. \tag{17.6}$$

We shall now find the solution of (17.4) or (17.5) without imposing any further restrictions. We shall temporarily introduce the following notation:

$$\begin{aligned} x + ct &= \xi, \\ x - ct &= \eta. \end{aligned} \tag{17.7}$$

We transform (17.5) to these independent variables. (17.5) can be rewritten symbolically as

$$\left(\frac{\partial}{\partial x} + \frac{1}{c}\frac{\partial}{\partial t}\right)\left(\frac{\partial}{\partial x} - \frac{1}{c}\frac{\partial}{\partial t}\right)\varphi = 0. \tag{17.8}$$

Then

$$\frac{\partial \varphi}{\partial x} = \frac{\partial \varphi}{\partial \xi}\frac{\partial \xi}{\partial x} + \frac{\partial \varphi}{\partial \eta}\frac{\partial \eta}{\partial x} = \frac{\partial \varphi}{\partial \xi} + \frac{\partial \varphi}{\partial \eta};$$

$$\frac{1}{c}\frac{\partial \varphi}{\partial t} = \frac{\partial \varphi}{\partial \xi}\frac{1}{c}\frac{\partial \xi}{\partial t} + \frac{\partial \varphi}{\partial \eta}\frac{1}{c}\frac{\partial \eta}{\partial t} = \frac{\partial \varphi}{\partial \xi} - \frac{\partial \varphi}{\partial \eta},$$

because, for constant t (i.e., $dt = 0$), $\frac{\partial \xi}{\partial x} = 1$ and $\frac{\partial \eta}{\partial x} = 1$, while for constant x (i.e., $dx = 0$), $\frac{1}{c}\frac{\partial \xi}{\partial t} = -\frac{1}{c}\frac{\partial \eta}{\partial t} = 1$ in accordance with the same equations. Thus, symbolically

$$\frac{\partial}{\partial x} + \frac{1}{c}\frac{\partial}{\partial t} = 2\frac{\partial}{\partial \xi}, \quad \frac{\partial}{\partial x} - \frac{1}{c}\frac{\partial}{\partial t} = 2\frac{\partial}{\partial \eta};$$

$$\left(\frac{\partial}{\partial x} + \frac{1}{c}\frac{\partial}{\partial t}\right)\left(\frac{\partial}{\partial x} - \frac{1}{c}\frac{\partial}{\partial t}\right)\varphi = 4\frac{\partial^2 \varphi}{\partial \xi\,\partial \eta} = 0.$$

Hence, wave equations (17.4) and (17.5) are written thus:

$$\frac{\partial^2 \mathbf{A}}{\partial \xi\,\partial \eta} = 0, \quad \frac{\partial^2 \varphi}{\partial \xi\,\partial \eta} = 0. \tag{17.9}$$

Integrating any of them with respect to ξ, we obtain

$$\frac{\partial \mathbf{A}}{\partial \eta} = \mathbf{C}(\eta), \frac{\partial \varphi}{\partial \eta} = C'(\eta). \tag{17.10}$$

It is not difficult now to integrate with respect to η:

$$\mathbf{A} = \int^{\eta} \mathbf{C}(\eta)\, d\eta + \mathbf{C}_1(\xi), \quad \varphi = \int^{\eta} C'(\eta)\, d\eta + C_1'(\xi).$$

Finally, the required solution is written as:

$$\mathbf{A} = \mathbf{A}_1(\eta) + \mathbf{A}_2(\xi), \quad \varphi = \varphi_1(\eta) + \varphi_2(\xi), \tag{17.11}$$

since the substitution of (17.11) into (17.9) gives zero identically. Passing to the variables x, t, we can write the solutions to (17.4), (17.5):

$$\mathbf{A} = \mathbf{A}_1(x - ct) + \mathbf{A}_2(x + ct), \quad \varphi = \varphi_1(x - ct) + \varphi_2(x + ct). \tag{17.12}$$

Plane travelling waves. The solution depending on $x + ct$ does not depend on the solution whose argument is $x - ct$; these are two linearly independent solutions. Therefore it is sufficient to consider one of them:

$$\mathbf{A} = \mathbf{A}(x - ct), \tag{17.13}$$

$$\varphi = \varphi(x - ct). \tag{17.14}$$

In order to satisfy the supplementary condition (17.6), we perform a gauge transformation:

$$\varphi(x - ct) = \varphi'(x - ct) - \frac{1}{c}\frac{\partial}{\partial t} f(x - ct) = \varphi' + \dot{f} \tag{17.15}$$

(the dot over f denotes differentiation with respect to the whole argument $\eta = x - ct$). But, if we put $\varphi' = -\dot{f}$, we obtain simply $\varphi = 0$. Then, from (17.6), we also obtain $A_x = 0$. Thus, for a solution of the form considered, depending on $x - ct$ only, the Lorenz condition is satisfied most simply by substituting $\varphi = 0, A_x = 0$.

The electric field component x is equal to zero:

$$E_x = -\frac{1}{c}\frac{\partial A_x}{\partial t} - \frac{\partial \varphi}{\partial x} = 0. \tag{17.16}$$

From the general result of Sec. 12, this property of E_x does not depend on a potential gauge transformation.

The magnetic field component x is also equal to zero:

$$H_x = \frac{\partial A_z}{\partial y} - \frac{\partial A_y}{\partial z} = 0. \tag{17.17}$$

We find the remaining field components:

$$\left.\begin{aligned} E_y &= -\frac{1}{c}\frac{\partial A_y}{\partial t} = \dot{A}_y, \quad E_z = -\frac{1}{c}\frac{\partial A_z}{\partial t} = \dot{A}_z, \\ H_y &= -\frac{\partial A_z}{\partial x} = -\dot{A}_z, \quad H_z = \frac{\partial A_y}{\partial x} = \dot{A}_y. \end{aligned}\right\} \tag{17.18}$$

From this equation it follows that **E** and **H** are perpendicular, because

$$\mathbf{EH} = E_y H_y + E_z H_z = 0. \tag{17.19}$$

They are equal in absolute magnitude, since $E = H = \sqrt{A_y^2 + A_z^2}$.

The solution of the form (17.13) has a simple physical meaning.

Let us take the value of **E** at an instant of time $t = 0$ on the plane $x = 0$. It is equal to **E** (0). It is clear that the **E** (0) will have the same value at the instant of time t on the plane $x = ct$, because **E** $(x - ct)$ = **E** (0) on that plane. We can also say that the plane on which the field **E** is equal to **E** (0) is translated in space through a distance ct in a time t, i.e., it moves with a velocity c. The same applies to any plane $x = x_0$, for which there was some value of field **E** (x_0) at the initial instant of time. To summarize, all planes with the given value of field are propagated in space with velocity c. Therefore, the solution **E** $(x - ct)$ is called a travelling plane wave.

We note that the form of the wave does not change as it moves; the distance between planes $x = x_1$ and $x = x_2$, for which **E** is equal to **E** (x_1) and **E** (x_2), is constant. This result holds for any arbitrary form of wave, provided it is travelling in free space.

Repeating, the velocity of propagation of a wave in empty space does not depend on its shape or amplitude and it is equal to a universal constant c.

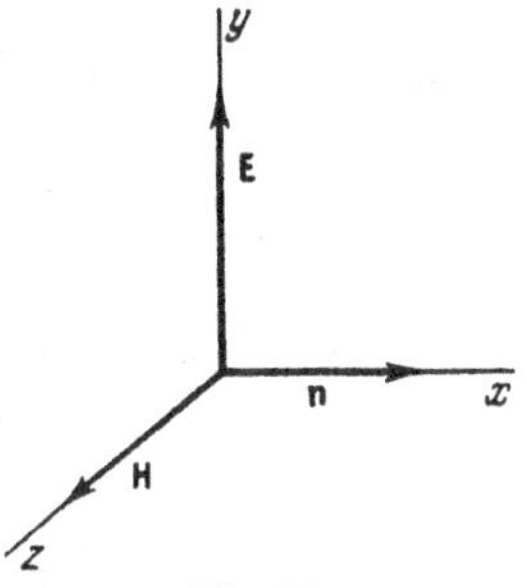

Fig. 24

The transverse nature of waves. The electric and magnetic fields, as we have seen from (17.19), are perpendicular to the direction of wave propagation, as well as to each other. This is why it is said that electromagnetic waves are transverse (as opposed to longitudinal sound waves in air, for which the oscillations occur in the direction of propagation). The direction of propagation, the electric field, and the magnetic field are shown in Fig. 24. In it, **n** is a unit vector along the x-axis.

In future it will be sufficient to take only one component of the electric field. For this it is necessary to take one of the coordinate axes, for example the y-axis, in the direction of the electric field, which in no way limits the generality. This is shown in Fig. 24.

The x-coordinate will be written in the form $x = \mathbf{rn}$, so that

$$E_y = \dot{A}_y(\mathbf{rn} - ct)\,. \tag{17.20}$$

But in this notation it is not necessary to relate the vector $\mathbf{n}$, in the direction of propagation to the x-axis. A solution with argument of the form (17.20) is applicable to any direction of n, provided, naturally, that $\mathbf{n}$, $\mathbf{E}$, and $\mathbf{H}$ are mutually perpendicular.

The momentum density of the wave [see (13.27)] is equal to

$$\frac{1}{4\pi c}[\mathbf{EH}] = \frac{1}{4\pi c}\dot{A}_y^2$$

and is directed along n. The energy density is

$$\frac{E^2 + H^2}{8\pi} = \frac{1}{4\pi}\dot{A}_y^2\,.$$

It differs from the momentum density by the factor c. This, as we shall see later, is very essential for the quantum theory of light.

Pressure of light. If a wave falls on an absorbing obstacle, for example, on a black wall, and is not reflected, then its momentum is transmitted to the wall in accordance with the conservation law. But momentum transmitted to a body in unit time is, by Newton's Second Law, nothing other than force. It follows that there is a force of $\frac{\dot{A}^2}{4\pi c}$ for every square centimetre of the absorbing barrier, upon which the wave is normally incident. Force referred to unit surface is, by definition, the *pressure* of the electromagnetic wave on the barrier. Consequently, electrodynamics predicts the existence of light pressure. This was observed and measured by P. N. Lebedev.

Harmonic waves. A special interest is attached to travelling waves for which the function $\mathbf{E}(x - ct)$ is harmonic. The most general harmonic solution is of the following form:

$$\mathbf{E} = \mathrm{Re}\left\{\mathbf{F}e^{-i\omega\left(t - \frac{\mathbf{rn}}{c}\right)}\right\}, \tag{17.21}$$

where the symbol Re {} denotes the real part of the expression inside the braces, $\mathbf{F}$ is a complex vector of the form $\mathbf{F}_1 + i\,F_2$ [Cf. (7.14c)], and ω is the wave frequency in the same sense as in equation (7.3). ω is the number of radians per second by which the argument of the exponential function changes.

The wave vector. The vector $\omega\,\frac{\mathbf{n}}{c}$ is called the wave vector. It is denoted by the letter $\mathbf{k}$:

$$\mathbf{k} \equiv \omega\frac{\mathbf{n}}{c}\,. \tag{17.22}$$

The geometric meaning of **k** is easy to explain. We define the wavelength, i.e., the distance $\Delta\,\mathbf{r}\cdot\mathbf{n}$ in space at which **E** assumes the same value. Let the required wavelength be λ. Then

$$e^{i\omega\frac{\lambda}{c}} = e^{i\omega\frac{\Delta\mathbf{r}\cdot\mathbf{n}}{c}} = e^{2\pi i}, \qquad (17.23)$$

because the period of the function e^{ix} is equal to $2\,\pi$. Hence,

$$\lambda = \frac{2\pi c}{\omega}. \qquad (17.24)$$

Comparing the wavelength with the wave vector, we obtain

$$\mathbf{k} = \frac{2\pi}{\lambda}\mathbf{n}\,, \; \lambda = \frac{2\pi}{k}. \qquad (17.25)$$

Polarization of a plane harmonic wave. Let us now study the nature of the oscillations of an electric field. To do this, we write the vector **F** in the form

$$\mathbf{F} = \mathbf{F}_1 + i\mathbf{F}_2 = (\mathbf{E}_1 - i\,\mathbf{E}_2)\,e^{i\alpha}. \qquad (17.26)$$

We choose the phase α so that the vectors $\mathbf{E}_1$ and $\mathbf{E}_2$ are mutually perpendicular. We multiply equation (17.26) by $e^{-i\alpha}$ and square. Then we obtain

$$(\mathbf{E}_1 - i\mathbf{E}_2)^2 = E_1^2 - E_2^2 = e^{-2i\alpha}(F_1^2 - F_2^2 + 2i\,(\mathbf{F}_1\,\mathbf{F}_2)). \qquad (17.27)$$

We have taken advantage of the fact that $\mathbf{E}_1$ and $\mathbf{E}_2$ are perpendicular. Because of this $(\mathbf{E}_1 - i\,\mathbf{E}_2)^2$ is a purely real quantity. Therefore, the imaginary part of the right-hand side of expression (17.27) must be put equal to zero. Representing $e^{-2i\alpha}$ as $\cos 2\,\alpha - i\,\sin 2\,\alpha$, we obtain

$$-(F_1^2 - F_2^2)\sin 2\alpha + 2\,(\mathbf{F}_1\,\mathbf{F}_2)\cos 2\alpha = 0$$

or

$$\operatorname{tg} 2\,\alpha = \frac{2\,(\mathbf{F}_1\,\mathbf{F}_2)}{F_1^2 - F_2^2}, \qquad (17.28)$$

whence the angle α is determined for the given solution (17.21).

It is now easy to express $\mathbf{E}_1$ and $\mathbf{E}_2$. Indeed, from (17.26), $\mathbf{E}_1 - \mathbf{E}_2 = (\mathbf{F}_1 + i\,\mathbf{F}_2)\,e^{-i\alpha} = \mathbf{F}_1\cos\alpha + \mathbf{F}_2\sin\alpha - i\,(\mathbf{F}_1\sin\alpha - \mathbf{F}_2\cos\alpha)$, so that

$$\left.\begin{aligned}\mathbf{E}_1 &= \mathbf{F}_1\cos\alpha + \mathbf{F}_2\sin\alpha\,,\\ \mathbf{E}_2 &= \mathbf{F}_1\sin\alpha - \mathbf{F}_2\cos\alpha\,.\end{aligned}\right\} \qquad (17.29)$$

We now include a constant phase in the exponent (17.21) and, for short, put

$$\alpha - \omega\left(t - \frac{\mathbf{r}\mathbf{n}}{c}\right) \equiv \psi. \qquad (17.30)$$

Then, in the most general case, the electric field for a plane harmonic wave will be

$$\mathbf{E} = \mathrm{Re}\,\{(\mathbf{E}_1 - i\,\mathbf{E}_2)\,e^{i\psi}\} = \mathbf{E}_1 \cos\psi + \mathbf{E}_2 \sin\psi\,. \tag{17.31}$$

Here, the vectors $\mathbf{E}_1$ and $\mathbf{E}_2$ are defined as perpendicular.

Let us assume that a wave is propagated along the x-axis. The y-axis is directed along $\mathbf{E}_1$, and the z-axis along $\mathbf{E}_2$. Hence, from (17.31), we obtain

$$E_y = E_1 \cos\psi\,, \quad E_z = E_2 \sin\psi\,. \tag{17.32}$$

Let us eliminate the phase ψ. We divide the first equation by E_1, the second by E_2, square and add. Then the phase is eliminated and an equation relating the field components remains:

$$\frac{E_y^2}{E_1^2} + \frac{E_z^2}{E_2^2} = 1\,. \tag{17.33}$$

It follows that the electric field vector describes an ellipse in the yz-plane moving along the x-axis with velocity c, and passes round the whole ellipse on one wavelength. Relative to a fixed coordinate system, the electric field vector describes a helix wound on an elliptic cylinder. The pitch of the helix is equal to the wavelength.

Such an electromagnetic wave is termed elliptically polarized. It represents the most general form of a plane harmonic wave (17.21).

If one of the components is equal to zero, for example $\mathbf{E}_1 = 0$ or $\mathbf{E}_2 = 0$, then the oscillations of $\mathbf{E}$ occur in one plane. Such a wave is termed plane polarized.

When E_1 is equal to E_2, the vector $\mathbf{E}$ describes a circle in the yz-plane. Depending on the sign of E_z, the rotation around the circle occurs in a clockwise or anticlockwise direction. Accordingly, the wave is termed right-handed or left-handed polarized. These waves are shown in Fig. 25. For the same value of phase ψ, the rotation occurs either in a clockwise or anticlockwise direction.

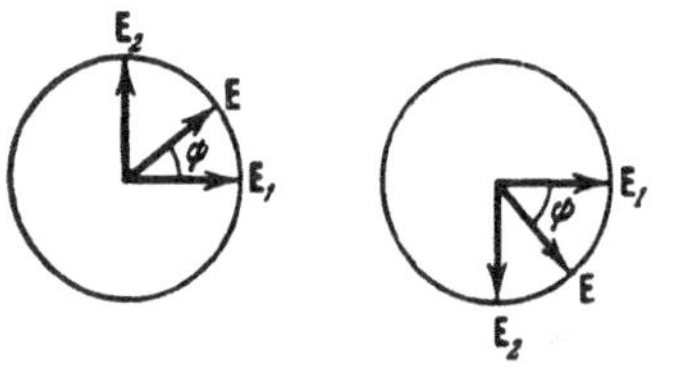

Fig. 25

The sum of two waves of equal amplitude, which are circularly polarized, gives a plane polarized wave. The relationship between their phases determines the plane of polarization. Thus, if the waves shown in Fig. 25 are added, the oscillations $\mathbf{E}_2$ and $-\mathbf{E}_2$ mutually cancel and only the plane polarized oscillation $\mathbf{E}_1$ remains.

In turn, a circularly polarized oscillation is resolved into two mutually perpendicular plane oscillations.

Certain crystals, for example tourmaline, are capable of polarizing light.

Unpolarized light. In nature, it is most common to observe unpolarized (natural) light. Naturally, such light cannot be strictly monochromatic (i.e., possessing strictly one frequency ω), for, as we have just shown, monochromatic light is always polarized in some way. But if we imagine that the components $\mathbf{E}_1$ and $\mathbf{E}_2$ in Fig. 25 are not related by a strict phase relationship (17.32), but randomly change their relative phases, then the resultant vector will also change its direction in a random manner. However, for this, it is necessary that the oscillation frequencies should vary in time within some interval $\Delta\omega$, since the difference of phase between two oscillations of strictly constant and identical frequency is constant.

The propagation of light in a medium. We shall now consider the question of the propagation of light in a material medium. At the end of the preceding section we said that the quantities ε and χ have meaning only for oscillations of a definite frequency ω. To simplify notation, we shall not use the symbol for a real part Re {}, remembering that the real part is always taken. Since all the quantities depend on time according to an $e^{-i\omega t}$ law, the derivative $\frac{\partial}{\partial t}$ reduces to a multiplication by $-i\omega$. Then the system of Maxwell's equations can be written in the following form:

$$\operatorname{rot} \mathbf{H} = -\frac{i\omega}{c}\varepsilon \mathbf{E}, \tag{17.34}$$

$$\operatorname{div} \mathbf{E} = 0, \tag{17.35}$$

$$\operatorname{rot} \mathbf{E} = \frac{i\omega}{c}\chi \mathbf{H}, \tag{17.36}$$

$$\operatorname{div} \mathbf{H} = 0. \tag{17.37}$$

Once again we look for a solution in the form of a plane wave. Since the time relationship is already eliminated, all the quantities depend only on one coordinate, for example, upon x. From (17.35) and (17.37), it follows that

$$\frac{\partial E_x}{\partial x} = 0, \quad \frac{\partial H_x}{\partial x} = 0$$

or

$$E_x = 0, \quad H_x = 0,$$

because a solution that is constant over all space does not represent any wave. Thus, the waves are transverse. Equations (17.34) to (17.37) are satisfied if we substitute $E_y = E(x)$, $E_z = 0$, $H_y = 0$, $H_z = H(x)$, or, in other words, if the electric field is directed along the y-axis and the magnetic field along the z-axis (a right-handed system).

Indeed, there then remain the following equations:

$$\frac{dH}{dx} = -\frac{i\omega}{c}\varepsilon E, \tag{17.38}$$

$$\frac{dE}{dx} = -\frac{i\omega}{c}\chi H. \tag{17.39}$$

Eliminating any of the quantities E or H, we obtain equations which are identical in form. For example

$$\frac{d^2 E}{dx^2} = -\frac{\omega^2}{c^2}\varepsilon\chi E, \tag{17.40}$$

whence

$$E = Fe^{i\frac{\omega}{c}\sqrt{\varepsilon\chi}\, x - i\omega t}. \tag{17.41}$$

If the wave is propagated in any arbitrary direction, and not along the x-axis, then the solution (17.41) is rewritten thus [here, the symbol Re {} is included for a comparison with (17.21)]:

$$E = \mathrm{Re}\left\{ Fe^{-i\omega\left(t - \frac{\mathbf{rn}}{c}\sqrt{\varepsilon\chi}\right)} \right\}. \tag{17.42}$$

And so, compared with (17.21), the wave velocity has been multiplied by $\frac{1}{\sqrt{\varepsilon\chi}}$. Accordingly, the wave vector will, instead of (17.22), formally satisfy the equation

$$\mathbf{k} = \frac{\omega}{c}\sqrt{\varepsilon\chi}\,\mathbf{n}. \tag{17.43}$$

However, $\frac{c}{\sqrt{\varepsilon\chi}}$ is the velocity of light in a medium, and (17.43) is the wave vector, provided ε and χ are real numbers. Then equation (17.43) will be fully analogous to (17.25). In this case, the solution (17.42) is periodic in space and in time and describes a plane wave travelling with velocity $\frac{c}{\sqrt{\varepsilon\chi}}$.

The ratio of the wave velocity in free space to that in a medium is called the refractive index of the medium for waves of given frequency ω. We note that for visible-light frequencies, $\varepsilon(\omega)$ has nothing in common with its static value. For example, water has a dielectric constant of 81, so that $\sqrt{\varepsilon} = \sqrt{81} = 9$, while the refractive index in the visible frequencies is approximately equal to 1.33 (χ can be considered equal to 1).

Absorption of light in a medium. We now consider a more general case of complex $\varepsilon = \varepsilon_1 + i\varepsilon_2$ (for simplicity we shall put $\chi = 1$). As was shown in exercise 18, Sec. 16, the imaginary part of ε accounts for absorption of light. We shall denote the root of the complex dielectric constant thus:

$$\sqrt{\varepsilon} = \sqrt{\varepsilon_1 + i\,\varepsilon_2} = \nu_1 + i\,\nu_2\,. \tag{17.44}$$

Let us substitute this expression into the exponent of equation (17.42), putting $n_x = 1$, $n_y = 0$, $n_z = 0$. Since $i^2 = -1$, we obtain

$$\mathbf{E} = \operatorname{Re}\{\mathbf{F}e^{-i\omega\left(t - \frac{x\nu_1}{c}\right)}\}\,e^{-\frac{x\nu_2\omega}{c}}\,. \tag{17.45}$$

Thus, the wave is damped in propagation.

Its amplitude diminishes e times at a distance $\frac{c}{\omega\nu_2}$. A solution of the form (17.45) cannot exist in a region which extends to infinity in all directions because $x = -\infty$ substituted into (17.45) yields $\mathbf{E} = \infty$. A solution which is damped in space can be used, for example, when an electromagnetic wave from free space is incident on an absorbing medium. And the x-axis in (17.45) must be considered as directed into the medium.

Exercises

1) Consider the reflection of a plane electromagnetic wave from the interface between two transparent (nonabsorbing) media a and b with refractive indexes $\nu_{1a} \equiv \nu_a$ and $\nu_{1b} \equiv \nu_b$ $(\nu_{2a} = \nu_{2b} = 0)$. Solve the problem in two cases: I) the electric vector lies in the plane drawn through the normal to the interface and through the wave vector $\mathbf{k}$, II) the electric vector is parallel

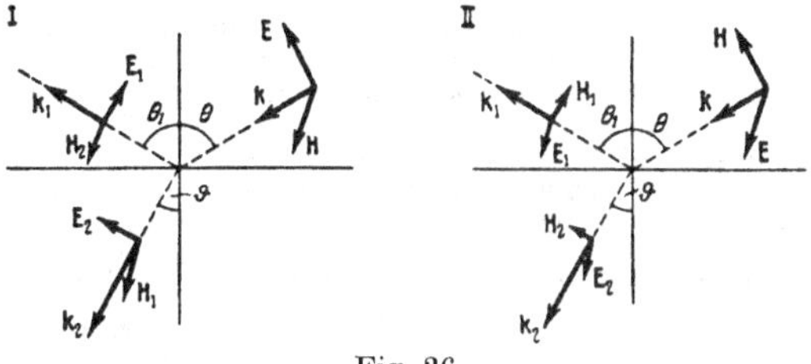

Fig. 26

to the interface between the two media. Calling the angle of incidence θ and the angle of refraction ϑ, find the ratio of the amplitudes of the incident and reflected waves for both cases, I and II (Fig. 26).

At the interface, the normal components of the inductions and the tangential components of the fields must be equal. In order to satisfy the conditions at the boundary, it is necessary to introduce a third wave, which is reflected from the interface. We shall take the equation of the interface to be $y = 0$. The phase of the incident wave at the interface is $\frac{\nu_a}{c} n_x x = \frac{\nu_a}{c} x \sin\theta$, that for the reflected wave $\frac{\nu_a}{c} x \sin\theta_1$, and that for the refracted wave $\frac{\nu_b}{c} x \sin\vartheta$. All three phases must coincide over the whole interface, whence

$$\nu_a \sin\theta = \nu_b \sin\vartheta \quad \text{(the law of refraction)},$$
$$\theta = \theta_1 \quad \text{(the law of reflection)}.$$

Taking into account that $H = \nu E$ [this is easily obtained from (17.39) and (17.41)], we write the boundary conditions (see exercise 1, Sec. 16):

$$\nu_a^2 (E \sin\theta - E_1 \sin\theta) = \nu_b^2 E_2 \sin\vartheta ,$$
$$E \cos\theta + E_1 \cos\theta = E_2 \cos\vartheta ,$$
$$\nu_a (E - E_1) = \nu_b E_2 ,$$

where E, E_1 and E_2 are the electric fields in the incident, reflected and refracted waves. We can see that, by eliminating E, E_1 and E_2 from these conditions, we again obtain the law of refraction. The ratio of the amplitudes is

$$\left| \frac{E_1}{E} \right| = \frac{\tan(\theta - \vartheta)}{\tan(\theta + \vartheta)} . \tag{I}$$

If $\theta + \vartheta = \frac{\pi}{2}$, then $E_1 = 0$ and reflection does not occur. (How can this be verified by double reflection ?)

In case II we must write down the boundary conditions and obtain the equation

$$\left| \frac{E_1}{E} \right| = \frac{\sin(\theta - \vartheta)}{\sin(\theta + \vartheta)} . \tag{II}$$

(I) and (II) are called the Fresnel equations.

2) In the case $\frac{\nu_a}{\nu_b} \sin\theta > 1$ (total internal reflection), show that instead of equation (I) and (II), a reflection coefficient of unity is obtained. Find at what depth the wave, passing in the medium b, is attenuated e times.

3) Find the frequency of electromagnetic oscillations in an infinite square prism with perfectly reflecting walls, assuming a longitudinal electric field constant along the length of the prism. Consider that the field inside the prism does not become zero.

We must consider that the tangential component of the electric field at the walls of the prism is equal to zero, so that the normal component of the Poynting vector **U** should become zero. The solution to Maxwell's equations can be obtained from the potential $A_x = A_0 \sin\frac{\pi z}{a} \sin\frac{\pi y}{a} e^{-i\omega t}$, $A_y = A_z = \varphi = 0$ (the x-coordinate is taken along the axis of the prism). It is of the form

$$E_x = E_0 \sin\frac{\pi z}{a} \sin\frac{\pi y}{a} e^{-i\omega t} ,$$

$$H_y = H_0 \cos\frac{\pi z}{a} \sin\frac{\pi y}{a} e^{-i\omega t} ,$$

$$H_z = - H_0 \sin\frac{\pi z}{a} \cos\frac{\pi y}{a} e^{-i\omega t}$$

on the condition that

$$E_0 = H_0 \text{ and } \omega^2 = \frac{2\pi^2}{a^2} c^2$$

(a is the side of the square).

4) Solve the same problem for a travelling wave in the prism (a waveguide). The field in a waveguide is not zero anywhere except at the walls.

The form of the vector potential in the previous problem suggests one of the following possible solutions:

$$A_x = A_{0x} e^{-i(\omega t - kx)} \sin \frac{\pi y}{a} \sin \frac{\pi z}{a},$$

$$A_y = A_{0y} e^{-i(\omega t - kx)} \cos \frac{\pi y}{a} \sin \frac{\pi z}{a},$$

$$A_z = \varphi = 0.$$

In order to satisfy the condition div $\mathbf{A} = 0$, we must demand that

$$ikA_{0x} - \frac{\pi}{a} A_{0y} = 0.$$

The normal (to the walls) components of the vector $\mathbf{U}$ are again equal to zero since $E_x = 0$ at the walls.

We determine the frequency from equation (12.37):

$$\omega^2 = c^2 k^2 + \frac{2\pi^2 c^2}{a^2}.$$

From here it can be seen that the frequency obtained in the example 3) is the smallest wave frequency that can be propagated in the prism. This wave corresponds to $\lambda = \infty$.

5) Show that when two waves, which are circularly polarized in opposite directions, and which have equal amplitudes (but with frequencies differing by a small quantity $\Delta\omega$) and are travelling in one direction, are combined, a wave is obtained whose polarization vector rotates to an extent depending on the distance of propagation.

Sec. 18. Transmission of Signals. Almost Plane Waves

The impossibility of transmitting a signal by means of a monochromatic wave. A plane monochromatic wave (17.42) extends without limit in all directions of space and in time. Nowhere, so to speak, does it have a beginning or an end. What is more, its properties are everywhere always the same; its frequency, amplitude and the distance between two travelling crests (i.e., the wavelength λ) are always constant. All this can be easily seen by considering a sinusoid or helix.

Let us now pose the problem of the possibility of transmitting an electromagnetic signal over a distance. In order to transmit the signal, an electromagnetic disturbance must be concentrated in a certain volume. By propagation, this disturbance can reach another region of space; detected by some means (for example, a radio receiver), it will transmit to the point of reception a signal about an event occurring at the point of transmission. Likewise, our visual perceptions are a continuous recording of electromagnetic (light) disturbances originating in surrounding objects.

A signal must somehow be bounded in time in order to give notice of the beginning and end of any event.

In order to transmit a signal the amplitude of the wave must, for a time, be somehow changed. For example, the amplitude of one of the waves of a sinusoid must be increased and we must wait until this increased amplitude arrives at the receiving device. A strictly monochromatic wave, i.e., a sinusoid, has the same amplitude everywhere and is therefore not suitable for the transmission of signals in time. In the same way, an ideal plane wave with a given wave vector cannot transmit the image of an object limited in space.

The propagation of a nonmonochromatic wave. We shall now consider what can be done by superimposing several sinusoids upon one another. Suppose that the frequencies of all these travelling waves are included within an interval $\omega_0 - \frac{\Delta\omega}{2} \leqslant \omega \leqslant \omega_0 + \frac{\Delta\omega}{2}$. We shall consider that the frequency interval $\Delta\omega$ is considerably smaller than the "carrier" frequency ω_0. The amplitudes of all the waves $E_0(\omega)$ will be assumed to be identical for any frequency within the chosen interval, and equal to zero outside that interval.

Then the resultant oscillation will be represented by the integral of all the partial oscillations:

$$E = \int E_0(\omega)\, e^{-i(\omega t - kx)} d\omega = E_0 \int\limits_{\omega_0 - \Delta\omega/2}^{\omega_0 + \Delta\omega/2} e^{-i(\omega t - kx)} d\omega. \qquad (18.1)$$

In this equation not only the frequency is variable, but also the absolute value of the wave vector k (the so-called wave number). According to (17.22), it is equal to $\frac{\omega}{c}$ in free space and, in a material medium, $k = \frac{\omega}{c}\nu$, where ν in turn is a function of frequency. In future, in this section, we shall always assume $\nu_2 = 0$, i.e., that there is no absorption.

Since the frequency lies within a small interval, k can be expanded in a series in powers of $\omega - \omega_0$:

$$k(\omega) = k(\omega_0) + (\omega - \omega_0)\left(\frac{dk}{d\omega}\right)_0. \qquad (18.2)$$

Substituting (18.2) in the integral (18.1), we obtain the following expression for the resultant field:

$$E = E_0 e^{-i(\omega_0 t - k_0 x)} \int\limits_{\omega_0 - \frac{\Delta\omega}{2}}^{\omega_0 + \frac{\Delta\omega}{2}} e^{-i(\omega - \omega_0)\left[t - \left(\frac{dk}{d\omega}\right)_0 x\right]} d\omega. \qquad (18.3)$$

We now introduce a new integration variable $\xi = \omega - \omega_0$. Then the integration can be easily performed and the field reduces to the following form:

$$E = E_0 e^{-i(\omega_0 t - k_0 x)} \int\limits_{-\Delta\omega/2}^{\Delta\omega/2} e^{i\xi\left[t - \left(\frac{dk}{d\omega}\right)_0 x\right]} d\xi =$$

$$= E_0 e^{-i(\omega_0 t - k_0 x)} \cdot \frac{2\sin\left\{\left[t - \left(\frac{dk}{d\omega}\right)_0 x\right]\frac{\Delta\omega}{2}\right\}}{t - \left(\frac{dk}{d\omega}\right)_0 x}. \qquad (18.4)$$

The shape of the signal. Let us now examine the expression obtained. It consists of two factors. The first of them, $E_0 e^{-i(\omega_0 t - k_0 x)}$, represents a travelling wave homogeneous in space with a mean "carrier" frequency ω_0. However, the amplitude of the resultant wave is no longer constant in space because of the second factor:

$$\frac{2\sin\left\{\left[t-\left(\frac{dk}{d\omega}\right)_0 x\right]\frac{\Delta\omega}{2}\right\}}{t-\left(\frac{dk}{d\omega}\right)_0 x} \equiv g\left\{\left[t-\left(\frac{dk}{d\omega}\right)_0 x\right]\frac{\Delta\omega}{2}\right\} \equiv g(\psi),$$

where the designation g and ψ are obvious from the equations. This factor has a greatest maximum at $x = \left(\frac{d\omega}{dk}\right)_0 t$, i.e., when the argument of the sine and the denominator are equal to zero. The other extremes will be the less the greater their number (Fig. 27). The greatest maximum is equal to $\Delta\omega$ (since $\frac{\sin\psi}{\psi} = 1$ for $\psi = 0$). This maximum is not situated at a fixed place, but moves in space with a velocity

$$v = \frac{d\omega}{dk}, \tag{18.5}$$

because, from the definition of the point of maximum $\psi = 0$, it follows that

$$x = \frac{d\omega}{dk} t = vt.$$

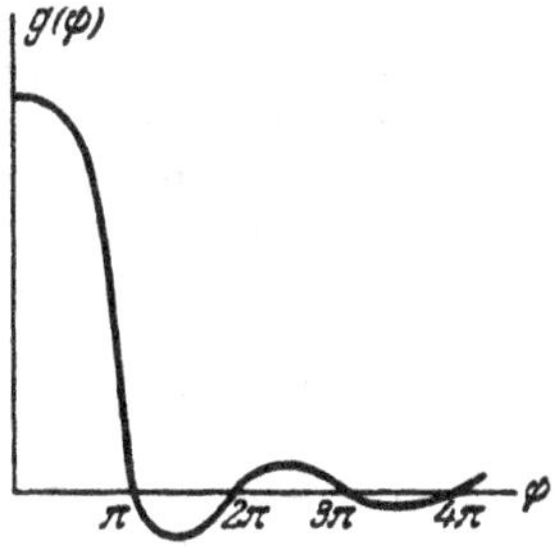

Fig. 27

As we indicated at the beginning of this section, a signal can be transmitted from one point of space to another by means of a displacement of the maximum, since this maximum is distinguished from other maxima.

A disturbance of this kind concentrated in space is called a *wave packet*.

The propagation of a signal of arbitrary shape. A wave packet need not necessarily have the form shown in Fig. 27. By choosing a relationship for $E_0(\omega)$ other than that in equation (18.1) (i.e., by choosing not a constant amplitude in the interval $\Delta\omega$, but a more complicated frequency function), the shape of $g(\psi)$ can be changed. For instance, the resultant amplitude may have the shape of a rectangle, so that the transmitted signal will resemble the dash in the Morse code. If the frequency ω_0 is within the radio-frequency range, then the signals can follow upon one another within audio frequency, in this way reproducing music or speech.

The frequency range and the duration of the signal. In order to transmit a signal, it is always necessary to choose a range of frequencies.

Let us determine this range. Suppose that the receiving device is situated at some point $x = \text{const}$. The width of the received signal can be seen from Fig. 27. In units of ψ, it is equal to π in order of magnitude. Therefore, the duration of the signal is determined from the equation

$$\Delta \psi = \frac{\Delta \omega}{2} \cdot \Delta t \sim \pi.$$

In other words, the duration of the signal Δt is related to the frequency interval $\Delta \omega$ necessary for its transmission by the expression

$$\Delta \omega \cdot \Delta t \sim 2 \pi. \tag{18.6}$$

It should be noted that this estimate refers only to the order of magnitude of $\Delta \omega$ and Δt. The determination of $\Delta \psi$ is, to some extent, arbitrary. In certain cases $\Delta \omega \cdot \Delta t \gg 2 \pi$, so that the estimate (18.6) is a lower figure.

If a radio station is required to transmit sounds audible to the human ear, then the quantity Δt is not greater than 0.5×10^{-4} sec, since the limit of audibility is 2×10^4 oscillations per sec. From this $\Delta \omega = 2 \pi \cdot 2.10^4$.

The range of $\Delta \omega$ is always less than the "carrier" frequency ω_0 which, even for the longest-wave transmitting stations, is not less than $1.5 \times 10^5 \times 2 \pi$. In practice, an interval of $\Delta \omega$ three or four times less than the value given is quite sufficient, since clipping off the very highest frequencies in music, singing or speech does not introduce any essential distortion.

Television transmissions require a considerably greater frequency interval, because an image must be reproduced 25 times every second; and, in turn, the image consists of tens of thousands of separate signals (points). As a result, the carrier frequency is about $2 \pi \times 6 \times 10^7$, corresponding to the metric band of radio waves. Such waves are propagated over a relatively small radius. They are screened by the curvature of the earth's surface like light. The relation (18.6) is always correct in order of magnitude; therefore, for distant television transmissions, it is necessary to have either relay stations, very high-placed transmitters, or cable lines.

Phase and group velocity. We shall now consider in more detail the velocity with which signals are transmitted. From (18.5), the velocity of a wave packet is

$$v = \frac{d \omega}{d k}.$$

It differs from the propagation velocity of the constant phase surface, which is expressed in terms of frequency and wave number as

$$u = \frac{\omega}{k}. \tag{18.7}$$

Indeed, the expression for a travelling monochromatic wave can be written in the following form:

$$E = E_0 e^{ik\left(x - \frac{\omega}{k} t\right)}.$$

Comparing this formula with the general expression for a travelling wave $E = E\,(x - ut)$, we arrive at (18.7). The velocity of the wave is u, and not c, because (18.7) is by no means necessarily related to the propagation of a plane wave in a vacuum.

$u = \frac{\omega}{k}$ is called the *phase velocity* of the wave; v is called the *group velocity* of the wave packet obtained by superimposing a group of waves. In a vacuum, v and u coincide because $\omega = ck$. However, if there is dispersion, i.e., a dependence of the refractive index on the frequency, then $\omega = \frac{c}{\nu} k$ so that $v \neq k$.

The group velocity may be regarded as the velocity of propagation of a signal only when it is less than the velocity of light in free space c. If the expression (18.5) formally gives $v > c$, we cannot avoid a more careful analysis that takes into account absorption. As a result, it turns out that an electromagnetic signal in the form of a very weak precursor is propagated with a velocity c, but the major portion of the wave energy arrives at the point of reception with a lesser velocity (see A. Sommerfeld, *Optik,* Wiesbaden, 1950).

As an example of the calculation of group velocity we shall take the dependence of frequency on the wave vector in the form:

$$\omega^2 = a^2 + c^2 k^2 .$$

This form is obtained for a waveguide (see exercise 4, Sec. 17, or exercise 19, Sec. 16, in the limiting case of extremely large frequencies). Whence the group velocity is

$$v = \frac{c^2 k}{\omega}$$

and since $ck < \omega$, we have $v < c$.

Here, the phase velocity proves to be greater than c:

$$u = \frac{\omega}{k} = c\,\frac{\omega}{ck} > c\,.$$

We note that $uv = c^2$.

In vector form the group velocity is defined as follows:

$$\mathbf{v} = \frac{\partial \omega}{\partial \mathbf{k}}\,. \tag{18.8}$$

If we make use of a more accurate dispersion law (obtained in exercise 19, Sec. 16), then for $\omega^2 \sim \omega_0^2$ there proves to be a frequency region for which $\varepsilon\,(\omega)$ is negative. For such frequencies the refractive index is purely imaginary and the expressions (18.5) and (18.8) become meaningless.

The form of a wave in space and the range of the wave vectors. An expression similar to (18.6) can also be obtained for the form of a wave in space at a definite instant of time. For this we must put $t=\text{const}$ and then, once again taking $\Delta\psi \sim \pi$, we obtain

$$\Delta\psi = \frac{\Delta\omega}{2}\frac{dk}{d\omega}\Delta x = \frac{\Delta k\,\Delta x}{2} \sim \pi$$

$$\Delta k \cdot \Delta x \sim 2\pi\,. \tag{18.9}$$

This means that if we want to limit the extent of an electromagnetic disturbance to a region Δx, we must perform a superposition of monochromatic waves in the interval of values k of order $\frac{2\pi}{\Delta x}$. In three dimensions (18.9) is rewritten thus:

$$\begin{aligned} \Delta k_x \cdot \Delta x &\sim 2\pi\,, \\ \Delta k_y \cdot \Delta y &\sim 2\pi\,, \\ \Delta k_z \cdot \Delta z &\sim 2\pi\,. \end{aligned} \tag{18.10}$$

The limiting accuracy of radiolocation. We shall explain the relations (18.10) by means of a graphic example. Let us suppose that an electromagnetic wave has, in some way, to be bounded on the sides, as in the case of a radiolocation (radar) beam. Let us find the greatest accuracy with which the locator can register the position of an object at a distance l. Obviously, this accuracy is given by the transverse diameter of the beam d at a distance l from the locator.

Let the frequency at which the locator works be equal to ω, then the corresponding wavelength is $\lambda = \frac{2\pi c}{\omega}$. If the electromagnetic wave were to be propagated in unbounded space it would have an accurately defined wave vector

$$\mathbf{k} = \frac{2\pi}{\lambda}\mathbf{n} \tag{18.11}$$

($\mathbf{n}$ is a unit vector in the direction of the beam). If the wave has a cross section d, then $\mathbf{k}$ can no longer be regarded as an accurately defined vector along $\mathbf{n}$. In order to write down an expression for the electromagnetic wave at any point in space occupied by the beam, it is necessary to take a group of plane waves whose vectors $\mathbf{k}$ lie inside a cone described by a certain angle of flare. The maximum deviation of the wave vectors of these plane waves from the mean vector $\mathbf{k}$, determined from (18.11), will be called $k_\perp$. Here, we do not have in mind a cone with a sharply bounded surface, but only an estimate of the angular flare of the beam. According to (18.10) $k_\perp$ is related to the whole cross section of the beam by the following relation:

$$d \cdot 2k_\perp \gtrsim 2\pi\,. \tag{18.12}$$

Here, we have put $\Delta x = d$, $\Delta K = 2\, K_\perp$ because the inaccuracy $k_\perp$ obtains on both sides of the axis of the beam.

The dimensions of the reflector of the locator itself can be ignored if the diameter of the beam is considered at a great distance; and this is of practical interest. In other words, d is determined only by the relationship (18.12) and is independent of the dimensions of the reflector.

The divergence of the beam of rays at every point is measured by the ratio $\frac{k_\perp}{k}$. For this reason, the ratio of the cross section of the beam d to the distance from the locator l cannot be less than the quantity $\frac{2k_\perp}{k}$:

$$\frac{d}{l} \gtrsim \frac{2k_\perp}{k}. \tag{18.13}$$

This relationship is shown in Fig. 28 for the limiting case of the equality. However, it must be borne in mind that it is not in reality an equality but an estimate of order of magnitude. (18.12) is also approximate and the symbol $\gtrsim$ must be written in it.

Thus, we have obtained two estimates for $k_\perp$:

$$k_\perp \gtrsim \frac{\pi}{d} \quad \text{(lower estimate)}$$

and, from (18.13),

$$k_\perp \leqslant \frac{kd}{2l} = \frac{\pi d}{\lambda l} \quad \text{(upper estimate)}.$$

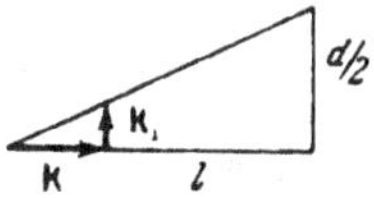

Fig. 28

Eliminating $k_\perp$ from these estimates, we obtain

$$\frac{\pi d}{l\lambda} \gtrsim \frac{\pi}{d}$$

or finally

$$d \gtrsim \sqrt{l\lambda}\,. \tag{18.14}$$

For example, if $l = 100$ km and $\lambda = 1$ m, then the position of the object cannot be determined with an accuracy exceeding 320 m. This is why the dimensions of the reflector could be neglected in the estimate.

The limit of applicability of the concept of a ray. Equations (18.10) indicate within what limits the concept of a ray is applicable in optics. Obviously, one can talk about a ray in a definite direction only when

$$\Delta k \ll k\,, \tag{18.15}$$

i.e., when the transverse broadening of the wave vector is considerably less than the wave vector itself. But $\Delta k \sim \frac{2\pi}{d}$ and $k \sim \frac{2\pi}{\lambda}$ so that (18.15) is equivalent to the condition

$$d \gg \lambda\,. \tag{18.16}$$

In other words, the dimensions of the region in which the concept of a light ray is defined must be considerably larger than the wavelength of the light wave. For example, a small circle in the wall of a camera-obscura of diameter, say, 1 mm is considerably greater than the wavelength of visible light, which is of an order of magnitude 0.5×10^{-4} cm. Therefore, the image obtained in a camera-obscura is formed with the aid of light rays.

The optics of light rays is called geometrical optics. A ray is defined only when its direction is given, i.e., the normal to the wave front. If we are given a beam of nonparallel (for example, converging) rays, then the wave front is curved. But the radius of its curvature at each point is considerably greater than the wavelength. Such a converging beam of rays represents a set of normals to an "almost plane" wave. The curvature of the wave front close to the focus of the rays may become comparable with the wavelength, and then there arise deviations from geometrical optics. Such deviations are called diffraction effects. They are also observed when a light wave falls on some opaque obstacle. In accordance with geometrical optics, we should have obtained a sharp shadow—a transition from a region where the field differs from zero to a region where it is equal to zero. But Maxwell's equations do not permit such solutions, which are discontinuous in free space (cf. the boundary conditions, exercise 1, Sec. 16). In actual fact, there always exists a transition zone between "light" and "shadow," in which the wave amplitude changes in a complicated oscillatory way.

Exercises

1) Find the limiting dimensions of an object which may be observed in a microscope using light of wavelength λ.

Denoting the semiangle of the cone of rays, drawn from the microscope objective to the object, by θ, we have $\Delta k = k \sin \theta$. Whence

$$\Delta x = \frac{2\pi}{\Delta k} = \frac{2\pi}{k \sin \theta} = \frac{\lambda}{\sin \theta}.$$

It is therefore convenient to use a beam of rays with large solid angles and small wavelengths.

2) Show that if the dispersion law of exercise 19, Sec. 16, is used, then $v < c$.

We write down an expression for the inverse of v:

$$\frac{1}{v} = \frac{\partial k}{\partial \omega} = \frac{1}{c}\left(\sqrt{\varepsilon} + \omega \frac{\partial \sqrt{\varepsilon}}{\partial \omega}\right).$$

This leads to the inequality

$$\varepsilon + \frac{\omega}{2} \frac{\partial \varepsilon}{\partial \omega} > \sqrt{\varepsilon}.$$

The derivative $\frac{\partial \varepsilon}{\partial \omega}$ is everywhere positive so that, for $\varepsilon > 1$, the inequality can be seen directly. When $\varepsilon < 1$ we have

$$\varepsilon + \frac{\omega}{2}\frac{\partial \varepsilon}{\partial \omega} = 1 + \frac{a^2 \omega^2}{(\omega_0^2 - \omega^2)^2}, \qquad \left(\frac{4\pi N e^2}{m} \equiv a^2\right).$$

Squaring both sides of the inequality, it is easy to see that this quantity is greater than $\sqrt{1 + \frac{a^2}{\omega_0^2 - \omega^2}} = \sqrt{\varepsilon}$.

Sec. 19. The Emission of Electromagnetic Waves

Basic equations and boundary conditions. So far we have considered electromagnetic waves irrespective of the charges producing them. In this section we shall consider the emission of waves by point charges moving in a vacuum. The basic system of equations in this case is (12.37) and (12.38) together with the Lorentz condition (12.36). We rewrite these equations anew:

$$\Delta \mathbf{A} - \frac{1}{c^2}\frac{\partial^2 \mathbf{A}}{\partial t^2} = -\frac{4\pi}{c}\mathbf{j}, \tag{19.1}$$

$$\Delta \varphi - \frac{1}{c^2}\frac{\partial^2 \varphi}{\partial t^2} = -4\pi\rho, \tag{19.2}$$

$$\operatorname{div} \mathbf{A} + \frac{1}{c}\frac{\partial \varphi}{\partial t} = 0. \tag{19.3}$$

We begin the solution with (19.2) proceeding in the following manner. We assume that ρ differs from zero only in an infinitesimal volume dV. We find the potential φ for such a "point" radiator. By virtue of the linearity of equation (19.2), the potential of the entire spatial distribution of charge appearing on the right-hand side of (19.2) is equal to the integral of the potentials due to infinitesimal small elements of charge $\delta e = \rho\, dV$.

In order to determine the solution uniquely, we must impose a certain boundary condition. It is assumed that the charges are situated in infinite space, i.e., that there are no conductors or dielectrics anywhere.

In free space, a boundary condition can be imposed only at an infinite distance away from the charges. In accordance with the posed problem of radiation, it is natural to suppose that there was no field for an infinitely large interval of time before the initiation of radiation at an infinitely large distance from the radiator:

$$\begin{aligned} \varphi\,(t \to -\infty\,, \quad r \to \infty) &= 0\,, \\ \mathbf{A}\,(t \to -\infty\,, \quad r \to \infty) &= 0\,. \end{aligned} \tag{19.4}$$

If no boundary conditions are imposed on the solution of an inhomogeneous equation, then any solution of the homogeneous equation can always be added to it so that a unique answer cannot be obtained.

The radiation of a small element of charge. Let us begin with an infinitesimal charge element $\delta e = \rho\, dV$. We place it at the coordinate origin. Then the solution of (19.2) will possess spherical symmetry. In Sec. 11, an expression for the Laplacian operator Δ was derived in spherical coordinates (11.46). As in the case of a static charge [equation (14.7)], we must retain only the term involving differentiation with respect to r and, this time, obviously, we must also differentiate with respect to time. For the time being we consider that the charge density at all points, except the origin, is equal to zero. Therefore, for all points for which $r \neq 0$, equation (19.20) is written thus:

$$\frac{1}{r^2}\frac{\partial}{\partial r}r^2\frac{\partial \varphi}{\partial r} - \frac{1}{c^2}\frac{\partial^2 \varphi}{\partial t^2} = 0. \tag{19.5}$$

Temporarily, we put

$$\varphi = \frac{\Phi(r, t)}{r}. \tag{19.6}$$

Then

$$\frac{\partial \varphi}{\partial r} = \frac{1}{r}\frac{\partial \Phi}{\partial r} - \frac{\Phi}{r^2}, \quad r^2\frac{\partial \varphi}{\partial r} = r\frac{\partial \Phi}{\partial r} - \Phi,$$

$$\frac{\partial}{\partial r}r^2\frac{\partial \varphi}{\partial r} = r\frac{\partial^2 \Phi}{\partial r^2} + \frac{\partial \Phi}{\partial r} - \frac{\partial \Phi}{\partial r} = r\frac{\partial^2 \Phi}{\partial r^2}.$$

Substituting this in (19.5) and multiplying by r (by convention, r is not equal to zero), we obtain

$$\frac{\partial^2 \Phi}{\partial r^2} - \frac{1}{c^2}\frac{\partial^2 \Phi}{\partial t^2} = 0. \tag{19.7}$$

But this is the equation, of the form (17.5), for the propagation of a wave. Its solution is similar to (17.12):

$$\Phi = \Phi_1\left(t + \frac{r}{c}\right) + \Phi_2\left(t - \frac{r}{c}\right). \tag{19.8}$$

The solution Φ_1 depends on the argument $t + \frac{r}{c}$, and the solution Φ_2 depends on the argument $t - \frac{r}{c}$. The first of these arguments, $t + \frac{r}{c}$, for $r \to \infty$, $t \to -\infty$ has a completely indeterminate form $\infty - \infty$, i.e., it is equal to anything. From the condition (19.4), the function Φ becomes zero when $r \to \infty$, $t \to -\infty$. Therefore Φ_1 becomes zero for any value of the argument, i.e., it is equal to zero everywhere. (The potential at infinity must tend to zero more strongly than $\frac{1}{r}$ so that there should be no radiation; see below in this section.) For the function Φ_2, condition (19.4) denotes that $\Phi_2(-\infty) = 0$. In other words, the function Φ_2 tends to zero at minus infinity. It does not follow from this, of course, that it is equal to zero everywhere. Thus,

$$\Phi = \Phi_2\left(t - \frac{r}{c}\right).$$

Omitting the index 2, we write the expression for φ as follows:

$$\varphi = \frac{1}{r}\Phi\left(t - \frac{r}{c}\right). \tag{19.9}$$

The function Φ is not as yet determined. From the form of its argument we conclude that it describes a travelling wave in the direction of increasing radii (because $t > 0$). Such a wave is termed diverging.

Retarded potential. The value of the function at $r = 0$, $t = 0$ is shifted to the point r in a time $t = \frac{r}{c}$ or, in other words, the potential at the point r and time t is determined by the charge, not at the instant of time t, but at an earlier instant $t - \frac{r}{c}$. The term $\frac{r}{c}$ is a measure of the retardation occurring as a result of the finite velocity of propagation of the wave.

But when the retardation $\frac{r}{c}$ becomes a very small quantity, the potential very close to the charge must be determined by the instantaneous value of the charge $\delta e\,(t)$. We know from Sec. 14 that the potential due to a point charge is equal to $\frac{\delta e}{r}$ (14.8), whence

$$\frac{\Phi\,(t)}{r} = \frac{\delta e\,(t)}{r} = \frac{\rho\,(t)\,d\,V}{r}.$$

Therefore,

$$\Phi\,(t) = \rho\,(t)\,d\,V, \tag{19.10}$$

and the retarded potential of a point charge $\varphi\left(t - \frac{r}{c}\right)$ is, in accordance with (19.9) and (19.10), equal to

$$\varphi\left(t - \frac{r}{c}\right) = \frac{\rho\left(t - \frac{r}{c}\right)}{r}\,dV. \tag{19.11}$$

Now, displacing the coordinate origin to another point, we obtain, like (14.9),

$$\varphi\left(t - \frac{|\mathbf{R} - \mathbf{r}|}{c}\right) = \frac{\rho\left(t - \frac{|\mathbf{R} - \mathbf{r}|}{c},\ \mathbf{r}\right)}{|\mathbf{R} - \mathbf{r}|}\,dV. \tag{19.12}$$

Here it is assumed that the charge density is given at the point $\mathbf{r}\,(x, y, z)$, and the potential is calculated at the point $\mathbf{R}\,(X, Y, Z)$, thus introducing the explicit dependence of ρ on the spatial argument $\mathbf{r}$. Finally, in order to obtain the complete solution to (19.2), we must integrate (19.12) over all the volume elements, i.e., with respect to $d\,V = dx\,dy\,dz$:

$$\varphi = \int \frac{\rho\left(t - \frac{|\mathbf{R} - \mathbf{r}|}{c},\ \mathbf{r}\right)}{|\mathbf{R} - \mathbf{r}|}\,dV. \tag{19.13}$$

For point charges, ρ denotes the special function that was defined in Sec. 12.

Equation (19.1) has exactly the same form as (19.2) and its solution satisfies the same boundary conditions. Therefore, the vector potential is written quite analogously to (19.13):

$$\mathbf{A}=\int \frac{\mathbf{j}\left(t-\frac{|\mathbf{R}-\mathbf{r}|}{c}, \mathbf{r}\right)}{c\,|\mathbf{R}-\mathbf{r}|}\,dV. \tag{19.14}$$

Comparing (19.14) with (15.10), which gives $\mathbf{A}$ for a stationary current, we see that $\mathbf{j}$ depends on the argument $\mathbf{r}$ in two ways: first, directly, in accordance with its spatial distribution and, secondly, via the time argument; since the system of currents is not infinitely small, but has finite dimensions, the retardation of a wave from various points of the system is different.

Retarded potential at a large distance from a system of charges. We shall now look for the form of the solutions of (19.12) and (19.13) at a great distance away from the radiating system. We note that the integrand depends on the argument R in both integrals in two ways: in the denominator and via the argument t. The function in the denominator depends very smoothly on R. Its expansion in terms of powers of R yields terms which tend to zero like $\frac{1}{R^n}$ at large distances away from the system. As will be shown later, they do not add anything to the radiation (for $n>1$). So we simply replace $\frac{1}{|\mathbf{R}-\mathbf{r}|}$ by $\frac{1}{R}$. At large distances from the system, the term $|\mathbf{R}-\mathbf{r}|$, appearing in the argument t of the numerator, looks like this:

$$|\mathbf{R}-\mathbf{r}|=R-\mathbf{r}\nabla R=R-\frac{\mathbf{r}\mathbf{R}}{R}=R-\mathbf{r}\mathbf{n}, \tag{19.15}$$

where $\mathbf{n}$ is a unit vector in the direction of $\mathbf{R}$. The subsequent terms of the expansion (19.15) contain R in the denominator and are insignificant. Thus, at a large distance from the radiating system, the potentials are:

$$\varphi=\frac{1}{R}\int \rho\left(t-\frac{R}{c}+\frac{\mathbf{r}\mathbf{n}}{c}, \mathbf{r}\right)dV, \tag{19.16}$$

$$\mathbf{A}=\frac{1}{cR}\int \mathbf{j}\left(t-\frac{R}{c}+\frac{\mathbf{r}\mathbf{n}}{c}, \mathbf{r}\right)dV. \tag{19.17}$$

An estimate of the retardation inside a system. The term $\frac{(\mathbf{r}\mathbf{n})}{c}$ in the arguments of the integrands of (19.16) and (19.17) indicates by how much an electromagnetic wave, coming from the more distant parts of the radiating system, is retarded in comparison with a wave radiated by the nearer parts of the system. In other words, the term $\frac{(\mathbf{r}\mathbf{n})}{c}$ determines the time that the electromagnetic wave takes in passing through the system of charges. If the velocity of the charges is equal to v, then, in that time, they are displaced through a distance $v\frac{(\mathbf{rn})}{c}$. The retardation inside the system is negligible when this distance is small in comparison with the size of the system r. Therefore, if $v\frac{(\mathbf{rn})}{c} \ll r$ (or, more simply, $v \ll c$), then the charges do not have time to change their *positions* noticeably during the time of propagation of the wave in the system.

However, in order that nothing should really change in the system, the charges must also maintain their velocities in that time, because the vector potential depends on the currents, i.e., on the particle velocities. This imposes a further condition which is formulated in the following manner. Let the charges oscillate and radiate light of frequency ω. The wavelength of the light is equal to $\lambda = \frac{2\pi c}{\omega}$. In the time $\frac{r}{c}$ the phase of the charge oscillations changes by $\omega\frac{r}{c}$. This change must be small in comparison with 2π, whence it follows that the size of the system must be small compared with the wavelength of the radiated light in order that the retardation inside the system should be insignificant. Thus, the term $\frac{(\mathbf{rn})}{c}$ in the argument of the integrand is small provided two inequalities are fulfilled: $v \ll c$, $r \ll \lambda$.

The vector potential to a dipole approximation. Let us assume that both the inequalities obtained have been fulfilled. We omit the term $\frac{\mathbf{rn}}{c}$ in the time argument in the expression for vector potential (19.17). Then the whole integrand will refer to the same instant of time $t-\frac{R}{c}$ and we obtain

$$\mathbf{A} = \frac{1}{Rc}\int \mathbf{j}\left(t - \frac{R}{c},\ \mathbf{r}\right) dV. \tag{19.18}$$

Recall now that $\mathbf{j} = \rho\mathbf{v}$ and that the charges are point charges. Then the integral (19.18) is reduced to a summation over separate charges:

$$\mathbf{A} = \frac{1}{Rc}\left\{\sum_i e_i \mathbf{v}^i\right\}_{t-\frac{R}{c}}. \tag{19.19}$$

Here, the lower index $t-\frac{R}{c}$ denotes that the whole sum must be taken at that instant of time. But $\mathbf{v}^i = \frac{d\mathbf{r}^i}{dt}$, so that

$$\mathbf{A} = \frac{1}{Rc}\frac{d}{dt}\left\{\sum_i e_i \mathbf{r}_i\right\}_{t-\frac{R}{c}} = \frac{\dot{\mathbf{d}}\left\{t-\frac{R}{c}\right\}}{Rc}. \tag{19.20}$$

Here we have used the definition for dipole moment (14.20). We note that (19.20) involves only a time derivative $\dot{\mathbf{d}}$. Therefore, the transformation (14.21), which corresponds to a change of coordinate origin, does not change $\dot{\mathbf{d}}$ either for a charged system or for a neutral system. In particular, (19.20) holds also for a single charge.

The approximation (19.20), in which **A** is expressed in terms of a derivative of the dipole moment of the system as a whole, is termed a *dipole approximation.*

The Lorentz condition to a dipole approximation. In Sec. 17, a potential gauge transformation for travelling plane waves was chosen such that the scalar potential became zero. We shall make the same gauge transformation for diverging spherical waves. To do this we must subject the vector potential to the following condition:

$$\operatorname{div} \mathbf{A} = 0, \tag{19.21}$$

which is obtained from (19.3) if we take $\varphi = 0$.

In condition (19.21) we should not differentiate **A** with respect to R in the denominator: each such differentiation increases the degree of R by unity, while the potential is determined at a large distance from the radiating system. Only terms inversely proportional to R contribute to the radiated energy (see below). The unit vector $\mathbf{n} = \frac{\mathbf{R}}{R}$, which will appear in differentiation, need not be differentiated a second time, since that would also give rise to superfluous degrees of R in the denominator.

We choose an arbitrary gauge function in the form

$$f = \frac{\mathbf{n}\,\mathbf{d}\left\{t - \frac{R}{c}\right\}}{R}. \tag{19.22}$$

Then, applying equation (11.37), it is easy to see that the condition (19.21) is fulfilled:

$$\operatorname{div}(\mathbf{A} + \nabla f) = -\frac{\mathbf{n}\,\ddot{\mathbf{d}}}{Rc^2} - \operatorname{div} \mathbf{n}\frac{(\mathbf{n}\,\dot{\mathbf{d}})}{Rc} = -\frac{\mathbf{n}\,\ddot{\mathbf{d}}}{Rc^2} + \mathbf{n}\cdot\mathbf{n}\frac{(\mathbf{n}\,\ddot{\mathbf{d}})}{Rc^2} = 0.$$

And the scalar potential is cancelled by $-\frac{1}{c}\frac{\partial f}{\partial t}$.

Field to a dipole approximation. Let us now calculate the electromagnetic field. We need to differentiate only inside the argument $t - \frac{R}{c}$. In calculating the magnetic field, we make use of (11.38) and of the fact that rot grad $f = 0$:

$$\mathbf{H} = \operatorname{rot} \mathbf{A} = \frac{1}{Rc} \operatorname{rot} \dot{\mathbf{d}}\left\{t - \frac{R}{c}\right\} = -\frac{1}{Rc^2}[\mathbf{n}\,\ddot{\mathbf{d}}]. \tag{19.23}$$

The electric field is

$$\mathbf{E} = -\frac{1}{c}\frac{\partial \mathbf{A}}{\partial t} - \frac{1}{c}\frac{\partial}{\partial t}\nabla f = -\frac{\ddot{\mathbf{d}}}{Rc^2} + \frac{1}{Rc^2}\mathbf{n}(\mathbf{n}\,\ddot{\mathbf{d}}) =$$
$$= \frac{1}{Rc^2}[\mathbf{n}[\mathbf{n}\,\ddot{\mathbf{d}}]] = [\mathbf{H}\,\mathbf{n}]. \tag{19.24}$$

From these equations it can be seen that the electric field, the magnetic field, and the vector $\mathbf{n}$ are mutually perpendicular. In addition, $H = E$, since $E^2 = [\mathbf{H}\mathbf{n}]^2 = H^2 - (\mathbf{H}\mathbf{n})^2$, and $\mathbf{H}\mathbf{n} = 0$, since $\mathbf{H}$ and $\mathbf{n}$ are perpendicular. Consequently, the wave at a point $\mathbf{R}$ at a great distance away from a radiating system is of the nature of a plane electromagnetic wave. This result was to be expected because the field is calculated far away from charges, where the wave front may be approximately regarded as plane and the solution becomes the same as obtained in Sec. 17.

Fig. 29 gives a general picture of the field. We situate the vector $\ddot{\mathbf{d}}$ at the centre of a sphere of large radius R so that $\ddot{\mathbf{d}}$ coincides with the polar axis or, in other words, is directed towards the "north pole." Let us draw the radius vector of some point. Through this point, we draw the meridian and the parallel. Then the electric field is tangential to the meridian and is directed "towards the south," while the magnetic field is tangential to the parallel and is directed "towards the east." It can be seen from equations (19.23) and (19.24) that the field becomes zero at the poles and maximum on the equator, i.e., on a plane perpendicular to $\ddot{\mathbf{d}}$. The field distribution in space does not possess spherical symmetry. We note that the transverse field cannot be spherically symmetrical for purely geometrical reasons. The zone, in which the field is calculated according to equations (19.23) and (19.24), is called a *wave zone.*

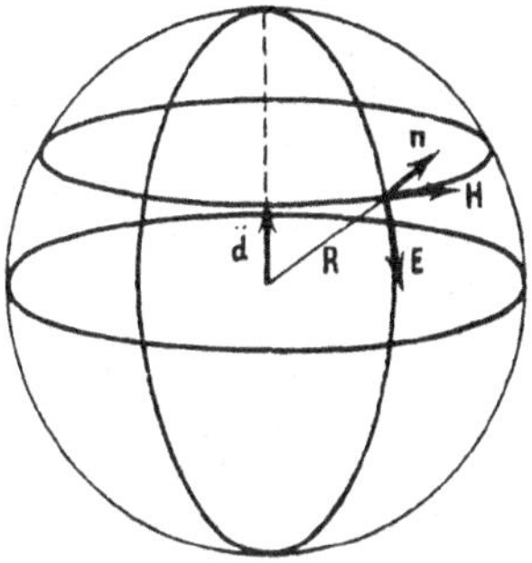

Fig. 29

The intensity of dipole radiation. Let us now find the energy lost by the system in radiation. We must calculate the energy flux crossing an infinitely distant surface. The energy flux density, or the Poynting vector, is

$$\mathbf{U} = \frac{c}{4\pi}[\mathbf{E}\mathbf{H}] = \frac{c}{4\pi}[[\mathbf{H}\mathbf{n}]\,\mathbf{n}] = \frac{c}{4\pi}\mathbf{n}H^2. \tag{19.25}$$

Hence, the energy flux is directed along a radius, as it should be in a wave zone. The total energy crossing a sphere of radius R in unit time is

$$\frac{d\mathscr{E}}{dt} = \int \mathbf{U}\,d\mathbf{s} = \frac{c}{4\pi}\int H^2\mathbf{n}\,d\mathbf{s} = \frac{c}{4\pi}\int H^2 ds, \tag{19.26}$$

because the vector $d\mathbf{s}$ is directed along $\mathbf{n}$. Further, $ds = R^2 \cdot \pi \sin\vartheta\, d\vartheta$, where ϑ is the polar angle. From (19.23)

$$H^2 = \frac{1}{R^2 c^4}\ddot{d}^2 \sin^2\vartheta. \tag{19.27}$$

By substituting (19.27) in (19.26), cancelling R^2, and integrating, we obtain an expression for the energy radiated in one second:

$$\frac{d\mathscr{E}}{dt} = \frac{2}{3}\frac{\ddot{d}^2}{c^3}. \tag{19.28}$$

We note that all the terms in the expression for fields containing R in the denominator to a higher degree than the first would not contribute anything to (19.28) for a sufficiently large R. It is for this reason that only first degree terms in R have been retained in the denominator.

The significance of equation (19.28). Equation (19.28) expresses a result of fundamental importance—energy is radiated whenever a charge is accelerated. Indeed, $\ddot{\mathbf{d}} = \sum_i e_i \ddot{\mathbf{r}}^i$. Hence it is necessary—so that $\ddot{\mathbf{d}}$ should differ from zero—that the charges should be in accelerated motion irrespective of the sign of the accelerations.

But then electrons moving in an atom should radiate energy continuously and should fall into the nucleus; every electron is in acceleration motion, otherwise its motion could not be finite (see Sec. 5). In actual fact atoms are obviously stable and the electrons do not fall into the nucleus.

Here, we realize that classical, Newtonian, mechanics can in no way be applied to the motion of an electron in an atom. In the third part of this book we shall explain the stability of atoms using quantum mechanics, where the very concept of motion differs qualitatively from that in classical mechanics.

Magnetic dipole and quadrupole radiation. We have indicated that charges must be in accelerated motion to radiate. But a simple example can be given when equation (19.28) yields zero even for accelerated charges. Let the system consist of two identical charged particles. According to Newton's Third Law their accelerations are equal and opposite in sign, so that $\ddot{\mathbf{d}} = \sum_i e_i \ddot{\mathbf{r}}^i = e(\ddot{\mathbf{r}}_1 + \ddot{\mathbf{r}}_2) = 0$. In this case this law is applicable because, to a dipole approximation, the retardation of electromagnetic interactions inside the system is considered as negligibly small and, hence, the interaction forces between charges are regarded as instantaneous. But there is then no need to take account of the momentum transmitted to the field and the total momentum of the particles is conserved, thereby leading to the condition $\ddot{\mathbf{r}}_1 = -\ddot{\mathbf{r}}_2$. For this case, approximation (19.18) does not describe the radiation and it becomes necessary to use higher-order approximations.

If, in the expansion in powers of $\frac{\mathbf{rn}}{c}$, a further term is retained in addition to the zero term, then a radiation is obtained which depends on the change in magnetic quadrupole and dipole moments of a system of charges. It is essential that this expansion be not in terms of inverse powers of R, as in electrostatics and magnetostatics, but in powers of the retardation inside the system.

We have already mentioned that the retardation inside the system is small when $v \ll c$ and $r \ll \lambda$. The ratio $\frac{v}{c}$ is involved in the magnetic moment of the system. Therefore, those terms in the expansion (involving powers of the retardation) which are proportional to $\frac{v}{c}$, account for magnetic dipole radiation. The quadrupole moment of the system involves an additional power of r compared with the dipole moment, and so quadrupole radiation is related to those terms of the expansion which are proportional to $\frac{r}{\lambda}$. Higher approximations are important in those cases for which lower-order approximations, for some reason or other, become zero, as is the case of the two identical charges.

The field due to a magnetic dipole radiator is similar to the field of a radiating electric dipole. Unlike the field represented in Fig. 29, the magnetic field, for magnetic dipole radiation, lies in the plane $\boldsymbol{\mu}$, (i.e., it is along a meridian), while the electric field is along a parallel. The equation for intensity is similar to (19.28), though it involves $(\ddot{\boldsymbol{\mu}})^2$ instead of $(\ddot{d})^2$. Since the magnetic moment is proportional to $\frac{v}{c}$, the intensity of magnetic dipole radiation is less than the intensity of electric dipole radiation in the ratio $\left(\frac{v}{c}\right)^2$.

The field of a radiating electric quadrupole has a more complicated configuration. The expression for the intensity of such radiation involves the square of the third derivative of the quadrupole moment of the system. In order of magnitude, the intensity of quadrupole radiation is less than the intensity of dipole radiation in the ratio $\left(\frac{r}{\lambda}\right)^2$

Exercises

1) Calculate the time that it takes a charge, moving in a circular orbit around a centre of attraction, to fall into the centre as a result of the radiation of electromagnetic waves. Regard the path as always approximately circular.

2) A particle with charge e and mass m passes, with velocity v, a fixed particle of charge e_1, at a distance ρ. Ignoring the distortion in the orbit of the oncoming particle, calculate the energy that this particle loses in radiation.

$$\text{Answer: } \Delta\mathscr{E} = \frac{2}{3}\frac{e^2}{m^2 c^3}\int_{-\infty}^{\infty} |\ddot{\mathbf{r}}|^2 \, dt = \frac{2}{3}\frac{e^4 e_1^2}{m^2 c^3}\int_{-\infty}^{\infty} \frac{dt}{(\rho^2 + v^2 t^2)^2} = \frac{\pi}{3}\frac{e^4 e_1^2}{m^2 c^3 \rho^3 v}.$$

3) Why is it that when two identical particles collide ($e_1 = e_2$, $m_1 = m_2$), magnetic dipole radiation does not result if the interaction is calculated according to the Coulomb law? The intensity of magnetic dipole radiation is $\frac{2}{3}\frac{\ddot{\mu}^2}{c^3}$.

4) A plane light wave falls on a free electron causing it to oscillate. The electron begins to radiate secondary waves, i.e., it scatters the radiation. Find the effective scattering cross section, defined as the ratio of the energy scattered in unit time to the flux density of the incident radiation.

We proceed from the fact that $\ddot{\mathbf{r}} = \frac{e\mathbf{E}}{m}$ and then determine $\frac{d\mathscr{E}}{dt}$ from (19.28). Dividing by the energy flux $\frac{cE^2}{4\pi}$, we obtain

$$\sigma = \frac{8\pi}{3} \frac{e^4}{m^2 c^4}.$$

Sec. 20. The Theory of Relativity

The law of addition of velocities and electrodynamics. In Sec. 15, the interactions of charges with a magnetic and an electric field was reviewed [see (15.34)]. But the motion of the charges was considered slow, in other words, the velocity satisfied the inequality $v \ll c$. Yet this inequality is by no means always satisfied. Electrons obtained in beta decay, particles in cosmic rays, and particles in accelerators move with velocities close to that of light. Hence, it is necessary to obtain the laws of mechanics for these ultra-highspeed charged particles.

If we attempt to apply Newtonian mechanical laws to these particles we will encounter an insurmountable contradiction—the law of addition of velocities cannot be applied in electrodynamics in its usual form (see Secs. 8 and 10).

The equations of Newtonian mechanics are of the same form for all inertial systems moving uniformly relative to each other. In such systems there are no inertial forces. Naturally, under no circumstances can the principle of the equivalence of inertial systems be violated, otherwise we should have to assume that there existed a reference system at absolute rest. We must also consider that the equations of electrodynamics appear the same for all inertial systems in free space—in the form (12.24)-(12.27). It follows from these equations that the velocity of propagation of electromagnetic disturbances is equal to c and is the same for all directions in space. If it turned out that in some inertial systems the velocity of light depended upon the direction of its propagation, then these systems would not be equivalent to a system in which the velocity of light is the same in all directions. In this system, the electrodynamical equations would admit of a solution in the form of a spherical wave—a solution similar to the one that was obtained in the preceding section. In all other inertial systems, the velocity of light would depend on the direction of the wave normal.

An analogy would arise with the propagation of sound in air: the velocity of sound in a system at rest relative to the air does not depend on direction, but in a system moving relative to the air the velocity of sound is less in the direction of motion and more in the opposite direction as a consequence of the law of velocity addition.

So far, it has been considered that light is transmitted in an elastic medium, "the ether," and it has been regarded as self-evident that the velocity of light must be governed by the same law of velocity addition as the velocity of sound in air. Then a reference system fixed in the "ether" would have to be regarded as being at absolute rest, while all the remaining systems, as in absolute motion. In these systems the velocity of light would depend on its direction of propagation, in accordance with the law

$$c' = c + v\,, \tag{20.1}$$

where, for simplicity, only that direction is taken which coincides with the relative velocity of the system.

Michelson's experiment. A direct experiment was performed which showed that the velocity of light cannot be combined with any other velocity and, in all reference systems, it is equal to a universal constant c. This was the famous Michelson experiment (1887) which we shall describe in brief. A ray of light falls on a half-silvered mirror SS (Fig. 30), where it is split up: a part of the light is reflected and falls on mirror A while the other part is transmitted and falls on mirror B. Let SA be perpendicular to the motion of the earth and let SB be parallel to the earth's motion. The light reflected from the mirrors A and B returns to the plate SS; the ray BS is reflected from it and falls on screen C while the ray AS is transmitted to the screen directly. Thus, both rays are entirely equivalent as regards transmissions and reflections, though in the sections AS and BS the light is propagated differently relative to the earth's motion.

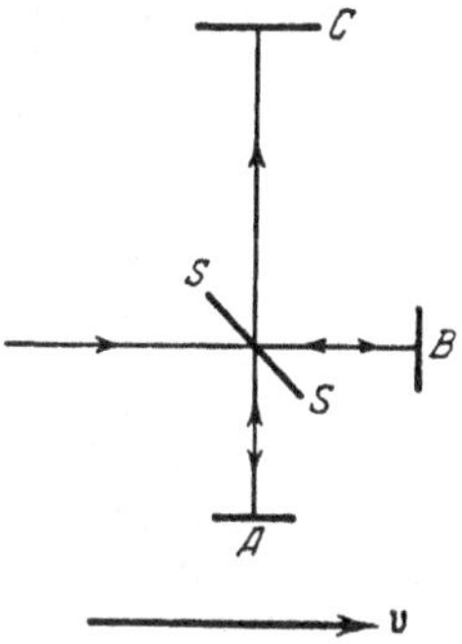

Fig. 30

Let us assume now that the velocity of light is combined with the motion of the earth according to the usual law of addition of velocities. Then, along the path SB, the velocity of light relative to the earth is equal to $c - V$, and $c + V$ along the return path, where V is the velocity of the earth. The time light takes to travel along the entire path SBS in both directions is

$$\frac{l}{c+V} + \frac{l}{c-V} = \frac{2lc}{c^2 - V^2} \cong \frac{2l}{c} + \frac{2lV^2}{c^3}\,,$$

where $l = SB$. We have used the fact that $V \ll c$. Along the section SA the velocity of light and the earth's velocity are perpendicular to each other (in a reference system fixed in the apparatus). If again the law of velocity addition holds, then the velocity of light relative to the apparatus, along the section SA, is equal to $\sqrt{c^2 - V^2}$ (c is the hypotenuse of the triangle, V and $\sqrt{c^2 - V^2}$ are the sides). The time taken by the light to travel along the whole path SAS, equal to $2\,l$, is

$$\frac{2l}{\sqrt{c^2 - V^2}} \cong \frac{2l}{c} + \frac{lV^2}{c^3}.$$

Thus, the difference in the passage times along the paths SBS and SAS is equal to $\frac{lV^2}{c^3}$. By means of repeated reflections, the path can be made sufficiently long (several tens of metres). Choosing it properly, we can arrange that the *proposed* time difference for the paths SAS and SBS is equal to a half-period of oscillation. The rays on the screen C should then mutually cancel. In order to be certain that the cancellation has occurred as a result of the combination of the velocity of light with the velocity of the earth, and not by accident, it is sufficient to rotate the apparatus through 45° so that the direction of the earth's velocity would be along the bisector of the angle ASB. Then the difference in time taken by the rays to travel along the paths SAS and SBS should at least become equal to zero; hence, if, in the previous position, the rays mutually cancelled as a result of the combination of the velocity of light with the earth's velocity then, in the new position, the rays would mutually reinforce each other. In other words, the interference bands on the screen would be displaced by half a wavelength.

Actually, no change in the path difference between the rays occurs when the apparatus is rotated, i.e., the expected effect is completely absent. An addition of the velocity of light and the earth's velocity does not occur.

The negative result of Michelson's experiment is completely understandable if we reject the postulate of an "ether." Nevertheless, at the time when the experiment was performed, no one as yet understood that electrodynamics does not require an "ether" in order to become as complete and clear a science as mechanics. The fact that the law of addition of velocities—a truism for physicists in the past—failed to hold, appeared as an inexplicable paradox.

In addition, Michelson's experiment apparently contradicted the phenomenon of astronomical abberation of light and Fizeau's experiment. (These will be considered later in this section in the light of the theory of relativity.)

Einstein's relativity principle. We shall not give the history of the painful attempts to explain this paradox but, instead, we will straightway present the correct solution to the problem as given by A. Einstein in 1905. It is known as the *special theory of relativity*. Despite the delusion widespread among laymen, this term by no means expresses the relativity of our physical knowledge. It expresses the mutual equivalence of all inertial systems moving relative to one another.

The equations of electrodynamics do not imply the presence of any elastic medium ("ether") for the transmission of electromagnetic disturbances. We have already discussed this in Sec. 12. The reality is the electric field itself. For this reason, the equations of electrodynamics are just as independent of the choice of inertial reference system as the equations of mechanics. Both sets of equations describe motion, i. e., the change of state with time directly. Mechanics describes the change of mass configuration, while electrodynamics describes changes of the electromagnetic field. The forms of the equations of motion cannot change as a result of the choice of inertial system.

This is why the result of Michelson's experiment does not contradict the notion of relativity of motion, but confirms it. Michelson's experiment shows that the velocity of light in free space is the same in all inertial systems. The velocity of propagation of interactions is a fundamental constant in the equations of electrodynamics. These equations are invariant to a transformation from one inertial system to another only when the velocity of propagation of the interactions in both systems is the same. And so the result of Michelson's experiment contradicts only the law of addition of velocities, i. e., Galilean transformations (8.1) and (8.2). This law of addition of velocities is confirmed experimentally only for relative velocities and for velocities of motion that are small compared with the velocity of light c. Obviously, it must be replaced by a more precise law for the region of high velocities. But this more precise law must also hold in mechanics for large particle velocities. This can be seen from the following reasoning.

Let the charges in a specific inertial system interact in some way with the electromagnetic field producing certain events (for example, collisions between the charges). They may be precalculated on the basis of the equations of mechanics and electrodynamics. In transforming to another inertial system, the equations of mechanics and electrodynamics must retain their form, otherwise, other consequences will follow from the transformed equations taken together; in particular, those events which were precalculated and occur in the first inertial system do not necessarily take place in another system. But events such as collisions, for example, are objective facts; they should be observed in all coordinate systems. Yet if we apply Galilean transformations (8.1) then the equations of Newtonian mechanics will not change, while the equations of electrodynamics will change, since the law of addition of velocities (10.12) is not applicable in electrodynamics. Therefore, we must find transformations to replace the Galilean transformations such as would leave both the equations of mechanics and the equations of electrodynamics invariant. But then it would become necessary to make the laws of Newtonian mechanics more precise, since they are correct only for low particle velocities.

A physical theory which cannot predict facts independently of the mode of their description is imperfect and contradictory. It is this that makes us reconsider the basic facts of mechanics, no matter how self-evident they seem to be in our everyday experience, which has to do with the motions of bodies at velocities that are small compared with that of light.

The Lorentz transformations. We look for transformations of a more general form than the Galilean transformations for passing from one inertial system to another. Like the Galilean transformations, they must satisfy certain requirements of a general nature. These requirements may be expressed as follows.

1) The transformation equations are symmetrical with respect to both systems. We shall denote the quantities that refer to one system by letters without primes (x, y, z, t), while those that refer to the other system will be primed (x', y', z', t'). We denote the velocity of the primed system with respect to the unprimed system by V. Then the mathematical form of the equations expressing unprimed quantities in terms of primed quantities (and the velocity V) is the same as that of the equations for the reverse transformation, if we change the sign of the velocity in them. This requirement is necessary for the equivalence of both systems.

2) The transformation must convert the finite points of one system to the finite points of the other, i.e., if (x, y, z, t) are finite, then a transformation with finite coefficients must leave (x', y', z', t') finite values.

Condition (1) greatly restricts the possible form of the transformations. For example, it can be seen that the transformation functions cannot be quadratic, because the inversion of a quadratic function leads to irrationality, just as that of the function of any degree other than the first. A linear-fractional transformation (i.e., the quotient of two linear expressions)—under certain limitations imposed on the coefficients—may be inverted retaining the same form. For example, for one variable the direct and the inverse linear-fractional functions look like

$$x' = \frac{ax + b}{ex + f}, \quad x = \frac{b - fx'}{ex' - a}.$$

But this function does not satisfy condition (2): if $x' = a/e$, x becomes infinite. Therefore, a linear function is the only possible one.

3) When the relative velocity of two systems tends to zero, the transformation equations yield an identity $(x' = x, y' = y, z' = z, t' = t)$.

4) A law of the addition of velocities is obtained from the transformation equations such that it leaves the velocity of light in free space invariant: $c' = c$.

Summarizing, we can say that the transformation equations: 1) maintain their form when inverted, 2) are linear, 3) become identities for small relative velocities, 4) leave the velocity of light in free space unchanged.

These four conditions are sufficient. The required equations can be obtained most simply if one of the coordinate axes (for example, the x-axis) is taken in the direction of the relative velocity. Then the other axes will not be affected by the transformation.

We return to Fig. 11 (page 68), but we will not make the arbitrary assumption that $t = t'$ (experiment supports this only for small relative velocities of both systems). Let us see what results from the conditions (1)-(4). If the velocity is along the x-axis, then, as has just been found, $y' = y$, $z' = z$. This can be seen simply from Fig. 11. In the most general form, linear transformations of x and t, appear thus:

$$x' = \alpha x + \beta t\,, \tag{20.2}$$

$$t' = \gamma x + \delta t\,. \tag{20.3}$$

The constant terms need not be written in these equations; they can be included in the definition of x or x' through choice of the coordinate origin.

Let us apply equation (20.2) to the origin of the primed system, $x' = 0$. This point moves with velocity V relative to the unprimed system. Hence, $x = Vt$. Substituting $x' = 0$, $x = Vt$ in (20.2), we obtain, after eliminating t,

$$\alpha V + \beta = 0\,. \tag{20.4}$$

We shall solve equations (20.2) and (20.3) with respect to x and t. Elementary algebraic computations give

$$x = \frac{\delta x' - \beta t'}{\alpha\delta - \beta\gamma}\,, \tag{20.5}$$

$$t = \frac{\gamma x' - \alpha t'}{\beta\gamma - \alpha\delta}\,. \tag{20.6}$$

Let us now apply condition (1). For this we note that the coefficients β and γ, which interrelate the coordinate and time, must change sign together with the velocity V. Otherwise, if the x and x' axes are turned in the opposite direction the equations will not preserve their form, and this is impermissible. Thus, the equations for the inverse transformation from unprimed quantities to primed have the same form as (20.2) and (20.3):

$$x = \alpha x' - \beta t'\,, \tag{20.7}$$

$$t = -\gamma x' + \delta t'\,. \tag{20.8}$$

Comparing (20.7) and (20.5), we obtain

$$\alpha = \frac{\delta}{\alpha\delta - \beta\gamma}\,, \tag{20.9}$$

$$-\beta = \frac{-\beta}{\alpha\delta - \beta\gamma}\,. \tag{20.10}$$

From (20.10) it follows that

$$\alpha\delta - \beta\gamma = 1 . \tag{20.11}$$

Then, from (20.9), we obtain

$$\alpha - \delta . \tag{20.12}$$

No other relationships are obtained from the comparison of the direct equations with the inverse equations.

We now use condition (4). We divide equation (20.2) by (20.3):

$$\frac{x'}{t'} = \frac{\alpha \dfrac{x}{t} + \beta}{\gamma \dfrac{x}{t} + \delta} . \tag{20.13}$$

Let x be a point occupied by a light signal emitted from the origin of the unprimed system at an initial instant of time $t=0$. Obviously, $\frac{x}{t}=c$. But in accordance with condition (4), $\frac{x'}{t'}=c$. Hence,

$$c = \frac{\alpha c + \beta}{\gamma c + \delta} . \tag{20.14}$$

We substitute the relations (20.4) and (20.12) into (20.14) in order to eliminate β and δ. There remains a relation between α and γ:

$$\gamma c^2 + \alpha c = \alpha c - \alpha V ,$$

whence

$$\gamma = -\alpha \frac{V}{c^2} . \tag{20.15}$$

We now substitute (20.15), (20.4) and (20.12) into (20.11) and obtain an equation for α:

$$\alpha^2 \left(1 - \frac{V^2}{c^2}\right) = 1 . \tag{20.16}$$

In extracting the square root, we must take the positive sign in accordance with condition (3), because then (20.3) becomes $t'=t$ for a small relative velocity. A minus sign would yield $t'=-t$, which is meaningless.

Now expressing all the coefficients α, β, γ, and δ in accordance with equations (20.16), (20.4), (20.15), and (20.12), respectively, and substituting (20.3) into (20.2), we arrive at the required transformations:

$$x' = \frac{x - Vt}{\sqrt{1 - \dfrac{V^2}{c^2}}} , \tag{20.17}$$

$$t' = \frac{t - \dfrac{Vx}{c^2}}{\sqrt{1 - \dfrac{V^2}{c^2}}} . \tag{20.18}$$

These equations are called Lorentz transformations. From (20.7) and (20.8), the inverse transformations are of the form

$$x = \frac{x' + Vt'}{\sqrt{1 - \frac{V^2}{c^2}}}, \tag{20.19}$$

$$t = \frac{t' + \frac{Vx'}{c^2}}{\sqrt{1 - \frac{V^2}{c^2}}}. \tag{20.20}$$

In order to explain the meaning of these equations we shall apply them to some special cases. Let a clock be situated at the origin $x' = 0$ of the primed system. It indicates a time t'. Then, from equation (20.20), it follows that

$$t = \frac{t'}{\sqrt{1 - \frac{V^2}{c^2}}}. \tag{20.21}$$

The clock which is at rest relative to its reference system we call the *observer's clock*. It can be seen from (20.21) that one observer, comparing his clock with that of another observer, will always observe that the latter clock is slow, i.e., that $t' \ll t$. If a clock is situated at the origin of the unprimed system (i.e., at the point $x = 0$), the transformation equation to the primed system is of the same form, since, from (20.18), we now obtain $t' = \frac{t}{\sqrt{1 - \frac{V^2}{c^2}}}$. This not only does not contradict (20.21), but expresses that very fact: a clock moving relative to an observer is slow compared with his own clock.

In the theory of relativity, a single universal time does not exist as in Newtonian mechanics. It is better to say that the absolute time of Newtonian mechanics is, in actual fact, an approximation, correct only for small relative velocities between clocks. The absoluteness of Newtonian time has sometimes given cause to regard it as an *a priori*, logical category independent of moving matter.

At any rate, Newton, by accepting instantaneous action at a distance, naturally had to consider time as universal; if we formally put $c = \infty$ in (20.18), we obtain $t' = t$. The instantaneous transmission of signals would allow us to synchronize clocks in all inertial systems independently of their relative velocities. In Newtonian mechanics, gravitational forces played the part of such instantaneous signals.

It is sometimes thought that, knowing the velocity of light c, we can introduce a correction into the readings of clocks in different inertial systems such that the rate of time will everywhere be the same. But it is precisely equation (20.21) that describes the relative passage of time in both reference systems *after* a correction has been introduced for the finite time of propagation of light. Time reduction, as has already been shown, is completely reciprocal. Consequently, it can in no way be accounted for by any change, resulting from motion, in the properties of clocks. The time reduction effect is purely kinematical.

It must also be added that in speaking about clocks we by no means necessarily have in mind clocks which have been made by human hands; any natural periodic process that gives a natural time scale will do as well, for example, the oscillations in a light wave. It is clear that the physical properties of a radiating atom cannot in the least depend upon the inertial system in which the atom is described. This is what gives us the right to assert that equation (20.21) refers to similar clocks.

At the same time we must remember that it is impossible to define time without relation to some periodic process, i.e., irrespective of motion.

Relativity and objectivity. The relativity of time by no means indicates a rejection of the objectivity of its measurement in any given system of reference. It is entirely of no consequence what observer is observing the clock. The relative character of time in inertial systems is the only thing that counts. We have all long since become accustomed to the relativity of the datum line of time measurement related to time zones (i.e., to the sphericity of the earth). The theory of relativity teaches us that the time scale is also relative.

The fact that an objective concept may be relative can be seen from the following example. In the Middle Ages it was thought that even direction in space was absolute, and it was then thought impossible to imagine that the earth was spherical since it would then follow that our antipodes would have to walk upside down! The concept of "up" or "down" was related not with the direction of a plumb-line at a given point on the globe, but with certain other categories characteristic of the ideology of the Middle Ages. The vertical directions in Moscow or Vladivostok form a substantial angle between each other, but nobody nowadays would think of arguing about which of them is the more "vertical." The concept of verticality is completely objective at every point on the globe, but is relative for different points. In the same way, time is *objective* in each inertial system, but is *relative* between them.

Contraction of the length scale. We shall now consider the question of the measurement of length. In order to find out the length of a moving body ("its scale"), we must simultaneously plot the coördinates of its ends in a fixed system. Obviously, a fixed observer has no fundamentally different means of measurement, for, otherwise, he would have to stop the motion of the scale (i.e., transfer it to his reference system). If the ends of the scale are fixed by a stationary observer at one time* we must put $t = 0$. From (20.17), there follows an expression for the length of a moving scale $\Delta x'$ measured by a fixed observer:

$$\Delta x' = \frac{\Delta x}{\sqrt{1 - \frac{V^2}{c^2}}}. \tag{20.22}$$

Like (20.21), this equation has a symmetrical inversion. If a "moving" observer measures a "stationary" scale, we must put $t' = 0$; it then turns out that $\Delta x = \frac{\Delta x'}{\sqrt{1 - \frac{V^2}{c^2}}}$. We conclude from (20.22) that a moving scale is shortened relative to a stationary observer. The contraction occurs in the direction of motion.

Lorentz supposed that this scale compression does not appear to both inertial systems, but, for some unknown reason, occurs when the scale moves relative to the "ether." Lorentz and others attempted in this way to explain the negative result of Michelson's experiment. Yet the very symmetry of the direct and inverse Lorentz transformations (20.17)-(20.20) (they were known before the advent of the theory of relativity), from which the contraction of length follows only as a special case, shows convincingly that there is no system at absolute rest relative to an "ether." It may be noted that by the beginning of the twentieth century, the "ether," which had been introduced by Huygens as a medium that transmitted light oscillations, remained in physics simply as a rudimentary concept. The discovery and confirmation of the electromagnetic nature of light made the hypothetical elastic medium quite superfluous (see Sec. 12). Only the theory of relativity disclosed the real meaning of the Lorentz transformations. But then, many concepts regarded as absolute in Newtonian mechanics, turned out to be related to the motion of inertial systems.

The formula for addition of velocities. We shall now find an equation for the addition of velocities arising from the Lorentz transformations. Differentiating (20.17) and (20.18) and dividing one by the other, we obtain

$$\frac{dx'}{dt'} \equiv v_x' = \frac{\frac{dx}{dt} - V}{1 - \frac{V}{c^2}\frac{dx}{dt}} \equiv \frac{v_x - V}{1 - \frac{V v_x}{c^2}}. \tag{20.23}$$

* The idea of the *simultaneity* of two operations performed in the same coordinate system may be uniquely defined with the aid of light signals. Indeed, observers at rest relative to each other at a given distance can always check their time with the aid of light signals by introducing a constant correction for the known time of propagation.

Noting that $dy' = dy$ and $dz' = dz$, we have a transformation of the velocity components perpendicular to V:

$$\frac{dy'}{dt'} \equiv v'_y = \frac{\frac{dy}{dt}\sqrt{1-\frac{V^2}{c^2}}}{1-\frac{V}{c^2}\frac{dx}{dt}} \equiv \frac{v_y\sqrt{1-\frac{V^2}{c^2}}}{1-\frac{Vv_x}{c^2}};\ v'_z = \frac{v_z\sqrt{1-\frac{V^2}{c^2}}}{1-\frac{Vv_x}{c^2}}. \quad (20.24)$$

For small velocities, (20.23) and (20.24) become the ordinary equations for addition of velocities. This can be seen if we let c tend to infinity, i.e., by putting $\frac{V}{c} = 0$.

It is easy to see that if $v = \sqrt{v_x^2 + v_y^2 + v_z^2} = c$, then likewise $v' = c$, i.e., the absolute value of the velocity of light does not change in passing from one inertial system to another. But the separate components of the velocity of light, which are less than c, may of course change; the direction of a light ray relative to different observers differs, since there is no absolute direction in space.

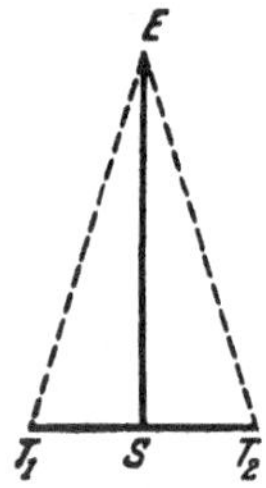

Fig. 31

Abberation of light. In this connection let us consider the phenomenon of the aberration of light. Astronomical aberration, or the deflection of light, consists in the fact that stars describe ellipses in the sky in the course of a year. Their origin is easy to explain: the velocity of the earth, in annual motion, combines differently with the velocity of the light emitted by the star (Fig. 31). If the velocity vector of the starlight relative to the sun is ES then the resultant direction of the velocity, for one position of the earth, is ET_1 and, in half a year's time, ET_2. These directions are projected on different points of the celestial sphere so that in the course of a year a star describes a closed ellipse. In angular units, the semi major axis of the ellipse is always equal to $\frac{V}{c}$ where V is the velocity of the earth. $\frac{V}{c} = \frac{1}{10,000} = 20''.25$.

We may ask the question: Why does not the velocity of light in Michelson's experiment combine with the earth's velocity but remains equal to c, while the phenomenon of aberration shows that velocities combine (we note that Michelson's experiment was also performed with an extra-terrestrial source of light). The explanation is that in Michelson's experiment it was the absolute value of the velocity of light c that was measured (from the path difference of the rays), while in the aberration of light there is a change in the *direction* of the velocity of light as a result of the combination of its components with the velocity of the earth. Considering that the velocity of light relative to the sun is perpendicular to the plane of the earth's orbit, we must put $v_x=0$, $v_y=c$, $v_z=0$ into (20.23) and (20.24). Then the components of the velocity of light relative to the earth are

$$v'_x = -V, \quad v'_y = c\sqrt{1-\frac{V^2}{c^2}}.$$

And, in accordance with Michelson's experiment, $v_x'^2+v_y'^2=c^2$. The direction of the projection of the velocity of light onto the plane of the earth's orbit (ecliptic) is reversed in the course of half a year, which is the reason why aberration occurs.

Similar equations are obtained in the more complicated case when the rays from the star are not perpendicular to the plane of the ecliptic. They coincide with the equations that follow from a simple addition of velocities if terms of the order $\frac{V^2}{c^2}$ are neglected.

Before the theory of relativity was put forward, it was wrongly supposed that the aberration of light contradicted Michelson's experiment.

Fizeau's experiment. Fizeau's experiment, which determined the velocity of light in a moving medium, was also believed to contradict Michelson's experiment. Fizeau's method was this. A beam of light was divided into two parts using a half-silvered mirror (Fig. 32). These beams were passed through tubes with flowing water; one beam in the direction of flow and the other in the opposite direction. For comparison, the same beams were passed through tubes in which the water was at rest. By subsequent reflections the beams once again combined and cancelled each other when the path difference between them was equal to an integral number of half wavelengths (i.e., when they were in opposite phase). Coherence between them was obtained due to the fact that they both came from the same source. In still water, the path difference was chosen so that the rays were reinforced, i.e., the phase difference was equal to an even number of half wavelengths. The path difference in flowing water was varied. Since the frequency of the light and the tube lengths remained unchanged, the change in path length indicated a change in the velocity of light relative to the tubes.

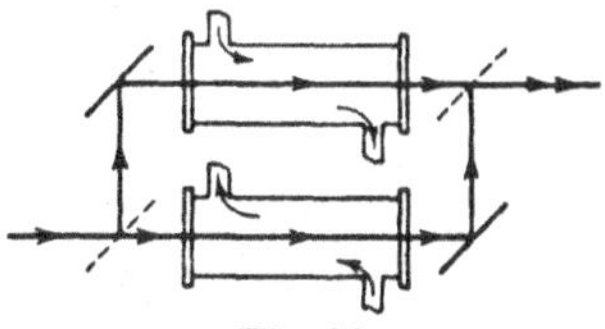

Fig. 32

First of all, we note that the result of Fizeau's experiment in no way contradicts the general ideas about the relativity of motion. A reference system fixed in flowing water is not equivalent to a system fixed in the tube, if we are studying the propagation of light in water.

Since the velocity of light in water is equal to $\frac{c}{\nu}$, where ν is the refractive index of the water, the general equation for the addition of velocities (20.23) shows that $\frac{c}{\nu}$ does not remain a constant quantity when passing to another coordinate system. At the same time, we cannot use the simple velocity-addition equation, because the denominator of equation (20.23) differs from unity by $\frac{V}{\nu c}$ (V is the velocity of the water). Considering that $V \ll c$ and expanding the denominator in a series up to the linear term inclusive, we find the change in the velocity of light in moving water (see exercise 1):

$$u = \frac{c}{\nu} \pm V\left(1 - \frac{1}{\nu^2}\right).$$

It was precisely this value, which differs from that given by the simple velocity-addition law, that was obtained by Fizeau. Since Michelson measured c and Fizeau measured $\frac{c}{\nu}$, there is no contradiction between them.

Interval. Despite the fact that x and t are changed separately by the Lorentz transformations we can construct a quantity which remains invariant (unchanged). It is easy to verify that this property is possessed by the difference $c^2t^2 - x^2$. Indeed,

$$c^2t^2 = \frac{c^2t'^2 + \frac{V^2x'^2}{c^2} + 2Vx't'}{1 - \frac{V^2}{c^2}},$$

$$x^2 = \frac{x'^2 + V^2t'^2 + 2Vx't'}{1 - \frac{V^2}{c^2}},$$

or

$$c^2t^2 - x^2 = c^2t'^2 - x'^2 \equiv s^2. \qquad (20.25)$$

The quantity s is called the *interval* between two events; that which occurred at the coordinate origin $x = 0$ and initial time $t = 0$, and another event that occurred at the point x and time t.

The word "event" may also be regarded in its most common everyday sense provided that its coordinates and time may be defined. If the first event is not related to the origin of the coordinate system and the initial instant, then

$$s^2 = c^2(t_2 - t_1)^2 - (x_2 - x_1)^2 = c^2(t_2' - t_1')^2 - (x_2' - x_1')^2. \qquad (20.26)$$

Considerable importance is attached to the interval between two infinitely close events:

$$ds^2 = c^2dt^2 - dx^2 . \tag{20.27}$$

It is not at all necessary to consider that both events occurred on the abscissa. Since $dy' = dy$ and $dz' = dz$, the interval is always invariant:

$$ds^2 = c^2dt^2 - dx^2 - dy^2 - dz^2 = c^2dt^2 - dl^2 = c^2dt'^2 - dl'^2 . \tag{20.28}$$

The interval, written in the form (20.28), is not related to any definite direction of velocity.

Space and time intervals. The interval provides for a very vivid way of studying various possible space-time relationships between two events. Let the spatial distance between the points at which the events occurred be taken along the abscissa, and the interval of time between them, along the ordinate axis (Fig. 33). To begin with, let $ct > l$, for example,

$$s^2 = c^2t^2 - l^2 \tag{20.29}$$

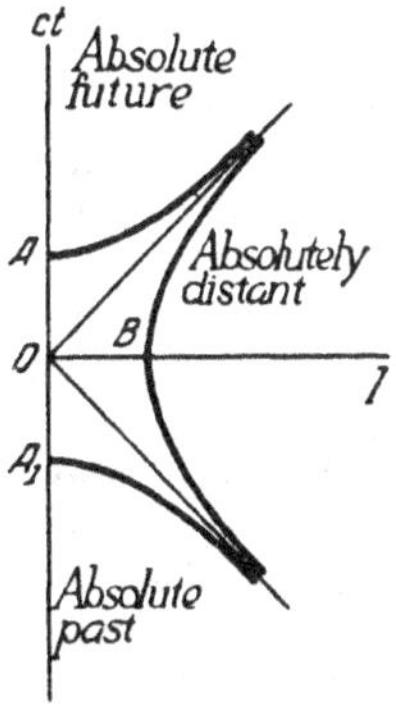

Fig. 33

We shall plot the values of ct and l corresponding to two definite events measured in quite different inertial systems. No matter what values are obtained as a result of these measurements of ct and l, the interval s (20.29) between the events is the same. It follows that the locus of the points, for all possible spatial distances l and time intervals ct, is an equilateral hyperbola $s^2 = c^2t^2 - l^2$. Two branches of the hyperbola are possible: one lies in the *past* relative to the event that occurred at $t = 0$, $x = 0$, while the other is completely in the *future*. It is easy to see that such a relationship *inevitably* results if the events are causally related. Let the events, in some reference system, be known to occur in the same place, for example, sowing and reaping. To this system there corresponds the point O (sowing) and the point A (reaping). But since all points of the given branch of the hyperbola lie at $t > 0$, the sowing in any reference system must occur earlier than reaping.

We can also proceed from causally related events which in our coordinate system do not occur at one point of space, such as firing and hitting the target (they occur at one point in a system fixed in the bullet). To the system fixed in the bullet there now corresponds a vertical section OA in Fig. 33, while to our system there corresponds some inclined line drawn from the origin to a point on the same upper hyperbola. Thus, here too, the second event—that of hitting the target—occurs after the shot in any reference system.

If the velocity of the bullet (or any material particle) is v, then $s^2 = c^2t^2 - l^2 > 0$. It necessarily follows from the inequality (20.29) that $l = vt < ct$ or $v < c$, if a reference system exists in which both events occurred at a single point in space. For this reason, the velocity of any material particle can only be less than the velocity of light c.

The region above the first asymptote is called "*absolute future*" relative to the initial event.

Although the example of a bullet hitting a target may appear to be a special case, in actual fact the foregoing reasoning may be used for all cases when an effect is related to a cause by some material-transfer process, either in the form of a particle or in the form of a wave packet. A reference system may be related to moving matter, and therefore any velocity of material transport satisfies the inequality $v \leqslant c$.

Consequently, the theory of relativity never contradicts the objective nature of causality. And it is precisely the sequence of cause and effect which determines the direction of time.

We can also consider other pairs of events. For example, let one event occur on the sun and the other, five minutes later, on the earth. Light travels from the sun to the earth in eight minutes; for the two given events $ct < l$, $s^2 = c^2t^2 - l^2 < 0$. For purely physical reasons, such events can in no way be related causally because the velocity of transport of matter does not exceed c. For such events, the interval term l^2 is not equal to zero in any system of reference. And so there is no coordinate system, relative to which events related by an imaginary ($s^2 < 0$) interval occur at a single point of space. Yet, on the other hand, their time sequence has not been defined: coordinate systems exist in which the first event occurs before the second, and there are systems in which the second event occurs before the first.* Thus, the theory of relativity denies the absolute nature of the simultaneity of two events occurring at different points in space and separated by an imaginary interval. O and B in Fig. 33 are such events. B lies on the hyperbola which, relative to O, belongs partly to the future and partly to the past. But O and B can in no way be related causally, since no interaction can arrive at B from O instantaneously.

Hence, the relativity of simultaneity does not contradict the absolute nature of causality.

The region between the asymptotes is called absolutely distant, relative to the coordinate origin.

The light cone. The asymptotes to the hyperbola $l = ct$ are of special interest. For them, $s = 0$.

---

* For example, the second event occurred earlier than the first relative to any system moving in the direction from the earth to the sun with a velocity exceeding $\frac{5}{8}$ c, as may easily be seen from (21.20).

The relationship $l=ct$ holds for two events related by an electromagnetic signal, for example, for the emission and absorption of a radio signal. For these two events, $s=0$ in all reference systems, because the velocity of light is invariant and we must always have that $l=ct$. Since the curve in Fig. 33 in actual fact corresponds to $3+1$ dimensions (three spatial and one time) instead of a plane, the locus of zero intervals is picturesquely called "a light cone."

Proper time. The concept of *proper time* of a particle is closely related to that of interval. This is the time measured in a coordinate system fixed to a particle. The displacement of the particle relative to this system is equal to zero by definition. Hence, the proper time that has elapsed between any two positions of the particle is proportional to the interval calculated for these two positions:

$$t_0=\frac{ds}{c}=\sqrt{dt^2-\frac{dl^2}{c^2}}=dt\sqrt{1-\frac{v^2}{c^2}}\,. \tag{20.30}$$

Here, v is the velocity of the particle relative to an arbitrarily chosen reference system in which the interval of time is equal to dt. From (20.30) and also (20.21), the proper time is always the shortest. For finite time intervals

$$t_0=\int dt\sqrt{1-\frac{v^2}{c^2}}\leqslant\int dt\,, \tag{20.31}$$

i.e.,

$$t_0\leqslant t\,. \tag{20.32}$$

From (20.32) there follows a consequence with which, at first sight, it is difficult to agree. Characteristic time is the time which determines the rhythm of life processes in the human organism. And for this reason if an imaginary traveller leaves the earth with a velocity close to that of light, and later returns to the earth, then, in accordance with (20.32), he will have grown less old than a person, initially of the same age, remaining on the earth.

The asymmetry between the traveller and person that remained on the earth is explained by the fact that the traveller was not moving inertially—he first travelled away and then returned. For this, it was necessary for him to turn about in some way, i.e., to lose the property of inertiality retained by observers on the earth. It may be noted that the time spent in turning may make up an infinitesimally small amount of the time of travel, if the journey itself is sufficiently long. This is why the turning operation cannot in any way re-establish equality between t and t_0. But this operation is necessary in order that the comparison between the ages of both observers can be performed, i.e., to return them to the same point of space and to a single coordinate system. Thus, disregarding the "striking" formulation of the traveller experiment, we may say that time in a noninertial system may differ to any extent from the time in an inertial system, even though the noninertial system may deviate from inertiality for an extremely short time.

The case of the human traveller is, of course, purely imaginary, technically speaking. But a relationship of the type (20.32) is observed in the decay of mesons in cosmic rays. The mean life of a positive π-meson, of mass 273 electronic masses, in decaying to a μ-meson, of mass 207 electronic masses, together with a neutral particle, is 2×10^{-8} sec (the negative π-meson is most often captured by nuclei). This time is measured for a π-meson stopped in the substance, i.e., it is proper time. The velocity of the meson, like that of any other particle, does not exceed c. If the relationship (20.32), expressing the relativity of time, did not exist, then a rapid π-meson with a velocity of the order of c would on the average travel through $c \times 2 \times 10^{-8}$ cm $= 600$ cm of air. Actually, the mean path of a π-meson is considerably greater due to the fact that its life-time, in a coordinate system fixed in the air, is considerably greater than its proper life-time.

Frequency and wave-vector transformations of electromagnetic waves. Proceeding from invariants, we may find out in what way the noninvariant quantities involved are transformed in passing from one reference system to another. We shall now show that the transformation properties of wave-vector components and frequency are those of coordinates and time.

In order to prove this, it is sufficient to note that the phase of a wave is invariant. Indeed, the phase characterizes some event, for example, the fading of the electric and magnetic fields at a certain instant of time and at a certain point in space. If we examine this wave in another coordinate system, then the coordinates and time (corresponding to this event) will have other values, though the event itself cannot change, of course. This is easy to understand if we imagine that the electric and magnetic fields are measured by the readings of some inertialess device. Two such devices, situated at the same point of space at a certain instant, but in motion relative to each other, must together indicate the zero value of the field. Otherwise, the coordinate system in which the electromagnetic field is equal to zero will in some way be distinguished from the others.

From (17.21) and (17.22), the expression for the phase of a wave is

$$\psi = x k_x + y k_y + z k_z - \omega t .$$

This quantity must be invariant in transforming to another coordinate system. Let us express x' and t' in accordance with (20.17) and (20.18) and substitute them into the condition of invariance of phase:

$$x k_x + y k_y + z k_z - \omega t = x' k'_x + y' k_y' + z' k_z' - \omega' t' =$$

$$= \frac{x - Vt}{\sqrt{1 - \frac{v^2}{c^2}}} k_x' + y' k_y' + z k_z' - \frac{t - \frac{Vx}{c^2}}{\sqrt{1 - \frac{v^2}{c^2}}} \omega'.$$

Now comparing the coefficients of x, y, z, and t, we obtain the transformation equations for the wave-vector components and frequency:

$$k_x = \frac{k_x' + \frac{\omega' V}{c^2}}{\sqrt{1 - \frac{V^2}{c^2}}}, \tag{20.33}$$

$$k_y = k_y', \tag{20.34}$$

$$k_z = k_z', \tag{20.35}$$

$$\omega = \frac{\omega' + k_x' V}{\sqrt{1 - \frac{V^2}{c^2}}}. \tag{20.36}$$

These are entirely analogous to equations (20.19) and (20.20).

The longitudinal and transverse Doppler effect. If the frequency of a source of light with respect to its own coordinate system is equal to ω', and the angle between its velocity and the line of sight is equal to ϑ', so that $k_x' = k' \cos \vartheta' = \frac{\omega'}{c} \cos \vartheta'$, then we obtain from (20.36)

$$\omega = \frac{\omega' \left(1 + \frac{V}{c} \cos \vartheta'\right)}{\sqrt{1 - \frac{V^2}{c^2}}}. \tag{20.37}$$

In particular, if the source moves along the line of sight (i.e., towards the observer),

$$\omega = \frac{\omega' \left(1 - \frac{V}{c}\right)}{\sqrt{1 - \frac{V^2}{c^2}}}. \tag{20.38}$$

These equations describe the well-known Doppler effect, by means of which the radial velocities of stars are measured. The square root in the denominator gives a correction introduced by the theory of relativity into the formula usually used.

If the velocity of the source is perpendicular to the ray, then, from the requirement that $k_x = 0$, we also obtain a change in frequency, although it is of second order with respect to $\frac{V}{c}$:

$$\omega' = \frac{\omega}{\sqrt{1 - \frac{V^2}{c^2}}}. \tag{20.39}$$

This transverse effect was observed by Ives in the radiation from moving ions (in canal rays) when the ratio $\frac{V}{c}$ was sufficient to detect the frequency displacement spectroscopically. This gives direct experimental proof of the contraction of the time scale in relative motion.

A comparison of inertial forces and the force of gravitation. Let us now investigate the transformation from inertial to noninertial systems. We define the latter as a system in which there are inertial forces.

All inertial forces have the common property that they are proportional to the mass of the body. Among the interaction forces, only one force is known which possesses that property—this is the force of Newtonian gravitation. The fact that gravitational force is proportional to the mass of the body is very well known, though very surprising all the same. The mass of a body may be defined from Newton's Second Law when any kind of force (electric, magnetic, elastic, etc.) acts on the body. It is therefore very difficult to understand why the force of interaction between bodies, namely, the force of gravitation, is proportional to that very mass involved in the expression for Newton's law [see (2.1)]. As is well known, all other interaction forces are independent of mass.

In addition, the very form of the gravitation law itself somewhat contradicts our physical intuition—in accordance with this law, gravitational forces are transmitted over any distance instantaneously.

Einstein called attention to the profound significance of the analogy between inertial forces and the force of gravitation. In certain cases these forces are indistinguishable in their action. For example, when an aeroplane performs a turn and, in doing so, inclines the plane of its wings, the passengers feel, as before, that the direction of gravity acting on them is perpendicular to the floor of the cabin. In this case, a resultant force consisting of gravity and a centrifugal force operates like the force of gravity—they are both proportional to the mass and act on all the bodies inside the aircraft in the same way. It is physically impossible to separate these two forces without considering objects outside the aircraft.

When a lift begins to rise the gravitational force is, as it were, increased—the force of inertia due to the acceleration of the lift is added to it.

In these examples (almost trivial), the inertial and gravitational forces are equivalent "on a small scale," that is, in certain small regions of space. In large regions, there is a certain essential difference between the behaviour of inertial forces and gravity. The latter diminishes with the distance from the centre of attraction, while inertial forces either remain constant or increase without limit. Thus, centrifugal force increases in proportion to the distance from the axis of rotation. The force of inertia in a coordinate system fixed in an accelerating lift is the same at any distance away from the lift.

The general theory of relativity. The basic idea of Einstein's gravitational theory is that motion in a gravitational field is the same sort of inertial motion as the accelerating of passengers relative to a braking carriage. It is precisely for this reason that acceleration due to the action of gravity does not depend on the mass of the body. In order to understand why the force of gravitation, as opposed to the well-known inertial forces, becomes zero at an infinite distance away from attracting bodies, we must assume that the space close to the attracting bodies does not have the geometrical properties of Euclidean space. In other words, we must take it that space and time obey non-Euclidean geometrical laws in the sense of the ideas first developed by Lobachevsky and later by Riemann. Free motion in such non-Euclidean (Riemannian) space is curvilinear. However, since it is precisely the properties of space itself that determine the curvature, acceleration of bodies does not depend upon their masses (if we can neglect the effect of the latter on the gravitational field) in the same sense as the field of a falling stone does not affect the gravitational field of the earth.

Thus, gravitational and inertial forces are indistinguishable in small regions of space. In such regions, a noninertial coordinate system is equivalent to an inertial system in which an additional gravitational field is operative, with the same acceleration of falling bodies which, in the noninertial system, is ascribed to inertial forces. For this reason, this theory of gravitation is also called the general theory of relativity in contrast to the special theory of relativity, which considers only inertial systems.

Since the equations of motion in the general theory of relativity (in the same way as all equations of motion) are formulated in differential form, equivalence on a small scale is quite sufficient for writing down the equations.

However, we must remember that a rotating coordinate system is not, as a whole, equivalent to a gravitational field. Indeed, a rotating system can, generally, only be determined for distances from the axis of rotation for which the velocity of rotation is less than that of light. This is why a rotating system is not equivalent to a nonrotating system, which has meaning in infinite space, too.

The mechanics of Einstein's general theory of relativity is considerably more complicated than Newtonian mechanics, which is included in this theory as a limiting case. But Einstein's theory is free from the gnosiological concepts, so alien to us, of the hypothesis of action at a distance. The properties of space and time in Einstein's theory are studied in inseparable unity with the motion of matter, and not only as a requisite for the motion of matter. Abstract space and time, which in Newtonian physics were sometimes regarded as almost belonging to logical, *a priori* categories, do not exist in Einstein's gravitational theory—in the general theory of relativity, space and time are endowed with physical properties.

The consequences of Einstein's gravitational theory. Einstein's refined gravitational theory leads to a series of results that may be verified by astronomical observation.

1) The perihelion of Mercury should rotate through 43″ per century. This is in excellent agreement with astronomical facts.

2) Rays of light from stars passing near the limb of the sun should be displaced towards it, since light is not propagated rectilinearly in non-Euclidean space. This result also agrees closely with accurate observations made during solar eclipses.

3) Spectral lines in heavy stars should be shifted towards the red end of the spectrum, and this, too, is found to be the case.

For the first time in the history of science, the general theory of relativity made it possible to pose the *cosmological* problem, i.e., the problem of the structure and development of the Universe. The present state of the cosmological problem is far from a solution due to insufficient astronomical data and to the mathematical difficulties associated with Einstein's gravitational equations. It should be noted that before the general theory of relativity, the cosmological problem was posed in a purely speculative way; Einstein's theory indicated the path for scientific investigation and has led to a series of important results.

Exercises

1) Calculate the change in the velocity of light propagated through flowing water in Fizeau's experiment.

$$u \pm = \frac{\frac{c}{\nu} \pm V}{1 \pm \frac{V}{\nu c}} \cong \left(\frac{c}{\nu} \pm V\right)\left(1 \mp \frac{V}{\nu c}\right) \cong \frac{c}{\nu} \pm V\left(1 - \frac{1}{\nu^2}\right).$$

Disregarding the theory of relativity, the result would be $u \pm = \frac{c}{\nu} \pm V$.

2) Obtain a precise equation for the aberration of light, with an arbitrary inclination ϑ of the ray of the ecliptic.

$$\text{Answer: } \cos\vartheta' = \frac{\cos\vartheta - \frac{V}{c}}{1 - \frac{V}{c}\cos\vartheta}.$$

3) Write down the equations for the Lorentz transformations for an arbitrary direction of the velocity **V** relative to a coordinate system.

In our equations $x = \frac{\mathbf{rV}}{V}$, $x' = \frac{\mathbf{r'V}}{V}$. The component perpendicular to the velocity is

$$\mathbf{r} - \frac{\mathbf{V}\,(\mathbf{rV})}{V^2} = \mathbf{r}' - \frac{\mathbf{V}\,(\mathbf{r'V})}{V^2}.$$

From (20.17)

$$\frac{\mathbf{r'V}}{V} = \frac{\frac{\mathbf{rV}}{V} - Vt}{\sqrt{1 - \frac{V^2}{c^2}}}.$$

Multiplying this equation by $\frac{\mathbf{V}}{V}$ and adding equations, we obtain

$$\mathbf{r}' = \mathbf{r} - \frac{\mathbf{V}\,(\mathbf{r}\,\mathbf{V})}{V^2} + \left(\frac{\mathbf{V}\,(\mathbf{r}\,\mathbf{V})}{V^2} - \mathbf{V}\,t\right)\left(1 - \frac{V^2}{c^2}\right)^{-1/2}.$$

4) Write down the Lorentz-transformation equations for the components of acceleration.

5) Show that the "four-dimensional volume element" $dx\,dy\,dz\,dt$ is invariant with respect to a Lorentz transformation.

$$dx = dx'\left(1 - \frac{V^2}{c^2}\right)^{1/2} \text{ from (20.22) and } dt = dt'\left(1 - \frac{V^2}{c^2}\right)^{-1/2}$$

from (20.21), whence the statement follows.

6) A light beam is within a solid-angle element $d\Omega$. Show that Lorentz transformations leave the quantity $\omega^2\,d\Omega$ invariant.

Use the result of exercise 2: $d\Omega = -2\pi d \cos\vartheta$.

Sec. 21. Relativistic Dynamics

Action for a particle in the theory of relativity. The adjective *relativistic* denotes invariance with respect to Lorentz transformations, which invariance satisfies the relativity principle. For example, Maxwell's equations in free space are relativistic.

In effect, the Lorentz transformation is derived from the requirement that the equations of electrodynamics remain invariant. Therefore, the proof of the relativistic invariance of Maxwell's equations, which proof will be given somewhat later in this section, is simply in the nature of a confirmation.

The situation with mechanics is altogether different. Newtonian mechanics satisfies only the Galilean relativity principle, which holds for velocities small compared with c. Therefore, it is necessary to find equations of mechanics such that they will be invariant with respect to Lorentz transformations.

In Sec. 10 it was shown how to develop a mechanics by proceeding from the principle of least action. And it was found possible to determine the form of the Lagrangian of a free particle by proceeding from two basic assumptions [see equations (10.11)-(10.13)]:

1) Action is invariant to Galilean transformations;

2) The Lagrangian of a free particle depends only on the absolute value of velocity; the velocity vector **v** cannot be involved in it because, in the absence of an external field, there are no distinguishable directions (in space) relative to which the vector **v** can be given.

In relativistic mechanics the first condition is replaced by the invariance to a Lorentz transformation, while the second condition remains unchanged. Both conditions are satisfied by an action function of the form

$$S=\int \alpha\, ds=\int \alpha c \sqrt{1-\frac{v^2}{c^2}}\, dt\,, \tag{21.1}$$

where we have used the relationship (20.30) between ds and dt. Agreement with the first condition can be seen from the fact that action is expressed in terms of interval only, while agreement with the second condition is obvious. No other invariant quantities can be constructed from dl and dt except the interval, whence the uniqueness of the choice (21.1).

The Lagrangian for a free particle. In order to define the constant α, we examine the limiting form of (21.1) for a small particle velocity. If $v \ll c$,

$$\sqrt{1-\frac{v^2}{c^2}} \cong 1-\frac{v^2}{2c^2}\,. \tag{21.2}$$

From the definition of the Lagrangian (10.2)

$$S=\int L\, dt \tag{21.3}$$

it follows that the Lagrangian is

$$L=\alpha c \sqrt{1-\frac{v^2}{c^2}} \cong \alpha c-\frac{\alpha v^2}{c^2}\,. \tag{21.4}$$

The first term in (21.4) is constant and can be omitted as not appearing in Lagrange's equation [see (10.8)]. The second term should be compared with the Lagrangian for a free particle in Newtonian mechanics:

$$L=\frac{mv^2}{2}\,. \tag{21.5}$$

Whence

$$\alpha=-mc\,. \tag{21.6}$$

The meaning of m here is the mass of the particle measured in a coordinate system in which the particle is at rest (or infinitely near rest). Thus, by its very definition, the quantity m is relativistically invariant. Finally, we have the Lagrangian in the form

$$L=-mc^2 \sqrt{1-\frac{v^2}{c^2}}\,. \tag{21.7}$$

Momentum in relativistic mechanics. From (21.17), we immediately obtain an expression for momentum in the theory of relativity:

$$\mathbf{p} = \frac{\partial L}{\partial \mathbf{v}} = \frac{m\mathbf{v}}{\sqrt{1-\frac{v^2}{c^2}}}. \tag{21.8}$$

As required, at small particle velocities it reduces to the Newtonian expression $\mathbf{p} = m\mathbf{v}$.

Sometimes the quantity $\frac{m}{\sqrt{1-v^2/c^2}}$ (i.e., the proportionality factor between velocity and momentum) is called the *mass of motion* of the particle, as opposed to the *rest mass* m. To avoid confusion we will not use the expression "mass of motion," and will take the term mass to mean the quantity m which is relativistically invariant by definition.

The limiting nature of the velocity of light. The limiting character of the velocity of light, about which we have already spoken in Sec. 20, can be seen from equation (21.8). As the velocity of a particle approaches the velocity of light, its momentum tends to infinity. The only exception is a particle whose mass is equal to zero. Its momentum, written in the form (21.8), gives the indeterminate form 0/0 for $v=c$ and can remain finite. But then the velocity of this particle must always equal c. This property, as we know, is relativistically invariant since the velocity of light is the same in all inertial systems. The momentum of such a particle must be given independently of its velocity [and not according to equation (21.8)], since the velocity is already determined and is equal to c. A velocity greater than c is utterly meaningless because it involves an imaginary quantity for momentum.

Energy in the theory of relativity. Let us now determine the energy of a particle. In accordance with the general definition for energy (4.4),

$$\mathscr{E} = \mathbf{v}\frac{\partial L}{\partial \mathbf{v}} - L = \frac{mv^2}{\sqrt{1-\frac{v^2}{c^2}}} + mc^2\sqrt{1-\frac{v^2}{c^2}} = \frac{mc^2}{\sqrt{1-\frac{v^2}{c^2}}}. \tag{21.9}$$

Equation (21.9) once again confirms the limiting nature of the velocity of light. When v tends to c, the energy of the particle $\mathscr{E}$ tends to infinity. In other words, an infinitely large quantity of work must be performed in order to impart to the particle a velocity equal to that of light.

Rest energy. From equation (21.9), the energy of a particle at rest is equal to mc^2. Let us apply this equation to a complex particle capable of spontaneously decaying into two or three particles. Many atomic nuclei and also unstable particles (mesons) are capable of such disintegration. In the disintegration, the energy must be conserved,

$$\mathscr{E} = \mathscr{E}_1 + \mathscr{E}_2, \tag{21.10}$$

because disintegration is spontaneous, caused not by any external interaction, but by some internal motion in the complex particle. The Lagrangian for this motion is not known explicitly, but in any case it cannot involve time. Therefore, the energy of a complex particle before disintegration is equal to the energy of the two particles formed after the disintegration, when there is no longer any interaction between them.

The energy of all these particles is expressed in accordance with equation (21.9), as applied to all free particles (whether simple or complex) when their motion is considered as a whole. The only possible form of the Lagrangian for such motion is (21.7), from which it follows that the energy is in the form (21.9). Substituting this expression in (21.10) and noting that the initial particle was at rest, we obtain

$$mc^2 = \frac{m_1 c^2}{\sqrt{1 - \frac{v_1^2}{c^2}}} + \frac{m_2 c^2}{\sqrt{1 - \frac{v_2^2}{c^2}}} . \tag{21.11}$$

But the terms $\mathscr{E}_1$ and $\mathscr{E}_2$ on the right are correspondingly greater than m_1c^2 and m_2c^2, whence we obtain the fundamental inequality

$$m \geqslant m_1 + m_2 . \tag{21.12}$$

Hence the mass of a complex particle capable of spontaneous disintegration is greater than the sum of the masses of its component particles. In Newtonian mechanics, the mass characterizing the motion of the system as a whole [see the last term of equation (4.17)] is equal to the sum of the masses of the component particles.

If we define the difference

$$T = \frac{mc^2}{\sqrt{1 - \frac{v^2}{c^2}}} - mc^2 \tag{21.13}$$

as the kinetic energy of a particle $\left(\text{for small energies it reduces to } T = \frac{mv^2}{2}\right)$ and call mc^2 the *rest energy*, then it can be seen from the law of conservation of energy (21.11) that part of the rest energy of a complex particle is converted into kinetic energy of the component particles, and part is converted into their rest energy. Only the total energies $\mathscr{E}$, and not the kinetic energies T, satisfy the conservation law because the kinetic energy of a complex particle as a whole is equal to zero before disintegration and cannot be equal to the essentially positive kinetic energy of the disintegration products.

In chemical reactions, the change in the rest masses of the reacting substances occurs in the order of 10^{-9} (and less) of the total mass.

In nuclear reactions, where the particle velocities are of the order $c/10$, the change in mass may approach one per cent.

When an electron and positron (a positive electron) are annihilated, their energy, including rest energy, is totally converted into the energy of electromagnetic radiation.

As we shall see from quantum theory, radiation is propagated in space in the form of separate particles—so-called light quanta (quantum mechanics teaches that this is compatible with the wave properties of radiation!). The velocity of a light quantum is equal to c so that its mass is identically equal to zero. For this reason, the total rest mass of the particles taking part in the annihilation process is $2\,mc^2$ before the annihilation and zero afterwards.

However, the change in the energy of the electromagnetic field is, of course, equal to $2\,mc^2$, provided the electron and positron did not have any additional kinetic energy. We could, by convention, call the energy of an electromagnetic field, divided by c^2, its mass. With such a definition of mass, the total "mass" would be conserved. But compared with the law of conservation of energy, such a law of conservation of "mass" does not contain anything new; it only repeats the law of conservation of energy in other units.

It is precisely the rest mass that is best to use in describing nuclear reactions, for a change in rest mass determines the energy which may be generated as a result of the reaction (in the form of kinetic energy of the disintegration products, or in the form of radiated energy).

There is no sense in calling the energy of a light quantum divided by the square of the velocity of light, its mass, because this quantity does not in any way characterize light quanta. This quantity has one value in one reference frame and another value in another frame, because the energy of any particle depends upon the reference system relative to which its motion is defined. Yet rest mass is a quantity that characterizes the particle. For example, the rest mass of an electron, involved in the expressions for all its mechanical integrals of motion, is equal to 9×10^{-28} gm. The corresponding quantity for a quantum is identically equal to zero, and, in this sense, characterizes a light quantum in the same way that the quantity 9×10^{-28} gm is characteristic of an electron.

The mass of a particle determines the relationship between the momentum and velocity of the particle in accordance with equation (21.8). It is impossible to determine the mass of a particle by its momentum alone, since particles with the same momenta can have quite different masses. For this reason, it is meaningless to state (though this is sometimes done) that the existence of light pressure (i.e., momentum of the electromagnetic field) proves that the light quantum has a finite mass.

It is sometimes said that a mass of one gramme is capable of releasing an energy of 9×10^{20} ergs (i.e., $1\,c^2$). However, if the substance consists of atoms the possibility of generating this energy is still questionable since up to now not a single process is known in which the total quantity of protons and neutrons (collectively called, nucleons) is changed.* This is why, the relative change in rest mass in nuclear reactions is always measured in fractions of one percent.

The possibilities of various reactions are also limited by the conservation of total charge.

The Hamiltonian for a free particle. We shall now express energy in terms of momentum. Squaring equation (21.9) and subtracting from it equation (21.8), after it has been squared and multiplied by c^2, we obtain

$$\mathscr{E}^2 - c^2 p^2 = m^2 c^4 . \tag{21.14}$$

We have called the energy expressed in terms of momentum the Hamiltonian [see (10.15)]. Hence,

$$\mathscr{E} = \mathscr{H} = \sqrt{m^2 c^4 + c^2 p^2} \tag{21.15}$$

Whence we obtain a relationship between the energy and momentum of a particle that has no rest mass:

$$\mathscr{E} = cp . \tag{21.16}$$

The Lorentz transformation for momentum and energy. We shall now find out how energy and momentum behave with respect to a Lorentz transformation. From equation (21.8) we get

$$\left.\begin{aligned} p_x &= \frac{m v_x}{\sqrt{1-\frac{v^2}{c^2}}} = \frac{m\,dx}{dt\sqrt{1-\frac{v^2}{c^2}}} = mc\frac{dx}{ds}, \\ p_y &= mc\frac{dy}{ds}, \quad p_z = mc\frac{dz}{ds}, \\ \mathscr{E} &= \frac{mc^2}{\sqrt{1-\frac{v^2}{c^2}}} = \frac{mc^2\,dt}{dt\sqrt{1-\frac{v^2}{c^2}}} = mc^3\frac{dt}{ds}. \end{aligned}\right\} \tag{21.17}$$

The quantities m, c and ds are invariant. Hence, the components p_x, p_y and p_z are transformed similar to dx, dy, and dz, i.e., similar to x, y, and z. In accordance with the last equation, energy transforms like time. We can make the following comparison: $p_x \sim x$, $p_y \sim y$, $p_z \sim z$, $\mathscr{E} \sim c^2 t$.

---

* In order to annihilate the whole mass we would have to first prepare "antimatter" (when ordinary matter interacts with antimatter they are mutually annihilated, cf. Sec 38). But this would require a like expenditure of energy.

Now substituting momentum and energy in the Lorentz transformation (20.17) and (20.18), we obtain

$$p'_x = \frac{p_x - \mathscr{E}\dfrac{V}{c^2}}{\sqrt{1 - \dfrac{V^2}{c^2}}}. \tag{21.18}$$

$$p'_y = p_y\,, \tag{21.19}$$

$$p'_z = p_z\,, \tag{21.20}$$

$$\mathscr{E}' = \frac{\mathscr{E} - p_x V}{\sqrt{1 - \dfrac{V^2}{c^2}}}. \tag{21.21}$$

We note that a correct transition from (21.18) to a nonrelativistic equation for the transformation of energy is obtained only when the rest energy mc^2 is substituted in place of $\mathscr{E}'$, for then $p'_x = p_x - mV$ (i.e., $v'_x = v_x - V$) in agreement with the Galilean law for addition of velocities.

Hence, if we demand that the Lorentz transformation yield the correct limiting transition to a Galilean transformation, it is necessary to include the rest energy of the particles in their total energy. Conversely, the kinetic energy T (21.13) does not give a correct limiting transition.

Further, we note that if we form the expression $\mathscr{E}^2 - c^2 p^2$ from equations (21.17) we obtain

$$\mathscr{E}^2 - c^2 p^2 = m^2 c^4 \left(\frac{c^2 dt^2 - dx^2 - dy^2 - dz^2}{ds^2}\right) = m^2 c^4$$

in accordance with (21.14)

The velocity of a system of particles in the theory of relativity. We shall now show how to determine the velocity of a system of particles in relativity theory. We shall consider two particles. Between the velocity, momentum, and energy of each particle there exists the relation

$$\mathbf{p} = \frac{\mathscr{E}\mathbf{v}}{c^2}. \tag{21.22}$$

It is obtained if we divide (21.8) by (21.9). The same equation can also be obtained somewhat differently. Let us determine, from (21.18), the velocity V of the coordinate system relative to which the momentum of the particle is equal to zero. Putting $p'_x = 0$ on the left-hand side of (21.18) we will have, on the right,

$$V = \frac{p_x c^2}{\mathscr{E}}$$

or, if the velocity is not along the x-axis at all, in accordance with (21.22) $\mathbf{V} = \mathbf{p}c^2/\mathscr{E} = \mathbf{v}$. As applied to a single particle, the statement

$\mathbf{v} = \mathbf{V}$ is trivial and simply denotes that the momentum of the particle, relative to a coordinate system moving with the same velocity as the particle itself, is equal to zero.

We now apply equation (21.18) to two particles in order to find the velocity of the coordinate system relative to which their total momentum is equal to zero. The total momentum in the primed system is $\mathbf{p}_1' + \mathbf{p}_2' = \mathbf{p}'$, and the total energy $\mathscr{E}_1' + \mathscr{E}_2' = \mathscr{E}'$. Let us take the x-axis along $\mathbf{p}'$. Since the Lorentz transformation is linear and homogeneous, it has the same form for the sum of two quantities as for each separately. Therefore, we immediately obtain an equation similar to (21.22):

$$\mathbf{V} = \frac{c^2 (\mathbf{p}_1' + \mathbf{p}_2')}{\mathscr{E}_1' + \mathscr{E}_2'} . \tag{21.23}$$

The primes may be omitted here. In order to obtain the limiting transition to the velocity of the centre of mass in Newtonian mechanics from (21.23), it is necessary to take $\mathbf{p}_1 = m_1 \mathbf{v}_1$, $\mathbf{p}_2 = m_2 \mathbf{v}_2$, $\mathscr{E}_1 = m_1 c^2$, $\mathscr{E}_2 = m_2 c^2$, i.e., the particle energies are replaced by their rest energies.

The quantity $\mathbf{V}$, expressed in terms of the particle velocities according to equations (21.8) and (21.9), does not have the form of a total derivative of any quantity with respect to time. Therefore, it is impossible in relativistic mechanics to determine its coordinates in terms of the velocity of the centre of mass. It is better to say that if we attempt to express the coordinates of the centre of mass by means of a classical (or some other) equation it is impossible to represent $\mathbf{V}$ in the form of a time derivative of these coordinates, except in the trivial case when $\mathbf{v}_1$ and $\mathbf{v}_2$ are constant. This is why the concept of centre of mass for particles moving in accelerated motion cannot be used.

As regards relative velocity, $\mathbf{v}_1 - \mathbf{v}_2$, it is meaningless in relativistic mechanics, since there is no simple law for the combination of velocities.

Action for particles in an electromagnetic field. Let us now turn to the equations of motion for a charged particle in an electromagnetic field. We already know the part of the action function which describes the interaction of charges and field. From equation (13.17), this is S_1. Since the variation of S_1 leads to Maxwell's equations, we can be certain of the relativistic invariance of S_1. As applied to point charges, we have already written S_1 in magnetostatics in equation (15.26). But action for free charges in the next equation, (15.27), was suitable only for small particle velocities.

We now know the Lagrangian for a fast particle in the absence of a field (21.7). Thus, the Lagrangian in an external field is equal to the sum of the relativistically invariant expressions (21.7) and (15.26):

$$P = -mc^2 \sqrt{1 - \frac{v^2}{c^2}} + \frac{e}{c} \mathbf{A} \mathbf{v} - e\varphi . \tag{21.24}$$

We now obtain an expression for momentum and energy. Momentum is

$$\mathbf{p} = \frac{\partial L}{\partial \mathbf{v}} = \frac{m\mathbf{v}}{\sqrt{1-\frac{v^2}{c^2}}} + \frac{e}{c}\mathbf{A} = \mathbf{p}_0 + \frac{e}{c}\mathbf{A}. \tag{21.25}$$

Here, $\mathbf{p}_0$ denotes momentum in the absence of a field.

From (4.6), the energy is

$$\mathscr{E} = \mathbf{v}\frac{\partial L}{\partial \mathbf{v}} - L = \mathbf{v}\mathbf{p}_0 + \frac{e}{c}\mathbf{v}\mathbf{A} + mc^2\sqrt{1-\frac{v^2}{c^2}} - \frac{e}{c}\mathbf{v}\mathbf{A} + e\varphi = \mathscr{E}_0 + e\varphi, \tag{21.26}$$

where $\mathscr{E}_0$ is the energy in the absence of an external field; according to (21.9), it is equal to

$$\mathscr{E}_0 = \mathbf{v}\mathbf{p}_0 + mc^2\sqrt{1-\frac{v^2}{c^2}} = \frac{mc^2}{\sqrt{1-\frac{v^2}{c^2}}}.$$

Thus, the linear term in velocity does not appear in the energy expressed in terms of momentum. It will be seen here that the Lagrangian is not of the form $T - U$ because it involves the linear term $\frac{e}{c}\mathbf{A}\mathbf{v}$.

The Hamiltonian for a charge in an external field. From (21.25) we obtain

$$\mathbf{p}_0 = \mathbf{p} - \frac{e}{c}\mathbf{A}, \tag{21.27}$$

and, from (21.26),

$$\mathscr{E}_0 = \mathscr{E} - e\varphi. \tag{21.28}$$

But we already know the expression for $\mathscr{E}_0$ in terms of $\mathbf{p}_0$ from equation (21.15), which relates to the energy and momentum of a free particle. Substituting $\mathbf{p}_0$ and $\mathscr{E}_0$ in (21.15) in accordance with the last equations, we obtain the Hamiltonian of a charge in a field:

$$\mathscr{H} = \sqrt{m^2c^4 + c^2\left(\mathbf{p} - \frac{e}{c}\mathbf{A}\right)^2} + e\varphi. \tag{21.29}$$

The equations of motion of a charge in an external field. From (21.29) we can obtain the equations of motion for a charge in an external field. However, it is simpler to make use of the Lagrangian (21.24). We know that Lagrange's equations are of the following form:

$$\frac{d}{dt}\frac{\partial L}{\partial \mathbf{v}} - \frac{\partial L}{\partial \mathbf{r}} = 0, \tag{21.30}$$

where equation (21.30) replaces three equations of the form (2.21) for the coordinates of $\mathbf{v}$ and $\mathbf{r}$.

The derivative $\frac{\partial L}{\partial \mathbf{v}}$ is equal to $\mathbf{p} = \mathbf{p}_0 + \frac{e}{c}\mathbf{A}$, so that its total time derivative is

$$\frac{d}{dt}\frac{\partial L}{\partial \mathbf{v}} = \frac{\partial \mathbf{p}_0}{dt} + \frac{e}{c}\frac{d\mathbf{A}}{dt}. \qquad (21.31)$$

In order to expand the expression $\frac{d\mathbf{A}}{dt}$, we first write it down for one component:

$$\frac{dA_x}{dt} = \frac{\partial A_x}{\partial t} + \frac{\partial A_x}{\partial x}\frac{dx}{dt} + \frac{\partial A_y}{\partial y}\frac{dy}{dt} + \frac{\partial A_z}{\partial z}\frac{dz}{dt} = \frac{\partial A_x}{\partial t} + (\mathbf{v}\nabla) A_x \qquad (21.32)$$

[see (11.31)], whence, going to a vector equation and substituting in (21.31), we have

$$\frac{d}{dt}\frac{\partial L}{\partial \mathbf{v}} = \frac{d\mathbf{p}_0}{dt} + \frac{e}{c}\left(\frac{\partial \mathbf{A}}{\partial t} + (\mathbf{v}\nabla)\mathbf{A}\right). \qquad (21.33)$$

Let us now calculate the right-hand side of (21.30). Instead of $\frac{\partial L}{\partial \mathbf{r}}$ we can write the completely equivalent expression ∇L:

$$\nabla L = \frac{e}{c}\nabla(\mathbf{A}\mathbf{v}) - e\nabla\varphi.$$

The gradient $\nabla(\mathbf{A}\mathbf{v})$ denotes coordinate differentiation, where only $\mathbf{A}$ and not $\mathbf{v}$ depends explicitly on the coordinates. Thus, applying equation (11.32), we find $\frac{\partial L}{\partial \mathbf{r}}$:

$$\frac{\partial L}{\partial \mathbf{r}} \equiv \nabla L = \frac{e}{c}(\mathbf{v}\nabla)\mathbf{A} + \frac{e}{c}[\mathbf{v}\,\mathrm{rot}\,\mathbf{A}] - e\varphi. \qquad (21.34)$$

Now, substituting (21.33) and (21.34) in (21.30) and taking all the terms involving potentials to the right-hand side, we obtain

$$\frac{d\mathbf{p}_0}{dt} = e\left(-\nabla\varphi - \frac{1}{c}\frac{\partial \mathbf{A}}{\partial t}\right) + \frac{e}{c}[\mathbf{v}\,\mathrm{rot}\,\mathbf{A}]. \qquad (21.35)$$

The right-hand side of (21.35) involves the electromagnetic fields in accordance with their definition in terms of potentials (12.28) and (12.29).

Hence, the equation of motion of a charge involves only the field and not the potential, as follows from the condition of gauge invariance. After substituting the fields the equation takes the form

$$\frac{d}{dt}\frac{m\mathbf{v}}{\sqrt{1-\frac{v^2}{c^2}}} = e\mathbf{E} + \frac{e}{c}[v\mathbf{H}]. \qquad (21.36)$$

The right-hand side of (21.36) is called the Lorentz force. In addition to the usual term, $e\mathbf{E}$, which we know from electrostatics, it involves a term similar to the Coriolis force. It is related to the part of the Lagrangian which is linear in velocity.

The magnetic part of the Lorentz force, $\frac{e}{c}[\mathbf{v}\mathbf{H}]$, is very similar to the expression for the force acting on a current in an external magnetic field and, naturally, can be obtained from it. We did not have to use this method of derivation because the part of the Lagrangian which describes the interaction between charges and field was already known from Sec. 13. And besides, the relativistic invariance of (21.36), which emerges obviously from derivation from the invariant Lagrangian function, is considerably more difficult to grasp from the elementary definition of a magnetic force acting on a current.

The work performed by a field on a charge. From equation (21.36), we can obtain an expression for the work done by an electromagnetic field on a charge. We know by definition that the work is equal to the change in kinetic energy. Let us multiply scalarly both parts of (21.36) by $\mathbf{v}$. We shall then have the expression $\frac{d\mathbf{p}_0}{dt}$ on the left-hand side. But $\mathbf{v} = \frac{\partial \mathscr{E}}{\partial \mathbf{p}}$ in accordance with Hamilton's equation (10.18), so that $\mathbf{v}\frac{d\mathbf{p}_0}{dt} = \frac{\partial \mathscr{E}_0}{\partial \mathbf{p}_0}\frac{d\mathbf{p}_0}{dt} = \frac{d\mathscr{E}_0}{dt} = \frac{dT}{dt}$, and on the left-hand side what we have is the required quantity for the change of kinetic energy in unit time. On the right-hand side the term $\mathbf{v}[\mathbf{v}\mathbf{H}] = [\mathbf{v}\mathbf{v}]\mathbf{H} = 0$ and there remains only the work done by the electric force:

$$\frac{d\mathscr{E}_0}{dt} = \frac{dT}{dt} = e(\mathbf{E}\mathbf{v}). \tag{21.37}$$

As was to be expected, the magnetic force $\frac{e}{c}[\mathbf{v}\mathbf{H}]$ does not perform work on the charge because it is perpendicular to the charge velocity at every given instant of time.

The Lorentz transformation for the field components. From (21.37) and equation (21.36), if we write it in terms of components, it is easy to obtain the Lorentz transformation equations for the field components. These equations must be written so that their form does not change in passing from one coordinate system to another. Let us take equation (21.36) for the component of momentum on x, and multiply it by $\frac{dt}{ds}$. We shall also multiply equation (21.37) by $\frac{dt}{ds}$ and also by $\frac{V}{c^2}$, where V is the relative velocity of the coordinate system. After this we subtract (21.37) from (21.36). Then on the left-hand side we have

$$\frac{d}{ds}\left(p_x - \frac{V\mathscr{E}}{c^2}\right) = \frac{dp'_x}{ds}\sqrt{1 - \frac{V^2}{c^2}}.$$

On the right-hand side we will have the expression

$$e\left(E_x\frac{dt}{ds}+\frac{1}{c}H_z\frac{dy}{ds}-\frac{1}{c}H_y\frac{dz}{ds}\right)-\frac{eV}{c^2}\left(E_x\frac{dx}{ds}+E_y\frac{dy}{ds}+E_z\frac{dz}{ds}\right)=$$
$$=eE_x\left(\frac{dt}{ds}-\frac{V}{c^2}\frac{dx}{ds}\right)+\frac{e}{c}\left(H_z-\frac{V}{c}E_y\right)\frac{dy}{ds}-\frac{e}{c}\left(H_y+\frac{V}{c}E_z\right)\frac{dz}{ds}.$$

But ds is invariant. Therefore, the quantities on the right-hand side must be transformed in accordance with the basic equations (20.17) and (20.18). Differentiating these equations, we obtain

$$\frac{dt}{ds}-\frac{V}{c^2}\frac{dx}{ds}=\frac{dt'}{ds}\sqrt{1-\frac{V^2}{c^2}},\quad \frac{dy}{ds}=\frac{dy'}{ds},\quad \frac{dz}{ds}=\frac{dz'}{ds}.$$

Now, dividing both sides of the equation by $\sqrt{1-\frac{V^2}{c^2}}$ and multiplying by $\frac{ds}{dt'}$, we will have an equation for the x-component of momentum in the new coordinate system:

$$\frac{dp'_x}{dt'}=eE_x+\frac{\left(H_z-\frac{V}{c}E_y\right)}{\sqrt{1-\frac{V^2}{c^2}}}\frac{e}{c}\frac{dy'}{dt'}-\frac{\left(H_y+\frac{V}{c}E_z\right)}{\sqrt{1-\frac{V^2}{c^2}}}\frac{e}{c}\frac{dz'}{dt'}.$$

In accordance with the principle of relativity, this equation must be written in the same way as for the unprimed coordinate system:

$$\frac{dp'_x}{dt'}=eE'_x+\frac{e}{c}H'_z\frac{dy'}{dt'}-\frac{e}{c}H'_y\frac{dz'}{dt'}.$$

Comparing the last two equations, we obtain the field transformation equations:

$$E'_x=E_x, \tag{21.38}$$

$$H'_y=\frac{H_y+\frac{V}{c}E_z}{\sqrt{1-\frac{V^2}{c^2}}}, \tag{21.39}$$

$$H'_z=\frac{H_z-\frac{V}{c}E_y}{\sqrt{1-\frac{V^2}{c^2}}}. \tag{21.40}$$

In the same way, though from other equations (21.36), is it easy to find other equations for field transformation:

$$H'_x=H_x, \tag{21.41}$$

$$E'_y=\frac{E_y-\frac{V}{c}H_z}{\sqrt{1-\frac{V^2}{c^2}}}, \tag{21.42}$$

$$E'_z = \frac{E_z + \frac{V}{c} H_y}{\sqrt{1 - \frac{V^2}{c^2}}}. \tag{21.43}$$

Consequently, in contrast to coordinates, it is not the longitudinal but the transverse components that are transformed in the field.

The change in field, in passing from one coordinate system to another, is verified to a nonrelativistic approximation $\left(\text{i.e., to the accuracy of terms of the order } \frac{V}{c}\right)$ in a unipolar induction experiment. A diagram of the experiment is shown in Fig. 34. The magnet NS rotates around its longitudinal axis. Two collectors connected by a fixed conductor are joined to the centre of the magnet and to its axis. When the magnet is rotated, an e.m.f. appears in the wire. This experiment is frequently interpreted as meaning that when the magnet is rotated the wire "cuts" its lines of force as if the lines were attached to the magnet like brushes.

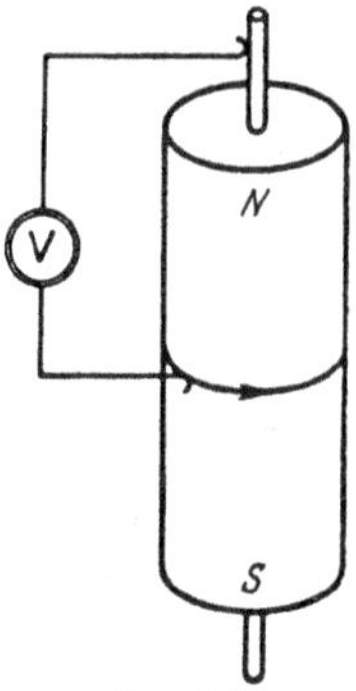

Fig. 34

Actually, unipolar induction must be understood as follows. There is only a magnetic field H in the coordinate system attached to the magnet, while the electric field is equal to zero. Hence, in a system fixed in the wire, relative to which the magnet moves, an electric field, too, should be observed in accordance with (21.42) or (21.43). This field is of an order of magnitude $\frac{V}{c} H$ and produces the e.m.f.

We note that a coordinate system is defined only when both the electric and magnetic fields are specified. It is insufficient to specifiy only one of them.

The invariants of an electromagnetic field. From equations (21.38)-(21.43), it is easy to obtain the following two invariants:

$$E'^2 - H'^2 = E^2 - H^2, \tag{21.44}$$

$$\mathbf{E}'\mathbf{H}' = \mathbf{E}\mathbf{H}. \tag{21.45}$$

From the invariance of these expressions it follows that the electromagnetic field of a plane wave appears similar in all systems. Indeed, in a plane wave, $E = H$ or $E^2 - H^2 = 0$. This property is invariant according to (21.44). Further, $\mathbf{E} \perp \mathbf{H}$, so that $(\mathbf{E}\,\mathbf{H}) = 0$. This property is invariant according to (21.45).

The quantity $(\mathbf{E}\,\mathbf{H})$ is invariant with respect to a Lorentz transformation. But with respect to a replacement of x, y, z by $-x$, $-y$, $-z$

(i.e., an inversion of the signs of the coordinates), it is not invariant, because in this case **E** changes sign while **H** does not (see Sec. 16). The quantity $E^2 - H^2$ is invariant even when the coordinate signs are inverted. But this quantity is the Lagrangian for a free electromagnetic field. Integrated over the invariant volume dx, dy, dz, dt (exercise 5, Sec. 20), it yields invariant action, as required, while the quantity (**EH**) does not give a real invariant.

The linearity of Maxwell's equations with respect to field. A real invariant can be formed from the quantity **EH** merely by squaring. It is, of course, not at all obvious beforehand why such a quantity as well as the square of the invariant $E^2 - H^2$ cannot appear in the Lagrangian for an electromagnetic field. The same can be said of higher-order terms which do not change sign in the substitution of x by $-x$, etc. But if some terms—other than quadratic with respect to field—are left in the Lagrangian, then Maxwell's equations will contain nonlinear terms.

The essential difference between nonlinear and linear equations is that the sum of two solutions of a nonlinear equation is not its solution. Indeed, if two electromagnetic waves are propagated in a vacuum they are simply combined, and in no way distort each other. In nonlinear theory, the velocity is a function of the wave amplitude, while in electrodynamics the velocity of light is a universal constant.

For this reason, the choice of the Lagrangian in the simplest form $E^2 - H^2$ expresses the experimental fact that the law of variation of any electromagnetic field in space and in time is in no way dependent upon whether another field is operative in that same charge-free region of space.

Actually the quantum electromagnetic field theory indicates the existence of certain nonlinear effects. In the range of phenomena for which classical electrodynamics is applicable, these effects are not essential.

Transformation of charge density and current density. From the definition of charge density one can find the law of its transformation. Since charge is an invariant quantity, we have

$$de = \rho\, dx\, dy\, dz = \rho_0\, dx_0\, dy_0\, dz_0\,, \tag{21.46}$$

where ρ_0 is the charge density in a system relative to which it is at rest and, hence, the quantity is also invariant by *definition*. Whence,

$$\rho = \rho_0 \frac{dx_0\, dy_0\, dz_0}{dx\, dy\, dz} = \rho_0 \frac{dt}{dt_0} = \rho_0 c \frac{dt}{ds}\,, \tag{21.47}$$

where we have used (20.30) and exercise 5, Sec. 20. The current density is

$$\left.\begin{aligned} j_x &= \rho v_x = \rho_0 c \frac{dx}{dt}\frac{dt}{ds} = \rho_0 c \frac{dx}{ds}, \\ j_y &= \rho_0 c \frac{dy}{ds}, \quad j_z = \rho_0 c \frac{dz}{ds}. \end{aligned}\right\} \tag{21.48}$$

From here it can be seen that the current components are transformed like coordinates, while charge density is transformed like time.

Let us consider a conductor in which a current is flowing. In the coordinate system in which the conductor is at rest, it remains neutral, but in other systems a charge density must appear on it. This fact does not contradict the invariance of total charge, but follows from it in accordance with (21.46)-(21.48).

The invariance of action for the field. Let us now verify that the action term (13.17) describing the interaction of field and charge is invariant. It follows from (21.27) and (21.28) that the vector potential transforms like momentum (i.e., like a radius vector), while the scalar potential transforms like energy (i.e., like time). For this reason the product

$$\mathbf{A}\mathbf{j} - \varphi\rho$$

behaves, with respect to a Lorentz transformation, like an interval; in other words, it remains invariant. Integrated over the invariant "four-dimensional" volume $dx\, dy\, dz\, dt$, it yields the invariant action term S_1. Hence, Maxwell's equations are obtained from the invariant action function S, so that they are also invariant themselves. This could also have been verified from equations (21.38)-(21.43).

Exercises

1) Find the scalar and vector potentials of a freely moving charge.

In its own coordinate system, the scalar potential is $\varphi_0 = \frac{e}{r_0}$ and the vector potential is equal to zero. Hence, in a system relative to which the charge moves, its scalar potential is

$$\varphi = \frac{\varphi_0}{\sqrt{1 - \frac{v^2}{c^2}}} = \frac{e}{r_0\sqrt{1 - \frac{v^2}{c^2}}},$$

and the vector potential is

$$\mathbf{A} = \frac{\varphi_0 \frac{\mathbf{v}}{c}}{\sqrt{1 - \frac{v^2}{c^2}}} = \frac{e\,\mathbf{v}}{r_0\, c\sqrt{1 - \frac{v^2}{c^2}}}.$$

Further, r_0 must be expressed in terms of coordinates in the fixed system

$$r_0 = \sqrt{x_0^2+y_0^2+z_0^2} = \sqrt{\frac{(x-vt)}{1-\frac{v^2}{c^2}} + y^2 + z^2}\,.$$

We can put ξ instead of vt, i.e., the abscissa of the moving charge. The electromagnetic disturbance arrives at the given point x, y, z from the point ξ', where the charge was situated earlier. We have

$$\frac{\xi-\xi'}{v} = \frac{\sqrt{(x-\xi')^2+y^2+z^2}}{c} = \frac{R'}{c}$$

from the definition of lag. Putting $\xi = vt$ in r_0, we obtain an expression for φ and $\mathbf{A}$ in terms of R'.

$$\varphi = e \Big/ \left(R' - \frac{\mathbf{v}\mathbf{R}'}{c}\right), \quad \mathbf{A} = e\,\mathbf{v} \Big/ \left(R' - \frac{\mathbf{v}\mathbf{R}'}{c}\right).$$

2) Find the motion of a charge in a constant uniform magnetic field.

If the field is in the direction of the z-axis, the equations of motion are of the following form:

$$\frac{dp_x}{dt} = \frac{e}{c}\frac{dy}{dt}H\,, \quad \frac{dp_y}{dt} = -\frac{e}{c}\frac{dx}{dt}H\,, \quad \frac{dp_z}{dt} = 0\,.$$

Further, $p^2 = \text{const}$, $p_z^2 = \text{const}$, $p_x^2 + p_y^2 = \text{const}$,

$$p_x = \frac{m v_x}{\sqrt{1-\frac{v^2}{c^2}}} = \frac{\mathcal{E} v_x}{c^2}\,, \qquad \mathcal{E} = \text{const}\,.$$

We look for the coordinates x and y in the form:

$$x = R\cos\omega t, \qquad y = R\sin\omega t\,.$$

For R and ω the following expressions result:

$$R = \frac{\mathcal{E} v}{ecH}\,, \qquad \omega = \frac{ecH}{\mathcal{E}}\,.$$

The particle moves along a helix. For small velocities, ω reduces to the constant value $\frac{eH}{mc}$.

3) Find the motion of a charge in a constant uniform electric field. The equations of motion are

$$\frac{dp_x}{dt} = eE\,, \quad \frac{dp_y}{dt} = 0\,, \quad \frac{dp_z}{dt} = 0\,, \quad \frac{d\mathcal{E}_0}{dt} = Ee\frac{dx}{dt}\,.$$

From the last equation we obtain

$$\sqrt{m^2c^4 + c^2(p_x^2+p_y^2+p_z^2)} - \sqrt{m^2c^4 + c^2(p_{x_0}^2+p_{y_0}^2+p_{z_0}^2)} = eEx\,.$$

From the first equation

$$p_x - p_{x_0} = eEt\,, \quad p_y - p_{y_0} = 0\,, \quad p_z - p_{z_0} = 0\,.$$

These equation integrals together give x as a function of t.

If $p_{z_0} = 0$, then dividing p_x by p_y, we have an expression for $\frac{dx}{dy}$ in terms of x (by eliminating t from the energy integral). The trajectory is of the form of a catenary.

4) Find the motion of a charge in a central attractive Coulomb field.

The energy integral is of the form

$$(\mathscr{E} - e\varphi)^2 - c^2\left(p_r^2 + \frac{M^2}{r^2}\right) = m^2 c^4, \qquad \varphi = -\frac{a}{r}.$$

Further, denoting the azimuth by ψ, we obtain

$$M = \frac{d\psi}{dt} \cdot \frac{m r^2}{\sqrt{1 - \frac{v^2}{c^2}}}, \qquad p_r = \frac{dr}{dt} \cdot \frac{m}{\sqrt{1 - \frac{v^2}{c^2}}}.$$

Whence

$$\frac{1}{r^2}\frac{dr}{d\psi} = \frac{p_r}{M}.$$

Substitution of p_r in the energy integral and separation of the variables r and ψ leads to an elementary quadrature. The trajectory for finite motion ($\mathscr{E} < mc^2$) is similar to an ellipse, but with a rotating perihelion.

5) Examine the collision of a travelling particle of zero mass with a particle of mass m at rest. Determine the energy of the incident particle after collision, if its angle of deflection ϑ is known.

$$\text{Answer: } \mathscr{E}' = \frac{\mathscr{E}}{1 + \frac{\mathscr{E}}{m c^2}(1 - \cos\vartheta)}.$$

6) Find the motion and radiation of a charge connected elastically to some point of space (with frequency ω_0) and situated in a uniform magnetic field $H_z = H$, $H_x = H_y = 0$.

The oscillations of the charge are governed by the following nonrelativistic equations:

$$m\ddot{x} = -m\omega_0^2 x + \frac{e}{c} H\dot{y},$$

$$m\ddot{y} = -m\omega_0^2 y - \frac{e}{c} H\dot{x},$$

$$m\ddot{z} = -m\omega_0^2 z.$$

The third equation does not depend on the first two. The first two equations are easily solved if we put $x = ae^{i\omega t}$, $y = be^{i\omega t}$. Then

$$a(\omega_0^2 - \omega^2) - i\omega\frac{eH}{mc} b = 0,$$

$$b(\omega_0^2 - \omega^2) + i\omega\frac{eH}{mc} a = 0.$$

Let us multiply the second equation by i and first subtract it from the first, and then add it. The combinations $a \pm ib$ then satisfy the equations

$$(a \pm ib)(\omega_0^2 - \omega^2) \mp (a \pm ib)\,\omega\frac{eH}{mc} = 0.$$

Cancelling $a \pm ib$, we arrive at the equations for frequencies

$$\omega^2 - \omega_0^2 \pm \frac{eH\omega}{mc} = 0 .$$

We regard $\frac{eH}{mc}$ as small compared with ω_0, replace ω by ω_0 in the term $\frac{eH\omega}{mc}$, and represent the difference $\omega_0^2 - \omega^2$ as $(\omega + \omega_0)(\omega - \omega_0)$, which is approximately equal to $2\omega_0(\omega - \omega_0)$. Then we obtain expressions for the frequencies of both oscillations:

$$\omega = \omega_0 \mp \frac{eH}{2mc} .$$

They differ from the undisplaced frequency by $\frac{eH}{2mc}$, i.e., by the Larmor frequency ω_L.

From the equations for a and b, after substituting $\omega = \omega_0 \mp \omega_L$, we have, to the same degree of approximation,

$$a = \pm i b .$$

If we represent the coordinates in real form, we obtain for both oscillations $x = a \cos(\omega_0 \mp \omega_L)\, t$; $y = \pm a \sin(\omega_0 \mp \omega_L)\, t$.

Thus, the radius vector of a particle performing oscillations with frequency $\omega_0 + \omega_L$ rotates in a clockwise direction, while for oscillations with frequency $\omega_0 - \omega_L$ it rotates in an anticlockwise direction. Thus, in accordance with Larmor's theorem, the frequency ω_L is added to the frequency ω_0 or subtracted from it, depending upon the direction in which the charge rotates (we note that the sign of ω_L is changed for a negative charge).

Let us consider the radiation of such a charge in a magnetic field. We know that the electric vector of the radiated electromagnetic wave lies in the same plane as the charge displacement vector. If radiation is observed to be due to the z-component of the dipole moment, its electric vector is along the z-axis and is proportional to $\ddot{z}$. Thus, the radiation is plane-polarized and is of frequency ω_0. The oscillation occurring along the field and having an undisplaced frequency radiates electromagnetic waves, which are polarized in the same plane as the magnetic field. This oscillation does not radiate at all in the direction of the magnetic field, but the oscillations with frequency $\omega_0 \pm \omega_L$ radiate circularly-polarized waves, and the electric-field vector rotates in the same direction as the charge displacement vector.

All three frequencies radiate in a direction perpendicular to the field. However, since the charge oscillations are viewed from one side in this position in circular rotation, the vector of electric-field oscillations lies in a plane perpendicular to the constant external magnetic field, so that waves with frequencies $\omega_0 \pm \omega_L$ are now also plane-polarized. In observations that are not at right angles to the field, we obtain elliptically-polarized oscillations and a plane-polarized oscillation of frequency ω_0.

The calculations set out here form the classical theory of the Zeeman effect. The line splitting that is actually observed for various values of magnetic field is correctly described only by quantum theory (Sec. 34).

PART III
QUANTUM MECHANICS

Sec. 22. The Inadequacy of Classical Mechanics. The Analogy Between Mechanics and Geometrical Optics

The instability of the atom according to the classical view. Rutherford's experiments in 1910 established that the atom consists of light negative electrons and a heavy positive nucleus of dimensions very small compared to the atom itself (see Sec. 6). For such a system to be stable, it is necessary that the electrons should revolve around the nucleus like planets about the sun, for unlike charges at rest would come together.

This stability condition of the atom is, nevertheless, insufficient. In the case of motion in an orbit, electrons will experience centripetal acceleration, but, as was shown in Sec. 19, a charged particle undergoing acceleration radiates electromagnetic waves, thereby transmitting its energy to the electromagnetic field. Thus, the energy of an electron moving around a nucleus should continuously diminish until the electron falls onto the nucleus. This statement is in striking contradiction to the obvious fact of the stability of atoms.

The Bohr theory. In 1913, N. Bohr suggested a compromise as a way out of this difficulty. According to Bohr, an atom has stable orbits such that an electron moving in them does not radiate electromagnetic waves. But in making a transition from an orbit of higher energy to one with lower energy, an electron radiates; the frequency of this radiation is related to the difference between the energies of the electron in these two orbits by the equation

$$h\omega = \mathscr{E}_1 - \mathscr{E}_2,$$

where h is a universal constant equal to 1.054×10^{-27} erg-sec.

Both of Bohr's principles were in the nature of postulates. But it was possible with their aid to explain, in excellent agreement with experiment, the observed spectrum of the hydrogen atom and also the spectra of a series of atoms and ions similar to the hydrogen atom (for example, the positive helium ion, which consists of a nucleus and one electron). Despite the fact that both of these, essentially quantum, postulates of Bohr were completely alien to classical physics and could in no way be explained on the basis of classical concepts, they represented an extraordinary step forward in the theory of the atom.

Indeed, the first postulate contains the statement that not every state of the atom is stationary, but only certain states. This statement, as we now know, derives from quantum mechanics just as directly as elliptical planetary orbits derive from Newtonian mechanics.

The Bohr theory was very successful in explaining the spectra of single-electron atoms. But the very next step, a two-electron atom such as the helium atom, did not yield to consistent calculation by the Bohr theory. The theory was even less capable of explaining the stability of the hydrogen molecule. For this reason, the situation in physics, notwithstanding a number of brilliant results of the Bohr theory, was completely unsatisfactory. Besides the particular difficulties that we have noted here, the Bohr theory was, on the whole, eclectic, since it was inconsistent in its combination of classical and quantum concepts.

Light quanta. The inadequacy of classical ideas appeared most obvious in the problem of the stability of the atom. But earlier there were many facts which classical (i.e., nonquantum) physics failed to explain. A case in point is the theory of an electromagnetic field in equilibrium with matter (for more detail, see Sec. 42). Here, classical theory leads to an absurd result—the total energy of an electromagnetic field in equilibrium with radiating matter is expressed in the form of a divergent, i.e., infinite, integral.

In order to give a satisfactory description of experimental facts, Planck, in 1900, postulated that sources of radiation emit and absorb energy of the electromagnetic field in *finite amounts*. These discrete quantities, or *quanta*, as they were called by Planck, are proportional to the frequency of the emitted or absorbed radiation. It is easy to see that the factor of proportionality must be the same as in Bohr's second postulate (actually, Planck introduced a quantity 2π times greater, but used a frequency $\nu = \frac{\omega}{2\pi}$, equal to the number of oscillations per second). Bohr's second postulate relates the properties of discreteness of stationary states of a radiating system (atom), occurring in line spectra, to the energy of the emitted quanta. Classically, it is just as impossible to explain this discreteness as it is to explain Planck's initial hypothesis.

The duality of electrodynamical concepts. At the beginning of the twentieth century, the classical theory of light also turned out to be incapable of explaining many facts without appealing to an additional hypothesis concerning light quanta. But at the same time, there was a whole range of phenomena, such as diffraction and interference of light, which appeared to be intimately bound up with its wave nature. It did not seem possible to explain these phenomena in terms of classical corpuscular concepts.

Another group of phenomena could be explained only on the basis of Planck's hypothesis concerning light quanta, and was in most obvious contradiction to the classical wave conceptions. Let us note two such phenomena.

I refer, firstly, to the so-called photoemissive effect, i.e., the emission of electrons from the surface of a metal in a vacuum when the metal is illuminated by ultraviolet rays. The energy of the photoelectrons depends only on the radiation frequency and is independent of its intensity. This can only be understood if we assumed that the energy of electromagnetic radiation is absorbed in the form of quanta, $\hbar\omega$. Then the kinetic energy of an electron will be equal to the energy of the quantum minus the electronic work function (the energy needed to remove the electron from the metal).

Einstein, to whom this explanation of the laws of the photoelectric effect belongs, went further and assumed that electromagnetic radiation is not only absorbed and emitted in the form of quanta, but is also *propagated* in that form.

Since the energy of a quantum is equal to $\hbar\omega$, and its velocity is c, it should possess a momentum $\frac{\hbar\omega}{c}$ (see Sec. 21). It follows that a quantum is a particle of zero mass. It will be noted that the energy and momentum of an electromagnetic wave are related in exactly this way (see Sec. 17).

The second phenomenon which exhibited the quantum properties of radiation provided for confirmation of Einstein's hypothesis concerning the momentum of light quanta. In the scattering of X-radiation by electrons, the latter may be regarded as free, since the characteristic frequencies of their motion in a substance are very small compared with the frequency of the incident radiation (we have considered such scattering in exercise 4, Sec. 19). It is essential that in accordance with classical theory the scattered radiation must be of the same frequency as the incident radiation. But experiment showed that the frequency of the scattered radiation is less than the frequency of the incident radiation, and depends upon the angle at which the scattering is observed (Compton effect). The displaced frequency can be calculated in relation to the scattering angle, if it is assumed that the act of scattering occurs as the collison of two free particles—a moving quantum and an electron at rest. A collision of this sort was considered in exercise 5, Sec. 21, especially for the case of an incident particle with zero mass. The equation obtained there gives a perfectly correct description of the frequency shift in the Compton effect, if we consider that the energy of the quantum is equal to $\mathscr{E} = \hbar\omega$ and its momentum $\mathbf{p} = \hbar\,\mathbf{k}$ (or $\hbar\,\omega/c$ in absolute magnitude).

Quantum mechanics. Thus, in the theory of light and the theory of the atom, a peculiar dualism arose: one and the same physical reality (the electromagnetic field or the atom) was described by two contradictory theories: classical and quantum. A way out of this situation was found in consistent quantum theory, where all motion possesses certain wave properties, which, however, cannot be detected in the motion of macroscopic bodies, but are essential in the description of the motion of such microscopic particles as quanta and electrons. The criterion to be used in order to determine whether it is necessary to take into account the wave properties of a given motion will be given in the following section. The only thing to note here is that it involves the constant $\hbar$.

The basic principles of quantum mechanics received direct experimental verification after the discovery of electron diffraction, whose laws are very similar to the diffraction laws of electromagnetic waves. All atomic phenomena are qualitatively and quantitatively fully accounted for by quantum mechanics.

The present state of nuclear theory. Nuclear phenomena are somewhat more complex. At the present time, we do not know the laws governing the interaction of nuclear particles. These laws are closely related to the properties of special nuclear fields, which properties differ in many respects from those of the electromagnetic field. At the present time, we still do not know the theory of nuclear fields. It may also be that there is still insufficient experimental data for the development of such a theory. Therefore, nuclear theory is, to date, considerably less developed than atomic theory, in which all interactions are of an electromagnetic nature and well known. In any case, the difficulties experienced by modern nuclear theory lie outside the region in which nonrelativistic quantum mechanics can be applied, and in no way affect its basis.

The correspondence between geometrical optics and classical mechanics. An essential role has been played in the formation of quantum theory by analogy between classical mechanics and wave optics. A correspondence will be established between them in the present section. Since geometrical optics is a limiting case of wave optics, an analogy between *geometrical* optics and classical mechanics permits of transition to the *wave* equations of quantum mechanics by means of a generalization. Let us establish an analogy between the equations of mechanics and geometrical optics, quite formally for the time being. Its meaning will be given later.

Surfaces of constant phase. Let us explain the significance of the wave phase in geometrical optics. To do this, we perform the limiting transition from wave optics to geometrical optics. We write the expression for the field in the form

$$\mathbf{E} = \mathbf{E}_0 (\mathbf{r}, t) \cos \frac{\chi(\mathbf{r}, t)}{\lambda} . \tag{22.1}$$

Here, λ is the wavelength, which is regarded as small compared with the linear dimensions of the region occupied by the field. In the limiting case of a plane wave the phase is

$$\frac{\chi}{\lambda} = -\omega t + \mathbf{k}\mathbf{r} \tag{22.2}$$

[Cf. (17.21), (17.22)]. Since $\omega = \frac{2\pi u}{\lambda}$, $\mathbf{k} = \frac{2\pi \mathbf{n}}{\lambda}$, where u is the phase velocity, it is convenient to obtain a relationship containing λ explicitly, describing the phase as $\varphi = \frac{\chi}{\lambda}$.

The expression for the field must be substituted into the wave equation

$$\Delta \mathbf{E} - \frac{1}{u^2} \frac{\partial^2 \mathbf{E}}{\partial t^2} = 0 . \tag{22.3}$$

In differentiating with respect to t and r, we retain only those terms containing the highest degree of λ in the denominator because λ is a small quantity. Hence,

$$\frac{\partial \mathbf{E}}{\partial t} \cong - \mathbf{E}_0 \frac{1}{\lambda} \frac{\partial \chi}{\partial t} \sin \frac{\chi}{\lambda} ,$$

$$\frac{\partial^2 \mathbf{E}}{\partial t^2} \cong - \mathbf{E}_0 \frac{1}{\lambda} \frac{\partial^2 \chi}{\partial t^2} \sin \frac{\chi}{\lambda} - \mathbf{E}_0 \frac{1}{\lambda^2} \left(\frac{\partial \chi}{\partial t} \right)^2 \cos \frac{\chi}{\lambda} .$$

As stated, the first term in the last expression must be discarded when $\lambda \to 0$. Whence,

$$\frac{\partial^2 \mathbf{E}}{\partial t^2} \cong - \mathbf{E}_0 \left(\frac{1}{\lambda} \frac{\partial \chi}{\partial t} \right)^2 \cos \frac{\chi}{\lambda}$$

and similarly

$$\Delta \mathbf{E} \cong - \mathbf{E}_0 \left(\frac{1}{\lambda} \nabla \chi \right)^2 \cos \frac{\chi}{\lambda} .$$

Substituting these expressions into the wave equation (22.3), we obtain a first-order differential equation for the phase $\varphi = \frac{\chi}{\lambda}$:

$$(\nabla \varphi)^2 - \frac{1}{u^2} \left(\frac{\partial \varphi}{\partial t} \right)^2 = 0 . \tag{22.4}$$

In the limiting case of a plane wave, it follows from (22.2) that

$$\mathbf{k} = \frac{\partial \varphi}{\partial \mathbf{r}} \equiv \nabla \varphi \tag{22.5}$$

and

$$\omega = - \frac{\partial \varphi}{\partial t} , \tag{22.6}$$

where

$$k^2 - \frac{\omega^2}{u^2} = 0 .$$

But, according to (22.4), this same equation is satisfied also by the quantities $\frac{\partial \varphi}{\partial \mathbf{r}}$, $\frac{\partial \varphi}{\partial t}$ in the expression of the almost plane wave (22.1). It follows that we can take equations (22.5) and (22.6) as *definitions* for the wave vector and frequency of an almost plane wave.

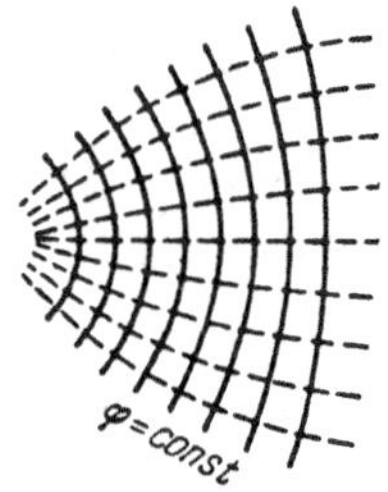

Fig. 35

The *wave vector* is directed along the normal to a surface of constant phase $\varphi = \text{const}$, i.e., along a light ray at a given point of space. The propagation of an *almost plane wave* may be represented as the displacement in space of a family of surfaces of constant phase. In Fig. 35, these surfaces are shown in cross section by solid lines, and the light rays are dashed lines.

At various instants of time t, a surface of definite value $\varphi = \varphi_0$ occupies various positions in space corresponding to the equation $\varphi(\mathbf{r}, t) = \varphi_0$. We determine the propagation velocity of this surface. To do this, we proceed from the condition obtained by differentiating $\varphi = \text{const}$:

$$d\varphi = \frac{\partial \varphi}{\partial t} dt + \frac{\partial \varphi}{\partial \mathbf{r}} d\mathbf{r} = 0 .$$

Let $d\mathbf{r}$ be a vector in a direction normal to the surface. Then $\left| \frac{d\varphi}{d\mathbf{r}} \right|$ is the absolute value of $\mathbf{k}$. From (22.5) and (22.6), we obtain

$$\left| \frac{d\mathbf{r}}{dt} \right| = \frac{\left| \frac{\partial \varphi}{\partial t} \right|}{\left| \frac{\partial \varphi}{\partial \mathbf{r}} \right|} = \frac{\omega}{k} = u ,$$

which coincides with the definition of phase velocity (18.7).

Phase and group velocities are, in general, different, since group velocity is equal to

$$\mathbf{v} = \frac{\partial \omega}{\partial \mathbf{k}}. \tag{22.7}$$

It is essential that for an almost plane wave, ω may be expressed as a function of $\mathbf{k}$, just as is done in the case of a plane wave.

Surfaces of constant action. We shall now consider a family of trajectories of identical particles moving in some force field. For example, these may be shrapnel particles formed when a shell is exploded (though not pieces of the shell itself having different masses!); it must be considered here that the shrapnel explosions occur at the same place continuously, so that the particles fly one after the other along each trajectory lagging only in time. It is not absolutely necessary to consider particles emerging from one point. Trajectories may be taken which are normal to some initial surface. Each particle emerging at $t = t_0$ has a definite trajectory depending on its initial coordinates and initial velocity. The value of the action S for each particle may be calculated along these trajectories from the equation [see (10.2)]

$$S = \int_{t_0}^{t} L dt . \tag{22.8}$$

Since at the instant $t = t_0$ the position of a particle is determined by its trajectory, and L is a known function of coordinates, velocities, and time, $L\{q(t), \dot{q}(t), t\}$, the action along each trajectory is also known as a function of time.

Let us join by surfaces the points of all the trajectories for which the value of S is constant. The equation of this surface is $S(\mathbf{r}, t) = S_0$. In accordance with (22.8), this surface coincides at the initial instant of time with the surface from which the particles emerged.

The relationship of momentum and energy with action. The surfaces of constant action are displaced in space and are orthogonal to the particle trajectories, because, from (10.23),

$$p_x = \frac{\partial S}{\partial x}, \quad p_y = \frac{\partial S}{\partial y}, \quad p_z = \frac{\partial S}{\partial z},$$

and, in general,

$$\mathbf{p} = \frac{\partial S}{\partial \mathbf{r}} \equiv \nabla S . \tag{22.9}$$

The partial derivatives are proportional to the direction cosines of the normal to the surface $S = S_0$.

Let us calculate, in addition, the partial derivative of S with respect to time. Since the action depends upon the coordinates and the time, its total derivative is equal to

$$\frac{\partial S}{\partial t} = \frac{\partial S}{\partial t} + \frac{\partial S}{\partial \mathbf{r}} \frac{d\mathbf{r}}{dt} . \tag{22.10}$$

But from (22.8)

$$\frac{dS}{dt} = L .$$

Substituting this into the left-hand side of (22.10), and $p = \frac{\partial S}{\partial \mathbf{r}}$ into the right-hand side of the same equation, we obtain

$$\frac{\partial S}{\partial t} = L - \mathbf{p}\mathbf{v} = -\mathscr{E} . \tag{22.11}$$

Similar quantities in an optical-mechanical analogy. Comparing equations (22.9) and (22.11) with (22.5) and (22.6), we conclude that a surface of constant action of a system of particles is propagated similar to a surface of constant phase: the *momentum of particles* is similar to the *wave vector*, while the *particle energy* is similar to the *frequency*. In accordance with Hamilton's equation (10.18), the particle velocity $\mathbf{v}$ analogous to the group velocity of a wave is

$$\mathbf{v} = \frac{\partial \mathscr{E}}{\partial \mathbf{p}}\,. \tag{22.12}$$

The velocity of a surface of constant action is

$$u = \frac{\left|\dfrac{\partial S}{\partial t}\right|}{\left|\dfrac{\partial S}{\partial \mathbf{r}}\right|} = \frac{\mathscr{E}}{p}\,. \tag{22.13}$$

This value by no means coincides with the particle velocity. Thus, for free particles

$$\mathscr{E} = \frac{mc^2}{\sqrt{1-\dfrac{v^2}{c^2}}}\,, \quad p = \frac{mv}{\sqrt{1-\dfrac{v^2}{c^2}}}\,,$$

so that $u = \frac{c^2}{v}$. The quantity u is analogous to the phase velocity of the waves.

The expression for group velocity (22.7) corresponds to Hamilton's equation (22.12). In Sec. 18 it was shown that the group velocity corresponds to the velocity of propagation for a wave packet, i.e., a disturbance concentrated within a certain region of space. Thus, the analogy between mechanics and geometrical optics establishes a correspondence between a particle and a wave packet.

The transition to geometrical optics provides for representation of the solution of the wave equation, in a certain region of space, in the form of a plane wave; however, the quantities defining this wave, such as the frequency and the wave vector, are themselves slowly varying functions of coordinates and time. The relationship between frequency and wave vector will be of the form $\omega = \omega(\mathbf{k}, \mathbf{r})$, where the quantities $\mathbf{k}$ and $\mathbf{r}$ that describe the wave propagation satisfy the same Hamiltonian equations as $\mathbf{p}$ and $\mathbf{r}$ of a particle moving along a trajectory. It is essential here that the vector $\mathbf{k}$ should not change much in magnitude and direction over a distance of one wavelength, and that the frequency ω should not change greatly in one oscillation period.

$\omega = ck$ for a plane wave in free space; this is completely analogous to the relationship (21.16) between the energy and momentum of a particle of zero mass.

If light is propagated in an inhomogeneous medium, the phase velocity u, appearing in equation (22.4), is a variable quantity. For example, when light is refracted at the boundary of two media, u has different values on both sides of the boundary. The propagation of light in an inhomogeneous medium is similar to the motion of a particle in a medium of variable potential energy.

The optical-mechanical analogy was established by Hamilton in 1825. However, up to the time of the formation of quantum mechanics (i.e., up to 1925), the physical significance of this analogy was not understood.

The law of transformation and the dimensions for similar quantities. The analogy between optical and mechanical quantities is relativistically invariant. Comparing formulae (20.33)-(20.36), for the transformation of wave vector and frequency, with (21.18)-(21.21), for the transformation of momentum and energy, we see that similar quantities are transformed in a similar manner.

The optical and mechanical quantities differ only in dimensions. Thus, phase has zero dimensions while action has the dimensions of $\int L\,dt$, i.e., gm. cm^2/sec. Accordingly, the wave vector and momentum, and the frequency and energy, likewise differ by the dimensions of action. As can be seen from a comparison of the Lorentz transformations for these quantities, the proportionality factor is an invariant quantity.

In the following section we shall show that the analogy between mechanics and geometrical optics emerges as a limiting relation from the precise wave equation of quantum mechanics.

Exercises

1) Formulate the equation of surfaces of constant action for a system of particles emerging from a single point in space $x = 0$, $z = 0$, in the plane $y = 0$ in a gravitational field. The absolute value of the particle velocities is v_0 and their direction is arbitrary.

We have

$$v_x = v_{0x}, \quad p_x = m v_{0x}, \quad x = v_{0x} t,$$

$$v_z = v_{0z} - g t, \quad p_z = m v_{0z} - m g t, \quad z = v_{0z} t - \frac{g t^2}{2}.$$

Eliminating the initial conditions for velocities, we obtain

$$p_x = m \frac{x}{t} = \frac{\partial S}{\partial x}, \quad p_z = \frac{m z}{t} - \frac{m g t}{2} = \frac{\partial S}{\partial z}, \quad S = \frac{m}{2}\left(\frac{x^2}{t} + \frac{z^2}{t} - g t z - \frac{g^2 t^3}{12}\right).$$

Indeed, from the expression for S we obtain

$$\mathscr{E} = -\frac{\partial S}{\partial t} = \frac{m}{2}\left[\left(\frac{x}{t}\right)^2 + \left(\frac{z}{t}\right)^2 + g z + \frac{g^2 t^2}{4}\right] = \frac{m v_0^2}{2}.$$

2) Proceeding from the fact that phase is analogous to action, show that light of given frequency is propagated along trajectories for which the propagation time of constant phase is least (Fermat principle).

At constant frequency

$$\varphi = \int \mathbf{k}\,d\mathbf{r} = \omega \int \frac{\mathbf{n}\,d\mathbf{r}}{u}.$$

But the product $\mathbf{n}\,d\mathbf{r}$ is equal to the displacement of the surface in a direction normal to it. It follows that $\frac{\mathbf{n}\,d\mathbf{r}}{u}$ is the propagation time dt. In accordance with the variational principle, which governs phase as well as action, the time must be least.

Sec. 23. Electron Diffraction

The essence of diffraction phenomena. Classical mechanics is analogous only to geometrical optics and by no means to wave optics. The difference between mechanics and wave optics is best of all illustrated by the example of diffraction phenomena.

Let us consider the following experiment. Let there be a screen with two small apertures. Let us assume that the distance between the apertures is of the same order of magnitude as the apertures themselves. Temporarily, we cover one of the apertures and direct a light wave on the screen. We shall observe the wave passing through the aperture by the intensity distribution on a second screen situated behind the first. Let us now cover the second aperture. The intensity distribution will be changed. Now let us open both apertures at once. An intensity distribution will be obtained which will not in any way represent the sum of the intensities due to each aperture separately. At the points of the screen, at which the waves from both apertures arrive in opposite phase, they will mutually cancel, while at those points at which the phase for both apertures is the same, they will re-inforce each other. In other words, it is not the intensities of light, i.e., the quadratic values, that are added, but the values of the fields themselves.

This type of diffraction can occur only because the wave passes through both apertures. Only then are definite phase differences obtained at points of the second screen for rays passing through each aperture.

We disregard here the diffraction effects associated with the passage through one aperture. These phenomena are due to the phase differences of the rays passing through various points of the aperture. Instead of examining such phase differences we consider that the phase of a wave passing through each aperture is constant, but we take into account the phase differences between waves passing through different apertures. Nothing is essentially changed by this simplification.

X-ray diffraction. In order to observe the diffraction of X-rays which are of considerably shorter wavelength than visible light, they can be made to scatter by correctly arranged atoms in a crystal lattice. Waves, scattered by different planes of the lattice, have constant phase differences 2π, 4π, $\ldots$, etc., for definite scattering directions. The distance between planes of the lattice, the scattering angles at which maxima are observed, and the wavelength are related by a simple formula (the Wolf-Bragg condition). From this equation we can determine the wavelength of X-radiation.

Diffraction by a crystal lattice is somewhat more complicated than in the experiment with two apertures, though, fundamentally, it occurs for the same reasons; the wave is scattered by all the atoms of the lattice, and the total amplitude of the scattered wave is the result of adding all the amplitudes of the waves scattered by all the atoms, with the path differences of the rays, i.e., the phase differences, taken into account.

Electron diffraction. The very same phenomena is observed when electrons (and also neutrons and other microparticles) are scattered by crystals. As we know, electrons act on a photographic plate or a luminescent screen in a way similar to X-rays; as a result, direct experiment shows that microparticles undergo diffraction that is governed by the same laws as the diffraction of electromagnetic waves.

However, for this, each electron must be scattered by all the atoms of the lattice, because the electrons travel entirely independently of each other; there can be no coherence (i.e., a constant phase difference) between them, nor can any arise. They may even pass through the crystal singly (see below). Only light waves which are originated from the same light source exhibit diffraction in such a manner: a stable diffraction pattern is obtained because the same wave passes through both apertures. If the waves passing through the apertures originated at different sources they could not cancel or reinforce each other at fixed points of the screen. The alternation of light and dark regions would depend on the relative phases with which the waves passed through the apertures; a constant phase difference cannot be maintained for light from different sources.

Electron diffraction demonstrates that the laws of motion in the microworld are wave-like in character; to obtain the same diffraction pattern for X-rays, each electron must be scattered by all the atoms of the lattice. This is clearly incompatible with the concept of a definite electron trajectory.

Diffraction phenomena prove that electron motion is associated with a phase of a certain magnitude.

The de Broglie wavelength. From diffraction experiments we can, without difficulty, determine the wavelength both for electrons and for X-rays. For electrons, it turns out to be very simply related to their velocities. Let us write down the expression for the wave vector obtained from experiment:

$$\mathbf{k} = \frac{\mathbf{p}}{h}. \tag{23.1}$$

Here, h is a universal constant having the dimensions of action spoken of in the preceding section (p. 229). It is equal to 1.054×10^{-27} erg-sec. Earlier, it was much more common to use a constant 2π times bigger. The value for h used in this book is frequently denoted by $\hbar$.

The relation (23.1) was suggested in 1923 by L. de Broglie, before the first experiments on electron diffraction. The quantity

$$\lambda = \frac{2\pi}{k} = \frac{2\pi h}{mv} \tag{23.2}$$

is called the de Broglie wavelength.

Equation (23.2) shows that a certain wavelength can be ascribed to the motion of each body, but in the motion of macroscopic bodies it is extremely small as a result of the smallness of h compared with the quantities which characterize the motion of macroscopic bodies, where the orders of magnitude correspond to the cgs system. Therefore, diffraction phenomena do not actually restrict the applicability of classical mechanics to macroscopic bodies.

The limits of applicability of classical concepts. The relationship between the quantities is entirely different when equation (23.2) is applied to the motion of an electron in an atom. The size of an atom is determined, for example, from experiments on X-ray diffraction or, more simply, by dividing the volume of one gram-atom of condensed material (solid or liquid) by the Avogadro number $N = 6.024 \times 10^{23}$. The atomic radius is of the order of 0.5×10^{-8} cm. From this it is easy to evaluate the velocity of an electron by equating the "centrifugal force" $\frac{mv^2}{r}$ to the force of attraction to the nucleus, equal to $\frac{e^2}{r^2}$ in a hydrogen atom. For the velocity, the value obtained is

$$v \sim \frac{e}{\sqrt{rm}},$$

whence the wavelength (23.2) is

$$\lambda \sim \frac{2\pi h}{e}\sqrt{\frac{r}{m}}.$$

Substituting the numerical values $m \sim 10^{-27}$ gm, $e \sim 4.8 \times 10^{-10}$ CGSE, we convince ourselves that the wavelength is approximately six times larger than r. In other words, a distance of the order of an atomic diameter can accommodate one third of a wave: this corresponds to dimensions which are characteristic of diffraction phenomena and completely "smears out" the trajectory over the atom.

Hence the motion of an electron in an atom is wave motion. Just as the concept of a ray has no place in optics, if the light is propagated in a region comparable with the wavelength, so the concept of an electron trajectory becomes meaningless for the motion of an electron in an atom.

The electron is not a wave, but a particle! It is necessary to warn the reader against certain common delusions. First of all, contrary to what is often written in the popular literature, the electron is never a wave even in quantum considerations. Without any doubt the electron remains a particle; for example, it is never possible to observe part of an electron. If a photographic plate replaces the second (rear) screen in the diffraction experiment, then at the point of incidence of each electron there will appear a single point of blackening. The point distribution characteristic of a diffraction pattern will also result when the electrons pass through a crystal singly.* Thus, it is not the electron that becomes a wave, but the laws of motion in the microworld that are wave-like in character.

It is clear that a diffraction pattern can in no way be obtained from a single electron. Since each electron gives a single point on the screen, we must have very many separate points in order to obtain the correct alternation of light and dark regions on the plate.

At the same time, diffraction would be utterly impossible if both screen apertures or all the crystal atoms did not actually participate in the passage of one and the same electron. In the diffraction experiment, electron trajectories simply do not exist. What is actually wave-like in the motion of a particle will be shown later.

In all its properties, the electron is a particle. Its mass and charge always belong to it and are never divided in any diffraction experiment or any other experiment that we know.

The incompatibility of trajectories and diffraction phenomena. Another common delusion is that the electron supposedly does possess a trajectory but that we are as yet unable to observe it due to imperfections in technical facilities, or to the inadequacy of our physical knowledge. In actual fact, the diffraction experiment shows that the electron definitely does not have a trajectory, just as diffracted light is not propagated in rays. To think that the development of physics will in future show the existence of an electron trajectory in the atom is just as unreasonable as to hope for the return of phlogiston in heat theory, or of a geocentric world system in astronomy.

Statistical regularity and the individual experiment. The absence of trajectories by no means signifies that we have lost all regularity. On the contrary, an identical diffraction experiment, performed, of course, with a large number of separate electrons of definite velocity, always yields an identical diffraction pattern. Thus, causal regularity undoubtedly exists. However, it is statistical in character appearing in a very large number of separate experiments, because each electron passage through a crystal may be regarded as a separate *independent* result.

Diffraction phenomena lead to a regular distribution of points on a photographic plate in the same way that a large number of shots at a target is subject to a law of dispersion. However, as opposed to bullets which fly along trajectories and therefore give a smooth distribution curve for the places where they strike the target, the blackened grains on a photographic plate caused by electrons are produced in a more intricate manner characteristic of wave motion. The distribution of bullet fits is due to indeterminacy in the initial firing conditions and becomes less in the case of better aiming, while the random character of electron

* In somewhat different form, this was shown by direct experiment by V. A. Fabrikant, N. G. Sushkin and L. M. Biberman, using currents of very low intensity.

behaviour presents a perfectly regular diffraction pattern and, for a given electron velocity, can in no way be reduced.

In addition, it may be noted that the statistical regularity in a diffraction experiment has nothing to do with the statistical regularities which govern the motion of a large ensemble of interacting particles. As has already been repeated several times, the same pattern is obtained completely independently of the way in which the electrons pass through the crystal—all at once, or singly. A certain phase governing the motion exists only because each electron interferes with itself.

Electron trajectories in a Wilson cloud chamber. It still remains to examine in more detail the question: In which cases do we, nevertheless, deal with the concept of an electron trajectory? In a cloud chamber, in a cathode-ray oscilloscope, and in many other instruments, electron trajectories can be precalculated very well from the laws of classical mechanics. In a cloud chamber, there even remains a cloud track along the line of motion of an electron.*

First, recall that under certain conditions light is also propagated along definite trajectories (rays). Geometrical optics is applicable when the inaccuracy in defining the wave vector Δk_x, subject to the inequality

$$\Delta k_x \cdot \Delta x \gtrsim 2\pi \tag{23.3}$$

[see (18.9)], is small in comparison with k. Substituting Δk_x for an electron in equation (23.1), we obtain the analogous expression of quantum theory:

$$\Delta p_x \cdot \Delta x \sim 2\pi h^{**}. \tag{23.4}$$

This is the so-called *uncertainty relation* of quantum mechanics. The concept of an electron trajectory has reasonable meaning if the uncertainties of all three momentum components Δp_x, Δp_y, Δp_z are small compared with the momentum itself:

$$\Delta p_x \ll p_x, \quad \Delta p_y \ll p_y, \quad \Delta p_z \ll p_z. \tag{23.5}$$

It may be pointed out that we have all along been saying "electron" simply to be specific. The same applies to a proton, neutron, meson and the like.

Let us suppose that the track of an electron in a cloud chamber is 0.01 cm wide and the electron energy is equal to 1.000 ev $= 1.6 \times 10^{-9}$ erg (1 electron-volt $= \frac{4.8 \times 10^{-10}}{300} = 1.6 \times 10^{-12}$ erg). According to (23.4)

---

* An electron passing through a gas ionizes the atoms in its path. The supersaturated vapour with which the chamber is filled condenses on the ions. Upon illumination, the droplets appear in the form of a cloud track.

** Sometimes, Δp_x, Δx are meant to signify not the "spreadings" p_x and x in themselves, but their mean square values. Then, $\Delta p_x \, \Delta x \gtrsim h$.

the component of momentum perpendicular to the trajectory has an indeterminacy

$$\frac{6.6 \times 10^{-27}}{10^{-2}} = 6.6 \times 10^{-25},$$

and the momentum itself is

$$p = \sqrt{2m\mathscr{E}} = \sqrt{2.9 \times 10^{-36}} = 1.7 \times 10^{-18}.$$

It follows that the relationships (23.5) are satisfied to an accuracy of up to four parts in ten million. The observation of a track in a cloud chamber does not allow us to determine the trajectory with accuracy sufficient to notice deviations (in electron motion) from the laws of classical mechanics.

The limitations of the concepts of classical mechanics. Thus, quantum mechanics does not abolish classical mechanics, but contains it as a limiting case, much the same as wave optics includes the geometrical optics of light rays as a limiting case. As we shall see later, quantum mechanics is concerned with the same quantities as classical mechanics, i.e., energy, momentum, coordinates, moment. But the finiteness of the quantum of action h imposes a limitation on the applicability of any two classical concepts (for example, coordinates and momentum) for one and the same motion.

The coordinate and momentum of an electron cannot simultaneously have precise values because the motion is wave-like. To attempt to define these precise values is just as meaningless physically as it is to seek precise trajectories for light rays in wave optics. In the same way that it is impossible to obtain, as a result of improvements in optical devices, a precise definition for light rays in wave optics, any progress in measuring techniques as applied to the electron will not allow us to determine its trajectory more precisely than indicated by the relation (23.4), since strictly speaking the trajectory *does not exist.*

Attempts are sometimes made to interpret relation (23.4) erroneously. It is taken that a trajectory cannot be determined because the precision of the initial conditions does not exceed Δp_x and Δx connected by relation (23.4). This would mean that some actual trajectory does exist but that it lies within a more or less narrow region of space and within a certain range of momenta. The "real" trajectory is likened to the imaginary trajectory from a gun to a target before firing. The path of the bullet is not precisely known beforehand, if only because strictly identical powder charges cannot be obtained. But this inaccuracy in the initial conditions for the bullet only leads to a smooth dispersion curve for the hits on the target, while the distribution of electrons indicates diffraction effects. The presence of diffraction shows that no "real-though-unknown-to-us" trajectory exists. As a matter of fact, relation (23.4) by no means indicates with what *error* certain quantities may be *measured* simultaneously, but to what extent these quantities have precise meaning in the given motion. It is this that the *uncertainty* principle of quantum mechanics expresses. The term "uncertainty" emphasizes the fact that what we are concerned with is not accidental errors of measurement or the imperfection of physical apparatus, but the fact of momentum and coordinate of a particle being actually nonexistent in the same state.

Sec. 24. The Wave Equation

The wave function. Diffraction of light occurs because the wave amplitudes are added. When the wave phases coincide the intensity (which is proportional to the square of the resultant amplitude) is maximum; when the phases are opposite the intensity is minimum. In the diffraction of electrons, a quantity similar to intensity is measured by the blackening of a plate, which blackening is proportional to the number of electrons incident on unit area. The distribution of the blackened grains on the plate obeys the same law as in the case of the diffraction of X-rays (in the sense of alternation of maxima and their relative positions). Thus, in order to explain the diffraction of electrons we must assume that with their motion there can be associated some wave function whose phase determines the diffraction pattern.

At the end of this section, we will show in the general case that such a wave function must be complex, since a real wave function cannot correspond to just any type of motion.

Probability density. Electrons move independently of one another and pass through a crystal singly, as it were. Therefore, the number of electrons in an element of volume dV is proportional to the probability of the appearance of one electron. Probability (a quantity similar to light intensity) must be quadratic with respect to the wave function, in the same way that light intensity is quadratic in wave amplitude. But since probability is a real quantity, it can only depend on the square of the modulus of the wave function. Let us put

$$dw = |\psi(x, y, z, t)|^2 dV. \tag{24.1}$$

Here dw is the probability of finding an electron in the volume dV at the instant t; then $|\psi|^2$ is the probability referred to unit volume or, otherwise, the probability density.

The linearity of the wave equation. Like in optics, where the laws of propagation of the wave itself are studied on the basis of Maxwell's equations, the intensity being found by squaring the wave amplitude, it is necessary in quantum mechanics to find the equation governing the quantity ψ and not the probability density. This equation must be linear. Indeed, two interfering waves, when combined, give a resultant wave. In order to obtain the same interference pattern as in optics, it is necessary to perform a simple algebraic addition of the functions; both the summands and the sum must satisfy the same wave equation. But only the solution of linear equations satisfy this requirement. The phases of the waves are very essential here since in order to formulate the laws of diffraction it is necessary to know not only the behaviour of the squares of the amplitude but also that of their phases.

In other words, we must have an equation for the wave function itself of the particle $\psi\ (x,\ y,\ z,\ t)$.

The wave function of a free particle. Proceeding from the analogy between geometrical optics and mechanics, it is easy to construct a wave corresponding to a free particle not subject to the action of external forces. We know that the state of a free particle is characterized by its momentum $\mathbf{p}$. But, in accordance with relationship (23.1), a wave with wave vector $\mathbf{k} = \frac{\mathbf{p}}{h}$ corresponds to a particle with momentum $\mathbf{p}$. It follows that the wave function of a free particle depends on coordinates in the following way (we write it in complex form):

$$e^{i\mathbf{kr}} = e^{i\frac{\mathbf{pr}}{h}}.$$

The time dependence of a wave function is also determined very simply, if we recall that the frequency of a wave corresponds to the particle energy (Sec. 22). The proportionality factor between them has the dimensions of action. As was shown at the end of Sec. 22, the factor of proportionality must be the same as between the wave vector and momentum; this follows from the condition of relativistic invariance of the correspondence between mechanical and optical quantities.* Hence

$$\omega = \frac{\mathscr{E}}{h}. \tag{24.2}$$

Whence we obtain the wave function of a free particle

$$\psi = e^{-i\omega t + i\mathbf{kr}} = e^{-i\frac{\mathscr{E}t}{h} + i\frac{\mathbf{pr}}{h}}. \tag{24.3}$$

The group velocity of the waves is

$$\mathbf{v} = \frac{\partial \omega}{\partial \mathbf{k}} = \frac{\partial \mathscr{E}}{\partial \mathbf{p}}. \tag{24.4}$$

Hence, it coincides with the particle velocity as it should in accordance with (22.12), thereby confirming (24.2).

The relation between wave function and action. We note that the wave function (24.3) may be written in the form

$$\psi = e^{i\frac{S}{h}}, \tag{24.5}$$

* Or at least from invariance to Galilean transformations.

where S is the action of the particle. Indeed, the action of a free particle is

$$S = -\mathscr{E}t + \mathbf{pr}, \tag{24.6}$$

because from this we obtain

$$\mathbf{p} = \frac{\partial S}{\partial \mathbf{r}} \equiv \nabla S, \quad \mathscr{E} = -\frac{\partial S}{\partial t},$$

as should be the case according to equations (22.9) and (22.11).

Equation (24.5) confirms the relationship established also in Sec. 22 between the wave phase and the action of a particle.

The wave equation for a free particle. The analogy between mechanics and optics by no means presupposes that the equations of mechanics are written in a relativistically invariant form. It is sufficient to recall that the analogy had already been established by Hamilton in 1825. The significance of the analogy consists in the fact that a correspondence is established between quantities: momentum and wave vector, energy and frequency, action and phase. In future, except in those cases where it is specifically stated otherwise, we shall proceed from the nonrelativistic form of the equations of mechanics.

Let us now find the differential equation satisfied by the wave function (24.3). We have

$$\frac{\partial \psi}{\partial t} = -\frac{i\mathscr{E}}{h}\psi, \tag{24.7}$$

$$\left.\begin{aligned} \frac{\partial \psi}{\partial x} &= \frac{\partial}{\partial x} e^{-\frac{i\mathscr{E}t}{h} + \frac{i(p_x x + p_y y + p_z z)}{h}} = \frac{ip_x}{h}\psi, \\ \frac{\partial^2 \psi}{\partial x^2} &= -\frac{p_x^2}{h^2}\psi . \end{aligned}\right\} \tag{24.8}$$

From (24.7) and (24.8) we obtain

$$-\frac{h}{i}\frac{\partial \psi}{\partial t} = \mathscr{E}\psi = -\frac{h^2}{2m}\left(\frac{\partial^2\psi}{\partial x^2} + \frac{\partial^2\psi}{\partial y^2} + \frac{\partial^2\psi}{\partial z^2}\right) = \frac{p^2}{2m}\psi \tag{24.9}$$

(here we are already using nonrelativistic expressions!), or in abbreviated form

$$-\frac{h}{i}\frac{\partial \psi}{\partial t} = -\frac{h^2}{2m}\Delta\psi, \tag{24.10}$$

where Δ s the Laplacian operator. Equation (24.10) holds because $\mathscr{E} = \frac{p^2}{2m}$, as can be seen from (24.9).

The Schrödinger equation. Let us now generalize equation (24.10) to the case of a particle moving in an external potential field U. In order to have a relation analogous to $\mathscr{E} = \frac{p^2}{2m} + U$ since $\mathscr{E} = \frac{p^2}{2m}$ for a free particle in (24.9), we must put

$$-\frac{h}{i}\frac{\partial \psi}{\partial t} = -\frac{h^2}{2m}\Delta\psi + U\psi. \tag{24.11}$$

E. Schrödinger formulated this equation in 1925, generalizing the de Broglie relations for free electrons to the case of bound electrons (this was also before the discovery of electron diffraction).

Equation (24.11) directly follows from (24.10) for the simplest case $U = \text{const}$, because then it is satisfied by the same substitution of (24.3), though with the momentum value $p = \sqrt{2m(\mathscr{E} - U)}$. From here it is only one step to generalization to the case of variable potential energy.

But this generalization must in no way be regarded as the derivation of the Schrödinger equation from the equations of prequantum physics, for it expresses a new physical law.

Its relationship to classical physics can be seen in the limiting transition, which is fully analogous to the transition from wave optics to geometrical optics.

The limiting transition to classical mechanics. Let us substitute the expression for the wave function (24.5) in (24.11). This expression must hold in the above limiting transition because then the wave phase surfaces $\varphi = \text{const}$ correspond to surfaces of constant action $S = \text{const}$ for particles.

$$\varphi = \frac{S}{h}. \tag{24.12}$$

Instead of the formal relationship considered in Sec. 22, we now have an *equality*, since we have introduced a new, universal physical constant h. Thus, we put

$$\psi = e^{i\frac{S}{h}}.$$

Whence

$$\frac{\partial \psi}{\partial t} = \frac{i}{h} \frac{\partial S}{\partial t} \psi,$$

$$\frac{\partial \psi}{\partial x} = \frac{i}{h} \frac{\partial S}{\partial x} \psi,$$

$$\frac{\partial^2 \psi}{\partial x^2} = \frac{i}{h} \frac{\partial^2 S}{\partial x^2} \psi - \frac{1}{h^2} \left(\frac{\partial S}{\partial x} \right)^2 \psi.$$

Let us substitute these derivatives in (24.11). Then, after eliminating ψ, we obtain the equality

$$-\frac{\partial S}{\partial t} = \frac{1}{2m} \left[\left(\frac{\partial S}{\partial x} \right)^2 + \left(\frac{\partial S}{\partial y} \right)^2 + \left(\frac{\partial S}{\partial z} \right)^2 \right] - \frac{ih}{2m} \Delta S + U. \tag{24.13}$$

The limiting transition from quantum to classical mechanics is attained by considering the de Broglie wavelength very small compared with the region in which motion occurs. Since the wavelength is proportional to the quantum of action h, the same limiting transition may be performed formally by considering that h tends to zero. This means that all the quantities having the dimensions of action are so large compared with the quantum h, that the latter can be neglected. In (24.13), passing to the limit $h=0$, we have

$$-\frac{\partial S}{\partial t}=\frac{(\nabla S)^2}{2m}+U \tag{24.14}$$

or $\mathscr{E}=\frac{p^2}{2m}+U$ from (22.9) and (22.11).

The limiting transition that we have performed here almost completely repeats the transition from wave optics to geometrical optics carried out in Sec. 22.

The correspondence between classical and quantum theory. We have seen that the Schrödinger wave equation does in fact give a correct limiting transition. This equation is, as it were, a fourth member in the following correspondence:

geometrical optics	→	classical mechanics
↓		↓
wave optics	→	quantum mechanics

The vertical arrows denote a transition from rays or trajectories to wave patterns, while the horizontal arrows denote a transition from waves to particles. The latter relates only to nonquantum electrodynamics because in the transition to quantum field equations the need arises for a corpuscular representation (see Sec. 27). Here, we consider only the analogy between quantum mechanics and classical wave optics.

The range of application of various theories. The regions in which quantum mechanics and wave optics can be applied do not overlap anywhere; in wave optics or, what is just the same, in electrodynamics, the velocity of light c is regarded as finite but the quantum of action h is considered arbitrarily small. In nonrelativistic quantum mechanics, c is considered arbitrarily large while h has a finite value. A quantum theory of the electromagnetic field, in which both h and c have finite values (i.e., the velocity ranges are comparable with c, and quantities with the dimensions of action are comparable with h), has, in essentials, also been completed at the present time. At any rate, any concrete problem requiring the application of quantum electrodynamics, may be uniquely solved to any required degree of precision, and the results agree with experiment. The existence of a light quantum as an independent particle is not a supplementary hypothesis which must be made in order to formulate quantum electrodynamical equations. The consistent quantization of electromagnetic field equations necessarily leads to the corpuscular aspect of the theory (for more detail see Sec. 27).

The *nonrelativistic* particle quantum mechanics (i.e., constructed on the relation $\mathscr{E} = \frac{p^2}{2m} + U$) is, in the region for which it is applicable, a theory which is just as consummate as Newtonian mechanics. Like the equations of Newtonian mechanics, the wave equation (24.11) is valid only for particle velocities small compared with the velocity of light. But still, in the region for which it is applicable, it is just as firmly established (in the same sense) as are the Newtonian laws for the motion of macroscopic bodies.

The grounds for this are absolutely the same—both nonrelativistic quantum theory and Newtonian mechanics agree with the widest range of experimental data, never contradicting them and providing for correct and unique predictions. In addition, they nowhere contain contradictory statements. The latter condition is, of course, not sufficient for a physical theory to be correct but it is at least necessary. The Bohr theory, or the old quantum mechanics (as it is otherwise known) did not satisfy this requirement; in addition to the classical concept of trajectory, it involved the quantum concept of discreteness of states. For this reason, it had always been clear that the Bohr formulation of quantum theory was not final and should be revised, no matter how wide the range of experimental facts that it explained.

Quantum mechanics permitted the construction of a consistent theory of atomic structure. The actual calculation of wave functions for electrons in complex atoms is a problem of enormous mathematical complexity.* However, it is, of course, by no means the purpose of quantum mechanics to calculate the spectra of complex atoms: the essential point is that quantum mechanics allows us to systematize atomic and molecular states in such a way that the very nature of the spectra is understood, whereas classical mechanics could not explain even the stability of the atom. Thanks to quantum mechanics such fundamentally important facts as the chemical affinity of atoms or Mendeleyev's periodic law are now understood.

In its domain, quantum mechanics will, of course, perfect methods of approach to various concrete problems. The correctness of its general principles will serve as a basis for such refinement.

The normalization condition for a wave function. Let us return to the wave equation (24.11). We shall write it for a wave function ψ and a conjugate function ψ^* (in the second equation we have to replace $-i$ by i):

$$-\frac{h}{i}\frac{\partial\psi}{\partial t} = -\frac{h^2}{2m}\Delta\psi + U\psi,$$

$$\frac{h}{i}\frac{\partial\psi^*}{\partial t} = -\frac{h^2}{2m}\Delta\psi^* + U\psi^*.$$

---

* It is considerably simplified thanks to approximation methods suggested by V. A. Fok.

Let us multiply the first equation by ψ^*, and the second by ψ, and subtract the second from the first. The term $\psi^* U \psi$ is eliminated and the remaining terms give

$$-\frac{h}{i}\left(\psi^* \frac{\partial \psi}{\partial t} + \psi \frac{\partial \psi^*}{\partial t}\right) = -\frac{h^2}{2m}(\psi^* \Delta \psi - \psi \Delta \psi^*). \qquad (24.15)$$

The left-hand side of the last equality is transformed to the form

$$-\frac{h}{i}\frac{\partial}{\partial t}\psi^* \psi = -\frac{h}{i}\frac{\partial}{\partial t}|\psi|^2.$$

We can write the right-hand side more fully thus:

$$-\frac{h^2}{2m}(\psi^* \Delta \psi - \psi \Delta \psi^*) = -\frac{h^2}{2m}(\psi^* \operatorname{div}\operatorname{grad}\psi - \psi \operatorname{div}\operatorname{grad}\psi^*) =$$
$$= -\frac{h^2}{2m}\operatorname{div}(\psi^* \operatorname{grad}\psi - \psi \operatorname{grad}\psi^*)$$

[see (11.27)]. Finally, we represent the equality obtained in the following form:

$$\frac{\partial}{\partial t}|\psi|^2 = -\operatorname{div}\left\{\frac{h}{2mi}(\psi^* \nabla \psi - \psi \nabla \psi^*)\right\}. \qquad (24.16)$$

The left-hand side of this equality is the time derivative of the probability density of finding a particle close to some point of space. Let us integrate (24.16) over the whole volume in which the particle might be situated. If this volume is finite then beyond its boundaries ψ and ψ^* must be equal to zero. But then, from the Gauss-Ostrogradsky theorem

$$\frac{\partial}{\partial t}\int |\psi|^2 dV = 0. \qquad (24.17)$$

It follows that the integral itself does not depend upon time. It is easy to see that it must be equal to unity because this is the probability of an electron being somewhere, i.e., the probability of a trustworthy event. The condition

$$\int |\psi|^2 dV = 1 \qquad (24.18)$$

is called the *normalization* condition of a wave function.

If the electron motion is infinite, i.e., ψ nowhere becomes zero, the normalization condition appears more complicated. However, in practice, it is always possible to consider that the volume in which the electron is situated is very large and finite, so that the condition (24.18) can be used. The physical results, naturally, do not depend upon the arbitrary choice of volume.

Probability-flux density. If we integrate (24.16) over any arbitrary volume, we obtain

$$\frac{\partial}{\partial t}\int|\psi|^2 dV = -\int \operatorname{div}\left\{\frac{h}{2mi}(\psi^*\nabla\psi - \psi\nabla\psi^*)\right\}dV =$$
$$= -\int \frac{h}{2mi}(\psi^*\nabla\psi - \psi\nabla\psi^*)\,dV. \qquad (24.19)$$

If on the left we have the change in the probability of finding an electron inside the given volume, then on the right-hand side we must have the flux of the probability of it passing through the boundary surface of the volume. According to (24.19) the density of the probability flux is equal to

$$\mathbf{j} = \frac{h}{2mi}(\psi^*\nabla\psi - \psi\nabla\psi^*). \qquad (24.20)$$

It follows that a real wave function gives $\mathbf{j}=0$, i.e., it cannot be used for describing the current of an electron. Therefore, in a general definition, the wave function ψ must be a complex quantity.

The equation for stationary states. Let us assume that potential energy does not depend explicitly upon time. Then classical mechanics leads to conservation of energy of the system. The action of such a system involves a term $-\mathscr{E}t$. But since $\psi = e^{i\frac{S}{h}}$ in quantum mechanics, too, we must seek a wave function in the form

$$\psi = e^{-i\frac{\mathscr{E}t}{h}}\psi_0(x, y, z). \qquad (24.21)$$

Substituting this in (24.11) and omitting the zero subscript, we obtain the equation

$$-\frac{h^2}{2m}\Delta\psi + U\psi = \mathscr{E}\psi. \qquad (24.22)$$

As we shall see in the next section, this equation has a solution which does not satisfy definite necessary conditions for all values of $\mathscr{E}$. Thus, it turns out that, in contrast to the energy in classical mechanics, the energy of a quantum system cannot always be arbitrarily given.

Exercise

Prove that if there are two solutions of (24.22) for different values of energy $\mathscr{E}$ and $\mathscr{E}'$, then

$$\int \psi^*(r, \mathscr{E})\,\psi(r, \mathscr{E}')\,dV = 0.$$

The functions $\psi^*(r, \mathscr{E})$ and $\psi(r, \mathscr{E}')$ satisfy the equations

$$-\frac{h^2}{2m}\Delta\psi^* + U\psi^* = \mathscr{E}\psi^*,$$
$$-\frac{h^2}{2m}\Delta\psi + U\psi = \mathscr{E}'\psi.$$

Let us multiply the first equation by ψ, the second by ψ^*, and subtract the second from the first. Integrating over the whole volume, similar to (24.19), and then transforming the volume integral on the left to a surface integral, we obtain the equation

$$(\mathscr{E} - \mathscr{E}') \int \psi^* (r, \mathscr{E}) \psi (r, \mathscr{E}') \, dV = 0 .$$

It follows that if $\mathscr{E} \neq \mathscr{E}'$, then the second factor is equal to zero as required. This is the so-called orthogonality property of wave functions. It will be shown in more general form in Sec. 30 because it forms one of the most important principles of quantum mechanics.

Sec. 25. Certain Problems of Quantum Mechanics

In this section we shall obtain solutions to the wave equation for certain cases which are partly illustrative and partly auxiliary. Nevertheless, many important laws are explained from these examples.

We have already obtained the solution of the wave equation for a free particle (24.3). We shall now examine the solutions for bound particles.

A particle in a one-dimensional, infinitely deep potential well. Let us suppose that a particle is constrained to move in one dimension remaining in an interval of length a, so that $0 \leqslant x \leqslant a$. We can imagine that at the points $x = 0$ and $x = a$, there are absolutely impenetrable walls which reflect the particle. A limitation of this type is represented with the aid of the potential energy curve shown in Fig. 36. $U = \infty$ at $x < 0$ and $x > a$. We put $U = 0$ at $0 \leqslant x \leqslant a$; this is the potential energy gauge. To leave the region $0 \leqslant x \leqslant a$, a particle would have to perform an infinitely large quantity of work. Thus, the probability for the particle to be at $x = 0$ or $x = a$ is equal to zero. With the aid of (24.1), we obtain

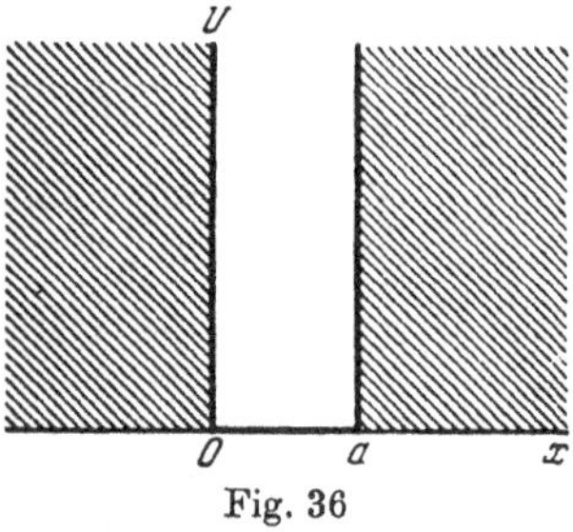

Fig. 36

$$\psi(0) = \psi(a) = 0 . \tag{25.1}$$

These boundary conditions may also be justified by means of a limiting transition from a well of finite depth. This will be done later.

Insofar as the potential energy is time independent, the wave equation must be written in the form of (24.22). The motion is one-dimensional and, therefore, we must take the total derivative $\frac{d^2}{dx^2}$ in place of Δ. From this we have

$$-\frac{h^2}{2m} \frac{d^2\psi}{dx^2} = \mathscr{E} \psi . \tag{25.2}$$

We introduce the shortened notation

$$\frac{2m\mathscr{E}}{h^2} \equiv \varkappa^2, \tag{25.3}$$

so that the wave equation will be of the form

$$\frac{d^2\psi}{dx^2} = -\varkappa^2\psi. \tag{25.4}$$

The solution to (25.4) is well known:

$$\psi = C_1 \sin \varkappa x + C_2 \cos \varkappa x. \tag{25.5}$$

But from (25.1) $\psi(0) = 0$ so that the cosine term must be omitted by putting $C_2 = 0$. There remains

$$\psi = C_1 \sin \varkappa x. \tag{25.6}$$

We now substitute the second boundary condition

$$\psi(a) = C_1 \sin \varkappa a = 0. \tag{25.7}$$

This is an equation in χ. It has an infinite number of solutions:

$$\varkappa a = (n+1)\pi, \tag{25.8}$$

where n is any integer equal to or greater than zero:

$$0 \leqslant n \leqslant \infty. \tag{25.9}$$

We discard the value $n = -1$ because, for $n = -1$, the wave function becomes zero everywhere, $\psi = \sin 0 = 0$. Hence, $|\psi|^2 = 0$, so that the particle simply does not exist anywhere (a "trivial solution"). Now substituting χ from the definition (25.3) and solving (25.8) with respect to energy, we find an expression for the energy

$$\mathscr{E}_n = \frac{\pi^2 h^2}{2ma^2}(n+1)^2. \tag{25.10}$$

Eigenvalues of energy and eigenfunctions. The boundary condition imposed on a wave function is, for a given problem, just as necessary as the wave equation itself. However, as can be seen from (25.10), the boundary condition is not satisfied for all values of the energy, but only for values belonging to a definite series of numbers by which the problem under consideration is given. It will be seen later on that, depending upon the conditions, these numbers may form a discrete series or a continuous sequence. They are called *eigenvalues* of the energy of a quantum mechanical system. The wave functions belonging to the energy eigenvalues are called *eigenfunctions*.

The foregoing example is the simplest, one in which the energy eigenvalues form a discrete series. The energy of a free particle forms a continuous series of eigenvalues. Indeed, the only condition which can be imposed on the wave function of a free particle consists in the fact that it must be finite everywhere, because the square of its modulus is the probability density of finding the particle at a given point in space. But the function (24.3) remains finite for all real values of p and $\mathscr{E}$.

The assembly of energy eigenvalues of a particle is termed its *energy spectrum*. The energy spectrum in an infinitely deep potential well is discrete, while the energy spectrum of a free particle is continuous.

The solution of Schrödinger's equation (24.22) for stationary states is always associated with finding the energy spectrum. In contrast to the Bohr theory, where the discreteness of the states appeared as a necessary but foreign appendage to classical motion, in quantum mechanics the very nature of the motion determines the energy spectrum. This will be seen especially clearly in the examples to follow.

The nodes or zeros of a wave function. The function ψ becomes zero n times in the interval $0 - a$ (except at its ends). The quantity of zeros ("nodes") of a wave function is equal to the number of the subscript of the energy eigenvalue.

This result is easily understood from the following considerations. In the interval $0 - a$ (for $n=0$) there is one sinusoidal half-wave; for $n=1$, there is one complete wave; for $n=2$, there is one and a half waves, etc. Thus, the greater n is, the less the de Broglie wavelength λ. But energy is proportional to the square of the momentum, i.e., it is inversely proportional to the square of λ, in accordance with (23.2). Therefore, the less λ is, the greater the energy. This conclusion holds, of course, for wave functions that are not of a purely sinusoidal shape, though not as a general quantitative relationship but, instead, qualitatively—the more zeros or "nodes" that the wave function has, the greater the energy. The least energy state has no zeros anywhere except at the limits of the interval $x=0$ and $x=a$. It is called the *ground* state, all the other states being termed *excited*.

Normalization of a wave function. It remains to determine the coefficient C_1 in order to define a wave function completely. We shall find it from the normalization condition (24.18):

$$\int_0^a |\psi|^2 dx = C_1^2 \int_0^a \sin^2 \varkappa x \, dx = C_1^2 \int_0^a \frac{1-\cos 2\varkappa x}{2} dx =$$

$$= C_1^2 \left(\frac{x}{2} - \frac{\sin 2\varkappa x}{4\varkappa} \right) \Bigg|_0^a = \frac{C_1^2 a}{2} = 1 .$$

The second term of the integrated expression becomes zero at both limits in accordance with (25.8). Thus,

$$C_1 = \sqrt{\frac{2}{a}} , \qquad (25.11)$$

$$\psi_n = \sqrt{\frac{2}{a}} \sin \frac{\pi(n+1)x}{a}. \qquad (25.12)$$

Real wave functions. The wave function (25.12) is real. Therefore, from (24.20), the current in this state is equal to zero. This can also be seen in the following way. The wave function (25.12) can be expanded into the sum of two exponentials. Each such exponential represents, together with a time factor, the wave function of a free particle (24.3), one of them corresponding to a momentum $p = h\varkappa$ and the other, to the same momentum but with opposite sign. Thus, a state with wave function (25.12) is represented as the superposition of two states with opposite momenta, these states having equal amplitudes. The mean momentum for a particle moving in a potential well according to classical mechanics is equal to zero; it changes sign for every reflection from the walls of the well. In this sense, we can say that the mean momentum is also equal to zero for quantum motion. The difference is that at every given instant classical momentum possesses a definite value, while the quantum momentum of a particle in a well never has a definite value; the wave function involves states with momenta of both signs. This corresponds to the uncertainty principle; since the particle coordinate is within the limits $0 \leqslant x \leqslant a$, the momentum cannot have an exact value.

In addition, we note that in this *particular* problem of a rectangular well the square of the momentum has a definite value since the uncertainty is extended only to the sign of the momentum. The square of the momentum in this case is proportional to the energy. The square of the momentum for a well of any arbitrary shape is also not fixed.

A particle in a three-dimensional infinitely deep potential well. Let us now suppose that a particle is contained in a box whose edges are a_1, a_2, a_3. Generalizing the boundary conditions (25.1), we conclude that the wave function becomes zero on all the sides of the box:

$$\psi(0, y, z) = \psi(x, 0, z) = \psi(x, y, 0) = \psi(a_1, y, z) =$$
$$= \psi(x, a_2, z) = \psi(x, y, a_3) = 0. \qquad (25.13)$$

The wave equation must now be written in three-dimensional form:

$$-\frac{h^2}{2m}\left(\frac{\partial^2 \psi}{\partial x^2} + \frac{\partial^2 \psi}{\partial y^2} + \frac{\partial^2 \psi}{\partial z^2}\right) = \mathscr{E}\psi. \qquad (25.14)$$

It is convenient to write the solution as follows:

$$\psi = C \sin \varkappa_1 x \cdot \sin \varkappa_2 y \cdot \sin \varkappa_3 z. \qquad (25.15)$$

It is written only in terms of sines and not cosines so as to satisfy the first line of the boundary conditions (25.13). We substitute (25.15) in (15.14) and, utilizing the fact that for every factor of (25.15) there exists an equality of the form,

$$\frac{\partial^2}{\partial x^2}\sin \varkappa_1 x = -\varkappa_1^2 \sin \varkappa_1 x\,; \tag{25.16}$$

this gives

$$\Delta \psi = -(\varkappa_1^2 + \varkappa_2^2 + \varkappa_3^2)\,\psi\,.$$

To satisfy equation (25.14) the energy must involve $\varkappa_1$, $\varkappa_2$ and $\varkappa_3$ in the following way:

$$\mathscr{E} = \frac{h^2}{2m}(\varkappa_1^2 + \varkappa_2^2 + \varkappa_3^2)\,. \tag{25.17}$$

The quantities $\varkappa_1$, $\varkappa_2$, $\varkappa_3$ are determined from the second line of the boundary conditions (25.13). The factors of (25.15) convert to zero either at $x=a_1$, or $y=a_2$ or $z=a_3$. In other words,

$$\begin{aligned} \sin \varkappa_1 a_1 &= 0\,, & \varkappa_1 a_1 &= n_1 \pi\,; \\ \sin \varkappa_2 a_2 &= 0\,, & \varkappa_2 a_2 &= n_2 \pi\,; \\ \sin \varkappa_3 a_3 &= 0\,, & \varkappa_3 a_3 &= n_3 \pi\,. \end{aligned} \tag{25.18}$$

Here n_1, n_2 and n_3 are integers of which none are equal to zero (otherwise ψ would be equal to zero over all the box).

Substituting $\varkappa_1$, $\varkappa_2$, $\varkappa_3$ from (25.18) in (25.17), we have the energy eigenvalues

$$\mathscr{E}_{n_1 n_2 n_3} = \frac{\pi^2 h^2}{2m}\left(\frac{n_1^2}{a_1^2} + \frac{n_2^2}{a_2^2} + \frac{n_3^2}{a_3^2}\right). \tag{25.19}$$

The least possible energy is

$$\mathscr{E}_{111} = \frac{\pi^2 h^2}{2m}\left(\frac{1}{a_1^2} + \frac{1}{a_2^2} + \frac{1}{a_3^2}\right). \tag{25.20}$$

It follows that the value $\mathscr{E}=0$ is impossible.

Calculating the number of possible states. To each value of the three numbers n_1, n_2, and n_3, there corresponds a single particle state. Let the numbers n_1, n_2, and n_3 be large in comparison with unity. Such numbers may be differentiated: the differential dn_1 denotes a number interval which is small compared with n_1, but still including many separate integral values of n_1. Then it stands to reason that there are exactly dn_1 possible integers, $1 \ll dn_1 \ll n_1$ included within the interval dn_1 (and similarly within the intervals dn_2 and dn_3). Let us plot n_1, n_2, and n_3 on coordinate axes. In this space we construct an infinitely small parallelepiped of volume $dn_1\,dn_2\,dn_3$. In accordance with the foregoing, there are $dn_1\,dn_2\,dn_3$ groups of three integers n_1, n_2, n_3 in this parallelepiped, each corresponding to one possible state of the particle in the box. Altogether, the number of such states in the examined interval of values n_1, n_2 and n_3 is

$$d\,N\,(n_1, n_2, n_3) = dn_1\,dn_2\,dn_3\,. \tag{25.21}$$

Substituting here $\varkappa_1$, $\varkappa_2$ and $\varkappa_3$ from (25.18), we obtain another expression for the number of states:

$$dN(\varkappa_1, \varkappa_2, \varkappa_3) = \frac{a_1 a_2 a_3 d\varkappa_1 d\varkappa_2 d\varkappa_3}{\pi^3} = \frac{V d\varkappa_1 d\varkappa_2 d\varkappa_3}{\pi^3}, \qquad (25.22)$$

where $V = a_1 a_2 a_3$ is the volume of the box. The numbers $\varkappa_1$, $\varkappa_2$ and $\varkappa_3$ take only positive values.

It was pointed out above that to each value of $\varkappa$ there correspond two values of the momentum projection, which are equal in magnitude and opposite in sign. Therefore, if we compare the number of states included in the intervals $d\varkappa_1$, and $\frac{dp_1}{h} \equiv \frac{dp_x}{h}$ then there are half as many states for the latter. Correspondingly, the number of states in the interval of values of momentum $dp_x\, dp_y\, dp_z$ is

$$dN(p_x, p_y, p_z) = \frac{V\, dp_x\, dp_y\, dp_z}{(2\pi h)^3}, \qquad (25.23)$$

where p_x, p_y and p_z assume all real values from $-\infty$ to ∞.

Equation (25.23) agrees with the uncertainty relation (23.4). If the motion is bounded along x by the interval a_1, then only those states differ physically for which the momentum projections differ by not less than $\frac{2\pi h}{a_1}$. Hence, there are $\frac{dp_x}{(2\pi h/a_1)} = \frac{a_1\, dp_x}{2\pi h}$ states within the interval dp_x. Multiplying $\frac{a_1\, dp_x}{2\pi h} \cdot \frac{a_2\, dp_y}{2\pi h} \cdot \frac{a_3\, dp_z}{2\pi h}$ we arrive at (25.23). In order to ensure coincidence of numerical coefficients with the results of rigorous derivation from the wave equation when evaluating the number of states from the uncertainty relation the quantity $2\pi h$ was selected on the right-hand side of (23.4) or 2π from (18.10).

We shall now consider the number of states after changing somewhat the independent variables. We plot the quantities $\varkappa_1$, $\varkappa_2$ and $\varkappa_3$ on the coordinate axes (Fig. 37). Let us construct in this "space" a sphere whose equation looks like

$$\varkappa_1^2 + \varkappa\ + \chi_3^2 = \mathbf{K}^2.$$

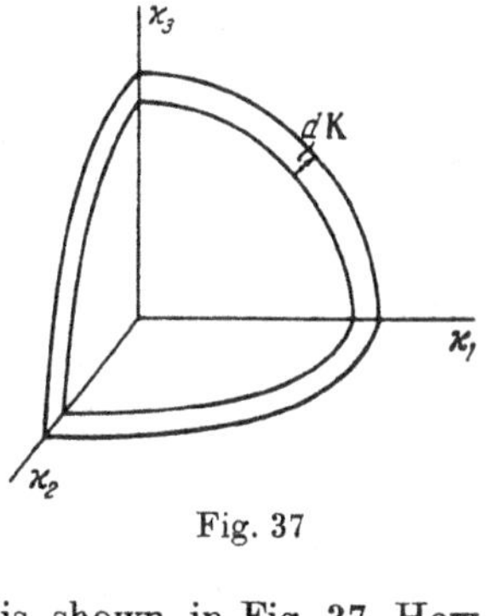

Fig. 37

The numbers $\varkappa_1$, $\varkappa_2$ and $\varkappa_3$ are positive so that we shall be interested only in one eighth of the sphere; this octant is shown in Fig. 37. How many states are included between the octants of two spheres with radii $\mathbf{K}$ and $\mathbf{K} + d\mathbf{K}$? The number of states is equal to the integral of (25.22) over the whole volume between the octants, or

$$dN(\mathbf{K}) = \int dN(\varkappa_1, \varkappa_2, \varkappa_3) = \frac{V.\, 4\pi \mathbf{K}^2 d\mathbf{K}}{8\pi^3} = \frac{V \mathbf{K}^2 d\mathbf{K}}{2\pi^2}. \quad (25.24)$$

This is evident simply from the fact that the volume is equal to the surface of the octant $\frac{4\pi \mathbf{K}^2}{8}$ multiplied by $d\mathbf{K}$. But from (25.17) $\mathbf{K}$ is very simply related to the energy of the particle:

$$\mathbf{K} = \frac{\sqrt{2m\mathscr{E}}}{h}.$$

Whence

$$dN(\mathscr{E}) = \frac{V m^{3/2} \mathscr{E}^{1/2} d\mathscr{E}}{2^{1/2} \pi^2 h^3}. \quad (25.25)$$

Thus, the number of states included between $\mathscr{E}$ and $\mathscr{E} + d\mathscr{E}$ increases in direct proportion to $\mathscr{E}^{1/2}$. In a one-dimensional potential well we would obtain $dN = dn = \frac{a d\varkappa}{\pi} = \frac{a m^{1/2} d\mathscr{E}}{2^{1/2} \pi h \mathscr{E}^{1/2}}$ Equation (25.25) has great significance in all that is to follow. We observe that this formula involves only the volume of the vessel V, irrespective of the ratio of the edges a_1, a_2, and a_3. In mathematical physics courses it is shown that the result (25.25) holds for energy eigenvalues which are sufficiently large compared with the energy of the ground state. The number of states is proportional to the volume of the vessel and is independent of its shape.

A one-dimensional potential well of finite depth. We shall now consider a one-dimensional potential well of finite depth. We specify it in the following way: $U = \infty$ at $-\infty \leqslant \leqslant x < 0$, $U = 0$ for $0 \leqslant x \leqslant a$ and $U = U_0$ for $a < x \leqslant \infty$. In other words, the potential energy for $x > 0$ is everywhere equal to U_0, except within a region of width a near the coordinate origin, which region we called the well. For $x < 0$ the potential energy is infinite (see Fig. 38).*

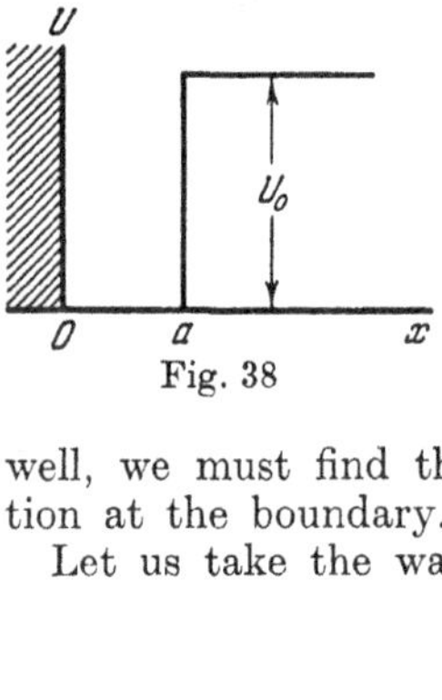

Fig. 38

Since the solution will be of different analytical form inside and outside the well, we must find the conjugation conditions for the wave function at the boundary.

Let us take the wave equation.

$$-\frac{h^2}{2m}\frac{d^2\psi}{dx^2} + U\psi = \mathscr{E}\psi \quad (25.26)$$

* It was shown in Sec. 19 that the three-dimensional wave equation can be reduced to a single-dimensional one, with the difference that the variable r must be positive by its very meaning. This may be attained formally by situating an infinitely high potential wall at $r = 0$. Fig. 38 actually refers to a spherical potential well with an angular-momentum value equal to zero, when there is no "centrifugal" term in the potential energy [see (5.8) and (31.5)].

and integrate both sides over a narrow region $a-\delta \leqslant x \leqslant a+\delta$, including the point of discontinuity of the potential energy $x=a$. The integration gives

$$-\frac{h^2}{2m}\left[\left(\frac{d\psi}{dx}\right)_{a+\delta}-\left(\frac{d\psi}{dx}\right)_{a-\delta}\right]=\int_{a-\delta}^{a+\delta}(\mathscr{E}-U)\,\psi\,dx. \qquad (25.27)$$

Even though U suffers a discontinuity at the boundary of the well, on the right it remains everywhere finite by arrangement. Therefore, when δ approaches zero, the integral on the right also approaches zero. It follows that the left-hand side of (25.27) is also zero. In other words,

$$\left(\frac{d\psi}{dx}\right)_{a+0}=\left(\frac{d\psi}{dx}\right)_{a-0}, \qquad (25.28)$$

the limit of the derivative on the right is equal to its limit on the left.

This argument would not hold in the problem of an infinitely deep well because then the integral in (25.27) would be indeterminate. Besides we notice that the derivative $\frac{d\psi}{dx}$ is finite at the points $x=a\pm\delta$ simply because the only solutions of equation (25.26) are those with finite derivatives (exponential function, sine or cosine).

We shall now show, by means of a limiting transition, that even the wave function itself does not suffer a discontinuity at the boundary. Let U initially have a finite discontinuity region of width δ and let the discontinuity of the function ψ be Δ. Before passing to the limit the derivative in the region of discontinuity is of the order of $\frac{d\psi}{dx}\sim\frac{\Delta}{\delta}$, so that when $\delta\to 0$, it diverges. Let us now multiply both sides of (25.26) by ψ and perform a transformation by parts:

$$\psi\frac{d^2\psi}{dx^2}=\frac{d}{dx}\left(\psi\frac{d\psi}{dx}\right)-\left(\frac{d\psi}{dx}\right)^2.$$

Let us integrate the transformed expression between $a-\delta$ and $a+\delta$. We then obtain

$$\left(\psi\frac{d\psi}{dx}\right)_{a+\delta}-\left(\psi\frac{d\psi}{dx}\right)_{a-\delta}-\int_{a-\delta}^{a+\delta}\left(\frac{d\psi}{dx}\right)^2dx=-\int_{a-\delta}^{a+\delta}\frac{2m}{h^2}(\mathscr{E}-U)\,\psi^2\,dx. \qquad (25.59)$$

We shall now perform the foregoing limiting process by indirect proof.

We may write the integrated terms thus:

$$\left(\psi\frac{d\psi}{dx}\right)_{a+\delta}-\left(\psi\frac{d\psi}{dx}\right)_{a-\delta}=[\psi(a+\delta)-\psi(a-\delta)]\left(\frac{d\psi}{dx}\right)_{a+\delta},$$

because the derivative $\frac{d\psi}{dx}$, as was shown, is not subject to a discontinuity. Within the assumed discontinuity region of the ψ-function,

$\frac{d\psi}{dx}$ is of the order of $\frac{\Delta}{\delta}$, but at the boundaries of the region it reverts to values which are independent of δ and are therefore finite in the limit. Hence, the whole integrated part on the left in (25.29) is of the order of $\Delta \cdot \left(\frac{d\psi}{dx}\right)_{a+\delta}$. The remaining integral is estimated as

$$\int_{a-\delta}^{a+\delta} \left(\frac{d\psi}{dx}\right)^2 dx \sim 2\delta \cdot \left(\frac{\Delta}{\delta}\right)^2 \sim \frac{2\Delta^2}{\delta}.$$

Hence it tends to infinity as δ tends to zero. The right-hand side of (25.29) is finite for $\delta \to 0$. Thus, by assuming that ψ has a finite discontinuity Δ we have arrived at a contradiction. It follows that ψ is continuous at the point a together with its first derivative.

Solutions in two regions. The wave equation for the region $0 \leqslant x \leqslant a$ (inside the well) is of the form

$$-\frac{h^2}{2m}\frac{d^2\psi}{dx^2} = \mathscr{E}\psi.$$

We take its solution

$$\psi = C_1 \sin \varkappa x, \tag{25.30}$$

where $\varkappa$ is defined from (25.3). The solution involving the sine only is taken because at the left-hand edge of the well, where the potential energy suffers an infinite discontinuity, ψ satisfies the boundary condition (25.1): $\varphi(0) = 0$.

The wave equation outside the well, when $x > a$, is

$$-\frac{h^2}{2m}\frac{d^2\psi}{dx^2} = (\mathscr{E} - U_0)\psi. \tag{25.31}$$

First of all let us take the case $\mathscr{E} > U_0$. Then, introducting the abbreviated notation

$$\frac{2m}{h^2}(\mathscr{E} - U_0) \equiv \varkappa_1^2, \tag{25.32}$$

we obtain (25.31) in the standard form (25.4)

$$\frac{d^2\psi}{dx^2} = -\varkappa_1^2\psi,$$

whence

$$\psi = C_2 \sin \varkappa_1 x + C_3 \cos \varkappa_1 x. \tag{25.33}$$

We must now satisfy the boundary conditions on the right-hand edge of the potential well where U suffers only a finite discontinuity. According to these conditions the wave function is itself continuous, i.e.,

$$C_1 \sin \varkappa a = C_2 \sin \varkappa_1 a + C_3 \cos \varkappa_1 a \tag{25.34}$$

and its derivative

$$\varkappa C_1 \cos \varkappa a = \varkappa_1 C_2 \cos \varkappa_1 a - \varkappa_1 C_3 \sin \varkappa_1 a. \tag{25.35}$$

From these equations we can determine C_2 and C_3 in terms of C_1, i. e., completely express the solution outside the well in terms of the solution inside the well. The equations (25.34) and (25.35) are linear with respect to C_2 and C_3 and have solutions for all values of coefficient.

$$C_2 = \frac{\varkappa_1 \sin \varkappa a \sin \varkappa_1 a + \varkappa \cos \varkappa a \cos \varkappa_1 a}{\varkappa_1} C_1 ,$$

$$C_3 = \frac{\varkappa_1 \sin \varkappa a \cos \varkappa_1 a - \varkappa \cos \varkappa a \sin \varkappa_1 a}{\varkappa_1} C_1 .$$

Therefore, the boundary conditions may be satisfied for any real values of $\varkappa$ and $\varkappa_1$. Thus, Schrödinger's equation is solvable for all $\mathscr{E}$. There is no discrete energy eigenvalues for $\mathscr{E} > U_0$.

We could adjust the potential energy in this problem to zero at infinity, i.e., consider it equal to zero for $x > a$ and equal to $-U$ for $a \geqslant x \geqslant 0$. Then the case which we have just considered would correspond to positive eigenvalues of the total energy.

Now let $\mathscr{E} < U_0$. We introduce the quantity

$$\frac{2m}{h^2}(U_0 - \mathscr{E}) \equiv \tilde{\varkappa}^2 . \qquad (25.36)$$

The wave equation is now written differently from that for $\mathscr{E} > U_0$, namely

$$\frac{d^2\psi}{dx^2} = \tilde{\varkappa}^2 \psi .$$

Its solution is expressed in terms of the exponential function

$$\psi = C_4 e^{\tilde{\varkappa} x} + C_5 e^{-\tilde{\varkappa} x} . \qquad (25.37)$$

But the exponential $e^{\tilde{\varkappa} x}$ tends to infinity as x increases. For $x = \infty$ it would give an infinite probability for finding the particle, and no finite value could be assigned to the integral $\int_0^\infty |\psi|^2 dx$. It follows that a physically meaningful solution exists only for $C_4 = 0$ and must be of the form

$$\psi = C_5 e^{-\tilde{\varkappa} x} \qquad (25.38)$$

Let us again try to satisfy the boundary conditions at $x = a$. This time they appear as follows:

$$C_1 \sin \varkappa a = C_5 e^{-\tilde{\varkappa} a} , \qquad (25.39)$$

$$\varkappa C_1 \cos \varkappa a = -\tilde{\varkappa} C_5 e^{-\tilde{\varkappa} a} \qquad (25.40)$$

Let us divide equation (25.40) by (25.39) in order to eliminate C_1 and C_5. We then obtain

$$\varkappa \cot \varkappa a = -\tilde{\varkappa} . \qquad (25.41)$$

From this equation we find the expression for $\sin \varkappa a$:

$$\sin \varkappa a = \pm \frac{1}{\sqrt{1 + \cot^2 \varkappa a}} = \pm \frac{1}{\sqrt{1 + \left(\frac{\tilde{\varkappa}}{\varkappa}\right)^2}} =$$

$$= \pm \frac{1}{\sqrt{1 + \frac{U_0 - \mathscr{E}}{\mathscr{E}}}} = \pm \sqrt{\frac{\mathscr{E}}{U_0}}. \tag{25.42}$$

Let us reduce this equation to a more convenient form. From (26.3)

$$\sqrt{\mathscr{E}} = \frac{h}{a\sqrt{2m}} \, . \, \varkappa a \, ,$$

so that

$$\sin \varkappa a = \pm \frac{h}{a\sqrt{2mU_0}} \, . \, \varkappa a \, , \tag{25.43}$$

only those solutions should be chosen for which ctg $\varkappa a$ is negative, in accordance with (25.41), i.e., $\varkappa a$ must lie in the second, fourth, sixth, eighth, etc., quadrants.

We shall solve this equation graphically (Fig. 39). The left-hand side of equation (25.43) is represented by a sinusoid, while the right-hand side is represented by two straight lines of slopes $\pm \frac{h}{a\sqrt{2mU_0}}$. If the absolute value of the slopes of the angle of inclination of these lines is less than $2/\pi$, they have one or several common points with the sinusoid in the quadrants corresponding to the roots of (25.41). The trivial point of intersection $\varkappa a = 0$ does not count because, for $\varkappa = 0$, the wave function is zero everywhere. Thus, in a well of finite depth of the form considered, there are only several energy eigenvalues.

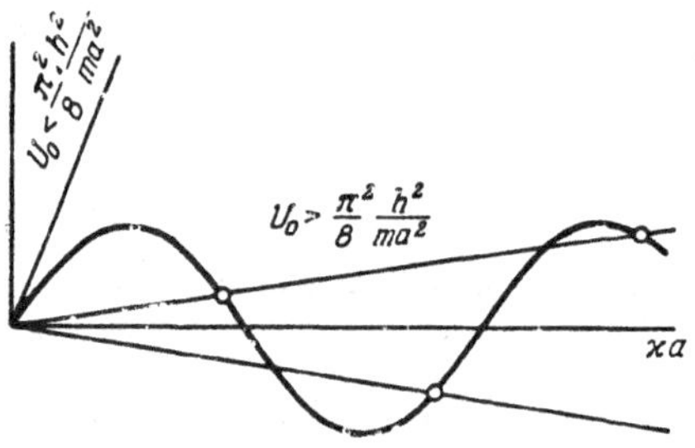

Fig. 39

If

$$\frac{h}{a\sqrt{2mU_0}} > \frac{2}{\pi}\, , \quad U_0 < \frac{\pi^2}{8}\frac{h^2}{ma^2}\, ,$$

there are in general no points of intersection of the straight lines with the sinusoid corresponding to energy eigenvalues. In Fig. 39 the points of intersection in the even quadrants are marked by small circles.

Finite and infinite motion. We shall now relate the shape of the energy spectrum to the type of motion. For $\mathscr{E} > U_0$ the solution outside the well is of the form (25.33). It remains finite also for an infinitely large x. Therefore, the integral $\int |\psi|^2 dx$ taken over the region of the well is infinitesimal compared with the same integral taken over all space. In other words, there is nothing to prevent the particle going to infinity. Such motion was termed infinite in Sec. 5. For $\mathscr{E} < U_0$, the solution (25.38), if it exists, is exponentially damped at infinity. Hence, the probability of the particle receding an infinite distance from the origin is equal to zero—the particle remains at a finite distance from the well all the time. This motion was termed finite in Sec. 5.

Thus, infinite motion has a continuous energy spectrum while finite motion has a discrete spectrum consisting of separate values. If the depth of the well is very small, the finite motion may be absent. It has no counterpart in classical mechanics. Finite motion is always possible in a potential well if $|\mathscr{E}| < U$.

The result that we have obtained does not only refer to a rectangular potential well. Indeed, if the potential energy is taken to be zero at infinity, then the solution with positive total energy is of the form (25.33) for sufficiently large x, while the solution with negative total energy is of the form (25.38). The latter contains only one arbitrary constant while (25.33) contains two constants. Both solutions must be extended to the coordinate origin in order that the condition $\psi(0) = 0$ can be satisfied at the origin (we consider that x is always greater than zero). Obviously, if we have two constants at our disposal we can always choose them so that the condition $\psi(0) = 0$ is satisfied.* Contrarily, a solution of the form (25.38) containing one constant becomes zero at the origin only for certain special values of $\varkappa$.

A continuous spectrum corresponding to infinite motion may be accounted for in the following way. A free particle moving in unbounded space has a continuous spectrum. The wave function of the particle in infinite motion differs from the wave function of a free particle only in the region of a potential well. But the probability of finding the particle in this region is infinitesimal if the whole region of motion is sufficiently large. Therefore, the wave function for infinite motion coincides with the wave function of a free particle in "almost" the entire space, i.e., in that region of space for which the probability of finding the particle is equal to unity, and the energy spectrum turns out to be the same as for a free particle.

The wave function in a region where the potential energy is greater than the total energy. If U_0 tends to infinity the function outside the

---

* If $\psi(0) = C_1\psi_1(0) + C_2\psi_2(0)$, then $\dfrac{C_1}{C_2} = -\dfrac{\psi_2(0)}{\psi_1(0)}$.

well very rapidly tends to zero. In the limit $U_0 \to \infty$, it tends to zero however close to the boundary $x = a$, thereby giving the boundary condition (25.1).

In the case of a finite U_0 the wave function outside the well does not become zero at once. Therefore a finite probability exists that the particle will be outside the well at a finite distance from it.

This would have been completely impossible in classical mechanics, as is obtained from (25.38) in the limiting transition $h \to 0$ for $\tilde{\varkappa} = \infty$ and ψ however small outside the well. This, naturally, should be the case: if the particle is situated outside the well its kinetic energy is $\mathscr{E} - U_0 < 0$. But the velocity of such a particle is an imaginary quantity. In classical mechanics it means that a given point of space is absolutely unattainable for the particle at the given value of its energy $\mathscr{E}$. In quantum mechanics, a coordinate and velocity never exist in the same states as precise quantities. Earlier, we interpreted this in terms of the uncertainty relation, i. e., we considered cases for which precision in the concept of velocity for a certain state was restricted by the limits $\frac{2\pi h}{m\Delta x}$. However, this is a lower limit and has to do with particles which are almost unaffected by forces. The appearance of an imaginary velocity in the equation for a bound particle shows that the very concept of velocity is not applicable to a region of space, however large, for which $U > \mathscr{E}$. We can express this differently by saying that, for $U > \mathscr{E}$, the uncertainty in the kinetic energy is always greater than the difference $U - \mathscr{E}$.

To summarize, in classical mechanics there is no counterpart to the motion of a bound particle outside a well.

Exercises

1) The potential energy is equal to zero for $x < 0$ and equal to U_0 for $x \geqslant 0$ (the potential threshold). Incident from the left are particles with energies $\mathscr{E} > U_0$. Find the reflection coefficient.

The wave function on the left is

$$C_1 e^{i\frac{px}{h}} + C_2 e^{-i\frac{px}{h}}.$$

On the right, above the threshold, the function is

$$C_3 e^{i\frac{p'x}{h}}, \text{ where } p' = \sqrt{2m(\mathscr{E} - U)}.$$

Find the ratio $|C^2|^2/|C_1|^2$ from the boundary conditions at $x = 0$, i.e., the ratio of the squares of the amplitudes of the reflected and incident waves. The ratio is equal to unity for $\mathscr{E} < U_0$.

2) The potential energy is equal to zero for $x < 0$ and for $x > a$. $U = U_0$ for $0 < x < a$ (a potential barrier). Particles are incident from the left with energies less than U_0. Find the coefficient of reflection.

The wave function to the left of the barrier is equal to $e^{ikx} + Ce^{-ikx}\left(k = \frac{p}{h}\right)$; under the barrier, i.e., for $0 \leqslant x \leqslant a$, the wave function is $C_1 e^{\varkappa x} \quad C_2 e^{-\varkappa x}$. We look for a wave function of the form $C_3 e^{ikx}$ beyond the barrier. This means that beyond the barrier the only wave is that travelling to the right (i.e., only the transmitted wave), while in front of the barrier we find both the incident wave and the reflected wave. The constants C, C_1, C_2, C_3 are determined from the continuity condition of the wave function and its first derivative at the boundaries of the barrier. The expressions for the constants C and C_3 are as follows:

$$C = \frac{2(\varkappa^2 + k^2)\,\mathrm{sh}\,\varkappa a}{(\varkappa + ik)^2 e^{-\varkappa a} - (\varkappa - ik)^2 e^{\varkappa a}},$$

$$C_3 = \frac{4ik\varkappa\, e^{-ika}}{(\varkappa + ik)^2 e^{-\varkappa a} - (\varkappa - ik)^2 e^{\varkappa a}}.$$

The particle flux on the left and right of the barrier is, respectively,

$$j = \frac{hk}{m}(1 - |C|^2), \quad j = \frac{hk}{m}|C_3|^2.$$

Substituting C and C_3 it is easy to see that both the expressions for flux coincide as expected.

If $\varkappa a \gg 1$, i.e., the barrier is transparent to a very small extent, we have

$$C \sim -1, \quad C_3 = -\frac{4ik}{\varkappa} e^{-ika} e^{-\varkappa a}.$$

Thus the flux diminishes exponentially with the thickness of the barrier.

It will also be noted that the total particle flux through the barrier is proportional to the particle density in front of the barrier, because the boundary conditions are linear and homogeneous with respect to the wave functions. By specifying the amplitude of the wave function on the left we determined the density and flux of the particles.

3) Verify the orthogonality property for wave functions (the exercise in the preceding section) for a particle in a box of finite and infinite depth.

Sec. 26. Harmonic Oscillatory Motion in Quantum Mechanics (Linear Harmonic Oscillator)

The wave equation for an oscillator. In Sec. 7 we considered harmonic oscillations with one degree of freedom. The Hamiltonian function of this system, called a linear harmonic oscillator, is of the form

$$\mathscr{H} = \frac{p^2}{2m} + \frac{m\omega^2 x^2}{2}. \tag{26.1}$$

Forming Hamilton's equations, we obtain

$$\dot{p} = -\frac{\partial \mathscr{H}}{\partial x} = -m\omega^2 x, \quad \dot{x} = \frac{\partial \mathscr{H}}{\partial p} = \frac{p}{m}.$$

Eliminating p, we arrive at the usual equation of harmonic oscillations (7.13):

$$\ddot{x} + \omega^2 x = 0.$$

In quantum mechanics the wave equation corresponding to this motion has the form [see (24.22)]

$$-\frac{h^2}{2m}\frac{d^2\psi}{dx^2}+\frac{m\omega^2x^2}{2}\psi=\mathscr{E}\psi. \tag{26.2}$$

Indeed, since the motion has only one degree of freedom, instead of the Laplacian Δ, we must simply write the second derivative. The potential energy is equal to $\frac{m\omega^2x^2}{2}$.

Let us now introduce other units of measurement, in particular, we shall take the unit of length equal to $\sqrt{\frac{h}{m\omega}}$, so that

$$x=\sqrt{\frac{h}{m\omega}}\,\xi. \tag{26.3}$$

The quantity ξ is dimensionless. The derivative $\frac{d\psi}{dx}$ is equal to

$$\frac{d\psi}{dx}=\sqrt{\frac{m\omega}{h}}\cdot\frac{d\psi}{d\xi}. \tag{26.4}$$

Further, we put

$$2\mathscr{E}=\varepsilon\cdot h\omega. \tag{26.5}$$

In terms of these dimensionless variables, equation (26.2) assumes the form

$$-\frac{d^2\psi}{d\xi^2}+\xi^2\psi=\varepsilon\psi. \tag{26.6}$$

Equation (26.6) does not contain any parameters of the problem, i.e., ω, m, and h. For this reason the eigenvalue ε can only be an abstract number. Comparing this with the expression for energy (26.5), we see that the energy eigenvalue of an harmonic oscillator is proportional to its frequency ω.

The transition to another dependent variable. It appears convenient to introduce a new dependent variable:

$$\psi=e^{-\frac{\xi^2}{2}}g(\xi). \tag{26.7}$$

Whence

$$\left.\begin{aligned}\frac{d\psi}{d\xi}&=-\xi e^{-\frac{\xi^2}{2}}g(\xi)+e^{-\frac{\xi^2}{2}}\frac{dg(\xi)}{d\xi},\\ \frac{d^2\psi}{d\xi^2}&=\xi^2e^{-\frac{\xi^2}{2}}g(\xi)-e^{-\frac{\xi^2}{2}}g(\xi)-2\xi e^{-\frac{\xi^2}{2}}\frac{dg(\xi)}{d\xi}+e^{-\frac{\xi^2}{2}}\frac{d^2g(\xi)}{d\xi^2}.\end{aligned}\right\} \tag{26.8}$$

We substitute (26.8) in (26.6) and perform the necessary rearrangement. The new dependent variable $g(\xi)$ having been introduced, the equation assumes the form

$$-\frac{d^2 g(\xi)}{d\xi^2} + 2\xi \frac{dg(\xi)}{d\xi} = (\varepsilon - 1) g(\xi). \tag{26.9}$$

Integration in the form of a series. It is possible to integrate equation (26.9) by expanding it in a power series of the form:

$$g(\xi) = g_0 + g_1\xi + g_2\xi^2 + g_3\xi^3 + \ldots = \sum_{n=0}^{\infty} g_n \xi^n. \tag{26.10}$$

In order to determine the coefficients of the expansion g_n, we must substitute the series (26.10) into equation (26.9), differentiate it by terms and compare the expressions for the same powers of ξ. The first derivative is

$$\frac{dg(\xi)}{d\xi} = g_1 + 2g_2\xi + 3g_3\xi^3 + \ldots = \sum_{n=1}^{\infty} n g_n \xi^{n-1},$$

so that

$$2\xi \frac{dg(\xi)}{d\xi} = 2g_1\xi + 4g_2\xi^2 + 6g_3\xi^3 + \ldots = \sum_{n=1}^{\infty} 2n g_n \xi^n. \tag{26.11}$$

The second derivative is

$$\frac{d^2 g(\xi)}{d\xi^2} = 2g_2 + 6g_3\xi + \ldots = \sum_{k=2}^{\infty} (k-1) k g_k \xi^{k-2}. \tag{26.12}$$

In the last summation we changed the summation index, denoting it by the letter k. We shall now revert to n, assuming $k-2=n$, $k=n+2$. Then

$$\frac{d^2 g(\xi)}{d\xi^2} = \sum_{n=0}^{\infty} (n+2)(n+1) g_{n+2} \xi^n. \tag{26.13}$$

Now substituting (26.13) and (26.11) into equation (26.9) and collecting coefficients of ξ^n, we obtain

$$\sum_{n=0}^{\infty} \xi^n [-(n+2)(n+1) g_{n+2} + 2n g_n - (\varepsilon - 1) g_n] = 0. \tag{26.14}$$

We know that for a power series to be equal to zero, all its coefficients must convert to zero. Thus,

$$g_{n+2} = g_n \frac{2n + 1 - \varepsilon}{(n+2)(n+1)}. \tag{26.15}$$

In this way the expansion proceeds in powers of (ξ^2) because the coefficients g_n go alternately.

Examining the series. Let us assume initially that $g_0 \neq 0$. Then, from equation (26.13), we find in turn $g_2, g_4, \ldots, g_{2k}$. Not a single odd coefficient will appear in the series if $g_1 = 0$. On the contrary, if $g_0 = 0$, $g_1 \neq 0$, then no even coefficients will appear in the series; for this reason it is sufficient to examine solutions containing only even or only odd powers of ξ. To be specific let us first take the series in even powers.

Let us examine the behaviour of the series (26.10) for large values of ξ. Terms involving high powers of ξ, i. e., large n are then predominant. But if n is a large number then, in equation (26.15), we can neglect all constant numbers where they appear in the sum or, difference with n. If n is large the equation will take on the form

$$g_{n+2} = \frac{2}{n} g_n . \tag{26.16}$$

Let $n = 2n'$ so that n' now changes by unity only. Putting this in (26.16), we obtain

$$g'_{n'+1} = \frac{1}{n'} g'_{n'}, \tag{26.17}$$

where we have introduced the notation $g'_{n'} \equiv g_{2n'} \equiv g_n$; $g'_{n'}$ are coefficients of the series in powers of (ξ^2).

If we now take a function containing only odd powers of ξ, then the terms involving $g_{2n'+1}$ will not differ, for large n', from the terms of a series in even powers of ξ because unity can be neglected in comparison with $2n'$. Therefore, the form of the coefficients for large n' is identical both for series in even and odd powers of ξ.

From (26.17) we find an expression for $g'_{n'+1}$:

$$g'_{n'+1} \cong \frac{g'_0}{n'(n'-1)(n'-2)\ldots 1} = \frac{g'_0}{n'!} . \tag{26.18}$$

Whence the expression for $g(\xi)$ in the case of large ξ is of the form:

$$g(\xi) \cong \sum_{n'=0}^{\infty} g'_{n'}(\xi^2)^{n'} = \sum_{n'=0}^{\infty} \frac{g'_0(\xi^2)^{n'}}{n'!} = g'_0 e^{\xi^2}. \tag{26.19}$$

Thus, the asymptotic expression for $g(\xi)$ is the exponential function e^{ξ^2}. But then, in accordance with the definition of $g(\xi)$ (26.7), the asymptotic form of $\psi(\xi)$ for large ξ is

$$\psi \cong e^{-\frac{\xi^2}{2}} \cdot e^{\xi^2} = e^{\frac{\xi^2}{2}} .$$

However, this form of ψ is not acceptable: the wave function must remain finite at infinity because its square is probability.

The condition for eigenvalues. There is only one possibility for obtaining a finite value of ψ at infinity. It is necessary that the series (26.10) should terminate at a certain n and that all the subsequent coefficients g_{n+2}, g_{n+4}, etc., be identically equal to zero. It can be seen from equation (26.15) that g_{n+2} becomes zero when

$$\varepsilon = 2n + 1, \tag{26.20}$$

where n is any integer or zero. Since g_{n+4} is linearly expressed in terms of g_{n+2}, it is sufficient for g_{n+2} to convert to zero to have the series terminate at g_n. It follows that when ε satisfies (26.20) the function $g(\xi)$ becomes a polynomial. The product of the polynomial $g(\xi)$ with the exponential $e^{-\frac{\xi^2}{2}}$ always tends to zero as ξ tends to infinity. Hence $\psi(\infty) = 0$. As was pointed out at the end of the last section, such motion is finite in the same sense as in classical mechanics: the probability that the particle will recede an infinite distance is equal to zero. To finite motion there corresponds a discrete energy spectrum; from (26.5) and (26.20)

$$\mathscr{E}_n = \frac{h\omega}{2}\varepsilon = h\omega\left(n + \frac{1}{2}\right). \tag{26.21}$$

The least possible value of energy $\mathscr{E}_0 = \frac{h\omega}{2}$. As we have already said in Sec. 25 a state with energy $\mathscr{E}_0$ is called the ground state. For this state the series $g(\xi)$ is already terminated at the zero term, because the number of the eigenvalue of energy determines the degree of the polynomial $g_n(\xi)$. The wave function of the ground state is of the especially simple form:

$$\psi_0(\xi) = g_0 e^{-\frac{\xi^2}{2}}. \tag{26.22}$$

This function does not have any zeros at a finite distance from the coordinate origin, which must be the case in the ground state. It may be noted that the state with zero energy would correspond to a particle at rest at the origin. However, such a state is not compatible with the uncertainty principle, since it has, simultaneously, a coordinate and velocity.

Oscillator wave functions. Let us also find the eigenfunctions for the first and second excited states. In the first state $\mathscr{E}_1 = h\omega\left(1 + \frac{1}{2}\right) = \frac{3}{2}h\omega$. The series will be terminated if we assume $g_0 = 0$, $g_1 \neq 0$. Then $\varepsilon = 3$ from (26.20) and $g_3 = g_5 = g_{2n+1} = 0$, etc., from (26.15). The even coefficients may be assumed at once equal to zero, for which purpose it is sufficient to take $g_0 = 0$.* In general, all functions with even n turn out even, while those with odd n are odd. In accordance with what has just been said, the wave function with $n = 1$ is

* If $g_0 \neq 0$ for $\varepsilon = 3$, then the series in even powers of ξ would have extended to infinity, which, as was shown, is impossible.

$$\psi_1 = g_1 \xi e^{-\frac{\xi^2}{2}}. \tag{26.23}$$

This function becomes zero for $\xi = 0$, i.e., it has one node.

In the same way it is easy to find ψ_2. Indeed, $\mathscr{E}_2 = \hbar\omega\left(2 + \frac{1}{2}\right) = \frac{5}{2}\hbar\omega$, $\varepsilon = 5$. From (26.15), the coefficient g_2 is

$$g_2 = g_0 \frac{1 - \varepsilon}{1 \cdot 2} = -2 g_0, \tag{26.24}$$

so that

$$\psi_2 = g_0 (1 - 2\xi^2) e^{-\frac{\xi^2}{2}}. \tag{26.25}$$

The nodes of this function are situated at the points $\xi = \pm \frac{1}{\sqrt{2}}$.

In general, the function ψ_n has n nodes. The functions for several small values of n are shown in Fig. 40.

We show the eigenvalue distribution and the potential energy curve in Fig. 41. It is very curious that the eigenvalues are separated by equal intervals. The oscillator problem qualitatively resembles that of an infinitely deep rectangular well (Sec. 25), but the energy level in the well increases in proportion to the square of the number.

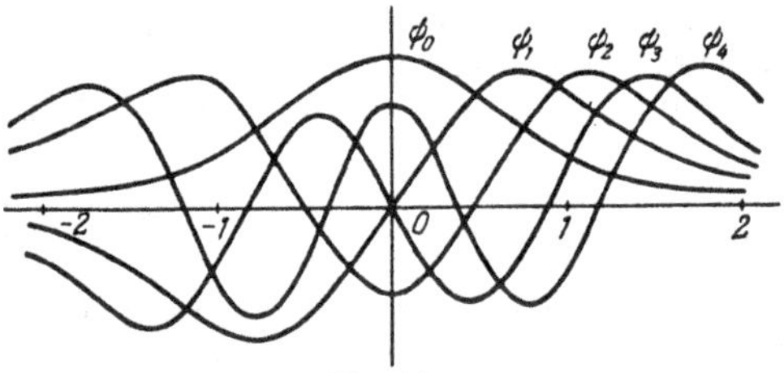

Fig. 40

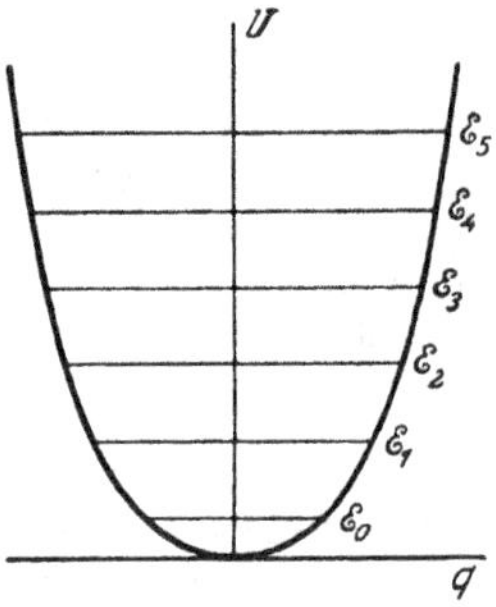

Fig. 41

Exercises

1) Show that, neglecting a constant factor, the function $g_n(\xi)$ may be written in the form

$$g_n(\xi) = e^{\xi^2}\left(\frac{d}{d\xi}\right)^n e^{-\xi^2}.$$

Verify this by substitution in equation (26.9), in which $\varepsilon = 2n + 1$.

2) Normalize the functions ψ_0 and ψ_1, taking advantage of the fact that

$$\int_{-\infty}^{\infty} e^{-\xi^2} d\xi = \sqrt{\pi}, \quad \int_{-\infty}^{\infty} \xi^2 e^{-\xi^2} d\xi = \frac{\sqrt{\pi}}{2}$$

(see exercises of Sec. 39).

Sec. 27. Quantization of the Electromagnetic Field

The electromagnetic field as a mechanical system. An electromagnetic field in a vacuum may be regarded as a mechanical system; this was shown in Sec. 13. It possesses a Lagrangian function, action, and so on. We are, therefore, justified in posing the problem of quantization of this system, i.e., applying quantum mechanics to it.

The basic difference between electrodynamics and the mechanics of point masses is that the degrees of freedom of an electromagnetic field are distributed continuously: in order to specify the field at a given instant of time, we must define its value at every point of space. In this sense electrodynamics resembles the mechanics of liquids or elastic bodies, if one regards them as continuous media ignoring the atomic structure of the substance. The degrees of freedom of a field are labelled by the coordinates of points in space, while the amplitude values of the potential are generalized coordinates [see (13.2)]. Potentials are usually chosen as generalized coordinates because they satisfy second-order equations in time, as do generalized coordinates in mechanics.

The potentials satisfy the Lorentz condition, which reduces to $\operatorname{div} \mathbf{A}=0$, provided the gauge transformations are chosen so as to eliminate the scalar potential.

The electromagnetic field coordinates defined in this way are not independent of one another. Indeed, the equations of electrodynamics involve coordinate derivatives, i.e., differences of field values at infinitely close points. In this sense, field equations resemble the equations for coupled oscillations: they are linear, but each one involves several generalized coordinates instead of one. The equations for coupled oscillations can be reduced to normal coordinates which are mutually independent. The same can be done with the wave equations of electrodynamics, thus separating the dependent variables therein. This considerably simplifies the application of quantum mechanics to radiation.

Clearly shown here is the generality of the methods of analytical mechanics: they permit determining generalized coordinates and momenta in such manner that quantum laws can then be applied uniquely.

The electromagnetic field in a closed volume. We must first of all represent the electromagnetic field as some kind of closed system, since quantum mechanics is most conveniently applied to such systems. We can assume, for example, that the radiation is contained in a box with mirror-type reflecting walls. At the walls of such an imaginary box ($x=0$ or $x=a_1$, $y=0$ or $y=a_2$, $z=0$ or $z=a_3$) the normal components of the Poynting vector $\mathbf{U}$ become zero. However, it is simpler to suppose that the field is periodic in space, and the lengths of the periods in three perpendicular directions are equal to the dimensions of the box; the period of the field along x is equal to a_1, that along y is equal to a_2, and that along z is equal to a_3. In other words,

$$\mathbf{A}(x, y, z) = \mathbf{A}(x + a_1, y, z) = \mathbf{A}(x, y + a_2, z) = \\ = \mathbf{A}(x, y, z + a_3). \tag{27.1}$$

We have, as it were, divided space into physically identical regions, after which it is sufficient to consider a single region.

The solution of equations describing a harmonic field in free space was found in Sec. 17 [see (17.21)]. Introducing a time dependence into the amplitude factor, we represent the potential in the following form:

$$\mathbf{A}(\mathbf{k}, \mathbf{r}, t) = \mathbf{A}_{\mathbf{k}}(t) e^{i\mathbf{kr}} + \mathbf{A}^*_{\mathbf{k}}(t) e^{-i\mathbf{kr}}, \tag{27.2}$$

where its reality is shown explicitly.

The potential satisfies the Lorentz condition which, for a plane wave, can be reduced to the form div $\mathbf{A} = 0$ (since $\varphi = 0$); whence, by (11.27), we obtain

$$\operatorname{div} \mathbf{A}(\mathbf{k}, \mathbf{r}, t) = \operatorname{div}(\mathbf{A}_{\mathbf{k}} e^{i\mathbf{kr}}) + \operatorname{div}(\mathbf{A}^*_{\mathbf{k}} e^{-i\mathbf{kr}}) = \\ = (\mathbf{A}_{\mathbf{k}} \nabla e^{i\mathbf{kr}}) + (\mathbf{A}^*_{\mathbf{k}} \nabla e^{-i\mathbf{kr}}) = i(\mathbf{k}\mathbf{A}_{\mathbf{k}}) e^{i\mathbf{kr}} - i(\mathbf{k}\mathbf{A}^*_{\mathbf{k}}) e^{-i\mathbf{kr}} = 0.$$

In order that this equation be satisfied for all $\mathbf{r}$, the coefficient of each exponential term must convert to zero. In other words, the vectors $\mathbf{A}_{\mathbf{k}}$ and $\mathbf{A}^*_{\mathbf{k}}$ are perpendicular to the wave vector $\mathbf{k}$:

$$(\mathbf{k}\mathbf{A}_{\mathbf{k}}) = 0, \quad (\mathbf{k}\mathbf{A}^*_{\mathbf{k}}) = 0. \tag{27.3}$$

For each $\mathbf{k}$ there exist two mutually perpendicular vectors $\mathbf{A}^{\sigma}_{\mathbf{k}}$ ($\sigma = 1, 2$) corresponding to two possible wave polarizations. Any vector in a plane perpendicular to $\mathbf{k}$ can be resolved into $\mathbf{A}^{(1)}_{\mathbf{k}}$ and $\mathbf{A}^{(2)}_{\mathbf{k}}$.

We shall now apply the periodicity condition (27.1) to each term of (27.2) separately. For the first term we obtain

$$\mathbf{A}_{\mathbf{k}} e^{i(k_x x + k_y y + k_z z)} = \mathbf{A}_{\mathbf{k}} e^{i[k_x(x + a_1) + k_y y + k_z z]} = \mathbf{A}_{\mathbf{k}} e^{i[k_x x + k_y(y + a_2) + k_z z]} = \\ = \mathbf{A}_{\mathbf{k}} e^{i[k_x x + k_y y + k_z(z + a_3)]},$$

whence it follows that

$$e^{ik_x a_1} = e^{ik_y a_2} = e^{ik_z a_3} = 1.$$

Therefore the components of the wave vector should be

$$k_x = \frac{2\pi n_1}{a_1}, \quad k_y = \frac{2\pi n_2}{a_2}, \quad k_z = \frac{2\pi n_3}{a_3}, \tag{27.4}$$

where n_1, n_2, n_3 are integers of any sign.

Consequently, each harmonic oscillation is given by three integers n_1, n_2, n_3 and a polarization σ, which can take two values. As indicated in Sec. 13, $\mathbf{A}^{\sigma}_{n_1 n_2 n_3}$ is a generalized coordinate. The number of such coordinates is infinite, but at least forms a denumerable set, and not a continuous set equivalent to the set of all points in space.

This, then, is the basic simplification introduced by the periodicity condition. This condition is, of course, only a mathematical convenience, there being no basic periods a_1, a_2, and a_3 in any final physical result.

An electromagnetic field is specified if its oscillation amplitudes are known for all values of n_1, n_2, n_3, and σ. The general solution is equal to the sum of partial solutions due to the linearity of electrodynamical equations (27.2):

$$\mathbf{A}(\mathbf{r}, t) = \sum_{\mathbf{k},\sigma} \mathbf{A}^{\sigma}(\mathbf{k}, \mathbf{r}, t) = \sum_{\mathbf{k},\sigma} (\mathbf{A}_{\mathbf{k}}^{\sigma} e^{i\mathbf{kr}} + \mathbf{A}_{\mathbf{k}}^{\sigma *} e^{-i\mathbf{kr}}). \tag{27.5}$$

Energy of the field. An electric field is calculated in accordance with (12.29) and (27.5):

$$\mathbf{E} = -\frac{1}{c}\frac{\partial \mathbf{A}}{\partial t} = -\frac{1}{c}\sum_{\mathbf{k},\sigma} (\dot{\mathbf{A}}_{\mathbf{k}}^{\sigma} e^{i\mathbf{kr}} + \dot{\mathbf{A}}_{\mathbf{k}}^{\sigma *} e^{-i\mathbf{kr}}). \tag{27.6}$$

The amplitudes of the field depend harmonically upon time, so that

$$\dot{\mathbf{A}}_{\mathbf{k}}^{\sigma} = -i\,\omega_{\mathbf{k}}\,\mathbf{A}_{\mathbf{k}}^{\sigma}, \quad \dot{\mathbf{A}}_{\mathbf{k}}^{\sigma *} = i\,\omega_{\mathbf{k}}\,\mathbf{A}_{\mathbf{k}}^{\sigma}. \tag{27.7}$$

Therefore

$$\mathbf{E} = \frac{i}{c}\sum_{\mathbf{k},\sigma} \omega_{\mathbf{k}} (\mathbf{A}_{\mathbf{k}}^{\sigma} e^{i\mathbf{kr}} - \mathbf{A}_{\mathbf{k}}^{\sigma *} e^{-i\mathbf{kr}}). \tag{27.8}$$

The magnetic field is determined from (21.28) and (12.28):

$$\mathbf{H} = \mathrm{rot}\,\mathbf{A} = \sum_{\mathbf{k},\sigma} ([\nabla e^{i\mathbf{kr}}, \mathbf{A}_{\mathbf{k}}^{\sigma}] + [\nabla e^{-i\mathbf{kr}}, \mathbf{A}_{\mathbf{k}}^{\sigma *}]) =$$

$$= i\sum_{\mathbf{k},\sigma} ([\mathbf{k}\mathbf{A}_{\mathbf{k}}^{\sigma}]\, e^{i\mathbf{kr}} - [\mathbf{k}\mathbf{A}_{\mathbf{k}}^{\sigma *}]\, e^{-i\mathbf{kr}}). \tag{27.9}$$

Let us now calculate the field energy. According to (13.21)

$$\mathscr{E} = \int \frac{E^2 + H^2}{8\pi}\, dV. \tag{27.10}$$

To obtain E^2 we perform summation over $\mathbf{k}$, $\mathbf{k}'$, σ, and σ':

$$E^2 = -\sum_{\mathbf{k},\mathbf{k}'\sigma,\sigma'} \frac{\omega_{\mathbf{k}}\,\omega_{\mathbf{k}'}}{c^2} (\mathbf{A}_{\mathbf{k}}^{\sigma}\mathbf{A}_{\mathbf{k}'}^{\sigma'} e^{i(\mathbf{k}+\mathbf{k}')\mathbf{r}} - \mathbf{A}_{\mathbf{k}}^{\sigma}\mathbf{A}_{\mathbf{k}'}^{\sigma' *} e^{i(\mathbf{k}-\mathbf{k}')\mathbf{r}} -$$

$$- \mathbf{A}_{\mathbf{k}}^{\sigma *}\mathbf{A}_{\mathbf{k}'}^{\sigma'} e^{-i(\mathbf{k}-\mathbf{k}')\mathbf{r}} + \mathbf{A}_{\mathbf{k}}^{\sigma *}\mathbf{A}_{\mathbf{k}'}^{\sigma' *} e^{-i(\mathbf{k}+\mathbf{k}')\mathbf{r}}). \tag{27.11}$$

It is expedient, when integrating E^2 over a volume, to change the order of summation and integration, each volume integral being reduced to the product of three integrals of the following type:

$$\int_0^{a_1} e^{i(k_x+k'_x)x}\,dx = \int_0^{a_1} e^{\frac{2\pi i}{a_1}(n_1+n'_1)}\,dx = \frac{a_1\left(e^{2\pi i(n_1+n'_1)}-1\right)}{2\pi i\,(n_1+n'_1)} = 0 \qquad (27.12)$$

$$\text{for } n_1+n'_1 \neq 0\,.$$

If $n_1+n'_1=0$ this integral is equal to a_1. Therefore the triple integral assumes one of two values:

$$\int e^{i(\mathbf{k}+\mathbf{k}')\mathbf{r}}\,d\,V = \begin{cases} a_1 a_2 a_3 = V & \text{for } \mathbf{k}' = -\mathbf{k}, \\ 0 & \text{for } \mathbf{k}' \neq -\mathbf{k}. \end{cases} \qquad (27.13)$$

It follows that in the expression for E^2 the double summation with respect to $\mathbf{k}$ and $\mathbf{k}'$ becomes a single one after integration, and we must replace $\mathbf{k}'$ by $-\mathbf{k}$ in the terms involving the product $\mathbf{A}_\mathbf{k}^\sigma\,\mathbf{A}_{\mathbf{k}'}^{\sigma'}$. In terms containing $\mathbf{A}_\mathbf{k}^\sigma\,\mathbf{A}_{\mathbf{k}'}^{\sigma'*}$ we replace k' by k due to the factor $e^{-i\mathbf{k}'\mathbf{r}}$. Thus,

$$\int E^2\,d\,V = -V\sum_{\mathbf{k},\sigma,\sigma'}\frac{\omega_\mathbf{k}^2}{c^2}\left(\mathbf{A}_\mathbf{k}^\sigma\mathbf{A}_{-\mathbf{k}}^{\sigma'} - \mathbf{A}_\mathbf{k}^\sigma\mathbf{A}_\mathbf{k}^{\sigma'*} - \mathbf{A}_\mathbf{k}^{\sigma*}\mathbf{A}_\mathbf{k}^{\sigma'} + \mathbf{A}_\mathbf{k}^{\sigma*}\mathbf{A}_{-\mathbf{k}}^{\sigma'*}\right). \quad (27.14)$$

But (given $\sigma\neq\sigma'$) $\mathbf{A}_\mathbf{k}^\sigma$ and $\mathbf{A}_{-\mathbf{k}}^{\sigma'}$, $\mathbf{A}_\mathbf{k}^{\sigma*}$ and $\mathbf{A}_\mathbf{k}^{\sigma'*}$ are orthogonal vectors. Therefore, instead of the double summation with respect to σ and σ', there also remains a single sum with respect to σ:

$$\int E^2\,d\,V = -\frac{V}{c^2}\sum_{\mathbf{k},\sigma}\omega_\mathbf{k}^2\left(\mathbf{A}_\mathbf{k}^\sigma\mathbf{A}_{-\mathbf{k}}^{\sigma} + \mathbf{A}_\mathbf{k}^{\sigma*}\mathbf{A}_{-\mathbf{k}}^{\sigma*} - 2\mathbf{A}_\mathbf{k}^\sigma\mathbf{A}_\mathbf{k}^{\sigma*}\right). \quad (27.15)$$

When calculating the integral of the square of the magnetic field we make use of the rule (27.13). But since the product $[\mathbf{k}'\,\mathbf{A}_{\mathbf{k}'}^{\sigma'}]$ is replaced by $-[\mathbf{k}\mathbf{A}_{-\mathbf{k}}^{\sigma'}]$ if $\mathbf{k}'=-\mathbf{k}$, we get

$$\int H^2 d\,V = V\sum_{\mathbf{k},\sigma,\sigma'}\left([\mathbf{k}\mathbf{A}_\mathbf{k}^\sigma]\,[\mathbf{k}\mathbf{A}_{-\mathbf{k}}^{\sigma'}] + [\mathbf{k}\mathbf{A}_\mathbf{k}^{\sigma*}]\,[\mathbf{k}\mathbf{A}_{-\mathbf{k}}^{\sigma'*}] + 2\,[\mathbf{k}\mathbf{A}_\mathbf{k}^{\sigma}]\,[\mathbf{k}\mathbf{A}_\mathbf{k}^{\sigma'}]\right). \quad (27.16)$$

The vector products may be expressed by known formulae:

$$[\mathbf{k}\mathbf{A}_\mathbf{k}^\sigma]\,[\mathbf{k}\mathbf{A}_\mathbf{k}^{\sigma'*}] = k^2\,\mathbf{A}_\mathbf{k}^\sigma\mathbf{A}_\mathbf{k}^{\sigma'*} - (\mathbf{k}\mathbf{A}_\mathbf{k}^\sigma)\,(\mathbf{k}\mathbf{A}_\mathbf{k}^{\sigma'*}) = k^2\mathbf{A}_\mathbf{k}^\sigma\mathbf{A}_\mathbf{k}^{\sigma'*}, \qquad (27.17)$$

where we have used the transversality condition (27.3). This expression becomes zero for $\sigma\neq\sigma'$.

Thus,

$$\int H^2\,d\,V = V\sum_{\mathbf{k},\sigma}k^2\left(\mathbf{A}_\mathbf{k}^\sigma\mathbf{A}_{-\mathbf{k}}^\sigma + \mathbf{A}_\mathbf{k}^{\sigma*}\mathbf{A}_{-\mathbf{k}}^{\sigma*} + 2\,\mathbf{A}_\mathbf{k}^\sigma\mathbf{A}_\mathbf{k}^{\sigma*}\right). \qquad (27.18)$$

Combining (27.14) and (27.18) and taking advantage of the fact that $\omega^2=c^2\,k^2$, we get

$$\mathscr{E} = \frac{1}{8\pi}\int (E^2 + H^2)\, dV = \frac{V}{2\pi c^2}\sum_{\mathbf{k},\sigma} \omega_{\mathbf{k}}^2 \mathbf{A}_{\mathbf{k}}^{\sigma} \mathbf{A}_{\mathbf{k}}^{\sigma *}. \tag{27.19}$$

Passing to real variables. In order to apply the usual equations of quantum mechanics to the electromagnetic field, it is convenient to pass to real variables

$$\mathbf{A}_{\mathbf{k}}^{\sigma} = \sqrt{\frac{\pi c^2}{V}}\left(Q_{\mathbf{k}}^{\sigma} + \frac{i P_{\mathbf{k}}^{\sigma}}{\omega_{\mathbf{k}}}\right)\mathbf{e}_{\mathbf{k}}^{\sigma}, \tag{27.20}$$

$$\mathbf{A}_{\mathbf{k}}^{\sigma *} = \sqrt{\frac{\pi c^2}{V}}\left(Q_{\mathbf{k}}^{\sigma} - \frac{i P_{\mathbf{k}}^{\sigma}}{\omega_{\mathbf{k}}}\right)\mathbf{e}_{\mathbf{k}}^{\sigma}, \tag{27.21}$$

where $\mathbf{e}_{\mathbf{k}}^{\sigma}$ is a unit vector in the direction of polarization of the wave. We get $\mathscr{E}$ expressed in terms of a sum of energies, or of the Hamiltonian functions for linear harmonic oscillators:

$$\mathscr{E} = \frac{1}{2}\sum_{\mathbf{k},\sigma} (P_{\mathbf{k}}^{\sigma 2} + \omega_{\mathbf{k}}^2 Q_{\mathbf{k}}^{\sigma 2}). \tag{27.22}$$

If we regard $P_{\mathbf{k}}^{\sigma}$ and $Q_{\mathbf{k}}^{\sigma}$ as ordinary classical dynamical variables, they satisfy the equations for a linear harmonic oscillator. Indeed, it follows from (27.22) and (10.18) that $\dot{P}_{\mathbf{k}}^{\sigma} = -\omega_{\mathbf{k}}^2 Q_{\mathbf{k}}^{\sigma}$ and $\dot{Q}_{\mathbf{k}}^{\sigma} = P_{\mathbf{k}}^{\sigma}$. This agrees with the harmonic time dependence of the amplitudes $\mathbf{A}_{\mathbf{k}}^{\sigma}$, $\mathbf{A}_{\mathbf{k}}^{\sigma *}$.

Each separate oscillator is characterized by four integers n_1, n_2, n_3, and σ, which label the independent degrees of freedom of the electromagnetic field. $Q_{\mathbf{k}}^{\sigma}$ are normal coordinates of the electromagnetic field [see (7.31)].

Quantization of the electromagnetic field. The result we have obtained (27.22) is of fundamental significance. It provides the most simple and vivid method of applying quantum mechanics to the electromagnetic field. Indeed, the equations of nonquantized oscillators are equivalent to the electrodynamical equations of a nonquantized electromagnetic field, the only difference being that they describe the field in other variables. But the oscillator problem in quantum mechanics having already been solved in Sec. 26, quantization can be performed in these variables as before. Quantization of the motion of oscillators representing a field *in vacuo* is just this quantization of the equations of electrodynamics performed in the appropriate system of variables. From (26.21), the energy of an oscillator in the Nth quantum state was

$$\mathscr{E} = \hbar\omega\left(N + \frac{1}{2}\right).$$

Therefore, with the aid of equation (27.22), the energy of the electromagnetic field is also reduced to the simple form:

$$\mathscr{E} = \sum_{n_1, n_2, n_3, \sigma} \hbar \omega_{n_1, n_2, n_3} \left(N_{n_1, n_2, n_3, \sigma} + \frac{1}{2} \right) \equiv \sum_{\mathbf{k}, \sigma} \hbar \omega_{\mathbf{k}} \left(N_{\mathbf{k}, \sigma} + \frac{1}{2} \right). \tag{27.23}$$

The numeral $N_{n_1, n_2, n_3, \sigma}$ gives the number of the quantum state of the oscillator, classified by the numbers n_1, n_2, n_3, and the polarization σ.

Quanta. We see that the energy of an oscillator can experience increments equal to $\hbar \omega_{n_1, n_2, n_3}$. This quantity of energy is called the energy quantum of the electromagnetic field. Disregarding the zero energy $\frac{\hbar \omega_{n_1, n_2, n_3}}{2}$ for the time being, we see that the field energy is equal to the sum of the energies of its quanta $\hbar \omega_{n_1, n_2, n_3}$. Thus is found a quantum expression for the energy of an electromagnetic field. It will be shown in exercise 1 of this section that the momentum of a field is equal to the sum of the momenta of its quanta, while the momentum of each quantum is found to be related to its energy by the expression

$$\mathbf{p} = \hbar \mathbf{k} = \frac{\hbar \omega}{c} \frac{\mathbf{k}}{k}. \tag{27.24}$$

Thus, a quantum possesses the properties of a particle of zero mass. The possibility of the existence of such particles was elucidated in connection with equation (21.16).

Polarization of quanta. A quantum has one more, so to say, internal degree of freedom, that of *polarization*. This peculiar degree of freedom corresponds to the "coordinate" σ, taking only two values $\sigma = 1$ and $\sigma = 2$. The energy does not depend upon σ. But, of course, in order to fully define the oscillation corresponding to a given quantum, we must indicate the number $\sigma = 1$, 2 as well as the three numbers n_1, n_2, n_3. We observe that these quanta relate only to the transverse field of electromagnetic waves and do not describe the Coulomb field.

The classical approximation of quanta and electrons. A quantum should by no means be regarded as the outcome of some mathematical sleight of hand that led us to equation (27.23). The quantum is an elementary particle just as real as the electron. For example, when X-rays are scattered by electrons, the energy of each *individual* quantum $\hbar \omega$ and its momentum $\hbar \mathbf{k}$ are involved in the general law of conservation of energy and momentum in collisions in the same way as for any other particle. In scattering, the frequency of a quantum diminishes, in direct proportion to its energy.

The essential difference between the quantum and the electron consists in the fact that in classical theory there is nothing to which the quantum corresponds; its energy $\mathscr{E}_\omega = \hbar \omega$ and its momentum $\mathbf{p} = \hbar \mathbf{k}$ tend to zero when $\hbar$ tends to zero. Yet the quantum mechanics of the electron admits of a classical approximation, since the quantum quantities for the electron become corresponding classical quantities, which do not tend to zero even when we take $h \to 0$. The relation (27.24) has a counterpart in classical electrodynamics; the energy density of an electromagnetic wave was shown in Sec. 17 to be related to its momentum by means of the factor c. In the limiting transition to classical theory, the energy of each quantum is regarded as infinitely small while their number $N_{\mathbf{k},\sigma}$ is infinitely large, so that the wave amplitude remains finite.

Occupation numbers. Passing from Cartesian coordinates to new independent variables (the components of the wave vector $\mathbf{k}$), we renumbered the radiation degrees of freedom, the quantities $Q_{\mathbf{k}}^{\sigma}$ now being the generalized coordinates. The state of a field is specified if all the "occupation" numbers $N_{\mathbf{k},\sigma}$ are known, because the number $N_{\mathbf{k},\sigma}$ defines the quantum state the given harmonic oscillator is in, i.e., the number of quanta in the state $\mathbf{k}$, σ. The numbers $N_{\mathbf{k},\sigma}$ may be regarded as the quantum variables of an electromagnetic field. When a field interacts with a radiator (for example, an atom) these numbers change. For example, if the number $N_{\mathbf{k},\sigma}$ has increased by unity this means that a quantum of corresponding frequency, direction, and polarization has been emitted.

The ground state of an electromagnetic field. Let us now examine expression (27.23) when $N_{\mathbf{k},\sigma}=0$. In other words, let us determine the ground state energy of the electromagnetic field. According to (27.23) it is

$$\mathscr{E}(0) = \frac{1}{2}\sum_{n_1, n_2, n_3, \sigma} h\,\omega_{n_1, n_2, n_3}\,. \tag{27.25}$$

But since the numbers n_1, n_2, and n_3 run through an infinite set of values, the sum (27.25) is infinitely large. It must be said that in this case the theory is not fundamentally defective because the zero energy itself (27.25), does not appear in any expression; the field energy is always measured from the ground state.

At the same time, the ground state ψ_0 of the quantum oscillators of an electromagnetic field leads to actually observable effects because the amplitude of a harmonic oscillator in the ground state is not equal to zero. It takes on all possible values, and, in accordance with (26.22), the probability of a certain value of Q is proportional to $|\psi_0(Q)|^2$.

In an electromagnetic field, the part of the oscillator coordinates is played by the generalized coordinates Q, in terms of which the field amplitudes are expressed linearly. Therefore, one can by no means assert that the field amplitudes are equal to zero in the ground state of an electromagnetic field (i.e., in the *absence of quanta*). The probabilities of definite amplitude values are given by the harmonic-oscillator wave functions. These functions are equal to $e^{-\frac{\omega Q^2}{2h}}$ in coordinates.

Electron-level shift produced by the ground state of a field. The ground state of an electromagnetic field affects observable quantities. One of the most important effects of this type consists in the following.

Let an electron move in the potential field of a nucleus. The value of the electromagnetic potentials of a field acting on the electron is usually chosen as $\mathbf{A}=0$, $\varphi=\frac{Ze}{r}$. Only the static Coulomb field has been taken into account. In actual fact, the potential of the radiating field must be added to the static potential, for example, in the form (27.5). As was indicated, this potential must not be considered equal to zero even when there are no quanta in the field. The field of radiation affects the energy eigenvalues of the electrons in an atom. They prove different from what they should be in a purely Coulomb nuclear field with $\mathbf{A}=0$, $\varphi=\frac{Ze}{r}$.

The solution of the problem of finding the energy eigenvalues for an atom, with account taken of the radiation field, encounters considerable inherent difficulties. First of all, this problem does not lend itself to a precise solution by means of mathematical analysis. It becomes necessary to solve it approximately, taking the correction to the energy produced by the radiating field as a small quantity. In actuality, however, a direct calculation of this correction leads to divergent integrals, i.e., to infinite expressions.

Nevertheless, it is possible to redetermine this correction so that a finite expression is obtained. To do this, one has to consider the analogous correction for the energy of a "free" electron not influenced by the external field of the nucleus, and consider the difference of two infinite integrals. If in doing so we take great care to follow the relativistic invariance of the expressions, the subtraction turns out to be a completely unique operation and does not contain any indeterminate quantity of the form $\infty - \infty$. The final correction does indeed turn out to be a small quantity compared with the binding energy of the electron in the atom in its ground state (4×10^{-6} ev for the hydrogen atom). The value of the correction is in excellent agreement with modern radiospectroscopic data.

Subtraction of infinities. The meaning of such a subtraction consists in the following. Physically, an electron is inseparable from its charge, i.e., from its radiation field. When we talk about a "free" electron, we always imply that the electron interacts with a radiation field, which cannot be regarded as equal to zero. The rest energy of an electron is equal to mc^2, where m denotes the observed value of the mass. In reality, this quantity encompasses the energy of all interactions of the electron, including interaction with the field of radiation.

Thus, in calculating the energy of an electron in a Coulomb field, the mass of the electron must be redefined so that, in the absence of any static external field, all the energy has a finite value mc^2. This redefinition operation or, as it is called, "renormalization" of mass allows us to find a finite quantity for the energy eigenvalue of an electron in an atom. Concerning the levels in a hydrogen atom, see Secs. 31 and 38.

The renormalization operation consists in the fact that the mass which appears formally in the equations of mechanics, together with the mass due to the interaction of the electron with radiation, is regarded as the observed finite quantity $m = 0.9106 \times 10^{-27}$ gm. Only this known quantity appears in the final result for any calculated effect.

Difficulties of the theory. The appearance of divergent (i.e., infinite) expressions in quantum electrodynamics is a defect of the theory and indicates a certain internal contradiction. The modern form of the theory as given by J. Schwinger, R. Feynman and F. Dyson, is apparently, not yet final.

It is all the more remarkable that despite this imperfection, quantum electrodynamics is capable—with the aid of renormalization—of yielding correct and unique answers when calculating concrete quantities observed in experiment.

Exercise

Calculate the electromagnetic field momentum in a vacuum in terms of the normal field coordinates $Q_{\mathbf{k}}^{\sigma}$

In the expression (13.27)

$$\mathbf{p} = \frac{1}{4\pi c}\int [\mathbf{E}\mathbf{H}]\, dV$$

we substitute the electric and magnetic fields from (27.8) and (27.9). After integration over the volume we obtain by (27.13)

$$\mathbf{p} = \frac{V}{4\pi c^2}\sum_{\mathbf{k},\sigma,\sigma'} \omega_{\mathbf{k}} \left([\mathbf{A}_{\mathbf{k}}^{\sigma}[\mathbf{k}\mathbf{A}_{-\mathbf{k}}^{\sigma'}]] + [\mathbf{A}_{\mathbf{k}}^{\sigma}[\mathbf{k}\mathbf{A}_{\mathbf{k}}^{\sigma'*}]] + [\mathbf{A}_{\mathbf{k}}^{\sigma*}[\mathbf{k}\mathbf{A}_{\mathbf{k}}^{\sigma'}]] + [\mathbf{A}_{\mathbf{k}}^{\sigma*}[\mathbf{k}\mathbf{A}_{-\mathbf{k}}^{\sigma'*}]]\right).$$

We rearrange the double vector products:

$$\mathbf{p} = \frac{V}{4\pi c^2}\sum_{\mathbf{k},\sigma} \omega_{\mathbf{k}} \left(\mathbf{k}\mathbf{A}_{\mathbf{k}}^{\sigma}\mathbf{A}_{-\mathbf{k}}^{\sigma} + 2\,\mathbf{k}\mathbf{A}_{\mathbf{k}}^{\sigma}\mathbf{A}_{\mathbf{k}}^{\sigma*} + \mathbf{k}\mathbf{A}_{\mathbf{k}}^{\sigma*}\mathbf{A}_{-\mathbf{k}}^{\sigma*}\right).$$

The quantities $\mathbf{k}\mathbf{A}_{\mathbf{k}}^{\sigma}\mathbf{A}_{-\mathbf{k}}^{\sigma}$ and $\mathbf{k}\mathbf{A}_{\mathbf{k}}^{\sigma*}\mathbf{A}_{-\mathbf{k}}^{\sigma*}$ are odd functions of $\mathbf{k}$, and disappear after summation over all $\mathbf{k}$. There remains only

$$\mathbf{p} = \frac{V}{2\pi c^2}\sum_{\mathbf{k},\sigma} \omega_{\mathbf{k}}\, \mathbf{k}\,\mathbf{A}_{\mathbf{k}}^{\sigma}\mathbf{A}_{\mathbf{k}}^{\sigma*}.$$

Substituting here the normal coordinates of the field from (27.20) and (27.21), we arrive at the expression

$$\mathbf{p} = \sum_{\mathbf{k},\sigma} \frac{\mathbf{k}}{\omega_{\mathbf{k}}}\,\frac{1}{2}\left(P_{\mathbf{k}}^{\sigma 2} + \omega_{\mathbf{k}}^2 Q_{\mathbf{k}}^{\sigma 2}\right) = \sum_{\mathbf{k},\sigma} \hbar\mathbf{k}\left(N_{\mathbf{k},\sigma} + \frac{1}{2}\right),$$

so that the momentum of each quantum is related to its energy by relation (27.24).

Sec. 28. Quasi-Classical Approximation

The classical limit of a wave function. It was shown in Sec. 24 that the limiting transition from quantum mechanics to classical mechanics is performed by means of the substitution of (24.5)

$$\psi = e^{i\frac{S}{h}}. \tag{28.1}$$

By substituting ψ into (24.11) (i.e., into the Schrödinger equation), eliminating $e^{i\frac{S}{h}}$, and formally making h tend to zero, we obtained the correct classical relation between energy and momentum. This limiting process signifies that the de Broglie wavelength $\frac{2\pi h}{mv}$ is very small compared with the region in which the motion occurs.

It is sometimes useful not to carry the limiting process to its end, namely in those cases when the asymptotic form of the almost classical wave function is important. If the wavelength is small compared with the region in which motion occurs, then the wave function has many nodes in this region and, as we know from Sec. 25, this corresponds to an energy eigenvalue with a large number. Thus, the passage from quantum laws of motion to classical laws is accomplished through a region in which the number of the eigenvalue is large compared with unity. If the energy eigenvalue is determined by several integers, as in the case with motion having several degrees of freedom (cf. the problem of the potential well), then all these numbers must be large compared with unity so that the motion should be close to the classical limit. The wave function in approximation (28.1) (where S is the classically calculated action) thus allows us to determine eigenvalues with large numbers [see (28.18)].

Solutions with real values of the exponent. The wave function does not convert to zero also for real values of the exponent in (28.1), i.e., for imaginary values of S. Imaginary values can only occur in those regions of space into which—for a given energy—the trajectories of classical motion cannot enter, because the potential energy would be greater in that case than the total energy, corresponding to a negative kinetic energy or an imaginary velocity. In these cases, the square of the modulus of the wave function determines the probability of a particle penetrating into a classically unattainable region of motion. Naturally, this probability does not convert to zero only anterior to the limiting transition, and not posterior to it, when h has already been eliminated from the equations. This is why approximation (28.1) is termed quasi-classical.

The quasi-classical approximation. Let us write the equations of transition to a quasi-classical approximation in a one-dimensional case. From (24.14), by substituting $\frac{\partial S}{\partial t} = -\mathscr{E}$ and $\nabla S = \frac{dS}{dx}$ into it (the one-dimensional case!); we have

$$\left(\frac{dS}{dx}\right)^2 = 2m(\mathscr{E} - U), \tag{28.2}$$

whence

$$S = \int \sqrt{2m(\mathscr{E} - U)}\, dx. \tag{28.3}$$

Unlike classical mechanics, equation (28.3) holds not only for $\mathscr{E} > U$, when the root is extracted from a positive quantity, but also when $\mathscr{E} < U$, when the action is imaginary.

In Sec. 25 we investigated a similar example of the precise wave equation for the problem of a potential well of finite depth. In the region of the well, the wave function was of the form $\psi = \sin \varkappa x$, while outside the well it approached zero exponentially like $e^{-\tilde{\varkappa} x}$. This resulted precisely when $\mathscr{E} < U$, i.e., in a classically unattainable region. Of course, an imaginary velocity signifies that a particle moving according to classical laws does not attain the given position. Therefore, the expression

$$\pm \frac{i}{h} \int \sqrt{2m(U - \mathscr{E})}\, dx \tag{28.4}$$

must not longer be understood as action, but simply as the exponent in the equation

$$\psi = e^{i\frac{S}{h}},$$

extended to the region $\mathscr{E} < U$, if there is such a region. When $h \to 0$, the wave function will be damped in this region infinitely quickly—like $e^{-\infty}$; and this denotes the unattainability of points where for classical motion $\mathscr{E} < U$.

The potential barrier. In the potential well problem, the region for which $U > \mathscr{E}$ extended rightward to infinity. Therefore, the wave function became zero at infinity. Considerable interest is attached to another problem, in which the potential energy at a certain distance away from the well again becomes less than the total energy. This is shown in Fig. 42. $U > \mathscr{E}$ for the region $x_1 \leqslant x \leqslant x_2$. Therefore, in classical mechanics a particle situated to the left of $x = x_1$ cannot under any circumstances attain the region $x > x_2$, from which it is separated by a potential barrier. In quantum mechanics, the wave function does not become zero between x_1 and x_2, since this region is finite (see exercise 2, Sec. 26).

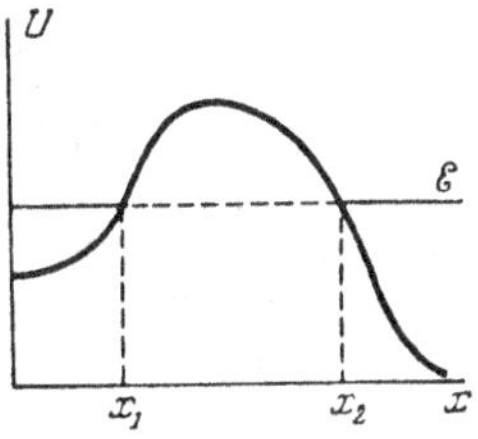

Fig. 42

In the approximation (28.1), the exponent in the equation is a real quantity when $x < x_1$. Therefore, the modulus of ψ remains equal to unity:

$$|\psi|^2 = \psi^*\psi = e^{-i\frac{S}{h}} e^{i\frac{S}{h}} = 1\,.$$

On the other hand, between x_1 and x_2, the modulus of ψ decreases according to the law

$$|\psi|^2 = \left(e^{-\frac{1}{h}\int\limits_{x_1}^{x}\sqrt{2m(U-\mathscr{E})}\,dx}\right)^2 = e^{-\frac{2}{h}\int\limits_{x_1}^{x}\sqrt{2m(U-\mathscr{E})}\,dx}\,. \tag{28.5}$$

At $x = x_2$, in comparison with the point $x = x_1$ it diminishes in the ratio

$$B = e^{-\frac{2}{h}\int\limits_{x_1}^{x_2}\sqrt{2m(U-\mathscr{E})}\,dx} \tag{28.6}$$

after which it again stops changing, since S becomes a real quantity.

Hence, the square of the modulus of the wave function diminishes between x_1 and x_2 according to the expression determined by the quantity B. A more precise theory provides a correction factor for B, though fundamentally, the function is determined by B alone. B is called the *barrier factor*. Somewhat later, it will be explained how the B factor is related to the penetration probability through the barrier in unit time.

The quasi-classical approximation is feasible only when the order of magnitude for the action is large compared with h. This was mentioned in Sec. 24, when the conditions for the limiting transition to classical mechanics were being determined. For this reason, equation (28.6) may be used only when the exponent is large compared with unity. If it is comparable with unity, the penetration probability through the barrier must be evaluated by means of precise wave functions.

The Mandelshtam analogy. In wave optics there is an analogy to the passage of a particle through a potential barrier. When a light wave falls on the boundary of a medium of small refractive index from a medium of larger refractive index at an angle whose sine is greater than the ratio of the indices of refraction, there occurs total internal reflection in accordance with the laws of geometrical optics. If the problem is solved in strict accord with wave optics, on the basis of Maxwell's equations (see exercise 2, Sec. 17), then it turns out that the wave penetrates somewhat into the second medium, but dies out exponentially in it. L. I. Mandelshtam took notice of this analogy between quantum mechanics and wave optics. It can be applied in the following manner.

Let us imagine two optically dense media separated by a layer optically less dense. Let a light ray fall on the interface of the media at an angle to the normal larger than the angle of total internal reflection. According to geometrical optics, the ray should be completely reflected by the layer, and it is absolutely immaterial whether or not there is a denser medium beyond the layer, or whether the reflection occurs from an infinitely thick, nondense medium. Similarly, a particle in classical mechanics is completely incapable of penetrating the barrier. According to the laws of wave optics, light penetrates into a nondense medium, but dies out in a thickness comparable with the wavelength. Therefore, if the second dense medium is situated closely enough, part of the light "seeps" into it.

The classical expression for the amplitude of a light wave may be regarded as the wave function in relation to a light quantum. The transition from quantum theory to classical electrodynamics consists in considering the occupation numbers $N_{\mathrm{k},\sigma}$ as large (see Sec. 28). Then the corresponding field amplitudes change to classical ones. For this reason the Mandelshtam analogy represents an example of quanta penentrating through a barrier. As has already been pointed out, the limiting transition for electrons occurs differently; it corresponds to the transition from wave optics to geometrical optics. Therefore, in the classical limit, electrons do not penetrate the barrier.

The existence of penetrations through a barrier clearly indicates that the concept of a trajectory is sometimes completely inapplicable in the case of quantum motion. A trajectory extended under the barrier would lead to imaginary velocity values.

Alpha disintegration. Passage through a potential barrier enables us to explain one of the most important facts of nuclear physics, that of alpha disintegration. The nuclear masses of heavy elements with atomic numbers greater than that of lead satisfy an inequality of the form (21.12):

$$m(A, Z) > m(A-4, Z-2) + m(4, 2). \tag{28.7}$$

Here A is the atomic weight and Z is the nuclear charge (i.e., the atomic number in the Mendeleyev table). Thus, $m(4,2)$ is the mass of a helium necleus with atomic weight 4 and atomic number 2. Such a nucleus emitted during alpha disintegration is called an alpha particle.

All that can be seen from equations (21.12) and (28.7) is that the spontaneous decay of a nucleus of mass $m(A, Z)$ is *possible*, though no indication is obtained about the time law of disintegration. The nuclei of certain elements have mean decay times of 10^{10} years while others have decay times of about 10^{-5} sec, which is a difference of 23 orders of magnitude. It will be noted that the energy of the alpha particles emitted differs here by a factor of only two. From experiment it turns out that the logarithm of the mean decay time of a nucleus is inversely proportional to the alpha-particle velocity. It is this logarithmic law that corresponds to the difference of 23 magnitudes. It is accounted for by the difference of barrier factors which depend exponentially upon the energy.

The potential-energy curve. At large distances from a nucleus, an alpha particle experiences a repulsive force of potential energy

$$U = \frac{2(Z-2)e^2}{r} \tag{28.8}$$

[cf. (3.4)]. At small distances, attractive forces must act because, otherwise, the nucleus (A, Z) could not exist at all. We do not know the *force law* (i. e., the shape of the potential-energy curve when the alpha particle is situated sufficiently close to the nucleus) and, therefore, in Fig. 43 we draw it at will. In this we must be guided by the following considerations. The special nuclear forces which hold the alpha particle in the nucleus before emission have a small radius of action, so the potential-energy curve has the form of a "potential well." Motion inside the nucleus corresponds to motion inside such a well. The transition region from the well to the Coulomb curve is not very essential for final results, i. e., it little affects the exponent of the barrier factor.

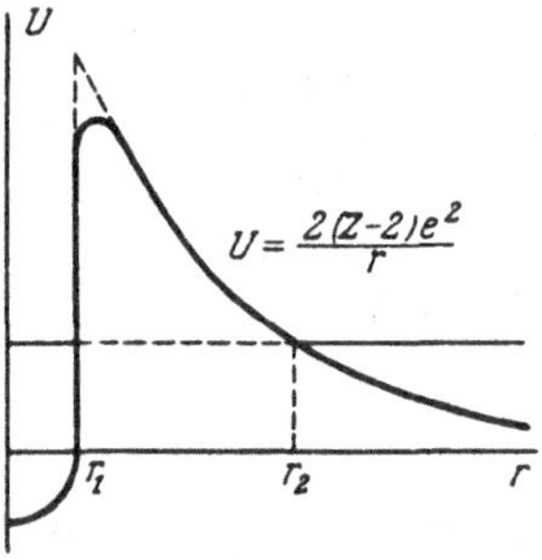

Fig. 43

The barrier factor for alpha disintegration. The energy of an alpha particle is positive at an infinite distance from the nucleus. It is this that signifies that the nucleus is capable of alpha disintegration, i.e., the alpha particle can move infinitely. In order to find the probability for alpha disintegration, we must calculate the barrier factor B in accordance with (28.6). Because nuclear forces are short-range forces, the transition region is small and we can extrapolate the Coulomb law, without sensible error in the integral, up to the point $r = r_1$ where $\mathscr{E}$ becomes greater than U. Point r_1 is the effective nuclear radius determined from alpha disintegration. Other data concerning the nucleus lead to somewhat different values for the respectively determined effective radius. This is understandable since r_1 is obtained on the particular assumption that the Coulomb law is valid up to the region for which the potential energy curve is taken in the form of a well with steep sides.

And so we determine the barrier factor according to the equation

$$B \doteq e^{-\frac{2}{h}\int\limits_{r_1}^{r_2}\sqrt{2m\left(\frac{2(Z-2)e^2}{r} - \mathscr{E}\right)}\,dr} \,. \tag{28.9}$$

The integral in the exponent can be easily calculated by the substitution

$$\frac{r\mathscr{E}}{2(Z-2)e^2} = \cos^2 x\,. \tag{28.10}$$

Then, after elementary treatment, it reduces to the form

$$\frac{2}{h}\sqrt{2m}\int_{r_1}^{r_2}\sqrt{\frac{2(Z-2)e^2}{r}-\mathscr{E}}\,dr=$$

$$=\sqrt{2m}\cdot\frac{2}{h}\cdot\frac{2(Z-2)e^2}{\sqrt{\mathscr{E}}}\left(\text{arc cos}\sqrt{\frac{\mathscr{E}r_1}{2(Z-2)e^2}}-\right.$$

$$\left.-\sqrt{\frac{\mathscr{E}r_1}{2(Z-2)e^2}}\sqrt{1-\frac{\mathscr{E}r_1}{2(Z-2)e^2}}\right). \tag{28.11}$$

The quantity $\frac{\mathscr{E}r_1}{2(Z-2)e^2}$ is the ratio of the alpha particle energy to the effective barrier height at the point r_1, taken according to equation (28.8). Let us evaluate this ratio. For heavy nuclei $2(Z-2)\simeq 180$, $r_1 \simeq 9\times 10^{-13}$ cm; we shall take $\mathscr{E}$ to be equal to 6 Mev, $e^2=23\times 10^{-20}$. Whence

$$\frac{\mathscr{E}r_1}{2(Z-2)e^2}=\frac{6\cdot 1.6\cdot 10^{-6}\cdot 9\cdot 10^{-13}}{180\cdot 23\cdot 10^{-20}}\simeq\frac{1}{5}.$$

To a first approximation, we shall consider this quantity as small.

Then, on the right-hand side of (28.11), we obtain the approximation

$$\frac{2\sqrt{2m}}{h}\cdot\frac{2(Z-2)e^2}{\sqrt{\mathscr{E}}}\left(\frac{\pi}{2}-2\sqrt{\frac{\mathscr{E}r_1}{2(Z-2)e^2}}\right)=$$

$$=\frac{2\pi e^2}{hv}\cdot 2(Z-2)-\frac{8}{h}\sqrt{mr_1e^2(Z-2)}. \tag{28.12}$$

It is easy to check the correctness of this expansion by a direct numerical substitution.

The time dependence of alpha disintegration. We shall now show how the expression for the barrier factor is related to the probability of alpha disintegration in unit time. In exercise 2, Sec. 25, it was found that the particle flux passing through a barrier is proportional to the particle density before the barrier. It is easy to see that the basic result obtained in the problem of a rectangular barrier coincides, in the limit, with the result of this section, if we go over to the quasi-classical approximation. Indeed, we, so to say, divide a barrier of arbitrary shape into separate, successive rectangular barriers. The penetration probability for each of them is $e^{-\frac{2}{h}\sqrt{2m(U-\mathscr{E})}\cdot\Delta x}=$ $=e^{-2\varkappa\Delta x}$, where Δx is the width of the rectangular barrier. The total penetration probability will be determined by reduction of the amplitude of the wave function over the whole width of the barrier. In other words, it will be proportional to the product $\Pi\, e^{-\frac{2}{h}\sqrt{2m(U-\mathscr{E})}\,\Delta x}$. This product can, obviously, also be represented like $e^{-\sum\frac{2}{h}\sqrt{2m(U-\mathscr{E})}\,\Delta x}$

or, in the limit, like $e^{-\frac{2}{h}\int\sqrt{2m(U-\mathscr{E})}\,dx}$ corresponding to (28.6). Thus, for any potential barrier, we can assert that the flux of transmitted particles is proportional to the particle density before the barrier multiplied by the barrier factor.

From this we can deduce the time dependence of alpha disintegration. The probability of an alpha particle existing inside the nucleus is equal to the integral of the square of the wave-function modulus over the volume of the nucleus, i.e., over the region $r<r_1$. As has just been indicated, the alpha-particle flux emitted from the nucleus is proportional to the probability density of their being in the nucleus, the constant of proportionality being basically determined by the barrier factor. It follows that the number of nuclei decaying in unit time is proportional to the total number of nuclei present that have not disintegrated by the given instant. The constant of proportionality depends upon the shape of the barrier and upon the state of the particles inside the nucleus, but it cannot be a function of time, as may be seen, for example, from the equations obtained in exercise 2, Sec. 25. Indeed, this constant is obtained from the solution of the wave equation with the time dependence eliminated, i.e., (24.22).

For this reason, the law for alpha disintegration is expressed by the equation

$$\frac{dN}{dt}=-\frac{\Gamma}{h}N, \quad N=N_0e^{-\frac{\Gamma t}{h}}, \tag{28.13}$$

where N is the number of nuclei that have not disintegrated at the given instant of time and N_0 is the initial number of nuclei. The quantity Γ has the dimensions of energy for convenience of comparison with other quantities of the same dimension. Every nucleus has the same probability of decaying in unit time, no matter how long it has been in existence. This probability is $\frac{\Gamma}{h}$ and does not depend upon time.

Equations (28.13) and (28.12) confirm the experimentally determined law which yields an inversely proportional relationship between the logarithm of the probability for alpha disintegration and the alpha-particle velocity v.

The nuclear wave function before decay. It is easy to obtain a time dependence for the wave function of a nucleus which has not emitted an alpha particle. It looks like

$$\psi=e^{-\frac{\Gamma t}{2h}}\cdot e^{-i\frac{\mathscr{E}_0 t}{h}}. \tag{28.14}$$

The first factor accounts for the exponential attenuation of amplitude according to a $e^{-\frac{\Gamma t}{2h}}$ law (since the probability, i.e., the square of the amplitude, diminishes according to $e^{-\frac{\Gamma t}{h}}$); the second factor is the usual wave-function time factor. Expression (28.14) is very similar to the well-known formula for damped oscillations, with the difference that in the given case it is the probability amplitude of the initial (not yet decayed) state of the nucleus that is damped.

The wave function (28.14) satisfies the initial condition $|\psi(0)|=1$. Since the wave-function equation is of the first order with respect to time, it is absolutely necessary to impose some initial condition. It was assumed that at the initial instant of time an alpha particle was definitely situated inside the nucleus. However, for this wave function, the right-hand side of equation (24.16) does not become zero: there is a finite flux for the probability of a particle being emitted from the nucleus. This is what leads to the exponential damping of the probability of the predecay state.

All nuclei before disintegration are described by exactly the same wave function (28.14), if at the instant $t=0$ they were in the initial state. Therefore, they all have a perfectly identical probability of decaying in unit time, and it is impossible to predict which one of them will decay earlier and which later. In exactly the same way, in the diffraction experiment it is impossible to say which part of the photographic plate will be hit by a given electron. The decay law is purely statistical, in the same way as the law for diffraction pattern.

In this sense, radioactive decay does not resemble the falling of ripe fruit from a tree—an alpha particle in a nucleus is always identically "ripe" for emission. This is suggested by the homogeneity of the decay law (28.13) with respect to time.

The indeterminacy of a nuclear energy level before decay. The wave function (28.14) does not belong to any eigenvalue of the energy $\mathscr{E}$, since the states with energy $\mathscr{E}$ have wave functions which are time-dependent according to $e^{-i\frac{\mathscr{E}t}{h}}$. Such states should exist for an unlimitedly long time because the amplitude of their probability does not fall off, while the probability for an undecayed nuclear state is reduced by e times in a time $\Delta t = \frac{h}{\Gamma}$. We can define the time interval $\frac{h}{\Gamma}$ as the characteristic (or mean) attenuation time.

Let us now suppose that the wave function (28.14) is represented as the sum of wave functions of states whose energies are determined accurately. We know that the time dependence of these functions is given by the factor $e^{-i\frac{\mathscr{E}t}{h}}$. In other words, we have to represent the "non-monochromatic" wave $e^{-\frac{\Gamma t}{2h}-i\frac{\mathscr{E}_0 t}{h}}$ as the sum of "monochromatic" waves $e^{-i\frac{\mathscr{E}t}{h}}$ *). In which energy interval $\Delta\mathscr{E}$ (by order of magnitude) will the amplitude of these monochromatic waves differ noticeably from zero?

We can answer this question by making use of relation (18.6). According to this relationship, a wave of duration Δt is represented by a group of monochromatic waves whose frequencies lie in the interval $\Delta\omega \gtrsim \frac{2\pi}{\Delta t} \sim \frac{2\pi\Gamma}{h}$. Substituting, in place of $\Delta\omega$, the equivalent (in this case) quantity $\frac{\Delta\mathscr{E}}{h}$, we arrive at the following estimate:

$$\Delta\mathscr{E}\,\Delta t \gtrsim 2\pi h\,. \tag{28.15}$$

This is the uncertainty relation for energy. The measure of uncertainty for the energy is the quantity Γ, i.e., the inverse value for the probability of decay in unit time. (28.15) should be formulated thus: an energy of a state existing during a limited time interval Δt is determined within the accuracy of the order of $\frac{2\pi h}{\Delta t}$. Only the energy of a state which exists an unlimitedly long time is fully determinate.

The meaning of the uncertainty relation for energy. The meaning of the uncertainty relation for the coordinate and the momentum (23.4) is not analogous to the meaning of (28.15). The estimate (23.4) expresses the fact that the coordinate and momentum do not exist in the same state; (28.15) signifies that if a state of the system has a finite duration Δt, then its energy at each instant of time within the interval Δt is not determined exactly, but is only contained within a region of the order of Γ.

The quantity Γ is termed the level width of the system. The concept of level width can be applied to any states of finite duration and not only to alpha disintegration. For example, the energy level of an atom in an excited state has a definite width, since an excited atom is capable of the spontaneous emission of a quantum.

Explanation of the level width. We shall now show how the level width of a nucleus capable of alpha disintegration can be found by considering the wave function variation under a potential barrier.

In was shown in Sec. 25 that infinite motion has a continuous spectrum. The motion of a system with a potential barrier is infinite because an alpha particle is capable of going to infinity. It follows, strictly speaking, that a nucleus capable of alpha disintegration should have a continuous spectrum.

---

* A wave with definite frequency is termed "monochromatic." A monochromatic wave corresponds to a single colour (chromos is the Greek for colour).

Let us now evaluate the energy level widths Γ for alpha-radioactive nuclei. From (28.15) it follows that even for a nucleus with a very short alpha-decay time ($t \sim 10^{-5}$ sec) $\Gamma \sim 10^{-22}$ erg $\sim 0.6 \times 10^{-10}$ ev. How is it possible to combine a continuous spectrum with such a narrow energy interval?

The solution to the wave equation between r_1 and r_2 is of the following form:

$$\psi = C_1 e^{-\frac{1}{h}\int_{r_1}^{r} \sqrt{2m(U-\mathscr{E})}\, dr} + C_2 e^{\frac{1}{h}\int_{r_1}^{r} \sqrt{2m(U-\mathscr{E})}\, dr} \qquad (28.16)$$

The first solution exponentially diminishes with r, while the second exponentially increases. It follows that if the barrier extended to infinity rightwards, a solution would exist only for $C_2 = 0$. The ratio C_2/C_1, determined from the boundary conditions at $r = r_1$, is a function of energy. It is the roots of equation $C_2(\mathscr{E}) = 0$ that give the possible energy eigenvalues for finite motion. The energy of a particle in a well of finite depth is obtained in just this way. A particle in a well, which was considered in Sec. 25, differs from a particle beyond the barrier in that the barrier is of finite width. Therefore, the second solution, proportional to C_2, need not be strictly equal zero, but may only be small compared with the first solution in any small interval of values $\mathscr{E}$ close to a root of equation $C_2(\mathscr{E}) = 0$. This region of values of $\mathscr{E}$ is what corresponds to the assumption that the modulus of the wave function *outside the nucleus* is small compared with the wave function *inside the nucleus*. In other words, if the energy of the nucleus is contained in a given region of values Γ, then we can say that the alpha particle is in some way bound in the nucleus, similar to the way that a particle can be bound in a real potential well. The higher or broader the potential barrier, the less the barrier factor B and the less the decay probability $\frac{\Gamma}{h}$ proportional to it. But then $\Delta\mathscr{E}$ is also correspondingly reduced, i.e., the continuous-spectrum state becomes closer to the discrete-spectrum state with an exact energy value $\mathscr{E}_0$. This is what explains the meaning of the uncertainty $\Delta\mathscr{E}$: it indicates how close the state is to a bound one, with an infinitely long life-time.

The uncertainty of $\Delta\mathscr{E}$ does not limit the applicability of the law of conservation of energy in any way; the total energy of a nucleus and alpha particle is constant. However, the state with a strictly defined energy relates simultaneously to disintegrated and nondisintegrated nuclei, while a nondisintegrated nucleus has an inexactly defined energy.

Any state capable of a spontaneous transition to another state with the same energy possesses a certain energy width. The energy is not defined exactly in each of these states separately, but a precisely defined energy corresponds to both states at once.

We can divide the total level width into partial widths related to the probabilities for various transitions. Thus, strongly excited nuclear states are capable of emitting neutrons of various energies and of radiating gamma quanta. Each possibility contributes its exponential in the term characterizing attenuation. The total attenuation is determined by the product of such exponentials. It follows that the total level width is equal to the sum of its widths in relation to all possibilities of disintegration.

The Bohr quantum conditions. Let us now apply the quasi-classical approximation to the finite motion of a particle in a potential well and find the energy levels. From (28.3) the wave function is

$$\psi = \sin\left(\frac{1}{h}\int_{x_1}^{x} dx\,\sqrt{2m\,(\mathscr{E} - U)} + \gamma\right). \qquad (28.17)$$

The real solution involving the sine is taken, since, in accordance with (24.20), it does not involve any particle flux outside the well, i.e., it corresponds to a stationary state.

Here x_1 is the left edge of the well for which $\mathscr{E} = U$. In changing x to x_1 to the right-hand edge x_2, the phase of the wave function can change by a whole multiple of π together with a certain addition, which we shall call β for the time being. In the whole length of the well, the sine changes its sign for a given value of the particle energy by as many times as π is added to the argument of the sine, i.e., the wave function (28.17) has as many nodes. However, the number of nodes is equal to the number of the energy eigenvalue n; therefore, an equality is obtained for the determination of $\mathscr{E}_n$:

$$\frac{1}{h}\int_{x_1}^{x_2} \sqrt{2m\,(\mathscr{E}_n - U)}\,dx = \pi n + \beta. \qquad (28.18)$$

If the edges of the well are not plumb at the points x_1 and x_2, then a very fine analysis shows that $\beta = \frac{\pi}{2}$. It turns out that $\beta = 0$ for plumb edges of the well, because the condition $\psi(x_1) = \psi(x_2) = 0$ is then imposed on the wave function, as in the problem of an infinitely deep rectangular well. In addition, we note that the points x_1 and x_2 are given by the result that $\mathscr{E} = U(x_1) = U(x_2)$, i.e., x_1 and x_2 depend upon $\mathscr{E}$.

It should be noted that, from its very meaning, equation (28.18) holds only for a large n, to a quasi-classical approximation. The integral is considerably greater than h for large n, and this signifies that $S(x_2) - S(x_1) \gg h$, whence, to a quasi-classical approximation, we obtain (28.17).

Equation (28.17) was postulated by Bohr in 1913 in determining the stationary orbits for a hydrogen atom (β was considered to be equal to zero). Bohr supposed that electrons in such orbits do not radiate light and do not fall onto the nucleus, while radiation occurs only in the case of a transition from one orbit to another.

Thus, it turns out that the Bohr quantum conditions emerge as a limiting case of quantum mechanics without any additional postulates.

Exercises

1) Determine the energy levels for a linear harmonic oscillator from equation (28.18).

$$\frac{1}{h}\int\limits_{-\sqrt{2\mathscr{E}/m\omega^2}}^{\sqrt{2\mathscr{E}/m\omega^2}}\sqrt{2m\left(\mathscr{E}-\frac{m\omega^2x^2}{2}\right)}\,dx=\pi\left(n+\frac{1}{2}\right).$$

From this $\mathscr{E}=h\omega\left(n+\frac{1}{2}\right)$ which is fortunately correct for all n and not only for $n \gg 1$.

2) Find the approximation that follows (28.3).

We look for S in the form $S = S_0 + hS_1$. Then

$$\psi = e^{i\frac{S_0}{h}+iS_1},$$

$$\frac{d^2\psi}{dx^2}=-\left(\frac{S_0'^2}{h^2}+\frac{2S_0'S_1'}{h}\right)\psi+\frac{i}{h}S_0''\psi=-\frac{2m}{h^2}(\mathscr{E}-U)\psi.$$

The zero approximation gives S_0 from (28.3). The first approximation yields

$$S_1'=\frac{i}{2}\frac{S_0''}{S_0'} \quad \text{or} \quad S_1 = i\ln\sqrt{S_0'},$$

so that $\psi=\frac{1}{\sqrt{S_0'}}e^{i\frac{S_0}{h}}$.

3) Find the factor B for a barrier of the form $U=0$ when $x<0$, $U=U_0-\alpha x$ when $x>0$, $\mathscr{E}<U_0$.

Sec. 29. Operators in Quantum Mechanics

Momentum eigenvalues. In a number of cases we are able to determine energy eigenvalues from the wave equation (24.22). However, it is very important to find the eigenvalues of other quantities too: linear momentum, angular momentum, etc. To do this, it is convenient to proceed from the form of ψ in the limiting transition to classical mechanics:

$$\varphi = e^{i\frac{S}{h}}. \tag{29.1}$$

Let us apply the operation $\frac{h}{i}\frac{\partial}{\partial x}$ to both sides of equation (29.1), i.e., we take the partial derivative with respect to x and multiply by $\frac{h}{i}$:

$$\frac{h}{i}\frac{\partial\psi}{\partial x} = \frac{\partial S}{\partial x} e^{i\frac{S}{h}} . \tag{29.2}$$

But in the classical limit S becomes the action of the particle, while $\frac{\partial S}{\partial x}$ becomes the component of momentum p_x [see (22.9)]. Therefore, the equation for the momentum eigenvalues that yields the correct transition to classical mechanics is of the following form:

$$\frac{h}{i}\frac{\partial\psi}{\partial x} = p_x\psi , \tag{29.3}$$

where p_x is the eigenvalue for the xth momentum projection.

Momentum and energy operators. Let us compare equation (29.3) with the wave equation (24.22):

$$\frac{1}{2m}\left[\left(\frac{h}{i}\frac{\partial}{\partial x}\right)^2 + \left(\frac{h}{i}\frac{\partial}{\partial y}\right)^2 + \left(\frac{h}{i}\frac{\partial}{\partial z}\right)^2\right]\psi + U\psi = \mathscr{E}\psi . \tag{29.4}$$

Here, the symbol $\left(\frac{\partial}{\partial x}\right)^2\psi$ denotes $\frac{\partial^2\psi}{\partial x^2}$ and similarly for $\left(\frac{\partial}{\partial y}\right)^2\psi$, $\left(\frac{\partial}{\partial z}\right)^2\psi$.

In order to find the energy and momentum eigenvalues we must perform a definite set of differential operations and multiplications by the function of coordinates in the left part of the equation. But these sets are connected in a very curious manner, as will now be shown. We shall call the symbol $\frac{\partial}{\partial x}$, multiplied by $\frac{h}{i}$, the momentum operator applied to a wave function. Instead of $\frac{h}{i}\frac{\partial}{\partial x}$ we will symbolically write $\hat{p}_x$. Then, it will be necessary to rewrite equation (29.3) as

$$\hat{p}_x\psi = p_x\psi . \tag{29.5}$$

This equation denotes exactly the same as (29.3), though the symbolic notation $\hat{p}_x$ should emphasize that the corresponding operation is applied in order to find the momentum eigenvalues.

The operation on the left-hand side of (29.4) we shall also symbolically call $\hat{\mathscr{H}}$. We write $\hat{\mathscr{H}}$ and not $\hat{\mathscr{E}}$ because the energy is assumed to be expressed in terms of momentum, similar to the Hamiltonian function $\mathscr{H}$. Then, in shorter notation, (29.4) appears as

$$\hat{\mathscr{H}}\psi = \mathscr{E}\psi . \tag{29.6}$$

$\hat{\mathscr{H}}$ is called the Hamiltonian operator, or the energy operator.

Comparing (29.4) and (29.3), we see that the momentum and energy operators are related by the same equations as the corresponding quantities:

$$\hat{\mathscr{H}} = \frac{1}{2m}(\hat{p}_x^2 + \hat{p}_y^2 + \hat{p}_z^2) + \hat{U}\,. \tag{29.7}$$

We have written $\hat{U}$ instead of simply U in order to emphasize that in this equation the expression $\hat{U}$ is not regarded as an independent quantity but, instead, as an operator acting upon ψ, i.e., a multiplication operator of ψ by U. Equation (29.7) is symbolic. It is understood that both sides are applied to ψ.

The meaning of operator symbolism. The usefulness of an abbreviated operator notation in quantum mechanics consists in the fact that the equations thus become more expressive. The relation between quantum laws of motion and classical laws, which are limiting cases with respect to the quantum ones, can be best of all seen in operator notation.

If in classical equations relating mechanical quantities we replace the momenta by their operators, we then obtain correct operator relationships of quantum mechanics. The limiting transition to classical mechanics restores the usual relationships between quantities. Indeed, in the limiting transition (29.1), the operator $\hat{\mathbf{p}} = \frac{h}{i}\nabla$ will give $\mathbf{p}\psi$. If we must perform the limiting transition for $\hat{p}^2$, then we need to differentiate only the exponential each time, because this yields the quantum of action in the denominator. For $h \to 0$, only terms with the highest degree of h in the denominator remain, and it is these very terms which are obtained in replacing the operator $\mathbf{p}$ by the quantity ∇S (i.e., by the classical momentum vector). We had an example of such a transition in Sec. 24 [see equations (24.13) and (24.14)].

The angular-momentum operator. It is now easy to define the angular-momentum operator. We shall begin with one component M_z. It is clear from Sec. 5 that the angular momentum M_z is at the same time a generalized momentum corresponding to the angle of rotation about the z-axis, i.e., $M_z = p_\varphi$. Then, from (10.23)

$$p_\varphi = \frac{\partial S}{\partial \varphi}\,. \tag{29.8}$$

Therefore, in quantum mechanics the operator p_φ must be of the form

$$\hat{p}_\varphi = \frac{h}{i}\frac{\partial}{\partial \varphi}\,. \tag{29.9}$$

At the same time, in accordance with classical mechanics, the projection M_z is related to the momentum projections thus:

$$M_z = x p_y - y p_x\,. \tag{29.10}$$

It follows that there must exist an operator relationship

$$\hat{p}_\varphi = \hat{M}_z = \frac{h}{i}\left(x\frac{\partial}{\partial y} - y\frac{\partial}{\partial x}\right) = x p_y - y p_x . \tag{29.11}$$

Let us check to see that the definitions (29.9) and (29.11) do, indeed, coincide. Let us pass to cylindrical coordinates:

$$x = r\cos\varphi , \tag{29.12}$$

$$y = r\sin\varphi . \tag{29.13}$$

From this we have

$$\frac{\partial\psi}{\partial x} = \frac{\partial\psi}{\partial r}\frac{\partial r}{\partial x} + \frac{\partial\psi}{\partial\varphi}\frac{\partial\varphi}{\partial x} , \tag{29.14}$$

$$\frac{\partial\psi}{\partial y} = \frac{\partial\psi}{\partial r}\frac{\partial r}{\partial y} + \frac{\partial\psi}{\partial\varphi}\frac{\partial\varphi}{\partial y} . \tag{29.15}$$

Solving (29.12) and (29.13) with respect to x and y, we have

$$r = \sqrt{x^2 + y^2} , \quad \varphi = \text{arc } \tan\frac{y}{x} ;$$

$$\frac{\partial r}{\partial x} = \frac{x}{\sqrt{x^2+y^2}} = \cos\varphi , \qquad \frac{\partial r}{\partial y} = \frac{y}{\sqrt{x^2+y^2}} = \sin\varphi ;$$

$$\frac{\partial\varphi}{\partial x} = -\frac{y}{x^2+y^2} = -\frac{\sin\varphi}{r} , \qquad \frac{\partial\varphi}{\partial y} = \frac{x}{x^2+y^2} = \frac{\cos\varphi}{r} .$$

Substituting all these expressions into equalities (29.14) and (29.15), and substituting the derivatives themselves into (29.11), we can see that both definitions for M_z (29.9) and (29.11) are identical.

Angular-momentum projection eigenvalues. Let us now find the eigenvalues for M_z. For this, it is necessary to solve the equation

$$\hat{M}_z\psi = M_z\psi , \tag{29.16}$$

i.e.,

$$\frac{h}{i}\frac{\partial\psi}{\partial\varphi} = M_z\psi . \tag{29.17}$$

This equation is very simply integrated:

$$\psi = e^{i\frac{M_z\varphi}{h}} . \tag{29.18}$$

As we know, a wave function is the probability amplitude. Function (29.18) is the amplitude of the probability that the particle possesses an azimuth angle φ, if the zth component of its angular momentum is equal to M_z. It is essential that not only the absolute value of the wave function has physical significance, but also its phase; this is indicated, for example, by the phenomenon of electron diffraction. For the phase of the wave function (29.18) to be determined, it must either not change at all or only by a multiple of 2π in rotating the coordinate system through 360°; this is because the position of the particle relative to such a system rotated through 360° does not change. If in this case the wave function were not returned to its initial value, it could not uniquely represent the probability amplitude. Thus,

$$\psi(\varphi + 2\pi) = \psi(\varphi),$$

i.e.,

$$e^{i\frac{M_z(\varphi + 2\pi)}{h}} = e^{i\frac{M_z \varphi}{h}}, \tag{29.19}$$

so that

$$e^{i\frac{2\pi M_z}{h}} = 1 = e^{2\pi i k}, \tag{29.20}$$

where k is an integer of any sign or zero. Whence we obtain the eigenvalues of M_z:

$$M_z = hk. \tag{29.21}$$

M_z is called the *orbital* angular-momentum projection for a particle. We shall see in Sec. 32 that a particle can have an angular momentum connected with its internal motion, which is not described by the wave function (29.18). Here, we have proven that the orbital angular-momentum components can only assume values which are whole multiples of h.

The Stern-Gerlach experiment. The discreteness of the angular-momentum spectrum is confirmed by direct experiment. The idea of the experiment consists in the following: a direct relationship exists between the orbital angular-momentum projection and the magnetic moment projection [see (15.25)]:

$$\mu_z = \frac{e}{2mc} M_z. \tag{29.22}$$

A narrow beam of vapour of the substance under investigation is passed between the poles of an electromagnet in a strongly inhomogeneous field; to achieve this, one of the poles may be made tapered. The particles—in the Stern-Gerlach experiment, they are atoms—enter the field parallel to the edge of the taper, i.e., they move in a direction perpendicular to the plane of the lines of force of the field. The plane of symmetry of the field passes through the edge of the taper and the initial direction of motion of the particle. We assume the z-axis to be perpendicular to the edge of the taper and to lie in the plane of symmetry of the field. If the mechanical moment of the electrons in the atoms has only discrete, integral projections on the z-axis, then the magnetic moment of the atoms is established in several definite ways. The deflecting force acting on the magnetic moment in a magnetic field is, by (15.40)

$$(\mu\nabla) H_z = \mu_z \frac{\partial H}{\partial z} = k \frac{eh}{2mc} \frac{\partial H}{\partial z}. \tag{29.23}$$

In the plane of symmetry of the field, H is directed along the z-direction and depends only upon z.

Since the angular momentum can only have a definite set of values, the deflecting force acting upon the atoms in the beam is also not arbitrary but has a very definite value for particles with respective angular-momentum projection $M_z = hk$. It can be seen from (29.23) that the force is a quantity which is a multiple of $\frac{eh}{2mc}\frac{\partial H}{\partial z}$. Therefore, the particles in the beam experience only those deflections in the magnetic field which correspond to the possible values of the force (29.23). In other words, the beam is split into several separate beams and does not proceed continuously, as would be the case for any nonintegral projections M_z.

Where the beam is formed, each particle was given a certain angular momentum. Motion in the magnetic field makes it possible to measure the projection of this angular momentum M_z in the direction of the field.

The impossibility of the simultaneous existence of two angular-momentum projections. From the fact that the angular-momentum projection on any axis is integral, it follows that the angular momentum does not have, simultaneously, projections on *two axes* in space.

Indeed, in the Stern-Gerlach experiment, the z-axis is absolutely arbitrary. We could have measured the angular-momentum projection on some axis in space and then pass the same beams through a magnetic field making a very small angle with the field in which the first measurement was performed. Both measurements will give only integral projections of the angular momentum. Both one and the same vector cannot simultaneously have integral projections on infinitely close, but otherwise arbitrary, directions; when the first measurement was performed, the angular momentum had a projection only on the first direction of the field, and, correspondingly, in the second measurement, it had projections only on the second direction of the field.

Similar to the way that coordinate and momentum do not exist simultaneously, it turns out that two angular-momentum projections do not exist in the same state.

The simultaneous existence of two physical quantities. We shall consider from a general point of view the question of which quantities of quantum mechanics can exist in the same state of a system. Let us suppose that in a certain state described by the wave function ψ there simultaneously exist two physical quantities λ and ν. This means that the wave function ψ is an eigenfunction of the two operators $\hat{\lambda}$ and $\hat{\nu}$. It satisfies two equations

$$\hat{\lambda}\psi = \lambda\psi \tag{29.24}$$

and

$$\hat{\nu}\psi = \nu\psi . \tag{29.25}$$

$\hat{\lambda}$ and $\hat{\nu}$ are, speaking generally, differential operators; λ and ν are numbers.

Let us apply the operator $\hat{\nu}$ to (29.24). Since there is a number λ on the right, it can be put on the left of the operator sign $\hat{\nu}$:

$$\hat{\nu}\hat{\lambda}\psi = \hat{\nu}\lambda\psi = \lambda\hat{\nu}\psi = \lambda\nu\psi . \qquad (29.26a)$$

In the last equation we made use of (29.25). We shall now apply the operator $\hat{\lambda}$ to (29.25):

$$\hat{\lambda}\hat{\nu}\psi = \hat{\lambda}\nu\psi = \nu\hat{\lambda}\psi = \nu\lambda\psi . \qquad (29.26b)$$

Let us subtract (29.26b) from (29.26a):

$$\hat{\nu}\hat{\lambda}\psi - \hat{\lambda}\hat{\nu}\psi = \lambda\nu\psi - \nu\lambda\psi = 0 . \qquad (29.27)$$

(29.27) can be symbolically written as an equality between operations.

$$\hat{\nu}\hat{\lambda} = \hat{\lambda}\hat{\nu} \text{ or } \hat{\nu}\hat{\lambda} - \hat{\lambda}\hat{\nu} = 0 . \qquad (29.28)$$

This symbolic equality means that the result of operating with $\hat{\nu}$ and $\hat{\lambda}$ should not depend upon the order of their actions, otherwise equations (29.24) and (29.25) cannot have a general solution.

We can also prove the inverse theorem: if two operators are commutative, i.e., the result of their action does not depend on their order, then it is possible to form a common eigenfunction ψ satisfying equations (29.24) and (29.25).

Commutations of certain operators. Let us now apply the obtained result to two quantities which definitely do not exist in the same state: the coordinate x and momentum p_x. We must calculate the commutation $\hat{p}_x\hat{x} - \hat{x}\hat{p}_x$.

Changing from symbolic notation to the usual one, we obtain

$$\frac{h}{i}\frac{\partial}{\partial x}x\psi - x\frac{h}{i}\frac{\partial\psi}{\partial x} = \frac{h}{i}\psi + x\frac{h}{i}\frac{\partial\psi}{\partial x} - x\frac{h}{i}\frac{\partial\psi}{\partial x} = \frac{h}{i}\psi . \quad (29.29)$$

Reverting to symbolic notation, we represent (29.29) in the following form:

$$\hat{p}_x\hat{x} - \hat{x}\hat{p}_x = \frac{h}{i} . \qquad (29.30)$$

Thus, the result of operating with $\hat{p}_x$ and $\hat{x}$ depends upon the order of their action; $\hat{p}_x$ and $\hat{x}$ are noncommutative. And this was to be expected because the quantities x and p_x do not exist simultaneously.

The eigenfunction of the operator $\hat{x}$ satisfies the equation $(\hat{x} - x')\psi = 0$. Consequently, it is equal to zero over the whole region where the coordinate x is not equal to the chosen eigenvalue x'. This function differs from zero only at one point $x = x'$. The eigenfunction of the momentum operator which satisfies equation (29.30) is $e^{i\frac{p_x x}{h}}$; it differs from zero over all the space. This example shows how great is the difference between the eigenfunctions of noncommuting operators.

The abreviated notation of (29.30) is a convenient representation of (29.29). In the more complex cases that will be examined later the convenience of this abbreviated notation is obvious. It must be borne in mind that the operator notation is simply a rationalization of mathematical symbolism, and there is nothing incomprehensible in the result that $\hat{p}_x\hat{x} - \hat{x}\hat{p}_x \neq 0$; such is the property of operator symbols. It will be recalled that in vector algebra there also exists noncommutative multiplication; and, what is more, not of symbols, but of quantities. In quantum mechanics the operator symbolism is most expedient.

The various momentum components are commutative:

$$\hat{p}_x\hat{p}_y - \hat{p}_y\hat{p}_x = 0 \tag{29.31}$$

(the word "operators," will be frequently omitted in future as being self-evident). The commutative relation (29.31) is obtained simply from the fact that the result of applying two partial derivatives does not depend upon the order of differentiation.

It is also obvious that

$$\hat{p}_y\hat{x} - \hat{x}\hat{p}_y = 0\,. \tag{29.32}$$

We now calculate the commutation of any two angular-momentum components. Let us take $\hat{M}_x$ and $\hat{M}_y$:

$$\hat{M}_x = \hat{y}\hat{p}_z - \hat{z}\hat{p}_y\,,$$
$$\hat{M}_y = \hat{z}\hat{p}_x - \hat{x}\hat{p}_z\,.$$

Let us first of all write the commutation without using the rules (29.30)-(29.32):

$$\hat{M}_x\hat{M}_y - \hat{M}_y\hat{M}_x = (\hat{y}\hat{p}_z - \hat{z}\hat{p}_y)(\hat{z}\hat{p}_x - \hat{x}\hat{p}_z) - (\hat{z}\hat{p}_x - \hat{x}\hat{p}_z)(\hat{y}\hat{p}_z - \hat{z}\hat{p}_y)\,.$$

We now group the terms here so that the order of coordinates and corresponding momenta is not disturbed:

$$\hat{M}_x\hat{M}_y - \hat{M}_y\hat{M}_x = \hat{y}\hat{p}_x(\hat{p}_z\hat{z} - \hat{z}\hat{p}_z) - \hat{x}\hat{p}_y(\hat{p}_z\hat{z} - \hat{z}\hat{p}_z)\,.$$

We substitute the commutation relation $\hat{p}_z\hat{z} - \hat{z}\hat{p}_z = \frac{h}{i}$ and then find the required result:

$$\hat{M}_x\hat{M}_y - \hat{M}_y\hat{M}_x = ih\,(\hat{x}\hat{p}_y - \hat{y}\hat{p}_x) = ih\,\hat{M}_z\,. \tag{29.33}$$

Now changing the indices x, y, z cyclically, we obtain the remaining commutation relations:

$$\hat{M}_y\hat{M}_z - \hat{M}_z\hat{M}_y = ih\,\hat{M}_x\,, \tag{29.34}$$
$$\hat{M}_z\hat{M}_x - \hat{M}_x\hat{M}_z = ih\,\hat{M}_y\,. \tag{29.35}$$

All three commutation relations can be easily remembered if we write them in contracted form thus:

$$[\mathbf{M}\mathbf{M}] = i\hbar\mathbf{M}. \tag{29.36}$$

Expanding this equality in components, we once again arrive at (29.33)-(29.35).

It will be noted that the vector product of an operator by itself cannot equal zero unless the operator components of the vector are noncommutating (but, for example, $[\hat{\mathbf{p}}\hat{\mathbf{p}}]=0$).

We have shown that there does not exist a state in which a system would possess two angular-momentum projections. Angular momentum has a projection only on one axis, in agreement with the Stern-Gerlach experiment. The only exception is when all three angular-momentum projections are equal to zero. The eigenfunction of such a state does not depend upon any angles at all [see (29.18)]. Therefore, as a result of applying differential operations of the type (29.9), where the differentiation is performed with respect to the angle of rotation about any arbitrary axis, this eigenfunction is multiplied by zero. The action of the operators of the angular-momentum components on such a function is commutative. This does not contradict the Stern-Gerlach experiment, because a vector can have a zero projection on two infinitely close, though arbitrarily orientated, axes, provided the vector itself is equal to zero. But if only one of the angular-momentum projections is not equal to zero, then the two others do not have definite values because, otherwise, there would be a contradiction in equality (29.36).

The square of the angular momentum. Let us now examine further properties of angular momentum. We shall show that even though two angular-momentum projections do not exist, a single angular-momentum projection exists together with its square

$$\hat{M}^2 = \hat{M}_x^2 + \hat{M}_y^2 + \hat{M}_z^2. \tag{29.37}$$

We shall verify this:

$$\hat{M}^2\hat{M}_z - \hat{M}_z\hat{M}^2 = \hat{M}_x^2\hat{M}_z - \hat{M}_z\hat{M}_x^2 + \hat{M}_y^2\hat{M}_z - \hat{M}_z\hat{M}_y^2,$$

because $\hat{M}_z$ and $\hat{M}_z^2$ are, of course, commutative. Let us add to the right-hand side of the last equality, and subtract from it the combinations $\hat{M}_x\hat{M}_z\hat{M}_x$ and $\hat{M}_y\hat{M}_z\hat{M}_y$; we take $\hat{M}_x$ and $\hat{M}_y$ outside the brackets, once on the right and another time on the left. Then we obtain

$$\begin{aligned}\hat{M}^2\hat{M}_z - \hat{M}_z\hat{M}^2 &= \hat{M}_x(\hat{M}_x\hat{M}_z - \hat{M}_z\hat{M}_x) + (\hat{M}_x\hat{M}_z - \hat{M}_z\hat{M}_x)\hat{M}_x + \\ &+ \hat{M}_y(\hat{M}_y\hat{M}_z - \hat{M}_z\hat{M}_y) + (\hat{M}_y\hat{M}_z - \hat{M}_z\hat{M}_y)\hat{M}_y = \\ &= -i\hbar\hat{M}_x\hat{M}_y - i\hbar\hat{M}_y\hat{M}_x + i\hbar\hat{M}_y\hat{M}_x + i\hbar\hat{M}_x\hat{M}_y = 0. \end{aligned} \tag{29.38}$$

Here we have made use of the commutation rules (29.33)-(29.35). The eigenvalue of M^2 will be found in the following section.

Exercises

1) Find the commutations, of

$$\hat{M}_x,\ \hat{p}_x;\ \hat{M}_x,\ \hat{p}_y;\ \hat{M}_x,\ \hat{p}_z;$$
$$\hat{M}_x,\ \hat{x};\ \ \hat{M}_x,\ \hat{y};\ \ \hat{M}_z,\ \hat{z};$$
$$\hat{M}^2,\ \hat{p}_x;\ \hat{M}^2,\ \hat{p}^2;\ \hat{M}^2,\ x;\ \ M^2,\ r^2.$$

2) Write down the Cartesian projections of momentum in spherical coordinates.

Spherical coordinates are expressed in terms of Cartesian coordinates in the following manner [see (3.5), (3,7), (3.8)]:

$$r=\sqrt{x^2+y^2+z^2},$$

$$\vartheta=\arccos\frac{z}{\sqrt{x^2+y^2+z^2}},\quad \varphi=\arctan\frac{y}{x}.$$

Whence we obtain the partial derivatives

$$\frac{\partial r}{\partial x}=\frac{x}{r}=\sin\vartheta\cos\varphi,\qquad \frac{\partial\vartheta}{\partial x}=\frac{xz}{r^2\sqrt{x^2+y^2}}=\frac{\cos\vartheta\cos\varphi}{r},$$

$$\frac{\partial r}{\partial y}=\frac{y}{r}=\sin\vartheta\sin\varphi,\qquad \frac{\partial\vartheta}{\partial y}=\frac{yz}{r^2\sqrt{x^2+y^2}}=\frac{\cos\vartheta\sin\varphi}{r},$$

$$\frac{\partial r}{\partial z}=\frac{z}{r}=\cos\vartheta,\qquad \frac{\partial\vartheta}{\partial z}=-\frac{\sqrt{x^2+y^2}}{r^2}=-\frac{\sin\vartheta}{r},$$

$$\frac{\partial\varphi}{\partial x}=-\frac{y}{x^2+y^2}=-\frac{\sin\varphi}{r\sin\vartheta},$$

$$\frac{\partial\varphi}{\partial y}=\frac{x}{x^2+y^2}=\frac{\cos\varphi}{r\sin\vartheta},$$

$$\frac{\partial\varphi}{\partial z}=0.$$

Further,

$$\frac{i}{h}\hat{p}_x=\frac{\partial}{\partial x}=\frac{\partial r}{\partial x}\frac{\partial}{\partial r}+\frac{\partial\vartheta}{\partial x}\frac{\partial}{\partial\vartheta}+\frac{\partial\varphi}{\partial x}\frac{\partial}{\partial\varphi}=\sin\vartheta\cos\varphi\frac{\partial}{\partial r}+\frac{\cos\varphi\cos\vartheta}{r}\frac{\partial}{\partial\vartheta}-$$
$$-\frac{\sin\varphi}{r\sin\vartheta}\frac{\partial}{\partial\varphi},$$

$$\frac{i}{h}\hat{p}_y=\frac{\partial}{\partial y}=\frac{\partial r}{\partial y}\frac{\partial}{\partial r}+\frac{\partial\vartheta}{\partial y}\frac{\partial}{\partial\vartheta}+\frac{\partial\varphi}{\partial y}\frac{\partial}{\partial\varphi}=\sin\vartheta\sin\varphi\frac{\partial}{\partial r}+\frac{\sin\varphi\cos\vartheta}{r}\frac{\partial}{\partial\vartheta}+$$
$$+\frac{\cos\varphi}{r\sin\vartheta}\frac{\partial}{\partial\varphi},$$

$$\frac{i}{h}\hat{p}_z=\frac{\partial}{\partial z}=\frac{\partial r}{\partial z}\frac{\partial}{\partial r}+\frac{\partial\vartheta}{\partial z}\frac{\partial}{\partial\vartheta}+\frac{\partial\varphi}{\partial z}\frac{\partial}{\partial\varphi}=\cos\vartheta\frac{\partial}{\partial r}-\frac{\sin\vartheta}{r}\frac{\partial}{\partial\vartheta}.$$

3) Write the angular-momentum projections on Cartesian axes in terms of spherical coordinates.

$$\hat{M}_x = \hat{y}\hat{p}_z - \hat{z}\hat{p}_y = \frac{h}{i}\left(-\sin\varphi\frac{\partial}{\partial\vartheta} - \cot\vartheta\cos\varphi\frac{\partial}{\partial\varphi}\right),$$

$$\hat{M}_y = \hat{z}\hat{p}_x - \hat{x}\hat{p}_z = \frac{h}{i}\left(\cos\varphi\frac{\partial}{\partial\vartheta} - \cot\vartheta\sin\varphi\frac{\partial}{\partial\varphi}\right),$$

$$\hat{M}_z = \hat{x}\hat{p}_y - \hat{y}\hat{p}_x = \frac{h}{i}\frac{\partial}{\partial\varphi}.$$

4) Write the expression for the square of the angular momentum in spherical coordinates:

$$\hat{M}^2 = \hat{M}_x^2 + \hat{M}_y^2 + \hat{M}_z^2 = (\hat{M}_x + i\hat{M}_y)(\hat{M}_x - i\hat{M}_y) - i(\hat{M}_y\hat{M}_x - \hat{M}_x\hat{M}_y) + \hat{M}_z^2 =$$
$$= (\hat{M}_x + i\hat{M}_y)(\hat{M}_x - i\hat{M}_y) - h\hat{M}_z + \hat{M}_z^2.$$

From exercise 3 we have

$$\hat{M}_x + i\hat{M}_y = \frac{h}{i}\left(ie^{i\varphi}\frac{\partial}{\partial\vartheta} - \cot\vartheta e^{i\varphi}\frac{\partial}{\partial\varphi}\right),$$

$$\hat{M}_x - i\hat{M}_y = \frac{h}{i}\left(-ie^{-i\varphi}\frac{\partial}{\partial\vartheta} - \cot\vartheta e^{-i\varphi}\frac{\partial}{\partial\varphi}\right).$$

Applying $\hat{M}_x + i\hat{M}_y$ to $\hat{M}_x - i\hat{M}_y$, we must observe the order of the "factors" $\frac{\partial}{\partial\varphi}$ and $e^{i\varphi}$, $\cot\vartheta$ and $\frac{\partial}{\partial\vartheta}$. We obtain

$$(\hat{M}_x + i\hat{M}_y)(\hat{M}_x - i\hat{M}_y) =$$
$$= -h^2\left(\frac{\partial^2}{\partial\vartheta^2} - i\frac{\partial}{\partial\vartheta}\cot\vartheta\frac{\partial}{\partial\varphi} + ie^{i\varphi}\cot\vartheta\frac{\partial}{\partial\varphi}e^{-i\varphi}\frac{\partial}{\partial\vartheta} + e^{i\varphi}\cot^2\vartheta\frac{\partial}{\partial\varphi}e^{-i\varphi}\frac{\partial}{\partial\varphi}\right) =$$
$$= -h^2\left(\frac{\partial^2}{\partial\vartheta^2} + i\csc^2\vartheta\frac{\partial}{\partial\varphi} + \cot\vartheta\frac{\partial}{\partial\vartheta} - i\cot^2\vartheta\frac{\partial}{\partial\varphi} + \cot^2\vartheta\frac{\partial^2}{\partial\varphi^2}\right).$$

Finally,

$$\hat{M}^2 = -h^2\left(\frac{1}{\sin\vartheta}\frac{\partial}{\partial\vartheta}\sin\vartheta\frac{\partial}{\partial\vartheta} + \frac{1}{\sin^2\vartheta}\frac{\partial^2}{\partial\varphi^2}\right).$$

This expression is obviously commutative with $M_z = \frac{h}{i}\frac{\partial}{\partial\varphi}$. This was shown in the present section in another way.

Sec. 30. Expansions into Wave Functions

The superposition principle. One of the most fundamental ideas of quantum mechanics consists in the fact that its equations are linear with respect to the wave function ψ. This result proceeds from the whole set of facts that confirm the correctness of quantum mechanics, in the same way as an analogous result in classical electrodynamics (see Sec. 21), which is also a generalization of experience.

For example, the diffraction of electrons shows that the amplitudes of wave functions are combined in the same simple way as the amplitudes of waves in optics; diffraction maxima and minima are situated at the same positions, which are determined only by the phase relationships, independently of the wave intensities. All this points to the linearity of wave equations; the solutions of nonlinear equations behave in an entirely different manner.

The sum of two solutions of a linear equation again satisfies the same equation. It follows from this that any solution of a wave equation can be represented in the form of a certain set of standard solutions, similar to the way that, in Sec. 18, a travelling nonperiodic wave was represented by a set of travelling harmonic waves (18.1).

The statement concerning the possibility of representing a single wave function in terms of the sum of other wave functions is called the *superposition principle.*

The Hermitian property of operators. Wave functions are usually represented with the aid of the sum of eigenfunctions of certain quantum-mechanical operators. In the present section it will be shown how such expansions are performed. First of all, however, it is necessary to establish certain general properties of the operators whose eigenvalues are physical quantities.

Obviously, these eigenvalues must be real numbers, although the operators themselves may depend explicitly upon $i=\sqrt{-1}$ [see (29.3), (29.10)]. We shall consider the equations for the eigenfunctions of the operator $\hat{\lambda}$ and another equation involving its conjugate:

$$\hat{\lambda}\psi = \lambda\psi\,, \tag{30.1a}$$

$$\hat{\lambda}^*\psi^* = \lambda^*\psi^*. \tag{30.1b}$$

We must find the condition for which the eigenvalues of the operator are real numbers, $\lambda^*=\lambda$.

To do this, we multiply (30.1a) by ψ^* and (30.1b) by ψ, integrate over the whole range of the variables x (upon which the operator $\hat{\lambda}$ depends), and subtract one from the other. Then we obtain

$$\int(\psi^*\hat{\lambda}\psi - \psi\hat{\lambda}^*\psi^*)\,dx = (\lambda-\lambda^*)\int\psi^*\psi\,dx\,.$$

But the integral of $\psi^*\psi=|\psi|^2$ cannot be equal to zero, since $|\psi|^2$ is an essentially positive quantity.

The eigenvalue λ is, by definition, real, i.e., $\lambda=\lambda^*$; therefore we arrive at the relation

$$\int(\psi^*\hat{\lambda}\psi - \psi\hat{\lambda}^*\psi^*)\,dx = 0\,. \tag{30.2}$$

Equation (30.2) can be regarded as a condition imposed upon the operator $\hat{\lambda}$. In fact, however, we must demand that the operator $\hat{\lambda}$ satisfy the equation (30.2) not only for its own eigenfunctions $\psi^*(\lambda, x)$ and $\psi(\lambda, x)$, but also for any pair of functions $\chi^*(x)$ and $\psi(x)$, provided these functions satisfy the same conditions of being finite, continuous, and single-valued as the eigenfunctions $\psi(\lambda, x)$:

$$\int (\chi^* \hat{\lambda} \psi - \psi \hat{\lambda}^* \chi^*)\, dx = 0 . \tag{30.3}$$

The necessity of such a condition will be explained later in this section. An operator for which equality (30.3) is satisfied is termed *Hermitian.*

In equation (30.3), dx is an abbreviated notation for $dV = dx\, dy\, dz$, if the integration is performed over a volume ($\hat{\lambda} = \hat{\mathscr{H}}$), or $d\varphi$, if $\hat{\lambda} = \hat{M}_z$, etc.

The Hermitian nature of the operators $\hat{p}_x$, $\hat{M}_z, \ldots$ is easily verified by integrating by parts. For example,

$$\int_0^{2\pi} \chi^* \hat{M}_z \psi\, d\varphi = \int_0^{2\pi} \chi^* \frac{h}{i} \frac{\partial \psi}{\partial \varphi}\, d\varphi = \frac{h}{i} \chi^* \psi \Big|_0^{2\pi} - \int_0^{2\pi} \psi \frac{h}{i} \frac{\partial \chi^*}{\partial \varphi}\, d\varphi .$$

The eigenfunctions of the operator $\hat{M}_z$ must satisfy the requirement of uniqueness (29.19); hence $\chi^*(0) = \chi^*(2\pi)$, $\psi(0) = \psi(2\pi)$ similar to the eigenfunctions of the operator $\hat{M}_z$. Therefore, the integrated quantity becomes zero. The operator $-\frac{h}{i}\frac{\partial}{\partial\varphi}$ is $\hat{M}_z^*$, so that

$$\int_0^{2\pi} \chi^* \hat{M}_z \psi\, d\varphi = \int_0^{2\pi} \psi M_z^* \chi^*\, d\varphi$$

in accordance with the general requirement (30.3).

The Hermitian nature of $\hat{\mathscr{H}}$ and $\hat{M}^2$ is proven by a double integration by parts.

The orthogonality of eigenfunctions. An important property of eigenfunctions follows from the Hermitian nature of operators. Let us consider the equations for two eigenvalues of the same operator $\hat{\lambda}$:

$$\hat{\lambda} \psi(\lambda, x) = \lambda \psi(\lambda, x) . \tag{30.4a}$$

$$\hat{\lambda}^* \psi^*(\lambda', x) = \lambda' \psi^*(\lambda', x) . \tag{30.4b}$$

We multiply (30.4a) by $\psi^*(\lambda', x)$ and (30.4b) by $\psi(\lambda, x)$, integrate with respect to dx, and subtract one from the other:

$$\int [\psi^*(\lambda', x) \hat{\lambda} \psi(\lambda, x) - \psi(\lambda, x) \hat{\lambda}^* \psi^*(\lambda', x)]\, dx = = (\lambda - \lambda') \int \psi^*(\lambda', x) \psi(\lambda, x)\, dx . \tag{30.5}$$

The left-hand side of this equation becomes zero in accordance with general requirements for Hermitian form (30.3). Therefore, if $\lambda' \neq \lambda$, the following integral must equal zero

$$\int \psi^*(\lambda', x) \psi(\lambda, x)\, dx = 0 \text{ for } \lambda \neq \lambda'. \tag{30.6}$$

This property was proved in exercise 1, Sec. 24, as applied to the eigenfunctions of the energy operator. It is called the property of *orthogonality*.

Several quantities, λ, ν, etc., may sometimes exist in the same state. For this it is necessary that the operators $\hat{\lambda}$, $\hat{\nu}$,... should be commutative. For example, for free motion there exist p_x, p_y, and p_z. Then we may form fonctions, which are eigenfunctions with respect to all the operators simultaneously:

$$\hat{\lambda}\psi(\lambda, \nu; x) = \lambda\psi(\lambda, \nu; x),$$

$$\hat{\nu}\psi(\lambda, \nu; x) = \nu\psi(\lambda, \nu; x);$$

the orthogonality condition for such functions is directly generalized in the form

$$\int \psi^*(\lambda', \nu'; x)\, \psi(\lambda, \nu; x)\, dx = 0, \tag{30.7}$$

if $\lambda' \neq \lambda$ or $\nu' \neq \nu$.

Expansion in eigenfunctions. Let us suppose that the eigenfunctions of a certain operator $\hat{\lambda}$ are known. These functions always satisfy (in addition to the equation $\hat{\lambda}\psi = \lambda\psi$) certain requirements: they are finite, continuous, single-valued, and so forth. Then, in accordance with the superposition principle, any function $\psi(x)$ which satisfies the same requirements may be represented as the sum of the eigenfunctions of the operator $\hat{\lambda}$:

$$\psi(x) = \sum_{\lambda'} c_{\lambda'}\, \psi(\lambda', x). \tag{30.8}$$

We shall show how to determine the expansion coefficients c_λ. To do this we multiply both sides of the equation by $\psi^*(\lambda, x)$ and integrate with respect to dx:

$$\int \psi^*(\lambda, x)\, \psi(x)\, dx = \sum_{\lambda'} c_{\lambda'} \int \psi^*(\lambda, x)\, \psi(\lambda', x)\, dx. \tag{30.9}$$

In accordance with the orthogonality condition all the integrals on the right-hand side of the equality (30.9) become zero except those for which $\lambda' = \lambda$. Consequently, there remains the equation

$$\int \psi^*(\lambda, x)\, \psi(x)\, dx = c_\lambda \int \psi^*(\lambda, x)\, \psi(\lambda, x)\, dx = c_\lambda \int |\psi(\lambda, x)|^2\, dx. \tag{30.10}$$

We shall consider that the eigenfunctions $\psi(\lambda, x)$ are normalized to unity, i.e., $\int |\psi|^2\, dx = 1$ [see (24.18)]. Then the expansion coefficient is

$$c_\lambda = \int \psi^*(\lambda, x)\, \psi(x)\, dx. \tag{30.11}$$

In the case when we have a system of commutative operators $\hat{\lambda}$, $\hat{\nu}$, equation (30.11) is directly generalized:

$$c_{\lambda,\nu} = \int \psi^* (\lambda, \nu; x)\, \psi (x)\, dx\,. \tag{30.12}$$

The meaning of the expansion coefficients. We have seen that a state $\psi (x)$ is represented as a superposition of states with definite values of the quantity λ. The component of the wave function ψ which corresponds to this value of λ is

$$c_\lambda \psi (\lambda, x)\,.$$

It represents the probability amplitude for the given value of the quantity λ in the state $\psi (x)$. In order to find the probability itself, w_λ, of the occurrence of quantity λ, we must eliminate the coordinate dependence, since λ and x do not exist in the same state.

To do this, let us integrate the probability density of the state with a given λ, i.e., $|c_\lambda|^2 |\psi (\lambda, x)^2$, over all x. From the normalization condition for eigenfunctions we obtain

$$w_\lambda = |c_\lambda|^2 \int |\psi (\lambda, x)|^2\, dx = |c_\lambda|^2\,. \tag{30.13}$$

The quantities $w_\lambda = |c_\lambda|^2$ have a basic property of probability: their sum is equal to unity, provided the function itself satisfies the condition of normalization (24.18). Indeed,

$$1 = \int |\psi (x)|^2\, dx = \int |\sum_\lambda c_\lambda \psi (\lambda, x)|^2\, dx =$$

$$= \sum_\lambda |c_\lambda|^2 \int |\psi (\lambda, x)|^2\, dx + \sum_\lambda \sum_{\lambda' \neq \lambda} c_\lambda \overset{*}{c}_{\lambda'} \int \psi^* (\lambda', x)\, \psi (\lambda, x)\, dx\,.$$

But on the orthogonality condition (30.6) a double summation is equal to zero. From this, in accordance with (24.18), it follows that

$$\sum |c_\lambda|^2 = \sum w_\lambda = 1\,. \tag{30.14}$$

Thus, the coefficient c_λ should be regarded as the probability amplitude, similar to $\psi (x)$. But $|\psi (x)|^2$ is the probability of detecting a particle with coordinate x independently of λ, while $|c_\lambda|^2$ is the probability of finding it with a given value of the quantity λ independently of x.

Expansion in angular-momentum projection eigenfunctions. The atomic beam in the Stern-Gerlach experiment is split into a certain number of separate beams, corresponding to the number of angular-momentum components along the magnetic field direction $M_z = hk$.

Let us denote the largest eigenvalue quantity k by the letter l. Then it is obvious that

$$l \geqslant k \geqslant -l, \tag{30.15}$$

i.e., k takes on $2l+1$ values.

The eigenfunction corresponding to $M_z = hk$ is

$$\psi(k) = \frac{1}{\sqrt{2\pi}} e^{ik\varphi} \tag{30.16}$$

(the factor $\frac{1}{\sqrt{2\pi}}$ is introduced for normalization, $\int_0^{2\pi} |\psi|^2 d\varphi = 1$).

If each of the separate beams is once again passed through a magnetic field parallel to the z-axis, there is no further splitting; this is because M_z in these beams has a single definite value and not the whole set of values in the range $hl \geqslant M_z \geqslant -hl$, as was the case in the initial beam. From this the meaning of the orthogonality of eigenfunctions is very well seen. If a particle is found in a beam corresponding to a given value of k, then the probability of finding it in a beam with a different value of the projection $M_z = hk' \neq hk$ is equal to zero. From the general rule, the probability is equal to the square of the modulus of the coefficient of the expansion c_k' of the function $\psi(k)$ in terms of the functions $\psi(k')$, i.e., in accordance with the general expression (30.11)

$$c_{k'} = \int_0^{2\pi} \psi^*(k)\, \psi(k')\, d\varphi .$$

From the orthogonality condition (30.6), the integral is naturally equal to zero if $k' \neq k$. Therefore the orthogonality condition is a necessary condition of particles being found in states with definite values of M_z or, as in the case of an arbitrary operator λ, in states with definite values of λ. But the orthogonality condition follows directly from the Hermitian nature of operators (30.3), while equation (30.2), concerning the functions with equal values of λ, is insufficient.

The Hermitian condition implies the reality of eigenvalues together with the possibility of "pure states," i.e., states with definite eigenvalues of quantities.

If the second magnetic field is along the x-axis, then splitting will again occur due to the component of angular momentum M_x, which does not exist simultaneously with M_z. The number of splitting components is again equal to $2l+1$, since it is determined by the maximum angular-momentum projection l. This quantity cannot depend upon the direction of the magnetic field, and is related only to the atomic states in the original beam.

The eigenfunctions of M_x are

$$\psi(k_1) = \frac{1}{\sqrt{2\pi}} e^{i k_1 \omega}, \tag{30.17}$$

where $l \geqslant k_1 \geqslant -l$, and ω is the angle of rotation about the x-axis. Functions (30.16) and (30.17) do not coincide, which is a natural consequence of their being functions of noncommutative operators.

As a result of magnetic splitting in a field directed along the x-axis, a beam with given value of k is split into $2l+1$ beams with definite values k_1. Hence, the function (30.16) will be represented as the superimposition of functions (30.17):

$$\psi(k) = \sum_{k_1=-l}^{l} c_{k_1} \psi(k_1). \tag{30.18}$$

The square of the modulus $|c_{k_1}|^2$ is proportional to the intensity of the beam of the given projection $M_x = hk_1$ obtained as a result of the secondary splitting of the beam with a given M_z.

Averages in quantum mechanics. Let us now find the average of λ in a state given by the wave function $\psi(x)$, represented in the form of a sum (30.8). *By definition* the mean value is

$$\bar{\lambda} = \sum_{\lambda} \lambda w_\lambda, \tag{30.19}$$

i.e., the sum of possible values of λ multiplied by the corresponding probabilities. Let us substitute here w_λ from (30.13) and c_λ^* from (30.11):

$$\bar{\lambda} = \sum_{\lambda} \lambda |c_\lambda|^2 = \sum_{\lambda} \lambda c_\lambda c_\lambda^* = \sum_{\lambda} \lambda c_\lambda \int \psi(x) \psi^*(\lambda, x)\, dx. \tag{30.20}$$

We shall now replace the product $\lambda\psi^*(\lambda, x)$ by $\hat{\lambda}^* \psi^*(\lambda, x)$ and we first sum and then integrate. Then we obtain

$$\bar{\lambda} = \int \psi(x) \sum_{\lambda} c_\lambda^* \hat{\lambda}^* \psi^*(\lambda, x)\, dx. \tag{30.21}$$

But the operator $\hat{\lambda}^*$ does not depend upon any definite value of λ $\left(\text{for example, if } \lambda = p_x, \text{ then } \hat{\lambda}^* = -\frac{h}{i}\frac{\partial}{\partial x}\right)$.
Therefore, $\hat{\lambda}^*$ stands outside the sign of the summation:

$$\bar{\lambda} = \int \psi(x) \hat{\lambda}^* \sum_{\lambda} c_\lambda^* \psi^*(\lambda, x)\, dx. \tag{30.22}$$

The sum $\sum_{\lambda} c_\lambda^* \psi^*(\lambda, x) = \psi^*(x)$, since this is an equation which is a conjugate complex of (30.8). Therefore,

$$\bar{\lambda} = \int \psi(x)\, \hat{\lambda}^* \psi^*(x)\, dx \tag{30.23}$$

or, from the Hermitian condition for the operator $\hat{\lambda}$ (30.2), (30.3),

$$\bar{\lambda} = \int \psi^*(x)\, \hat{\lambda} \psi(x)\, dx\,. \tag{30.24}$$

Thus, in order to calculate the mean value of λ in a state $\psi(x)$, it is not necessary to know the eigenvalues of λ, since it is sufficient to calculate the integral (30.24).

The eigenvalues of the square of the angular momentum. If $\psi(x)$ is one of the eigenfunctions of the operator $\hat{\lambda}$, then the mean value $\bar{\lambda}$ is simply reduced to this eigenvalue. Indeed, then

$$\bar{\lambda} = \int \psi^*(\lambda, x)\, \hat{\lambda} \psi(\lambda, x)\, dx = \lambda \int |\psi(\lambda, x)|^2 dx = \lambda\,.$$

Taking advantage of the foregoing remark, it is easy to calculate the mean value of the square of the angular momentum.

First of all it may be noted that in the Stern-Gerlach experiment the mean values of the squares of all three angular-momentum projections must be the same, because it is absolutely immaterial what the notation of the coordinate axis is along which the magnetic field is directed:

$$\overline{M_x^2} = \overline{M_y^2} = \overline{M_z^2}\,. \tag{30.25}$$

It follows that the mean value $\overline{M^2}$ is equal to three times the mean value $\overline{M_z^2}$:

$$\overline{M^2} = \overline{M_x^2} + \overline{M_y^2} + \overline{M_z^2} = 3\overline{M_z^2}\,. \tag{30.26}$$

In the original beam all values of $M_z = hk$ from $-hl$ to hl are equally probable. This means that $\overline{M_z^2}$ is equal to

$$\overline{M_z^2} = h^2 \frac{\sum\limits_{k=-l}^{l} k^2}{2l+1} = h^2 \frac{l(l+1)(2l+1)}{3 \cdot (2l+1)} = \frac{h^2 l(l+1)}{3}\,, \tag{30.27}$$

whence

$$\overline{M^2} = h^2 l(l+1)\,. \tag{30.28}$$

But it was shown in Sec. 29 that $\hat{M}^2$ is commutative with $\hat{M}_z$, so that M^2 and M_z exist in one and the same state. In the Stern-Gerlach experiment the atoms in the beam occur predominantly in the ground state. This state is characterized by a certain absolute value of the angular momentum. Therefore, the mean value of the angular momentum in such a state is equal to its eigenvalue

$$M^2 = \overline{M^2} = h^2 l(l+1)\,. \tag{30.29}$$

The result (30.29) may appear somewhat surprising because the eigenvalue of the square of the angular momentum is equal not to the square of its greatest projection $h^2 l^2$, but to some greater amount. However, if M_z^2 were equal to $h^2 l^2$, i.e., its greatest value and $M^2 = h^2 l^2$, then for the remaining projections there would remain an identical zero. The other projections cannot have any definite values, including zero values, at the same time as $M_z \neq 0$. Therefore, the square of the angular momentum is somewhat greater than the square of the maximum value of any of its projections. The only exception is when all three projections are equal to zero (see Sec. 29).

Composition of angular momenta. Knowing the absolute value of the angular momentum, we can now indicate a rule for the composition of the angular momenta of two mechanical systems. Let the greatest angular-momentum projection of one system equal hl_1 and that of the other system hl_2; and also let $l_1 \geqslant l_2$. Then the projection of the smaller angular momentum in the direction of the larger one is contained between hl_2 and $-hl_2$, which, when added to the larger angular momentum, yields values ranging from $h(l_1+l_2)$ to $h(l_1-l_2)$. It follows that the greatest projection of the resultant angular momentum upon any arbitrary direction in space is equal, in units of h, to

$$l = l_1 + l_2,\ l_1 + l_2 - 1,\ l_1 + l_2 - 2, \ldots, l_1 - l_2. \tag{30.30}$$

The eigenvalues of the square of the sum of the angular momenta are

$$h^2 (l_1 + l_2)(l_1 + l_2 + 1),\ h^2 (l_1 + l_2 - 1)(l_1 + l_2), \ldots,$$
$$h^2 (l_1 - l_2)(l_1 - l_2 + 1).$$

The rule for composition of angular momenta formulated here agrees with the result that the value of a vector sum is contained between the sum and the difference of the absolute values of the vectors.

Quantum equations of motion. Let us suppose that a certain operator $\hat{\lambda}$ is given. It is required to find the operator form of its total time derivative, i.e., $\dot{\hat{\lambda}}$. We shall first of all determine the total derivative of the mean value $\bar{\lambda}$. In accordance with (30.24), for any state with wave function ψ, this derivative is

$$\dot{\bar{\lambda}} = \frac{\partial}{\partial t}\int \psi^* \hat{\lambda} \psi \, dx = \int \frac{\partial \psi^*}{\partial t} \hat{\lambda} \psi \, dx + \int \psi^* \frac{\partial \hat{\lambda}}{\partial t} \psi \, dx + \int \psi^* \hat{\lambda} \frac{\partial \psi}{\partial t} \, dx.$$

Let us substitute here the derivatives $\frac{\partial \psi^*}{\partial t}$ and $\frac{\partial \psi}{\partial t}$ from the Schrödinger equation (24.11), whose right-hand side we shall represent as $\hat{\mathscr{H}} \psi$, where $\hat{\mathscr{H}}$ is the Hamiltonian operator [see (29.7)]. From this,

$$\dot{\bar{\lambda}} = \int \frac{i}{h} \cdot (\hat{\mathscr{H}}^* \psi^*) \hat{\lambda} \psi \, dx + \int \psi^* \frac{\partial \hat{\lambda}}{\partial t} \psi \, dx - \int \frac{i}{h} \psi^* \hat{\lambda} \hat{\mathscr{H}} \psi \, dx.$$

We transform the first integral on the right-hand side in accordance with the Hermitian condition for $\hat{\mathscr{H}}$, namely

$$\int (\hat{\mathscr{H}}^* \psi^*)(\hat{\lambda}\psi)\, dx = \int \psi^* \hat{\mathscr{H}} \hat{\lambda} \psi \, dx .$$

We now combine all three integrals and obtain

$$\dot{\bar{\lambda}} = \int \psi^* \left(\frac{\partial \hat{\lambda}}{\partial t} + \frac{i}{h} [\hat{\mathscr{H}} \hat{\lambda} - \hat{\lambda} \hat{\mathscr{H}}] \right) \psi \, dx . \tag{30.31}$$

If we now define the operator $\hat{\dot{\lambda}}$ by the equality

$$\dot{\bar{\lambda}} = \int \psi^* \hat{\dot{\lambda}} \psi \, dx , \tag{30.32}$$

then we obtain the equation

$$\hat{\dot{\lambda}} = \frac{\partial \hat{\lambda}}{\partial t} + \frac{i}{h} [\hat{\mathscr{H}} \hat{\lambda} - \hat{\lambda} \hat{\mathscr{H}}] . \tag{30.33}$$

The operators of linear momentum, angular momentum, and coordinate that have been employed up till now do not depend upon time explicitly. For them, only the second term of (30.33) remains:

$$\hat{\dot{\lambda}} = \frac{i}{h} [\hat{\mathscr{H}} \hat{\lambda} - \hat{\lambda} \hat{\mathscr{H}}] . \tag{30.34}$$

Thus, if a given operator commutes with the Hamiltonian operator $\hat{\mathscr{H}}$, then $\hat{\dot{\lambda}} = 0$. It is then natural to call the quantity λ a quantum integral of motion. In accordance with the general result of Sec. 29, quantum integrals of motion have a common state with energy, since their operators are commutative.

We shall now find the equations of motion for the $\hat{x}$ and $\hat{p}_x$ operators. From (29.7), the energy operator $\hat{\mathscr{H}}$ is equal to $\frac{\hat{p}^2}{2m} + \hat{U}$. Here, only $\hat{p}_x^2$ is noncommutative with $\hat{x}$; for $\hat{p}_x^2$ we find

$$\begin{aligned} \hat{p}_x^2 \hat{x} - \hat{x} \hat{p}_x^2 &= \hat{p}_x^2 \hat{x} - \hat{p}_x \hat{x} \hat{p}_x + \hat{p}_x \hat{x} \hat{p}_x - \hat{x} \hat{p}_x^2 = \\ &= \hat{p}_x (\hat{p}_x \hat{x} - \hat{x} \hat{p}_x) + (\hat{p}_x \hat{x} - \hat{x} \hat{p}_x) \hat{p}_x = \frac{2h}{i} \hat{p}_x . \end{aligned}$$

It follows that

$$\hat{\dot{x}} = \frac{\hat{p}_x}{m} , \tag{30.35}$$

i.e., the operator $\hat{\dot{x}}$ is related to the operator $\hat{p}_x$ by the same expression as the quantities $\dot{x} = v_x$ and p_x in classical mechanics.

Let us now find $\hat{\dot{p}}_x$. p_x does not commute with $\hat{U}$. The commutator of $\hat{U}$ and p_x is easily evaluated:

$$(\hat{U} \hat{p}_x - \hat{p}_x \hat{U}) \psi = \frac{h}{i} \left(\hat{U} \frac{\partial \psi}{\partial x} - \frac{\partial}{\partial x} \hat{U} \psi \right) = - \frac{h}{i} \frac{\partial \hat{U}}{\partial x} \cdot \psi ,$$

whence, symbolically,

$$\hat{U}\hat{p}_x - \hat{p}_x\hat{U} = -\frac{h}{i}\frac{\partial \hat{U}}{\partial x} \tag{30.36}$$

Hence,

$$\dot{\hat{p}}_x = -\frac{\partial \hat{U}}{\partial x}, \tag{30.37}$$

which is completely analogous to the classical relationship between the momentum derivative and the force.

The quantum equations of motion (30.34) were the starting point for W. Heisenberg, who arrived at quantum mechanics independently of Schrödinger. The equivalence of both approaches was shown somewhat later.

The wave function and measurement of quantities. The probability amplitudes characterize the properties of a system in relation to the results of measuring certain quantities. If a system occurs in a state with wave function $\psi(x)$, and the quantity λ is measured, then the probability of obtaining the given value of λ is [see (30.11), (30.13)]

$$|c_\lambda|^2 = \left|\int \psi^*(x)\,\psi(\lambda, x)\,dx\right|^2.$$

For example, in the Stern-Gerlach experiment, the particles in the original beam have all angular-momentum projections between $-hl$ and hl. Measurement results yield $2l+1$ beams, each of them corresponding to the zth angular-momentum projection given by a definite value hk. However, the same measurement in a field directed along the x-axis of the original system would split the beam according to the xth angular-momentum projections. Both angular-momentum projections do not exist simultaneously, and the initial states of the particles in the beam were identical. It follows that, as a *result of measurement*, the particles occur either with a definite zth, or with a definite xth angular-momentum projection.

A measurement of a microscopic entity essentially changes the state of the latter. This is the fundamental difference between the concepts of measurement in classical and in quantum physics: a classical measurement has an infinitesimally small effect on the object being measured.

As a result of measurement, the angular-momentum projections in the original beam acquire $2l+1$ values, no matter how the measurement is performed. The state of these particles *after measurement* is essentially different and depends upon how the measurement was performed. But by performing measurements of a large number of identical entities, we can find out in what state they were *before measurement*, quite independently of the method of measurement. For this reason, a quantum measurement yields physical results which are just as objective as those given by a classical measurement though, obviously, within the limits permitted by the uncertainty principle.

Thus, in the Stern-Gerlach experiment it appears that the particles had an absolute angular-momentum value $M^2 = h^2 l\,(l+1)$, while the direction of the angular momentum in space was arbitrary (a non-polarized beam).

The repeated measurement of the zth angular-momentum projection, in the beams which had passed earlier through a field directed along the z-axis, gives a definite value of $M^2 = h^2 l\,(l+1)$ *and* a definite value of $M_z = hk$.

Exercises

1) Expand the function $\psi = \frac{1}{\sqrt{a}}$, in an infinitely deep rectangular potential well, in terms of functions (25.12).

$$c_n = \int_0^a \psi\,\psi_n\,dx = \frac{\sqrt{2}}{a}\int_0^a \sin\frac{\pi(n+1)x}{a}\,dx =$$

$$= \frac{\sqrt{2}}{\pi(n+1)}\left[-\cos\pi(n+1)+1\right] = \frac{\sqrt{2}}{\pi(n+1)}\left[1-(-1)^{n+1}\right].$$

2) Find the energy eigenvalues for a symmetrical quantum top. The energy of the symmetrical top is

$$\mathcal{E} = \frac{1}{2J_1}(M_1^2 + M_2^2) + \frac{1}{2J_3}M_3^2.$$

Introducing M^2, we have

$$\mathcal{E} = \frac{1}{2J_1}(M^2 - M_3^2) + \frac{1}{2J_3}M_3^2.$$

Substituting the eigenvalues for the angular momentum and its projections, we at last find

$$\mathcal{E} = \frac{h^2}{2J_1}\left[l(l+1) - k^2\right] + \frac{h^2k^2}{2J_3} = \frac{h^2}{2J_1}l(l+1) + h^2k^2\left(\frac{1}{2J_3} - \frac{1}{2J_1}\right).$$

Sec. 31. Motion in a Central Field

The motion of an electron in a central attractive field is the principal problem in the quantum mechanics of the atom. And it is not necessary to regard the field as strictly Coulomb in character. For example, in alkali-metal atoms, an outer electron which is bound relatively weakly to the nucleus moves in the field of the nucleus and the so-called atomic residue (i. e., all the other electrons). The charge-density distribution for these electrons possesses spherical symmetry and therefore produces a central field. We shall suppose that the potential energy of the electron is equal to $U(r)$, where r is the distance from the nucleus.

The energy operator and the angular-momentum integral. The equation for the energy eigenvalues of an atom (24.22) is, as usual,

$$-\frac{h^2}{2m}\Delta\psi + U\psi = \mathcal{E}\psi. \tag{31.1}$$

Here, m is the reduced mass of the nucleus and the electron, which mass is very close to the mass of the electron. Since the field is central we must pass to spherical coordinates. The Laplacian operator in spherical coordinates was obtained in Sec. 11 (11.46). Using this expression, we rewrite (31.1) explicitly:

$$-\frac{h^2}{2m}\left[\frac{1}{r^2}\frac{\partial}{\partial r}r^2\frac{\partial\psi}{\partial r}+\frac{1}{r^2}\left(\frac{1}{\sin\vartheta}\frac{\partial}{\partial\vartheta}\sin\vartheta\frac{\partial\psi}{\partial\vartheta}+\frac{1}{\sin^2\vartheta}\frac{\partial^2\psi}{\partial\varphi^2}\right)\right]+ \\ +U(r)\psi=\mathscr{E}\psi. \quad (31.2)$$

The operator involving angular differentiation is simply the square of the angular momentum introduced by us in exercise 4, Sec. 29. Therefore, equation (31.2) can also be rewritten as

$$-\frac{h^2}{2m}\frac{1}{r^2}\frac{\partial}{\partial r}r^2\frac{\partial\psi}{\partial r}+\frac{\hat{M}^2}{2mr^2}\psi+U(r)\psi=\mathscr{E}\psi. \quad (31.3)$$

It follows that the Hamiltonian operator $\hat{\mathscr{H}}$ [see (29.6)] is related to the angular-momentum operator in the following way:

$$\hat{\mathscr{H}}=-\frac{h^2}{2m}\frac{1}{r^2}\frac{\partial}{\partial r}r^2\frac{\partial}{\partial r}+\frac{\hat{M}^2}{2mr^2}+\hat{U}(r). \quad (31.4)$$

Reducing to an ordinary differential equation. The operator $\hat{M}^2$ involves only the angles ϑ and φ and derivatives with respect to them. All the derivatives with respect to angles in the operator $\hat{\mathscr{H}}$ are contained in the one term $\hat{M}^2$, while all the remaining terms involve only r and the derivative with respect to r. Consequently, the operators $\hat{\mathscr{H}}$ and $\hat{M}^2$ are commutative, since $\hat{M}^2$ commutes with any function of r and, of course, with r itself. Commutative operators have eigenvalues in the same state. Therefore, in a central field, the square of the angular momentum and one of its projections have (together with energy) eigenvalues, which, in accordance with (30.34), are quantum integrals of motion. All the other quantities which are not integrals of motion do not exist in the same energy state (in classical mechanics they, naturally, exist but are not conserved).

Thus, in equations (31.3) and (31.4), we can substitute in place of $\hat{M}^2$ its eigenvalue $h^2l(l+1)$ from (30.29). Then any angular dependence will be eliminated from equation (31.3) and, in place of the partial derivative with respect to r, we will get the total derivative:

$$-\frac{h^2}{2m}\frac{1}{r^2}\frac{d}{dr}r^2\frac{d\psi}{dr}+\frac{h^2l(l+1)}{2mr^2}\psi+U\psi=\mathscr{E}\psi. \quad (31.5)$$

It is considerably more simple to solve this equation than the partial differential equation (31.2). The form of (31.5) corresponds to (5.6) in classical mechanics, where it was also possible to eliminate all variables except r with the aid of the angular-momentum integral.

Reduction to one-dimensional form. It is convenient to reduce equation (31.5) to a one-dimensional form. To do this—the treatment is similar to that used in the problem of the propagation of spherical waves [cf. (19.6)]—we introduce the function

$$\chi = r\psi, \quad \psi = \frac{\chi}{r}. \tag{31.6}$$

Without repeating the computations by means of which the one-dimensional form (19.6) was obtained, we write down the analogous equation for χ:

$$-\frac{h^2}{2m}\frac{d^2\chi}{dr^2} + \frac{h^2 l(l+1)}{2mr^2}\chi + U\chi = \mathscr{E}\chi. \tag{31.7}$$

The wave function at large and small distances from the nucleus. As long as the form of $U(r)$ has not yet been made definite, we can consider (31.7) only in two limiting cases: for very large and for very small distances from the nucleus.

The field of the atomic residue is not effective at very small distances from the nucleus, and there remains only the Coulomb relationship $U = -\frac{Ze^2}{r}$ (Z is the atomic number of the element). However, if r is very small then the term $\frac{h^2 l(l+1)}{2mr^2}\psi$ is, in any case, larger than the term $U\psi$, which involves r in the denominator only in the first degree, and all the more greater than $\mathscr{E}\psi$. Hence, in direct proximity to the nucleus, the wave equation is of very simple form:

$$\frac{d^2\chi}{dr^2} = l(l+1)\frac{\chi}{r^2}. \tag{31.8}$$

In this form it is solved by the substitution

$$\chi = r^\alpha, \tag{31.9}$$

so that

$$\alpha(\alpha-1) = l(l+1). \tag{31.10}$$

This equation has two roots:

$$\alpha = l+1 \text{ and } \alpha = -l. \tag{31.11}$$

But the second root gives $\psi = r^{-l-1}$ from (31.7); at the point $r=0$, this function of ψ becomes infinite for all l. Therefore, we must discard the root $\alpha = -l$ and take the relationship between ψ and r for small r in the form

$$\psi = \frac{\chi}{r} = r^l\,*. \tag{31.12}$$

---

* The result (31.12) is true for $l = 0$ as well, even though the term $\frac{h^2 l(l+1)}{2mr^2}\psi$ in this case does not exist at all and cannot exceed $U(r)\psi$.

The greater the angular momentum, the higher the order of the wave-function zero at the coordinate origin. Only for $l=0$ does it remain finite close to the nucleus. This can be understood by analogy with classical mechanics: angular momentum is the product of momentum by the "arm," i.e., by the distance from the origin; $l=0$ corresponds to a zero "arm" and a zero angular momentum. Therefore, there is a nonzero probability of finding the electron at the origin. In the old version of quantum mechanics (due to Bohr), the electron orbit with zero angular momentum passed through the nucleus. The larger angular-momentum values correspond to larger "arms" and, correspondingly, in quantum mechanics, to a smaller probability of finding an electron close to the nucleus.

The behaviour of the wave function close to the origin can also be explained as follows. A centrifugal repulsive force acts on the particle; to this force there corresponds an effective potential energy $\frac{h^2 l(l+1)}{2mr^2}$. This energy limits the classically possible region of motion for small r. In quantum mechanics the particle penetrates the centrifugal barrier, though more weakly the greater r, i.e., the higher the barrier. There is no barrier for $l=0$ and there is nothing to prevent finding the particle at the origin.

The terms $\frac{h^2 l(l+1)}{2mr^2}\psi$ and $U\psi$ must be discarded for large r in the wave equation, because $U(r)$ is assumed to be zero at infinity, $U(\infty)\neq 0$. Then the equation is also greatly simplified:

$$\frac{d^2\chi}{dr^2} = -\frac{2m\mathscr{E}}{h^2}\chi. \tag{31.13}$$

Its general solution appears thus:

$$\chi = C_1 e^{\frac{\sqrt{-2m\mathscr{E}}}{h}r} + C_2 e^{\frac{\sqrt{-2m\mathscr{E}}}{h}r}. \tag{31.14}$$

Positive and negative energy values. We consider two cases. Let the energy be positive, $\mathscr{E}>0$. Here, χ appears as follows:

$$\chi = C_1 e^{i\frac{r\sqrt{2m\mathscr{E}}}{h}} + C_2 e^{-i\frac{r\sqrt{2m\mathscr{E}}}{h}}. \tag{31.15}$$

Both terms remain finite for any value of r. Therefore, two constants, C_1 and C_2, must be retained in the solution. We came across the same situation in considering the solution of wave equation (25.33) for a potential well of finite depth.

Any general solution of a second-order differential equation involves two arbitrary constants. Let us suppose that the solution (31.12), which holds for small r only, is continued into the region of large r, where it is not of the simple form r^l, but nevertheless satisfies the precise equation (31.7). A certain integral curve is obtained for this equation. But any integral curve can be represented by properly choosing the constants in the general solution. As r tends to infinity this solution acquires its asymptotic form (31.15) if $\mathscr{E}>0$. The expression (31.15) remains finite when $r \to \infty$ for any constants C_1 and C_2. It follows that, for a positive energy, the wave equation always has a finite solution for any values of r. Therefore, the values for $\mathscr{E}>0$ correspond to a continuous energy spectrum, since the wave function satisfies the required conditions at zero and at infinity for any $\mathscr{E}>0$. In accordance with (31.15), the probability of finding an electron at infinity for $r \to \infty$ does not become zero; i.e., this case corresponds to infinite motion, as in the classical problem considered in Sec. 5 (see also Sec. 25).

Thus, the general rule has been confirmed that infinite motion possesses a continuous energy spectrum.

Now let $\mathscr{E}<0$ or $\mathscr{E}=-|\mathscr{E}|$. Then (31.14) must be represented thus:

$$\chi = C_1 e^{\frac{r\sqrt{2m|\mathscr{E}|}}{h}} + C_2 e^{-\frac{r\sqrt{2m|\mathscr{E}|}}{h}}. \tag{31.16}$$

Here the first solution tends to infinity together with r and we must therefore put $C_1=0$, so that χ will involve one instead of two arbitrary constants:

$$\chi = C_2 e^{\frac{r\sqrt{2m|\mathscr{E}|}}{h}}. \tag{31.17}$$

The condition for eigenvalues. If we now draw an integral curve from the coordinate origin, proceeding from (31.12), then, as a rule, for large r it will not be reduced to the form (31.17). For all negative energy values, except certain ones, the integral curve is represented in the form (31.16) at infinity when $\mathscr{E}<0$ and, hence, does not satisfy the boundary condition imposed on the wave function. Only for those energy values for which it turns out that

$$C_1(\mathscr{E}) = 0 \tag{31.18}$$

does the wave equation have a solution. This corresponds to a discrete energy spectrum. At the same time, $\chi(\infty)$ becomes zero, so that the finite motion has a discrete energy spectrum, as expected.

The Coulomb field. The transition to atomic units. We shall now find this spectrum for an electron in a purely Coulomb field:

$$U(r) = -\frac{Ze^2}{r}. \tag{31.19}$$

This occurs in a hydrogen atom (though not in a molecule!), in singly ionized helium, doubly ionized lithium, etc. Z, as usual, denotes the atomic number of the nucleus.

The wave equation (31.7) is now written as

$$-\frac{h^2}{2m}\frac{d^2\chi}{dr^2}+\frac{h^2 l(l+1)}{2mr^2}\chi-\frac{Ze^2}{r}\chi=-|\mathscr{E}|\chi. \tag{31.20}$$

We have straightway taken the case of negative energies that leads to a discrete spectrum.

It is convenient here to change the units of length and energy similar to the way it was done in the problem of the harmonic oscillator (Sec. 26). In place of the CGS system (where the basic units are the arbitrary quantities centimetre, gram, second) we take the following units: the elementary charge e, the mass of the electron m, and the quantum of action h. From these quantities we form the unit of length

$$\frac{h^2}{me^2}=5.2917\times 10^{-9}\text{ cm}$$

and energy

$$\frac{me^4}{h^2}=27\text{ ev}.$$

Hence, if we put $e=1$, $m=1$, $h=1$ in equation (31.20), then length and energy will be measured in these units. Let us call this length ξ:

$$r=\frac{h^2}{me^2}\xi, \tag{31.21}$$

and energy ε:

$$|\mathscr{E}|=\frac{me^4}{h^2}\varepsilon, \tag{31.22}$$

so that, of the constants, the wave equation will involve only the atomic number Z:

$$-\frac{d^2\chi}{d\xi^2}+\frac{l(l+1)}{\xi^2}\chi-\frac{2Z}{\xi}\chi=-2\varepsilon\chi. \tag{31.23}$$

Solution by the series-expansion method. We look for the solution of this equation in the form of a series expansion. We shall proceed here from the solutions obtained for large and small values of ξ (i.e., r).

In accordance with equations (31.12) and (31.17), we write χ in following form:

$$\chi=\xi^{l+1}e^{-\xi\sqrt{2\varepsilon}}(\chi_0+\chi_1\xi+\chi_2\xi^2+\dots)=$$
$$=\xi^{l+1}e^{-\xi\sqrt{2\varepsilon}}\sum_{n=0}^{\infty}\chi_n\xi^n=e^{-\xi\sqrt{2\varepsilon}}\sum_{n=0}^{\infty}\chi_n\xi^{n+l+1}. \tag{31.24}$$

The first factor determined the form of χ for $\xi\to 0$, the second factor should basically correspond to the form of χ for large ξ, and the series interpolates, as it were, between the limiting values.

Differentiating (31.24) twice, we obtain

$$\frac{d^2\chi}{d\xi^2} = 2\varepsilon e^{-\xi\sqrt{2\varepsilon}} \sum_{n=0}^{\infty} \chi_n \xi^{n+l+1} - 2\sqrt{2\varepsilon}\, e^{-\xi\sqrt{2\varepsilon}} \sum_{n=0}^{\infty} (n+l+1)\chi_n \xi^{n+l} +$$

$$+ e^{-\xi\sqrt{2\varepsilon}} \sum_{n=0}^{\infty} (n+l+1)(n+l)\chi_n \xi^{n+l-1}. \qquad (31.25)$$

The first term on the right is simply $-2\varepsilon\chi$. Hence, it cancels with the same term in (31.23). We group the remaining terms so that in one of them the degree of ξ is everywhere less by unity than in (31.24) and, in the other, less by two units. In addition we eliminate the common factor $e^{-\xi\sqrt{2\varepsilon}}$. We shall now have an equality between two such series:

$$\sum_{n=0}^{\infty} [l(l+1) - (n+l+1)(n+l)]\chi_n \xi^{n+l-1} =$$

$$= \sum_{n=0}^{\infty} [2Z - 2\sqrt{2\varepsilon}\,(n+l+1)]\chi_n \xi^{n+l}. \qquad (31.26)$$

An equality between series is possible only when the coefficients of the same powers of ξ coincide. On the left-hand side the power ξ^{n+l} will have a coefficient involving χ_{n+1}.

Hence

$$\chi_{n+1} = \chi_n \frac{2[Z - (n+l+1)\sqrt{2\varepsilon}]}{l(l+1) - (n+l+1)(n+l+2)}. \qquad (31.27)$$

Examining the series and the condition for eigenvalues. From the relationship (31.27), all the coefficients χ_n are determined consecutively. We must neglect the constant numbers l and Z in equation (31.27) when n are large; there then remains the limit

$$\chi_{n+1} = \frac{2\sqrt{2\varepsilon}\,\chi_n}{n}. \qquad (31.28)$$

We met with a similar expression in the problem of the harmonic oscillator (26.16). In the case of large ξ it reduces the whole series to an exponential form:

$$\sum \chi_n \xi^n \cong e^{2\xi\sqrt{2\varepsilon}} \qquad (31.29)$$

But such a series cannot give a correct solution to the wave equation because, if we substitute (31.29) in (31.24), we obtain $\psi(\infty) = \infty$ despite the boundary condition. However, if all the coefficients become zero from a certain χ_{n+1} onwards, the series (31.29) degenerates to a polynomial. Then, being multiplied by $e^{-\xi\sqrt{2\varepsilon}}$, it gives $\psi(\infty) = 0$, as expected. It can be seen from (31.27) that χ_{n+1} is equal to zero if

$$Z-(n+l+1)\sqrt{2\varepsilon}=0, \tag{31.30}$$

i.e.,

$$\varepsilon=\frac{Z^2}{2(n+l+1)^2}. \tag{31.31}$$

Finally, going over to conventional units and taking into account the sign of the energy, we obtain the required spectrum:

$$\mathscr{E}=-\frac{Z^2 m e^4}{2h^2(n+l+1)^2}. \tag{31.32}$$

Quantum numbers. The number n is the degree of the polynomial $\sum_n \chi'_n \xi^n$ (it is called the Sonin-Laguerre polynomial). A more detailed analysis shows that this polynomial becomes zero exactly n times, corresponding to its degree. Therefore, if we examine the dependence of the wave function on radius, it has n zeros or "nodes," not counting the zero at $r=\infty$ and at $r=0$, which all functions with $l\neq 0$ have. The term *node* instead of *zero* is given by analogy with the nodes of a vibrating string fixed at both ends. In future we shall call n_r the degree of the polynomial and denote by the letter n the whole sum

$$n\equiv n_r+l+1. \tag{31.33}$$

It is convenient to use these quantities also in the more complex cases of many-electron atoms. Even though the energy in such a case does not have the simple form (31.32), the numbers n, n_r, and l are convenient for classification of the states.

l is called the *azimuthal* quantum number. As we know, it defines the angular momentum of an electron. The following system of notation is used in spectroscopy: the electron state with $l=0$ is called the s-state and, corresponding to $l=1, 2, 3$, we have the p-, d- and f-states. There are no greater values of l in nonexcited atoms. Combining the angular momenta of separate electrons according to the rule of vector addition (30.30), we obtain the angular momentum L of the atom as a whole. The states with $L=0, 1, 2, 3$ are termed S, P, D, F, while states with greater L are named by subsequent letters of the Latin alphabet.

k, [see Eq. (29.21)], i.e., the angular-momentum projection on some axis in units of h, is called the *magnetic* quantum number, since the external magnetic field is usually directed along this axis.

n_r is the number of wave-function zeros as related to the radius (for $r\neq 0$ and $r\neq\infty$) and is called the *radial* quantum number.

Finally, the sum (31.33) is called the *principal* quantum number. In accordance with (31.32), the binding energy of an electron in a hydrogen atom is

$$\mathscr{E}_n = -\frac{m e^4}{2 \hbar^2 n^2} = -\frac{13{,}5}{n^2}\ \text{ev}. \tag{31.34}$$

An analogous expression is obtained also for the positive helium ion. Apart from the difference of $Z^2 = 4$ times, there is a more subtle difference due to the fact that the reduced mass of the helium atom differs somewhat from the reduced mass of a hydrogen atom as a result of a difference in the nuclear masses.

The state with $n = 1$ is the ground state. The atom cannot emit light in this state because it is impossible to make a transition to a lower state. For more detail about radiation, see Sec. 34.

The parity of a state. The state of an electron in an atom is characterized by one more property, which (as opposed to energy and angular momentum) does not correspond to any classical analogue. This is the *parity* of a wave function with respect to coordinates.

To begin with let us consider the wave function of a separate electron. The wave equation (31.1) does not change its form if we substitute

$$x = -x' \quad y, = -y', \quad z = -z'. \tag{31.35}$$

This transformation is termed inversion: it transforms a right-handed coordinate system to a left-handed one. No rotation in space can make these systems coincide (like left-hand and right-hand gloves) (see Sec. 16).

The wave equation (31.1) is linear. Therefore, if it has not changed its form, then its solution (determined by the boundary conditions within the accuracy of the constant factor) can acquire only a certain additional factor:

$$\psi(x, y, z) = C\,\psi(x', y', z'). \tag{31.36}$$

But, in principle, the primed left-handed system differs in no way from the unprimed, right-handed system. For this reason, the transformation of inversion must involve the *same* transformation factor C:

$$\psi(x', y', z') = C\,\psi(x, y, z). \tag{31.37}$$

Substituting this in (31.36), we obtain

$$\psi(x, y, z) = C^2\,\psi(x, y, z),$$

whence

$$C^2 = 1, \quad C = \pm 1. \tag{31.38}$$

The function is termed *even* for $C = 1$ and *odd* for $C = -1$. The eigenfunctions of a linear harmonic oscillator possessed an analogous property; here the energy operator was also even, $\hat{\mathscr{H}}(x) = \hat{\mathscr{H}}(-x)$, while the wave functions alternated depending upon the eigenvalue number n (i.e., they were either even or odd).

Parity and orbital angular momentum. Let us now find out what it is that determines the parity of a wave function in a central field.

To do this, it is convenient to utilize its form near the coordinate origin:

$$\psi = r^l. \tag{31.39}$$

In order to find the angular dependence of the wave function as well, it is sufficient to investigate it to the approximation that yields equation (31.39) since the terms U and $\mathscr{E}$, which do not depend upon the angles, are thereby discarded. The angular dependence of the solutions of the precise and shortened equation is the same. This shortened equation is, obviously, simply the Laplace equation

$$\Delta\psi = 0. \tag{31.40}$$

Equation (31.40) is satisfied by a homogeneous polynomial in x, y, z

$$\psi = x^l + ax^{l-1}y + \ldots + bx^{l-k-m}y^k z^m + \ldots \tag{31.41}$$

of degree l for certain relationships between its coefficients a, . . ., b, ... It is clear that the degree l of this polynomial is equal to the degree l in equation (31.39). But the degree l of (31.41) defines the even or odd nature of ψ with respect to the inversion (31.35). It follows that wave functions with even orbital angular momenta l are even, and those with odd orbital angular momenta l are odd. In a multi-electron atom, the total parity of the wave function is determined by the parity of all the wave functions for the separate electrons (this by no means signifies that the wave function of an atom is equal to the product of the wave functions of the separate electrons!). Therefore, the parity of the total wave function is equal to the parity of the number $\sum_i l_i$ where l_i are the orbital quantum numbers of the electrons. As we know, the total angular momentum of the atom is equal to the vector sum of the angular momenta of its electrons.

Parity as an integral of motion. We shall explain the significance of the parity of a wave function. To begin with, let us point out that the inversion (31.35) can be represented by an operator $\hat{G}$ such that

$$\hat{G}\psi(x, y, z) = \psi(-x, -y, -z). \tag{31.42}$$

Since the Hamiltonian operator in an atom $\hat{\mathscr{H}}$ is an even function of coordinates, we can write

$$\hat{G}\hat{\mathscr{H}} = \hat{\mathscr{H}}. \tag{31.43}$$

Whence it follows that the parity operator is commutative with the Hamiltonian operator

$$\hat{G}\hat{\mathscr{H}}\psi = \hat{\mathscr{H}}\hat{G}\psi. \tag{31.44}$$

The eigenvalues of the operator $\hat{G}$ are the numbers $C = \pm 1$ (31.38), because

$$\hat{G}\psi = \psi(-x, -y, -z) = C\psi. \tag{31.45}$$

According to (31.44) and (29.28) these numbers exist simultaneously with the energy eigenvalues.

We shall now consider what limitations can be imposed, by the law of conservation of parity, on possible transitions in the atom. Suppose we have an excited multi-electron atom with total angular momentum $L=0$, i. e., in the S-state. Then let there be in this atom s-electrons and an odd number of p-electrons. Consequently, the atom is in an odd state. Let the excitation energy be sufficient for the atom to emit one of the p-electrons, so that after the rearrangement of the electron cloud the atom remains in the S-state with $L=0$. Since angular momenta are combined vectorially, such a state may result both for an odd and an even number of p-electrons. According to the law of conservation of total angular momentum, an electron may be emitted only with an angular momentum equal to zero because, according to assumption, the angular momentum of the rest of the system is equal to zero before and after the transition. It follows that the electron can be emitted only in an even s-state.

After the emission of the electron, an even number of p-electrons remains in the ion, and the emitted electron is also found to be in an even state. But this is impossible since the initial state was odd and the final state was even, the total energy being constant. Hence, the laws of conservation of parity and angular momentum may exclude transitions which are permissible energetically. We have considered a typical case of a transition which is "forbidden" by parity selection rules ($L=0$ into $L=0$ with changed parity).

The law of conservation of parity by no means follows from the law of conservation of angular momentum, since parity depends upon the arithmetic sum of l while total angular momentum depends upon the vector sum.

The law of conservation of angular momentum in quantum mechanics must always be used together with the law of conservation of parity. In origin, these laws have a common basis: they both follow from the invariance of equations with respect to the orientation of coordinate axes in space. But all possible orientations are not exhausted by axis rotations alone: an additional transformation is inversion which is not reduced to any rotation. It is this that yields the parity conservation law in addition to the law of conservation of angular momentum.

In this form the parity conservation law can be unconditionally applied to those systems in which electromagnetic interactions occur. The considerably weaker interactions which occur in certain elementary-particle conversions probably satisfy a modified parity conservation law (see Sec. 38).

Hydrogen-like atoms. Alkali-metal atoms somewhat resemble the hydrogen atom. The outer electron in these atoms is relatively weakly bound to the atomic residue, which consists of the nucleus and all the remaining electrons. The wave functions for electrons of the atomic residue differ from zero at smaller distances from the nucleus than the wave function for the outer electron, so that the residue screens, as it were, the nuclear charge. The field in which the outer electron moves is approximately Coulomb, provided only that it is not situated in the region of the residue. It is for this reason that the spectra of alkali-metal atoms resemble the hydrogen-atom spectrum. The energy levels of these atoms, which are due to excitation of the outer electron, are given by the equation

$$\mathscr{E}_{n,l} = -\frac{me^4}{2h^2}\frac{1}{[n+\Delta(l)]^2}, \qquad (31.46)$$

where the correction $\Delta(l)$ depends upon the azimuthal quantum number. It accounts for the deviation of the field from a purely Coulomb one at small distances from the nucleus.

Thus, the energy levels in alkali metals—like the energy levels of all atoms—depend upon n and l. An exception is the hydrogen atom, where the energy depends only upon n; this is a special property of a purely Coulomb field. For example, when $n=2$, the azimuthal quantum number can take on two values: $l=0$ and $l=1$, while the corresponding energy levels of the hydrogen atom are close to each other (the splitting of these levels is due to relativistic corrections to the wave equation).

Exercise

Construct and normalize the wave functions in a hydrogen atom with $l=0$, 1, 2 and $n=1$, 2, 3. Take advantage of the fact that $\int\limits_0^\infty e^{-x}x^n\,dx = n!$

Sec. 32. Electron Spin

The insufficiency of three quantum numbers for the electron in an atom. From equation (31.34) the ground state of a hydrogen atom has a principal quantum number n equal to unity. For $n=1$ the azimuthal quantum number l and the radial quantum number n_r must be equal to zero, since $n=n_r+l+1$, and n_r and l can in no way be less than zero. The ground state of a hydrogen atom is the s-state. The orbital motion of an s-electron does not produce a magnetic moment because the magnetic moment is proportional to the mechanical moment. Yet, if the Stern-Gerlach experiment is performed for atomic hydrogen, the atomic beam will split, but only into two parts. However, when $l=0$, as we have already said, there should be no splitting due to orbital angular momentum, while for $l=1$, the beam should split into $2l+1=3$ beams corresponding to the number of projections k of the angular momentum l $(-1, 0, 1)$.

The same results if, instead of hydrogen, we take an alkali metal. The electron cloud of any alkali metal consists of an atomic residue in an S-state, i.e., one lacking an orbital angular momentum and one electron in the s-state. In this sense, aklali-metal atoms resemble the hydrogen atom. For this reason, the state of the atom is not described by the three quantum numbers n, l, and k.

Intrinsic angular momentum or electron spin. Splitting into two beams can be accounted for only by an angular momentum whose greatest projection is equal to $\hbar/2$. Then it has only two projections $\hbar/2$ and $-\hbar/2$.

The Stern-Gerlach experiment was given only as an example. In fact, not only this experiment, but the whole enormous aggregate of knowledge about the atom indicates that the electron possesses a mechanical moment $\hbar/2$ that is not related to its orbital motion. This mechanical moment is termed the *spin*. It can be said that an electron is somewhat reminiscent of a planet which has an angular momentum due not only to its revolution about the sun, but also to rotation on its own axis.

The analogy with a planet is not far-reaching since the angular momentum of rotating rigid body can be made equal to any value, while the spin of an electron always has projections $\pm\hbar/2$ and no others. Therefore, spin is a purely quantum property of the electron; in the limiting transition to classical mechanics it becomes zero. We must not take the word "spin" too literally, for the electron actually does not resemble a rigid body like a top or a spindle.

Spin degree of freedom. The analogy between an electron and a top consists in the fact that their motion is not described by their position in space alone, but possesses an internal rotational degree of freedom.

There is a certain analogy between the electron and the light quantum. As was shown in Sec. 28, in addition to its wave vector, the state of a quantum is described by a polarization variable which takes on two values. Similarly, the electron has, in addition to its spatial coordinates, a spin variable σ which assumes two values (since spin has only two projections).

Spin operators. When we write $\psi(x)$ we have in mind the whole group of values of the wave function for all x, i.e., ψ at all points of space. The action of an operator on $\psi(x)$ denotes a linear transformation of ψ in the whole space, since, in accordance with the superposition principle, all operators in quantum mechanics are linear. Taking into account the spin variable, one has to write $\psi(x, \sigma)$, where σ takes on only two values. The action of the spin operator on $\psi(x, \sigma)$ denotes the replacement of $\psi(x, 1)$ by some linear combination of $\psi(x, 1)$ and $\psi(x, 2)$; the action of the operator on $\psi(x, 2)$ is determined analoguously. Linear operators depending upon σ can denote nothing other than a linear substitution as applied to a function of "two points" $\sigma = 1$ and $\sigma = 2$.

We shall try to determine, in explicit form, how spin angular-momentum projection operators should act upon functions of the spin variable σ. The following requirements are to be imposed on them.

1) The eigenvalues of all three spin projections must be equal to $\pm h/2$.

2) The same commutation rules (29.33)-(29.35) must exist for them as for the components of orbital angular momentum, otherwise the sum of the orbital and spin angular-momentum operators will not possess the property of angular momentum.

3) For the same reason we must require that the spin-projection operators should be Hermitian.

4) In coordinate-system rotations, spin-projection operators must behave in the same way as vector components so that the commutation rules for these operators, in a rotated system, should not differ from the rules of the original system in which the operators were defined.

Corresponding to these requirements, W. Pauli found the required operators, which we shall now form.

We shall write the group of functions $\psi(x, \sigma)$ in columns instead of rows; the meaning of $\psi(x, \sigma)$ does not thereby change, of course. In addition, for brevity, we shall omit the coordinate dependence contained in the argument x. Thus, $\psi(\sigma)$ denotes the column

$$\begin{pmatrix} \psi(1) \\ \psi(2) \end{pmatrix}. \tag{32.1}$$

Here, each component satisfies Schrödinger's coordinate equation (24.22). In the most general case the action of a linear operator on the function (32.1) reduces it to the form

$$\begin{pmatrix} \alpha\psi(1) + \beta\psi(2) \\ \gamma\psi(1) + \delta\psi(2) \end{pmatrix}.$$

As we know, one of the angular-momentum projections can always exist together with the square of the total angular momentum, since the substitution rules for spin components are the same as for orbital angular momentum (condition 2). To be specific, we shall consider that there exists the zth projection σ_z. If the operator $\hat{\sigma}_z$ has an eigenvalue in the given state, its application leads to the multiplication of the function (32.1) by some number without mixing of the components. This number is equal to $\pm h/2$, depending upon what the sign of the spin projection σ_z is in the given state. Let the function $\psi(1)$ be multiplied by $+h/2$ and the function $\psi(2)$ by $-h/2$, so that

$$\hat{\sigma}_z\psi = \hat{\sigma}_z\begin{pmatrix} \psi(1) \\ \psi(2) \end{pmatrix} = \frac{h}{2}\begin{pmatrix} \psi(1) \\ -\psi(2) \end{pmatrix}. \tag{32.2}$$

The equality sign between columns denotes a line by line equality of the expressions, i.e., $\hat{\sigma}_z \psi(1) = \frac{h}{2}\psi(1)$, $\hat{\sigma}_z(2) = -\frac{h}{2}\psi(2)$. The form of the functions corresponding to various spin projections can immediately be seen: to the projection $h/2$ there corresponds a function $\begin{pmatrix}\psi(1)\\ 0\end{pmatrix}$, while the function $\begin{pmatrix}0\\ \psi(2)\end{pmatrix}$ corresponds to the projection $-h/2$. The first of them, if we substitute it in (32.2), is entirely multiplied by $h/2$, while the second is multiplied by $-h/2$, since the change of sign for the zero component of the function does not signify anything.

The operators $\hat{\sigma}_x$ and $\hat{\sigma}_y$ cannot have eigenvalues in these states. It follows that they must in any case also interchange the components of the wave function that we have defined, and not merely multiply them by numbers. Simple multiplication operators would be commutative with $\hat{\sigma}_z$. Once the form $\hat{\sigma}_z$ is given, we can also determine the operators for the other two components.

Let us temporarily go over to atomic units (see Sec. 31), i.e., we put $h = 1$. We shall look for the operator $\hat{\sigma}_x$ (acting on the two functions) in the most general possible form*:

$$\hat{\sigma}_x \psi = \begin{pmatrix}\alpha\psi_1 + \beta\psi_2\\ \gamma\psi_1 + \delta\psi_2\end{pmatrix}. \tag{32.3}$$

In other words, we suppose that it replaces ψ_1 by $\alpha\psi_1 + \beta\psi_2$ and ψ_2 by $\gamma\psi_1 + \delta\psi_2$. We act on $\hat{\sigma}_x\psi$ with $\hat{\sigma}_z$. Then, by definition of σ_z (32.2), we must change the sign in the lower row and divide both rows by two:

$$\hat{\sigma}_z \hat{\sigma}_x \psi = \hat{\sigma}_z \begin{pmatrix}\alpha\psi_1 + \beta\psi_2\\ \gamma\psi_1 + \delta\psi_2\end{pmatrix} = \frac{1}{2}\begin{pmatrix}\alpha\psi_1 + \beta\psi_2\\ -\gamma\psi_1 - \delta\psi_2\end{pmatrix}.$$

If we act upon $\hat{\sigma}_z\psi$ with $\hat{\sigma}_x$, then we must first of all put a minus sign in front of ψ_2 and divide both components by two, and then substitute them in (32.3). This will yield

$$\hat{\sigma}_x \hat{\sigma}_z \psi = \hat{\sigma}_x \frac{1}{2}\begin{pmatrix}\psi_1\\ -\psi_2\end{pmatrix} = \frac{1}{2}\begin{pmatrix}\alpha\psi_1 - \beta\psi_2\\ \gamma\psi_1 - \delta\psi_2\end{pmatrix}.$$

From (29.35), the difference $\hat{\sigma}_z\hat{\sigma}_x - \hat{\sigma}_x\hat{\sigma}_z$ must be equal to $i\sigma_y$ in atomic units:

$$(\hat{\sigma}_z \hat{\sigma}_x - \hat{\sigma}_x \hat{\sigma}_z)\psi = \begin{pmatrix}\beta\psi_2\\ -\gamma\psi_1\end{pmatrix} = i\hat{\sigma}_y\psi. \tag{32.4}$$

Thus, the operator $\hat{\sigma}_y$ interchanges the functions ψ_1 and ψ_2 and multiplies them by β and $-\gamma$, where β and γ appeared in the definition (32.3) for $\hat{\sigma}_x$.

* The argument $\sigma = 1$ and $\sigma = 2$ will in future be replaced by the index.

But if we proceed from $\hat{\sigma}_y$, defining it analogously to (32.3), it will turn out that $\hat{\sigma}_x$ also interchanges functions, i.e., it does not contain the coefficients α and δ. Therefore we obtain

$$\hat{\sigma}_x\psi=\begin{pmatrix}\beta\psi_2\\ \gamma\psi_1\end{pmatrix},\quad \hat{\sigma}_y\psi=\begin{pmatrix}-i\beta\psi_2\\ i\gamma\psi_1\end{pmatrix}.$$

We now form the difference $\hat{\sigma}_x\,\hat{\sigma}_y - \hat{\sigma}_y\,\hat{\sigma}_x$. First acting with $\hat{\sigma}_x$ upon $\hat{\sigma}_y\psi$ and then with $\hat{\sigma}_y$ upon $\hat{\sigma}_x\psi$, we equate their difference to $i\,\hat{\sigma}_z\psi$ [see (29.33)]:

$$(\hat{\sigma}_x\hat{\sigma}_y-\hat{\sigma}_y\hat{\sigma}_x)\psi=\begin{pmatrix}2i\beta\gamma\psi_1\\ -2i\beta\gamma\psi_2\end{pmatrix}=i\,\hat{\sigma}_z\psi=\frac{i}{2}\begin{pmatrix}\psi_1\\ -\psi_2\end{pmatrix},$$

whence it follows that

$$\gamma=\frac{1}{4\beta}. \tag{32.5}$$

Hence, conditions (29.33)-(29.35) lead to the following form for $\hat{\sigma}_x$ and $\hat{\sigma}_y$:

$$\hat{\sigma}_x\psi=\begin{pmatrix}\beta\psi_2\\ \dfrac{1}{4\beta}\psi_1\end{pmatrix};\quad \hat{\sigma}_y\psi=\begin{pmatrix}-i\beta\psi_2\\ \dfrac{i}{4\beta}\psi_1\end{pmatrix}. \tag{32.6}$$

The operators $\hat{\sigma}_x$, $\hat{\sigma}_y$ and $\hat{\sigma}_z$ must be Hermitian. We once again deduce the Hermitian condition (30.3) for the operator $\hat{\sigma}_x$, insofar as the result of Sec. 30 related to a continuous variable x, while here we consider a discrete variable $\hat{\sigma}$. Let us write similarly to (30.1a) and (30.1b), with spin dependence in explicit form:

$$\hat{\sigma}_x\psi=\begin{pmatrix}\beta\psi_2\\ \dfrac{1}{4\beta}\psi_1\end{pmatrix}=\sigma_x\begin{pmatrix}\psi_1\\ \psi_2\end{pmatrix};\quad \hat{\sigma}_x^*\psi^*=\begin{pmatrix}\beta^*\psi_2^*\\ \dfrac{1}{4\beta^*}\psi_1^*\end{pmatrix}=\sigma_x\begin{pmatrix}\psi_1^*\\ \psi_2^*\end{pmatrix}.$$

Summation with respect to σ now corresponds to integration with respect to x. We multiply the first two equations by ψ_1^* and ψ_2^*, respectively, and the second two by ψ_1 and ψ_2, sum with respect to σ, and equate the results utilizing the fact that σ_x is a real number:

$$\psi_1^*\beta\psi_2+\psi_2^*\frac{1}{4\beta}\psi_1=\psi_1\beta^*\psi_2^*+\psi_2\frac{1}{4\beta^*}\psi_1^*.$$

As was shown in Sec. 30, this condition must be identically satisfied for any two functions χ^* and ψ. From this we obtain

$$\chi_1^*\beta\psi_2+\chi_2^*\frac{1}{4\beta}\psi_1=\psi_1\beta^*\chi_2^*+\psi_2\frac{1}{4\beta^*}\chi_1^*. \tag{32.7}$$

But this equation can hold only if

$$\beta=\frac{1}{4\beta^*},\ |\beta|^2=\frac{1}{4},$$

i.e., $\beta = \frac{1}{2} e^{i\nu}$. There remains an arbitrary phase factor $e^{i\nu}$ which we choose equal to unity. Thus,

$$\hat{\sigma}_x \psi = \frac{1}{2} \begin{pmatrix} \psi_2 \\ \psi_1 \end{pmatrix}, \tag{32.8}$$

$$\hat{\sigma}_y \psi = \frac{1}{2} \begin{pmatrix} -i\psi_2 \\ i\psi_1 \end{pmatrix}. \tag{32.9}$$

We note three operator relations

$$\begin{gathered} \hat{\sigma}_x \hat{\sigma}_y = -\hat{\sigma}_y \sigma_x = \frac{i}{2} \hat{\sigma}_z; \quad \hat{\sigma}_z \hat{\sigma}_x = -\hat{\sigma}_x \hat{\sigma}_z = \frac{i}{2} \hat{\sigma}_y; \\ \hat{\sigma}_y \hat{\sigma}_z = -\hat{\sigma}_z \hat{\sigma}_y = \frac{i}{2} \hat{\sigma}_x, \end{gathered} \tag{32.10}$$

which are directly verified by substitution, and also expressions for the squares $\hat{\sigma}_x^2$, $\hat{\sigma}_y^2$, and $\hat{\sigma}_z^2$, which are obtained when they act twice upon ψ:

$$\begin{gathered} \hat{\sigma}_x^2 \psi = \frac{1}{2} \hat{\sigma}_x \begin{pmatrix} \psi_2 \\ \psi_1 \end{pmatrix} = \frac{1}{4} \begin{pmatrix} \psi_1 \\ \psi_2 \end{pmatrix} = \frac{1}{4} \psi; \\ \hat{\sigma}_y^2 \psi = \frac{1}{4} \psi; \quad \hat{\sigma}_z^2 \psi = \frac{1}{4} \psi. \end{gathered} \tag{32.11}$$

Hence, in accordance with condition (1), the eigenvalues of $\hat{\sigma}_x^2$, $\hat{\sigma}_y^2$, and $\hat{\sigma}_z^2$ are $\frac{1}{4}$. Since each operator is commutative with its square, we see that the eigenvalues of $\hat{\sigma}_x$, $\hat{\sigma}_y$, and $\hat{\sigma}_z$ should equal the square roots of the eigenvalues of their squares, i.e., $\pm 1/2$. Naturally, these eigenvalues of $\hat{\sigma}_x$, $\hat{\sigma}_y$, and $\hat{\sigma}_z$ only exist separately and by no means simultaneously.

The vector properties of spin operators. In order to prove finally that the operators $\hat{\sigma}_x$, $\hat{\sigma}_y$, and $\hat{\sigma}_z$ possess the properties of angular-momentum components, we must be convinced that in coordinate rotations they transform like vector projections, i.e., we must verify that condition (4) is satisfied.

Let us suppose that a rotation occurs around the z-axis through an angle ω. Then we must prove that the operators

$$\begin{aligned} \hat{\sigma}_x' &= \hat{\sigma}_x \cos\omega + \hat{\sigma}_y \sin\omega, \\ \hat{\sigma}_y' &= -\hat{\sigma}_x \sin\omega + \hat{\sigma}_y \cos\omega, \end{aligned} \tag{32.12}$$

formed by analogy with the vector projections on rotated coordinate axes possess the same properties (1) and (2) as the original operators $\hat{\sigma}_x$ and $\hat{\sigma}_y$. First of all we have

$$\begin{aligned} \hat{\sigma}'^2_x &= (\hat{\sigma}_x \cos\omega + \hat{\sigma}_y \sin\omega)(\hat{\sigma}_x \cos\omega + \hat{\sigma}_y \sin\omega) = \\ &= \hat{\sigma}_x^2 \cos^2\omega + \hat{\sigma}_y^2 \sin^2\omega + (\hat{\sigma}_x \hat{\sigma}_y + \hat{\sigma}_y \hat{\sigma}_x) \sin\omega \cos\omega. \end{aligned}$$

It can be seen from (32.10) that

$$\hat{\sigma}_x\hat{\sigma}_y + \hat{\sigma}_y\hat{\sigma}_x = 0 \tag{32.13}$$

(and analogously for any pair of components). Further, $\hat{\sigma}_x^2$ and $\hat{\sigma}_y^2$, operating on ψ-functions, act like numbers, i.e., they simply multiply it by 1/4. But then this also means that $\hat{\sigma}'^2_x = \frac{1}{4}(\cos^2\omega + \sin^2\omega) = \frac{1}{4}$. Thus, the first property of $\hat{\sigma}'_x$ and $\hat{\sigma}'_y$ is retained under the rotation of the components.

We now form the difference $\hat{\sigma}'_x\hat{\sigma}'_y - \hat{\sigma}'_y\hat{\sigma}'_x$:

$$\begin{aligned}\hat{\sigma}'_x\hat{\sigma}'_y - \hat{\sigma}'_y\hat{\sigma}'_x = &-\hat{\sigma}_x^2\cos\omega\sin\omega + \hat{\sigma}_y^2\sin\omega\cos\omega + \hat{\sigma}_x\hat{\sigma}_y\cos^2\omega - \\ &-\hat{\sigma}_y\hat{\sigma}_x\sin^2\omega + \hat{\sigma}_x^2\cos\omega\sin\omega - \hat{\sigma}_y^2\sin\omega\cos\omega + \hat{\sigma}_x\hat{\sigma}_y\sin^2\omega - \\ &-\hat{\sigma}_y\hat{\sigma}_x\cos^2\omega = \hat{\sigma}_x\hat{\sigma}_y - \hat{\sigma}_y\hat{\sigma}_x = i\hat{\sigma}_z.\end{aligned} \tag{32.14}$$

But $\hat{\sigma}_z = \hat{\sigma}'_z$ since the rotation occurs about the z-axis.

Any rotation in space may be obtained by successive rotations about three axes. Therefore, it was sufficient to show that the basic properties of the operators are preserved under rotations about any one of the three axes.

The total angular-momentum operator. If we now form the sum of the operators

$$\hat{j}_x = \hat{M}_x + \hat{\sigma}_x; \quad \hat{j}_y = \hat{M}_y + \hat{\sigma}_y; \quad \hat{j}_z = \hat{M}_z + \hat{\sigma}_z, \tag{32.15}$$

then it will possess all the properties of an angular-momentum operator. Naturally, we could not have added the components of $\boldsymbol{\sigma}$ to the components of $\mathbf{M}$ if they both did not transform identically under rotations of the coordinate system, since, otherwise, equations (32.15) would be noninvariant with respect to the choice of the system of axes.

The vector $\mathbf{j}$ is called the *total angular momentum of an electron*. If the orbital angular momentum of the electron has a greatest projection l, then the greatest projection of $\mathbf{j}$ can equal $l + \frac{1}{2}$ or $l - \frac{1}{2}$. In the first case, we say that the spin and the orbital angular momentum are *parallel*; in the second case, we say that they are *antiparallel*.

Spin magnetic moment. The spin of an electron, similar to its orbital angular momentum, is associated with a definite magnetic moment. But experiment shows that the ratio of spin magnetic moment to mechanical moment is twice as great as for orbital angular momentum. There is nothing paradoxical in this because the result (15.25) cannot be applied to spin. At the same time we can deduce the spin magnetic moment from the Dirac relativistic wave equation for an electron (Sec. 38); in agreement with experiment, it is found to be

$$\boldsymbol{\mu}_\sigma = \frac{e}{mc}\boldsymbol{\sigma}. \tag{32.16}$$

Hence, the projection of μ_σ on any axis is

$$(\mu_\sigma)_z = \pm \frac{eh}{2mc} \equiv \pm \mu_0. \tag{32.17}$$

The quantity μ_0 is termed the *Bohr magneton.* This is a natural unit of magnetic moment.

The ratio $\frac{(\mu_\sigma)_z}{\sigma_z} = \frac{e}{mc}$ is called the *spin gyromagnetic ratio.* It was first discovered in determining the mechanical moment caused by magnetization of iron rods (the Einstein-de Hass experiment). Spin was not known at that time and it appeared strange that the gyromagnetic ratio was not equal to $\frac{e}{2mc}$ as follows from (15.25). It is now known that magnetism in iron is connected with the spin of certain of its electrons.

The fine structure of atomic levels. Spin magnetic moment interacts with the magnetic moment of orbital motion and with the spin angular momenta or other electrons, if the atom is of the multi-electron type. This interaction is proportional to the magnitude of both magnetic moments, i.e., it involves the product of gyromagnetic ratios. The latter is inversely proportional to c^2 and, hence, is an essentially relativistic effect.

Electron velocities in atoms are everywhere small compared with the velocity of light, with the exception of the internal regions of the atoms of heavy elements. Therefore, a quantity involving c^2 in the denominator is usually small compared with other quantities on the atomic scale; the interaction energy for magnetic moments is less than the distance—due to electrostatic interaction—between energy levels. As a result of the interaction between spin and orbital angular momenta, the energy level of a separate electron corresponding to a total electron angular momentum $j = l + \frac{1}{2}$ differs a little (when placed in a central field) from a level with a total angular momentum $j = l - \frac{1}{2}$*; this is because the angular momenta are parallel in the first case and antiparallel in the second. But the energies of two parallel and antiparallel angular momenta differ.

The only scalar quantity which is linear with respect to each of two pseudovectors μ_1 and μ_2 is $\mu_1 \mu_2$. Therefore, to the lowest approximation, the interaction energy of two magnetic moments is proportional to $\mu_1 \mu_2$.

The spacing between the levels $j = l + \frac{1}{2}$ and $j = l - \frac{1}{2}$ is small compared with that between electron levels with different l. Therefore, a magnetic interaction contributes only a small splitting of the

---

* The angular momentum is determined by means of its greatest projection.

electron level with a given l into two levels. This splitting is called the fine structure of the level.

We note that such a simple splitting into two levels takes place for a separate electron in a central field, for example, for the outer electron in an alkali-metal atom.

Isotopic (isobaric) spin. The splitting of an atomic level into two levels with $j=l+1/2$ and $j=l-1/2$ is due to weak magnetic interactions between spin and orbital magnetic moments. Since each magnetic moment contains c in the denominator [see (32.16)], such interaction is relativistic in nature and must vanish if the electrostatic forces alone are taken into account. This means that the energies of two states with spins parallel and antiparallel to the magnetic moment coincide if magnetic forces are completely neglected.

An analogous situation exists in the domain of nuclear interaction. The nuclear forces which hold nuclear particles (neutrons and protons) together are not of electromagnetic origin. At least we have no indications that both types of force—nuclear and electromagnetic—can be deduced in a unique manner from some first principle. As yet, no experiment suggests that such derivation is at all possible. On the contrary, there are many facts proving that nuclear interactions are independent of the electrical properties of particles.

First, we have the so-called mirror nuclei. These are pairs of nuclei which have all the neutrons interchanged with all the protons, and vice versa. For example, H^3 consists of one proton and two neutrons, and He^3, of two protons and one neutron. All the main properties of such nuclei are similar both qualitatively and quantitatively, and the small differences that still do exist can readily be explained by the difference in charge and magnetic moment of the neutron and proton. We can therefore say that the substitution, in a nucleus, of all protons by neutrons and all neutrons by protons leaves the nuclear interactions invariant, i.e., the nuclear interaction between two protons and two neutrons is the same if we neglect electromagnetic forces.

Second, the scattering of neutrons and protons on protons indicates that the elementary nuclear interactions neutron-proton and proton-proton are also the same. This is a stronger statement than the previous one, because the interaction between two unlike particles is also taken into account.

Comparing this situation with that in the atom, we can say that there is no splitting of nuclear states if the strongest interactions alone are considered; the actual splitting is due to the much weaker electromagnetic interactions.

Let us, therefore, neglect for a time the weakest interactions. We can then consider the neutron and the proton as two states of a single particle—the nucleon. These states do not differ in energy like those of an electron with two different spin projections in the absence of a magnetic field. If such a field is switched off, both states of the electron fall together in energy; if all the electromagnetic interactions are switched off, certain states of nucleon pairs fall together, too.

We have said that the spin can be considered as an internal degree of freedom of the electron. It is reasonable to say that the electric charge is the internal degree of freedom in the nucleon. Both degrees of freedom assume only two values with a dichotomic variable corresponding to them. It will be shown that there exists a far-reaching formal analogy between these degrees of freedom. Let us say that the nucleon possesses (besides its usual, nucleon, spin) another "spin" variable, which defines its "charge state." Like mechanical spin, this variable assumes only two values. It is called the isotopic spin (sometimes, and more consistently, the isobaric spin). We shall say that the projection on some imaginable axis of isotopic spin is equal to $+1/2$ which corresponds to a proton, the opposite projection corresponding to a neutron. Some years ago, the reverse convention was used, but this is immaterial. Now let us consider three nucleon pairs: proton-proton, neutron-proton, and neutron-neutron. According to what we have already said, the first pair corresponds to a resultant projection 1 of the isotopic spin, the second pair to a projection 0, and the third, to -1. In the absence of electromagnetic forces, none of the three projections split in energy.

But if these states coincide in energy they can be considered as having a resultant spin 1 and differing in their projections only. Spin angular momentum 1 can assume just three projections; and here we can say that the resultant isotopic spin angular momentum 1 has three different projections on some imaginable z-axis. In the absence of electromagnetic forces, the physical choice of such a "z-axis" is unimportant. Note, for comparison, that if no magnetic field is applied to the electron, any direction in space is preferred (for example, the z-axis).

Changes in projections of ordinary spin can be due simply to rotations of the coordinate frame. If no preferred directions in space exist, such rotations are unlimited. Now we can consider different isotopic spin projection as due to "rotation" of some frame also. But this rotation is purely formal in nature and has nothing in common with the rotation of geometrical space, except their mathematical expressions. If the isotopic spin vector $\hat{\vec{\tau}}$ with components $\hat{\tau}_x$, $\hat{\tau}_y$, $\hat{\tau}_z$ is introduced, then its rotations are described exactly by the same formulae as (32.12). The corresponding angle of rotation has no more geometrical meaning than the axes which rotate.

The formulae for isotopic spin rotation are deducible from the dichotomic nature of that variable and from the similarity of three different two-nucleon states, so there is no reason to abolish the vivid geometrical terminology of "projections" and "rotations."

Let us now formulate the situation in a quantum-mechanical fashion. For several nucleons, it is possible to define their resulting isotopic spin operator.

$$\hat{\vec{\tau}} = \sum_i \hat{\vec{\tau}}_i \qquad (32.18)$$

Its different components do not commute. But its square, $\hat{\tau}^2$, commutes with one projection, say $\hat{\tau}_z$, which defines the resulting charge of the given system. The Hamiltonian of the nuclear interactions (with electromagnetic interactions neglected) commutes with both $\hat{\tau}^2$ and $\hat{\tau}_z$, just like the Hamiltonian for an electron in a central field commutes (to a nonrelativistic approximation) with $\hat{\mu}^2$ and μ_z. It follows that τ^2 and τ_z exist in nuclear states with a given energy. In other words, nuclear states can be distinguished by their τ^2 and τ_z values. This ascribing of τ^2, τ_z to nuclear levels is approximate, like the distinguishing of atomic levels by n, l, k, the difference being that no account is taken of the magnetic properties of spin.

In heavy nuclei, electrostatic interactions are very important because they increase in proportion to the square of the atomic number. Nuclear interactions increase linearly with the number of nucleons, as the mass defect of nuclei does. So in heavy nuclei both types of interaction are of an equal order of magnitude and the neglect of electrical interaction has no meaning. No definite isotopic spin values can be attributed even approximately to the levels of heavy nuclei.

The isotopic spin variables are very important in the classifying of elementary particles.

Exercises

1) Write down the transformation of $\hat{\sigma}_x$, $\hat{\sigma}_y$, $\hat{\sigma}_z$ for any arbitrary rotation in space and prove that the properties of the operators do not change.

The general expression for the transformation of vector component is the following (see Sec. 9):

$$\hat{\sigma}_i' = \alpha_{ik}\hat{\sigma}_k\,; \quad \alpha_{ik} = \cos\left(\angle\, x'_i x_k\right),$$

where the coefficients α_{ik} satisfy the conditions

$$\alpha_{in}\alpha_{nk} = \begin{cases} 0 \text{ for } i \neq k\,, \\ 1 \text{ for } i = k\,, \end{cases}$$

where, according to the summation convention, n runs from 1 to 3.

2) Find the eigenvalues of the scalar product $\hat{\sigma}_1\hat{\sigma}_2$ of two electrons with parallel and antiparallel spins.

We begin with the equation

$$(\hat{\boldsymbol{\sigma}}_1 + \hat{\boldsymbol{\sigma}}_2)^2 = \hat{\sigma}_1^2 + \hat{\sigma}_2^2 + 2\hat{\boldsymbol{\sigma}}_1\hat{\boldsymbol{\sigma}}_2\,.$$

But $\hat{\sigma}_1^2 = \hat{\sigma}_2^2 = \hat{\sigma}_x^2 + \hat{\sigma}_y^2 + \hat{\sigma}_z^2 = \dfrac{3}{4}$ [see (32.11)]; the maximum projection of $\sigma_1 + \sigma_2$ is equal to zero for antiparallel spins, and is equal to unity for parallel spins. From this, $(\sigma_1+\sigma_2)^2=0$ in the first case and is equal to $1\cdot 2=2$ in the second case. This gives

$$\boldsymbol{\sigma}_1\boldsymbol{\sigma}_2=\frac{0-\frac{3}{2}}{2}=-\frac{3}{4}\ \text{(antiparallel spins)},$$

$$\boldsymbol{\sigma}_1\boldsymbol{\sigma}_2=\frac{2-\frac{3}{2}}{2}=\frac{1}{4}\ \text{(parallel spins)}.$$

3) Write down the eigenfunctions for $\hat{\sigma}_x$, $\hat{\sigma}_y$, and $\hat{\sigma}_z$.

$$\text{for } \sigma_x=\frac{1}{2},\ \psi=\begin{pmatrix}1\\1\end{pmatrix}, \qquad \text{for } \sigma_x=-\frac{1}{2},\ \psi=\begin{pmatrix}1\\-1\end{pmatrix}.$$

$$\text{for } \sigma_y=\frac{1}{2},\ \psi=\begin{pmatrix}1\\i\end{pmatrix}, \qquad \text{for } \sigma_y=-\frac{1}{2},\ \psi=\begin{pmatrix}1\\-i\end{pmatrix}.$$

$$\text{for } \sigma_z=\frac{1}{2},\ \psi=\begin{pmatrix}1\\0\end{pmatrix}, \qquad \text{for } \sigma_z=-\frac{1}{2},\ \psi=\begin{pmatrix}0\\1\end{pmatrix}.$$

Thus, the eigenfunctions of all three noncommutative operators differ.

4) Express the scalar product $\mathbf{j}_1, \mathbf{j}_2$ in terms of the resultant angular momentum j, j_1, and j_2.

By definition $j^2=j_1^2+j_2^2+2\mathbf{j}_1\mathbf{j}_2$. Substituting here the squares of the angular momenta, we obtain

$$\mathbf{j}_1\mathbf{j}_2=\frac{1}{2}\{j(j+1)-j_1(j_1+1)-j_2(j_2+1)\}.$$

Sec. 33. Many-Electron Systems

The Mendeleyev periodic law. Long before the atom became an object of physical study its properties were investigated in chemistry. And chemistry discovered and studied such properties of the atom as were utterly alien to prequantum physics. In this category belongs, first of all, valency or the chemical affinity of atoms. On the basis of a vast quantity of experimental material accumulated in chemistry, Mendeleyev constructed a generalizing and systematic periodic law. This was a new law that allowed Mendeleyev to predict the existence of many elements, which were discovered later. And what is more, the basic chemical and many physical properties of these elements were correctly predicted. At the present time, too, the Mendeleyev law guides scientific investigation into the study of the periodic structure of nuclear shells.*

The Pauli principle. The wave equation for a single particle is inadequate for an explanation of Mendeleyev's periodic law. It is ne-

* We have in mind the quantum-mechanical theory of nuclear shells, and not the rather widespread speculative constructions which are mainly based on the arithmetic relationships between the atomic numbers and atomic weights of the elements.

cessary to introduce a new principle concerning many-electron systems—the so-called Pauli principle. We shall first of all formulate it in such a way that it can be conveniently used to investigate the electron shells of an atom, to wit, an atom cannot have more than one electron with a given group of four quantum numbers: the principal quantum number n, the azimuthal quantum number l, the magnetic quantum number k_l and the spin quantum number k_σ. The spin quantum number is a measure of the spin projection onto the same axis onto which the orbital angular momentum is projected.

The Pauli principle is substantiated by relativistic quantum mechanics (see Sec. 38). Here we shall simply use it as a supplementary principle of quantum mechanics.

The addition of angular momenta of two electrons with identical n and l. We shall first of all show how the Pauli principle is applied in adding the angular momenta of two electrons for which the principal and azimuthal quantum numbers are the same. The accepted practice is to say that these electrons belong to the same shell. Usually (though not always) electrons in different shells possess quite different energies—a fact which justifies the classification by shells.

Let us take the simplest case when $n=1$. Then, in accordance with the definition of n (31.33), $l=0$. But, for l equal to zero, the magnetic quantum number k_l is also equal to zero. Hence, three quantum numbers are the same for the electrons and, according to the Pauli principle, the fourth number k_σ must differ. However, k_σ can only have two values, $+\frac{1}{2}$ and $-\frac{1}{2}$, and each of its values can only have one electron for given n, l, and k_l. Thus, an atom can have only two electrons with $n=1$. Their spins are antiparallel and therefore the resultant spin S is equal to zero. The resultant orbital angular momentum L is also equal to zero.

Let us now take two electrons in the p-state, i.e., with $l=1$ and with the same principal quantum numbers. Either the magnetic or the spin quantum numbers, or both, must differ. The p-electron can be in six states, which we list writing the magnetic quantum number first and the spin projection second:

$$A:1,\frac{1}{2}\,;\ B:0,\frac{1}{2}\,;\ C:-1,\frac{1}{2}\,;\ D:1,-\frac{1}{2};\ E:0,-\frac{1}{2}\,;\ F:-1,-\frac{1}{2}\,.$$

It follows that two electrons can occupy any two different states of the six. As is known, the number of combinations of six things, two at a time, is equal to $C_6^2=\frac{6\times 5}{2}=15$. These fifteen states differ by the total orbital angular momentum L and the total spin S, as well as by their projections. The latter depend upon the choice of coordinate axes and will interest us only insofar as they characterize the relative directions of **L** and **S**.

We shall first of all find those states which correspond to the greatest projections of **L** and **S**, because they determine the possible eigenvalues of L and S. In any case, of the fifteen states only those must be taken, for which the total spin projections and orbital angular momentum are non-negative, since it is obvious that negative projections cannot be greatest in relative value. The states with positive projections number eight out of fifteen, and from these eight we take only those which possess the greatest projections. We rewrite all eight states:

AD: 2, 0; BD: 1, 0; CD: 0, 0; AB: 1, 1; AC: 0, 1; AE: 1, 0; BE: 0, 0; AF: 0, 0.

The state with maximum orbital angular-momentum projection is AD. Hence, a state exists for which the orbital angular momentum is equal to two and the spin angular momentum is zero (here, and in future, the angular momentum is characterized by the greatest projection). The indicated state also yields projections 1, 0 and 0, 0. Such states are, for example, the BD and CD states; therefore they need not be considered, since they do not define the vector sum. The AB state has the maximum spin angular momentum. It follows that a state exists for which the orbital and spin angular momenta are equal to unity. Their projection can be 1, 0; 0, 1; and 0, 0. These are AC, AE, and BE, which, like BD and CD, no longer interest us. There remains one more state with projections 0, 0.

Thus, only three states are possible:

$$L=2,\ S=0;\ L=1,\ S=1;\ L=0,\ S=0.$$

Composition of angular momenta for three electrons with identical n and l. For three p-electrons we obtain the following seven states with positive projections:

$$ABC: 0, \tfrac{3}{2};\ ACE: 0, \tfrac{1}{2};\ ABD: 2, \tfrac{1}{2};\ ABE: 1, \tfrac{1}{2};\ ABF: 0, \tfrac{1}{2};$$

$$ACD: 1, \tfrac{1}{2};\ ABE: 0, \tfrac{1}{2}.$$

The maximum spin projection is 3/2 for a zero orbital angular-momentum projection. The maximum orbital angular-momentum projection is 2 with a spin projection $\frac{1}{2}$. These two states, together with their projections, are listed in order from ABC to ABF. ACD and ABE remain, to which there correspond $L=1$ and $S=\frac{1}{2}$. In all, we have $L=0$, $S=3/2$; $L=2$, $S=\frac{1}{2}$; $L=1$, $S=\frac{1}{2}$.

Normal coupling. States with different values of the total orbital and spin angular momenta L and S, and with the same principal quantum numbers for the electrons, differ in energy. This difference occurs as a result of the electrostatic, and not magnetic, interaction between electrons. In order to explain why the resultant orbital angular momentum affects the interaction energy, we examine two p-electrons. The sum of their orbital angular momenta can yield two, unity or zero. If two is obtained, then the angular dependence for the wave functions of both electrons is the same (not only do the azimuthal quantum numbers coincide, but also the magnetic quantum numbers, that is why at least the spin projections must differ). We shall call the wave functions of both electrons $\psi_{1,1}(\mathbf{r}_1)$ and $\psi_{1,1}(\mathbf{r}_2)$, where $\mathbf{r}_1$ and $\mathbf{r}_2$ are the radius vectors of both electrons and the indices refer to the quantum numbers l and k_1.

The interaction energy between electrons is approximately

$$e^2\int \frac{|\psi_{1,1}(\mathbf{r}_1)|^2 |\psi_{1,1}(\mathbf{r}_2)|^2}{|\mathbf{r}_1 - \mathbf{r}_2|} dV_1 dV_2,$$

because $e|\psi_{1,1}(\mathbf{r}_1)|^2$ and $e|\psi_{1,1}(\mathbf{r}_2)|^2$ represent the densities of the charge distribution. The approximation consists in the fact that the effect of the interaction on the wave functions and the so-called "exchange" [see (33.32)] has not been taken into account.

If the resultant moment is unity, we correspondingly obtain the other estimation:

$$e^2\int \frac{|\psi_{1,1}(\mathbf{r}_1)|^2 |\psi_{1,0}(\mathbf{r}_2)|^2}{|\mathbf{r}_1 - \mathbf{r}_2|} dV_1 dV_2.$$

Here the magnetic quantum numbers are equal to zero or unity.

This integral is clearly different from the previous one. Thus, there appears to be interaction between the orbital angular momenta when they are to form the total angular momentum; this interaction does not involve c^2 in the denominator, i.e., it is electrostatic in character.

In multi-electron atoms, Pauli's principle imposes definite conditions on the choice of spatial wave functions for given spins. As an example, let us consider the state with spin 3/2, which, as was just established, is possible in a system with three p-electrons. In accordance with the Pauli principle, three different spatial wave functions for the separate electrons having $k_l = 1$, 0 and -1 correspond to this state. The corresponding electron densities coincide in space less than, for example, in a state with magnetic quantum numbers 1, 1, 0, to which, according to the Pauli principle, there must correspond spin projections $\frac{1}{2}, -\frac{1}{2}, \frac{1}{2}$ in order that all three pairs of k_l, k_σ should be different. But the less the electron wave functions coincide in space, the less the Coulomb repulsion energy between the electrons, because the mean distance between like charges is greater. For this reason, the state to which the

Pauli principle assigns the greatest possible spin possesses the least repulsion energy.

There are three p-electrons in the ground state of nitrogen. As was just indicated, the state with the least energy occurs when all three spins are parallel. The next state, for which the orbital angular momentum is equal to 2 and the spin is equal to $\frac{1}{2}$, lies approximately 2.2 ev higher, while the state with orbital angular momentum 1 and spin $\frac{1}{2}$ lies 3.8 ev higher.

We can explain why a lesser energy corresponds to a greater resultant orbital angular momentum. Wave functions, for which the orbital angular-momentum projections differ only in sign, are closer to each other than functions for which the angular-momentum projections differ in absolute value. But those functions which correspond to a closer spatial electron density distribution lead to a larger repulsion energy, while angular momenta in opposite directions, when summed, yield a lesser resultant angular momentum than the angular momenta whose projections differ in magnitude also. Thus, the state with the greatest spin possesses the least energy and, for a given spin, it is the state with the greatest orbital angular momentum that has the least energy (Hund's first rule).

This is the way the orbital and spin angular momenta are combined. In calculating the electrostatic energy only, the state of the atom is defined by the absolute values of L and S. But a magnetic interaction takes place between the resultant orbital angular momentum and the resultant spin angular momentum of a system of electrons, analogous to that of a separate electron (see Sec. 32). To a first approximation, this interaction is described by the scalar product $A\boldsymbol{\mu}_L\,\boldsymbol{\mu}_s$, where $\boldsymbol{\mu}_L$ and $\boldsymbol{\mu}_s$ are the magnetic moments for the orbital and spin motion of the system, and A is a factor of proportionality.

The scalar product of two angular momenta assumes as many values as are possessed by the resultant angular momentum for a given absolute value of the component angular momenta (see exercise 4, Sec. 32). This is clearly shown with the aid of a so-called vector model: a triangle is constructed on the vectors $\mathbf{L}$, $\mathbf{S}$ and $\mathbf{J}=\mathbf{L}+\mathbf{S}$. In accordance with the law of composition of angular momenta (30.30), the side J can equal $L+S$, $L+S-1$, $\ldots$, $|L-S|$. The energy level of an atom with given values of L and S is split into as many fine structure levels as can be assumed by J, i.e., $2S+1$ levels if S is less than L, and $2L+1$ levels if L is less than S.

The system of levels described here occurs with the so-called Russel-Saunders normal coupling (of orbital and spin angular momenta): the energy states with different L and S differ considerably more than the energy states with given L and S but different J. The group of energy levels differing only in total angular momentum J is termed a *multiplet*.

In heavy elements, where the spin-orbital interaction for separate electrons is great, the spin of each electron in a shell is combined with its orbital angular momentum to form a resultant angular momentum j [see (32.15)]; only then do the angular momenta j of separate electrons combine. This may be accounted for by the fact that, the relativistic effect of magnetic-moment interactions is not small compared with the energy of electrostatic repulsion between electrons in the inner regions of the atoms of heavy elements, where the electron velocity is close to the velocity of light. The type of coupling which occurs when the j of separate electrons are added, is termed j-j coupling. j-j coupling also occurs between nuclear particles as a result of the large spin-orbital interaction characteristic of nuclear forces.

The spectroscopic notation for levels. In general form, the spectroscopic notation for the resultant state of an atom is written thus:

$$^{2S+1}L_J^{g,u}.$$

The main symbol is L, i.e., the letters S, P, D, F, etc., depending upon what L is equal to: 0, 1, 2, 3, ... As a left superscript we put $2S+1$. As a right subscript we put J, i.e., the vector sum of L and S from the number of the fine-structure components. Finally, the right superscripts denote an odd (u) or even (g) state, respectively.

For example, the ground state of a nitrogen atom has $L=0$, $S=3/2$ and is formed by three p-electrons. Hence, its spectroscopic designation is $^4S^u_{3/2}$ because the total angular momentum can only equal the spin angular momentum ($L=0$), and $\sum l = 3$ is an odd number.

The notations for the next two states of nitrogen are

$$^2D^u \text{ and } ^2P^u,$$

or, if the multiplet splitting is taken into account, then

$$^2D^u_{5/2} \text{ or } ^2D^u_{3/2} \text{ and } ^2P^u_{3/2} \text{ or } ^2P^u_{1/2},$$

depending upon the resultant angular momentum J.

If the ground state of the atom has L and S not equal to zero, the resultant angular momentum is determined by Hund's second (empirical) rule: when there are less than half the possible number of electrons in a shell, the least energy corresponds to a multiplet level for which $J=|L-S|$, and to that of $J=L+S$ when there is more than half the possible number. Since the electron angular-momentum l can have $2l+1$ projections, and there are two values k_σ for each projection k_l, then there can be in all $2(2l+1)$ electrons in a shell with given values of l and n. The total number of electrons in an atom with a given principal quantum number n is

$$\sum_{l=0}^{n-1} 2(2l+1) = 2n^2. \tag{33.1}$$

The electron configuration corresponding to the least energy occurs in the ground state. It is determined by Hund's first and second rules.

The dependence of energy on the azimuthal quantum number. Before we can go over to a description of the Mendeleyev periodic system we have to remark on the dependence of the energy of an electron on the azimuthal quantum number. The energy of an electron in all atoms, except the hydrogen atom, depends upon l as well as upon n. For large l the electron is situated comparatively far away from the nucleus; in other words, it is more weakly bound to the nucleus than for small l. For a given n, the energy of an electron is greater, the larger l. When the field greatly differs from a Coulomb field, the dependence of energy upon l is so strong that an increase in the principal quantum number n, with a simultaneous decrease in l, leads to a smaller energy increase than the increase of l for a given n. In other words, the state with quantum numbers $n+1$, 0 can have a lower energy than the state with quantum numbers n, l. This will become clear in the later examples.

Filling the first shells. As was mentioned, the shell with $n=1$ is filled by two electrons in the $1s$-state (the 1 in front denotes the quantity n). Hydrogen has one electron in this shell and helium has two. The helium shell is completely filled and has a $^1S_0^g$ state. The electron configuration for the ground state of a helium atom is so stable that if any other atom approaches close to it the total energy can only increase, so that repulsion forces are produced. Helium is completely inert chemically. The forces between helium atoms are small as a result of the symmetry and stability of their electron shells. Therefore, helium gas is liquified at an extremely low temperature.*

After helium, the shell structure with $n=2$ begins. The first electron of this shell, i.e., a $2s$-electron, appears in lithium. The two inner $1s$-electrons occurring in the helium configuration strongly screen the nuclear charge and, consequently, the outer electron is weakly

* The condensation of helium into a liquid at low temperatures is due to the so-called Van der Waals forces, which arise out of the mutual electrostatic polarization of approaching atoms. These forces act at larger distances than the forces of chemical affinity, and are very small compared with them.

bound. Such is the alkali-metal electron configuration in the case of lithium, and analogous electron configurations subsequently result each time (Na, K, Rb, Cs) from the addition of an *s*-electron to a nucleus surrounded by a noble-gas electronic cloud. The next 2 *s*-electron has an energy which is comparatively close to the energy of a 2 *p*-electron: the energy of the electron is still weakly dependent on the azimuthal quantum number since the field is approximately Coulomb. A large energy is needed for an electron to go from a 1 *s*-shell to a 2 *s*- or a 2 *p*-shell, while a small energy is needed for the transition from a 2 *s*- to a 2 *p*-shell. For this reason, the beryllium electronic configuration, having two 2 *s*-electrons, is not very stable with respect to an electron transition to the 2 *p*-shell. In other words, filling the 2 *s*-shell does not give the electron configuration of a noble gas. Indeed, as we know, beryllium is a metal.

After beryllium, the 2 *p*-shell fills up, and is completely filled for the noble gas neon. Neon follows fluorine, which requires one electron for the shell to be filled. The energy required for an electron to be added to the fluorine 2 *p*-shell, to fill the shell of neon, is large. This explains the chemical activity of fluorine and the other halogens, which are similarly situated with respect to the noble gases.

There can be eight electrons in a shell for which $n=2$. This is the first group of the Mendeleyev system. The shell with $n=3$ is then filled, though initially only the first two subshells: 3 *s* and 3 *p*. The elements of the second group have an outer electron-shell structure similar to the elements of the first group. The chemical properties of atoms are basically determined by the outer shells. This explains the similarity of chemical properties, on the basis of which Mendeleyev formulated his law. Argon has a filled shell, i.e., still another group of eight elements is completed. The noble-gas configuration is obtained for argon because the 3 *p*-state, on the one hand, and the 3 *d* and 4 *s* states, on the other hand, differ considerably in energy.

By considering the possible states of shells which, to be filled, lack less than half the possible number of electrons, we can consider that unfilled states behave like electrons. For example, if there are two of the six electrons wanting in a 2 *p*-shell, then we can combine the states of the two "holes," similar to the way that the states of two 2 *p*-electrons were combined at the beginning of this section. In doing so, correct results are always obtained, provided that Hund's second rule is used in finding the total angular momentum J of the ground state, i.e., that we take $J=L+S$. It is easy to see that four electrons in the shell are equivalent to two holes by applying the Pauli principle first to an electron and then to a "hole," (see exercise 2).

Let us now give, in one table, the scheme for building up the first eighteen places in the periodic system of elements; this table shows the number of electrons having given quantum numbers.

Element	$n = 1$, $l = 0$	$n = 2$, $l = 0$	$n = 2$, $l = 1$	$n = 3$, $l = 0$	$n = 3$, $l = 1$	Ground state
H	1					$^2S^g_{1/2}$
He	2					$^1S^g_0$
Li	2	1				$^2S^g_{1/2}$
Be	2	2				$^1S^g_0$
B	2	2	1			$^2P^u_{1/2}$
C	2	2	2			$^3P^g_0$
N	2	2	3			$^4S^u_{3/2}$
O	2	2	4			$^3P^g_2$
F	2	2	5			$^2P^u_{3/2}$
Ne	2	2	6			$^1S^g_0$
Na	2	2	6	1		$^2S^g_{1/2}$
Mg	2	2	6	2		$^1S^g_0$
Al	2	2	6	2	1	$^2P^u_{1/2}$
Si	2	2	6	2	2	$^3P^g_0$
P	2	2	6	2	3	$^4S^u_{3/2}$
S	2	2	6	2	4	$^3P^g_2$
Cl	2	2	6	2	5	$^2P^u_{3/2}$
Ar	2	2	6	2	6	$^1S^g_0$

The filling order after the 3*p*-shell. After argon, the 4*s*-shell begins to fill instead of the 3*d*-shell. The new group begins with the alkali metal, potassium. The sum $n+l$ is the same for the 3*p*- and 4*s*-shells and is equal to 4, while it is already greater by unity in the 3*d*-shell. The 4*p*-shell is filled after the 3*d*-shell, with the same value of the sum $n+l=5$, and then the 5*s*-shell. It is seen that this rule is observed later on, too; the filling of the shells with the same sum $n+l$ proceeds in order of increasing n. But there are certain deviations from this rule during the filling of the *d*- and *f*-shells.

In the shells with $n=1, 2, 3$, there are altogether $2.1^2+2.2^2+ +2.3^2=2+8+18 = 28$ electrons. There are a further eight electrons in the 4*s*- and 4*p*-states, and another two electrons in the 5*s*-state. The 5*s*-state is followed by electrons with $n+l=6$, where we begin with the least n, i.e., with 4*d*. There are $2\ (4+1)=10$ more of these electrons. The 4*d*-electrons are followed by 5*p*-electrons, of which there are six, and then by the same simple rule we get the 6*s*-state.

Rare-earth elements. The next value of $n+l=7$, the least being $n=4$. Hence, beginning with the 57th place (in actuality, with the

58th place) the 4*f*-shell can begin to fill acquiring at once two 4*f*-electrons. This shell is already inside the atom as a result of the form of the potential distribution within the atom. The screening of the nuclear charge by atomic electrons leads, at large distances away from the nucleus, to the potential decreasing like $\frac{1}{r^4}$ instead of $\frac{1}{r}$ (see Sec. 44). If we combine the potential energy of an electron, calculated with allowance for screening with the centrifugal energy, it turns out that the *d*- and *f*-states possess a minimum resultant effective potential energy deep inside the atom (see Sec. 5, footnote to p. 45). Indeed, the centrifugal energy is greater than the potential energy both close to the nucleus (with allowance made for screening) as well as far away from the nucleus. Therefore, the effective potential energy U_M is positive for large r as well as for small r. In other words, for the *d*- and *f*-states, the U_M curve goes *higher* for large r than for the *s*- and *p*-states, and it turns out that the effective potential well for *d*- and *f*-electrons is situated *closer* to the nucleus than to the boundaries of the *s*- and *p*-electron shells. Thus, the *d*- and *f*-shells are, as it were, filled inside the atom. But the chemical properties of atoms depend mainly upon the outer electrons which, in filling the 4*f*-shell, change very little. This is how the group of $2(2\cdot 3+1)=14$ chemically similar elements, termed *rare-earth* elements, originates.

It should be pointed out that the *d*- and *f*-shells are not filled successively as a result of "competition" with outer shells: for example, there are three *d*-electrons and two *s*-electrons in V_{23}, five *d*-electrons and one *s*-electron in the next element Cr_{24}, while Mn_{25} also has five *d*-electrons but two *s*-electrons.

The statistical theory of the atom, which will be set out in Sec. 44, permits us, in rough outline, to find the potential distribution inside an atom. It becomes possible, from this distribution, to predict rather accurately the places in the periodic system where elements with $l=2$ and 3 appear.

The 5*f*-shell fills up (beginning with thorium) in a whole group of elements similar to the rare earths. A large part of this group consists of the artificially produced *transuranium* elements.

The wave equation of a two-electron system. We shall now formulate Pauli's principle using wave functions. The simplest way to do this is to consider a two-electron system. The wave equation for two electrons may be written thus:

$$\hat{\mathscr{H}}\Phi = \left[-\frac{\hbar^2}{2m}\Delta_1 - \frac{\hbar^2}{2m}\Delta_2 = U(\mathbf{r}_1, \mathbf{r}_2)\right]\Phi = \mathscr{E}\Phi. \qquad (33.2)$$

Here Δ_1 and Δ_2 are the Laplacian operators with respect to variables of the first and second electrons, $U(\mathbf{r}_1, \mathbf{r}_2)$ is the potential energy of their interaction with the external field and with each other. For example, in a helium atom

$$U(\mathbf{r}_1, \mathbf{r}_2) = -\frac{2e^2}{r_1} - \frac{2e^2}{r_2} + \frac{e^2}{|\mathbf{r}_1 - \mathbf{r}_2|}. \tag{33.3}$$

The wave function depends upon the spatial and spin variables of both particles:

$$\Phi = \Phi(\mathbf{r}_1, \sigma_1; \mathbf{r}_2, \sigma_2). \tag{33.4}$$

The interaction of spin magnetic moment with orbital motion is weak. Therefore, to a first approximation, the spin-orbital interaction can be neglected in the potential energy operator. This corresponds to $U(\mathbf{r}_1, \mathbf{r}_2)$ in equation (33.3). If the effect of spin motion upon orbital motion is small then the probability of a certain value of spin and coordinate is equal to the product of the probabilities of both values, and the probability amplitude Φ also divides into a product of amplitudes

$$\Phi(\mathbf{r}_1, \sigma_1; \mathbf{r}_2, \sigma_2) = \Psi(\mathbf{r}_1, \mathbf{r}_2)\,\chi(\sigma_1, \sigma_2). \tag{33.5}$$

The probability amplitude of orbital motion satisfies equation (33.2), provided it does not involve the spin operator. But even when the system is placed in an external homogeneous magnetic field $\mathbf{H}$, the following operator is added to $\hat{\mathscr{H}}$

$$\hat{U}_{mag} = \frac{e}{mc}[(\hat{\boldsymbol{\sigma}}_1 \mathbf{H}) + (\hat{\boldsymbol{\sigma}}_2 \mathbf{H})] = \frac{eH}{mc}(\hat{\sigma}_{z1} + \hat{\sigma}_{z2}), \tag{33.6}$$

where it is taken that the z-axis coincides with the direction of the field (the minus sign is replaced by a plus sign because the electronic charge is negative). The action of the operator $\hat{\sigma}_{z1} + \hat{\sigma}_{z2}$ on the spin function simply gives the total spin projection of the system. For this reason, in the presence of an external homogeneous magnetic field, $\hat{U}_{\text{mag}}$ is replaced by a number which is added to the total energy of the system.

The symmetry of the $\hat{\mathscr{H}}$ operator with respect to particle interchange. When examining the $\hat{\mathscr{H}}$ operator in (33.2) we see that it is completely symmetrical with respect to a coordinate interchange for both particles, i.e., it does not change its form if the first electron is called the second, and the second electron the first:

$$\hat{\mathscr{H}}(\mathbf{r}_1, \boldsymbol{\sigma}_1; \mathbf{r}_2, \boldsymbol{\sigma}_2) = \hat{\mathscr{H}}(\mathbf{r}_2, \boldsymbol{\sigma}_2; \mathbf{r}_1, \boldsymbol{\sigma}_1). \tag{33.7}$$

But equation (33.2) is linear. Therefore, if the form does not change due to operation (33.7), then the wave function can only be multiplied by some constant number P:

$$\Phi(\mathbf{r}_1, \sigma_1; \mathbf{r}_2, \sigma_2) = P\Phi(\mathbf{r}_2, \sigma_2; \mathbf{r}_1, \sigma_1). \tag{33.8}$$

Because $\mathbf{r}_1$, σ_1 and $\mathbf{r}_2$, σ_2 are involved in the same way in all the equations, we can interchange them in (33.8) obtaining

$$\Phi(\mathbf{r}_2, \sigma_2; \mathbf{r}_1, \sigma_1) = P\Phi(\mathbf{r}_1, \sigma_1; \mathbf{r}_2, \sigma_2). \tag{33.9}$$

Substituting (33.9) in (33.8), we shall have

$$\Phi(\mathbf{r}_2, \sigma_2; \mathbf{r}_1, \sigma_1) = P^2\Phi(\mathbf{r}_2, \sigma_2; \mathbf{r}_1, \sigma_1)$$

or

$$P^2 = 1, \quad P = \pm 1. \tag{33.10}$$

In this comparatively simple case, when there are only two particles, the transformation is similar to the symmetry transformation for a wave function under reflection [see (31.38)].

The commutative operator for coordinates and spin variables. We can define a coordinate commutative operator for electrons $\hat{P}_r$ such that

$$\hat{P}_r\Psi(\mathbf{r}_1, \mathbf{r}_2) = \Psi(\mathbf{r}_2, \mathbf{r}_1). \tag{33.11}$$

If the wave equation is symmetrical with respect to interchange of $\mathbf{r}_1$ and $\mathbf{r}_2$ (without interchanging the spin variables σ_1 and σ_2) then, by repeating the foregoing argument, we see that the eigenvalues of P_r for two electrons are equal to ± 1.

An analogous operator is also defined for spin

$$\hat{P}_\sigma\chi(\sigma_1, \sigma_2) = \chi(\sigma_2, \sigma_1), \tag{33.12}$$

where the eigenvalues of $\hat{P}_\sigma$ are likewise equal to ± 1.

We denote the set of orbital quantum numbers of the first electron by the letter n_1 (in place of n_1, l_1, k_{l_1}), and those of the second electron by the letter n_2. Then the orbital wave function Ψ is written in more detail as

$$\Psi = \Psi(n_1, \mathbf{r}_1; n_2, \mathbf{r}_2).$$

It follows from requirement (33.10) that

$$\hat{P}_r\Psi(n_1, \mathbf{r}_1; n_2, \mathbf{r}_2) = \Psi(n_1, \mathbf{r}_2; n_2, \mathbf{r}_1) = \pm\Psi(n_1, \mathbf{r}_1; n_2, \mathbf{r}_2). \tag{33.13}$$

The function in (33.13) with an upper sign is termed symmetric; with a lower sign, antisymmetric.

The wave function of a two-electron system. Introducing, in addition, the spin quantum numbers k_{σ_1} and k_{σ_2}, which determine the form of the spin wave functions (see exercise 3, Sec. 32), we write the total wave function of a two-electron system as

$$\Phi(n_1, k_{\sigma_1}, \mathbf{r}_1, \sigma_1; n_2, k_{\sigma_2}, \mathbf{r}_2, \sigma_2).$$

The total permutation of spin and spatial coordinates in this function occurs as a result of the action of the $\hat{P}$ operator, which is

$$\hat{P} = \hat{P}_r \hat{P}_\sigma . \tag{33.14}$$

Operating with (33.14) on the function Φ, we have

$$\hat{P}\Phi(n_1, k_{\sigma_1}, \mathbf{r}_1, \sigma_1; n_2, k_{\sigma_2}, \mathbf{r}_2, \sigma_2) = \Phi(n_1, k_{\sigma_1}, \mathbf{r}_2, \sigma_2; n_2, k_{\sigma_2}, \mathbf{r}_1, \sigma_1) . \tag{33.15}$$

According to (33.10) this function is also either symmetric or antisymmetric. But it can now be seen immediately that only the antisymmetric function satisfies the Pauli principle. Indeed, let the states of both electrons be identical, i.e., $n_1 = n_2$, $k_{\sigma_1} = k_{\sigma_2}$. Then, if the function Φ is antisymmetric, we obtain

$$\begin{aligned} \hat{P}\Phi(n_1, k_{\sigma_1}, \mathbf{r}_1, \sigma_1; n_1, k_{\sigma_1}, \mathbf{r}_2, \sigma_2) &= \Phi(n_1, k_{\sigma_1}, \mathbf{r}_2, \sigma_2; n_1, k_{\sigma_1}, \mathbf{r}_1, \sigma_1) = \\ &= -\Phi(n_1, k_{\sigma_1}, \mathbf{r}_1, \sigma_1; n_1, k_{\sigma_1}, \mathbf{r}_2, \sigma_2) = \\ &= \Phi(n_1, k_{\sigma_1}, \mathbf{r}_1, \sigma_1; n_1, k_{\sigma_1}, \mathbf{r}_2, \sigma_2) . \end{aligned} \tag{33.16}$$

By definition, the operator P interchanges only the variables $\mathbf{r}$ and σ, and by no means the quantum numbers n, k_σ. The first equation of (33.16) denotes the result of a $\hat{P}$ operation, the second takes into account the antisymmetry of the wave function, while the third is obtained from the first by permutation of all four arguments relating to the electrons. The possibility of such a permutation for any function is obvious, since it does not matter which particle is considered first and which second when writing down the wave function; the interchange of the four values n_1, k_{σ_1}, $\mathbf{r}_1$ σ_1 and n_1, k_{σ_1}, $\mathbf{r}_2$, σ_2 in the last equality of (33.16) simply does not denote anything: it is immaterial which arguments are written first—those relating to the first electron, i.e., n_1, k_{σ_1}, $\mathbf{r}_1$, σ_1, or those relating to the second electron n_1, k_{σ_1}, $\mathbf{r}_2$, σ_2. Hence, the function $\Phi(n_1, k_{\sigma_1}, \mathbf{r}_1, \sigma_1; n_1, k_{\sigma_1}, \mathbf{r}_2, \sigma_2)$ is equal to itself with the sign reversed, i.e., it becomes zero identically.

This property is possessed only by an antisymmetric function and not by a symmetric function; the latter would become identically equal to itself. But if the antisymmetric wave function of two electrons occurring in identical states is identically equal to zero, the probability amplitude of this state of a system of two electrons is equal to zero for any values of the variables $\mathbf{r}_1$, $\mathbf{r}_2$, σ_1, σ_2. Only an antisymmetric function is compatible with the Pauli principle.

The same applies to the wave function for a many-electron system: it is antisymmetric with respect to a simultaneous permutation of spatial and spin variables for any electron pair. This is the generalized formulation of the Pauli principle.

Particles with half-integral spin. Experiment shows that all elementary particles with half-integral spin obey the Pauli principle: protons, neutrons, electrons, and positrons. Complex particles, consisting of an even number of elementary particles with half-integral spin, have a symmetric wave function because, for a complete interchange of all variables relating to such a complex particle, we must make an even number of permutations of the elementary particle variables it consists of. But by changing the sign an even number of times we do not change it at all. For this reason, nuclei with even atomic weights (for example, D^2, He^4, N^{14}, O^{16}, etc.) and, therefore, having symmetric wave functions are not subject as units to the Pauli principle, while He^3, Li^7, etc., have an antisymmetric wave function, that is to say, they are subject to the Pauli principle.

Elementary particles not subject to the Pauli principle. Light quanta do not obey the Pauli principle since there can be an unlimited number of quanta in a state with a given wave vector **k** and given polarization. All particles with integral spin possess a wave function which is symmetric with respect to a complete permutation of the variables relating to any pair of particles.

The Pauli principle and the limiting transition to classical theory. The Pauli principle enables us to understand why the wave properties of light quanta are conserved in the limiting transition to classical theory, while the wave properties of electrons are not.

We shall consider quanta in definite states, i.e., having a certain polarization and wave vector. The number of such quanta can be infinitely large, since quanta are not subject to the Pauli principle. We note that this was not introduced as a supplementary hypothesis concerning the properties of light quanta, but was directly obtained in Sec. 27 in the quantization of electromagnetic field equations: the number of quanta in a state $N_{k,\sigma}$ is the quantum number for the corresponding oscillator. If this quantum number is large then the motion of the oscillator becomes classical, and, as we know, its oscillation amplitude is proportional to the amplitude of a field with a given polarization and wave vector. Thus, the limiting transition yields a classical wave pattern.

In accordance with the Pauli principle, there cannot be more than one electron in each state. Therefore, the probability-amplitude absolute values are always limited by the normalization to unity and, consequently, do not pass to wave amplitudes which can be defined classically.

The ortho- and para-states of two electrons. Let us now return to the case for which the wave function can be represented in the form (33.5). Since the whole product is antisymmetric, one of its factors must be symmetric and the other antisymmetric. This simple result refers only to the two-electron problem.

Let us consider the wave function for two electrons. Since the spin of each electron is equal to $\frac{1}{2}$ (in atomic units), the resultant spin can only be equal to zero or unity. Both these states of a system of two electrons have special names. The state with spin unity is termed the *ortho-state*, while that with spin equal to zero is called the *para-state*.

As has already been said, the magnetic interaction of spins is small. If we can neglect it, then it is easy to write down the spin wave functions for the ortho- and para-states. Let $\chi\,(k_{\sigma_1}, \sigma_1)$ be a function of the spin variable for the first particle σ_1, assuming, as we know, only two values $\sigma_1 = 1$ and $\sigma_1 = 2$. k_{σ_1} denotes the eigenvalue of the spin projection. Depending upon whether k_{σ_1} is equal to $\frac{1}{2}$ or $-\frac{1}{2}$, the function χ has the form shown in exercise 3, Sec. 32.

Without assigning a definite form to χ, we write down the spin wave function for two particles which do not have a spin magnetic interaction:

$$\chi\,(k_{\sigma_1}, \sigma_1 ; k_{\sigma_2}, \sigma_2) = \chi\,(k_{\sigma_1}, \sigma_1) \cdot \chi\,(k_{\sigma_2}, \sigma_2)\,, \tag{33.17}$$

i.e., the probability amplitude for both particles breaks up into the product of the amplitudes for each particle separately. However, it must be taken into account that this function must be either symmetric or antisymmetric. If $k_{\sigma_1} = k_{\sigma_2}$, then (33.17) is symmetric by itself:

$$\chi\,(k_{\sigma_1}, \sigma_1 ; k_{\sigma_2}, \sigma_2) = \begin{cases} \chi\left(\frac{1}{2}, \sigma_1\right) \chi\left(\frac{1}{2}, \sigma_2\right), \\ \chi\left(-\frac{1}{2}, \sigma_1\right) \chi\left(-\frac{1}{2}, \sigma_2\right). \end{cases} \tag{33.18}$$

For $k_{\sigma_1} \neq k_{\sigma_2}$ we must form either a symmetric or an antisymmetric combination of χ:

$$\chi_{sym} = \frac{1}{\sqrt{2}} \left[\chi\left(\frac{1}{2}, \sigma_1\right) \chi\left(-\frac{1}{2}, \sigma_2\right) + \chi\left(\frac{1}{2}, \sigma_2\right) \chi\left(-\frac{1}{2}, \sigma_1\right)\right], \tag{33.19}$$

$$\chi_{antisym} = \frac{1}{\sqrt{2}} \left[\chi\left(\frac{1}{2}, \sigma_1\right) \chi\left(-\frac{1}{2}, \sigma_2\right) - \chi\left(\frac{1}{2}, \sigma_2\right) \chi\left(-\frac{1}{2}, \sigma_1\right)\right]; \tag{33.20}$$

$\frac{1}{\sqrt{2}}$ is introduced for normalization.

The magnitude of the spin projection, i.e., 0, 1 or -1, depends upon the choice of the z-axis. But the symmetry or antisymmetry of a wave function is an internal property and cannot depend upon the choice of coordinate axes. Therefore, the state (33.19) must be regarded as one state together with (33.18), if we judge by the total spin value. They are distinguished by the spin projections, and the total number of these states is three, as is required for a total spin equal to 1. The upper line of (33.18), as is evident, corresponds to a total projection 1, the lower line to a projection -1, and (33.19) to a projection 0. (33.20) corresponds to a total spin of zero. In the accepted terminology, the state with unity spin is to be regarded as the ortho-state while that with zero spin, the para-state.

This definition of ortho- and para-states from the symmetry of the spin wave function also holds for particles with spin other than $\frac{1}{2}$. But the resultant spin in the ortho- and para-states turns out, in this case, to be ambiguously related to the symmetry of the function. The example of deuterons with spin 1 will be examined in Sec. 41.

Ortho- and para-states of helium. The two electrons in the helium atom can occur either in the ortho-state or the para-state. In the first case, the atom has spin unity, in the second case, zero. The symmetric and antisymmetric spin functions are the eigenfunctions of the spin-commutation operator $\hat{P}_\sigma$: when operated upon by the operator $\hat{P}_\sigma$ they give ± 1, i.e., the eigenvalues of $\hat{P}_\sigma$. To the approximation (33.2)-(33.3), the Hamiltonian* is commutative with $\hat{P}_\sigma$, so that $\hat{P}_\sigma$ is an integral of motion. Therefore, transitions between the ortho- and para-states, during which the total spin is not conserved, are far less probable than transitions with conservation of spin.

Only when spin-orbital interaction is taken into account, when the wave function cannot be expressed as a product of the form (33.5), is $\hat{P}_\sigma$ not an independent integral of motion. But the corresponding terms in the Hamiltonian,** which describe the spin-orbit interaction, are inversely proportional to c^2. To this approximation, it is only the total permutation operator of the spin and spatial variables for both electrons $\hat{P}$ that is an integral of motion, because the total wave function of the two electrons is always antisymmetric in accordance with the Pauli principle.

The eigenfunction of a hydrogen molecule in a zero approximation. Concluding this section we shall consider the quantum mechanical explanation for the homopolar chemical bond. Such a bond occurs, for example, between the two atoms in a hydrogen molecule. It was first considered by Heitler and London.

We assume that the atoms are independent in the zero approximation. Each electron is situated close to its own nucleus. We shall denote the nuclei by the letters a and b, and the electrons by the numbers 1 and 2. In the initial approximation, the interaction

* The Hamiltonian means the Hamiltonian operator.

** See A. I. Akhiezer and V. B. Berestetsky, *Quantum Electrodynamics*, GTTI, 1953, equation (37.10). [English translation by Consultants Bureau, Inc. New York, N. Y., 1957.]

between atoms is not taken into account. But this does not mean that the wave functions of two electrons can be taken in the form

$$\Psi = \psi(r_{a_1})\,\psi(r_{b_2}),$$

because this function is neither symmetric nor antisymmetric with respect to the interchange of the electron coordinates. Neglecting the spin-orbital interaction, we must write the spatial wave function in one of the two forms:

$$\Psi = \psi(r_{a_1})\,\psi(r_{b_2}) + \psi(r_{a_2})\,\psi(r_{b_1}), \tag{33.21}$$

or

$$\Psi = \psi(r_{a_1})\,\psi(r_{b_2}) - \psi(r_{a_2})\,\psi(r_{b_1}), \tag{33.22}$$

assuming that the total wave function Φ is obtained by multiplication of the spatial function by the spin function of opposite symmetry. In this form the wave functions are compatible with Pauli's principle. We write r_{a_1} and r_{b_2} in scalar form because the wave functions of a hydrogen atom in the ground state do not depend upon the angle.

The wave equation for a hydrogen molecule has the following form:

$$\left(-\frac{h^2}{2m_p}\Delta_a - \frac{h^2}{2m_p}\Delta_b - \frac{h^2}{2m}\Delta_1 - \frac{h^2}{2m}\Delta_2 - \frac{e^2}{r_{a_1}} - \frac{e^2}{r_{b_2}} - \right.$$
$$\left. - \frac{e^2}{r_{a_2}} - \frac{e^2}{r_{b_1}} + \frac{e^2}{r_{12}} + \frac{e^2}{r_{ab}}\right)\Psi = \mathscr{E}\Psi. \tag{33.23}$$

The first two terms describe the motion of the nuclei of the molecule. They involve the mass of the proton m_p in the denominator, and are therefore exceedingly small compared with the terms describing the motion of the electrons. Physically, this means that the nuclei move considerably slower than the electrons, so that we can find the electron wave function for a fixed distance between the nuclei. Then $\mathscr{E}$ is a function of the distance between the nuclei. If this function has a minimum, corresponding to a stable equilibrium for a given electronic state, then it becomes possible for atoms to form a molecule. We shall not in future write the terms corresponding to the nuclear kinetic energy: they must be taken into account when we consider the vibrational, rotational or translational motion of molecules, though the very position of stable equilibrium, which is determined by the electronic motion, can be found without allowance for $-\dfrac{h^2}{2m_p}\Delta_a$ and $-\dfrac{h^2}{2m_p}\Delta_b$.

The terms of the Hamiltonian appearing in the first line of (33.23) refer to separated atoms. We shall call this part $\left(\text{without } -\dfrac{h^2}{2m_p}\Delta_a \text{ and } -\dfrac{h^2}{2m_p}\Delta_b\right)$ $\mathscr{H}_0$, where the index 0 indicates the degree of the approximation. The second line involves terms due to atomic interactions: the attraction of electrons to "alien" nuclei, and the Coulomb repulsion between electrons and nuclei. We shall call this part $\hat{\mathcal{H}}_1$ and consider it a perturbation: this is true, strictly speaking, only in a qualitative sense.

Perturbation method. Using the notation $\hat{\mathcal{H}}_0$ and $\hat{\mathcal{H}}_1$, the wave equation is written as

$$\hat{\mathcal{H}}\Psi = (\hat{\mathcal{H}}_0 + \hat{\mathcal{H}}_1)\Psi = \mathscr{E}\Psi. \qquad (33.24)$$

The energy eigenvalue can be expanded as the sum of the energies of the noninteracting hydrogen atoms [from (31.34)] and the interaction energy of the atoms:

$$\mathscr{E} = \mathscr{E}_0 + \mathscr{E}_1. \qquad (33.25)$$

We consider the operator $\hat{\mathcal{H}}_1$ and the energy $\mathscr{E}_1$ as corrections. Accordingly, we separate the wave function into a zero approximation function given by one of the expressions (33.21) or (33.22), and a correction Ψ_1:

$$\Psi = \Psi_0 + \Psi_1. \qquad (33.26)$$

We shall neglect the quantities $\mathcal{H}_1\Psi_1$ and $\mathscr{E}_1\Psi_1$ because, in our approximation, they can be considered as being of the second order of magnitude.

Substituting (33.25) and (33.26) in (33.24) and omitting these small terms, we obtain

$$\hat{\mathcal{H}}_0\Psi_0 + \hat{\mathcal{H}}_1\Psi_0 + \hat{\mathcal{H}}_0\Psi_1 = \mathscr{E}_0\Psi_0 + \mathscr{E}_1\Psi_0 + \mathscr{E}_0\Psi_1. \qquad (33.27)$$

But from the definition of Ψ_0 we have

$$\hat{\mathcal{H}}_0\Psi_0 = \mathscr{E}_0\Psi_0, \qquad (33.28)$$

since $\mathscr{E}_0$ is the zero approximation energy and Ψ_0 is an eigenfunction of $\hat{\mathcal{H}}_0$.

We multiply the remaining terms by Ψ_0 and integrate over the volume $dV_1\,dV_2$ of both electrons*:

$$\int\Psi_0\hat{\mathcal{H}}_0\Psi_1 dV_1 dV_2 + \int\Psi_0\hat{\mathcal{H}}_1\Psi_0 dV_1 dV_2 = $$
$$= \mathscr{E}_1\int\Psi_0^2\, dV_1 dV_2 + \mathscr{E}_0\int\Psi_0\Psi_1 dV_1 dV_2. \qquad (33.29)$$

We can transform the first term on the left making use of the fact that $\hat{\mathcal{H}}_0$ is an Hermitian operator (Ψ_0 is a real wave function):

$$\int\Psi_0\hat{\mathcal{H}}_0\Psi_1 dV_1 dV_2 = \int\Psi_1\hat{\mathcal{H}}_0\Psi_0 dV_1 dV_2 = \mathscr{E}_0\int\Psi_0\Psi_1 dV_1 dV_2, \qquad (33.30)$$

* For simplicity, we consider here the real Hamiltonian and the real wave functions; in the more general case, we must multiply the left by Ψ_0^*.

so that it cancels with the last term on the left, in accordance with (33.28). The same could have been proved by using a definite form for $\hat{\mathscr{H}}_0$ in the given case.

Finally, from (33.29) we obtain

$$\mathscr{E}_1 = \frac{\int \Psi_0 \hat{\mathscr{H}}_1 \Psi_0 \, dV_1 \, dV_2}{\int \Psi_0^2 \, dV_1 \, dV_2} . \tag{33.31}$$

The denominator of this expression is the square of the normalization factor of the function Ψ_0, which square appears because the expressions (33.21) and (33.22) are not normalized to unity. Changing to a normalized wave function,

$$\Psi_0' = \frac{\Psi_0}{\sqrt{\int \Psi_0^2 \, dV_1 \, dV_2}},$$

we represent the energy correction to a first approximation in the form

$$\mathscr{E}_1 = \int \Psi_0' \hat{\mathscr{H}}_1 \Psi_0' \, dV_1 \, dV_2 . \tag{33.31 a}$$

This expression is equal to the average of the perturbation energy over nonperturbed motion [see (30.24)].

It can be seen from the very deduction of equation (33.32), in which only the general Hermitian property for operators is used, that the result is of a general nature.

Bound state of hydrogen molecule. From (33.21) or (33.22), we can substitute either a symmetric or an antisymmetric coordinate function in the formula for the energy of a hydrogen molecule. Evaluating the integral shows that $\mathscr{E}_1(r_{ab})$ has a minimum for the symmetric form of the spatial wave function only. The depth of this minimum corresponds approximately to the binding energy of a hydrogen molecule. We cannot expect very good agreement with experiment here, since the approximation used was more qualitative than quantitative in nature.

In calculating the energy, the following integral is of essential importance

$$e^2 \int \psi(r_{a_1}) \psi(r_{b_2}) \frac{1}{r_{12}} \psi(r_{a_2}) \psi(r_{b_1}) \, dV_1 \, dV_2 , \tag{33.32}$$

and is termed the *exchange* integral. It cannot be correlated with any classical quantity because it involves only probability amplitudes and not densities.

We note that the antisymmetric spatial function (33.22) possesses a nodal surface between the nuclei because it changes sign when the places of the nuclei a und b are exchanged. A symmetric function does not have nodes and therefore corresponds to a smaller total energy, i.e., a more stable state. This function is multiplied by an antisymmetric spin function so that a stable molecular state has a zero resultant spin. Thus, the homopolar bond of two hydrogen atoms forming a molecule is related to a "saturation," as it were, of spins. There is no longer a stable equilibrium position for a third hydrogen atom near a hydrogen molecule.

The tendency towards spin saturation of electron pairs is pronounced in *homopolar* bonds.

Exercises

1) Find the possible states of a system of two d-electrons with the same principal quantum numbers.

Each electron can occur in ten states:

$$A: 2, \frac{1}{2};\ B: 1, \frac{1}{2};\ C: 0, \frac{1}{2};\ D: -1, \frac{1}{2};\ E: -2, \frac{1}{2};\ F: 2, -\frac{1}{2};\ G: 1, -\frac{1}{2};$$

$$H: 0, -\frac{1}{2};\ I: -1, -\frac{1}{2};\ J: -2, -\frac{1}{2}.$$

The states with positive projections of spin and orbital moment are

AB: 3,1; AC: 2,1; AD: 1,1; AE: 0,1; AF: 4,0; AG: 3,0; AH: 2,0; AI: 1,0; AJ: 0,0; BC: 1,1; BD: 0,1; BF: 2,0; BG: 2,0; BH: 1,0; BI: 0,0; CF: 2,0; CG: 1,0; CH: 0,0; DF: 1,0; DG: 0,0; EF: 0,0.

Choosing the states with maximum angular-momentum projections, we obtain three resultant states with zero spin:

$$^1S,\ ^1D,\ ^1G \text{ or } ^1S^g_0,\ ^1D^g_2,\ ^1G^g_4,$$

and two states with unity spin:

$$^3P,\ ^3F \text{ or } ^3P^g_2,\ ^3P^g_1,\ ^3P^g_0 \text{ and } ^3F^g_4,\ ^3F^g_3,\ ^3F^g_2.$$

2) Show that, in a system of four p-electrons with the same principal numbers, the states are the same as in a system of two p-electrons; in other words, that two electrons have the same states as two "holes."

Sec. 34. The Quantum Theory of Radiation

In this section we shall find the probability of an excited atom emitting a light quantum in unit time, and we shall compare the probabilities of such radiation transitions as correspond to various changes in the atomic states. But first we must deduce a general formula for the probability of quantum transitions (this formula will also be used in Sec. 37).

Transitions between states with the same energy. Let us suppose that a system has two states corresponding to the same energy but different in some other respect. For example, in this section we will consider an excited atom having energy excess $\mathscr{E}_1 - \mathscr{E}_0$ above the ground state. This atom is capable of emitting a light quantum with energy $\hbar\omega = \mathscr{E}_1 - \mathscr{E}_0$, in which case the atom will go to the ground state. Strictly speaking, there is only one state consisting of an atom and electromagnetic field with a certain energy $\mathscr{E}_1$ (if there are no other quanta in the field). The energy of such a system is rigorously determined, though the state is not defined in more detail.

Also possible is the following approach to the problem. Let an atom at the initial instant of time be in an excited state but capable of the spontaneous emission of a quantum. Then the energy of the atom is no longer specified with full rigour, but lies in some narrow interval $\Delta\mathscr{E}$, where $\Delta\mathscr{E} \sim \frac{2\pi\hbar}{\Delta t}$ and Δt is the mean lifetime of the atom in the excited state before radiation [see (28.15)]. If the mean lifetime of the atom in the excited state Δt is such that $\frac{2\pi\hbar}{\Delta t}$ is considerably less than the energy level spacing of the atom, then the energy uncertainty can be neglected to a first approximation, assuming that the atom initially occurred in a state with an accurate energy value $\mathscr{E}_1$; it is also necessary to calculate the probability that, in a certain interval of time t, the atom will go to the ground state, and a quantum with energy $\hbar\omega = \mathscr{E}_1 - \mathscr{E}_0$ will appear in the electromagnetic field.

The reason for the transition is interaction with the electromagnetic field. Here the lifetime of the atom in the excited state Δt is so great that $\Delta\mathscr{E} \ll \mathscr{E}_1 - \mathscr{E}_0$. For this reason, the interaction of the atom with the electromagnetic field can be interpreted as a small perturbation superimposed on the excited atom with energy $\mathscr{E}_1$.

The same type of problem concerning the transition probability due to a perturbation can also be formulated for other transitions. For example, if the total excitation energy of an atom is greater than its ionization energy, it is possible for an electron to be emitted from the atom without radiation. In this case the excited state of the atom and the ion+electron state belong to the same energy. Each of them separately does not have a strictly defined energy.

Transition probability. A radiation transition with the emission of a quantum is caused by interaction between an atom and an electromagnetic field. We shall suppose for the time being that this interaction is "switched off"; then the energy of the atom and field separately becomes an exact integral of motion. We shall call its eigenvalue in the initial state $\mathscr{E}_1$. Then, if the interaction is "turned on," a finite probability exists of the system making a transition to some state which, energetically, is very close to $\mathscr{E}_1$, but otherwise very different from the initial state; for example, the atom was excited in the initial state and there were no quanta in the field, while in the final state the atom went to the ground state and a quantum appeared in the field.

Let us divide the Hamiltonian of the system into two terms: $\hat{\mathscr{H}} = \hat{\mathscr{H}}^{(0)} + \hat{\mathscr{H}}^{(1)}$, where $\hat{\mathscr{H}}^{(0)}$ corresponds to the separated atom and field while $\hat{\mathscr{H}}^{(1)}$ describes the interaction. We then deduce a general formula for the transition probability, and apply it to a radiation. We shall therefore call $\hat{\mathscr{H}}^{(0)}$ the Hamiltonian of the unperturbed system, and regard $\hat{\mathscr{H}}^{(1)}$ as a small perturbation causing the transition. The eigenfunctions and eigenvalues of the operator $\hat{\mathscr{H}}^{(0)}$ are determined from the equation

$$-\frac{h}{i}\frac{\partial \psi_m^{(0)}}{\partial t} = \hat{\mathscr{H}}^{(0)} \psi_m^{(0)}. \qquad (34.1)$$

Allowing for perturbation, the wave function satisfies the equation

$$-\frac{h}{i}\frac{\partial \psi}{\partial t} = (\hat{\mathscr{H}}^{(0)} + \hat{\mathscr{H}}^{(1)}) \psi. \qquad (34.2)$$

Considering that $\hat{\mathscr{H}}^{(1)}$ is a small perturbation, we represent the wave function in the form

$$\psi = \psi_1^{(0)} + \psi^{(1)}, \qquad (34.3)$$

the "product" $\hat{\mathscr{H}}^{(1)} \psi^{(1)}$ will be neglected as being of the second order. Then, for $\psi^{(1)}$ we obtain the nonhomogeneous equation

$$-\frac{h}{i}\frac{\partial \psi^{(1)}}{\partial t} - \hat{\mathscr{H}}^{(0)} \psi^{(1)} = \hat{\mathscr{H}}^{(1)} \psi_1^{(0)}. \qquad (34.4)$$

We shall look for $\psi^{(1)}$ in the form of an eigenfunction expansion of the operator $\psi^{(0)}$:

$$\psi^{(1)} = \sum_m c_m(t)\, \psi_m^{(0)}, \qquad (34.5)$$

each of the functions $\psi_m^{(0)}$ satisfying the homogeneous equation (34.1). Substituting the series (34.5) in the nonhomogeneous equation and using the indicated property of the function $\psi_m^{(0)}$, we arrive at the following equality:

$$-\frac{h}{i}\sum_m \frac{\partial c_m}{\partial t} \psi_m^{(0)} = \hat{\mathscr{H}}^{(1)} \psi_1^{(0)}. \qquad (34.6)$$

The coefficients c_m can be determined therefrom by taking advantage of the orthogonality property of the eigenfunctions $\psi_m^{(0)}$ (30.6). For this it is necessary to multiply both sides of (34.6) by $\psi_n^{(0)*}$ and integrate over a volume. Then only the term $-\frac{h}{i}\frac{\partial c_n}{\partial t}$ remains on the left, while on the right a certain integral is obtained which is characteristic of the perturbation method set out here:

$$-\frac{h}{i}\frac{\partial c_n}{\partial t}=\int\psi_n^{(0)*}\hat{\mathscr{H}}^{(1)}\psi_1^{(0)}\,dV. \tag{34.7}$$

In order to integrate this equation we must determine the time dependence of the right-hand side. It involves wave functions [that satisfy equation (34.1)] together with their time factor in the form (24.21). It is assumed that the operator $\hat{\mathscr{H}}^{(1)}$ does not depend explicitly on time. Then

$$\int\psi_n^{(0)*}\hat{\mathscr{H}}_1\psi_1^{(0)}\,dV=e^{-i\frac{(\mathscr{E}_1-\mathscr{E}_n)t}{h}}\int\psi_{0n}^{(0)*}\hat{\mathscr{H}}^{(1)}\psi_{01}^{(0)}\,dV, \tag{34.8}$$

where the integral multiplied by the exponent does not depend upon time. We have supposed that at the initial instant of time the system was in a state with energy $\mathscr{E}_1$: in other words, that $|c_1(0)|=1$, $c_{n\neq 1}(0)=0$. Therefore, equation (34.7) is integrated thus:

$$\frac{h}{i}c_n=\frac{h\left(e^{-i\frac{(\mathscr{E}_1-\mathscr{E}_n)t}{h}}-1\right)}{i(\mathscr{E}_1-\mathscr{E}_n)}\int\psi_{0n}^{(0)*}\hat{\mathscr{H}}^{(1)}\psi_{01}^{(0)}\,dV, \tag{34.9}$$

or, once again introducing the exponential factor under the integral sign, i.e., reverting to the functions $\psi_n^{(0)*}$, $\psi_1^{(0)}$ we obtain

$$c_n(t)=\frac{1-e^{\frac{i(\mathscr{E}_1-\mathscr{E}_n)t}{h}}}{(\mathscr{E}_1-\mathscr{E}_n)}\int\psi_n^{(0)*}\hat{\mathscr{H}}^{(1)}\psi_1^{(0)}\,dV. \tag{34.10}$$

Consequently, the probability that at the instant of time t the system will be in a state with wave function $\psi_n^{(0)}$ is, from (30.13), equal to

$$w_n(t)=|c_n(t)|^2=\frac{\left(1-e^{i\frac{(\mathscr{E}_1-\mathscr{E}_n)t}{h}}\right)\left(1-e^{-i\frac{(\mathscr{E}_1-\mathscr{E}_n)t}{h}}\right)}{(\mathscr{E}_1-\mathscr{E}_n)^2}\left|\int\psi_n^{(0)*}\hat{\mathscr{H}}^{(1)}\psi_1^{(0)}\,dV\right|^2=$$

$$=\left(2-2\cos\frac{(\mathscr{E}_1-\mathscr{E}_n)t}{h}\right)\frac{\left|\int\psi_n^{(0)*}\hat{\mathscr{H}}^{(1)}\psi_1^{(0)}\,dV\right|^2}{|\mathscr{E}_1-\mathscr{E}_n|^2}=$$

$$=4\sin^2\frac{(\mathscr{E}_1-\mathscr{E}_n)t}{2h}\,\frac{\left|\int\psi_n^{(0)*}\hat{\mathscr{H}}^{(1)}\psi_1^{(0)}\,dV\right|^2}{|\mathscr{E}_1-\mathscr{E}_n|^2}. \tag{34.11}$$

Matrix elements. We shall be concerned with the expression (34.11) somewhat later. First of all let us introduce a system of notation which, in general, is very convenient in quantum mechanics. The integral (34.7) for any pair of eigenfunctions ψ_n, ψ_k^* and for an arbitrary operator $\hat{\lambda}$ is denoted thus:

$$\lambda_{nk}\equiv\int\psi_k^*\hat{\lambda}\,\psi_n\,dV, \tag{34.12}$$

where the integration is performed over all the independent variables involved in the Hamiltonian $\hat{\mathscr{H}}$.

The quantities λ_{nk} form a square table so that the index n will designate the rows, and k the columns:

$$\begin{array}{l} \lambda_{11}, \lambda_{12}, \lambda_{13}, \ldots, \lambda_{1k}, \ldots, \\ \lambda_{21}, \lambda_{22}, \lambda_{23}, \ldots, \lambda_{2k}, \ldots, \\ \lambda_{31}, \lambda_{32}, \lambda_{33}, \ldots, \lambda_{3k}, \ldots, \\ \cdot\ \cdot\ \cdot\ \cdot\ \cdot\ \cdot\ \cdot\ \cdot\ \cdot\ \cdot\ \cdot\ \cdot\ \cdot \\ \lambda_{n1}, \lambda_{n2}, \lambda_{n3}, \ldots, \lambda_{nk}, \ldots, \\ \cdot\ \cdot\ \cdot\ \cdot\ \cdot\ \cdot\ \cdot\ \cdot\ \cdot\ \cdot\ \cdot\ \cdot\ \cdot \end{array} \tag{34.13}$$

Such a table is termed a *matrix* in mathematics, while the separate quantities λ_{nk} are called *matrix elements*. The right-hand side of (34.7) contains the matrix element $\hat{\mathscr{H}}^{(1)}$.

We note an important property of the matrix elements of Hermitian operators. In accordance with the Hermitian condition (30.3)

$$\int \psi_k^* \hat{\lambda}\, \psi_n \, dV = \int \psi_n \hat{\lambda}^* \psi_k^* \, dV = \left(\int \psi_n^* \hat{\lambda}\, \psi_k \, dV \right)^*, \tag{34.14}$$

where the conjugate sign* on the right refers to the whole integral. Proceeding now from the definition (34.12), we write:

$$\lambda_{nk} = \lambda_{kn}^* . \tag{34.15}$$

A matrix whose elements satisfy equation (34.15) is termed Hermitian.

The relationship between matrix elements of different quantities. Let us take the matrix elements of both sides of the operator equalities (30.35) and (30.37), and put the time derivative before the integral:

$$\frac{d}{dt} x_{nk} = p_{nk} , \tag{34.16}$$

$$\frac{d}{dt} p_{nk} = -\left(\frac{\partial U}{\partial x} \right)_{nk} . \tag{34.17}$$

The time dependence of the matrix elements was found in equation (34.8), namely

$$x_{nk} = e^{-\frac{i(\mathscr{E}_n - \mathscr{E}_k)t}{h}} \int \psi_{0k}^* \hat{x}\, \psi_{0n} \, dV,$$

$$p_{nk} = e^{-\frac{i(\mathscr{E}_n - \mathscr{E}_k)t}{h}} \int \psi_{0k}^* \hat{p}\, \psi_{0n} \, dV.$$

Therefore,

$$\frac{d}{dt} x_{nk} = i \frac{\mathscr{E}_k - \mathscr{E}_n}{h} x_{nk} , \tag{34.18}$$

$$\frac{d}{dt} p_{nk} = i \frac{\mathscr{E}_k - \mathscr{E}_n}{h} p_{nk} . \tag{34.19}$$

Thus, matrix elements depend harmonically upon time.

The energy difference $\mathscr{E}_k - \mathscr{E}_n$ can be conveniently represented by means of the Bohr frequency condition (see beginning of Sec. 23):

$$\frac{\mathscr{E}_k - \mathscr{E}_n}{h} \equiv \omega_{kn}. \tag{34.20}$$

Therefore, the operator relationships (34.16) and (34.17), rewritten for matrix elements, appear thus:

$$p_{nk} = i m \omega_{kn} x_{nk}, \tag{34.21}$$

$$-\left(\frac{\partial U}{\partial x}\right)_{nk} = i \omega_{kn} p_{nk}. \tag{34.22}$$

The matrix form of the equations of quantum mechanics was found by Heisenberg.

The probability for transition to a continuous spectrum. Let us now investigate the expression for the transition probability (34.11), rewriting it in matrix notation:

$$w_n(t) = 4 \sin^2 \frac{(\mathscr{E}_1 - \mathscr{E}_n) t}{2h} \cdot \frac{|\mathscr{H}^{(1)}_{1n}|^2}{(\mathscr{E}_1 - \mathscr{E}_n)^2}. \tag{34.23}$$

In the examples dealing with radiation and ionization that we have mentioned, the final state of the system belonged to a continuous energy spectrum. Indeed, the energy spectrum of an electromagnetic field is continuous since the field can contain quanta of any frequency ω. In the ionization example, the spectrum of the electron emitted from the atom was continuous because the motion of the electron was infinite.

If the state n belongs to a continuous spectrum, it is more interesting to determine the total probability of transition to any of the states with energy $\mathscr{E}_n$, i.e., to find the integral of (34.23) with respect to $d\mathscr{E}_n$. Now it is not advisable to state the final energy $\mathscr{E}_n$ since it varies continuously; it is better simply to write $\mathscr{E}$. Then there are $dN(\mathscr{E})$ states contained within the energy interval between $\mathscr{E}$ and $\mathscr{E} + d\mathscr{E}$. The example of $dN(\mathscr{E})$ was given in Sec. 25, equation (25.25). (The transition to a continuous spectrum in Sec. 25 is achieved simply by means of an infinite increase in the box dimensions, which are not contained in any physical result. The distance between neighbouring levels is then infinitely reduced.)

So let

$$dN(\mathscr{E}) = z(\mathscr{E})\, d\mathscr{E}. \tag{34.24}$$

The total probability of transition to the continuous spectrum is

$$W = \int w(\mathscr{E}_1, \mathscr{E})\, dN(\mathscr{E}) = \int \frac{4 \sin^2 \dfrac{(\mathscr{E} - \mathscr{E}_1) t}{2h}}{(\mathscr{E} - \mathscr{E}_1)^2} |\mathscr{H}^{(1)}(\mathscr{E}_1, \mathscr{E})|^2 z(\mathscr{E})\, d\mathscr{E}. \tag{34.25}$$

For clarity in notation we shall put the indices $\mathscr{E}_1$, $\mathscr{E}$ of $\mathscr{H}^{(1)}$ in brackets and not as subscripts, treating them as the arguments of the function, which in fact they are with respect to $\mathscr{H}^{(1)}$.

We shall denote the argument of the sine by the letter ξ:

$$\frac{(\mathscr{E}-\mathscr{E}_1)\,t}{2h}=\xi.$$

Passing to the integration variable ξ we obtain

$$W=\frac{2t}{h}\int\frac{\sin^2\xi}{\xi^2}\left|\mathscr{H}^{(1)}\left(\mathscr{E}_1,\mathscr{E}_1+\frac{2h\xi}{t}\right)\right|^2 z\left(\mathscr{E}_1+\frac{2h\xi}{t}\right)d\xi. \quad (34.26)$$

The function $\frac{\sin^2\xi}{\xi^2}$ has a principal maximum for $\xi=0$. Its next maximum is already twenty times smaller. For this reason, in the integral (34.26), the main part is played by the values of ξ of the order unity. But then the instant of time t can always be chosen so that $\frac{2h\xi}{t}$ is considerably less than $\mathscr{E}_1$. In other words, it is permissible, in the arguments of the functions $\mathscr{H}^{(1)}(\mathscr{E}_1,\mathscr{E})$ and $z(\mathscr{E})$, to replace $\mathscr{E}$ simply by $\mathscr{E}_1$ and to take out the functions $|\mathscr{H}^{(1)}(\mathscr{E}_1, \mathscr{E}=\mathscr{E}_1)|^2$ and $z(\mathscr{E}=\mathscr{E}_1)$ from under the integral sign. It is shown thereby that if the time t is sufficiently long, the energies of the initial and final states $\mathscr{E}_1$ and $\mathscr{E}$ are defined so accurately that they can be considered simply equal to one another, in accordance with the law of conservation of energy in the transition. Naturally, the law of conservation of energy holds always, but, for sufficiently small values of t, it is impossible to determine the energy of the final state, for the uncertainty relation (28.15) for the given case is of the form $(\mathscr{E}-\mathscr{E}_1)\,t\sim 2\pi h$. Hence, if t tends to infinity, the precise equality $\mathscr{E}=\mathscr{E}_1$ is obtained.

Since the function $\frac{\sin^2\xi}{\xi^2}$ decreases rapidly with increasing ξ, the integration should be extended from $-\infty$ to ∞. Since the remaining values have been taken out from under the integral sign, the integral itself can be evaluated. It is

$$\int_{-\infty}^{\infty}\frac{\sin^2\xi}{\xi^2}\,d\xi=\pi. \quad (34.27)$$

From this

$$W=\frac{2\pi}{h}\,|\mathscr{H}^{(1)}(\mathscr{E}_1,\mathscr{E}=\mathscr{E}_1)|^2 z(\mathscr{E}_1)\cdot t. \quad (34.28)$$

Then the transition probability in unit time is

$$\frac{dW}{dt}=\frac{2\pi}{h}\,|\mathscr{H}^{(1)}(\mathscr{E}_1,\mathscr{E}=\mathscr{E}_1)|^2 z(\mathscr{E}_1). \quad (34.29)$$

We write the second argument $\mathscr{E}=\mathscr{E}_1$ to emphasize that the state $\psi(\mathscr{E})$ coincides with $\psi(\mathscr{E}_1)$ only with respect to energy. The formula (34.29) has very many applications.

The matrix element corresponding to the emission of a quantum. With the aid of expression (34.29) it is possible to obtain rigorously an expression for radiation intensity. This result is based on quantization of the electromagnetic field performed in Sec. 27. We shall not give the other, less rigorous, result based on the analogy between classical equations and the equations for matrix elements.

In order to simplify subsequent computations, we shall, from the start, take advantage of the law of conservation of energy for radiation. In considering transitions of an atom from a state with energy $\mathscr{E}_1$ to a state with energy $\mathscr{E}_0$, we take only those quanta which satisfy the energy conservation law in accordance with the Bohr frequency condition $\mathscr{E}_1 - \mathscr{E}_0 = \hbar\omega$. Further, we shall first of all consider quanta with a definite direction of the wave vector **k**, and a definite polarization $\boldsymbol{\sigma}$. In addition, we assume that there were no quanta of this type in the field initially, i.e., in the initial state $N_{\mathbf{k}\sigma}=0$.

In this case the perturbation energy operator is the product of two operators:

$$\hat{\mathscr{H}}^{(1)} = -\frac{e}{c}\,(\hat{\mathbf{v}}\,\hat{\mathbf{A}})\,. \tag{34.30}$$

This expression is obtained from (15.32), if the term that is linear in $\hat{\mathbf{A}}$ is retained in the equation, and if we put $\varphi=0$.

The wave function of the atom and the field in a nonperturbed state, i.e., with the interaction between them "switched off," is expressed as the product of the wave functions of the atom and the field. The wave function of the field is represented as the product of the wave functions of separate oscillators with different **k**, $\boldsymbol{\sigma}$. All these functions are orthogonal and normalized. Therefore, in calculating the matrix element of the quantity $\mathbf{A}_{\mathbf{k}}^{\sigma}$ or of the coordinate of the oscillator with given **k**, $\boldsymbol{\sigma}$, we must take the wave functions corresponding to the given oscillator. In accordance with the normalization condition, the integral over all the remaining coordinates of the field gives unity.

In the vector potential we take the term relating to **k** and $\boldsymbol{\sigma}$:

$$\hat{\mathbf{A}}_{k,\sigma} = \sqrt{\frac{\pi c^2}{V}}\,\mathbf{e}_{\mathbf{k}}^{\sigma}\left[\left(\hat{Q}_{\mathbf{k}}^{\sigma} + \frac{i\hat{P}_{\mathbf{k}}^{\sigma}}{\omega_{\mathbf{k}}}\right)e^{i\mathbf{kr}} + \left(\hat{Q}_{\mathbf{k}}^{\sigma} - \frac{i\hat{P}_{\mathbf{k}}^{\sigma}}{\omega_{\mathbf{k}}}\right)e^{-i\mathbf{kr}}\right], \tag{34.31}$$

where we have used equations (27.20) and (27.21) that express the field amplitudes in terms of real variables.

We must calculate the matrix element describing the transition between two states, with no quantum described by wave vector **k** and polarization $\boldsymbol{\sigma}$ in the first one, and with only one such quantum in the second one. We shall write these states with subscripts 0 and 1. Then, from (34.21),

$$(P_{\mathbf{k}}^{\sigma})_{01} = i\,\omega_{10}\,(Q_{\mathbf{k}}^{\sigma})_{01} = i\,\omega_{\mathbf{k}}\,(Q_{\mathbf{k}}^{\sigma})_{01}\,, \tag{34.32}$$

since ω_{10} is equal to the energy difference between the initial and final states of the field divided by h, i.e., just equal to the frequency of the emitted quantum ω_k. Substituting this into the expression for the matrix element $(\mathbf{A}_{\mathbf{k}}^{\sigma})_{01}$ we find that

$$(\mathbf{A}_{\mathbf{k}}^{\sigma})_{01} = 2\sqrt{\frac{\pi c^2}{V}}\,\mathbf{c}_{\mathbf{k}}^{\sigma}\,(Q_{\mathbf{k}}^{\sigma})_{01}\,e^{-i\mathbf{k}\mathbf{r}}\,, \tag{34.33}$$

since the coefficient of $e^{i\mathbf{k}\mathbf{r}}$ becomes zero.

For simplicity we shall temporarily omit the indices $\mathbf{k}$ and $\boldsymbol{\sigma}$. We must evaluate the integral

$$Q_{01} = \int \varphi_1\,Q\,\varphi_0\,dQ\,. \tag{34.34}$$

Here, the field-oscillator wave functions are denoted by φ_0 and φ_1 in order not to confuse them with the atom wave functions.

We know the functions φ_0 and φ_1 from Sec. 26. From equations (26.22) and (26.23) we have

$$\varphi_0 = g_0\,e^{-\frac{1}{2}\frac{\omega Q^2}{h}}\,, \quad \varphi_1 = g_1\,e^{-\frac{1}{2}\frac{\omega Q^2}{h}}\sqrt{\frac{\omega}{h}}\,Q\,. \tag{34.35}$$

The coefficients g_0 and g_1 are found from the normalization condition (see exercise 1, Sec. 26):

$$1 = g_0^2\int_{-\infty}^{\infty} dQ\left(e^{-\frac{\omega Q^2}{2h}}\right)^2 = g_0^2\sqrt{\frac{h}{\omega}}\int_{-\infty}^{\infty} dx\,e^{-x^2} = g_0^2\sqrt{\frac{h\pi}{\omega}}\;;\; g_0 = \sqrt[4]{\frac{\omega}{h\pi}}\;;$$

$$1 = g_1^2\int_{-\infty}^{\infty} dQ\left(e^{-\frac{1}{2}\frac{\omega Q^2}{h}}\sqrt{\frac{\omega}{h}}\,Q\right)^2 = g_1^2\sqrt{\frac{h}{\omega}}\int_{-\infty}^{\infty} x^2 e^{-x^2}\,dx = g_1^2\sqrt{\frac{h\pi}{4\omega}}\;;$$

$$g_1 = \sqrt[4]{\frac{4\omega}{h\pi}}\,.$$

We note that the product $Q\varphi_0$ is proportional to φ_1. Hence, the integral (34.34) would have vanished due to the orthogonality condition, if any other function, $\varphi_2, \varphi_3, \ldots, \varphi_n$ had been substituted, in place φ_1, into the integral. Therefore, only one quantum can be emitted with a given frequency, direction, and polarization. The same can also be shown for any arbitrary initial stage of the field. Absorption of quanta also occurs singly. For Q_{01} we have

$$Q_{01} = \sqrt{\frac{2}{\pi}}\,\frac{\omega}{h}\int_{-\infty}^{\infty} Q^2 e^{-\frac{\omega Q^2}{h}}\,dQ = \sqrt{\frac{h}{2\omega}}\,. \tag{34.36}$$

The dipole approximation. It is now necessary to calculate the matrix element given by two atomic states:

$$(\mathscr{H}^{(1)})_{10} = \frac{e}{c}\int \psi_0^* (\hat{\mathbf{v}}\,\mathbf{A}_{01})\,\psi_1\,dV. \tag{34.37}$$

Substituting $\mathbf{A}_{01}$ from (34.33) and (34.36), we reduce this matrix element to the form

$$(\mathscr{H}^{(1)})_{10} = e\sqrt{\frac{2\pi h}{V\omega}}\int \psi_0^* (\mathbf{e}_{\mathbf{k}}^{\sigma}\,\hat{\mathbf{v}})\,e^{-i\mathbf{k}\mathbf{r}}\,\psi_1\,dV. \tag{34.38}$$

The wave function of the discrete spectrum of an atom differs from zero in the region near 10^{-8} cm, i.e., of the order of atomic dimensions. The wavelength of visible light is about 0.5×10^{-4} cm, i.e., several thousand times greater. Therefore, we can consider that the phase of a wave changes very little in the region of the atom, and we remove the exponential factor from the integration, taking it at some mean point (for example, the nucleus).

This corresponds to a dipole approximation defined in Sec. 19, the wavelength being considerably greater than the dimensions of the radiating system. The other condition concerning the applicability of the dipole approximation is that the electron velocity must be considerably less than the velocity of light—a thing that occurs in atoms of small and medium atomic weight.

To the dipole approximation we have

$$(H^{(1)})_{10} = e\sqrt{\frac{2\pi h}{V\omega}}\,e^{-i\mathbf{k}\mathbf{r}}\,\mathbf{e}_{\mathbf{k}}^{\sigma}\int \psi_0^*\,\hat{\mathbf{v}}\,\psi_1\,dV. \tag{34.39}$$

From (34.18) the velocity matrix element is directly expressed in terms of the coordinate matrix element

$$\mathbf{v}_{10} = i\,\omega_{01}\,\mathbf{r}_{10} = -\,i\,\omega_{\mathbf{k}}\,\mathbf{r}_{10}, \tag{34.40}$$

because, according to the law of conservation of energy, $\omega_{10} = \dfrac{\mathscr{E}_1 - \mathscr{E}_0}{h}$ is equal to the frequency of the radiated light.

The square of the modulus of the matrix element is

$$|\mathscr{H}_{10}^{(1)}|^2 = e^2\,\frac{2\pi h\omega}{V}\,|e_{\mathbf{k}}^{\sigma}\mathbf{r}_{10}|^2. \tag{34.41}$$

We shall now take into account the fact that an emitted quantum can have two different polarizations. If we are not specially interested in the probability of quantum emission with a given polarization, then the probability must be summed over the polarizations, i.e., over σ. To begin with, let us assume the vector $\mathbf{k}$ to be in the direction of the z-axis. Then the unit vector $\mathbf{e}_{\mathbf{k}}^{\sigma}$ can have two directions: along the x-axis and along the y-axis. Accordingly,

$$\sum_{\sigma} \left| \mathbf{e}_{\mathbf{k}}^{\sigma} \mathbf{r}_{10} \right|^2 = |x_{10}|^2 + |y_{10}|^2. \tag{34.42}$$

Let us find the average of this expression over all possible directions of quantum emission. It is then obvious that

$$\overline{|x_{10}|^2} = \overline{|y_{10}|^2} = \overline{|z_{10}|^2} = \frac{1}{3} |\mathbf{r}_{10}|^2. \tag{34.43}$$

By performing this averaging after summation with respect to σ, we obtain

$$\sum_{\sigma} \left| \mathscr{H}_{12}^{(1)} \right|^2 = e \cdot \frac{2\pi h\omega}{V} \cdot \frac{2}{3} |\mathbf{r}_{10}|^2. \tag{34.44}$$

In order to find the probability of emission of a quantum in unit time, we must multiply (34.44) by $\frac{2\pi}{h} z(\mathscr{E})$. $z(\mathscr{E})$ is found from equation (25.24), where we must put $k = \frac{\omega}{c}$:

$$z(\omega) = \frac{dN(\omega)}{d(h\omega)} = \frac{\omega^2}{2\pi^2 c^3 h}. \tag{34.45}$$

Finally, from equation (34.29), we find the expression for the probability to a dipole approximation:

$$\frac{dW}{dt} = \frac{4}{3} \frac{\omega^3}{hc^3} e^2 |\mathbf{r}_{10}|^2. \tag{34.46}$$

We can write the product $e\mathbf{r}_{10}$ as $\mathbf{d}_{10}$, i.e., as the dipole moment matrix element.

The intensity of radiation is equal to the radiation probability in unit time multiplied by the energy of the quantum:

$$\frac{d\mathscr{E}}{dt} = \frac{4}{3} \frac{\omega^4}{c^3} |\mathbf{d}_{10}|^2. \tag{34.47}$$

This equation greatly resembles the classical formula (19.28). However, we have the square of the modulus $\omega^4 |\mathbf{d}_{10}|^2$ in place of the square of the second derivative of dipole moment $\ddot{\mathbf{d}}^2$. The correspondence between classical and quantum theory displayed here may be demonstrated by means of the matrix equations (34.16) and (34.17) too. Directly applied to electrodynamics, it leads to equation (34.47). We have given a more rigorous deduction, based on the quantization of electrodynamical equations, in order to illustrate the generality of the methods of quantum theory.

Compared with the classical formula (19.28), the quantum expression contains an extra factor of two (4/3 instead of 2/3). This is explained in the following way. We represent a classical dipole moment, varying harmonically, in the following manner:

$$\mathbf{d} = \mathbf{d}_1 e^{i\omega t} + \mathbf{d}_1^* e^{-i\omega t}, \quad \ddot{\mathbf{d}} = -\omega^2 (\mathbf{d}_1 e^{i\omega t} + \mathbf{d}_1^* e^{-i\omega t}). \qquad (34.48)$$

The terms $\mathbf{d}_1 e^{i\omega t}$ and $\mathbf{d}_1^* e^{-i\omega t}$ depend upon time like the matrix elements $\mathbf{d}_{10}$ and $\mathbf{d}_{01}$. Let us form the time average of $(\ddot{\mathbf{d}})^2$. The terms involving $e^{2i\omega t}$ and $e^{-2i\omega t}$ drop out in averaging, and there remains

$$\overline{(\ddot{\mathbf{d}})^2} = \omega^4 (2\mathbf{d}_1 \mathbf{d}_1^*) = 2\omega^4 |\mathbf{d}_1|^2.$$

But it is the quantum formula which corresponds to the mean radiation intensity $\left(\hbar\omega \cdot \frac{dW}{dt}\right)$, so that the factor of two is due to the time averaging of the square of the dipole moment, given in the form (34.48).

The expression (34.47) confirms what was said in Sec. 22 about an atom being stable in quantum theory: radiation is always associated with a transition of the atom from one state to another. But no atomic state exists for which the energy is less than the energy of the ground state, that is why the atom can exist in the ground state for an indefinitely, long time.

The selection rules for the magnetic quantum number. It follows from equation (34.41) that, if $\mathbf{r}_{10} = 0$, the intensity is equal to zero, at any rate to a dipole approximation. We shall now find the conditions for which $\mathbf{r}_{10}$ differs from zero. First of all we notice that if the vector defining the polarization direction is equal to $\mathbf{e}_\mathbf{k}^\sigma$ then, to a dipole approximation, radiation of such a quantum is possible provided the matrix element for the projection of the electron radius vector along the direction of quantum polarization differs from zero. Let us assume that the quantum polarization is along the z-axis. Let the magnetic quantum number of the electron be equal to k before the transition, and k' after the transition. Then the dependence of the wave function upon azimuth ψ is given by the equations

$$\psi_1 = f_1(r, \vartheta)\, e^{ik\varphi}, \quad \psi_0^* = f_0^*(r, \vartheta)\, e^{-ik'\varphi},$$

because $e^{ik\varphi}$ and $e^{ik'\varphi}$ are eigenfunctions of the angular-momentum z-projection. Hence

$$z_{10} = \int f_0^*(r, \vartheta)\, r\cos\vartheta \cdot f_1(r\vartheta)\, r^2 \sin\vartheta\, dr\, d\vartheta \int_0^{2\pi} e^{i\varphi(k-k')}\, d\varphi;$$

$$\int_0^{2\pi} e^{i\varphi(k-k')}\, d\varphi = \left.\frac{e^{i\varphi(k-k')}}{i(k-k')}\right|_0^{2\pi} = \begin{cases} 0 \text{ for } k \neq k', \\ 2\pi \text{ for } k = k'. \end{cases}$$

$z = r\cos\vartheta$ in polar coordinates and, therefore, does not depend upon φ. For this reason, the matrix element z_{10} differs from zero only when $k' = k$.

Instead of considering plane polarized radiation along x or y, let us take circularly polarized radiation in the xy-plane. In such radiation there is a constant phase shift between the xth and the yth components equal to $\frac{\pi}{2}$ (see sec. 17, Fig. 25). Consequently, we must determine the matrix elements of the quantities

$$(x + e^{\pm i \frac{\pi}{2}} y)_{10} = (x \pm iy)_{10} = (r \sin \vartheta \, e^{\pm i\varphi})_{10} .$$

Substituting the expressions for the wave functions involving φ explicitly, we find

$$\int_0^{2\pi} e^{i(k-k' \pm 1)\varphi} d\varphi = \frac{e^{i(k-k' \pm 1)}}{k - k' \pm 1} = 0, \text{ if } k' \neq k \pm 1 . \qquad (34.49)$$

Hence, radiation which is circularly polarized in the xy-plane can only be emitted if the magnetic quantum number changes by ± 1.

The rules that determine what change of quantum number governs the emission of a given radiation are called *selection rules.*

The selection rules for dipole radiation with respect to the magnetic quantum number forbid the changing of k by more than unity.

The selection rules for the azimuthal quantum number and parity. The magnetic quantum number is the angular-momentum projection. Since the angular-momentum projection does not change by more than unity, the angular momentum itself (i.e., the azimuthal quantum number) cannot change by more than unity.

But l for a separate electron cannot remain unchanged, because then the functions ψ_1 and ψ_0 must have the same parity. Here, the product $\psi_0^* z \psi_1$ will turn out to be an odd function while its integral, i.e., the matrix element $\int \psi_0^* z \psi_1 \, dV$, will become identically zero. In exactly the same way $\int \psi_0^* x \psi_1 dV$ and $\int \psi_0^* y \psi_1 \, dV$ will also become zero. This is why, for a dipole transition of one electron in the atom, the azimuthal quantum number changes by ± 1.

Angular momentum and parity of a light quantum. As was indicated in Sec. 13, an electromagnetic field possesses angular momentum. If from equation (13.28) we determine the angular momentum of a quantum emitted during dipole radiation, it comes out equal to unity. And the state of the quantum is odd because it is determined by the parity of the dipole-moment vector components **d**, which, obviously, change sign for the interchange $x \to -x$, $y \to -y$, $z \to -z$. Hence, the selection rules for the azimuthal quantum number and parity of the state of the atom must be interpreted as the conservation laws of total angular momentum and total parity of the atom + quantum system in radiation. Clearly, if the angular momentum of a quantum is equal to unity, the angular momentum of the atom cannot change by greater than unity during radiation.

The selection rules for spin and total angular momentum. If the spin is in no way related to the orbital motion, the spin functions for the initial and final states must be the same, otherwise the transition dipole moment is equal to zero due to the orthogonality of spin functions that correspond to different spin eigenvalues.

This selection rule is approximate in character and is valid for light atoms. Taking into account the spin-orbital interaction, we must consider the selection rules for the total angular momentum $\mathbf{j} = \mathbf{M} \pm \boldsymbol{\sigma}$ [see (22.15)]. Since the angular momentum of a dipole quantum is equal to unity, we obtain the condition for j'—j: $j' = j$ or $j' = j \pm 1$. Here, the parity of the state must change. However, since the parity is not directly related to j but only to l, the transition $j' = j$ is also possible. But the transition from $j = 0$ to $j' = 0$ is forbidden, because in this transition the quantum cannot acquire the angular momentum. It is necessary to note that the angular momenta for quanta of higher multipole order than dipole can only be greater than unity, so that the transition from $j = 0$ to $j' = 0$ is forbidden for all approximations, and not only to the dipole approximation.

The selection rules for many-electron atoms. By considering a light quantum as a particle with unity angular momentum, it is easy to obtain the selection rules also for cases when the states of more than one electron change. Neglecting the spin-orbital interaction, the selection rules are the following: $S' = S$, $L' = L$ or $L' = L \pm 1$ and the parity is reversed. The transition $L' = L$ is possible here, because, in a many-electron system, parity is not related to total angular momentum.

Magnetic dipole radiation. A system of charges may radiate as a magnetic dipole as well as an electric dipole. Magnetic dipole radiation is usually related to a change of spin projection k_σ. Since the spin of an electron is one-half, the angular momentum of an atom changes by unity for a "flip" of the spin of an electron and for an unchanged orbital angular momentum. The moment of a magnetic dipole quantum is equal to unity just like the moment of an electric dipole quantum. But the parities of the electric and magnetic quanta are reversed. Indeed, the components of electric dipole moment change signs in an inversion of the coordinate system (31.35), while the magnetic-moment components do not change signs because the magnetic moment, like the angular momentum, is a pseudovector (see Sec. 16).

As was pointed out in Sec. 19, the intensity of magnetic dipole radiation is less than the intensity of electric dipole radiation, their ratio being $\left(\frac{v}{c}\right)^2$, where v is the charge velocity and c is the velocity of light. This ratio is about 10^{-5} for light elements.

Quadrupole radiation. In Sec. 19 it was shown that radiation is possible due to the change of quadrupole moment for the system. Here, electric quadrupole quanta occurring in an even state are radiated because the electric quadrupole moment is an even coordinate function. The angular momentum of a quadrupole quantum is equal to two.

Quadrupole radiation can occur when dipole radiation is forbidden by the selection rules. From Sec. 19, quadrupole radiation is obtained when taking into account the retardation inside the system. The order of magnitude of this retardation is determined by the ratio of the dimensions of the system to the wavelength of the emitted light. Therefore, the probability of quadrupole radiation is less than the probability of dipole radiation in the ratio $\left(\frac{r}{\lambda}\right)^2$, where r is the size of the system.

$\lambda \sim 0.5 \times 10^{-4}$ cm for visible light while the atomic dimensions are $r \sim 0.5 \times 10^{-8}$ cm. Therefore, in order of magnitude, the probability of a quadrupole transition is 10^8 times less than the probability of a dipole transition.

Metastable atoms. If an atom can go from an excited state to the ground state only by means of a transition which is forbidden in dipole radiation, it remains excited considerably longer than for a dipole transition. For a strong forbiddence it may remain excited for a very long time (even on the ordinary scale, and not the atomic scale). Such an atom is termed *metastable*. Usually, in gases which are not highly rarefied a metastable atom gives up its excitation energy to other atoms in collisions and not by means of radiation. Radiation will then not be observed. But in highly rarefied matter, for example in the solar corona or in a gaseous nebula, the spectral lines due to the de-excitation of metastable atoms are very bright. For example, in the spectra of nebulae, there occurs an intense magnetic dipole line of doubly-ionized oxygen atoms.

Nuclear isomerism. Transitions with very large Δj (up to 5) are observed in nuclei. For small excitation energies, of the order of several tens of kilovolts, metastable nuclei have very large de-excitation times—days or months. Such nuclei are called *isomers* with respect to the basic unexcited state of the nucleus. The phenomenon of nuclear isomerism in artificially radioactive nuclei was first discovered by I. V. Kurchatov and L. I. Rusinov (in Br^{80}).

The totally forbidden transition. The transition from $j=0$ to $j'=0$, with an energy of 1,414 kev is observed in the RaC nucleus. Since the radiation in this case is completely forbidden, the nucleus simply ejects an electron from the inner atomic shell by means of an electrostatic interaction; this may be explained as follows.

If an internal nuclear rearrangement occurs, the charge distribution inside it somehow changes. For a $0 \rightarrow 0$—transition, one spherically symmetrical charge distribution is rearranged into another, which is also symmetrical, but with a different radial dependence. Therefore, in accordance with Gauss' theorem, only the electric field *inside* the nucleus is changed. The field outside the nucleus cannot change; for instance, it cannot radiate quanta. The wave functions of the s-states of the electrons differ from zero in the nucleus. It follows that a change of field inside the nucleus is capable of influencing an electron and imparting to it an energy sufficient for ejection from the atom. In accordance with the law of conservation of energy, the electron, upon ejection from the atom, will have an energy equal to the energy difference of both spherically symmetrical states of the nucleus minus the binding energy in the atom.

It may be stated generally that the ejection of electrons from an atom shortens the lifetime of metastable isometric nuclei, since it makes transitions more possible.

Sec. 35. The Atom in a Constant External Field

A classical analogue. In considering the behaviour of a system of charges situated in an external magnetic field, it is very convenient to proceed from the idea of the Larmor precession of magnetic moment around the field. The only component of the angular momentum conserved in such precession is that directed along the field, both transversal components averaged over the precessional motion being zero.

The situation in quantum mechanics is analogous, with the difference that the projections perpendicular to the field do not exist as physical quantities. In this way a simple correspondence is established between the integrals of classical and quantum mechanics. The angular-momentum projection on the magnetic field is such a corresponding quantity; it can be called a *quantum integral of motion.*

An external magnetic field superimposed on an atom perturbs its state in a definite way. The Hamiltonian operator for such an atom may be divided into the operator $\hat{\mathscr{H}}^{(0)}$ for the unperturbed atom and the perturbation operator $\hat{\mathscr{H}}^{(1)}$ due to the magnetic field.

Addition of magnetic moments. Let us first of all write down the operator $\hat{\mathscr{H}}^{(1)}$ explicitly. It was shown in Sec. 32 that spin motion does not produce the same magnetic motion as orbital motion, namely, the magnetic moment for orbital motion is

$$\boldsymbol{\mu}_{orb} = \frac{e}{2mc}\mathbf{L}, \tag{35.1}$$

and the spin magnetic moment

$$\boldsymbol{\mu}_{sp} = \frac{e}{mc}\mathbf{S}. \tag{35.2}$$

Therefore the total magnetic moment is

$$\boldsymbol{\mu} = \boldsymbol{\mu}_{orb} + \boldsymbol{\mu}_{sp} = \frac{e}{2mc}(\mathbf{L} + 2\mathbf{S}). \tag{35.3}$$

Hence, the magnetic moments are not combined according to the same law as mechanical moments:

$$\mathbf{J} = \mathbf{L} + \mathbf{S}. \tag{35.4}$$

Comparing (35.3) and (35.4), we see that the magnetic moment of an atom is not proportional to its mechanical moment.

In accordance with (15.35), the perturbation energy caused by the magnetic field is equal to (the moments are expressed in h units)

$$\hat{\mathscr{H}}^{(1)} = -(\boldsymbol{\mu}\mathbf{H}) = \frac{eh\mathbf{H}}{2mc}(\mathbf{L} + 2\mathbf{S}) = \mu_0\mathbf{H}(\mathbf{J} + \mathbf{S}). \tag{35.5}$$

Here $\mu_0 = \frac{eh}{2mc}$ is the Bohr magneton. The plus sign resulted because the charge of the electron is $-e$. We note that the magnetic energy in expression (15.35) was defined as a correction to the Hamiltonian, i.e., to the energy expressed in terms of momenta. Therefore, in quantum theory it is directly interpreted in terms of operators.

The vector model of the atom. To a first approximation, the energy correction is equal to the mean value of the perturbing energy taken over unperturbed motion [see (33.31)]. Therefore, we first find the unperturbed state of the atom without the superimposition of a magnetic field. Let us suppose that a normal coupling exists in the atom between the total spin and the total orbital angular momentum (see Sec. 33), i.e., all the orbital angular momenta of the electrons are combined in one resultant orbital angular momentum **L**, and all the spin angular momenta are combined in one resultant spin angular momentum **S**. Examples of such orbital and spin angular-momentum composition were given in Sec. 33 (in the text and in the exercises). For example, in combining the angular momenta of two np-electrons, the following states are obtained: ^1D, ^3P, and ^1S. All these states are formed in accordance with the Pauli principle, and possess spatial wave functions of different forms. For this reason, in all three states, the energy for purely electrostatic electron interactions differs by magnitudes of the order of an atomic unit, i.e., by several electron-volts.

Let us choose the ground state of these states. In accordance with Hund's first rule, this is the ^3P state. We have not written the subscript J here because it can have three values: $J = 2$, $J = 1$, and $J = 0$. Accordingly, we have written 3 on the upper left. The states which differ only in J, for identical L and S, are considerably closer to each other than the three states with differing S or L listed above.

Let us estimate the order of magnitude for multiplet level splitting, i.e., the spacing of levels with different J. A magnetic field of moment

μ is of the order $\frac{\mu}{r^3}$, so that the interaction energy of two moments is $\frac{\mu^2}{r^3}$. To evaluate the order of magnitude we put one Bohr magneton in place of μ, i.e., 10^{-20}, and $r \sim 0.5 \times 10^{-8}$. This results in an interaction energy of the order 10^{-15}, i.e., thousandths of an electron-volt (in practice, greater multiplet splitting is observed due to larger μ and smaller effective radius values). In any case, the levels $^3P_2$, $^3P_1$, and $^3P_0$, which are comparatively different from the other two levels ^1D and ^1S, occur close to one another. The ^3P level is split into three fine-structure levels which, in the given case, corresponds to the superscript 3. If $L < S$, the number of components of the multiplet splitting is determined by L.

Each of the levels with a given J corresponds to a definite configuration of the vectors **L** and **S**. In classical theory, we would say that **L** and **S** are parallel in the state with $J=2$, antiparallel for $J=0$, and perpendicular for $J=1$. Of course, the latter one is not at all meaningful in quantum theory because only one angular-momentum projection exists. The projection of S on L is equal to zero for $J=1$, and the other projections do not exist.

At the beginning of this section we indicated that, in the classical analogy, those components which are not conserved are, in some way, averaged over the Larmor precession of angular momenta and yield zero. In this case, we are not concerned with precession in an external magnetic field, but with that in the internal field of the magnetic moments themselves. Since **J** is an exact integral of motion we can, in a visual demonstration, consider that the direction of **J** is fixed in space, while the triangle consisting of the vectors **L**, **S**, and **J** precesses about **J** in space. In the cases for which $J=2$ and $J=0$ the triangle degenerates to a straight line. Thus, to each of the multiplet levels there corresponds a definite vector model given by **L**, **S**, and **J**. We note that this refers to normal coupling.

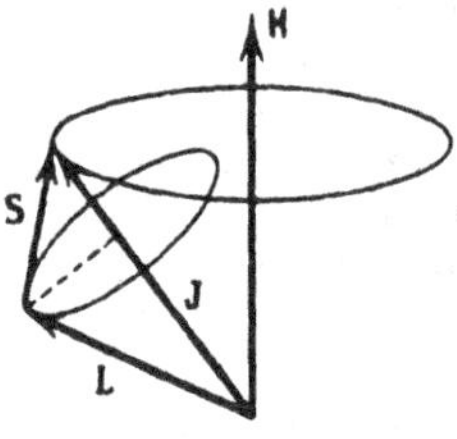

Fig. 44

An external magnetic field **H** causes the vectors **L**, **S** to precess about its direction. It is most simple here to consider the two opposing limiting cases. We shall examine them.

A weak external field. Let the external field be weak compared with the effective internal field that the multiplet level splitting is due to. Since the Larmor precession frequency is proportional to the magnetic field, the triangle **LSJ** in this case rotates about the side **J** considerably faster than the precession about **H**. During the time of one rotation about **H**, the triangle can rotate very many times about **J**. Therefore, the coupling of the vectors **L**, **S**, and **J** in the triangle is not disrupted, as it were, due to the internal magnetic forces forming the triangle being large compared with the external magnetic force. We have shown this idea in Fig. 44.

Let us now find the correction due to the magnetic field. In calculating the mean value of the perturbation energy from the unperturbed motion, it is convenient to make use of the Larmor precession model. In this case, two forms of precession must be considered: the triangle **LSJ** about **J** and the precession of **J** about the magnetic field.

Equation (35.5) involves the vectors **J** and **S**. It is very simple to average **J**: we must take its projection on the magnetic field J_z. We shall consider that the z-axis coincides with the direction of **H**. The projection of **S** upon **H** is not meaningful because the vector **S** together with the triangle **LSJ** rotates considerably more rapidly about **J** than about **H**. The component **S**, perpendicular to **J**, is averaged by the precession in motion unperturbed by the external field. There remains the projection parallel to **J** and equal to

$$\mathbf{S}_J = \frac{\mathbf{J}(\mathbf{S}\mathbf{J})}{J^2}\,. \tag{35.6}$$

Obviously, the projection of this vector upon **H** is equal to $J_z \frac{(\mathbf{S}\mathbf{J})}{J^2}$.

Thus, the mean value of $\hat{\mathscr{H}}^{(1)}$ is proportional to J_z and is equal to

$$\mathscr{E}^{(1)} = \bar{\hat{\mathscr{H}}}_1 = \mu_0 H J_z \left(1 + \frac{(\mathbf{S}\mathbf{J})}{J^2}\right). \tag{35.7}$$

It is now necessary to give a quantum meaning to the product $(\mathbf{SJ})$. From the definition of **J** (35.4) we have

$$\mathbf{L} = \mathbf{J} - \mathbf{S}\,. \tag{35.8}$$

Squaring this equation, we get

$$\mathbf{L}^2 = \mathbf{J}^2 + \mathbf{S}^2 - 2(\mathbf{S}\mathbf{J})\,. \tag{35.9}$$

Expressing the square of the angular momentum in h units, in accordance with (30.29), we have

$$L^2 = L(L+1)\,.$$

Let us make similar substitutions for $\mathbf{J}^2$ and $\mathbf{S}^2$. Therefore

$$\frac{(\mathbf{S}\mathbf{J})}{J^2} = \frac{J(J+1) + S(S+1) - L(L+1)}{2J(J+1)}\,. \tag{35.10}$$

Substituting (35.10) in (35.7) we obtain finally

$$\mathscr{E}^{(1)} = \bar{\hat{\mathscr{H}}}^{(1)} = \mu_0 H J_z \left(1 + \frac{J(J+1) + S(S+1) - L(L+1)}{2J(J+1)}\right). \tag{35.11}$$

Thus, the fine-structure level with a given J is split into as many levels as there are different projections of **J** on the magnetic field, i.e., $2J+1$ levels. For given L and S, the following definite factor corresponds to each value of J:

$$g = 1 + \frac{J(J+1) + S(S+1) - L(L+1)}{2J(J+1)}, \tag{35.12}$$

It is called the Landé factor.

For example, for $L=S=J=1$, we obtain

$$g = 1 + \frac{1}{2}\,\frac{2+2-2}{2} = \frac{3}{2}.$$

Analogously, for $J=2$, $S=L=1$

$$g = 1 + \frac{1}{2}\,\frac{6+2-2}{6} = \frac{3}{2}.$$

The level with $J=0$ does not split.

Splitting in a strong field. The representation of splitting set out here corresponds to reality only as long as the magnetic field is so weak that the spacing between the $2J+1$ levels of $\mu_0 g H J_z$ in the magnetic field is small compared with that between the unsplit multiplet levels themselves with differing J. When the splitting in the magnetic field is comparable with that of the multiplet itself, or is somewhat greater, the pattern becomes more complicated, but in a strong field it once again becomes very simple.

Therefore we shall consider the opposite extreme case, when the external field is strong compared with the internal field, so that the coupling between the vectors **L**, **S**, **J** in the triangle is disrupted. The necessity for this disruption in a sufficiently strong magnetic field can be explained by the fact that **S** precesses twice as fast as **L**. Then, from the classical analogy, each of the vectors **S** and **L** precesses independently about the magnetic field, so that the correction to the energy is given by a different expression from (35.11):

$$\mathscr{E}^{(1)} = \hat{\mathscr{H}}^{(1)} = \mu_0 e H (L_z + 2S_z). \tag{35.13}$$

Here L_z is the projection of the orbital angular momentum upon the z-axis, and S_z is the total-spin projection of the atom upon the same axis (in $\hbar$ units). Naturally, the total values of L and S are not changed by the magnetic field, though the distribution of levels in a strong magnetic field is not related to the multiplet structure, as was the case in a weak field, but only with the possible projections of **L** and **S** on the magnetic field. The vectors **L** and **S** precess about the field far more rapidly than they precess about **J** without the field. This is why the coupling in the triangle is disrupted.

The projections S_z and L_z are changed by unity, therefore all the levels $\mathscr{E}^{(1)}$ in expression (35.13) are equidistant. Of course, certain values of $\mathscr{E}^{(1)}$ may be repeated several times if the sum $L_z + 2S_z$ assumes the same value in several ways. For example, if $L=1$, $S=1$, then we get the following range of values of the sum: $1+2=3$, $0+2=2$, $1+0=1$, $-1+2=1$, $0+0=0$, $-1+0=-1$, $1-2=-1$, $0-2=-2$, $-1-2=-3$; there are in all seven equidistant values, and 1 and -1 are obtained in two ways (i.e., each of them from the confluence of two levels), so that there are nine states in all. We note that in a weak field the same multiplet split thus: $J=2$ into 5 levels, $J=1$ into 3 levels and $J=0$ did not split. As was to be expected, the total number of different states in the strong and weak fields is the same.

The radiation spectrum for level splitting in a strong field. Let us now see what spectral lines appear when light is emitted from an atom situated in a magnetic field. To begin with, let us consider a strong field, because the pattern of the splitting of spectral lines is simpler in this case than in that of a weak field. Both levels, upper and lower, resulting from two multiplets are split into a certain number of equidistant levels in accordance with formula (35.13).

Let the radiation be observed in a direction perpendicular to the magnetic field. The radiation polarization vector is perpendicular to the direction of propagation, i.e., it is either directed along the magnetic field or in a third perpendicular direction, say along the x-axis (the magnetic field is along the z-axis). The selection rules for radiation polarized along z and along x are different. For polarization along the z-axis, the orbital magnetic quantum number must be conserved. S_z is also conserved for all polarization in which the spin-orbital interaction is neglected. Therefore, all lines polarized along the z-axis, i.e., along the magnetic field, have the same frequency, which corresponds to the energy difference of the two initial levels $\mathscr{E}_1 - \mathscr{E}_0$ prior to splitting in the magnetic field: the correction (35.13) is cancelled in calculating the difference $\mathscr{E}_1^{(1)} - \mathscr{E}_0^{(1)}$. A wave polarized along the x-axis can be represented as the sum of two waves circularly polarized with opposite directions of polarization. The selection rule for these lines is that k can change only by ± 1. Consequently, the radiation polarized along the x-axis has a frequency that differs from the initial frequency by $\pm \frac{eH}{2mc}$. In observing the spectral lines emitted perpendicularly to a strong magnetic field, the original line is thus split into three lines separated by an interval which is equal to the Larmor frequency for the given field.

If we drill a hole in the shoe of an electromagnet it is possible to observe radiation propagated along the magnetic field. It is circularly polarized in the xy-plane. The selection rules for right- and left-hand circular polarization correspond to a change of k by ± 1, so that there will be observed two lines spaced from the centre by $\pm \frac{eH}{2mc}$. Thus, when the field is switched on, the original line will split into two lines separated by an interval equal to twice the Larmor frequency.

Exactly the same picture is found in the classical oscillatory motion of a charge situated in a magnetic field. This problem was considered in exercise 6, Sec. 21.

The effect of the splitting of spectral lines in a magnetic field was discovered by Zeeman before the quantum theory of the atom appeared. Therefore, the then accepted theoretical explanation of the Zeeman effect corresponded to the classical problem, where it was considered that the charge performs an oscillatory motion. However, in observing spectra, this classical picture applies only in strong magnetic fields such that the splitting of lines obtained is considerably greater than the spacing between multiplet levels. Under these conditions the Zeeman effect is termed *normal*, because outwardly it corresponds to the theoretical ideas of the time at which it was discovered. It may be noticed that a field which is strong for one multiplet can still be weak for another.

Spectral-line splitting in a weak magnetic field. The Zeeman effect in a weak magnetic field is termed *anomalous*. A spectral pattern is obtained which is entirely different from the classical. First of all, the number of splitting components can differ from the normal. The distances between them are also quite different.

As an example let us consider the anomalous Zeeman effect in the so-called D-line doublet of sodium. This line is double without an external magnetic field. It corresponds to the two transitions ${}^2P_{\frac{1}{2}} \rightarrow {}^2S_{\frac{1}{2}}$ and ${}^2P_{\frac{3}{2}} \rightarrow {}^2S_{\frac{1}{2}}$. The ${}^2P$ level has an orbital angular momentum 1 and spin $\frac{1}{2}$. Therefore, the resultant value of the total angular momentum J can be $1+\frac{1}{2}=\frac{3}{2}$ and $1-\frac{1}{2}=\frac{1}{2}$. This is where we get the fine doublet structure of the ${}^2P$ level in the absence of an external field. The ${}^2S$ level cannot split without a field because it has an orbital angular momentum of zero. The double D-line in the sodium spectrum arises in the transition from the doublet level to the single. According to our rough estimate of the fine-structure splitting, the difference in frequency between its components amounts to about one thousandth of the mean frequency of the doublet. The ${}^2P_{\frac{1}{2}}$ level is lower than the ${}^2P_{\frac{3}{2}}$ level.

Let us now calculate the Landé factor for three levels.

1) ${}^2P_{3/2}$: $J = 3/2, \quad L = 1, \quad S = 1/2,$

$$g = 1 + \frac{1}{2}\,\frac{3/2 \cdot 5/2 + 1/2 \cdot 3/2 - 1 \cdot 2}{3/2 \cdot 5/2} = \frac{4}{3}.$$

2) ${}^2P_{1/2}$: $J = 1/2, \quad L = 1, \quad S = 1/2,$

$$g = 1 + \frac{1}{2}\,\frac{1/2 \cdot 3/2 + 1/2 \cdot 3/2 - 1 \cdot 2}{1/2 \cdot 3/2} = \frac{2}{3}.$$

3) $^2S_{1/2}$: $J = 1/2$, $L = 0$, $S = 1/2$,

$$g = 1 + \frac{1}{2}\frac{1/2 \cdot 3/2 + 1/2 \cdot 3/2}{1/2 \cdot 3/2} = 2 .$$

In accordance with equation (35.11), we have an expression for the energy of the $^2P_{\frac{3}{2}}$ state in a magnetic field. For conciseness we shall denote the quantity $\mu_0 H$ by the single letter β. Then

$$\mathscr{E}^{(1)} = \frac{4}{3}\beta J_z .$$

But J_z takes on four values: 3/2, 1/2, —1/2, —3/2. Hence, in a field, the $^3P_{\frac{3}{2}}$ level splits into four levels, whose energy differences from the central, unperturbed, state are

$$\mathscr{E}^{(1)}(-3/2) = -2\beta , \quad \mathscr{E}^{(1)}(-1/2) = -\frac{2}{3}\beta ,$$

$$\mathscr{E}^{(1)}(+1/2) = \frac{2}{3}\beta , \quad \mathscr{E}^{(1)}(+3/2) = 2\beta$$

respectively.

We obtain two energy values for the $^2P_{\frac{1}{2}}$ fine-structure level:

$$\mathscr{E}^{(1)}(-1/2) = -\frac{1}{3}\beta ; \quad \mathscr{E}^{(1)}(+1/2) = \frac{1}{3}\beta .$$

And, finally, for the lower $^2S_{\frac{1}{2}}$ level, we get

$$\mathscr{E}^{(1)}(-1/2) = -\beta , \ \mathscr{E}^{(1)}(+1/2) = \beta .$$

Let us now find the spectral pattern. We start with the $^2P_{\frac{1}{2}} \to {}^2S_{\frac{1}{2}}$ transitions. The oscillations polarized along the field obey the selection rule $\Delta J_z = 0$. Hence, their frequencies are shifted relative to the central position by

$$\mathscr{E}^{(1)}(^2P_{1/2}, -1/2) - \mathscr{E}^{(1)}(^2S_{1/2} -1/2) = -\frac{1}{3}\beta + \beta = \frac{2}{3}\beta$$

and by

$$\mathscr{E}^{(1)}(^2P_{1/2}, 1/2) - \mathscr{E}^{(1)}(^2S_{1/2}, 1/2) = \frac{1}{3}\beta - \beta = -\frac{2}{3}\beta .$$

Unlike the normal Zeeman effect, a double line has been obtained also for radiation polarized along the magnetic field.

For perpendicular polarizations we have

$$\mathscr{E}^{(1)}(^2P_{1/2}, -1/2) - \mathscr{E}^{(1)}(^2S_{1/2}, 1/2) = -\frac{1}{3}\beta - \beta = -\frac{4}{3}\beta \quad \text{(right-handed polarization)},$$

$$\mathscr{E}^{(1)}(^2P_{1/2}, 1/2) - \mathscr{E}^{(1)}(^2S_{1/2}, -1/2) = \frac{1}{3}\beta + \beta = \frac{4}{3}\beta \quad \text{(left-handed polarization)}.$$

Let us take the transition $^2P_{\frac{3}{2}} \to {}^2S_{\frac{1}{2}}$. If the oscillation is polarized along the field we once again have, of course, two lines, though with other spacing:

$$\mathscr{E}^{(1)}(^2P_{3/2}, -1/2) - \mathscr{E}^{(1)}(^2S_{1/2}, -1/2) = -\frac{2}{3}\beta + \beta = \frac{1}{3}\beta,$$

$$\mathscr{E}^{(1)}(^2P_{3/2}, 1/2) - \mathscr{E}^{(1)}(^2S_{1/2}, 1/2) = \frac{2}{3}\beta - \beta = -\frac{1}{3}\beta.$$

We have, for both circular polarizations:

$$\mathscr{E}^{(1)}(^2P_{3/2}, -3/2) - \mathscr{E}^{(1)}(^2S_{1/2}, -1/2) = -2\beta + \beta = -\beta,$$

$$\mathscr{E}^{(1)}(^2P_{3/2}, -1/2) - \mathscr{E}^{(1)}(^2S_{1/2}, 1/2) = -\frac{2}{3}\beta - \beta = -\frac{5}{3}\beta.$$

These are the results for right-handed polarization. The corresponding splitting for left-handed polarization is β and $\frac{5}{3}\beta$.

Thus, one component of the D-line is split into six Zeeman components, and the other into four.

In the given case, the Zeeman effect remains anomalous as long as β is negligibly small compared with one thousandth of a volt, or the magnetic field is very much less than 5,000 CGSE units.

A diagram of the splitting is shown in Fig. 45.

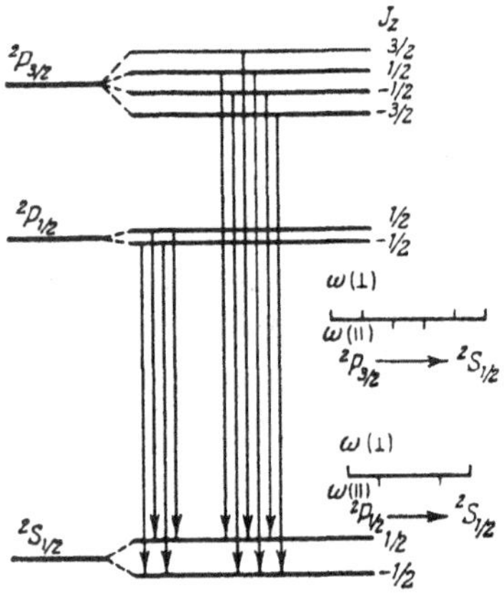

Fig. 45

The atom in an electric field (Stark effect). The multiplet levels for a certain total angular momentum J split in an electric field, too. We shall consider first of all the case of a weak field, when the level shift caused by the field is small compared with natural multiplet splitting.

First of all, we must bear in mind that the angular-momentum projection on the electric field is determined only within the accuracy of the sign, because the angular momentum is a pseudovector while the electric field is a real vector. In reversing all the coordinate signs, the angular-momentum components change sign while the electric-field components do not change sign. But since the choice of right-handed or left-handed coordinate system is arbitrary, the projections of the angular momenta on the electric field are physically determined only to the accuracy of the sign. If J is an integer, the number of its projections which differ in absolute value is equal to $J+1$ $(0, 1, \ldots, J)$, while if J is a half-integral number then the total number of projections is: $J+\frac{1}{2}\left(\frac{1}{2}, \frac{3}{2}, \ldots, J\right)$. For example, if $J=\frac{1}{2}$, there is only one nonnegative projection. Therefore, the state with angular momentum $\frac{1}{2}$ is not split by an electric field, at any rate as long as the coupling between **L** and **S** is not disrupted. For comparison we note that the magnetic field splits the state with $J=\frac{1}{2}$ into two states because the magnetic field, like the angular momentum, is a pseudovector.

In a stronger electric field the coupling between L and S is disrupted. In this case the scheme of splitting is the following. The vector **L** is integral. It has $L+1$ projections on the electric field. We must project **S** onto its projection. But since **L** and **S** are both pseudovectors, the number of projections of **S** upon **L** is already equal to $2S+1$. The only exception is when the projection of **L** upon the field is equal to zero. This level splits into $S+1$ or $S+\frac{1}{2}$ levels according to S.

The square-law Stark effect. The amount of splitting is determined by the relative shift of neighbouring levels. As was shown in Sec. 33 (33.31 a), the shift of an energy level is equal to the average of the perturbation energy for unperturbed motion. Proceeding from (14.28), we have the following expression for the perturbation energy in a homogeneous electric field

$$\hat{\mathscr{H}}^{(1)} = -(\hat{\mathbf{d}}\mathbf{E}) . \qquad (35.14)$$

But it is easy to see that the average of this quantity is equal to zero. Indeed, the wave function of an atomic state with given J is always odd or even (with the exception of hydrogen, see below). Therefore, the product $\psi_J^* \psi_J$ must be even. From (30.24), the average of $\hat{\mathscr{H}}^{(1)}$ is equal to

$$\overline{\hat{\mathscr{H}}^{(1)}} = -e\mathbf{E}\int \psi_J^* \mathbf{r} \psi_J dV . \qquad (35.15)$$

But the integrand is an odd function, so that its integral is identically equal to zero.

Level splitting is obtained only to a second approximation, if into (35.15) we substitute wave functions which have already been perturbed by the external field. This splitting is governed by a square-law field dependence.

The linear Stark effect. In a hydrogen atom the electron energy depends only upon the principal quantum number n and does not depend upon l. Therefore, the state with $\mathscr{E}=\mathscr{E}_n$ is represented as a superposition of states with l varying from 0 to $n-1$. But the wave function is even for even l, and odd for odd l. Hence, the function with $\mathscr{E} = \mathscr{E}_n$ does not have a definite parity, so that the integral (35.15) does not become zero. Therefore, in the hydrogen atom we observe line splitting which depends linearly upon the electric field.*

Highly excited atomic states always more or less resemble hydrogen-atom states, because the nucleus and the atomic residue act upon an electron, which has receded far from the nucleus, in a way similar to a point charge. The energies of these states depend upon l in accordance with the expression (31.46). These states give a linear Stark effect if the perturbation produced by the field shifts the levels more strongly than they are split in l.

Ionization of the atom by a constant field. A constant electric field not only shifts the energy levels of an atom, but also qualitatively changes its whole state.

Let us write down the potential energy of an electron in an atom situated in an external electric field E which is directed along the z-axis:

$$U = U_0(r) + eEz\,. \tag{35.16}$$

For a sufficiently large and negative z the potential energy far away from the atom is less than in the atom. The potential well in the atom is separated from the region of large negative z (where the potential energy can be still less) by a potential barrier. But there is always the probability of a spontaneous electron transition through the potential barrier into the free state. Transitions of this type were considered in Sec. 28 as applied to alpha disintegration.

Any state of an atom put in a constant electric field may be ionized, but, naturally, if the field is weak the probability of ionization becomes vanishingly small. In a strong field the potential barrier becomes transparent, especially for highly excited atomic states. If the time for the spontaneous ejection of an electron in such a state turns out to be less than the radiation time, the corresponding line in the spectrum disappears.

Thus, a weak perturbation inside an atom (the atomic unit of field intensity $E = \frac{m^2 e^5}{\hbar^4} = 5.13 \cdot 10^9$ v/cm, so that the external field is always small compared with the atomic field) essentially affects the state since the conditions at infinity change. But if the broadening of the atomic levels is still small compared with the distance between them, they can be regarded, as before, as discrete.

Exercise

Construct a diagram for the splitting of the multiplet $^3P \rightarrow {}^3S$ and the transitions in a strong and weak magnetic field.

---

* The relativistic expression (38.28) for the energy of a hydrogen atom involves n and j. The orbital angular momentum $l = j \pm {}^1/_2$ for a given j, so that a state with given n and j (in the same way as to a nonrelativistic approximation) does not have definite parity and yields a linear Stark effect.

Sec. 36. Quantum Theory of Dispersion

The classical theory of dispersion, a brief outline of which was given in exercise 19, Sec. 16, proceeds from the concept of a charge elastically bound in an atom. The forced oscillations of these charges under the action of a sinusoidally varying field lead to an electrical polarization of the medium proportional to the field. Whence the dielectric constant can be easily calculated as a function of the frequency.

The classical theory of dispersion is in good agreement with experiment. Yet, at the present time it is well known that the charges in atoms are by no means bound by elastic forces. For this reason, the success of classical dispersion theory may appear incomprehensible.

Even though the charges are not bound by elastic forces, there exist quantities, relating to the motion of the charges, which vary harmonically with time: these are the coordinate matrix elements [see (34.18)]. Similar harmonic oscillations occur, as is well known, in the classical mechanics of elastically bound particles. The dipole moment of an atom, induced by an external alternating field, is expressed in terms of the dipole-moment matrix elements directly related to the coordinate matrix elements. In the present section, a quantum theory of dispersion will be formulated which will lead to the same expression for dielectric constant as classical theory; it will also indicate which quantities should correspond to each other in both theories.

The wave equation for an atom in a given field of radiation. In order to calculate the dipole moment induced by a field, we must first of all determine the wave function of the atom in the external field. In contrast to the previous section, where the behaviour of an atom in a constant external field was studied, we shall here consider the interaction of an atom with an alternating external field which varies according to the law

$$\mathbf{E} = \mathbf{E}_0 \cos \omega t \,. \tag{36.1}$$

It turns out to be more convenient here to write down the field, straightway in real form instead of taking the real part of the final result in order to have a real Hamiltonian.

The wavelength of a light ray is rightly considered large compared with atomic dimensions (this was confirmed by the estimate in Sec. 34), so that the field **E** may be considered homogeneous: its phase is constant over the whole atom.

We determined the energy for a system of charges in an external homogeneous field in Sec. 14 [see (14.28) and (35.14)]. The correction to the Hamiltonian—due to a homogeneous electric field—looks like

$$\hat{\mathscr{H}}^{(1)} = -(\hat{\mathbf{d}}\mathbf{E}) \,. \tag{36.2}$$

If we call the Hamiltonian of an unperturbed system $\hat{\mathscr{H}}^{(0)}$, then Schrödinger's equation will be of the form

$$-\frac{h}{i}\frac{\partial\psi}{\partial t}=\hat{\mathscr{H}}^{(0)}\psi+\hat{\mathscr{H}}^{(1)}\psi\,. \tag{36.3}$$

Separating the wave function into an unperturbed part $\psi^{(0)}$ and a perturbation $\psi^{(1)}$, and regarding the perturbation as relatively small, we obtain an equation which we have already used in Sec. 34:

$$-\frac{h}{i}\frac{\partial\psi^{(1)}}{\partial t}-\hat{\mathscr{H}}^{(0)}\psi^{(1)}=\hat{\mathscr{H}}^{(1)}\psi^{(0)}\,. \tag{36.4}$$

Expansion in eigenfunctions. We seek the unknown function $\psi^{(1)}$ in the form of a wave function expansion with time-dependent coefficients:

$$\psi^{(1)}=\sum_n c_n\psi_n^{(0)}\,. \tag{36.5}$$

We obtained an equation in (34.7), Sec. 34, for the expansion coefficients

$$-\frac{h}{i}\frac{dc_n}{dt}=\int\psi_n^{(0)*}\,\hat{\mathscr{H}}^{(1)}\psi_0^{(0)}\,dV\,. \tag{36.6}$$

The right-hand side of this equation depends upon time in a different way from that in equation (34.7), because the perturbation operator $\hat{\mathscr{H}}^{(1)}$ involves time explicitly [see (36.1)]. Let us consider that the unperturbed state of the atom is its ground state, which we shall write with a subscript 0, i.e., $\psi_0{}^{(0)}$. Then there will simply be the matrix element $\mathbf{d}_{0n}$ on the right-hand side of (36.6) multiplied by $-\mathbf{E}_0\cos\omega t$. The time dependence for the matrix element was found in Sec. 34. Using the notation (34.20) we can write

$$\mathbf{d}_{0n}=e^{i\,\omega_{n0}}\,\mathbf{d}'_{0n}\,. \tag{36.7}$$

The representation of a matrix element together with its time dependence is termed the *Heisenberg* representation, and that without the dependence, in the form $\mathbf{d}'_{0n}$, is the *Schrödinger* representation. Substituting (36.7) in (36.6) we obtain

$$-\frac{h}{i}\frac{dc_n}{dt}=-\frac{1}{2}\left(e^{i\,(\omega+\omega_{n0})\,t}+e^{i\,(\omega_{n0}-\omega)\,t}\right)(\mathbf{E}_0\,\mathbf{d}'_{0n})\,. \tag{36.8}$$

In order to integrate this equation we must impose a certain initial condition upon c_n. It is natural to suppose that the external field acts for a sufficiently long time so that all the transition processes related to turning on the field do not affect the states. We can assume, for example, that the external field depends upon time according to the law:

$$\begin{aligned} \mathbf{E} &= \mathbf{E}_0 e^{\alpha t} \cos \omega t \quad \text{for} \quad t < 0\,, \\ \mathbf{E} &= \mathbf{E}_0 \cos \omega t \quad \text{for} \quad t \geqslant 0\,, \end{aligned} \tag{36.9}$$

i.e., the amplitude gradually rises with time to the value $\mathbf{E}_0$. This law for the change of the field must be substituted into (36.8), integration performed from $-\infty$ to any t, and α must tend to zero. After this, at each instant of time ($t < 0$ or $t \geqslant 0$) there will be a single dependence of c_n upon t:

$$c_n = \frac{1}{2h}\left(\frac{e^{i(\omega_{n0}-\omega)t}}{\omega_{n0}-\omega} + \frac{e^{i(\omega_{n0}+\omega)t}}{\omega_{n0}+\omega}\right)(\mathbf{E}_0\,\mathbf{d}'_{0n})\,. \tag{36.10}$$

Induced dipole moment. The mean value of the dipole moment is calculated according to the general formula (30.24) for mean values:

$$\bar{\mathbf{d}} = \int \psi^* \mathbf{d} \psi \, dV = \int (\psi^{(0)*} + \psi^{(1)*})\, \mathbf{d}\, (\psi^{(0)} + \psi^{(1)})\, dV\,. \tag{36.11}$$

The quadratic term in $\psi^{(1)}$ must, of course, be discarded, since the calculations are performed to the accuracy of terms proportional to $\mathbf{E}$ in the first degree. In addition, the term $\int \psi^{(0)*}\, \mathbf{d}\, \psi^{(0)}\, dV$ does not depend at all upon $\mathbf{E}$ and, therefore, is irrelevant to the problem of polarization produced by an external field. Also, this term is usually equal to zero, as indicated in the previous section in connection with the expression (35.15). Hence, the mean dipole moment responsible for dispersion is

$$\bar{\mathbf{d}} = \int (\psi^{(0)*}\, \mathbf{d}\, \psi^{(1)} + \psi^{(1)*}\, \mathbf{d}\, \psi^{(0)})\, dV\,. \tag{36.12}$$

We shall substitute here the expansion (36.5) and integrate the series term by term:

$$\bar{\mathbf{d}} = \sum_n \left(c_n \int \psi_0^{(0)*}\, \mathbf{d}\, \psi_n^{(0)} dV + c_n^* \int \psi_n^{(0)*}\, \mathbf{d}\, \psi_0^{(0)}\, dV\right). \tag{36.13}$$

The integrals involved here are once again dipole-moment matrix elements. Substituting their expressions from (36.7), we write the mean dipole moment as

$$\bar{\mathbf{d}} = \sum_n (c_n e^{i\omega_{0n}t}\, \mathbf{d}'_{n0} + c_n^* e^{-i\omega_{0n}t}\, \mathbf{d}'_{0n})\,. \tag{36.14}$$

With the aid of expression (36.10), we finally obtain for c_n:

$$\begin{aligned} \bar{\mathbf{d}} = \frac{1}{2h} \sum_n \Bigg[&\left(\frac{e^{-i\omega t}}{\omega_{n0} - \omega} + \frac{e^{i\omega t}}{\omega_{n0} + \omega}\right) \mathbf{d}_{n0}\,(\mathbf{E}_0\, \mathbf{d}_{0n}) + \\ &+ \left(\frac{e^{i\omega t}}{\omega_{n0} - \omega} + \frac{e^{-i\omega t}}{\omega_{n0} + \omega}\right) \mathbf{d}_{0n}\,(\mathbf{E}_0\, \mathbf{d}_{n0}) \Bigg]. \end{aligned} \tag{36.15}$$

Here we could already have written $\mathbf{d}_{n0}$ in place of $\mathbf{d}'_{n0}$, because the time factors of $\mathbf{d}_{n0}$ and $\mathbf{d}_{0n}$ cancel.

Polarization. In order to calculate the polarization of atoms by a light-wave field, it is sufficient to know only the dipole moment projection on the field. If, for example, the electric field of an incident wave is directed along the x-axis, then the expression (36.15) involves only angular-momentum transition components directed along the x-axis, i.e., the matrix elements of x:

$$\overline{\mathbf{d}} = \frac{e^2}{2h}\sum_n\left[\left(\frac{e^{-i\omega t}}{\omega_{n0}-\omega} + \frac{e^{i\omega t}}{\omega_{n0}+\omega}\right)\mathbf{E}_0\,|x_{0n}|^2 + \right.$$

$$\left. + \left(\frac{e^{i\omega t}}{\omega_{n0}-\omega} + \frac{e^{-i\omega t}}{\omega_{n0}+\omega}\right)\mathbf{E}_0\,|x_{0n}|^2\right]. \tag{36.16}$$

We have made use of the Hermitian nature of matrix elements expressed by the relationship (34.15). In other words, we have put $|x_{0n}|^2$ in place of $x_{0n}\,x_{n0}$.

Now, by performing a simple algebraical transformation and introducing the electric field E itself instead of its amplitude E_0, we have:

$$\overline{\mathbf{d}} = \sum_n \frac{2\,\omega_{n0}\,e^2\,|x_{0n}|^2}{h\,(\omega_{n0}^2 - \omega^2)}\,\mathbf{E}. \tag{36.17}$$

The dispersion formula. Let us consider the polarization of a medium $\mathbf{P} = N\overline{\mathbf{d}}$, where N is the number of atoms in unit volume.* The electric induction is related to the electric field and polarization by the relationship (16.23) which, in the given case, is of the form

$$\mathbf{D} = \mathbf{E} + 4\pi\mathbf{P} = \left(1 + \frac{4\pi N e^2}{h}\sum_n \frac{2\,\omega_{n0}\,|x_{0n}|^2}{\omega_{n0}^2 - \omega^2}\right)\mathbf{E}. \tag{36.18}$$

But $\mathbf{D} = \varepsilon\mathbf{E}$ from the definition of dielectric constant, so that

$$\varepsilon = 1 + \frac{4\pi N e^2}{h}\sum_n \frac{2\,\omega_{n0}\,|x_{0n}|^2}{\omega_{n0}^2 - \omega^2}. \tag{36.19}$$

We note that this expression is correct only when the frequency of the incident radiation is not close to one of the natural frequencies of the atom ω_{n0}. Otherwise the denominator in (36.19), and correspondingly in all the previous equations up to (36.10), can become zero; in any case, it becomes small. But then the perturbation caused by the field is large, and it cannot be regarded as weak, utilizing the expansion $\psi = \psi^{(0)} + \psi^{(1)}$ and neglecting $|\psi^{(1)}|^2$. Physically, this means that if the frequency of the incident light is close to the frequency of one of the absorption lines (or, what is just the same, the emission lines), we must take into account the damping in the amplitude of the excited atomic states due to radiation. In other

---

* Such additivity of dipole moments is true only for gases.

words, the amplitudes of the excited atomic states must not be taken with a purely imaginary time dependence, but of the form (28.14):

$$\psi_n^{(0)} = e^{-\frac{\Gamma_n t}{2h} - i\frac{\mathscr{E}_n t}{h}}. \tag{36.20}$$

The quantum evaluation of Γ_n for radiation damping is rather complicated, and we shall not deal with it.

A comparison of classical and quantum dispersion formulae. We shall now go over to a comparison of the quantum formula of dispersion (36.19) and the classical formula. We shall write the latter for the case when there are many types of oscillation whose frequencies are ω_{n0} (instead of a single frequency ω_0). We shall separate the total number of oscillators N into parts corresponding to each separate frequency ω_{n0}:

$$N = \sum_n N_n. \tag{36.21}$$

If we introduce the relative fractions of each oscillation by the formula

$$f_n \equiv \frac{N_n}{N}, \tag{36.22}$$

then it is obvious that

$$\sum_n f_n = 1. \tag{36.23}$$

Let us suppose that the frequency of the incident radiation is not close to any one of the natural frequencies ω_{n0}. Then the classical dispersion formula is generalized to the case of many frequencies in the following way:

$$\varepsilon = 1 + \frac{4\pi Ne^2}{m}\sum_n \frac{f_n}{\omega_{n0}^2 - \omega^2}. \tag{36.24}$$

Comparing it with the quantum formula of dispersion (36.19), we see that both formulae become identical if we put

$$f_n = \frac{2m\,\omega_{n0}}{h}\,|\,x_{0n}\,|^2. \tag{36.25}$$

But to make this equation meaningful we must, in accordance with (36.23), impose upon the right-hand side of (36.25) the condition

$$\sum_n \frac{2m\,\omega_{n0}}{h}\,|\,x_{0n}\,|^2 = 1. \tag{36.26}$$

Let us prove that this is actually what occurs. To do so we write the commutation relation between $\hat{p}_x$ and $\hat{x}$ [see (29.3)]

$$\hat{p}_x\hat{x} - \hat{x}\hat{p}_x = \frac{h}{i}. \tag{36.27}$$

We multiply it by ψ_0^* on the left and ψ_0 on the right (we omit the upper index 0) and integrate over the whole volume. We can take advantage of the normalization condition on the right-hand side of the equality, i.e., of the fact that $\int |\psi_0|^2 dV = 1$:

$$\int \psi_0^* \hat{p}_x \hat{x} \psi_0 dV - \int \psi_0^* \hat{x} \hat{p}_x \psi_0 dV = \frac{h}{i}. \tag{36.28}$$

We expand the products $\hat{x}\psi_0$ and $\psi_0^*\hat{x}$ in an eigenfunction series ψ_n, according to the general formula (30.8):

$$\hat{x}\psi_0 = \sum_n a_n \psi_n \; ; \; \psi_0^* \hat{x} = \sum_n a_n^* \psi_n^* . \tag{36.29}$$

The expansion coefficients are determined from (30.11):

$$a_n = \int \psi_n^* \hat{x} \psi_0 dV \; ; \; a_n^* = \int \psi_0^* \hat{x} \psi_n dV . \tag{36.30}$$

In other words, they are equal to the matrix elements x_{0n}. Now, substituting the expansion (36.29) into (36.28), we obtain

$$\sum_n \left(x_{0n} \int \psi_0^* \hat{p}_x \psi_n dV - \int \psi_n^* \hat{p}_x \psi_0 dV \cdot x_{n0} \right) = \frac{h}{i}. \tag{36.31}$$

But this expression contains the momentum matrix elements, which can be replaced by coordinate matrix elements by equation (34.21):

$$\int \psi_0^* \hat{p}_x \psi_n dV = (p_x)_{n0} = i m \omega_{0n} x_{n0} ,$$

$$\int \psi_n^* \hat{p}_x \psi_0 dV = (p_x)_{0n} = i m \omega_{n0} x_{0n} .$$

After this substitution, equality (36.31) can be easily reduced to the required form (36.26) if we take advantage of the fact that $x_{n0} = x_{0n}^*$ and $\omega_{0n} = -\omega_{n0}$.

We note that the oscillator fractions f_n (they are also called "oscillator forces") are proportional to the same matrix elements as are involved in the probabilities of radiation or absorption of the appropriate quanta. Therefore, the dispersion properties of a substance may be associated with the intensity of the spectral lines emitted by it.

Incoherent scattering. In addition to the dipole moment $\overline{\mathbf{d}}$ determined by equation (36.11), we can also calculate the transition moments corresponding to radiation with a frequency which is less than that of the incident light. In other words, we can calculate the intensity of light scattering with a change of frequency. Such scattering is termed incoherent.

A very important case is when the radiation energy, which remains in the substance upon incoherent scattering, contributes to exciting the oscillatory motion of the molecules. This phenomenon was discovered by L. I. Mandelshtam together with G. S. Landsberg and, independently, by Raman. It is frequently accompanied by the excitation of oscillations which do not manifest themselves in the direct absorption of quanta as a result of the appropriate selection rules for molecular oscillations. In this case, incoherent scattering yields important information concerning the molecular structure of substances.

Sec. 37. Quantum Theory of Scattering

The effective cross-section concept in quantum theory. The concept of an effective scattering cross-section of particles, which was defined in Sec. 6 in terms of classical mechanics, is directly extended to quantum mechanics. Indeed, the differential effective scattering cross-section of the particles inside a given solid angle is the ratio of the number of scattered particles in this element of angle to the flux density of the incident particles. Since flux and flux density can be defined quantum-mechanically, the effective cross-section has the same sense in quantum theory as it has in classical theory.

In practice, however, it is very difficult to calculate the effective cross-section. Therefore, we shall consider certain special cases in which the solution to the problem is comparatively simple.

The Born approximation. Let us suppose that a particle with energy $\mathscr{E}$ is scattered in a given potential field U. We shall first consider the case of $\mathscr{E} \gg U$. Then, the change in the wave vector of the particle in the field is of the order

$$\frac{\sqrt{2m(\mathscr{E} - U)}}{h} - \frac{\sqrt{2m\mathscr{E}}}{h} \sim \sqrt{\frac{m}{2\mathscr{E}}}\,\frac{U}{h}.$$

If the dimensions of the region in which the field acts are of the order a then the total phase change of the wave function in the scattering field is estimated as

$$\sqrt{\frac{m}{2\mathscr{E}}}\,\frac{Ua}{h}.$$

This quantity must be considerably smaller than unity in order that the perturbation produced by the field may be regarded as weak. In the case when $U \gg \mathscr{E}$, the wave number is estimated as $\sqrt{\frac{2mU}{h}} \gg \sqrt{\frac{2m\mathscr{E}}{h}}$. It follows then that the criterion of smallness for a phase change is $\frac{a\sqrt{2mU}}{h} \ll 1$ (upper estimate).

Under these conditions the action of the field U must be regarded as a weak perturbation imposed upon the wave function.

We shall proceed from the general formula (34.29) for the transition probability. Let the initial momentum of the incident particle equal $\mathbf{p}$

prior to scattering in a centre-of-mass system (see Sec. 6), and $\mathbf{p}'$ after scattering. We consider the scattering to be elastic, so that $p=p'$. To a zero approximation we choose the wave functions $\psi^{(0)}(\mathbf{p})$ and $\psi^{(0)}(\mathbf{p}')$ in the form of plane waves, which corresponds to free motion. We write them as

$$\psi^{(0)}(\mathbf{p})=\frac{1}{\sqrt{V}}e^{\frac{i\mathbf{pr}}{h}};\ \psi^{(0)*}(\mathbf{p}')=\frac{1}{\sqrt{V}}e^{-\frac{i\mathbf{p}'\mathbf{r}}{h}}. \tag{37.1}$$

These functions are normalized to unity in the volume V (which, of course, falls out of the final result). The approximation (37.1) for $\psi^{(0)}(\mathbf{p})$, $\psi^{(0)*}(\mathbf{p}')$ is called a *Born approximation.*

The function $\psi^{(0)}(\mathbf{p})$ corresponds to a flux density $\frac{\mathbf{v}}{V}$; this is immediately seen from (24.20):

$$\mathbf{j}=\frac{h}{2mi V}\left(e^{-i\frac{\mathbf{pr}}{h}}\nabla e^{i\frac{\mathbf{pr}}{h}}-e^{i\frac{\mathbf{pr}}{h}}\nabla e^{-i\frac{\mathbf{pr}}{h}}\right)=\frac{\mathbf{p}}{mV}=\frac{\mathbf{v}}{V}. \tag{37.2}$$

From (37.1), the matrix element for the transition probability that appears in (34.29) is

$$\mathscr{H}^{(1)}(\mathbf{p},\mathbf{p}')=\frac{1}{V}\int e^{i\frac{(\mathbf{p}-\mathbf{p}')\mathbf{r}}{h}}U(\mathbf{r})\,dV. \tag{37.3}$$

In order to find the scattering probability, we must multiply the square modulus of (37.3) by $\frac{2\pi}{h}$, by the number of finite states $z(\mathscr{E})$ in a unit energy interval, and by the element of solid angle $d\Omega$. This number can be determined directly from (25.25) if we take into account that the fraction of the states corresponding to a solid angle element $d\Omega$ is $\frac{d\Omega}{4\pi}$.

Therefore,

$$z(\mathscr{E})=\frac{dN(\mathscr{E})}{d\mathscr{E}}\cdot\frac{d\Omega}{4\pi}=\frac{Vm^{3/2}\mathscr{E}^{1/2}}{2^{1/2}\pi^2h^3}d\Omega. \tag{37.4}$$

The differential effective scattering cross-section inside a solid-angle element $d\Omega$ is equal to the scattering probability in unit time [defined from (34.29)], divided by the incident flux $\frac{v}{V}=\frac{1}{V}\sqrt{\frac{2\mathscr{E}}{m}}$. Therefore,

$$d\sigma=\left|\int e^{i\frac{(\mathbf{p}-\mathbf{p}')\mathbf{r}}{h}}U(\mathbf{r})\,dV\right|^2\frac{m^2}{4\pi^2h^4}d\Omega. \tag{37.5}$$

The integral appearing here is the matrix element calculated for two functions $\psi^{(0)}(\mathbf{p})$, $\psi^{(0)*}(\mathbf{p}')$ with normalization over unit volume, $V=1$. Therefore, introducing $\mathbf{k}\equiv\frac{\mathbf{p}}{h}$, $\mathbf{k}'\equiv\frac{\mathbf{p}'}{h}$, we write

$$U_{\mathbf{kk}'} = \int e^{i(\mathbf{k}-\mathbf{k}')\mathbf{r}}\, U(\mathbf{r})\, dV, \tag{37.6}$$

$$d\sigma = \frac{m^2}{4\pi^2 h^4} \left| U_{\mathbf{kk}'} \right|^2 d\Omega. \tag{37.7}$$

Scattering by a central field. Simplifications appear in expression (37.5) if the field U is central, i.e., if it depends only upon r. Let us calculate $U_{\mathbf{kk}'}$ for this case. In defining the polar angle ϑ we choose the direction of the vector $\mathbf{k} - \mathbf{k}'$ as the polar axis. Then

$$U_{\mathbf{kk}'} = \int e^{i(\mathbf{k}-\mathbf{k}')\mathbf{r}}\, U(r)\, dV = 2\pi \int_0^\infty r^2\, dr\, U(r) \int_0^\pi e^{i|\mathbf{k}-\mathbf{k}'|r\cos\vartheta} \sin\vartheta\, d\vartheta. \tag{37.8}$$

Bearing in mind that $\sin\vartheta\, d\vartheta = -d\cos\vartheta$, we can integrate with respect to ϑ immediately, obtaining

$$U_{\mathbf{kk}'} = 2\pi \int_0^\infty r^2\, dr\, U(r) \left(\frac{e^{i|\mathbf{k}-\mathbf{k}'|r} - e^{-i|\mathbf{k}-\mathbf{k}'|r}}{i\,|\mathbf{k}-\mathbf{k}'|\,r} \right) =$$

$$= \frac{4\pi}{|\mathbf{k}-\mathbf{k}'|} \int_0^\infty r\, U(r) \sin(|\mathbf{k}-\mathbf{k}'|\, r)\, dr. \tag{37.9}$$

As we have already said, $k = k'$. Therefore, the vector difference is easily expressed in terms of the deflection angle θ for the particle:

$$|\mathbf{k}-\mathbf{k}'|^2 = 2k^2 - 2(\mathbf{kk}') = 2k^2(1-\cos\theta) = 4k^2 \sin^2\frac{\theta}{2}. \tag{37.10}$$

This can also be seen from a geometrical construction. We have

$$U_{\mathbf{kk}'} = \frac{2\pi}{k\sin\frac{\theta}{2}} \int_0^\infty r\, U(r) \sin\left(2kr\sin\frac{\theta}{2}\right) dr. \tag{37.11}$$

Thus, a calculation of $U_{\mathbf{kk}'}$ reduces to calculation of a single integral (37.11).

Rutherford's formula. For the case of a Coulomb field, $U = \pm \dfrac{Ze^2}{r}$. The integral $U_{\mathbf{kk}'}$ is found in the following artificial manner. We define the integral $\int_0^\infty \sin x\, dx$ thus

$$\lim_{a\to 0} \int_0^\infty e^{-ax} \sin x\, dx = \lim_{a\to 0} \frac{1}{a^2+1} = 1.$$

Then

$$\int_0^\infty \sin a x\, d x = \frac{1}{a}$$

and, finally,

$$U_{\mathbf{k}\mathbf{k}'} = \pm \frac{2\pi Z e^2}{k \sin \frac{\theta}{2}} \int_0^\infty \sin\left(2 k r \sin \frac{\theta}{2}\right) d r = \pm \frac{\pi Z e^2}{k^2 \sin^2 \frac{\theta}{2}}. \tag{37.12}$$

Substituting this in (37.7), we obtain a final expression for the differential effective scattering cross-section:

$$d\sigma = \frac{Z^2 e^4 d\Omega}{4 m^2 v^4 \sin^4 \frac{\theta}{2}}, \tag{37.13}$$

where we have taken advantage of the fact that $p = hk = mv$. This result curiously agrees with the precise classical Rutherford formula (6.19).

It turns out that the result (37.13) is also obtained from a precise solution of the wave equation for the case of a Coulomb field. Thus, Rutherford's formula is extended to quantum mechanics unchanged.

The Born approximation in the theory of scattering by a Coulomb field can be regarded as a series expansion in square powers of the charge, or, more precisely, Ze^2. But since the precise formula does not involve powers higher than $(Ze^2)^2$, the result of the Born approximation coincided with the precise result.

We shall now estimate the limits of applicability of the method under consideration for the Coulomb field. To do this, we make use of the first criterion established at the beginning of the section for the applicability of this method. Since the product Ua in this case is equal to Ze^2, we arrive at the following condition:

$$\sqrt{\frac{m}{2\mathscr{E}}}\,\frac{Ua}{h} \sim \frac{Ze^2}{hv} \ll 1. \tag{37.14}$$

The quantity $e^2/hc = 1/137$. Therefore, we write (37.14) otherwise thus:

$$\frac{Z}{137}\frac{c}{v} \ll 1. \tag{37.15}$$

But $Z \sim 90$ for heavy elements, so that (37.15) is not satisfied in general. Of course, Rutherford's formula is applicable to nonrelativistic particles in this case too, because it is exact; but in calculating a correction, for example, arising from a distortion of the nuclear field by the field of atomic electrons, the Born approximation yields an incorrect result.

With the condition (37.15), formula (37.13) is applicable also to the scattering of relativistic particles at small angles provided m is replaced by $\frac{m}{\sqrt{1-\frac{v^2}{c^2}}}$.

The collision parameter (aiming distance) and angular momentum. The Born approximation cannot be used when large forces act upon a particle, even if they are concentrated in a small region. First of all let us define what is meant by a "small" region.

It is convenient here to compare the classical aiming distance of ρ (see Sec. 6) with the angular-momentum eigenvalue in quantum mechanics. For large angular-momentum eigenvalues, when the quasi classical approximation is applicable, we can give the following estimate:

$$hl \sim mv\rho. \tag{37.16}$$

Whence

$$\rho \sim \frac{hl}{mv} = \frac{\lambda l}{2\pi}, \tag{37.17}$$

where λ is the de Broglie wavelength.

It can be seen from here that to a change in the angular momentum by unity there corresponds an increment of $\frac{\lambda}{2\pi} = \frac{1}{k}$ in the aiming distance. Accordingly, the smallest collision parameter is given by $l=0$ and $\rho \sim \frac{\lambda}{2\pi}$. Here the particle is scattered in the s-state.

Let us consider the case when the radius of action of the forces is less than $\frac{\lambda}{2\pi}$. Then a particle with an angular momentum other than zero hardly at all experiences scattering. We have shown [see (31.12)] that the wave function for a particle with angular momentum l becomes zero, like r^l, at the origin. Therefore, the probability of finding a particle with $l>0$ in the region of action of the forces is very small if the radius of action of the forces is much less than $\frac{\lambda}{2\pi}$.

Separating the wave function with zero angular momentum. Let us take the term, corresponding to $l=0$, out of the wave function. To do this, it is necessary to expand the function (37.1), i.e., a plane wave, in a series of eigenfunctions of the operator $\hat{M}^2$. The function corresponding to $l=0$ is especially simple: it does not depend upon the angle. Indeed, the operator $\hat{M}^2$ involves only angular derivatives. Operating upon a function which does not depend on the angle is equivalent to multiplying the function by zero. Normalizing the angular function of the s-state to unity, we find

$$\varphi_0 = \frac{1}{\sqrt{4\pi}},$$

then $\int |\varphi_0|^2 \, d\Omega = 1$.

The expansion coefficient for this function is, according to the general formula (30.11),

$$c(r) = \int \varphi_0 \psi^{(0)}(\mathbf{p})\, dV = \frac{1}{\sqrt{4\pi}\, V}\int e^{i\,\mathbf{kr}} d\Omega =$$

$$= \frac{2\pi}{\sqrt{4\pi}\, V}\int_0^\pi e^{ikr\cos\vartheta}\sin\vartheta\, d\vartheta = 2\sqrt{\frac{\pi}{V}}\,\frac{\sin kr}{kr}. \qquad (37.18)$$

$c(r)$ satisfies the radial wave equation for a free particle (31.5), if we put $U=0$, $l=0$ in it. In actual fact, equation (31.7), which is obtained from (31.5) by substituting $\psi = \frac{\chi}{r}$ when $l=0$, $U=0$, has the solution

$$\chi = \sin\frac{\sqrt{2m\mathscr{E}}}{h}\, r.$$

But $\frac{\sqrt{2m\mathscr{E}}}{h} = k$, so that the function $\chi = rc(r)$.

$c(r)$ tends to a finite limit when $r=0$. This corresponds to the boundary condition for a radial ψ-function at the coordinate origin.

Let there now be, close to the origin, a scattering field $U(r)$, which diminishes so rapidly with distance that $U(r)=0$ when $r \sim \frac{\lambda}{2\pi}$. Then, in its radial dependence, the s-state wave function satisfies, as before, equations (31.5)-(31.7) for $r > \frac{\lambda}{2\pi}$, because no forces act upon the particle in this region. The solution to (31.5), which is more general than (37.18), is

$$c'(r) = A\cdot 2\sqrt{\frac{\pi}{V}}\,\frac{\sin(kr+\delta)}{kr}, \qquad (37.19)$$

where δ is some phase shift depending upon the definite form of the potential $U(r)$. Naturally, the solution (31.19) cannot be extended to $r < \frac{\lambda}{2\pi}$, because the particle in this region is no longer free from the action of forces.

We shall show that the effective scattering cross-section is expressed in terms of δ. An example of determining δ is given in exercises 2 and 3.

Determining the scattered wave. Let us suppose that, in some way, an exact solution ψ' of the wave equation (31.12) has become known in a given scattering field $U(r)$. This solution must satisfy the same conditions at infinity as a plane wave, because these conditions correspond to the particle scattering problem. We represent the function ψ' in the form of the sum

$$\psi' = \psi^0(\mathbf{p}) + \psi_{scat} \qquad (37.20)$$

This equation separates the function ψ^0 (**p**), which corresponds to the incident wave, from the complete solution of the wave equation. The second term $\psi_{scat}=\psi'-\psi^0$ (**p**) describes the scattered particles.

If we now expand ψ' and ψ^0 (**p**) in a series of eigenfunctions of the square of the angular momentum, it turns out that a short-range force field distorts only the term ψ', which corresponds to $l=0$. All the remaining terms of both functions are the same. As a result, when $r > \frac{\lambda}{2\pi}$ we obtain

$$\psi_{scat} = [c'(r) - c(r)]\,\varphi_0 = 2\sqrt{\frac{\pi}{V}}\left[\frac{A\sin(kr+\delta)-\sin kr}{kr}\right]\frac{1}{\sqrt{4\pi}}\,. \quad (37.21)$$

But for large r the scattered particles can move only away from the scatterer. This means that ψ_{scat} involves only the function $\frac{e^{ikr}}{r}$ and does not contain the function $\frac{e^{-ikr}}{r}$. Indeed, if we put $\psi = \frac{e^{ikr}}{r}$ in (24.20), then the flux j will acquire a positive sign, while if we substitute $\frac{e^{-ikr}}{r}$ the flux will be negative, i.e., incoming.

We write ψ_{scat} in complex form:

$$\psi_{scat} = \frac{1}{2i\sqrt{V}\cdot kr}\,[A\,(e^{i(kr+\delta)} - e^{-i(kr+\delta)}) - (e^{ikr} - e^{-ikr})]\,. \quad (37.22)$$

To exclude the incoming wave e^{-ikr}, we put $Ae^{-i\delta}=1$, whence

$$A = e^{i\delta}\,, \quad (37.23)$$

$$\psi_{scat} = \frac{1}{2i\sqrt{V}\cdot kr}\,(e^{2i\delta}-1)\,e^{ikr}\,. \quad (37.24)$$

The effective scattering cross-section. From (24.20), the flux density of the scattered particles at infinity is

$$j = \frac{1}{4Vk^2r^2}\,|\,e^{2i\delta}-1\,|^2\,\frac{hk}{m}\,. \quad (37.25)$$

The total effective scattering cross-section σ is equal to the whole flux $4\pi r^2 j$ divided by the flux density of the incident particles $\frac{v}{V} = \frac{hk}{mV}$:

$$\sigma = \frac{\pi}{k^2}\,|\,e^{2i\delta}-1\,|^2\,. \quad (37.26)$$

Passing to real quantities we write

$$|\,e^{2i\delta}-1\,|^2 = (e^{2i\delta}-1)(e^{-2i\delta}-1) = 2-2\cos 2\delta = 4\sin^2\delta\,,$$

so that

$$\sigma = \frac{4\pi}{k^2}\sin^2\delta\,. \quad (37.27)$$

It can be seen from here that the greatest value of σ is for the scattering of particles in the s-state $\frac{4\pi}{k^2}$. This formula has many applications in nuclear physics since nuclear forces are large and short-range.

Since scattered particles occur in the s-state, their distribution in a centre-of-mass system is isotropic, i.e., it does not depend upon the scattering angle. This agrees with the statement made at the end of Sec. 6.

Exercises

1) Find the effective scattering cross-section of fast particles by hydrogen atoms in the ground state.

The wave function for the ground state of a hydrogen atom with $n = 1$, $l = 0$ is

$$\psi_0 = Be^{-\xi\sqrt{2\varepsilon_0}} = Be^{-\xi},$$

because, of the polynomial χ there remains only the first term, which does not depend upon ξ, while $\sqrt{2\varepsilon_0} = 1$. The coefficient B is found from the normalization condition

$$\left(\frac{h^2}{me^2}\right)^3 4\pi B^2 \int_0^\infty e^{-2\xi}\xi^2 d\xi = 1,$$

whence

$$B = \sqrt{\frac{1}{\pi}\left(\frac{me^2}{h^2}\right)^3}.$$

The potential interaction energy of the charge e with the atom is

$$U = -\frac{e^2}{r} + \int \frac{e^2\psi_0^2(r')}{|\mathbf{r}-\mathbf{r}'|} dV'.$$

The first term of the integral $U_{\mathbf{kk}'}$ was found in the text. It is

$$-\frac{\pi e^2}{k^2 \sin^2\frac{\theta}{2}}.$$

We integrate the second term in the following way:

$$e^2 \int e^{i(\mathbf{k}-\mathbf{k}')\mathbf{r}} dV \int \frac{\psi_0^2(r')\,dV'}{|\mathbf{r}-\mathbf{r}'|} =$$
$$= e^2 \int \psi_0^2(r')\, e^{i(\mathbf{k}-\mathbf{k}')\mathbf{r}'} dV' \int \frac{e^{i(\mathbf{k}-\mathbf{k}')(\mathbf{r}-\mathbf{r}')}}{|\mathbf{r}-\mathbf{r}'|} dV.$$

In the last integral it is necessary to take the origin at the point $\mathbf{r}'$, so that it reduces to the same form as (37.12):

$$\int \frac{e^{i(\mathbf{k}-\mathbf{k}')\mathbf{r}}}{r} dV = \frac{\pi}{k^2\sin^2\frac{\theta}{2}}.$$

Hence,

$$U_{\mathbf{kk}'} = -\frac{\pi e^2}{k^2\sin^2\frac{\theta}{2}}\left(1 - \int \psi_0^2(r')\, e^{i(\mathbf{k}-\mathbf{k}')\mathbf{r}'} dV'\right).$$

The quantity inside the brackets is called the screening factor. Evaluating it in the same manner as (37.8), we obtain

$$\int \psi_0^2(r')\, e^{i(\mathbf{k}-\mathbf{k}')\mathbf{r}'}\, dV' = \frac{2\pi}{k\sin\frac{\theta}{2}} \int_0^\infty r'\,\psi_0^2(r')\,\sin\left(2kr'\sin\frac{\theta}{2}\right) dr'.$$

The integral reduces to the form

$$\int_0^\infty x\sin ax\, e^{-bx}\,dx = -\frac{\partial}{\partial b}\int_0^\infty \sin ax\, e^{-bx}\,dx = -\frac{\partial}{\partial b}\frac{a}{a^2+b^2} = \frac{2ab}{(a^2+b^2)^2}.$$

Here, $a = 2k\sin\frac{\theta}{2}$, $b = \frac{2me^2}{h^2}$, so that the screening factor is

$$1 - \frac{1}{\pi}\left(\frac{me^2}{h^2}\right)\frac{2\pi}{k\sin\frac{\theta}{2}}\,\frac{2\cdot 2k\sin\frac{\theta}{2}\cdot\frac{2me^2}{h^2}}{\left[4k^2\sin^2\frac{\theta}{2}+\left(\frac{2me^2}{h^2}\right)^2\right]^2} = 1 - \frac{1}{\left[1+\frac{h^4k^2}{m^2e^4}\sin^2\frac{\theta}{2}\right]^2} =$$

$$= 1 - \frac{1}{\left[1+\left(\frac{hv}{e^2}\right)^2\sin^2\frac{\theta}{2}\right]^2}.$$

It was assumed in the last transformation that the scattered particle is an electron. Then, strictly speaking, we should have formed a function which is antisymmetrical together with the function of the atomic electron; this we did not do. The final formula for the effective scattering cross-section differs from (37.13) by the square of the screening factor. We note that this factor is correctly obtained only in the Born approximation, in contrast to Rutherford's formula (37.13), which is exact.

For $\theta = 0$, the effective cross-section turns out to be finite, because $\theta = 0$ corresponds to large aiming distances, when the nuclear charge is screened by the charge of an electron.

2) Calculate the effective scattering cross-section for a particle by an impermeable sphere of radius a, which is very much less than $\frac{\lambda}{2\pi} = \frac{1}{k}$.

In accordance with (25.1), the wave function at the surface of the impermeable sphere becomes zero. Hence, the solution (37.19) has the form

$$c'(r) = A\cdot 2\sqrt{\frac{\pi}{V}}\,\frac{\sin k(r-a)}{kr}.$$

From this, $\delta = -ka$, while

$$\sigma = \frac{4\pi}{k^2}\sin^2 ka.$$

But $ka \ll 1$ from the conditions, so that $\sin ka \sim ka$.

Finally, $\sigma = 4\pi a^2$, i.e., the effective scattering radius is twice the radius of the sphere. In classical theory $\sigma = \pi a^2$ (see exercise 1, Sec. 6).

3) Examine the scattering of particles with energy $\mathscr{E}$ in the s-state by a spherical potential well of constant depth $|U_0|$ and radius a; consider that there exist the following relations:

$$U_0 = -\frac{\pi^2}{8}\frac{h^2}{ma^2}(1+\varepsilon), \qquad \frac{\mathscr{E}}{|U_0|} \ll \varepsilon \ll 1$$

(see Sec. 25). For $\varepsilon > 0$, there exists in the well a bound state of a particle of energy $\mathscr{E}_0$ close to the upper boundary of the well. Unlike Fig. 38, it is assumed here that U becomes zero for $r > a$. Express the cross-section in terms of the energy level $\mathscr{E}_0$.

The conjugation condition for the wave functions with $r = a$ is of the form

$$\frac{k \cos(ka+\delta)}{\sin(ka+\delta)} = \frac{\varkappa \cos \varkappa a}{\sin \varkappa a},$$

where $\varkappa = \frac{\sqrt{2m(\mathscr{E} + |U_0|)}}{h}$, $k = \frac{\sqrt{2m\mathscr{E}}}{h}$. Neglecting ka, we obtain an expression for the effective cross-section:

$$\sigma = \frac{4\pi}{k^2} \sin^2 \delta = \frac{4\pi}{k^2(1+\cot^2\delta)} = \frac{4\pi}{k^2 + \varkappa^2 \cot^2 \varkappa a}.$$

In accordance with the conditions imposed upon $E/|U_0|$ we have, approximately,

$$\cot \varkappa a \cong \frac{\pi\varepsilon}{4}.$$

From (25.41) the condition for finding the level $\mathscr{E}_0$ is of the form

$$\varkappa_0 \cot \varkappa_0 a = -\tilde{\varkappa}$$

Supposing that $\left|\frac{\mathscr{E}_0}{U_0}\right| \ll \varepsilon$, we obtain $\left(\frac{\pi\varepsilon}{4}\right)^2 = \left|\frac{\mathscr{E}_0}{U_0}\right|$. This confirms the assumption concerning the order of smallness $\left|\frac{\mathscr{E}_0}{U_0}\right|$, since $\sqrt{\varepsilon} \gg \varepsilon$. For the cross-section we finally have

$$\sigma = \frac{4\pi}{k^2 + k_0^2},$$

where $k_0 = \frac{\sqrt{2m|\mathscr{E}_0|}}{h}$. We note that the formula obtained also holds for $\varepsilon < 0$, when in actuality there is no level at all in the well. In this case we talk about a "virtual" level. The straight line in Fig. 39 intersects the first half-cycle of the sinusoid just before $\varkappa a = \frac{\pi}{2}$.

A similar case occurs when neutrons are scattered by protons with antiparallel particle spins.

Sec. 38. The Relativistic Wave Equation for an Electron

The equation for a spinless particle. Schrödinger's equation (24.11) is formed on the basis of the nonrelativistic relationship between energy and momentum

$$\mathscr{E} = \frac{p^2}{2m} + U.$$

Therefore, it can be applied only to electrons whose velocity is considerably less than that of light, and whose kinetic energy is considerably less than the rest energy:

$$T = \frac{mc^2}{\sqrt{1 - \frac{v^2}{c^2}}} - mc^2 \ll mc^2 .$$

Immediately after Schrödinger obtained the nonrelativistic equation, the first attempts were made to build a relativistic wave equation (Fock, Klein, Gordon). In formula (21.30)

$$(\mathscr{H} - e\varphi)^2 = c^2\left(\mathbf{p} - \frac{e}{c}\mathbf{A}\right)^2 + m^2c^2$$

$-\frac{h}{i}\frac{\partial}{\partial t}$ was substituted in place of $\mathscr{H}$, as is usual in quantum mechanics, and in place of $\mathbf{p}$, the operator $\frac{h}{i}\nabla$. In this way a wave equation was obtained in relativistically invariant form

$$\left(\frac{h}{i}\frac{\partial}{\partial t} + e\varphi\right)\left(\frac{h}{i}\frac{\partial}{\partial t} + e\varphi\right)\psi =$$
$$= c^2\left(\frac{h}{i}\nabla - \frac{e}{c}\mathbf{A}\right)\left(\frac{h}{i}\nabla - \frac{e}{c}\mathbf{A}\right)\psi + m^2c^4\psi , \qquad (38.1)$$

which equation, however, is not applicable to electrons. The fact of the matter is that equation (38.1) does not take into account the spin of the electron, because it involves only a single wave function. Yet in Sec. 32 we saw that a particle with spin 1/2 must be described by at least two wave functions. These two functions could be introduced into the nonrelativistic equation purely formally, assuming that each of them satisfies it. But the interaction of spin and orbit is a relativistic effect; therefore, a correct equation for fast electrons must take it into account automatically, without any additional hypotheses concerning spin magnetic moment. This equation must involve operators which act upon the spin degree of freedom.

The inapplicability of equation (38.1) to the electron was very quickly seen: the fine structure of the levels of the hydrogen atom obtained from this equation was incorrect. A nonspin equation cannot explain, first of all, the number of splitting components; this is decisively against it.

Charged particles without spin—mesons—take part in nuclear interactions. Equation (38.1) can be applied to them, at least if it is shown that such mesons can be regarded, to some sort of approximation, separately from protons and neutrons.

But for electrons one has to form a relativistic wave equation that takes spin into account. Such an equation was obtained by Dirac.

The Dirac equation. Following the line of Dirac's argument, we begin with the equation for a free electron. The starting relationship is

$$\mathscr{E} = \sqrt{c^2(p_x^2 + p_y^2 + p_z^2) + m^2c^4} . \qquad (38.2)$$

Instead of $\mathcal{E}$ and $\mathbf{p}$ we must substitute the derivatives $-\frac{h}{i}\frac{\partial}{\partial t}$ and $\frac{h}{i}\nabla$. However, to do this it is necessary to define the meaning of the square root of an operator. Dirac supposed that, in the operator sense, a root is equal to an expression like

$$\sqrt{c^2(\hat{p}_x^2+\hat{p}_y^2+\hat{p}_z^2)+m^2c^4} = c\,(\hat{\alpha}_x\hat{p}_x+\hat{\alpha}_y\hat{p}_y+\hat{\alpha}_z\hat{p}_z)+\hat{\beta}mc^2, \tag{38.3}$$

where $\hat{\alpha}_x$, $\hat{\alpha}_y$, $\hat{\alpha}_z$, and $\hat{\beta}$ act on the internal degrees of freedom of the electron such as, for example, the spin degree of freedom.

Let us square both sides of the equation and attempt to choose the operators $\hat{\alpha}_x$, $\hat{\alpha}_y$, $\hat{\alpha}_z$, and $\hat{\beta}$ in such a way as to obtain an identity, i.e., so as to eliminate terms of the type $\hat{p}_x\hat{p}_y, \ldots, mc^3\hat{p}_x, \ldots$:

$$\begin{aligned} c^2(\hat{p}_x^2+\hat{p}_y^2+\hat{p}_z^2)+m^2c^4 &= c^2(\hat{\alpha}_x^2\hat{p}_x^2+\hat{\alpha}_y^2\hat{p}_y^2+\hat{\alpha}_z^2\hat{p}_z^2)+\hat{\beta}^2m^2c^4+ \\ &+ c^2(\hat{\alpha}_x\hat{\alpha}_y+\hat{\alpha}_y\hat{\alpha}_x)\hat{p}_x\hat{p}_y+c^2(\hat{\alpha}_x\hat{\alpha}_z+\hat{\alpha}_z\hat{\alpha}_x)\hat{p}_x\hat{p}_z+ \\ &+ c^2(\hat{\alpha}_y\hat{\alpha}_z+\hat{\alpha}_z\hat{\alpha}_y)\hat{p}_y\hat{p}_z+mc^3(\hat{\alpha}_x\hat{\beta}+\hat{\beta}\hat{\alpha}_x)\hat{p}_x+ \\ &+ mc^3(\hat{\alpha}_y\hat{\beta}+\hat{\beta}\hat{\alpha}_y)\hat{p}_y+mc^3(\hat{\alpha}_z\hat{\beta}+\hat{\beta}\hat{\alpha}_z)\hat{p}_z\,. \end{aligned}$$

Hence, the operators must be subject to the conditions

$$\begin{aligned} \hat{\alpha}_x^2=\hat{\alpha}_y^2=\hat{\alpha}_z^2=\hat{\beta}^2=1\,, \\ \hat{\alpha}_x\hat{\alpha}_y+\hat{\alpha}_y\hat{\alpha}_x=\hat{\alpha}_x\hat{\alpha}_z+\hat{\alpha}_z\hat{\alpha}_x=\hat{\alpha}_y\hat{\alpha}_z+\hat{\alpha}_z\hat{\alpha}_y= \\ =\hat{\alpha}_x\hat{\beta}+\hat{\beta}\hat{\alpha}_x=\hat{\alpha}_y\hat{\beta}+\hat{\beta}\hat{\alpha}_y=\hat{\alpha}_z\hat{\beta}+\hat{\beta}\hat{\alpha}_z=0\,. \end{aligned} \tag{38.4}$$

These operator equalities greatly resemble the spin operator relationships (32.10), (32.11). It can already be seen from this that the operators $\hat{\alpha}_x$, $\hat{\alpha}_y$, $\hat{\alpha}_z$, and $\hat{\beta}$ at least act upon the spin degree of freedom of an electron. To the accuracy of the factor 1/4, the relations (38.4) agree with (34.11) and (32.13) for $\hat{\sigma}_x$, $\hat{\sigma}_y$, and $\hat{\sigma}_z$.

But the operators $\boldsymbol{\alpha}$ and $\boldsymbol{\sigma}$ are not identical. This can easily be seen by proceeding from the opposite: assume that $\hat{\alpha}_x=\hat{\sigma}_x$, $\hat{\alpha}_y=\hat{\sigma}_y$, $\hat{\alpha}_z=\hat{\sigma}_z$. In order to obtain the wave equation we must equate the right-hand side of (38.3) to $-\frac{h}{i}\frac{\partial\psi}{\partial t}$. Now let us perform an inversion of the coordinate system. All momentum components will change sign so that the sign of $\hat{p}_z$ will also change. But the operator $\hat{\alpha}_z$ in front of $\hat{p}_z$, if it equals $\hat{\sigma}_z$, should not interchange the wave-function components. Therefore, the equality between the left- and right-hand sides of the wave equation breaks down when the coordinate system is inverted; but this should not be. Therefore, $\hat{\boldsymbol{\alpha}}\neq\hat{\boldsymbol{\sigma}}$.

The necessity for a four-component wave function. We shall be confronted with the same difficulty if we consider that the operators $\hat{\alpha}_x$, $\hat{\alpha}_y$, $\hat{\alpha}_z$ act upon the same wave functions as those appearing with

$-\frac{h}{i}\frac{\partial}{\partial t}$ on the other side of the equation. Therefore, we must assume that coordinate differentiation, on the one hand, and time differentiation, on the other, are applied to different pairs of functions, of which one pair changes sign in inversion of the coordinate system while the other does not. This is sufficient to ensure invariance of the equation with respect to inversion.

Thus, we shall say that the wave functions depend not only upon the spin variable σ, but also upon some other internal variable ρ, which also takes on two values.

Let us define the operators which are completely analogous to spin operators and which act upon the variable $\boldsymbol{\sigma}$ and upon the variable ρ. Bearing in mind that the factor 1/2, with which the spin operators were multiplied by in Sec. 32, is no longer needed, we shall write $\hat{\boldsymbol{\sigma}}_1$, $\hat{\boldsymbol{\sigma}}_2$, $\hat{\boldsymbol{\sigma}}_3$ and, correspondingly, $\hat{\rho}_1$, $\hat{\rho}_2$, and $\hat{\rho}_3$ for the variable ρ. These operate upon the wave function analogously:

$$\hat{\sigma}_1\begin{pmatrix}\psi(1,\rho)\\ \psi(2,\rho)\end{pmatrix}=\begin{pmatrix}\psi(2,\rho)\\ \psi(1,\rho)\end{pmatrix};\quad \hat{\sigma}_2\begin{pmatrix}\psi(1,\rho)\\ \psi(2,\rho)\end{pmatrix}=\begin{pmatrix}-i\psi(2,\rho)\\ i\psi(1,\rho)\end{pmatrix};$$
$$\hat{\sigma}_3\begin{pmatrix}\psi(1,\rho)\\ \psi(2,\rho)\end{pmatrix}=\begin{pmatrix}\psi(1,\rho)\\ -\psi(2,\rho)\end{pmatrix}, \tag{38.5}$$

$$\hat{\rho}_1\begin{pmatrix}\psi(\sigma,1)\\ \psi(\sigma,2)\end{pmatrix}=\begin{pmatrix}\psi(\sigma,2)\\ \psi(\sigma,1)\end{pmatrix};\quad \hat{\rho}_2\begin{pmatrix}\psi(\sigma,1)\\ \psi(\sigma,2)\end{pmatrix}=\begin{pmatrix}-i\psi(\sigma,2)\\ i\psi(\sigma,1)\end{pmatrix};$$
$$\hat{\rho}_3\begin{pmatrix}\psi(\sigma,1)\\ \psi(\sigma,2)\end{pmatrix}=\begin{pmatrix}\psi(\sigma,1)\\ -\psi(\sigma,2)\end{pmatrix}. \tag{38.6}$$

From formulae (32.10) and (32.11) we find the basic relations for the operators $\hat{\boldsymbol{\sigma}}$ and $\hat{\rho}$:

$$\left.\begin{aligned}\hat{\sigma}_1^2=\hat{\sigma}_2^2=\hat{\sigma}_3^2=1;\ \hat{\sigma}_1\hat{\sigma}_2=i\hat{\sigma}_3;\ \hat{\sigma}_3\hat{\sigma}_1=i\hat{\sigma}_2;\ \hat{\sigma}_2\hat{\sigma}_3=i\hat{\sigma}_1;\\ \hat{\sigma}_1\hat{\sigma}_2=-\hat{\sigma}_2\hat{\sigma}_1;\ \hat{\sigma}_1\hat{\sigma}_3=-\hat{\sigma}_3\hat{\sigma}_1;\ \hat{\sigma}_2\hat{\sigma}_3=-\hat{\sigma}_3\hat{\sigma}_2\end{aligned}\right\} \tag{38.7}$$

and similarly for $\hat{\rho}$.

All the operators $\hat{\boldsymbol{\sigma}}$ are commutative with all the operators $\hat{\boldsymbol{\rho}}$ because they operate upon different variables. In order that the operators $\hat{\alpha}_x$, $\hat{\alpha}_y$, $\hat{\alpha}_z$ together should satisfy the "anticommutative" relations with the operator $\hat{\beta}$, as in the third line of (38.4), we arrange that all three components of $\hat{\alpha}$ are proportional to one operator of the components of $\hat{\rho}$, for example, $\hat{\rho}_1$ and $\hat{\beta}$ is simply $\hat{\rho}_3$. We notice that $\hat{\rho}_1$ interchanges the functions while $\hat{\rho}_3$ does not. To have the operators $\hat{\alpha}_x$, $\hat{\alpha}_y$, and $\hat{\alpha}_z$ anticommutative, as in the second line of (38.4), we put them proportional to $\hat{\boldsymbol{\sigma}}_1$, $\hat{\boldsymbol{\sigma}}_2$, $\hat{\boldsymbol{\sigma}}_3$, respectively. Thus,

$$\hat{\alpha}_x=\hat{\rho}_1\hat{\sigma}_1;\ \hat{\alpha}_y=\hat{\rho}_1\hat{\sigma}_2,\ \hat{\alpha}_z=\hat{\rho}_1\hat{\sigma}_3,\ \hat{\beta}=\hat{\rho}_3. \tag{38.8}$$

It is obvious that, as a result of the commutative nature of $\hat{\rho}$ and $\hat{\boldsymbol{\sigma}}$ and of the definition of the operators (38.5) and (38.6), all the operators formed in (38.8) satisfy the conditions (38.4).

Thus, the wave function in the Dirac equation has four components, according to the number of values for $\boldsymbol{\sigma}$ and ρ ($\boldsymbol{\sigma}=1, 2$; $\rho=1, 2$). For convenience in future we shall number the symbols $\boldsymbol{\sigma}$, ρ from one to four, putting

$$\psi_1 = \psi(1, 1),\ \psi_2 = \psi(2, 1),\ \psi_3 = \psi(1, 2),\ \psi_4 = \psi(2, 2).$$

It is convenient to write the functions ψ_1, ψ_2, ψ_3, ψ_4 in columns, as in Sec. 32. Now using (38.5) and (38.6), we find out how the operators $\hat{\alpha}_x$, $\hat{\alpha}_y$, $\hat{\alpha}_z$, and $\hat{\beta}$ act upon the four-component wave function

$$\hat{\alpha}_x\psi = \hat{\rho}_1\hat{\sigma}_1\begin{pmatrix}\psi(1, 1)\\ \psi(2, 1)\\ \psi(1, 2)\\ \psi(2, 2)\end{pmatrix} = \hat{\rho}_1\begin{pmatrix}\psi(2, 1)\\ \psi(1, 1)\\ \psi(2, 2)\\ \psi(1, 2)\end{pmatrix} = \begin{pmatrix}\psi(2, 2)\\ \psi(1, 2)\\ \psi(2, 1)\\ \psi(1, 1)\end{pmatrix} = \begin{pmatrix}\psi_4\\ \psi_3\\ \psi_2\\ \psi_1\end{pmatrix};$$

$$\hat{\alpha}_y\psi = \begin{pmatrix}-i\psi_4\\ i\psi_3\\ -i\psi_2\\ i\psi_1\end{pmatrix};\ \hat{\alpha}_z\psi = \begin{pmatrix}\psi_3\\ -\psi_4\\ \psi_1\\ -\psi_2\end{pmatrix};\ \hat{\beta}\psi = \begin{pmatrix}\psi_1\\ \psi_2\\ -\psi_3\\ -\psi_4\end{pmatrix}. \tag{38.9}$$

The choice of operators (38.8) and (38.9) is not unique. It is possible to form other operators with the same properties. For example, one could have chosen ρ_2 instead of ρ_1. We shall examine below (exercise 4) the implications of this fact.

The Dirac equation in expanded form. Summarizing, according to (38.2), (38.3) and (38.8) Dirac's equation can be written as

$$-\frac{h}{i}\frac{\partial\psi}{\partial t} = c(\hat{\alpha}_x\hat{p}_x + \hat{\alpha}_y\hat{p}_y + \hat{\alpha}_z\hat{p}_z)\psi + \hat{\beta}mc^2\psi = c(\hat{\boldsymbol{\alpha}}\hat{\mathbf{p}})\psi + mc^2\hat{\beta}\psi. \tag{38.10}$$

In accordance with (38.9), this equation must be understood as a system of four equations, which we shall write explicitly, first of all replacing $-\frac{h}{i}\frac{\partial\psi}{\partial t}$ by $\mathscr{E}\psi$ (because ψ is proportional to the factor $e^{-\frac{iEt}{h}}$):

$$\left.\begin{aligned}\mathscr{E}\psi_1 &= c(\hat{p}_x\psi_4 - i\hat{p}_y\psi_4 + \hat{p}_z\psi_3) + mc^2\psi_1,\\ \mathscr{E}\psi_2 &= c(\hat{p}_x\psi_3 + i\hat{p}_y\psi_3 - \hat{p}_z\psi_4) + mc^2\psi_2,\\ \mathscr{E}\psi_3 &= c(\hat{p}_x\psi_2 - i\hat{p}_y\psi_2 + \hat{p}_z\psi_1) - mc^2\psi_3,\\ \mathscr{E}\psi_4 &= c(\hat{p}_x\psi_1 + i\hat{p}_y\psi_1 - \hat{p}_z\psi_2) - mc^2\psi_1.\end{aligned}\right\} \tag{38.11}$$

As usual, here $\hat{p}_x$, $\hat{p}_y$, $\hat{p}_z$ are the components of the vector $\frac{h}{i}\nabla$, i.e., $\frac{h}{i}\frac{\partial}{\partial x}$, $\frac{h}{i}\frac{\partial}{\partial y}$, $\frac{h}{i}\frac{\partial}{\partial z}$. The system of equations (38.11) is applied to a free electron since it does not involve the scalar potential and components of the vector potential. Therefore, the coordinate dependence of the wave function is determined by the factor

$$e^{i\frac{\mathbf{pr}}{\hbar}},$$

so that the whole group of four ψ has the form

$$\psi = \begin{pmatrix} a_1 \\ a_2 \\ a_3 \\ a_4 \end{pmatrix} e^{i\frac{\mathbf{pr}}{\hbar}} \tag{38.12}$$

where the amplitudes a_1, a_2, a_3, and a_4 do not depend upon coordinates. The action of the operators $\hat{p}_x$, $\hat{p}_y$, $\hat{p}_z$ on this group of four ψ leads simply to a multiplication of its components by p_x, p_y, p_z. Consequently, the differential equation (38.10) for a free electron leads to the algebraic equation

$$\mathscr{E} a = c\,(\hat{\boldsymbol{\alpha}}\mathbf{p})\,a + mc^2\hat{\beta}\,a, \tag{38.13}$$

where by a is understood the whole column

$$a = \begin{pmatrix} a_1 \\ a_2 \\ a_3 \\ a_4 \end{pmatrix}.$$

Here, operator properties are preserved only by $\hat{\alpha}_x$, $\hat{\alpha}_y$, $\hat{\alpha}_z$, and $\hat{\beta}$, which rearrange the amplitudes a in the same way as the functions ψ. In other words, the amplitudes depend upon the internal variables σ and ρ.

Energy eigenvalues. We apply the operator

$$\hat{\mathscr{E}} = c^2\,(\hat{\boldsymbol{\alpha}}\hat{\mathbf{p}}) + mc^2\,\hat{\beta} \tag{38.14}$$

to both sides of equality (38.13). As a result of the anti-commutation properties of the Dirac operators (38.4), only their squares, equal to unity, remain on the right. Hence, the following equation will result:

$$\mathscr{E}^2 a = c^2\,p^2\,a + m^2\,c^4\,a. \tag{38.15}$$

Here, the components of a are no longer interchanged because there are numbers in front of a, and not operators. All four equations (38.15) have the same form $\mathscr{E}^2\,a_1 = c^2\,p^2\,a_1 + m^2\,c^2\,a_1$, etc. For these equalities to be satisfied we must subject the energy to the usual relativistic relationship, i.e., we cancel a_1:

$$\mathscr{E}^2 = c^2\,p^2 + m^2\,c^4. \tag{38.16}$$

The last equality refers simply to magnitudes. Thus the energy eigenvalue for a free electron, determined from the Dirac equation, is

$$\mathscr{E} = \pm\sqrt{m^2 c^4 + c^2 p^2}. \tag{38.17}$$

The two signs in the equality correspond to the internal degree of freedom which an electron possesses in addition to spin. Only the plus sign is taken in classical mechanics, since free electrons do not have negative energy. The square root in (38.17) is not less than mc^2 in absolute value, so that a region of width $2\,mc^2$ exists in which the energy cannot occur. But all the quantities in classical equations vary continuously; therefore, once the energy has been defined with a positive sign, it cannot jump across this forbidden region of width $2\,mc^2$, and remains positive all the time. In other words, energy which is defined as positive in the initial conditions remains positive from the equations of motion.

Negative kinetic energies in the Dirac theory. In quantum theory it is shown that discontinuous transitions, too, are possible between different states. For example, an electron with energy greater than mc^2 could emit a light quantum and remain with an energy less than $-mc^2$. But such electrons with negative energy and mass are not observed in nature. Their properties would be very strange: upon radiating light, they would each time reduce their energy and, as it were, drop into a state with $\mathscr{E} = -\infty$. All the electrons in the universe would rather quickly fall into this state; but, as we see, this has not happened!

Thus, Dirac's equation admits of the possibility of states, which cannot simply be excluded, because electrons may transfer to them from other observable states. But, on the other hand, there are no electrons in nature with negative energy $\mathscr{E} = -\sqrt{m^2 c^4 + c^2 p^2}$. At the same time, the Dirac equation describes quite correctly a great assemblage of electron properties: as we shall soon see, it yields a relationship between the spin and magnetic moment of an electron that agrees with experiment, it leads to an accurate formula for the fine-structure levels of the hydrogen atom, etc. In addition, mathematical investigations show that there is no essentially different relativistically invariant wave equation for a particle with spin one half and mass differing from zero. Therefore, one should not simply reject the Dirac equation: it is better to attempt to supplement it with some kind of hypothesis.

Vacuum in the Dirac theory. Dirac suggested that vacuum should be redefined. Earlier, vacuum was understood as being a state of matter in which there are no charges, say electrons. A *vacuum* must now be called that state in which all negative energy levels are occupied by electrons. That this redefinition is not verbal but has a physical meaning will be seen very soon from what follows.

If all the negative energy levels are occupied, then, in accordance with the Pauli principle, no electrons can transfer to them from positive energy states. Thus, the Pauli principle is necessary for relativistic quantum theory to be able, in general, to describe the properties of electrons. This is the basic reason why the Pauli principle is necessary as an element of quantum mechanics. In order to avoid misunderstanding we shall give a somewhat fuller definition of a "vacuum" in field theory, which has a rather different meaning from that in experimental physics. In field theory, *vacuum* signifies the ground state of the field; for an electromagnetic field, for example, it is the state of the field in which there are no quanta. We saw in Sec. 27 that this state is endowed with observable physical properties.

In exactly the same way, if all the negative energy states are occupied, then all the remaining electrons can no longer reduce their energy by making a transition to negative states. When there are no electrons with positive energy, electrons can no longer reduce their energy in any way if all the negative energy states are occupied. This explains the definition of vacuum as a ground state.

Pair production. All observed phenomena occur, so to speak, on the background of a state in which the negative kinetic-energy levels are filled.

However, this "background" can manifest its existence in a real physical process. Close to a nucleus, a quantum with energy greater than mc^2 is capable of effecting the emission of an electron from a negative-energy to a positive-energy state. Proximity to the nucleus is necessary so as to satisfy the law of conservation of momentum. For the proof of this simple statement, see exercise 1.

But after an electron has been removed from a negative energy state there will remain a "hole," i.e., an unoccupied level. In an electric field, electrons with negative mass (mass has the same sign as energy) do not move against the field, towards the anode, but along the field, to the cathode, against the applied force. And with them moves the hole which thus behaves like an electron of positive charge and mass.

Experiment will show that as a result of the ejection of an electron from a negative energy state, two charges have appeared: negative and positive. Such a positive electron, or positron, was discovered by Anderson after Dirac had formulated his theory of the background. The attitude towards the Dirac equation was somewhat suspicious before the discovery of the positron, while the idea of background was considered far-fetched and intended only to hide the defects of the theory.

In actual fact Dirac's theory is an unusual example of scientific foresight. The discovery of the "antiproton" (i.e., a proton with negative charge) by Segrè once again confirmed the generality of Dirac's conceptions concerning particles with spin 1/2.

Pair annihilation. When a positron and electron meet they can annihilate each other if the electron transfers to an unoccupied level belonging to a negative energy state. Its energy will be imparted to electromagnetic radiation in the form of two or three quanta. A single quantum cannot result when annihilation occurs in free space because, in this case, momentum is not conserved, in the same way that a single quantum cannot form a pair in free space. Single-quantum annihilation, too, is possible in a nuclear field.

The Dirac equation and quantum electrodynamics. Modern quantum electrodynamics is based upon the quantum theory of the electromagnetic field (Sec. 27) and the Dirac electron theory, with account taken of direct and reverse transitions from negative energy to positive energy. It actually describes an electron and a positron in a completely symmetrical way: the nonsymmetry of the charge that has appeared in our terminology as a result of the fact that the positron was defined as a "hole," is only apparent. The background of electrons with negative energy may be, as it were, subtracted from the equations without altering the physical content of the theory, and so the equations become symmetrical with respect to the sign of the charge. The concept of particles and antiparticles is extended also to particles without spin (for example, π^+- and π^--mesons).

It is characteristic that the relativistic quantum theory describes change in the number of particles: electrodynamics treats of absorption and emission of quanta, electron theory is concerned with the creation and annihilation of pairs.

The electron-positron-photon field. Electrons together with a field form a sort of unified electron-positron-electromagnetic field. Insofar as interaction exists between electrons and positrons, on the one hand, and quanta, on the other, the division of a unified field into charges and quanta becomes, in a certain sense, artificial and, in any case, approximate. The strength of interaction between field and charge is determined by the dimensionless parameter

$$\frac{e^2}{\hbar c} = \frac{1}{137}.$$

Since this parameter is a rather small number, the approximation arising from the separated charges and field seems satisfactory.

As was mentioned in Sec. 27, the theory still has certain difficulties. The most difficult problem—even the approach to which is not known—consists in an explanation of the number $\frac{1}{137}$: since it is an abstract number, the theory should, in principle, derive it from certain general physical principles. But these principles have not yet been formulated.

Nevertheless, all problems in which quantum electrodynamics is used in calculating experimental quantities have complete and unique solutions. Therefore, despite certain imperfections, the quantum theory of the electromagnetic field possesses the essential features of a correct theory: agreement with experiment and a very specific mechanism for calculation.

The quantum theory for other fields is quite a different matter. The only other comparatively satisfactory situation is in the theory of the field responsible for the beta disintegration of nuclei and other so-called "weak interactions," like π-μ-meson decay, etc. (see below).

As regards the theory of nuclear forces, all that is known from a series of unsuccessful attempts is what form the theory cannot have. However, there are many experimental facts that permit of a conclusion concerning the physical nature of nuclear forces. These forces are undoubtedly related, at least in part, with the so-called π-mesons—particles with mass 273 times that of the electron mass. These mesons play the part of quanta in the field of nuclear forces. But they interact with nucleons (i.e., with protons and neutrons) so intensely that it is doubtful whether there is any sense in a separate consideration of nucleons and π-mesons as opposed to electrodynamics, where, to an initial approximation, electrons and quanta may be considered separately.

In addition to π-mesons, there are heavy K-mesons, which disintegrate into three, and sometimes only into two, π-mesons.

Upon decay, π-mesons give μ-mesons, which interact weakly with nuclei. The role played by such weakly interacting particles in the general scheme of nuclear forces is mysterious in the extreme.

An analysis of experimental data shows that the three basic types of elementary interactions differ essentially as to their strength:

1) The strongest are nuclear interactions. These include, for example, interactions between π-mesons and nucleons.

2) Electromagnetic interactions between quanta and charged elementary particles are approximately one hundred times weaker than nuclear interactions.

3) Interactions which are related to beta disintegration or, for example, to the decay of heavy mesons of mass 960 electron masses into 2 or 3 π-mesons are weaker than nuclear interactions by a factor of 10^{11}.

Landau, and also Lee and Yang, have shown that the laws for weak interactions cannot be invariant with respect to a simple transformation from a right-handed to a left-handed coordinate system (inversion). For the interaction to remain invariant, it is necessary, simultaneously with inversion, to transfer from particles to antiparticles, i.e., from electrons to positrons, from protons to antiprotons, from π^+- to π^--mesons, etc.

Thus, the simple law of conservation of parity, which obtains for nuclear and electromagnetic interactions, changes its form for weak interactions. Starting with the principle of "combined parity,"

R. Feynman and R. Gell-Mann (and, independently, R. Marshak and Snolarshan) have succeeded in constructing a universal Hamiltonian for all the weak interactions. The original form was suggested by Fermi. It contains a new universal constant of the order of 10^{-49} erg·cm^3.

Classification of elementary particles. It is rather difficult to define exactly just what an "elementary" particle is. At the beginning of this century, the atoms of elements were considered to be elementary particles, since they were thought to be indivisible. Now we know that atoms consist of electrons, protons, and neutrons, and we are quite sure that these latter do not consist of still other particles of a more "elementary" nature in the same sense as atoms do.

Several elementary particles transform into one another. In some cases we can precalculate the laws of such transformations, for instance, for electrons, positrons, and photons, or, to a less extent, for beta transformations; but we know very little about strong nuclear interactions, which are undoubtedly also connected with some conversion of elementary particles.

To visualize this situation better let us consider the following process: a neutron emits a negative π-meson, and a proton absorbs it. The neutron converts into a proton, and the proton into a neutron, and the whole process of emission and reabsorption can be treated as an "exchange" interaction. This sort of interaction is to some extent analogous to electromagnetic interaction, where a photon is emitted by one electron and absorbed by another, but unlike the electromagnetic forces, we can describe the mechanism of nuclear forces only in words. All attempts to do more have as yet failed.

Despite the lack of a theory of elementary particles, it is now possible to bring them into some order. This classification is due to R. Gell-Mann.

We shall not describe the experimental proofs of the existence of all the elementary particles listed below; such proof can be found elsewhere. Let us be satisfied in stating that their actual existence is quite definite, in contrast to the "existence" of the ephemeral elementary particles which disappear from the pages of scientific papers after a careful investigation.

The Gell-Mann classification is based primarily on particle interaction. First, a particle exists which is capable of electromagnetic interaction only. This is the photon, or the quantum of electromagnetic radiation. Another group of particles is not capable of strong nuclear interactions: the μ-meson, the electron, and the neutrino—the so-called "leptons" (light particles). Still another group of particles, consisting of π- and K-mesons, is capable of nuclear interactions. The masses of these particles are intermediate between those of nucleons and leptons. π- and K-mesons can appear and disappear in the nuclear transformations of other highly energetic particles; no conservation law for their number exists, but the charge conservation law is never violated. The name baryon (heavy particles) has been given to a fourth group of elementary particles. This group consists of stable nucleons (neutron and proton), and unstable hyperons: Λ, $\sum$, and Ξ, which transform spontaneously into nucleons. A conservation law concerning the baryon number exists, and is as strong as the charge conservation law.

All the particles listed above, except the photon and the π°-meson have counterparts—antiparticles. It is very important that a particle need not to be electrically charged to be able to have an antiparticle, as witness the antineutron. The process of particle-antiparticle interaction is one of annihilation; thus, the antineutron and the neutron are mutually annihilated, and in the process create π-mesons. True neutral particles are only those that do not possess antiparticles, or, in other words, such that physically coincide with them in a transformation from the ordinary world into the "antiworld." This means a mathematical transformation of all the wave equations interchanging positrons and electrons, negative antiprotons and protons, etc. (some think that antinebulae actually exist in the universe).

If the physical laws governing the antiworld are the same as those that govern our ordinary world, then the Hamiltonian of electromagnetic interaction must be invariant under transformation from one to the other. The sign of the charge evidently changes in such a transformation, and the charge enters the Hamiltonian multiplied by the amplitude of the electromagnetic field, the vector potential A, so the latter must change its sign simultaneously. We conclude that the amplitude of the photon is odd with respect to a transformation from the ordinary world to the antiworld. The π°-meson decays into two photons; consequently its amplitude must be even.

Such parity of true neutral particles is one of their very important characteristics.

We now pass to a nontrivial point in the Gell-Mann classification. We first consider the decay of the electrically neutral Λ particle

$$\Lambda \to \pi^\circ + n \quad \text{or} \quad \Lambda \to \pi^- + p$$

(both are possible). The mean time of such decay is of the order of 10^{-10} sec, but the Λ particle itself was created in a nuclear collision which lasted less than 10^{-21} sec. It appears rather puzzling that a particle that can be created so quickly should disappear so slowly. It seems to contradict the general reversibility of physical laws in time. This was why the Λ particle came to be called "strange." The only explanation is that both processes are of a totally different nature. The generation of a particle is due to a strong interaction, and the decay, to a weak interaction. It is enough to suppose that the Λ particle is always created from a nucleon accompanied by a

K-meson. This has been verified experimentally, though indirectly. Such a process does not violate the baryon conservation law. If Λ and K particles take part in the interaction simultaneously, it is strong, and if only one of them transforms into something else, the interaction is weak.

To distinguish between these two types of interaction, Gell-Mann introduced a new characteristic of mesons and of baryons—their "strangeness," S. It is defined as follows:

$$\frac{Q}{e} = \frac{n}{2} + \tau_z + \frac{S}{2}.$$

Here, Q is the charge of the baryon, e is the magnitude of the elementary charge, n is the difference between the number of baryons and antibaryons (1 for baryons, —1 for antibaryons, and 0 for mesons), and τ_z is the z-component of the isotopic spin (see Sec. 32). As all the heavy particles take part in strong interactions, a definite value of τ_z can be ascribed to each one of them. n must be taken equal to 1 for each baryon. For nucleons, τ_z is $+\frac{1}{2}$, as they can have only two values of charge: 1 and 0. Λ has no charge and can have only $\tau_z = 0$. $\sum$ has three values of charge: 1, 0, —1, and equal values of τ_z. Lastly, Ξ is either neutral ($\tau_z = 1/2$) or negative ($\tau_z = -1/2$). Substituting these values of charge and τ_z into the definition of strangeness, we find that nucleons have $S = 0$, Λ and $\sum$ hyperons have $S = -1$ and Ξ has $S = -2$. It is noteworthy that $\sum^+$ and $\sum^-$ are not particle and antiparticle, as both have $n = 1$. Each of them has an antibaryon, for which $n = -1$.

For K- and π-mesons, $n = 0$, since they are not baryons. This gives $S = 0$ for π-mesons and $S = 1$ for K-mesons. Unlike $\sum^\pm$, $\pi^\pm$ are related as particle and antiparticle.

Then a selection rule is defined: the given interaction is strong only when the resulting strangeness of all particles entering into the reaction is conserved. For instance, every interaction of nucleons and π-mesons only is strong (if it does not violate any other conservation laws, except strangeness).

The simultaneous creation of Λ and K particles belongs to strong interactions also, since one of them has $S = 1$, and the other $S = -1$. But the spontaneous decay of the Λ particle into a nucleon and a π-meson is due to a weak interaction, because, here, the strangeness is not conserved.

Transitions with $\Delta S = 2$ are forbidden more strongly than with $\Delta S = 1$. That is why the Ξ particle must decay first into a Λ or $\sum$

particle, which in turn decays into nucleons and π-mesons. These statements agree with the cascade nature of Ξ decay.

There is no reason as yet to attribute definite values of τ_z and S to leptons, since they do not take part in strong interactions.

The transition to the nonrelativistic wave equation. It is instructive, in comparing the relativistic wave equation with Schrödinger's equation, to perform the limiting transition. We shall consider that the energy of the electron is positive and that its velocity v is considerably less than the velocity of light. Then $\mathscr{E}$ differs from mc^2 by a small quantity $\frac{mv^2}{2}$. If we take $mc^2\psi_1$ and $mc^2\psi_2$ to the left-hand side in the first and second equations (38.11), the components ψ_1 and ψ_2 are multiplied by the quantity $\mathscr{E} - mc^2$, i. e., by $\frac{mv^2}{2}$. The wave-function components ψ_3 and ψ_4 appear on the right, multiplied by cp_x or by cp_y. It follows that, for a positive energy, the components ψ_3 and ψ_4 are less than ψ_1 and ψ_2 in the ratio $\frac{v}{c}$. The same follows from the second two equations (38.11): the components ψ_3 and ψ_4 appear on the right multiplied by $\sim 2\,mc^2$, and ψ_1 and ψ_2 are on the left with a factor of the order cp.

For negative energies the components ψ_3 and ψ_4 are large, while ψ_1 and ψ_2 are less in the ratio $\frac{v}{c}$.

Consequently, in the nonrelativistic limit Dirac's equation is reduced to two-component form (as is required according to Sec. 32) for a description of particles with spin $\frac{1}{2}$.

Spin magnetic moment. We shall now show how spin magnetic moment is obtained. It is first of all necessary to write down the Dirac equation in an electromagnetic field. We know that for a transition from the equation for a free particle to an equation for a particle in a field it is necessary to replace the momentum $\mathbf{p}$ by $\mathbf{p} - \frac{e}{c}\mathbf{A}$, where $\mathbf{A}$ is the vector potential, and the energy $\mathscr{E}$ by $\mathscr{E} - e\varphi$ (see Sec. 21). Thus, the Dirac equation in the presence of an electromagnetic field is of the form

$$(\mathscr{E} - e\varphi)\,\psi = c\left(\hat{\mathbf{p}} - \frac{e}{c}\mathbf{A}\right)\hat{\boldsymbol{\alpha}}\,\psi + \hat{\beta}\,mc^2\,\psi. \tag{38.18}$$

As was pointed out, the relations between the wave-function components in the nonrelativistic limit without field appear as

$$\psi_3 = \frac{1}{2\,mc}\left[(\hat{p}_x - i\,\hat{p}_y)\,\psi_2 + \hat{p}_z\,\psi_1\right],$$

$$\psi_4 = \frac{1}{2\,mc}\left[(\hat{p}_x + i\,\hat{p}_y)\,\psi_1 - \hat{p}_z\,\psi_2\right]. \tag{38.19}$$

It is convenient to write down these relations, according to Sec. 32, with the aid of the operators $\boldsymbol{\sigma}$, the definition of which involves the factor $\frac{1}{2}$ [see (32.2), (32.8), (32.9)]:

$$\psi' = \frac{1}{mc}\left(\hat{\boldsymbol{\sigma}}, \hat{\mathbf{p}} - \frac{e}{c}\mathbf{A}\right)\psi. \tag{38.20}$$

Here, the two small wave-function components ψ_3 and ψ_4 are for short denoted by ψ', and the two large components ψ_1 and ψ_2 are called ψ. In addition, the momentum $\mathbf{p}$ is replaced by $\mathbf{p} - \frac{e}{c}\mathbf{A}$.

We shall also call $\mathscr{E} = mc^2 + \mathscr{E}'$, where $\mathscr{E}'$ is the energy value which appears in the nonrelativistic theory. Then, after replacing $\hat{\mathbf{p}}$ by $\hat{\mathbf{p}} - \frac{e}{c}\mathbf{A}$, we obtain from the first two equations of (38.11)

$$(\mathscr{E}' - e\varphi)\psi = 2c\left(\hat{\boldsymbol{\sigma}}, \hat{\mathbf{p}} - \frac{e}{c}\mathbf{A}\right)\psi'. \tag{38.21}$$

We can now eliminate ψ' from the equations, so that only the relations for the large components ψ remain.

Substituting ψ' in (38.21) from (38.20), we shall have

$$(\mathscr{E}' - e\varphi)\psi = \frac{2}{m}\left(\hat{\boldsymbol{\sigma}}, \hat{\mathbf{p}} - \frac{e}{c}\mathbf{A}\right)\left(\hat{\boldsymbol{\sigma}}, \hat{\mathbf{p}} - \frac{e}{c}\mathbf{A}\right)\psi. \tag{38.22}$$

When squaring the operator $\left(\hat{\boldsymbol{\sigma}}, \hat{\mathbf{p}} - \frac{e}{c}\mathbf{A}\right)$ we must take into account the commutation relations between the components of $\hat{\boldsymbol{\sigma}}$ and also between $\mathbf{p}$ and $\mathbf{A}$. Taking advantage of the fact that $\hat{\boldsymbol{\sigma}}_x^2 = \hat{\boldsymbol{\sigma}}_y^2 = \hat{\boldsymbol{\sigma}}_z^2 = \frac{1}{4}$, and calling $\hat{p}_x - \frac{e}{c}A_x \equiv P_x \ldots$, and analoguously for P_y and P_z, we shall first of all have

$$(\mathscr{E}' - e\varphi)\psi = \left[\frac{1}{2m}\left(\hat{\mathbf{p}} - \frac{e}{c}\mathbf{A}\right)^2 + \frac{2}{m}(\hat{\boldsymbol{\sigma}}_x\hat{\boldsymbol{\sigma}}_y\hat{P}_x\hat{P}_y + \hat{\boldsymbol{\sigma}}_y\hat{\boldsymbol{\sigma}}_x\hat{P}_y\hat{P}_x) + \ldots\right]\psi. \tag{38.23}$$

Further, we utilize the fact that $\hat{\boldsymbol{\sigma}}_x\hat{\boldsymbol{\sigma}}_y = -\hat{\boldsymbol{\sigma}}_y\hat{\boldsymbol{\sigma}}_x = \frac{i\hat{\sigma}_z}{2}$ [see (32.13)]. and also the commutation relations of the form

$$\left.\begin{aligned} \hat{P}_x\hat{P}_y - \hat{P}_y\hat{P}_x &= -\frac{e}{c}(\hat{p}_x A_y + A_x\hat{p}_y - \hat{p}_y A_x - A_y\hat{p}_x) = \\ &= -\frac{e}{c}(\hat{p}_x A_y - A_y\hat{p}_x) + \frac{e}{c}(\hat{p}_y A_x - A_x\hat{p}_y) = \\ &= \frac{eh}{ci}\left(\frac{\partial A_x}{\partial y} - \frac{\partial A_y}{\partial x}\right) = -\frac{eh}{ic}H_z \end{aligned}\right\} \tag{38.24}$$

[cf. (30.36)]. Substituting this in (38.23), we obtain an equation for the components

$$(\mathscr{E}' - e\varphi)\psi = \left[\frac{1}{2m}\left(\hat{\mathbf{p}} - \frac{e}{c}\mathbf{A}\right)^2 - \frac{eh}{mc}(\hat{\sigma}_x H_x + \hat{\sigma}_y H_y + \hat{\sigma}_z H_z)\right]\psi.$$

Going over to vector notation, we arrive at the nonrelativistic wave equation

$$\mathscr{E}'\psi = \frac{1}{2m}\left(\hat{\mathbf{p}} - \frac{e}{c}\mathbf{A}\right)^2\psi + e\varphi\psi - \frac{eh}{mc}(\hat{\boldsymbol{\sigma}}\mathbf{H})\psi. \quad (38.25)$$

Compared with the Hamiltonian operator for a spinless particle, the Hamiltonian of an electron involves an additional term:

$$\mathscr{H}_\sigma = -\frac{eh}{mc}(\hat{\sigma}\mathbf{H}) \equiv -(\hat{\boldsymbol{\mu}}_\sigma\mathbf{H}). \quad (38.26)$$

But since $\hat{\boldsymbol{\sigma}}$ is an additional mechanical moment, we see that the electron has an additional magnetic moment

$$\hat{\boldsymbol{\mu}}_\sigma = \frac{eh}{mc}\hat{\boldsymbol{\sigma}} \quad (38.27)$$

in accordance with what was affirmed in (32.17) ($\hat{\boldsymbol{\sigma}}$ here is a dimensionless operator). Spin differs from orbital angular momentum in that its magnetic moment does not contain a factor 2 in the denominator. Thus, the so-called spin magnetic anomaly follows naturally from the Dirac equation.

The radiation-field correction to the magnetic moment. Equation (38.27) is, of course, correct only in the nonrelativistic limit. But even in this limit it is not completely exact. As was indicated in Sec. 27, the state of an electromagnetic field in which there are no quanta interacts with charged particles. Strictly speaking, insofar as there is an interaction between the charges and the field, the state of each separately cannot be defined with complete precision. It is therefore not surprising that any state of a field is in some way perturbed by the presence of charges, and any state of the charges is perturbed by the field. As a result of this, the magnetic moment of an electron, as is shown by the rather exact calculations of Schwinger and others, is greater than one magneton by a very small quantity, whose relative fraction is $\frac{e^2}{2\pi hc}$. This result is in complete agreement with experiment.

The magnetic moments of the proton and neutron do not at all agree with the Dirac theory. For instance, on Dirac's theory, a neutral particle (the neutron) should not have a magnetic moment at all. In actual fact, the neutron possesses a magnetic moment directed opposite to the spin.

This is usually explained by the strong interaction between the nucleon and the nuclear force field or, as it is sometimes called, the meson field. There is a certain analogy here with the correction to the magnetic moment of the electron. This correction is small because the interaction constant $\frac{e^2}{\hbar c}$ is small. Nuclear interaction is very strong, and so the result is a large "correction," if one can use that expression for a quantity which, in the case of a proton, is twice as great as the basic magnetic moment given by the Dirac theory.

At the present time we are unable to calculate the magnetic moment of a nucleon, since no theory of nuclear forces exists.

Nevertheless, the nucleon is undoubtedly to some extend a Dirac particle, as confirmed by the existence of the antiproton.

Energy eigenvalues of a hydrogen atom. In accordance with the Dirac equation, the energy eigenvalues of a hydrogen atom, or of any single-electron atom, are calculated in the following way:

$$\frac{\mathscr{E}}{mc^2} = \frac{1}{\sqrt{1 + \left[\frac{\alpha Z}{n - \left(j + \frac{1}{2}\right) + \sqrt{\left(j + \frac{1}{2}\right)^2 - \alpha^2 Z^2}}\right]^2}} - 1, \qquad (38.28)$$

where n is the principal quantum number, $j = l \pm \frac{1}{2}$, i.e., j is the total electron angular momentum, $\alpha = \frac{e^2}{\hbar c} = \frac{1}{137}$. If we regard αZ as small compared with unity, then the nonrelativistic formula (31.34) results. It follows from formula (38.28) that the states $2p_{1/2}$ and $2s_{1/2}$, with the same $n = 2$ and $j = \frac{1}{2}$, have the same energy. In practice, these states of the atom are somewhat split as a result of the interaction of light quanta with the ground state of the field. The calculated splitting agrees with experiment with considerable accuracy.

Exercises

1) Prove that a quantum cannot give rise to an electron-positron pair in free space in the absence of an additional external field.

The conservation laws in the absence of a field are written thus:

$$-\sqrt{m^2 c^4 + c^2 p^2} + \hbar\omega = \sqrt{m^2 c^4 + c^2 p_1^2}, \quad \mathbf{p} + \frac{\hbar\omega}{c}\mathbf{n} = \mathbf{p}_1 .$$

Here, $\mathbf{p}$ is the electron momentum in a negative energy state, $\mathbf{n}$ is a unit vector in the direction of the quantum momentum, $\mathbf{p}_1$ is the electron momentum in a positive energy state. Substituting $\mathbf{p}_1$ in the first equality and squaring the left- and right-hand sides, it is easy to see that this equation is not satisfied.

Another method of proof is based on simple reasoning. A transition to another inertial system can always make the energy of a quantum less than $2\,mc^2$. A quantum cannot give rise to a pair in such a system, simply because it has insufficient energy. But what is impossible in one reference system is impossible in all systems, because the possibility or impossibility of an event does not depend upon the choice of the reference system.

The preceding argument no longer holds if pair production is considered close to a nucleus. Here, the nucleus is at rest in one reference system and in motion in another. Where the energy of the quantum is less than $2\,mc^2$, the moving nucleus will "help" it to give rise to a pair. Naturally, it is in no way possible for a quantum to give rise to a pair if its energy in the *rest system of the nucleus* is less than $2\,mc^2$.

2) Obtain the solution to the Dirac equation for a free electron.

Let us equate ψ_1 to zero. Then the first equation of (38.11) is satisfied if we take $\psi_3 = Ac\,(p_x - i\,p_y)$, $\psi_4 = -Ac\,p_z$. The second equation of (38.11) gives

$$\psi_2 = \frac{Ac^2\,(p_x^2 + p_y^2 + p_z^2}{\mathscr{E} - mc^2} = \frac{A\,(\mathscr{E}^2 - m^2c^4)}{\mathscr{E} - mc^2} = A(\mathscr{E} + mc^2)\,.$$

The third equation of (38.11) reduces to the identity

$$(\mathscr{E} + mc^2)\,\psi_3 = Ac\,(\mathscr{E} + mc^2)\,(p_x - ip_y) = c(p_x - ip_y)\,\psi_2 = Ac\,(p_x - ip_y)\,(\mathscr{E} + mc^2)\,.$$

The fourth equation of (38.11) also reduces to an identity. The number A is determined from the normalization condition

$$|\,\psi_1\,|^2 + |\,\psi_2\,|^2 + |\,\psi_3\,|^2 + |\,\psi_4\,|^2 = 1$$

or

$$A^2\,[(\mathscr{E} + mc^2)^2 + c^2\,p_x^2 + c^2\,p_y^2 + c^2\,p_z^2] = 1\,,$$

$$A = \frac{1}{\sqrt{2\,\mathscr{E}\,(\mathscr{E} + mc^2)}}\,.$$

The components ψ_3 and ψ_4 are small compared with ψ_2 if $v \ll c$. Therefore the solution corresponds to positive energy. Another solution with positive energy is obtained if we take $\psi_2 = 0$. Negative energy solutions are obtained if we choose $\psi_3 = 0$ or $\psi_4 = 0$.

3) Show that from Dirac's equation there follows a charge-conservation equation which is analogous to (24.16):

$$\frac{\partial}{\partial t}\,|\,\psi\,|^2 = -\,\mathrm{div}\,(\psi^*\,c\hat{\boldsymbol{\alpha}}\,\psi),$$

where $|\,\psi\,|^2 = |\,\psi_1\,|^2 + |\,\psi_2\,|^2 + |\,\psi_3\,|^2 + |\,\psi_4\,|^2$.

Write down equation (38.18) and its complex conjugate; multiply the first by ψ^* and the second by ψ; subtract the second from the first and utilize the Hermitian nature of the operators $\hat{\boldsymbol{\alpha}}$ and $\hat{\beta}$.

4) Show that if ψ is a solution with positive energy $\mathscr{E}$, then $\rho_2\,\psi$ is a solution with negative energy $-\,\mathscr{E}$.

The equation for ψ is

$$\mathscr{E}\,\psi = c\,(\hat{\boldsymbol{\alpha}}\,\hat{\mathbf{p}})\,\psi + mc^2\,\beta\,\psi\,.$$

Whence

$$\mathscr{E}\,\rho_2\,\psi = c\,\rho_2\,(\hat{\boldsymbol{\alpha}}\,\hat{\mathbf{p}})\,\psi + mc^2\,\rho_2\,\hat{\beta}\,\psi = -\,[c\,(\hat{\boldsymbol{\alpha}}\,\hat{\mathbf{p}}) + mc^2\,\hat{\beta}]\,\rho_2\,\psi\,.$$

This proves that a negative energy solution cannot be avoided.

5) Prove that the operators

$$\hat{\sigma}_x = \frac{h}{2i}\,\hat{\alpha}_y\,\hat{\alpha}_z,\quad \hat{\sigma}_y = \frac{h}{2i}\,\hat{\alpha}_z\,\hat{\alpha}_x,\quad \hat{\sigma}_z = \frac{h}{2i}\,\hat{\alpha}_x\,\hat{\alpha}_y,$$

acting upon *four-component* functions, are spin operators.

We have

$$\hat{\sigma}_x^2 = -\frac{h^2}{4}\hat{\alpha}_y\hat{\alpha}_z\hat{\alpha}_y\hat{\alpha}_z = \frac{h^2}{4}\hat{\alpha}_y^2 = \frac{h^2}{4},$$

$$\hat{\sigma}_x\hat{\sigma}_y = -\frac{h^2}{4}\hat{\alpha}_y\hat{\alpha}_z\hat{\alpha}_z\hat{\alpha}_x = -\frac{h^2}{4}\hat{\alpha}_y\hat{\alpha}_x = i\frac{h}{2}\hat{\sigma}_z,$$

$$\hat{\sigma}_y\hat{\sigma}_x = -i\frac{h}{2}\hat{\sigma}_z,$$

so that the spin operators determined here possess all the required properties (see Sec. 32). This can also be seen from the definition of $\hat{\boldsymbol{\alpha}}$ in terms of $\hat{\boldsymbol{\sigma}}$ and $\hat{\boldsymbol{\rho}}$. We notice that the spin operators do not permute functions of the first pair ψ_1, ψ_2 and of the second pair ψ_3, ψ_4, but make permutations only inside each pair.

6) Show that according to the Dirac equation only the sum of the orbital and spin angular momenta and not each angular momentum separately satisfies the angular-momentum conservation law.

The total angular momentum is defined as

$$\hat{\mathbf{J}} = \hat{\mathbf{M}} + \hat{\boldsymbol{\sigma}} = [\mathbf{r}\hat{\mathbf{p}}] + \hat{\boldsymbol{\sigma}},$$

$$\hat{\mathbf{J}}_z = x\hat{p}_y - y\hat{p}_x + \frac{h}{2i}\hat{\alpha}_x\hat{\alpha}_y.$$

We calculate the commutator with the Hamiltonian:

$$\hat{\mathscr{H}}\hat{J}_z - \hat{J}_z\hat{\mathscr{H}} = [c(\hat{\alpha}_x\hat{p}_x + \hat{\alpha}_y\hat{p}_y + \hat{\alpha}_z\hat{p}_z) + \hat{\beta}mc^2]\left(x\hat{p}_y - y\hat{p}_x + \frac{h}{2i}\hat{\alpha}_x\hat{\alpha}_y\right) -$$

$$-\left(x\hat{p}_y - y\hat{p}_x + \frac{h}{2i}\hat{\alpha}_x\hat{\alpha}_y\right)[c(\hat{\alpha}_x\hat{p}_x + \hat{\alpha}_y\hat{p}_y + \hat{\alpha}_z\hat{p}_z) + \hat{\beta}mc^2] =$$

$$= c\hat{\alpha}_x\hat{p}_y(\hat{p}_x x - x\hat{p}_x) - c\hat{\alpha}_y\hat{p}_x(\hat{p}_y y - y\hat{p}_y) + \frac{hc}{2i}\hat{p}_x(\hat{\alpha}_x\hat{\alpha}_x\hat{\alpha}_y - \hat{\alpha}_x\hat{\alpha}_y\hat{\alpha}_x) +$$

$$+\frac{hc}{2i}\hat{p}_y(\hat{\alpha}_y\hat{\alpha}_x\hat{\alpha}_y - \hat{\alpha}_x\hat{\alpha}_y\hat{\alpha}_y) = \frac{hc}{i}(\hat{\alpha}_x\hat{p}_y - \hat{\alpha}_y\hat{p}_x + \hat{p}_x\hat{\alpha}_y - \hat{p}_y\hat{\alpha}_x) = 0.$$

The Hamiltonian is commutative also with the square of the total angular momentum $\hat{J}^2 = \hat{J}_x^2 + \hat{J}_y^2 + \hat{J}_z^2$. The integrals of motion are $\hat{J}^2$ and $\hat{J}_z$, and not $\hat{M}^2$, $\hat{\sigma}^2$ and $\hat{M}_z$, $\hat{\sigma}_z$ separately.

PART IV

STATISTICAL PHYSICS

Sec. 39. The Equilibrium Distribution of Molecules in an Ideal Gas

The subject of statistical physics. The methods of quantum mechanics set out in the third part make it possible, in principle, to describe any assembly of electrons, atoms, and molecules comprising a macroscopic body.

In practice, however, even the problem of an atom with two electrons presents such great mathematical difficulties that nobody, so far, has solved it completely. It is all the more impossible not only to solve but even to write down the wave equation for a macroscopic body consisting, for example, of 10^{23} atoms with their electrons.

Yet in large systems, we encounter certain general laws of motion for which it is not necessary to know the wave function of the system to describe them. Let us give one very simple example of such a law. We shall suppose that there is only one molecule contained in a large, completely empty vessel. If the motion of this molecule is not defined beforehand, the probability of finding it in any half of the vessel is equal to 1/2. If there are two molecules in the same vessel, the probability of finding them in the same half of the vessel simultaneously is equal to $\left(\frac{1}{2}\right)^2 = 1/4$. The probability of finding all of a gas, consisting of N particles, in the same half of the vessel (if the vessel is filled with gas) is $(1/2)^N$, i.e., an unimaginably small number. On the average, there will always be an approximately equal number of molecules in each half of the vessel. The greater the number of molecules forming the gas, the closer to unity will be the ratio of the number of molecules in both halves of the vessel, no matter at what time they are observed.

This approximate equality for the number of molecules in equal volumes of the same vessel gives an almost obvious example of a statistical law applicable only to a large assembly of objects. In addition to a spatial distribution, molecules possess a definite velocity distribution, which, however, can in no way be uniform (if only because the probability of an infinitely large velocity is equal to zero).

Statistical physics studies the laws governing the motion of large assemblies of electrons, atoms, quanta, molecules, etc. The problem of the velocity distribution of gas molecules is one of the simplest that is solved by the methods of statistical physics.

Statistical physics introduces a series of new quantities, which cannot be defined in terms of single-body dynamics or the dynamics of a small number of bodies. An example of such a statistical quantity is temperature, which turns out to be closely related to the mean energy of a gas molecule. If a gas is confined only to one half of a vessel, and the barrier dividing the vessel is then removed, the gas will itself uniformly fill both halves. Similarly, if the velocity distribution of the molecules is disturbed in some way, then, as a result of collisions between the molecules there will be established a very definite statistical distribution, which, for constant external conditions, will be maintained approximately for an indefinitely long time. This example involving collisions shows that regularity in statistics arises not only because a large assembly of objects is taken, but also because they interact.

The statistical law in quantum mechanics. Quantum mechanics also describes statistical regularities, but relating to a separate object. Here, the statistical regularity manifests itself in a very large number of identical experiments with identical objects, and is in no way related to the interaction of these objects. For example, the electrons in a diffraction experiment may pass through a crystal with arbitrary time intervals and nevertheless give exactly the same statistical picture for the blackening of a photographic plate as if they had passed through the crystal simultaneously.

Regularities in alpha disintegration cannot be accounted for by the fact that there are a very large number of nuclei: since there is practically no interaction between nuclei inducing the process, the statistical character predicted by quantum mechanics is only *manifested* for a large number of identical objects; it is by no means *due* to their number. In this connection, a description of phenomena in quantum mechanics involves the concept of probability phase, similar to the concept of the phase of a light wave.

In principle, the wave equation can also be applied to systems consisting of a large number of particles. The solution of such an equation represents a detailed quantum-mechanical description of the state of the system. Let us suppose that as a result of the solution of the wave equation we have obtained a certain spectrum of energy eigenvalues of the system

$$\mathscr{E} = \mathscr{E}_0, \mathscr{E}_1, \mathscr{E}_2, \ldots, \mathscr{E}_n, \ldots \tag{39.1}$$

in states with wave functions

$$\psi_0, \psi_1, \psi_2, \ldots, \psi_n, \ldots$$

Then the wave function for any state, as was shown in Sec. 30, can be represented in the form of a sum of ψ-functions of states with definite energy values:

$$\psi = \sum_n c_n \psi_n. \tag{39.2}$$

The square of the modulus

$$w_n = |c_n|^2 \tag{39.3}$$

gives the probability (when the energy of a system in the state ψ is measured) that the result will be the nth value.

The expansion (39.2) makes it possible to determine not only amplitudes, but also relative probability phases corresponding to a detailed quantum-mechanical description of the system.

The methods of statistical physics make it possible straightway to determine approximately the quantities $w_n = |c_n|^2$, i.e., the probabilities themselves omitting their phases. For this reason, the wave function of the system cannot be determined from them, although it is possible to find the practically important mean values of quantities that characterize macroscopic bodies (for example, their mean energy).

In this section we shall consider how to calculate the probability w_n as applied to an ideal gas.

Ideal gases. An ideal gas is a system of particles whose interaction can be neglected. The interaction resulting from collisions between molecules is essential only when the statistical distribution w_n is in the process of being established. When this distribution becomes established the effect of interaction is very slight.

As regards condensed (i.e., solid and liquid) bodies, the molecules are all the time in vigorous interaction, so that the statistical distribution depends essentially upon the forces acting between the bodies.

But even in a gas the particles must not be regarded as absolutely independent. For example, Pauli's principle imposes essential limitations on the possible quantum states of a gas. We shall take these limitations into account when calculating probabilities.

The states of separate particles of a gas. In order to distinguish the states of separate particles from the state of the gas as a whole we shall denote their energies by the letter ε and the energy of the whole gas by $\mathscr{E}$. Thus, for example, if the gas is contained in a rectangular potential well (see Sec. 25), then the energy values for each particle are calculated according to equation (25.19).

Let ε take on the following series of values:

$$\varepsilon = \varepsilon_0, \varepsilon_1, \varepsilon_2, \ldots, \varepsilon_k, \ldots, \tag{39.4}$$

where there are n_0 particles in the state with energy ε_0 and in general there are n_k particles with energy ε_k in the gas. Then the total energy of the gas is

$$\mathscr{E} = \sum_k n_k \varepsilon_k. \tag{39.5}$$

By giving different combinations of numbers n_k, we will obtain the total energy values forming the series (39.1).

We have repeatedly seen that the energy value ε_k does not yet define the particle states. For example, the energy of a hydrogen atom depends only upon the principal quantum number n,* so that the atom can have $2n^2$ states for a given energy [see (33.1)]. This number, $2n^2$, is called the *weight* of the state with energy ε_n. But it is also possible to place the system under such conditions that the energy value defines the energy in principle uniquely. We note, first of all, that in all atoms except hydrogen the energy depends not only on n, but on l also, i.e., on the azimuthal quantum number. Further, account of the interaction between spin and orbit shows that there is a dependence of the energy upon the total angular-momentum j and, finally, if the atom is placed in an external magnetic field, the energy also depends upon the projection of angular momentum on the magnetic field. Thus, one energy value mutually corresponds uniquely to one state of the atom.

In a magnetic field all the $2n^2$ states with the same principal quantum number are split. We now consider how the states of a gas in a closed vessel are split. We shall suppose that the vessel is of the form of a box with incommensurable squares of the sides a_1^2, a_2^2, a_3^2. Then, in accordance with equation (25.19), the energy of the particles is proportional to the quantity $\frac{n_1^2}{a_1^2} + \frac{n_2^2}{a_2^2} + \frac{n_3^2}{a_3^2}$ where n_1, n_2, n_3 are positive integers. Any combination of these integers gives one and only one number for the incommensurable values a_1^2, a_2^2, a_3^2. Therefore, specification of the energy defines all three *integers* n_1, n_2, n_3. If the particles possess an intrinsic angular momentum, we can, so to speak, remove the degeneracy by placing the gas in a magnetic field (an energy eigenvalue to which there correspond several states of a system is termed *degenerate*). We shall first consider only completely removed degeneracy.

States of an ideally closed system. We shall now consider the energy spectrum of a gas completely isolated from possible external influences and consisting of absolutely noninteracting particles. For simplicity, we shall assume that one value of energy corresponds to each state of the system as a whole, and, conversely, one state corresponds to each energy value. This assumption is true if all the energy eigenvalues for each particle are incommensurable numbers).* We shall

* Not to be confused with n_k!

* In a rectangular box the state $\varepsilon(n_1, n_2, n_3)$ has an energy which is commensurable with the energy of state $\varepsilon(2n_1, 2n_2, 2n_3)$. Therefore, the energy of all states can be incommensurable only in a box of more complex form than rectangular.

call these numbers ε_k. Then, if there are n_k particles in the k-th state, the total energy of the gas is equal to $\mathscr{E} = \sum_k n_k \varepsilon_k$. But, for incommensurable ε_k, it is possible in principle to determine all n_k from this equation, provided $\mathscr{E}$ is precisely specified. It is clear, however, that the energy of a gas consisting of a sufficiently large number of particles must be specified with trully exceptional accuracy for it to be possible to really find all n_k from $\mathscr{E}$.

It is not a question of determining the state of an individual particle from its energy $\mathscr{E}$, but of finding the state of the whole gas from the sum of the energies of all of its particles. Every interval of values $d\mathscr{E}$, even very small (though not infinitely small), will include very many eigenvalues $\mathscr{E}$. Each of them corresponds to its own set of values n_k, i.e., to a definite state of the system as a whole.

States of a nonideally closed system. Energy is an exact integral of motion only in an ideally closed system. The state of such a system is maintained for an indefinitely long time, and the conservation of the quantity $\mathscr{E}$ provides for the constancy of all n_k. But nature does not (and cannot) have ideally closed systems. Every system interacts in some way with the surrounding medium. We will regard this interaction as weak and will determine how it affects the behaviour of the system.

Let us assume that the interaction with the medium does not noticeably disturb the quantum levels of separate particles. Nevertheless, every level ε_k ceases being a precise number and receives a small, though finite, width $\Delta\mathscr{E}_k$. This is sufficient for the meaning of the equation $\mathscr{E} = \sum_k n_k \varepsilon_k$ to change in a most radical way: in a system consisting of a large number of particles, the equation containing approximate quantities ε_k no longer defines the number n_k.

In other words, an interaction with the surrounding medium, no matter how weak, makes impossible an accurate determination of the state from the total energy $\mathscr{E}$.

Transitions between close-energy states. In an ideally closed system, transitions were forbidden for all states corresponding to an energy interval $d\mathscr{E}$ because the energy conservation law held strictly. For weak interaction with the medium, all transitions that do not change the total energy of the gas as a whole are possible to an accuracy which is, in general, compatible with the determination of the energy of a nonideally closed system.

Let us suppose that the interaction with the medium is so weak that, for some small interval of time, it is possible, in principle, to determine all the quantities n_k and thus to give the total energy of the gas $\mathscr{E} = \sum_k n_k \varepsilon_k$. But over a large interval of time the state of the gas can now vary within the limits of that interval of total energy which is given by the inaccuracy in the energy of separate states $\Delta\varepsilon_k$. All transitions will occur that are compatible with the approximate equation $\mathscr{E} = \sum_k n_k (\varepsilon_k \pm \Delta\varepsilon_k)$. Naturally, the state in which all $\Delta\varepsilon_k$ are of one sign is extremely improbable; this is why the double symbol $\pm$ is written. We must find the state that is formed as a result of all possible transitions in the interval $d\mathscr{E}$.

The probabilities of direct and reverse transitions. A very important relation exists between the probabilities for a direct and reverse transition. Let us first of all consider this relation on the basis of equation (34.29), which is obtained as a first approximation in perturbation theory. Let there be two states A and B in the system, with wave functions ψ_A and ψ_B. The same value of energy corresponds to these states, within the limits of inaccuracy $d\mathscr{E}$ given by the interaction of the system with the medium. In the interval $d\mathscr{E}$, both states may be regarded as belonging to a continuous spectrum. Then, from (34.29), the probability of a transition from A to B in unit time is equal to $\frac{2\pi}{h}\left|\mathscr{H}_{AB}\right|^2 g_B$, and from B to A, $\frac{2\pi}{g}\left|\mathscr{H}_{BA}\right|^2 g_A$, where

$$\mathscr{H}_{AB} = \int \psi_B^* \hat{\mathscr{H}} \psi_A \, dV,$$

$$\mathscr{H}_{BA} = \int \psi_A^* \hat{\mathscr{H}} \psi_B \, dV$$

(the weights of the states are denoted by g_A, g_B). But, if $g_A = g_B$, then the probabilities for direct and reverse transitions, which we shall call W_{AB} and W_{BA}, are equal because $|\mathscr{H}_{AB}|^2 = |\mathscr{H}_{BA}|^2$. Naturally, the transition is only possible because the energies $\mathscr{E}_A$ and $\mathscr{E}_B$ are not defined with complete accuracy, and a small interval $d\mathscr{E}$ is given in which the energy spectrum is continuous. In an ideally closed system we would have $\mathscr{E}_A \neq \mathscr{E}_B$.

The relationship we have found only holds to a first approximation in the perturbation method. However, there is also an accurate relationship that can be deduced from the general principles of quantum mechanics. In accordance with the accurate relationship, the probabilities for transitions from A to B and from B^* to A^* are equal; here A^* and B^* differ from A and B in the signs of all the linear- and angular-momentum components.

The equal probability for states with the same energy. We have seen that, due to interaction with the medium, transitions will occur, in the system, between all kinds of states A, B, $C, \ldots$, belonging to the same energy interval $d\mathscr{E}$. If we wait long enough the system will pass equal intervals of time in the states A, B, $C, \ldots$. This is most easily proven indirectly, supposing first of all that the probabilities for direct and reverse transitions are simply equal: $W_{AB}=W_{BA}$. The refinement $W_{AB}=W_{B^*A^*}$ does not make any essential change.

For simplicity we shall consider only two states such that $W_{AB}=W_{BA}$. We at first assume that t_A is greater than t_B, so that the system will more frequently change from A to B than from B to A. But this cannot continue over an indefinitely long time, because if the ratio t_A: t_B increases, the system will finally be constantly in A despite the fact that a transition is possible from A to B. Only the equality $t_A=t_B$ can hold for an indefinite time (on the average) on account of the fact that direct and reverse transitions occur on the average with equal frequency. The same argument shows that if there are many states for which direct and reverse transitions are equally probable, then over a sufficiently long period of time the system will, on the average, spend the same time in each state.

We can suppose that $t_A=t_{A^*}$, because the states A and A^* differ only in the signs of all linear and angular momenta (and also the sign of the external magnetic field, which must also be changed so that the magnetic energy of all particles is the same in states A and A^*). If we proceed from the natural assumption that $t_A=t_{A^*}$, then all the precinged argument can be extended to the case when $W_{AB}=W_{B^*A^*}$.

We have thus seen that the system spends the same time in all states (with the same weight) that belong to the same total energy interval $d\mathscr{E}$.

The probability of a separate state. We will call the limit of the ratio t_A/t, when t increases indefinitely, the *probability* of the state q_A. The equality of all t_A implies that corresponding states are equally probable. But this allows us to define the probability of each state directly. Indeed, if there are P states, then $\sum_{A=1}^{P} q_A=1$, because $\sum_{A=1}^{P} t_A = t$. But since the states are, as proven, equiprobable, we find that $q_A=1/P$. Similarly, the probability that a tossed coin will fall heads is equal to 1/2, since the occurrence of heads or tails is practically equiprobable.

Hence, the problem of finding probability is reduced to that of combinatorial analysis. But in order to use this analysis we must determine which states of the system can be regarded as physically different. When computing the total number P we must take each such state once.

Specification of gas states in statistics. If a gas consists of identical particles, for example, electrons, helium atoms, etc., then its state is precisely given if we know how many particles occur in each one of the states. It is not meaningful to ask which particles occur in a certain state, since identical particles cannot, in principle, be distinguished from one another. If the spin of the particles is half-integral then Pauli's exclusion principle must hold and in each state there will occur either one particle or none at all.

As an illustration for calculating the number of states of a system as a whole, let us suppose that there are only two particles and that each particle can have only two states a and b ($\varepsilon_a = \varepsilon_b$), each with weight unity. In all, three different states of the system are conceivable:

1) both particles in state a; state b is unoccupied;
2) the same in state b; a is unoccupied;
3) one particle in each state.

In view of the indistinguishability of the particles, state (3) must be counted once because the interchange of identical particles between states does not have meaning. If, in addition, the particles are subject to the exclusion principle, then only the third state is possible.

Thus, if the exclusion principle is applicable to the particles then the system can have only one state and, if it is not, then three states are possible. Pauli exclusion greatly reduces the number of possible states of a system. A system of two different particles, for example, an electron and proton, would have four states because these particles can obviously be distinguished.

Let us further consider the example of three particles occupying three states. If Pauli exclusion is operative, then one, and only one, state of the system as a whole is possible; one particle occurs in each separate state. If there is no exclusion, then the indistinguishable particles can be distributed thus: one in each quantum state, or two particles in one state and the third in one of the two remaining states (this gives six states for the system as a whole), and all three particles in any quantum state. Thus we have obtained $1 + 3 + 6 = 10$ states for the system as a whole.

If these three particles differed, for example, if they were π^+-, π^0-, and π^--mesons of zero spin, then each of them could have any one of three states independently of the others. Consequently, a system of three such particles could, as a whole, have $3^3 = 27$ states. Later on we shall derive a general formula for calculating the number of states.

Particles not subject to the Pauli exclusion principle. There is no sense, for future deductions, in considering that each state of a particle of given energy has unit weight. We shall denote the weight of a state with energy ε_k by the letter g_k. In other words, g_k different states of a particle have the same energy ε_k. For every particle these states are equally probable.

Let us assume that n_k particles have energy ε_k and are not subject to Pauli exclusion. It is required to calculate the number of ways that these n_k particles can be distributed in g_k states. We shall call this number $P_{g_k n_k}$. In accordance with what we have proved above, the probability for each arrangement as a whole is $\frac{1}{P_{g_k n_k}}$.

In order to calculate $P_{g_k n_k}$ we will, as is usual in combinatorial analysis, call the state a "box" and the particle a "ball." The problem is: how many ways are there of placing n_k balls in g_k boxes without numbering the balls, i.e., without desiring to know which ball lies in which box. If the particles are not subject to the Pauli exclusion principle then we must suppose that each box can accommodate any number of balls.

Let us mix up all the boxes and all the balls so that we obtain n_k+g_k objects. From these objects we take any box and put it aside. The n_k+g_k-1 objects which remain are then randomly taken from the common pile, irrespective of whether they are box or ball, and placed in one row with this box from left to right. The following series may be obtained:

bx, bl, bl, bx, bx, bl, bl, bl, bx, bl, bx, bl, bl, bx, bx, bx.

Since a box must appear on the left, the remaining objects can be distributed among themselves $(n_k+g_k-1)!$ ways.

We now throw each ball into the closest box on the left. In the distribution we have used there will be two balls in the first box, none in the second box, three in the third, one in the fourth, and so on. There are $(n_k+g_k-1)!$ distributions in all, but they are not all distinguishable. Indeed, if we place the second ball in the position of the first or any other one, nothing will change in the series shown. There are $n_k!$ permutations between the balls. In exactly the same way the boxes can be interchanged with the boxes because it does not matter in which order these boxes appear. Only we must not touch the first box, because it always appears on the left by convention. In all, there are $(g_k-1)!$ permutations of the boxes. It follows that, of all the possible $(n_k+g_k-1)!$ arrangements in the series, only the following set of arrangements is different:

$$P_{g_k n_k} = \frac{(n_k + g_k - 1)!}{n_k!\,(g_k - 1)!}. \tag{39.6}$$

If, for example, $n_k=3$, $g_k=3$ then $P_{g_k n_k} = \frac{5!}{3!\,2!} = 10$, which is what we have already seen from direct computation.

Particles subject to Pauli exclusion. In the case of particles subject to Pauli exclusion the calculation of $P_{g_k n_k}$ is still simpler. Indeed, here we always have the inequality $n_k \leqslant g_k$, because not more than one particle occurs in each state. Therefore, of the total number of g_k states n_k are occupied.

The number of ways in which we can choose n_k states is equal to the combination of g_k things n_k at a time:

$$P_{g_k n_k} = C_{g_k}^{n_k} = \frac{g_k!}{n_k!(g_k - n_k)!}\,. \tag{39.7}$$

There are as many possible different states in the case of $n_k \leqslant g_k$, and there is one particle or none at all in any of the g_k states.

The most probable distribution of particles among states. The numbers g_k and n_k refer to a single definite energy. The total number of states of a gas is equal to the product of the numbers $P_{g_k n_k}$ for all the states separately:

$$P = \prod_k P_{g_k n_k}\,. \tag{39.8}$$

So far we have only used combinatorial analysis. And besides it has been shown that all separate states taken separately are equally probable. The quantity P depends upon the distribution of particles among the states. It can be seen that, in fact, a gas is always close to a state where the distribution of separate particles among the states corresponds to the maximum value of P possible for a given total energy $\mathscr{E}$ and for a given total number of particles.

We shall explain this statement by a simple example from gambling, as is usually done in probability theory (most easily seen here is the manifestation of large-number laws in a game of chance). Let a coin be tossed N times. The probability that it will fall heads once is equal to $1/2$. The probability that it will fall heads all N times is equal to $(1/2)^N$. The probability that it will fall N—1 times heads, and once tails, is equal to $(1/2)^{N-1} \times 1/2 \times N$, because this single occasion can turn out to be anyone, from the first to the last, and the probabilities for mutually exclusive events are additive. The probability for a double tails is equal to $\left(\frac{1}{2}\right)^N \frac{N(N-1)}{2}$.

The last factor shows how many ways two events can be chosen from the total number N (the number of combinations of N two at a time). In general, the probability that the coin will fall tails k times is

$$q_k = \left(\frac{1}{2}\right)^{N-k} \left(\frac{1}{2}\right)^k \frac{N!}{k!(N-k)!}\,.$$

The sum of all probabilities is, of course, unity:

$$\sum q_k = \left(\frac{1}{2}\right)^N \left(1 + N + \frac{N(N-1)}{1\cdot 2} + \frac{N(N-1)(N-2)}{1\cdot 2\cdot 3} + \dots\right) = \\ = \left(\frac{1}{2}\right)^N 2^N = 1\,,$$

because the sum of binomial coefficients is equal to 2^N.

Considering the series q_k, we can see that q_k increases up to the middle of the sum, i.e., as far as $k = N/2$, and then decreases symmetrically with respect to the middle of the sum. Indeed, the kth term is obtained from the $(k-1)$th term multiplied by $\frac{N-k+1}{k}$ so that the terms increase as long as $N/2 > k$.

Every separate series for tails appearing is in every way equally probable with all other series. The probability for any given series is equal to $(1/2)^N$. But if we are not interested in the sequence of heads and tails, but only in their total number, then the probabilities will be equal to the numbers q_k. For $N \gg 1$, the function q_k has a very sharp maximum at $k = N/2$ and rapidly falls away on both sides of $N/2$. If we call the total number of N tosses a "game," then in the overwhelming majority of games, heads will be obtained approximately $N/2$ times (if N is large). The probability maximum will be sharper, the greater N is. We will not, here, refine this as applied to the game of pitch and toss (see exercise 1), but will return to the calculation of the number of states of a gas.

On the basis of the equal probability for the direct and reverse processes between any pair of states, we have shown that any previously defined distribution of particles among states has the same probability of being established for a given total energy. In the same way, every separate sequence of heads and tails in each separate game is of equal probability. But, if we do not specify the states of a gas by denoting which of the g_k states with a given energy are filled, and give only the total number of particles in a state with energy ε_k, then we obtain a probability distribution with a maximum similar to the probability distribution of games according to the *total* number of occurrences of tails irrespective of their sequence. The only difference is that in the example of pitch and toss the probability depends upon one parameter k, and the probability for the distribution of gas particles among states depends upon all n_k.

Our problem is to find this distribution for particles with integral and half-integral spins.

It is most convenient to look for the maximum of the logarithm of P rather than P itself. $\ln P$ is a monotonic function of the argument and assumes a maximum value at the same time as the argument P.

Stirling's formula. In calculations we shall require logarithms of factorials. For the factorials of large numbers, there is a convenient approximate formula which we shall deduce here.

It is obvious that

$$\ln n! = \ln(1 \cdot 2 \cdot 3 \cdot 4 \ldots n) = \sum_{k=1}^{n} \ln k .$$

The logarithms of large numbers vary rather slowly since the difference $\ln(n+1) - \ln n$ is inversely proportional to n. Therefore, the sum can be replaced by an integral:

$$\ln n! = \sum_{k=1}^{n} \ln k \cong \int_0^n \ln k \, dk = n \ln n - n = n \ln \frac{n}{e} . \qquad (39.9)$$

This is the well-known mathematical formula of Stirling in somewhat simplified form. It becomes more accurate the greater n is.

Additional maximum conditions. And so we must look for the values of the numbers n_k for which the quantity

$$S = \ln P = \ln \prod_k P_{g_k n_k} \qquad (39.10)$$

is a maximum at a given total energy

$$\mathscr{E} = \sum_k n_k \varepsilon_k \qquad (39.11)$$

and for a given total number of particles

$$N = \sum_k n_k . \qquad (39.12)$$

This kind of extremal condition is termed bound, because additional conditions (39.11) and (39.12) are imposed upon it.

We shall first of all find n_k for particles which are not subject to Pauli exclusion, i.e., those with integral spin. To do this we must substitute the expression for P from (39.6) in (39.10), take the differential dS with respect to all n_k, and equate it to zero. We have

$$S = \ln P = \ln \prod_k \frac{(g_k + n_k - 1)!}{n_k! \, (g_k - 1)!} = \sum_k \ln \frac{(g_k + n_k - 1)!}{n_k! \, (g_k - 1)!} . \qquad (39.13)$$

We substitute here the expression for factorials according to Stirling's formula (39.9):

$$S = \sum_k \left[(g_k + n_k - 1) \ln \frac{g_k + n_k - 1}{e} - n_k \ln \frac{n_k}{e} - (g_k - 1) \ln \frac{g_k - 1}{e} \right] . \qquad (39.14)$$

Since g_k is a large number, unity can naturally be neglected everywhere. We must, of course, differentiate with respect to n_k in formula (39.14), because g_k is the given number of all states. Then

$$dS = \sum_k dn_k \, [\ln (g_k + n_k) - \ln n_k = \sum_k dn_k \ln \frac{g_k + n_k}{n_k} = 0 . \qquad (39.15)$$

It must not be concluded from this equation that the coefficients of every dn_k are equal to zero, because n_k are dependent quantities.

The relationship between them is given by the two equations (39.11) and (39.12) and, in differential form, are as follows:

$$d\mathscr{E} = \sum_k \varepsilon_k \, dn_k = 0\,, \tag{39.16}$$

$$dN = \sum_k dn_k = 0\,. \tag{39.17}$$

From these equations, we could eliminate any two of the numbers dn_k, substitute them in (39.15), and afterwards regard the remaining dn_k as independent quantities. Then their coefficients may be regarded as equal to zero.

The method of undetermined coefficients. The elimination of dependent quantities is most conveniently achieved by the method of undetermined coefficients. This makes it possible to preserve the symmetry between all n_k. Let us multiply equation (39.16) by an indefinite coefficient which we denote by $1/\theta$; the meaning of this notation will be explained later. We multiply the second equation (39.17) by a coefficient which we denote μ/θ, so that we have introduced, as is required from the number of supplementary conditions, two quantities, θ and μ. After this we combine all three equations (39.15)-(39.17) and regard all dn_k as independent, and θ, and μ as unknown values which should be determined from equations (39.11) and (39.12). The maximum condition is now written as

$$dS - \frac{d\mathscr{E}}{\theta} + \frac{\mu \, dN}{\theta} = 0\,. \tag{39.18}$$

We look for the extremum of one quantity $S - \frac{\mathscr{E}}{\theta} + \frac{\mu N}{\theta}$, and then choose θ and μ so that the energy and number of particles equal the given values. But if the extremum is determined for one function without conditions, then all its arguments become mutually independent, and we are entitled to equate any differential dn_k to zero regardless of the other differentials.

Equation (39.18), written in terms of dn_k, has the following form:

$$dS - \frac{d\mathscr{E}}{\theta} + \frac{\mu \, dN}{\theta} = \sum_k dn_k \left(\ln \frac{g_k + n_k}{n_k} - \frac{\varepsilon_k}{\theta} + \frac{\mu}{\theta} \right) = 0\,. \tag{39.19}$$

Bose-Einstein distribution. Let us now put all the differentials except dn_k equal to zero. According to what we have just said this is justified. Then, for equation (39.19) to hold, we must put the coefficient of dn_k equal to zero:

$$\ln \frac{g_k + n_k}{n_k} + \frac{\mu}{\theta} - \frac{\varepsilon_k}{\theta} = 0\,.$$

Naturally, this equation holds for all k. Solving it with respect to n_k, we arrive at the required most probable distribution of the number of particles according to state:

$$n_k = \frac{g_k}{e^{\frac{\varepsilon_k - \mu}{\theta}} - 1}. \tag{39.21}$$

This formula is called the Bose-Einstein distribution. As to particles for which this distribution is applicable, they are said to obey Bose-Einstein statistics or, for short, Bose statistics. They have either integral or zero spin. The unknown quantities θ and μ, i.e., the parameters in the distribution, are given by equations as functions of N and $\mathscr{E}$:

$$\sum_k \frac{\varepsilon_k g_k}{e^{\frac{\varepsilon_k - \mu}{\theta}} - 1} = \mathscr{E}, \tag{39.22}$$

$$\sum_k \frac{g_k}{e^{\frac{\varepsilon_k - \mu}{\theta}} - 1} = N, \tag{39.23}$$

so that the problem posed of finding the most probable values of n_k is, in principle, solved.

Fermi-Dirac distribution. We shall now find the quantities n_k for the case when the particles are subject to Pauli exclusion. In accordance with (39.7) and Stirling's formula we have for the quantity S:

$$S = \ln \prod_k \frac{g_k!}{n_k!\,(g_k - n_k)!} =$$
$$= \sum_k \left[g_k \ln \frac{g_k}{e} - n_k \ln \frac{n_k}{e} - (g_k - n_k) \ln \frac{g_k - n_k}{e} \right]. \tag{39.24}$$

Differentiating S and substituting into equation (39.18), we obtain

$$dS - \frac{d\mathscr{E}}{\theta} + \frac{\mu\, dN}{\theta} = \sum_k dn_k \left(\ln \frac{g_k - n_k}{n_k} - \frac{\varepsilon_k}{\theta} + \frac{\mu}{\theta} \right) = 0, \tag{39.25}$$

whence, by the same method, we arrive at the extremum condition:

$$\ln \frac{g_k - n_k}{n_k} - \frac{\varepsilon_k}{\theta} + \frac{\mu}{\theta} = 0,$$

and the required distribution appears thus·

$$n_k = \frac{g_k}{e^{\frac{\varepsilon_k - \mu}{\theta}} + 1}. \tag{39.26}$$

Here, $n_k \leqslant g_k$ as is the case for particles subject to Pauli exclusion. For such particles, formula (29.36) is called a Fermi-Dirac distribution. The parameters θ and μ are determined analogously to (39.22) and (39.23):

$$\sum_k \frac{\varepsilon_k g_k}{e^{\frac{\varepsilon_k - \mu}{\theta}} + 1} = \mathscr{E}\,, \tag{39.27}$$

$$\sum_k \frac{g_k}{e^{\frac{\varepsilon_k - \mu}{\theta}} + 1} = N\,. \tag{39.28}$$

Concerning the parameters θ and μ. The parameter θ is an essentially positive quantity, because otherwise it would be impossible to satisfy equations (39.22), (39.23) and (39.27), (39.28). Indeed, there is no upper limit to the energy spectrum of gas particles. For an infinitely large ε and $\theta < 0$, we would obtain $e^{\frac{\varepsilon_k}{\theta}} = 0$, so that, by itself, a Bose distribution would lead to the absurd result $n_k < 0$. In (39.23), on the left, we would have the negative infinite quantity $\sim -\sum g_k$, which can in no way equal N. Similarly, a Fermi distribution would lead to infinite positive quantities on the left-hand sides of (39.27) and (39.28); and this is impossible for finite N and $\mathscr{E}$ on the right. Therefore,

$$\theta \geqslant 0\,. \tag{39.29}$$

In the following section it will be shown that the quantity θ is proportional to the absolute temperature of the gas.

The quantity μ is very important in the theory of chemical and phase equilibria. These applications will be considered later (see the end of Sec. 46 and the succeeding sections of the book).

The weight of a state. Here we give a few more formulae for the weight of a state of an ideal gas particle. The weight of a state with energy between ε and $\varepsilon + d\varepsilon$ is given by the formula (25.25), whose left-hand side we shall denote now by $dg\,(\varepsilon)$. In addition we assume that the particles have an eigenmoment j, so that we must take into account the number of possible projections of $\mathbf{j}$, equal to $2j+1$:

$$dg\,(\varepsilon) = (2j+1)\,\frac{V m^{3/2} \varepsilon^{1/2} d\varepsilon}{2^{1/2} \pi^2 h^3}\,. \tag{39.30}$$

For electrons $j = 1/2$, so that $2j + 1 = 2$.

For light quanta we must use formula (25.24), replacing K in it by ω/c and multiplying by two, according to the number of possible polarizations of the quantum:

$$dg\,(\omega) = \frac{V \omega^2 d\omega}{\pi^2 c^3}\,. \tag{39.31}$$

It is also useful to know the weight of a state whose linear momentum is between p_x and $p_x + dp_x$, p_y and $p_y + dp_y$, p_z and $p_z + dp_z$. It is determined in accordance with (25.23), also with account taken of the factor $2j+1$. Thus, for electrons, we obtain

$$dg(\mathbf{p}) = 2\,\frac{V\,dp_x\,dp_y\,dp_z}{(2\pi\hbar)^3}\,. \tag{39.32}$$

Exercises

1) Write down the formula for the probability that heads are obtained k times for large N, where k is close to the maximum q_k.

The general formula for probability is of the form:

$$q_k = \frac{N!}{(N-k)!\,k!}\,2^{-N}.$$

We shall consider the numbers N and k as large. It is more convenient here to use Stirling's formula in a somewhat more exact form than (39.9), namely:

$$\ln N! = N\ln\frac{N}{e} + \frac{1}{2}\ln 2\pi N.$$

We put $k = \frac{N}{2} + x$, $N-k = \frac{N}{2} - x$, where x is a quantity small compared with $\frac{N}{2}$. Then, in the correction terms of Stirling's formula $1/2\ \ln 2\pi k$ and $1/2\ \ln 2\,\pi\,(N-k)$, the quantity x can be neglected. We expand the denominator in a series up to x^2:

$$\ln(N-k)! = \ln\left(\frac{N}{2} - x\right)! = \frac{N}{2}\ln\frac{N}{2e} - x\ln\frac{N}{2} + \frac{x^2}{N} + \frac{1}{2}\ln 2\pi\frac{N}{2},$$

$$\ln k! = \ln\left(\frac{N}{2} + x\right)! = \frac{N}{2}\ln\frac{N}{2e} + x\ln\frac{N}{2} + \frac{x^2}{N} + \frac{1}{2}\ln 2\pi\frac{N}{2}.$$

The correction terms are

$$\frac{1}{2}\left(\ln 2\pi N - 2\ln 2\pi\frac{N}{2}\right) = \frac{1}{2}\ln\frac{2}{\pi N}\,.$$

Substituting this in the expression for q_k and taking antilogarithms we arrive at the required formula:

$$q = \sqrt{\frac{2}{\pi N}}\;e^{-\frac{2x^2}{N}}\,.$$

q has a maximum at $x = 0$ and dies away on both sides. q is reduced e times in the interval $x_e = \sqrt{\frac{N}{2}}$, characterizing the sharpness of the maximum. Compared with the whole interval of variation x, the interval x_e comprises $\frac{x_e}{N/2} = \sqrt{\frac{2}{N}}$. For example, for $N = 1{,}000$, the maximum is approximately equal to 1/40. The ratio $\frac{x_e}{N}$ is about 2% so that, basically, the heads fall between 475 to 525 times. The probability that heads (or tails) will fall 400 times out of a thousand is equal to $1/40\cdot e^{-\frac{2\cdot 10{,}000}{1{,}000}} = 1/40\,e^{-20}$. In other words, it is

e^{+20}, i.e., several hundreds of millions of times less than the probability for heads appearing five hundred times.

2) Verify that the probability q has been normalized, i.e., that $\int q(x)\,dx = 1$.

Since the probability decreases very rapidly with increase in the absolute value of x, the integration can be extended from $-\infty$ to ∞ without noticeable error. Then

$$\int_{-\infty}^{\infty} q(x)\,dx = \sqrt{\frac{2}{\pi N}} \int_{-\infty}^{\infty} e^{-\frac{2x^2}{N}}\,dx = \frac{1}{\sqrt{\pi}} \int_{-\infty}^{\infty} e^{-\xi^2}\,d\xi .$$

We shall now show that the integral appearing here is indeed equal to $\sqrt{\pi}$ We shall call it I:

$$I = \int_{-\infty}^{\infty} e^{-\xi^2}\,d\xi .$$

Squaring, we get

$$I^2 = \int_{-\infty}^{\infty} e^{-\xi^2}\,d\xi \int_{-\infty}^{\infty} e^{-\eta^2}\,d\eta = \int_{-\infty}^{\infty}\int_{-\infty}^{\infty} e^{-(\xi^2+\eta^2)}\,d\xi\,d\eta .$$

The integration spreads over the whole $\xi\eta$-plane. Let us go over to polar coordinates:

$$\xi = \rho\cos\varphi , \quad \eta = \rho\sin\varphi .$$

Instead of $d\xi\,d\eta$ we must put $\rho\,d\rho\,d\varphi$, as is usual for polar coordinates. Then,

$$I^2 = \int_0^{\infty} \rho e^{-\rho^2}\,d\rho \int_0^{2\pi} d\varphi = -2\pi\cdot\frac{1}{2}e^{-\rho}\Big|_0^{\infty} = \pi ,$$

or

$$I = \sqrt{\pi},$$

so that

$$\int_{-\infty}^{\infty} q(x)\,dx = 1 .$$

3) Find the mean square deviation, for the occurrence of heads, from the most probable number, i.e., $\overline{x^2}$.

We have

$$\overline{x}^2 = \int_{-\infty}^{\infty} x^2 q(x)\,dx = \sqrt{\frac{2}{\pi N}} \int_{-\infty}^{\infty} x^2 e^{-\frac{2x^2}{N}}\,dx = \frac{N}{2}\frac{1}{\sqrt{\pi}} \int_{-\infty}^{\infty} \xi^2 e^{-\xi^2}\,d\xi .$$

In order to calculate the integral we make use of the result of exercise 2, writing $\xi^2\alpha$ in the exponent instead of ξ^2.

$$\int_{-\infty}^{\infty} e^{-\alpha\xi^2}\,d\xi = \frac{\sqrt{\pi}}{\sqrt{\alpha}} .$$

We differentiate both sides of this equation with respect to α:

$$-\int_{-\infty}^{\infty} \xi^2 e^{-\alpha\xi^2} d\xi = -\frac{\sqrt{\pi}}{2\alpha^{3/2}}.$$

Putting $\alpha = 1$ we arrive at the required formula:

$$\int_{-\infty}^{\infty} \xi^2 e^{-\xi^2} d\xi = \frac{\sqrt{\pi}}{2}.$$

We notice, incidentally, that

$$\int_{-\infty}^{\infty} \xi^4 e^{-\xi^2} d\xi = \frac{3\sqrt{\pi}}{4}$$

and in general

$$\int_{-\infty}^{\infty} \xi^{2n} e^{-\xi^2} d\xi = \frac{1 \cdot 2 \cdot 3 \ldots (2n-1)}{2^n} \sqrt{\pi}.$$

For $\overline{x^2}$, we obtain

$$\overline{x^2} = \frac{N}{4}.$$

Expressing N in terms of $\overline{x^2}$, we can write down the probability-distribution law:

$$q(x)\,dx = \frac{1}{\sqrt{2\pi\overline{x^2}}} e^{-\frac{x^2}{2\overline{x^2}}}.$$

Thus, the width of the distribution is very simply related to the mean square deviation of the quantity from its most probable value*:

$$x_e = \sqrt{2\overline{x^2}}.$$

Of course, this relationship holds only for an exponential distribution of the type obtained in exercise 1.

Sec. 40. Boltzmann Statistics (Translational Motion of a Molecule. Gas in an External Field)

Boltzmann distribution. Long before the Bose and Fermi quantum distribution formulae (39.21) and (39.26) had been obtained, Boltzmann derived a classical energy distribution law for the molecules of an ideal gas. This law is obtained from both quantum laws by means of a limiting process. We shall perform this transition purely formally at first, and then decide which real conditions it corresponds to.

* For $x = x_e$, the probability $q(x)$ decreases e times compared with $q(0)$; it is for this reason that x_e characterizes the width of the distribution.

Let ε be measured from zero, and let the ratio μ/θ be large in absolute value and negative. Then

$$e^{-\frac{\mu}{\theta}+\frac{\varepsilon}{\theta}}$$

is considerably greater than unity for all ε. Here, unity in the denominator of both distribution formulae can be neglected as compared with the exponent, and both the Bose and Fermi formulae take on the same limiting form

$$n_k = g_k e^{\frac{\mu-\varepsilon_k}{\theta}}. \tag{40.1}$$

This is the Boltzmann distribution. Let us now determine the constant μ from the normalization condition for the distribution:

$$\sum_k n_k = N. \tag{40.2}$$

Let us suppose that the gas molecules possess some internal degrees of freedom (in addition to the external transport degrees of freedom) that may be related to electron excitation, the vibration of nuclei with respect to each other, and the rotation of the molecule in space. The energy of all these degrees of freedom is quantized. Without defining it more exactly for the time being, we can write the total energy of a molecule ε in the form of a sum of the energies of translational and internal motion:

$$\varepsilon = \frac{p^2}{2m} + \varepsilon^{(i)}. \tag{40.3}$$

Accordingly, the weight of a state of given energy is also represented as the product of two weights: one relates to translational motion and is given by the formula (39.32), while the other we denote simply by $g_{(i)}$ (we also agree to include in it the factor $2j+1$):

$$g = \frac{V\,dp_x\,dp_y\,dp_z}{(2\pi h)^3}\,g_{(i)}. \tag{40.4}$$

Therefore, formula (40.2) can be written thus:

$$\frac{V}{(2\pi h)^3}e^{\frac{\mu}{\theta}}\sum_i g_{(i)}e^{-\frac{\varepsilon(i)}{\theta}}\int_{-\infty}^{\infty}\int_{-\infty}^{\infty}\int_{-\infty}^{\infty}e^{-\frac{p^2}{2m\theta}}\,dp_x\,dp_y\,dp_z = N. \tag{40.5}$$

Expanding the translational-motion energy into $\frac{p_x^2+p_y^2+p_z^2}{2m}$, we see that the momentum integral is represented as the product of three integrals of the form

$$\int_{-\infty}^{\infty} e^{-\frac{p_x^2}{2m\theta}}\,dp_x.$$

These integrals are easily calculated from the second formula of exercise 3, Sec. 39. Each of them is equal to $\sqrt{2\pi m\theta}$, so that condition (40.5) reduces to the form

$$e^{-\frac{\mu}{\theta}} = \frac{V}{N}\left(\frac{m\theta}{2\pi}\right)^{3/2}\frac{1}{h^3}\sum g_{(i)}\,e^{-\frac{\varepsilon^{(i)}}{\theta}}. \qquad (40.6)$$

If the gas is monatomic then the quantities $\varepsilon^{(i)}$ refer to electron excitations. Therefore, if $\varepsilon^{(1)} \gg \theta$, then, actually, only the zero term appears in the summation over the states*. But since the energy is measured from $\varepsilon^{(o)}$ as from zero, the whole summation, actually, reduces only to the zero term $g_{(o)}$. It is of the order of unity. For example, when the ground state has angular momentum 1/2, $g_{(o)} = 2$. We then obtain the condition for the applicability of Boltzmann statistics in the form

$$-\frac{\mu}{\theta} = \ln\left[\frac{g_{(o)}}{h^3}\frac{V}{N}\left(\frac{m\theta}{2\pi}\right)^{3/2}\right] \gg 1. \qquad (40.7)$$

For the inequality (40.7) to be satisfied, it is sufficient to satisfy one of two conditions:

1) the density of the gas is very small, i.e., the volume occupied by the gas at a given temperature θ is large;

2) the temperature θ for a given volume V is very high.

In the case when the gas is not monatomic, these conditions are quantitatively changed somewhat because $\sum_i g_{(i)}\,e^{-\frac{\varepsilon^{(i)}}{\theta}}$ is also some function of θ. But qualitatively, the conditions of applicability of Boltzmann statistics still hold.

Classical and quantum statistics. We have seen that for small densities or high temperatures the quantum distribution laws for a gas pass into the classical Boltzmann law. From now on we shall agree to call the Bose and Fermi statistics *quantum* statistics and the Boltzmann statistics, *classical*, regardless of wheather the energy spectrum is discrete or continuous. Those statistics will be termed quantum for which the indistinguishability of separate particles is taken into account. In other words, a quantum definition of the state of a system lies at the basis of quantum statistics: the *number* of particles in all quantum states must be given. The classical definition of the state of a system indicates *which* particles are found in the given states. The Boltzmann formula (40.1) can be obtained from this classical definition.

Maxwell distribution. In this section we will not be concerned with the statistics of the internal motion of molecules, and will consider

* The relation between θ and temperature is given by formula (40.25).

only their translational motion. In accordance with (40.3), the energy of the translational motion of molecules is separable from their internal energy. Therefore, the Boltzmann distribution breaks up into the product of two factors. We are not interested in the first of the two factors, but the second, relating to translational motion, is of the form

$$e^{-\frac{p^2}{2m\theta}}.$$

The weight of a state relating to a given absolute value p is obtained by changing to polar coordinates in formula (39.32):

$$dg(p)\,\frac{Vp^2\,dp}{2\pi^2 h^3} \tag{40.8}$$

[cf. (25.24)].

Thus, the distribution according to the kinetic energies of translational motion is written in the form:

$$dn(p) = Ae^{-\frac{p^2}{2m\theta}}p^2\,dp. \tag{40.9}$$

It is applicable both to monatomic and polyatomic gases if m is the mass of a molecule as a whole.

The constant A is found from the normalization condition

$$A\int_0^\infty p^2 e^{-\frac{p^2}{2m\theta}}\,dp = N. \tag{40.10}$$

The value of the integral was found in exercise 3, Sec. 39. From this we obtain

$$A = N\frac{\sqrt{2}}{\sqrt{\pi(m\theta)^3}}. \tag{40.11}$$

In place of the momentum distribution of molecules, it is sometimes useful to have their velocity distribution. For this it is sufficient to substitute $p = mv$ in the distribution (40.9):

$$dn(v) = N\frac{\sqrt{2m^3}}{\sqrt{\pi\theta^3}}e^{-\frac{mv^2}{2\theta}}v^2\,dv. \tag{40.12}$$

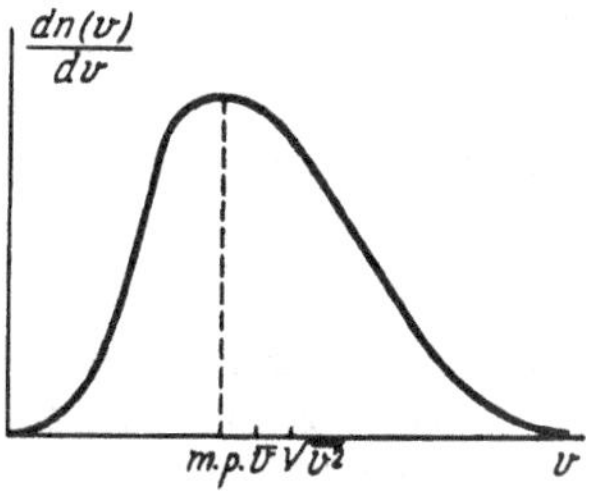

Fig. 46

This distribution had already been deduced by Maxwell, before Boltzmann, and for this reason it is called the Maxwell distribution.

In Fig. 46 we have plotted the ratio $\frac{dn(v)}{dv}$ on the ordinate. For small v, this quantity is close to zero because of the factor v^2 in the equation for the weight of a state; after the zero point it reaches a maximum and exponentially decreases to zero again for large velocities. We thus see that a gas contains molecules with every possible velocity value.

The velocities of gas molecules. The greatest number of molecules have a velocity corresponding to the maximum of the distribution curve shown in Fig. 46. This maximum is determined from equation (40.12). The corresponding velocity is termed the most probable; it is

$$v_{\text{m.p.}} = \sqrt{\frac{2\,\theta}{m}}. \tag{40.13}$$

We find the mean velocity by calculating the integral (we omit the factor N, because the mean value of velocity relates to a single molecule):

$$\overline{v} = \sqrt{\frac{2\,m^3}{\pi\,\theta^3}}\int\limits_0^\infty e^{-\frac{m v^2}{2\,\theta}} v^3 dv = \frac{1}{2}\sqrt{\frac{2}{\pi}\cdot\frac{m^3}{\theta^3}}\left(\frac{2\,\theta}{m}\right)^2 = \sqrt{\frac{8\,\theta}{\pi\,m}}. \tag{40.14}$$

The mean square velocity is also interesting

$$\overline{v^2} = \sqrt{\frac{2\,m^3}{\pi\,\theta^3}}\int\limits_0^\infty e^{-\frac{m v^2}{2\,\theta}} v^4 dv = \sqrt{\frac{2}{\pi}\,\frac{m^3}{\theta^3}}\sqrt{\left(\frac{2\,\theta}{m}\right)^5}\cdot\frac{3}{8}\sqrt{\pi} = \frac{3\,\theta}{m}. \tag{40.15}$$

(the result of exercise 3, Sec. 39 is used in the derivation).

The ratio $\sqrt{\overline{v^2}} : \overline{v} : v_{\text{m.p.}} = \sqrt{3} : \sqrt{\frac{8}{\pi}} : \sqrt{2}$.

The mean energy of a single molecule is equal to

$$\varepsilon = \frac{m\overline{v^2}}{2} = \frac{3}{2}\,\theta, \tag{40.16}$$

and the mean energy for the whole gas is N times greater:

$$\overline{\mathscr{E}} = N\varepsilon = \frac{3}{2}\,N\,\theta. \tag{40.17}$$

This result relates to the energy of translational motion of the molecules. Numerical evaluations of velocity will be performed below.

The relationship between energy density and pressure. We shall now derive a very important relationship between the density of kinetic energy of a gas and its pressure. This relationship holds for any statistics and depends only upon the form of the expression for energy in terms of momentum.

The pressure of a gas is defined as the force with which the gas acts upon unit area perpendicular to its direction. This force is equal to the normal (to the surface) component of momentum transmitted by the gas molecules in unit time. Let the direction of the normal to the surface coincide with the x-axis. We first choose those molecules which have a velocity component along the x-axis equal to v_x. They will reach the surface of a volume in unit time if they initially were situated in a layer of width v_x. Let us cut out a cylinder from this layer with base of unit area and height equal to v_x. The volume of this cylinder is v_x. If $dn\,(v_x)$ is now the number of molecules whose velocity component normal to the surface is v_x, then the density of these molecules is $\frac{dn\,(v_x)}{V}$. There are $v_x \frac{dn\,(v_x)}{V}$ such molecules in a cylinder of volume v_x. Each of them, upon elastically colliding with a wall, will reverse its normal velocity component, and the wall will receive a momentum

$$m v_x - (-m v_x) = 2 m v_x. \tag{40.18}$$

Thus, all the gas molecules having a velocity v_x, transfer to the wall in unit time a momentum

$$2 m v_x \frac{dn\,(v_x)}{V} \cdot v_x = 2 m v_x^2 \frac{dn\,(v_x)}{V}. \tag{40.19}$$

In order to obtain the gas pressure on the wall we must integrate (40.19) over all v_x from 0 to ∞, and not from $-\infty$ to ∞, because molecules moving away from the wall will not strike it. Thus, the pressure of the gas on the wall is

$$p = \frac{2m}{V} \int_0^{\infty} v_x^2 \, dn\,(v_x) = \frac{m}{V} \int_{-\infty}^{\infty} v_x^2 \, dn\,(v_x). \tag{40.20}$$

On the other hand, the mean kinetic energy of the gas is

$$\overline{\mathscr{E}} = \frac{m}{2} \int_{-\infty}^{\infty} v_x^2 dn\,(v_x) + \frac{m}{2} \int_{-\infty}^{\infty} v_y^2 dn\,(v_y) + \frac{m}{2} \int_{-\infty}^{\infty} v_z^2 \, dn\,(v_z) =$$

$$= \frac{3m}{2} \int_{-\infty}^{\infty} v_x^2 \, dn\,(v_x), \tag{40.21}$$

because the mean values of the squares of all the velocity components are identical.

Comparing now (40.20) and (40.21) we find that the gas pressure is equal to two thirds of the density of its kinetic energy:

$$p = \frac{2}{3} \frac{\mathscr{E}}{V}. \tag{40.22}$$

This result was published by D. Bernoulli, as early as 1738, a century and a half before statistical physics began to develop as an independent science.

Only two assumptions have been used in the derivation of (40.22): identical values of the three velocity projections are equiprobable and the kinetic energy is equal to $\frac{mv^2}{2}$. The concrete form of the distribution function is not essential.

The Clapeyron equation. If a gas is subject to Boltzmann statistics, then, in accordance with (40.17), the mean kinetic energy $\overline{\mathscr{E}}$ is equal to $\frac{3N\theta}{2}$. Substituting this in (40.22) we obtain

$$pV = N\theta. \tag{40.23}$$

But from the definition of absolute temperature

$$pV = RT. \tag{40.24}$$

From this we obtain the relationship between "statistical" temperature θ, measured in ergs, and the temperature T, measured in degrees Centigrade:

$$\theta = \frac{R}{N} T = \frac{8.314 \times 10^7\, T}{6.024 \times 10^{23}} = 1.38 \times 10^{-16}\, T. \tag{40.25}$$

The ratio $k = \frac{R}{N}$ is called Boltzmann's constant. It is equal to 1.38×10^{-16}. The temperature can also be measured in electron-volts, one electron-volt being equal to 1.59×10^{-12} erg. Translating ergs into degrees with the aid of Boltzmann's constant, we find that 1 ev $=$ $= 11{,}600°$.

As is known, the specific heat of an ideal monatomic gas is equal to $\frac{3}{2} R$, thus corresponding to an energy $\frac{3}{2} RT$. Replacing RT by $N\theta$, we find $\overline{\mathscr{E}} = \frac{3}{2} N\theta$ in agreement with (40.17).

The relationship (40.25) allows us to calculate the mean velocity of gas molecules without using the Avogadro number N. Indeed,

$$\bar{v} = \sqrt{\frac{8\theta}{\pi m}} = \sqrt{\frac{8RT}{\pi N m}} = \sqrt{\frac{8RT}{\pi M}},$$

where M is the molecular weight of the gas. For example, the mean velocity of hydrogen molecules at a temperature of 300° K is

$$\bar{v} = \sqrt{\frac{8 \cdot 8.3 \cdot 10^7 \cdot 300}{3.14 \cdot 2}} \simeq 1{,}800\, m/sec.$$

This value is comparable with the exit velocity of a gas into a vacuum or with the velocity of sound [see (47.30)].

The thermonuclear reaction. When nuclei collide reactions are possible between them that proceed with the release of energy. For example, in a deuteron-deuteron collision one of two reactions can occur (besides elastic scattering):

$$D_1^2 + D_1^2 = \begin{cases} He_2^3 + n_0^1, \\ H_1^3 + H_1^1. \end{cases}$$

Here H_1^3 is tritium and n_0^1 a neutron. Another example is

$$Li_3^6 + D_1^2 = 2\,He_2^4.$$

In order that charged nuclei may be able effectively to collide, they must overcome the potential barrier of Coulomb repulsion, which was considered in Sec. 28. The dependence upon energy for the probability of passing through the potential barrier is basically determined by the barrier factor [see the first term on the right in (28.12)]:

$$e^{-\frac{2\pi Z_1 Z_2 e^2}{h v_{\parallel}}}. \tag{40.26}$$

Here, $Z_1 e$ and $Z_2 e$ are the charges of the colliding nuclei and $v_{\parallel}$ is the relative velocity along their joining line [recall that (28.12) refers to *one-dimensional* motion].

The reaction can be produced by accelerating the particles in a discharge tube. But charged particles, striking a substance, mainly spend their energy on ionization and excitation of the atoms. And since, according to (40.26), the probability of the reaction at small energy is vanishingly small, the majority of incident particles do not cause a reaction. Of the total number of particles it turns out that 10^{-5}—10^{-6} are effective. Therefore, the energy yield of the reaction is considerably less than the total energy spent in accelerating the beam of particles.

The situation is different if the substance used for the reaction is at a very high temperature, of the order of 10^7 degrees. At this temperature, the nuclei of the heated substance already react at a sufficiently high rate, and transmission of energy to electrons does not occur because their mean energy is the same as that of the nuclei.

Let us calculate the rate of a nuclear reaction occurring under such conditions. It is termed *thermonuclear*.

Let the effective cross-section for the reaction between nuclei with relative velocity $v_{\parallel}$ be $\sigma(v_{\parallel})$. We assume that different nuclei react: we shall call them 1 and 2. Let us construct on each nucleus 2 a cylinder with base area $\sigma(v_{\parallel})$ and height numerically equal to $v_{\parallel}$. Then, by definition of $\sigma(v_{\parallel})$, all those nuclei 1 which occur in the volume of these cylinders and which have velocity $v_{\parallel}$ relative to nuclei 2 will be involved in the reaction in unit time.

The number of such events in unit volume and unit time is equal to the product of

$$n_1 \cdot v_{\parallel}\,\sigma(v_{\parallel})\,n_2\,dq(v_{\parallel}), \tag{40.27}$$

where n_1 and n_2 are the numbers of nuclei 1 and 2 in unit volume, and $dq(v_{\parallel})$ is the probability that the relative velocity is equal to $v_{\parallel}$.

Indeed, a cylinder of volume $v_\parallel\ \sigma\ (v_\parallel)$ can be constructed on each nucleus 2, and there will be $n_1\ \sigma\ (v_\parallel)\ v_\parallel$ nuclei 1 in each cylinder. The velocity distribution of the nuclei is taken into account by multiplying by $dq\ (v_\parallel)$. If 1 and 2 are identical nuclei, then expression (40.27) must be halved so that each reaction is not taken into account twice. We indicate this by the factor (2) in the denominator of expression (40.28).

Let us now determine the probability factor $dq\ (v_\parallel)$. The absolute-velocity distribution is given by the product of two Maxwellian factors of the form

$$e^{-\frac{m_1 v_1^2}{2\theta}}\ e^{-\frac{m_2 v_2^2}{2\theta}}\ .$$

In the exponent of this expression is the sum of the energies of both nuclei. In accordance with formula (3.17), it can be split into the kinetic energy of the motion of the centre of mass of the nuclei and the kinetic energy of their relative motion. Hence, in the product a factor is separated that gives the relative-velocity distribution:

$$e^{-\frac{m_1 m_2 v^2}{2(m_1+m_2)\theta}} \equiv e^{-\frac{m v^2}{2\theta}} = e^{-\frac{m v_\parallel^2}{2\theta}}\ e^{-\frac{m v_\perp^2}{2\theta}}\ ,$$

$$\text{where} \quad m \equiv \frac{m_1 m_2}{m_1 + m_2} \quad [\text{see } (3.20)]\ , \quad v^2 \equiv v_\parallel^2 + v_\perp^2\ .$$

Normalizing the distribution over $v_\parallel$ to unity and passing to the reduced mass m, we obtain an expression for the probability that the value of relative velocity along the line joining the nuclei will be $v_\parallel$.

$$dq\ (v_\parallel) = \sqrt{\frac{m}{2\pi\theta}}\ e^{-\frac{m v_\parallel^2}{2\theta}}\ dv_\parallel\ .$$

The barrier factor (40.26) depends upon $v_\parallel$.

Thus, the overall rate of a thermonuclear reaction is

$$r = \frac{n_1 n_2}{(2)} \int_0^\infty \sigma\ (v_\parallel)\ v_\parallel\ dq\ (v_\parallel)\ \frac{\text{events}}{cm^3 \cdot sec}\ . \tag{40.28}$$

Taking into account the barrier factor, we write the dependence of effective cross-section upon the rate as

$$\sigma\ (v_\parallel) = \sigma_0\ (v_\parallel)\ e^{-\frac{2\pi Z_1 Z_2 e^2}{h v_\parallel}}\ .$$

The factor σ_0 here depends considerably less upon the rate than the exponential function.

The integral in (40.28) now reduces to the form:

$$\int_0^\infty \sigma_0(v_\parallel)\, v_\parallel\, e^{-\frac{2\pi Z_1 Z_2 e^2}{h v_\parallel} - \frac{m v_\parallel^2}{2\theta}}\, dv_\parallel . \tag{40.29}$$

It can be calculated, to a good approximation in the case when the temperature is so low, that the greater part of the reaction proceeds at the "tail" of the Maxwellian distribution at a rate greater than the mean. Let us show how this calculation is done.

We denote the argument of the exponential in the integrand thus:

$$f(v_\parallel) \equiv \frac{2\pi Z_1 Z_2 e^2}{h v_\parallel} + \frac{m v_\parallel^2}{2\theta} \equiv \frac{a}{v_\parallel} + \frac{b}{2} v_\parallel^2,$$

$$a \equiv \frac{2\pi Z_1 Z_2 e^2}{h}, \quad b \equiv \frac{m}{\theta}.$$

We find the minimum of the function $f(v_\parallel)$ from the condition

$$\frac{df}{dv_\parallel} = -\frac{a}{v_\parallel^2} + b v_\parallel = 0; \quad v_\parallel^0 = \sqrt[3]{\frac{a}{b}}. \tag{40.30}$$

We shall see that the basic contribution to the integral is given by values of $v_\parallel$ close to $v_\parallel^0$. Near the minimum, $f(v_\parallel)$ can be represented in the form

$$f(v_\parallel) = f(v_\parallel^0) + \frac{1}{2}(v_\parallel - v_\parallel^0)^2 \left(\frac{d^2 f}{dv_\parallel}\right)_0 =$$

$$= \frac{3}{2}\sqrt[3]{a^2 b} + \frac{3}{2} b (v_\parallel - v_\parallel^0)^2, \tag{40.31}$$

and the integral (40.29) is written as

$$\int_0^\infty \sigma_0(v_\parallel)\, v_\parallel\, e^{-f(v_\parallel^0) - \frac{3}{2} b (v_\parallel - v_\parallel^0)^2}\, dv_\parallel . \tag{40.32}$$

The minimum of $f(v_\parallel)$ corresponds to the rate $v_\parallel^0$ at which the greatest number of reactions occur. The ratio of the rate $v_\parallel^0$ to the mean relative rate $\overline{|v_\parallel|}$ is, according to (40.30),

$$\frac{v_\parallel^0}{\overline{|v_\parallel|}} = \sqrt{\frac{\pi}{2}}\sqrt[3]{a}\sqrt[6]{b} = \left(\frac{\pi}{2}\right)^{1/2}\left(\frac{2\pi Z_1 Z_2 e^2}{h}\right)^{1/3}\left(\frac{m}{\theta}\right)^{1/6}, \tag{40.33}$$

since

$$\overline{|v_\parallel|} = 2\sqrt{\frac{m}{2\pi\theta}} \int_0^\infty v_\parallel e^{-\frac{m v_\parallel^2}{2\theta}}\, dv_\parallel = \sqrt{\frac{2\theta}{\pi m}} = \sqrt{\frac{2}{\pi b}}.$$

We shall call the temperature low if the ratio $\frac{v_\parallel^0}{\overline{|v_\parallel|}}$ is several times greater than unity. Then the maximum of the integrand of (40.32) is very sharp at the point $v_{\|}=v_{\|}^0$, because it decreases e times when $v_{\|}$ deviates from $v_{\|}^0$ by an amount $\sqrt{\frac{2}{3b}}$, which is considerably less than $v_{\|}^0$.

It was therefore justifiable to terminate the expansion in (40.31) with the second term. In addition, the quantities $\sigma_0\,(v_{\|})$ and $v_{\|}$ can be taken outside the integral sign when $v_{\|}=v_{\|}^0$. The error in both approximations is of the order $\frac{\overline{|V_{\|}|}}{V_{\|}^0}$. The integration can be taken from $-\infty$ to ∞ because the integrand rapidly decreases as $v_{\|}$ recedes from $v_{\|}^0$, so that the error is exponentially small.

Thus,

$$\int_0^\infty \sigma\,(v_{\|})\,v_{\|}\,e^{-f(v_{\|}^0)-\frac{3}{2}b(v_{\|}-v_{\|}^0)^2}\,d v_{\|}\cong$$

$$\cong\sigma_0\,(v_{\|}^0)\,v_{\|}^0\,e^{-f(v_{\|}^0)}\int_{-\infty}^{\infty} e^{-\frac{3b}{2}(v_{\|}-v_{\|}^0)^2}\,d v_{\|}=$$

$$=\sigma_0\,(v_{\|}^0)\,v_{\|}^0\sqrt{\frac{2\pi}{3b}}\;e^{-f(v_{\|}^0)}\,. \tag{40.34}$$

Substituting the values a and b and using (40.28), we find the expression for the rate of a thermonuclear reaction

$$\mathbf{r}=\frac{n_1 n_2}{(2)\sqrt{3}}\,v_{\|}^0\sigma_0(v_{\|}^0)\,e^{-\frac{3}{2}\sqrt[3]{\left(\frac{2\pi Z_1 Z_2 e^2}{h}\right)^2\frac{m}{\theta}}}\;;\quad v_{\|}^0=\sqrt[3]{\frac{2\,\pi Z_1 Z_2 e^2\,\theta}{hm}}\,. \tag{40.35}$$

The exponential factor depends very strongly upon the temperature. For example, for a reaction in deuterium, this factor changes 3,600 times when the temperature is increased from 100 to 200 ev.

Thermonuclear reactions are the source of stellar energy and for this reason play as important a part in our life as chemical reactions!

Ideal gas in an external field. We shall now consider an ideal gas acted upon by an external field with potential U. The potential energy can depend both upon the position of the molecules in space as well as their orientation (if the gas is not monatomic).

The total energy of a molecule is

$$\varepsilon=\frac{p^2}{2m}+\varepsilon^{(i)}+U\,. \tag{40.36}$$

If U depends upon the position of a molecule in space, i.e., $U=U\,(x, y, z)$, then we must pass from a finite volume V, in the weight factor (40.4), to an infinitely small volume $dV=dx\,dy\,dz$. Then part of the distribution function that depends upon coordinates x, y, z can be separated, and a formula is obtained defining the dependence of gas density upon coordinates:

$$dn(x,y,z) = n_0 e^{-\frac{U(x,y,z)}{\theta}} dx\,dy\,dz\,. \qquad (40.37)$$

Here, the potential energy calibration is U (0, 0, 0) = 0, and the gas density at this point is equal to n_0. For example, in a gravitational field, $U = mgz$, so that

$$dn(z) = n_0 e^{-\frac{mgz}{\theta}} dz\,. \qquad (40.38)$$

It should be noted that in the earth's atmosphere the "barometric" formula (40.38) is rather more applicable qualitatively because air temperature is not constant with height.

In addition, the "barometric" formula indicates that the composition of the air must vary with height as a result of the different molecular weights of nitrogen, oxygen, and other gases. Actually, the air composition is almost uniform vertically because of vigourous mixing processes.

The nonequilibrium state of planetary atmospheres. In place of the approximate expression for the potential energy in a gravitational field, let us substitute its exact expression (3.4). Let us first of all express the constant a in formula (3.4) in terms of more convenient quantities. The force of gravity at the earth's surface is $-mg$ and, from the general gravitational law, it is equal to $-\frac{a}{r_0^2}$, where r_0 is the radius of the earth. From this $a = mgr_0^2$, so that $U = -\frac{mgr_0^2}{r}$. Therefore, the gas density must vary with height according to the law

$$n = n_\infty e^{\frac{mgr_0^2}{2r\theta}}\,. \qquad (40.39)$$

This quantity remains finite even at an infinite distance from the earth, and since the exponent is equal to unity at infinity we have called the proportionality factor n_∞.

Near the earth, where $r = r_0$, the density is greater than n_∞ by as many times as the quantity

$$e^{\frac{mgr_0}{\theta}} = e^{\frac{Mgr_0}{RT}}$$

is greater than unity.

The radius of the earth $r_0 \approx 6.4 \times 10^8$ cm, $g \approx 10^3$ cm/sec^2. From this we obtain for oxygen

$$\frac{Mgr_0^2}{RT} \cong \frac{32 \cdot 10^3 \cdot 6.4 \cdot 10^8}{8.3 \cdot 10^7 \cdot 300} \approx 800\,.$$

In actual fact the density of the earth's atmosphere at infinity is equal to zero. Therefore, it follows from formula (40.39) that the atmosphere cannot arrive at the most probable state when in the earth's gravitational field, and is gradually dispersed into space. The most probable state of a gas is called statistical equilibrium (see Sec. 45). The equilibrium density of the atmosphere at infinity is e^{800} times less than at the earth's surface. Therefore the present state of the atmosphere is very close to equilibrium. For the moon, equilibrium has been reached: its atmosphere has completely escaped!

A kinetic interpretation of the dispersion of planetary atmospheres. It is easy to understand the reason for the recession of gases to infinity. Any particle whose velocity exceeds 11.5 km/sec is capable of overcoming the earth's attraction: its motion is infinite. In accordance with the Maxwell distribution (40.12) a gas will always have molecules with every possible velocity. In literal notation, the velocity of molecules capable of going to infinity is defined by the equation

$$\frac{mv^2}{2} = mgr_0. \tag{40.40}$$

Taking v^2 from this equation and substituting into the Maxwell distribution, we once again obtain the exponential $e^{-\frac{mgr_0}{\theta}}$ for the fraction of molecules capable of leaving the atmosphere. It is easy to estimate the number of such molecules in the atmosphere at any instant of time. The earth's surface is 5×10^{18} cm^2. There is about 1,030 gm of air above every square centimetre, i.e., about 35 moles. Hence, the total number of molecules in the atmosphere is $5 \times 10^{18} \times$ $\times 35 \times 6 \times 10^{23} = 10^{44}$, and the fraction of molecules of velocity greater than 11.5 km/sec is $e^{-800} = 10^{-344}$. Therefore the mean number of molecules capable of leaving the earth at each instant is only 10^{-300}. Of course, those molecules close to the earth's surface will not be able to "carry" their energy to the upper layers of the atmosphere because of collisions with other molecules.

The dielectric constant of a gas. We shall now consider a gas whose molecules have a constant dipole moment in a constant and uniform electric field. Those molecules can have characteristic dipole moments for which there is some preferred direction: NO, CO, H_2O (along the altitude of the triangle passing through O), NH_3 (along the axis of symmetry of a three-sided pyramid). The more symmetrical molecules do not possess moments: H_2, O_2, CH_3 (tetrahedron), CO_2 (this proves that the CO_2 molecule has the form of a rod with the C atom in the centre).

Rotational motion is quantized. In the next section it will be shown that for all gases except hydrogen, at a temperature of several tens of degrees (from absolute zero), the states with large quantum numbers are already excited. In these states the motion may be rega ded as classical. Then the total rotational energy of a molecule simply breaks down into the kinetic energy of rotation (see Sec. 9) and potential energy, which depends upon the orientation of the dipole moment relative to the external electric field:

$$U = -(\mathbf{d}\mathbf{E}) = -d \cdot E \cos\vartheta$$

[see (14.28)]. In classical motion, the potential and kinetic energies may be regarded as quantities instead of operators. Therefore, in the Boltzmann distribution the factor that depends only upon potential energy is split off:

$$dn(\vartheta) = A e^{\frac{d \cdot E \cos\vartheta}{\theta}} \sin\vartheta \, d\vartheta. \tag{40.41}$$

Here, $\sin\vartheta \, d\vartheta$ is proportional to the element of solid angle in which the vector $\mathbf{d}$ lies [cf. (6.15)].

Let us now determine the electric polarization of a gas in an external field. For this we must calculate the mean projection of the dipole moment upon the electric field, i.e.,

$$\overline{d \cdot \cos\vartheta} = \frac{d \cdot \int\limits_0^\pi e^{\frac{d \cdot E \cos\vartheta}{\theta}} \cos\vartheta \sin\vartheta \, d\vartheta}{\int\limits_0^\pi e^{\frac{d \cdot E \cos\vartheta}{\theta}} \sin\vartheta \, d\vartheta}. \tag{40.42}$$

It turns out that it is sufficient merely to find an expression for the integral in the denominator, because equation (40.42) can be rewritten thus:

$$\overline{d \cdot \cos\vartheta} = \theta \frac{\partial}{\partial E} \ln\left(\int\limits_0^\pi e^{\frac{d \cdot E \cos\vartheta}{\theta}} \sin\vartheta \, d\vartheta \right). \tag{40.43}$$

Indeed, differentiating the integral with respect to the parameter E, we revert to (40.42). The integral can be calculated using the fact that $\sin\vartheta \, d\vartheta = -d(\cos\vartheta)$:

$$\int\limits_0^\pi e^{\frac{d \cdot E \cos\vartheta}{\theta}} \sin\vartheta \, d\vartheta = -\frac{\theta}{Ed} e^{\frac{d \cdot E \cos\vartheta}{\theta}} \Bigg|_0^\pi = \frac{2\theta}{Ed} \sinh \frac{Ed}{\theta}. \tag{40.44}$$

The integral in formula (40.43) is called a *statistical integral.* For quantized energy levels, it is replaced in the general case by a statistical sum. The expression for the summation and integral will be met with many times again. It is very convenient in calculating mean values.

Substituting (40.44) in (40.43), we obtain an expression for the mean projection of the dipole moment on the electric field

$$\overline{d\cdot\cos\vartheta} = d\cdot\left(\coth\frac{Ed}{\theta} - \frac{\theta}{Ed}\right). \qquad (40.45)$$

This expression was obtained by Langevin. Let us investigate the right-hand side in two limiting cases: $E \ll \frac{\theta}{d}$ (weak field) and $E \gg \frac{\theta}{d}$ (strong field).

If the field is weak then we can use the expansion of coth x in terms of x:

$$\coth x = \frac{1}{x} + \frac{x}{3},$$

whence

$$\overline{d\cdot\cos\vartheta} = \frac{Ed^2}{3\theta}. \qquad (40.46)$$

The polarization of the gas is

$$P = N\overline{d\cdot\cos\vartheta} = \frac{NEd^2}{3\theta}, \qquad (40.47)$$

and the dielectric constant is calculated from the definitions of induction (16.23) and (16.29):

$$D = E + 4\pi P = E\left(1 + \frac{4\pi Nd^2}{3\theta}\right) \equiv \varepsilon E. \qquad (40.48)$$

In a strong field $\coth\frac{Ed}{\theta}$ tends to unity and $\frac{\theta}{Ed}$ tends to zero. Therefore, $\overline{\cos\vartheta}$ tends to unity. This means that all the dipoles are orientated along the external field and saturation sets in. Then $D = E + 4\pi Nd$. We notice that, for $T = 300°$ K, this case would correspond to a field $E \gg 10^7$ v/cm, which is considerably greater than the breakdown potential.

Paramagnetism of gases. We shall now find how the magnetic permeability χ is calculated. Here we must take into account the fact that the magnetic moment is related to the mechanical moment of electrons, and the latter is a quantized quantity, i.e., it takes on a discrete series of values. Usually an electronic mechanical moment does not have a value greater than several units, so that the limiting transition to classical theory cannot be performed. An atom can also have a magnetic moment (as opposed to an electric dipole moment). Therefore, let us determine the magnetic susceptibility arising from the orientation of atomic magnetic moments in an external field **H**.

Let us suppose that an atom in the ground state possesses an orbital angular momentum L, a spin angular momentum S and a total angular momentum J. In other words, the ground state is a multiplet state. Let the multiplet splitting (fine structure) be defined by the energy interval Δ, so that the level with the closest value $J \pm 1$ differs from the ground level by the quantity Δ. If the energy of the ground level is ε_0, then the closest level has an energy $\varepsilon_0 + \Delta$. The ratio of the number of atoms in the ground state to the number of atoms in the closest state, belonging to the given multiplet, is, according to (40.1)

$$\frac{2J+1}{2(J \pm 1)+1} \cdot \frac{e^{-\frac{\varepsilon_0}{\theta}}}{e^{-\frac{(\varepsilon_0+\Delta)}{\theta}}} = \frac{2J+1}{2(J \pm 1)+1} e^{\frac{\Delta}{\theta}}. \tag{40.49}$$

Thus, if the multiplet splitting Δ is considerably greater than θ, the majority of the atoms are in the ground state. If they are placed in an external magnetic field then each of the multiplet levels is split into $2J+1$ levels, corresponding to its value of J. Suppose that the field corresponds to the anomalous Zeeman effect in the sense that the splitting of each multiplet level in the magnetic field is considerably less than the fine-structure splitting Δ as defined in Sec. 35.

Then, from (35.11), the energy of an atom in the ground state is

$$\varepsilon = -\frac{eHh}{2mc} g_L J_z = H \mu_0 g_L J_z, \tag{40.50}$$

where g_L is the Lande factor [see (35.12)] and μ_0 is the Bohr magneton.

The number of such atoms is given by the Boltzmann distribution

$$n(J_z) = A e^{-\frac{\mu_0 g_L H J_z}{\theta}}. \tag{40.51}$$

We must again determine the mean value of the magnetic moment projection on the field:

$$-\mu_0 g_L \overline{J_z} = \frac{-\sum\limits_{-J}^{J} \mu_0 g_L J_z e^{-\frac{\mu_0 g_L H J_z}{\theta}}}{\sum\limits_{-J}^{J} e^{-\frac{\mu_0 g_L H J_z}{\theta}}} = \theta \frac{\partial}{\partial H} \ln \left[\sum_{-J}^{J} e^{-\frac{\mu_0 g_L H J_z}{\theta}} \right]. \tag{40.52}$$

We have put the minus sign on the left because the electronic charge is negative. Formula (40.52) involves a statistical sum. The summation is performed only over those levels which are obtained when the ground state of the multiplet is split in the magnetic field, since the number of atoms in an excited state is small.

Summing the geometric progression, it is not difficult here to obtain a general formula similar to the Langevin formula (40.45). But we shall confine ourselves to the case of a weak field, when the exponential function is expanded in a series. The expansion must be taken up to the second term inclusive because the sum of the terms which are linear in J_z is equal to zero:

$$\sum_{-J}^{J} e^{-\frac{\mu_0 g_L H J_z}{\theta}} = \sum_{-J}^{J}\left[1 - \frac{\mu_0 g_L H J_z}{\theta} + \frac{1}{2}\left(\frac{\mu_0 g_L H J_z}{\theta}\right)^2\right] =$$

$$= 2J + 1 + \frac{(\mu_0 g_L H)^2}{2\theta^2}\sum_{-J}^{J} J_z^2.$$

We calculated the sum of J_z^2 in Sec. 30 [see (30.27)]. Using the value for the sum then obtained, we write the required mean moment thus:

$$-\mu_0 g_L \overline{J}_z = \theta\frac{\partial}{\partial H}\ln\left(2J + 1 + \frac{1}{6}\frac{\mu_0^2 g_L^2 H^2 J(J+1)(2J+1)}{\theta^2}\right) =$$

$$= \frac{1}{3}\frac{\mu_0^2 g_L^2 H J(J+1)}{\theta}, \tag{40.53}$$

where we have once again neglected terms of higher order in H.

Formula (40.53) is completely analogous to (40.46) for the electric moment of dipole molecules produced by a field. The characteristic magnetic moment is represented by the quantity $\mu_0 g_L \sqrt{J(J+1)}$ so that the Lande factor g_L takes into account the spin magnetic anomaly. Thus, magnetic susceptibility can be calculated from data obtained from spectroscopic observations.

Paramagnetism of rare earths. There are almost no elements for which we can completely verify formula (40.53) as applied to the gaseous state. But in rare earths the moment of the electronic cloud is due to the 4 f-shell, which occurs, as mentioned in Sec. 33, deep inside the atom. When such an atom is part of a crystal lattice, the 4 f-shell is but slightly subjected to the action of the electric field of the neighbouring atoms so that its state may be regarded as being almost the same as for a free atom of a rare-earth element. Therefore, (40.53) is applicable to those chemical compounds of rare-earth elements where other elements do not possess a characteristic magnetic moment. Its agreement with experiment is very satisfactory for almost all the elements of the rare-earth group.

Exercises

1) Find the mean relative velocity of two molecules of different gases occurring in a mixture.

The relative velocity distribution is given by a formula similar to the $v_{\parallel}$ distribution but written for all three velocity components. This formula is similar to (40.12), but it involves the reduced mass $m = \frac{m_1 m_2}{m_1 + m_2}$ instead of the mass of a single molecule. Hence, like (40.14), the mean relative velocity turns out equal to

$$\bar{v} = \sqrt{\frac{8\,\theta}{\pi\, m}} = \sqrt{\frac{8\,\theta\,(m_1 + m_2)}{\pi\, m_1\, m_2}}\,.$$

If the molecules are identical, their mean relative velocity is $\sqrt{2}$ times the mean absolute velocity.

2) Calculate the velocity of a bimolecular reaction r', if the effective cross-section depends upon the velocity component (along the line joining the nuclei) in the following way:

$$\sigma(v_{\parallel}) = \begin{cases} 0 & v_{\parallel} < \sqrt{\dfrac{2A}{m}}, \\ \sigma_0 & v_{\parallel} \geqslant \sqrt{\dfrac{2A}{m}}. \end{cases}$$

Then, from the general formula (40.28), we find

$$r' = \frac{n_1 n_2}{(2)} \sqrt{\frac{m}{2\pi\theta}} \int\limits_{\sqrt{\frac{2A}{m}}}^{\infty} \sigma_0\, v_{\parallel}\, e^{-\frac{m v_{\parallel}^2}{2\theta}}\, d v_{\parallel} = \frac{n_1 n_2}{(2)} \sqrt{\frac{\theta}{2\pi m}}\, \sigma_0\, e^{-\frac{A}{\theta}}.$$

The decisive quantity in this result is the exponential factor $e^{-\frac{A}{\theta}}$. The quantity A is called the activation energy. It is equal to the height of the potential barrier over which the colliding particles must pass in order that the reaction may occur. Unlike a thermonuclear reaction, it is assumed here that the motion of the reacting particles is classical. Transitions below the barrier make a vanishingly small contribution in chemical reactions.

Sec. 41. Boltzmann Statistics (Vibrational and Rotational Molecular Motion)

Molecular energy levels. In order to apply statistics to gases consisting of molecules, we must classify the energy levels of the molecules. The fact that a nucleus is considerably heavier than an electron, and, therefore, moves much slower, is very helpful here. We have used this in Sec. 33, when considering the question of the binding energy of two hydrogen atoms in a hydrogen molecule. The eigenfunction can be found for any relative positions of the nuclei. In a diatomic molecule the position of the nuclei is defined by a single parameter—the distance between them. The energy eigenvalue of the electrons depends upon this distance. Adding the energy of Coulomb repulsion of the nuclei to the electron energy, we obtain, for a given electron wave-function, the energy of the molecule as a function of the distance between the nuclei. For example, in a hydrogen molecule, the curves for this relationship are of different form in the case of parallel and antiparallel spin orientations (Fig. 47). The lower curve refers to the state with a symmetric spatial wave function and antiparallel spins, while the upper curve relates to the states with an antisymmetric spatial function and parallel spins. The lower curve has a minimum at $r = r_e$, so that hydrogen atoms may form a molecule only in a definite electron state.

In the general case, several different electronic states can have a minimum. The distances between corresponding potentional curves are defined from a wave equation of the type (33.23). In this equation we can neglect terms involving the masses of the nuclei in the denominator. Therefore, the energy scale separating different electronic states of the molecules is the same as for an atom, i.e., from one to ten electron-volts.

Close to the minimum of potential energy, nuclei may perform small oscillations. To a first approximation, these oscillations are harmonic so that their energy is given by the general formula (26.21):

$$\varepsilon_v = \hbar \omega \left(v + \frac{1}{2}\right). \tag{41.1}$$

Here, v is called the vibrational quantum number of the molecule. This number is, naturally, integral. Fig. 47 shows a more general dependence of energy upon v, taking into account that the potential energy curve is not a parabola as in Fig. 41. However, practically, the deviations from formula (41.1) affect but little the statistical quantities, because dissociation occurs when oscillations with large v are excited (see Sec. 51).

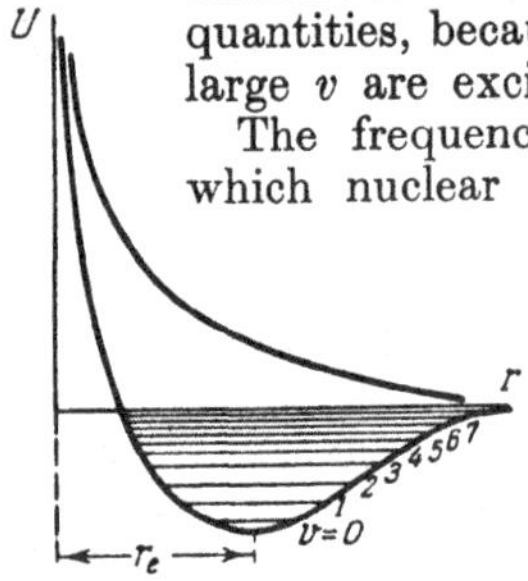

Fig. 47

The frequency ω depends upon the electronic state in which nuclear oscillations occur. In accordance with the general formulae for frequency (7.10)-(7.12), $\omega = \sqrt{\frac{1}{m}\left(\frac{\partial^2 U}{\partial r^2}\right)_{r=r_e}}$, so that the frequency is inversely proportional to the square root of the reduced mass of the nuclei. Therefore, the vibrational quantum is considerably less than the distance between electronic levels, which is independent of the nuclear mass. It is of the order of tenths of an electron-volt.

In addition to vibrational motion, a molecule with two atoms may also perform rotational motion. Rotation is most simply taken into account when the resultant spin of the electrons is equal to zero and the total orbital angular-momentum projection of the electrons on a line joining the nuclei is also equal to zero. These conditions are satisfied in the electronic ground state for nearly all molecules, with the exception of O_2, the resultant spin of which is equal to unity (but the projection of the electronic moment on the axis is zero), and NO, where the spin is one half (and the orbital angular-momentum projection of the electrons on the axis is zero). Disregarding these exceptions, we may consider a molecule of two atoms as a solid rotator, i.e., a system of two point masses at a fixed distance r_e corresponding to the minimum of the lower potential curve in Fig. 47 (see exercise 2, Sec. 30); (our case corresponds to $J_3=0$ and $k=0$, so that the closest excited level in k with $k=1$ is moved to infinity. The rotational moment of the rotator is perpendicular to the line joining the nuclei since its projection on this line is equal to zero).

As we know from Sec. 5 [see (5.6)] the rotational energy of two particles is

$$\varepsilon_r = \frac{M^2}{2\,m r_e^2},$$

where m is the reduced mass and mr_e^2 is equal to the moment of inertia of the rotator J_1. Going over to the quantum formula, we substitute the angular-momentum eigenvalue. It is usual to denote it by the letter K, so that

$$\varepsilon_r = \frac{h^2 K(K+1)}{2\,m r_e^2}. \tag{41.2}$$

This formula corresponds to the energy of a symmetric top with $k=0$ (cf. exercise 2, Sec. 30). It involves the mass of the nuclei in the denominator. Therefore, the distances between neighbouring rotational levels are of the order of a thousandth of an electron-volt and less.

Thus, to a good approximation, the total energy of a two-atom molecule can be written in the form of a sum with three terms:

$$\varepsilon = \varepsilon_e + \varepsilon_v + \varepsilon_r = \varepsilon_e + h\omega\left(v + \frac{1}{2}\right) + \frac{h^2 K(K+1)}{2\,m r_e^2}, \tag{41.3}$$

where $\varepsilon_e \sim \frac{1}{m^0}$ (i.e., it is independent of the nuclear mass m), $\varepsilon_v \sim \frac{1}{m^{1/2}}$, $\varepsilon_r \sim \frac{1}{m}$.

The excitation of electronic levels. If we substitute the expression (41.3) in the Boltzmann distribution, the latter separates into the product of three distributions involving electronic, rotational, and vibrational states. Let us suppose that a gas is considered with temperature not exceeding several thousand degrees, for example, 2,000-3,000°. Then if the energy of electronic excitation is several electron-volts (1 ev = 11,600°, since the temperature can be defined in energy units), the fraction of molecules in excited electronic states is a very small number: $e^{-\frac{\varepsilon_e}{\theta}}$. In those cases when there are very low electron levels, the Boltzmann factor may also be other than a small quantity. But, as a rule, dissociation of the molecules sets in earlier than excitation of their electronic levels (see Sec. 51).

Excitation of vibrational levels. Let us examine the vibrational states. For generality we may consider not only molecules with two atoms, but also polyatomic molecules. If the oscillations of such molecules are harmonic, we can make the transition to normal coordinates, as was shown in Sec. 7. Then the vibrational energy assumes the form of a sum of the energies of independent harmonic oscillators. The energy levels for each such harmonic oscillator are given by a formula of the form (41.1) with a frequency ω corresponding to a given normal oscillation.

Molecular oscillations are basically divided into two types: "valent," in which the distances between neighbouring nuclei mutually change, and "deformational," where only the angles between the "valence directions" change. For example, in a CO_2 molecule, having a straight-line equilibrium form $O=C=O$, valent oscillations alter the distance between the carbon nucleus and the oxygen nuclei, while deformational oscillations move the C nucleus out of the straight-line configuration. The frequencies of deformational vibrations are several times less than those of valent oscillations. The estimation $\hbar\omega \sim 0.1$ ev related to valent oscillations.

In any case, if the vibrational energy breaks up into the sum of energies of separate independent oscillations, then the distribution function also splits into the product of distribution functions for each separate oscillation.

Let us calculate the mean energy for one normal oscillation:

$$\bar{\varepsilon}_v = \frac{\sum\limits_{v=0}^{\infty} \hbar\omega\left(v+\frac{1}{2}\right) e^{-\frac{\hbar\omega\left(v+\frac{1}{2}\right)}{\theta}}}{\sum\limits_{v=0}^{\infty} e^{-\frac{\hbar\omega\left(v+\frac{1}{2}\right)}{\theta}}} = \theta^2 \frac{\partial}{\partial\theta} \ln\left(\sum_{v=0}^{\infty} e^{-\frac{\hbar\omega\left(v+\frac{1}{2}\right)}{\theta}}\right); \qquad (41.4)$$

here we have used the same transformation as in the derivation of (40.43) and (40.52). Formula (41.4) involves the statistical sum for a harmonic oscillator. The sum of the geometric progression inside the logarithm sign is very easily calculated. Indeed,

$$\sum_{v=0}^{\infty} e^{-\frac{\hbar\omega\left(v+\frac{1}{2}\right)}{\theta}} = e^{-\frac{\hbar\omega}{2\theta}} \sum_{v=0}^{\infty}\left(e^{-\frac{\hbar\omega}{\theta}}\right)^v = \frac{e^{-\frac{\hbar\omega}{2\theta}}}{1-e^{-\frac{\hbar\omega}{\theta}}}. \qquad (41.5)$$

Substituting this in (41.4) and differentiating, we get

$$\bar{\varepsilon}_v = \frac{\hbar\omega}{2} + \frac{\hbar\omega}{e^{\frac{\hbar\omega}{\theta}}-1}. \qquad (41.6)$$

The first term in (41.6) simply denotes the zero energy of an oscillation of given frequency. The oscillation possesses this energy at absolute zero because then the second term in (41.6) does not contribute anything. The second term has a very simple meaning. If we write the mean energy in terms of the mean vibrational quantum number $\bar{v}$

$$\bar{\varepsilon}_\nu = \frac{h\omega}{2} + h\omega\bar{v}, \tag{41.7}$$

then it is obvious that

$$\bar{v} = \frac{1}{e^{\frac{h\omega}{\theta}} - 1}. \tag{41.8}$$

For this reason, the factor $\left(e^{\frac{h\omega}{\theta}} - 1\right)^{-1}$ signifies the mean number of quanta possessed by a vibration at a temperature $\theta = kT$. At a low temperature, $\bar{v}$ is close to zero. For example, for oxygen and nitrogen, $h\omega$ is about 0.2 ev, or 2,000-3,000°. Therefore, at room temperature oxygen and nitrogen occur in the ground vibrational state. In the case of hydrogen the reduced mass of a molecule is 14 times less than that of nitrogen. Its vibrational quantum is close to 6,000°. In polyatomic molecules, where deformational vibrations occur, such oscillations can be excited at temperatures of the order of 300-600°.

Vibrational energy at high temperatures. If the temperature is very high compared with $h\omega$, then $e^{\frac{h\omega}{\theta}}$ can be replaced by the expansion $1 + \frac{h\omega}{\theta}$. Substituting this in (41.6), we obtain

$$\bar{\varepsilon}_\nu = \frac{h\omega}{2} + \theta. \tag{41.9}$$

The first term does not relate to thermal excitation. Besides, it is considerably less than θ. Thus it turns out that at a sufficiently high temperature the mean energy per oscillation is equal to θ, irrespective of the frequency. The same can be obtained by proceeding from the nonquantized expression for the energy of a harmonic oscillator:

$$\varepsilon_\omega = \frac{p^2}{2m} + \frac{m\omega^2 q^2}{2}. \tag{41.10}$$

Substituting this in the Boltzmann distribution and calculating the mean energy, we have

$$\bar{\varepsilon}_\omega = \frac{\int\limits_{-\infty}^{\infty} dp \int\limits_{-\infty}^{\infty} dq\, \varepsilon_\omega e^{-\frac{\varepsilon_\omega}{\theta}}}{\int\limits_{-\infty}^{\infty} dp \int\limits_{-\infty}^{\infty} dq e^{-\frac{\varepsilon_\omega}{\theta}}} = \theta^2 \frac{\partial}{\partial\theta} \ln\left(\int\limits_{-\infty}^{\infty} dp \int\limits_{-\infty}^{\infty} dq e^{-\frac{\varepsilon_\omega}{\theta}} \right). \tag{41.11}$$

The statistical integral inside the logarithm is calculated in the usual way:

$$\int_{-\infty}^{\infty} e^{-\frac{p^2}{2m\theta}} dp \int_{-\infty}^{\infty} e^{-\frac{m\omega^2 q^2}{2\theta}} dq = \sqrt{2\pi m\theta}.\sqrt{\frac{2\pi\theta}{m\omega^2}} = \frac{2\pi}{\omega}\theta. \tag{41.12}$$

Whence $\overline{\varepsilon_\omega} = \theta$. Then the total vibrational energy of a gas occurring at a frequency ω is

$$\overline{\mathscr{E}}_\omega = N\theta = kT, \tag{41.13}$$

and its contribution to the specific heat is correspondingly equal to R [see (40.17)]. Thus, at a high temperature the specific heat due to vibrational degrees of freedom tends to a constant limit.

The excitation of rotational levels.* Let us now consider rotational energy. The weight of a state with a given value of moment K is, as usual, equal to $2K+1$, in accordance with the number of possible projections of K. Especially interesting is the case when a diatomic molecule consists of two identical nuclei. In classifying the states of such a molecule it is necessary to take nuclear spin into account. Indeed, the wave equation for a molecule consisting of identical atoms does not change form when the nuclei are interchanged. Therefore, if the nuclei have half-integral spin, the wave function must be antisymmetric with respect to the interchange of both nuclei, while if they have integral spin it must be symmetric. The symmetry of the eigenfunction of a molecule is determined by the symmetry of its factors [in the approximation (41.3) it is separated into factors]: electronic, vibrational, rotational, and nuclear spin. The electronic term of most molecules does not change when the nuclei are interchanged. The vibrational function depends only upon the absolute value of the distance between the nuclei and therefore does not change either. The rotational eigenfunction is even with respect to this permutation in the case of even K, and odd for odd K. Therefore, if the nuclear spin is half-integral, then the spin function must be antisymmetric for even K and symmetric for odd K, so that the resultant wave function may always be antisymmetric. If the nuclear spin is integral, the position is reversed, and if it is equal to zero, then odd K are in general excluded because then the spin factor simply does not exist.

Rotational energy of para- and ortho-hydrogen. We shall now consider the rotational states of a hydrogen molecule. The total nuclear spin for hydrogen can equal unity (the ortho-state) and zero (the para-state). The weight of a state with spin 1 is 3 and that with spin 0 is 1. The state with $K=0$ is even in the rotational wave function. Hence, it must be odd in the spin function, i.e., it must have spin 0, (see Sec. 33). But the state with zero moment possesses the least rotational

---

* The hypothesis that the rotation of molecules participates in the thermal motion of a gas was put forward by M. V. Lomonosov as far back as 1745.

energy. Therefore, only para-hydrogen is stable close to absolute zero.

At a temperature other than zero all those states, for which the Boltzmann factor $e^{-\frac{h^2 K(K+1)}{2mr_e^2\theta}}$ is of the order unity, are excited. Taking the moment of inertia of a hydrogen molecule to be equal to $0.45 \times \times 10^{-40}$, we can see that already at $T = 300°$ K the summation over all odd moments

$$\sum_{K=1,3,5\ldots} (2K+1)\, e^{-\frac{h^2 K(K+1)}{2mr_e^2\theta}}$$

differs from the summation over even moments by several thousandths. But since the states with even moments are, for hydrogen, nuclear-spin ortho-states, each state with even moment has an additional weight factor 3 according to the number of projections of spin 1. Thus, at room temperature, 3/4 of hydrogen is ortho-hydrogen and 1/4 para-hydrogen. If hydrogen is rapidly cooled the ratio 3:1 is retained for a long time because the ortho-para-transition proceeds slowly. Such a state is obviously nonequilibrium since all the hydrogen in an equilibrium state, at a temperature close to absolute zero, must be in the para-state.

One of the methods of obtaining pure para-hydrogen is to adsorb hydrogen onto any substance that disrupts the molecular bonds during adsorption, for example, activated carbon. When desorbing the hydrogen by pressure reduction at low temperature, the change is that to the para-state. If the hydrogen is then heated to room temperature it stays in the para-state for quite a long time.

Let us now write down the formulae for the mean rotational energy of ortho- and para-hydrogen. For simplicity we shall denote the factor $\frac{h^2}{2mr_e^2}$ in the rotational energy by the letter B. Then

$$\bar{\varepsilon}_{\text{para}} = \frac{\sum\limits_{K=0,2,4\ldots} (2K+1)\, e^{-\frac{B}{\theta}K(K+1)} \cdot BK(K+1)}{\sum\limits_{K=0,2,4,\ldots} (2K+1)\, e^{-\frac{B}{\theta}K(K+1)}} =$$

$$= \theta^2 \frac{\partial}{\partial \theta} \ln \left[\sum_{K=0,2,4,\ldots} (2K+1)\, e^{-\frac{B}{\theta}K(K+1)} \right]. \tag{41.14}$$

The difference between $\bar{\varepsilon}_{\text{ortho}}$ and $\bar{\varepsilon}_{\text{para}}$ is that the summation is performed over odd K. For a mixture at room temperature

$$\bar{\varepsilon}_r = \frac{1}{4}\bar{\varepsilon}_{\text{para}} + \frac{3}{4}\bar{\varepsilon}_{\text{ortho}}. \tag{41.15}$$

At very low temperature, it is sufficient to retain only the term with $K=2$ in the summation (41.14), so that

$$\overline{\varepsilon}_{\text{para}} = \theta^2 \frac{\partial}{\partial\theta} \ln\left(1 + 5e^{-\frac{6B}{\theta}}\right) \cong 30Be^{-\frac{6B}{\theta}}. \qquad (41.16)$$

For ortho-hydrogen we obtain

$$\overline{\varepsilon}_{\text{ortho}} = \theta^2 \frac{\partial}{\partial\theta} \ln\left(3e^{-\frac{2B}{\theta}} + 7e^{-\frac{12B}{\theta}}\right) \cong$$

$$\cong B\frac{6e^{-\frac{2B}{\theta}} + 84e^{-\frac{12B}{\theta}}}{3e^{-\frac{2B}{\theta}}} \cong 2B\left(1 + 14e^{-\frac{10B}{\theta}}\right). \qquad (41.17)$$

The determination of nuclear spins from rotational specific heat. The rotational specific heat of hydrogen makes it possible to determine the spin of a proton. Let us consider formula (41.17). In it, the first term is a constant. It is due to the fact that a molecule of ortho-hydrogen would have a rotational energy $2B$ even at absolute zero. This energy does not contribute to the specific heat because it does not depend upon the temperature. Defining specific heat as the derivative $\frac{\partial \varepsilon}{\partial \theta}$, we see that for a sufficiently low temperature the ratio of the specific heats of ortho- to para-hydrogen tends to zero as

$$e^{-\frac{4B}{\theta}}.$$

Therefore, if ordinary hydrogen is rapidly cooled to a low temperature, its rotational specific heat will be determined by a quarter of its molecules in the para-state. It will be four times less than the rotational specific heat of pure para-hydrogen at the same temperature.

Thus, by measuring the specific heat of the equilibrium state of hydrogen at low temperature (i.e., the para-state) and of rapidly cooled hydrogen, we can determine the spin of a proton or, knowing the spin from other data, we can show that protons are subject to Pauli exclusion because they possess an antisymmetric wave function.

The rotational specific heat of molecules consisting of different atoms. Diatomic molecules that do not consist of identical atoms possess equal nuclear-spin weights for states with odd and even K. Therefore, their mean rotational energy is expressed thus:

$$\overline{\varepsilon}_r = \theta^2 \frac{\partial}{\partial\theta}\left[\ln \sum_{K=0}^{\infty} (2K+1)e^{-\frac{BK(K+1)}{\theta}}\right]. \qquad (41.18)$$

The sum inside the logarithm cannot be written in finite form, but it is easily tabulated. Let us evaluate the temperature at which use of an integral as a substitute for the summation is justified. Thus, for hydrogen

$$B = \frac{h^2}{2\,m r_e^2} \cong \frac{1.11 \cdot 10^{-54}}{1.67 \cdot 10^{-24}\,(0.74)^2 \cdot 10^{-16}} = 1.2 \cdot 10^{-14}\,\text{erg},$$

which corresponds to a temperature of 87° K.

Here, m is the reduced mass of two protons, equal to half the proton mass; $r_e \sim 0.74 \times 10^{-8}$ cm (where we obtained the moment of inertia used above). For other gases B is of the order of several degrees so that for all temperatures at which these gases are not in the liquid state the ratio B/θ is a small quantity. To a good approximation, the summation in (41.18) may be replaced by an integral. If we take

$$K\,(K+1) = x,$$

then

$$(2\,K+1)\,dK = 2\,K + 1 = dx \quad (dK = 1)$$

and

$$\sum (2\,K+1)\,e^{-\frac{B K (K+1)}{\theta}} \cong \int_0^\infty e^{-\frac{B x}{\theta}}\,dx = \frac{\theta}{B}. \tag{41.19}$$

Substituting this in (41.18), we have an expression for the rotational energy of a diatomic molecule or any linear molecule

$$\bar{\varepsilon}_r = \theta = \frac{RT}{N}. \tag{41.20}$$

We note that the concepts of "high" temperature for vibrations and rotations do not coincide in the least. With respect to the rotational specific heat of oxygen, the temperature must be higher than 10° K to be regarded as high, while with respect to vibrational specific heat, it must be above 2,000° K. Therefore, in a very wide range of temperatures, in particular at room temperature, the specific heats of diatomic gases are constant, and consist of a translational part $\frac{3}{2}\,R$ and a rotational part equal to R, so that the total specific heat is $\frac{5}{2}\,R$.

It may be seen by numerical computation that the rotational specific heat does not tend to a constant limit monotonically, but passes through a maximum at $\theta = 0.81\;B$, equal to $1.1\;R$.

The rotational energy for a polyatomic molecule will be calculated in Sec. 47.

Exercise

Find the rotational energy of para- and ortho-deuterium.

Particles with integral spin have a symmetric wave function. Let us now consider a system of two particles with integral spin, for example, a deuterium molecule. For comparison we shall also take two particles with spin zero. The spin function of the latter is identically equal to unity; therefore their orbital wave function can be only symmetrical. With respect to the rotational function, interchange of the nuclei is equivalent to a reflection at the coordinate origin. Hence, if the spin of a deuteron were zero then the spectrum of molecular deuterium would show the lines, corresponding to odd rotational quantum numbers, to be absent. In actual fact they exist in the deuterium spectrum, and the weight of states with even K is twice as great as for those with odd K. This is seen from the relative intensity of spectral lines that correspond to transitions from the appropriate states.

We shall show that for a deuteron spin of unity, the weight of the ortho-states turns out twice the weight of the para-states. A spin projection of unity takes on three values: 1, 0, —1. We denote the spin wave functions (of both deuterons) that correspond to these projections as $\psi_1(1)$, $\psi_1(0)$, $\psi_1(-1)$ and $\psi_2(1)$, $\psi_2(0)$, $\psi_2(-1)$. Let us form all the spin wave functions of deuterium that correspond to a total spin projection 0; we shall only take symmetric and antisymmetric combinations:

Symmetric functions	*Antisymmetric functions*
$\psi_1(1)\psi_2(-1)+\psi_1(-1)\psi_2(1)$,	$\psi_1(1)\psi_2(-1)-\psi_1(-1)\psi_2(1)$.
$\psi_1(0)\psi_2(0)$,	

For the total spin projection ± 1, we obtain

$\psi_1(1)\psi_2(0)+\psi_1(0)\psi_2(1)$,	$\psi_1(1)\psi_2(0)-\psi_1(0)\psi_2(1)$,
$\psi_1(-1)\psi_2(0)+\psi_1(0)\psi_2(-1)$,	$\psi_1(-1)\psi_2(0)-\psi_2(-1)\psi_1(0)$.

And for a total projection ± 2 we have

$$\psi_1(1)\psi_2(1),$$
$$\psi_1(-1)\psi_2(-1).$$

The symmetric state has a maximum spin projection of two. Hence, the state for which the spins are parallel is symmetric. But there are six symmetric spin wave-function projections in all, and spin 2 has $2\cdot 2+1=5$ projections. Hence, of the functions with zero resultant projection, we can construct one function corresponding to a zero projection of spin 2. The other function with zero resultant projection corresponds to a resultant spin 0.

In all, deuterium has six ortho-states with a symmetric spin wave function. A spin unity has states given by an antisymmetric spin function because the maximum spin projection in these states is equal to unity. Thus, there are three para-states. An even rotational function of a deuterium molecule corresponds to the ortho-states, and an odd rotational function corresponds to the para-states. Then the total function is symmetric, as the case should be for integral particle spins. The weight (due to spin) for the ortho-states is six and for the para-states it is three. Therefore, the statistical sum of ortho-deuterium is

$$6\sum_{K=0,2,4,\ldots}(2K+1)\,e^{-\frac{BK(K+1)}{\theta}},$$

and for para-deuterium it is equal to

$$3\sum_{K=1,3,5,\ldots}(2K+1)\,e^{-\frac{BK(K+1)}{\theta}}$$

Here the equilibrium state at absolute zero is the ortho-state. The energies of both states [see (41.16) and (41.17)] are

$$\bar{\varepsilon}_{\text{ortho}} \cong 30\, B e^{-\frac{6B}{\theta}} ;$$

$$\bar{\varepsilon}_{\text{para}} \cong B \left(2 + 28\, e^{-\frac{10B}{\theta}}\right).$$

Compared with hydrogen, the ortho- and para-states are interchanged here. Close to absolute zero, the basic contribution to the specific heat is given only by the ortho-state. Two thirds of all the molecules in equilibrium deuterium occur in this state at room temperature. Therefore, the rotational specific heat of rapidly cooled deuterium is less than that of equilibrium deuterium at the same temperature in the ratio 2/3. Thus, by measuring this ratio we can show that the spin of a deuteron is equal to unity and not zero.

Sec. 42. The Application of Statistics to the Electromagnetic Field and to Crystalline Bodies

The statistical equilibrium of matter and radiation. In this section we shall first of all consider radiation in a state of statistical equilibrium with matter. The conditions for such equilibrium are achieved inside a closed cavity in an opaque body. The walls of the opaque cavity absorb radiation of all frequencies and hence they also radiate all frequencies: if a direct quantum transition is permissible, then the reverse transition is also permissible. Therefore, radiation arrives at a statistical equilibrium with matter, that is, in unit time there is an equal amount of absorbed and emitted energy of electromagnetic radiation per unit surface of the cavity for every direction, frequency, and polarization.

An equilibrium density of radiation energy is thus set up in the cavity. It can be shown that in this case temperature of radiation is equal to the temperature of the walls. The necessity of this will be especially clearly seen in the sections dealing with the fundamentals of thermodynamics (Sec. 45 and 46); for the time being we shall merely note that it is natural to regard the temperatures of systems in equilibrium as identical.

The absolutely black body. Equilibrium radiation can be experimentally studied by making a small aperture in the wall of the cavity: if it is of sufficiently small dimensions the equilibrium state will not be noticeably changed. Radiation incident on such an aperture from outside the cavity is absorbed in it and does not get outside. In this sense the aperture resembles a black body which does not reflect light rays. For this reason it is called an "*absolute black body,*" and the equilibrium radiation coming from the aperture is called "*black-body radiation.*"

This term is somewhat paradoxical since it contradicts the obvious picture. Indeed, an absolutely black body in equilibrium radiates more than a nonblack body because it absorbs more, and in equilibrium the radiation and absorption are equal. If a body having a cavity and aperture is brought to an incandescent state, the aperture will exhibit the brightest glow.

The statistics of an oscillator field representation. Planck's formula. In this section we shall consider the application of statistics to equilibrium radiation. For this it is necessary to quantize the radiation. Unlike the statistics of a gas, the statistics of radiation does not permit a limiting transition to equations, with the quantum of action being eliminated entirely. This will become clear a little later.

In quantizing the field, a double approach is possible. Firstly, a field may be represented as a set of linear harmonic oscillators by characterizing each oscillator with a definite wave vector **k** and polarization σ ($\sigma = 1, 2$). It is obvious that all these oscillators are different (as to their k and σ). The quantum properties of such oscillators are not apparent in calculating the number of states of the field; their only manifestation is that the energy of each of them cannot be equated to an arbitrary number, but belongs to an oscillator-energy spectrum; i.e., equal to $\hbar\omega\left(n + \frac{1}{2}\right)$ where n is an integer.

When an oscillator is in thermal equilibrium, the mean number of its vibrational quanta is given by a formula similar to (41.8):

$$\overline{n} = \frac{1}{e^{\frac{\hbar\omega}{\theta}} - 1}. \tag{42.1}$$

The energy of each quantum is equal to $\hbar\omega$ and the number of oscillations with frequency ω is, according to (25.24),

$$dg(\omega) = \frac{V\omega^2\, d\omega}{\pi^2 c^3}. \tag{42.2}$$

Here, in contrast to formula (25.24), both possible polarizations of oscillation with a given frequency are taken into account, and $K = \omega/c$ has been substituted. Hence the energy of an electromagnetic field in the frequency interval $d\omega$ is

$$d\mathscr{E}(\omega) = \frac{V\hbar\omega^3\, d\omega}{\pi^2 c^3\left(e^{\frac{\hbar\omega}{\theta}} - 1\right)}. \tag{42.3}$$

The radiation spectrum of the sun is close to this frequency distribution.

The statistics of light quanta. Let us now approach formula (42.1) from another direction. We have said that the electromagnetic field is viewed as an assemblage of elementary particles—light quanta.

Quanta of the same frequency, direction, and polarization are indistinguishable from one another. Therefore quantum statistics are applicable to them as to particles. At the same time quanta have integral angular momenta; this was mentioned in Sec. 34. Therefore they are not subject to Pauli exclusion, and possess a Bose and not Fermi distribution. But, as opposed to gas molecules, which are subject to a Bose distribution, the number of quanta is not a constant quantity, since quanta may be absorbed and radiated. This is why the supplementary condition (39.12) does not apply to quanta.

It is easy to pass from the general Bose distribution to a special case, when condition (39.12) is not imposed; for this it is sufficient to put equal to zero the parameter μ, by which equation (39.12) is multiplied (μ was introduced to satisfy the condition $N = \text{const}$). Then the Bose distribution is simplified:

$$n = \frac{1}{e^{\frac{\varepsilon}{\theta}} - 1}. \tag{42.4}$$

Taking into account that for a quantum $\varepsilon = \hbar\omega$, we once again obtain (42.1). Thus, formula (42.1) denotes either the mean vibrational quantum number of an oscillator in an assembly subject to Boltzmann statistics, or the mean number of light quanta subject to Bose statistics. As we have already said, certain oscillators obey Boltzmann statistics: they are differentiated by the numbers n_1, n_2, n_3, σ (see Sec. 27), while the statistics of distinguishable particles is nonquantum. Let it be recalled that we differentiate between quantum and nonquantum statistics according as the particles are distinguishable or not.

The impossibility of the limiting transition $\hbar \to 0$ in the statistics of the electromagnetic field. Let us now turn, for a time, to the oscillator picture. On classical theory, the mean energy of an oscillator is equal to θ [see (41.11)-(41.13)]. If we multiply it by $dg\,(\omega)$, the classical Rayleigh-Jeans formula for the energy of equilibrium radiation results.

$$d\mathscr{E}\,(\omega)_{\text{class}} = \frac{V\omega^2\, d\omega}{\pi^2 c^3}\,\theta. \tag{42.5}$$

But this formula is obviously inadequate for large frequencies: upon integration with respect to ω it gives an infinite total energy. It was precisely here, in statistics, that the classical representations first so obviously failed. Therefore, in 1900, Planck proposed formula (42.3): it was here that the quantum of action appeared for the first time in physics.

Formula (42.5) is correct only for frequencies that satisfy the inequality $\hbar\omega \ll \theta$.

The total energy of equilibrium radiation. It is easy to find the total energy of equilibrium electromagnetic radiation from formula (42.3). Integrating with respect to ω, we obtain

$$\mathscr{E} = \frac{Vh}{\pi^2 c^3} \int_0^\infty \frac{\omega^3 \, d\omega}{e^{\frac{h\omega}{\theta}} - 1} = \frac{Vh}{\pi^2 c^3} \frac{\theta^4}{h^4} \int_0^\infty \frac{x^3 \, dx}{e^x - 1}. \tag{42.6}$$

The integral in (42.6) is merely an abstract number, equal to $\frac{\pi^4}{15}$ (see Appendix, p. 586), so that the required energy is proportional to the fourth power of the absolute temperature (the Stefan-Boltzmann law).

Radiation from an absolutely black body. The result (42.6) can be verified from the emissivity of an "absolutely black body." It is easy to relate it to the energy $\mathscr{E}$. For this it is sufficient to calculate how many quanta fall from inside in unit time upon unit surface of a cavity, normal to the surface. We have indicated that if we take away a small section of the wall, radiation will pass through the aperture with the same composition as that falling on the wall.

The velocity of each quantum is c, so that its normal component is equal to $c \cos \vartheta$, where ϑ is the angle with the normal. In unit time these quanta will strike a square centimetre of the wall from the whole volume of a cylinder with base 1 cm² and height $c \cos \vartheta$. The energy included in the volume of this cylinder is equal to $\frac{\mathscr{E}}{V} \cdot c \cos \vartheta$.

The fraction of quanta flying in unit solid angle is equal to $\frac{1}{4\pi}$, so that the total energy falling on a square centimetre of the wall in unit time is

$$\frac{1}{4\pi} \int_0^{2\pi} d\varphi \int_0^{\frac{\pi}{2}} \sin\vartheta \, d\vartheta \cdot c \cos\vartheta \, \frac{\mathscr{E}}{V} = \frac{c}{4} \frac{\mathscr{E}}{V} = \frac{\pi^2 \theta^4}{60 c^2 h^3} = \frac{\pi^2 k^4}{60 c^2 h^3} T^4. \tag{42.7}$$

The constant in front of T^4 is equal to 5.67×10^{-5} erg/cm² sec · deg⁴. Formula (42.7) cannot be directly applied to an incandescent solid body without ascertaining to what extent it may be regarded as black.

Due to the fact that the sun's luminous shell (chromosphere) is nearly opaque to radiation, the spectrum it emits is close to the equilibrium spectrum (42.3), even though it does not exactly coincide with it. The temperature of the chromosphere, as determined from (42.3), is approximately 5,700°.

The pressure of equilibrium radiation. It is also easy to calculate the pressure of equilibrium radiation. It is convenient in doing so to apply the same reasoning that led to formula (42.7). Now, however, instead of calculating the number of quanta, it is necessary to calculate their normal component of momentum transmitted through a square centimetre of surface. This component is equal to the quantum energy $\hbar\omega$ divided by c and multiplied by $\cos\vartheta$. Therefore, unlike formula (42.7), we must integrate $\cos^2\vartheta$ instead of $\cos\vartheta$. In addition, for every incident quantum in the equilibrium state there is a similar quantum radiated in the reverse direction, so that the transferred momentum is doubled. Whence the pressure is

$$p = \frac{\mathscr{E}}{cV} \cdot \frac{1}{4\pi} \cdot 2c\int_0^{2\pi} d\varphi \int_0^{\pi/2} \cos^2\vartheta \sin\vartheta\, d\vartheta = \frac{\mathscr{E}}{3V}, \tag{42.8}$$

i.e., one third of the energy density. The same would be obtained from the derivation of equation (40.22) if the momentum were put equal to ε/c instead of mv. We note that in Lebedev's experiments, where the pressure of a directed beam was measured, and not of light arriving uniformly from all directions, $p = \mathscr{E}/V$; the pressure of the directed beam is equal to the energy density without the factor 1/3 (see Sec. 17).

From (42.8) and (42.6), the pressure of electromagnetic radiation increases in proportion to the fourth power of the temperature while the gas pressure is proportional to the first power. Therefore, radiation pressure will always predominate at a sufficiently high temperature.

At high temperatures the pressure of a substance can always be calculated from the ideal-gas formula, because the interaction energy between particles becomes small compared with their kinetic energy. Hence,

$$p = \frac{N\theta}{V}.$$

By considering that atoms are dissociated into nuclei and electrons, it is easy to express the ratio N/V in terms of the mass density. Let us suppose that the substance consists of hydrogen. Then for every proton there is one electron. If the density of the substance is ρ, then the ratio N/V is $2\rho/m$, where m is the mass of a proton and the factor 2 takes into account the electron. This gives

$$p_m = \frac{2\rho}{m}\,\theta. \tag{42.9}$$

From (42.8) and (42.6), the radiation pressure p_r is

$$p_r = \frac{\pi^2\theta^4}{45(\hbar c)^3}. \tag{42.10}$$

From this we obtain the relationship between density and temperature when the radiation pressure becomes equal to the gas pressure:

$$\rho = \frac{\pi^2}{90} \cdot \frac{m}{(hc)^3} \theta^3 = 1.5 \times 10^{-23} T^3.$$

For example, for a density $\rho = 1$ gm/cm³, both pressures become equal if the temperature is equal to 4×10^7 deg. Radiation pressure is important in the interiors of certain classes of stars.

The frequency corresponding to the maximum radiation-energy density in a spectral interval $d\omega$. The maximum energy in the distribution occurs at a frequency determined from the equation

$$\left(\frac{d}{d\omega} \frac{\omega^3}{e^{\frac{h\omega}{\theta}} - 1} \right)_{\omega = \omega_0} = 0. \tag{42.11}$$

Performing the differentiation, we have

$$1 - e^{-\frac{h\omega_0}{\theta}} = \frac{h\omega_0}{3\theta}. \tag{42.12}$$

This equation has a single solution with respect to $\frac{h\omega_0}{\theta}$.

$$\frac{h\omega_0}{\theta} = 2.822. \tag{42.13}$$

Thus, the frequency corresponding to maximum energy in the spectrum of black-body radiation is directly proportional to the absolute temperature (Wien's law):

$$\omega_0 = \frac{2.822\,\theta}{h}. \tag{42.14}$$

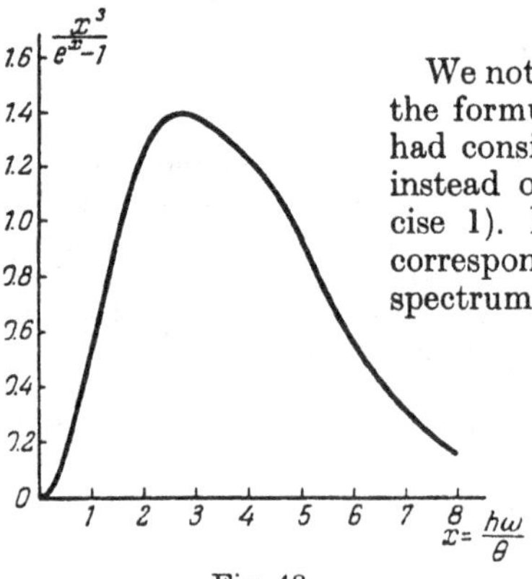

Fig. 48

We notice that the numerical coefficient in the formula would have been different if we had considered the wavelength distribution instead of frequency distribution (see exercise 1). It is interesting to note that the corresponding wavelength λ_0 in the solar spectrum is very close to that for the maximum sensitivity of the human eye. The curve of the distribution $\frac{(h\omega)^3/\theta^3}{e^{\frac{h\omega}{\theta}} - 1}$ is shown in Fig. 48.

Spontaneous and forced emission of quanta. At the beginning of this section we pointed out that thermal equilibrium between atoms and radiation is attained in a closed cavity. The presence of atoms capable of radiation and absorption is necessary in general in order that the radiation may arrive at equilibrium; this is because separate oscillators, corresponding to normal oscillations of the electromagnetic field, are completely independent of one another, and any initial nonequilibrium distribution is maintained until there is an exchange of quanta via absorbing atoms.

In Sec. 34 we derived an expression for the probability of light emission by an atom. According to (34.46), the radiation probability in unit time is

$$W_{10} = \frac{4}{3}\frac{\omega_{10}^3}{\hbar c^3}\,|\mathbf{d}_{10}|^2. \qquad (42.15)$$

We shall now consider atoms which are in thermal equilibrium with matter. Let the frequency ω_{10} satisfy the relationship $\hbar\omega_{10}=\varepsilon_1-\varepsilon_0$, where ε_1 and ε_0 are the energies of two atomic states. In equilibrium, atoms with energy ε_1 radiate as many quanta with frequency ω_{10} as are absorbed by atoms with energy ε_0.

In accordance with the principle of detailed balance, the probabilities for direct and reverse transitions are connected by the following relation:

$$g_1 W_{10} = g_0 W_{01}\,. \qquad (42.16)$$

Indeed, the first-approximation formula of perturbation theory (34.29) is applicable to radiation and absorption processes, since the interaction of matter with radiation may be regarded as weak. From this formula, the probabilities for the transitions $1\to 0$ and $0\to 1$ are, respectively,

$$W_{10} = \frac{2\pi}{\hbar}\,|\mathscr{H}_{10}|^2 g_0;\; W_{01} = \frac{2\pi}{\hbar}\,|\mathscr{H}_{01}|^2 g_1\,. \qquad (42.17)$$

But according to the Hermitian condition (34.15), the squares of the moduli of the matrix elements $|\mathscr{H}_{01}|^2$ and $|\mathscr{H}_{10}|^2$ are the same so that if we multiply expressions (42.17) by the weights of the initial states, the result will be equation (42.16).

The formula for the probability of absorption related to the case when a single quantum of frequency ω_{10} existed in the field before absorption. If there were $n\,(\omega_{10})$ such quanta before absorption, then it is natural to assume that the probability of absorbing one of them in unit time is $n\,(\omega_{10})$ times greater. This assumption is justified in electromagnetic-field quantum theory.

We shall therefore assume the probability of absorbing in unit time one of the $n\,(\omega_{10})$ identical quanta in the field to be equal to $n\,(\omega_{10})$ $g_0 W_{01}$. In accordance with the principle of detailed balance we must have the same probability for the reverse transition, i. e., the emission of a quantum by an atom occurring in state 1 when there are $n\,(\omega_{10})-1$ such quanta in the field; this is because the transition is reversed with respect to the one just considered. We represent both transitions thus:

1st state of the system:	quantum absorption →	2nd state of the system:
atom with energy ε_0		atom with energy ε_1
n quanta with frequency ω_{10}	← quantum radiation	$n-1$ quanta with frequency ω_{10}

Thus, in accordance with the principle of detailed balance, the probability of emission of a quantum must likewise be $n g_1 W_{10}$, which can also be represented as $[(n-1)+1]\, g_1 W_{10}$. Because of equation (42.16) the probabilities for both direct and reverse transitions will be equal. Hence, if $n-1$ quanta exist in the field, then the probability of emission is proportional to n, i.e., to the number of quanta increased by unity. If, for example, there were no quanta in the field before emission, this factor of proportionality is equal to unity. In this case the emission is termed spontaneous. But when there are quanta in the field, they stimulate, as it were, further emission of quanta with the same frequency, direction of propagation, and polarization. The emission produced by them is called forced. The existence of forced emission can also be proved by means of quantum field theory, just as the proportionality factor n in the absorption probability. The idea of forced emission was introduced by Einstein.

The derivation of Planck's formula from the relationship between the quantum emission and absorption probabilities. Let us now consider atoms in thermal equilibrium with an electromagnetic field. Let the quantity $n(\omega_{10})$ denote the equilibrium number of quanta. The condition of statistical equilibrium is that atoms occurring in state 0 absorb as many quanta with frequency ω_{10} in unit time as are emitted by atoms in state 1. Then the number $n(\omega_{10})$ does not change with time, i.e., equilibrium is attained.

The number of acts of absorption by all the atoms in unit time (from state 0 in which there are N_0 atoms) is equal to

$$N_0 W_{01}\, n(\omega_{10})\,. \qquad (42.18\text{a})$$

The number of acts of emission by all atoms in state 1 in unit time is

$$N_1 W_{10}\,[n(\omega_{10})+1]\,, \qquad (42.18\text{b})$$

because, as we have seen, it involves the number of quanta increased by unity, i.e., $n(\omega_{10})+1$.

Naturally, expressions (42.18a) and (42.18b) no longer denote the probabilities for direct and reverse transitions, but the probabilities for transitions from a state with the same number of quanta $n(\omega_{10})$, which probabilities lead to a reduction or to an increase by unity of the same number. The condition for thermal equilibrium is that these probabilities are equal:

$$N_0 W_{01}\, n(\omega_{10}) = N_1 W_{10}\,[n(\omega_{10})+1]\,. \qquad (42.19)$$

Here, we substitute N_0 and N_1 from the Boltzmann distribution (40.1):

$$e^{\frac{\mu-\varepsilon_0}{\theta}} g_0 W_{01} n(\omega_{10}) = e^{\frac{\mu-\varepsilon_1}{\theta}} g_1 W_{10} [n(\omega_{10}) + 1]. \qquad (42.20)$$

We now take advantage of the fact that $h\omega_{10} = \varepsilon_1 - \varepsilon_0$, and also of the relationship (42.16). Then there remains an equation for the equilibrium number of quanta $n(\omega_{10})$

$$e^{\frac{h\omega_{10}}{\theta}} \cdot n(\omega_{10}) = n(\omega_{10}) + 1. \qquad (42.21)$$

Whence Planck's formula is immediately obtained

$$n(\omega_{10}) = \frac{1}{e^{\frac{h\omega_{10}}{\theta}} - 1}. \qquad (42.22)$$

Thus, the idea of forced emission leads to a correct frequency distribution of quanta. Note that the part of forced emission is the more important the greater n is compared with unity. But large n correspond to the classical limit; it follows that forced or induced emission is by nature classical and spontaneous emission is a quantum effect.

We notice that the theory of a field consisting of Bose particles always leads to the concept of forced emission, provided the principle of detailed balance is used.

The probability of the appearance in the field, of the $(n+1)$th particle is proportional to $n+1$, while the probability of the particle disappearing is proportional to n. Naturally, if the Bose particles are charged (as, for example, π-mesons) then only those transitions are possible which are compatible with the conservation of total charge of the system.

As regards Fermi particles, we must bear in mind that a transition to a filled level is impossible. Therefore, if the probability that a level is filled is f, then the number of transitions to this level in unit time is proportional to $1-f$.

The oscillation spectrum for the lattice of a solid body. Let us now apply statistics to the crystal lattice of a solid body. As applied to the crystal lattice, statistics is in many ways similar to the theory of equilibrium radiation.

The vibrations of atoms in a lattice may be described in normal coordinates, after which their energy is reduced to approximately the same form as (27.22): it consists of the sum of the energies of separate oscillators. To each oscillator, there corresponds a travelling wave (in the lattice) of the displacements of atoms from their equilibrium positions. An example of such a wave, travelling along a chain of atoms, will be given in exercise 4.

However, there exist the following differences between the set of oscillators for an electromagnetic field and those for a solid crystalline body.

1) The number of degrees of freedom for an electromagnetic field is infinite, so that it always contains all frequencies from zero to ∞. A solid body has a finite number of degrees of freedom equal to $3N$, where N is the number of atoms. Therefore, the range of vibrational frequencies extends from zero to some maximum frequency ω_{max}.

2) The dependence of frequency upon the wave vector of an electromagnetic field is defined by the simple law $\omega = ck$. In the oscillations of a solid body, the frequency depends upon the wave vector in a very complex manner. Only in the limit do the atomic vibrations become elastic vibrations of a continuous medium for very long waves (i.e., for small k), so that the atomic structure of the crystal can be ignored.

In a continuous medium the frequency is proportional to the wave vector $\omega = u_\sigma k$, where the index σ must denote that the wave velocity u depends upon its polarization. Here, as opposed to the electromagnetic field in an elastic body, there are three wave polarizations for each $\mathbf{k}$ (when taking into account the atomic structure of a crystal, waves of another type besides elastic sometimes occur: see Fig. 49). The directions of polarization depend upon the elastic properties of crystals and upon $\mathbf{k}$.

In an isotropic elastic body two of these polarizations are transverse, for which the velocity is equal to u_t, and one is longitudinal with velocity u_l, so that σ ranges through three values as in the crystal.

The number of oscillations occurring in a given interval of values $\mathbf{k}$ can be obtained, as usual, by proceeding from the relationship between the oscillation number and the wave vector. Each oscillation is defined by three integers (n_1, n_2, n_3) and σ. The wave vector has components proportional to n_1, n_2, n_3:

$$k_x = \frac{2\pi n_1}{a_1}; \quad k_y = \frac{2\pi n_2}{a_2}; \quad k_z = \frac{2\pi n_3}{a_3} \tag{42.23}$$

(a_1, a_2, a_3 are the crystal dimensions). From this,

$$dg_{\mathbf{k}} = dn_1\, dn_2\, dn_3 = \frac{a_1 a_2 a_3\, dk_x\, dk_y\, dk_z}{(2\pi)^3} = \frac{V\, dk_x\, dk_y\, dk_z}{(2\pi)^3}. \tag{42.24}$$

Comparing this formula with (25.22), we notice that now the denominator is $(2\pi)^3$, while before it was simply π_3. The difference is due to the fact that here the numbers n_1, n_2, n_3 are found from the periodicity condition, as for an electromagnetic field [see (27.4)], and the expansion is performed for travelling waves instead of standing waves. For this reason, an additional factor 2^3 appears in the denominator of (42.24) as compared with (25.18). But here the numbers n_1, n_2, n_3 range through all values from $-\infty$ to ∞, while in Sec. 25 they varied only from zero to ∞, filling one octant. Thus, the total number of states turns out to be the same, irrespective of the method used to count them—by travelling waves or by standing waves. And this is the way it should be [cf. (25.23)].

The energy of a solid body. It is now easy to write down an expression for the energy for an interval dk_x, dk_y, dk_z and for a given polarization of oscillations σ. Like in (42.3) we have

$$d\mathscr{E}(\mathbf{k}, \sigma) = \frac{V\hbar\omega_\sigma\, dk_x\, dk_y\, dk_z}{8\pi^3\left(e^{\frac{\hbar\omega_\sigma}{\theta}} - 1\right)}. \tag{42.25}$$

In order to find the total crystal energy, we must integrate this expression over all $dk_x\, dk_y\, dk_z$ and sum over σ. Unlike the case of an electromagnetic field, here the integration must not be performed to infinity, but only between limits such that the total number of oscillations equals the number of degrees of freedom $3N$ (because there are N atoms in the lattice and each one has three vibrational degrees of freedom):

$$V\sum_\sigma \iiint \frac{dk_x\, dk_y\, dk_z}{(2\pi)^3} = 3N. \tag{42.26}$$

In order to find out what is meant here by summation over σ, let us consider the possible types of vibration of a crystal lattice composed of atoms.

Two types of lattice vibration. If we confine ourselves only to crystal lattices of elements where there is only a single atom in an elementary cell, then the index σ will indeed range in value from 1 to 3. In reality, lattices are sometimes of more complex form, and the possible types of vibration become correspondingly more complicated. This can be illustrated by a simple example. Let there be two atoms in a cell, as shown in Fig. 49 by solid and open circles. The length d corresponds to one constant of the lattice. Let us imagine a vibration of some definite wavelength λ. In Fig. 49 $\lambda = 4\,d$ (a single half-wave is shown). This vibration may be effected in two ways: both atoms in the elementary cell are either displaced to one side (Fig. 49a), or in opposite directions (Fig. 49b). The second vibration corresponds to a greater frequency for a given wavelength than the first because the restoring force for the second vibration is greater.

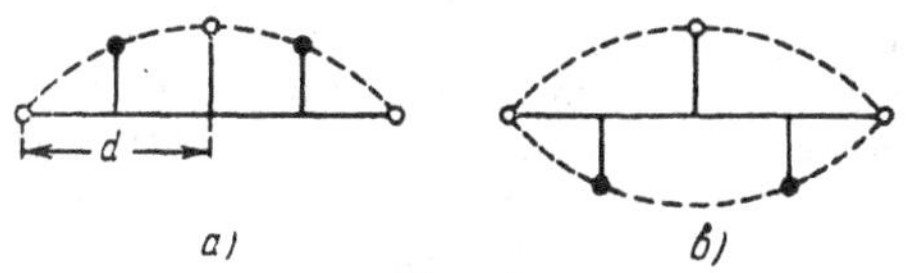

Fig. 49

If there are i atoms in an elementary cell then $3\,i$ types of vibration exist in the three-dimensional case. Three types correspond to the case (a) in Fig. 49, when all the i atoms are displaced in the same direction, and, in the limiting case of long wavelengths, the whole lattice vibrates like a continuous medium.

The total number of crystal vibrations is equal to $3\,iN' = 3\,N$, where N' is the number of elementary cells. Obviously, the number of vibrations is equal to the number of degrees of freedom, i.e., three times the number of atoms in the lattice.

Calculating the energy of a crystal lattice. (42.25) cannot be integrated in general form because the dependence of frequency upon **k** and σ is different for different lattices and for different types of vibration. We must therefore confine ourselves to two cases.

a) The temperature θ is considerably greater than the limiting-frequency quantum $\hbar\,\omega_{max}$. It is then all the more so greater than the other quanta, so that we can neglect all terms, except the first, in the exponential series

$$e^{\frac{\hbar\,\omega_\sigma}{\theta}} - 1 = \frac{\hbar\,\omega_\sigma}{\theta}.$$

Substituting this in (42.25), we obtain a simple expression for the lattice energy:

$$\overline{\mathscr{E}} = V\theta \sum_\sigma \int\int\int \frac{d\,k_x\,d\,k_y\,d\,k_z}{8\,\pi^3} = 3\,N\,\theta = 3\,R\,T. \qquad (42.27)$$

Here we have made use of the fact that the total number of lattice vibrations is equal to the number of its degrees of freedom $3N$. Hence, the specific heat of the lattice is equal to $3R$ and is the same for all elements in molar units. This law is well satisfied for very many elements already at room temperature (the Dulong and Petit law). Exceptions are, for example, diamond and beryllium, for which the large frequency ω_{max} is due to a relatively small atomic weight, since frequency is proportional to $M^{-\frac{1}{2}}$.

Expression (42.27) fits the general law for the temperature dependence of vibrational energy at high temperatures (41.13).

Very frequently, in a crystal lattice we can distinguish the molecules of the substance of which it is formed. We cannot, of course, draw a really strict distinction between atomic and molecular crystals but, qualitatively, this distinction is fully meaningful. In molecular crystals we can separately consider the motion of atoms inside molecules (purely vibrational, in the given case) and the motion of molecules as a whole relative to their equilibrium positions in the crystal. The latter correspond not only to definite centre-of-mass coordinates of the molecules in the lattice, but also to certain distinct orientations in space. Usually all the degrees of freedom of the motion of molecules in a crystal are vibrational. Solid hydrogen forms an exception where the molecules rotate almost freely (this rotation is similar to the rotation of a pendulum if its total energy is sufficient for transition through its upper position). The frequency spectrum for all vibrations, translational and rotational, consists of very many dispersion curves with different σ (according to the number of modes of vibration) and with its frequency dependent upon the wave vector. This spectrum is complicated by the vibrations of atoms inside molecules, similar to case (b) in Fig. 49. Since all the possible vibrations are excited by temperature increases, the dependence of specific heat upon temperature is of very complex form in molecular crystals.

b) The temperature is considerably less than $h\,\omega_{max}$. Then the factor

$$\left(e^{\frac{h\omega_{max}}{\theta}} - 1\right)^{-1}$$

is so small that integration can be taken to infinity without any essential error, because only the small frequencies, for which the quantum is of the order θ, contribute noticeably, i.e., $h\omega \sim \theta$. For large frequencies, the Planck factor $\left(e^{\frac{h\omega}{\theta}} - 1\right)^{-1}$ cancels the contributions of the corresponding vibrations.

However, in the case of small frequencies the lattice vibrations pass into the vibrations of a continuous medium, for which vibrations the frequency is related to the wave vector by the simple formula

$$\omega_\sigma = u_\sigma\left(\frac{\mathbf{k}}{k}\right) \cdot k. \tag{42.28}$$

The propagation velocity of such waves depends upon the direction of propagation and upon polarization, but does not depend upon the absolute value of k. The remaining types of vibration, whose frequency does not become zero for small k, are not excited at low temperature since the corresponding quanta are comparable with $h\,\omega_{max}$.

It is expedient to transform the volume element $dk_x\,dk_y\,dk_z$ to spherical coordinates, i.e., to replace it by the expression $k^2 dk\,d\Omega$, where $d\Omega$ is an element of solid angle for the directions $\mathbf{k}$. Here, in accordance with what has just been said, the integration with respect to k must be taken to infinity.

Thus, we obtain a formula for the total crystal energy at low temperature

$$\overline{\mathscr{E}} = \frac{Vh}{8\pi^3} \sum_{\sigma=1}^{3} \int u_\sigma\,d\Omega \int_0^\infty \frac{k^3\,dk}{e^{\frac{h u_\sigma k}{\theta}} - 1}. \tag{42.29}$$

The inner integral is taken in the same way as in the formula for the energy of an electromagnetic field (42.6), so that

$$\overline{\mathscr{E}} = \frac{\pi V}{120} \frac{\theta^4}{h^3} \int \left(\sum_\sigma \frac{1}{u_\sigma^3}\right) d\Omega. \tag{42.30}$$

Thus the energy of a crystal lattice is proportional to the fourth power of the absolute temperature, while the specific heat is proportional to the third power. This refers to temperatures considerably smaller than $h\,\omega_{max}$.

The Debye interpolation formula. P. Debye—the author of the theory of crystal specific heats at low temperatures which is set out here—proposed an interpolation formula for intermediate temperatures when the results (42.30) and (42.28) do not hold. The Debye formula reduces to both these formulae in the limiting cases of high and low temperatures. The intermediate interval is described qualitatively, but in certain agreement with experiment. In order to obtain the Debye formula, we suppose that the law

$$\omega = u_\sigma k$$

holds for all **k**, where u_σ is the usual propagation velocity of elastic waves. We may even take $u_1 = u_2 = u_t$, $u_3 = u_l$, where u_t and u_l are the velocities of transverse and longitudinal waves in a given substance in the polycrystalline state, which velocities are independent of the direction of propagation of the wave. We define the upper frequency limit ω_{max} from the condition that the total number of vibrations is equal to $3N$. For this we must go over to spherical coordinates in (46.26):

$$\frac{V}{2\pi^2}\sum_\sigma \frac{1}{u_\sigma^3}\int_0^{\omega_{max}} \omega^2\, d\omega = 3N, \tag{42.31}$$

or, changing to u_l, u_t, we have

$$\omega_{max} = \left(\frac{18\pi^2 N}{V\left(\frac{2}{u_t^3} + \frac{1}{u_l^3}\right)}\right)^{1/3}. \tag{42.32}$$

Condition (42.31) is selected so that at high temperatures the correct law $\overline{\mathscr{E}} = 3N\theta$ is automatically obtained. At medium temperatures $\theta \sim h\,\omega_{max}$; $k = \frac{\omega_{max}}{u_\sigma}$ is substituted as the upper limit in the integral (42.29) in place of ∞, so that the energy expression has the form

$$\overline{\mathscr{E}} = \frac{Vh}{2\pi^2}\left(\frac{2}{u_t^3} + \frac{1}{u_l^3}\right)\int_0^{\omega_{max}} \frac{\omega^2\, d\omega}{e^{\frac{h\omega}{\theta}} - 1}. \tag{42.33}$$

Changing to the integration variable $x = \frac{h\omega}{\theta}$ and denoting $h\omega_{max} = \theta_D$, we can rewrite the lattice energy thus:

$$\overline{\mathscr{E}} = \frac{V}{2\pi^2}\left(\frac{2}{u_t^3} + \frac{1}{u_l^3}\right)\frac{\theta^4}{h^3}\int_0^{\frac{\theta_D}{\theta}}\frac{x^3\,dx}{e^x - 1}. \tag{42.34}$$

At low temperature $\theta_D \gg \theta$, so that the upper limit in the integral is replaced by infinity. Then the integral is equal to $\frac{\pi^4}{15}$, and for the energy we have

$$\overline{\mathscr{E}} = \frac{V\pi^2}{30}\frac{\theta^4}{h^3}\left(\frac{2}{u_t^3} + \frac{1}{u_l^3}\right). \tag{42.35}$$

The exact formula (42.30) assumes the same form if we replace u_σ in it by u_t and u_l, which are independent of direction.

We shall now show how to determine θ_D from experimental data on specific heat and, independently, from elastic constants. The following values of specific heat C are known for tungsten (from the data of F. F. Lange): $T = 26.2°$K, C$=0.21$ cal/mol · deg.; T $= 38.9°$K, C$=0.75$ cal/mol · deg. The cube of the temperature ratio is equal to 3.37, and the ratio of specific heats is 3.58. We may assume that in the given temperature range the T^3 law for specific heat holds. Substituting $\theta_D = h\,\omega_{max}$ in formula (42.35), we determine ω_{max} with the aid of (42.32). This gives

$$C = \frac{12}{5}\pi^4 N\left(\frac{\theta}{\theta_D}\right)^3.$$

Converting this to heat units, we write

$$C = 232\,R\left(\frac{\theta}{\theta_D}\right)^3.$$

Here $R = 1.96$ cal/mol · deg. Substituting the specific heat at the lowest temperature, we find $T_D = 340°$.

We now determine T_D by proceeding from the elastic constants for tungsten. We have to give, without derivation, the formulae which connect u_t and u_l with the shear modulus and the bulk modulus for tungsten (see L. D. Landau and E. M. Lifshits, *The Mechanics of Continuous Media*, Gostekhizdat, 1953, p. 744 or A. Love. *A Treatise on the Mathematical Theory of Elasticity*, Ch. XIII, Cambridge, 1927).

$$u_l = \sqrt{\frac{K + \frac{4G}{3}}{\rho}}; \quad u_t = \sqrt{\frac{G}{\rho}}.$$

Here, K is the bulk modulus, which, for tungsten, is about 3.14×10^{12} dyne/cm^2 at low temperature. G is the shear modulus equal to 1.35×10^{12} dyne/cm^2. The density of tungsten is $\rho = 19.3$ gm/cm^2. Hence, $u_l = 5 \times 10^5$ cm/sec, $u_t = 2.64 \times 10^5$ cm/sec. For tungsten the ratio N/V is equal to 0.635×10^{23}. Whence, if we calculate it from (42.32), ω_{max} is equal to 4.61×10^{13} sec^{-1} and $T_D = \frac{4.61 \times 10^{13} \times 1.05 \times 10^{-27}}{1.38 \times 10^{-16}} = 352°$.

The agreement with what was obtained from specific heat turns out to be even better than could have been expected, because the elastic constants do not strictly refer to the temperature at which the specific heat was determined, and also because tungsten is a crystalline substance and its elastic properties are characterized by three moduli of elasticity instead of two (see Landau and Lifshits, *loc. cit.*, p. 675). For a number of substances we have the following values of Debye temperature T_D: Pb — 88°, Na — 172°, Cu — 315°, Fe — 453°, Be — 1,000°, diamond — 1,860° (all from absolute zero).

At high temperature, $\theta \gg \theta_D$, we must put $e^x - 1 \simeq x$, so that

$$\mathscr{E} = \frac{V}{6\pi^2}\left(\frac{2}{u_t^3} + \frac{1}{u_l^3}\right)\frac{\theta \cdot \theta_D^3}{h^3} = \frac{V}{6\pi^2}\left(\frac{2}{u_t^3} + \frac{1}{u_l^3}\right)\omega_{\max}^3\,\theta = 3N\theta\,, \qquad (42.36)$$

which is what we demanded.

For $\theta \approx \theta_D$, formula (42.34) agrees with experiment qualitatively. We note that we must not expect complete agreement, because the initial assumptions made in deriving this formula are not quantitative in character. It is not worth the attempt to make formula (42.34) more accurate, without taking into account the exact form of the dependence of ω upon $\mathbf{k}$. The attempts at correcting this formula, which are sometimes made, are simply in the nature of adjustments.

Exercises

1) Write down the formula for the wavelength distribution of black-body radiation energy. Proceeding from the fact that $\omega = \frac{2\pi c}{\lambda}$, we have

$$d\mathscr{E}(\lambda) = \frac{16\pi^2 hc\, V d\lambda}{\lambda^5\left(e^{\frac{2\pi hc}{\lambda\theta}} - 1\right)}\,.$$

The maximum is defined by the equation

$$\frac{2\pi hc}{\lambda_0\,\theta} = 4.965\,.$$

2) Show that if Bose particles interact with a Boltzmann gas the probability of a particle appearing in a certain state is proportional to $n + 1$, where n is the number of particles already in that state, and the probability of a particle disappearing is n.

Let the energy of a Boltzmann particle be ε and that of a Bose particle, η. Let us consider the process in which there occurs the transition

$$\varepsilon + \eta \rightarrow \varepsilon' + \eta'$$

i.e., the interaction of these particles changes their initial state with energies ε, η to a state with energies ε' and η'. In statistical equilibrium we must observe the balance

$$W_{\varepsilon\varepsilon'} N_\varepsilon n_\eta (1 + n_{\eta'}) = W_{\varepsilon'\varepsilon} N_{\varepsilon'} (1 + n_\eta) n_{\eta'},$$

where $W_{\varepsilon\varepsilon'}$ is the probability of direct transition and $W_{\varepsilon'\varepsilon}$ is the reverse-transition probability. Putting

$$N_\varepsilon = g_\varepsilon e^{\frac{\mu - \varepsilon}{\theta}}, \qquad N_{\varepsilon'} = g_{\varepsilon'} e^{\frac{\mu - \varepsilon'}{\theta}},$$

$$n_\eta = \left(e^{\frac{\eta - \mu_1}{\theta}} - 1\right)^{-1}, \qquad n_{\eta'} = \left(e^{\frac{\eta' - \mu_1}{\theta}} - 1\right)^{-1},$$

we see that the balance equation is satisfied if $W_{\varepsilon\varepsilon'} = W_{\varepsilon'\varepsilon}$. For simplicity we have put $g_\varepsilon = g_{\varepsilon'}$. The presence of spontaneous emission is due to the Bose distribution.

3) Find the total number of quanta in black-body radiation at a given temperature

$$N = \frac{V}{\pi^2 c^3} \int_0^\infty \frac{\omega^2 \, d\omega}{e^{\frac{h\omega}{\theta}} - 1} = \frac{V\theta^3}{\pi^2 h^3 c^3} \int_0^\infty \frac{x^2 \, dx}{e^x - 1}.$$

Further (see Appendix),

$$\frac{1}{e^x - 1} = \sum_{n=1}^\infty e^{-nx}.$$

Hence,

$$\int_0^\infty \frac{x^2 \, dx}{e^x - 1} = \sum_{n=1}^\infty \int_0^\infty e^{-nx} x^2 \, dx = \sum_{n=1}^\infty \frac{1}{n^3} \int_0^\infty e^{-y} y^2 \, dy = 2 \sum_{n=1}^\infty \frac{1}{n^3}.$$

The sum is approximately 1.2, so that

$$N = \frac{2.4}{\pi^2} \frac{V\theta^3}{h^3 c^3}.$$

4) The atoms are situated in the form of a linear chain. We shall denote the displacement of the nth atom by a_n. The force acting between the nth and $(n+1)$th atoms is equal to $\alpha(a_{n+1} - a_n)$. Find the equations for the vibrations of the chain. Ignore the interaction between the more distant neighbours.

The vibration equation for the nth atom is

$$m\ddot{a}_n = \alpha(a_{n+1} + a_{n-1} - 2a_n).$$

We look for a_n in the form

$$a_n = b(t) e^{ifn}.$$

Substituting this in the initial equation, we find, after cancelling e^{ifn}

$$m\ddot{b}(t) = \alpha b(t)(e^{if} + e^{-if} - 2) = 2\alpha b(t)(\cos f - 1) = -4\alpha \sin^2 \frac{f}{2} \cdot b(t),$$

so that the oscillation frequency for a given value of f is

$$\omega_f = 2\sqrt{\frac{\alpha}{m}} \left| \sin \frac{f}{2} \right|.$$

If the distance between the atoms is d then $n = \frac{x}{d}$, where x is the equilibrium position of the nth atom. Putting $\frac{f}{d} = k$, we have $e^{ifn} = e^{ikx}$, so that f can be called the wave vector, considering that the length is measured in units of d. For small f, as was asserted, the frequency is proportional to f:

$$\omega_f = \sqrt{\frac{\alpha}{d}} \cdot f.$$

Sec. 43. Bose Distribution

The choice of sign of μ. The Bose distribution has very peculiar properties at low temperatures. We shall suppose that the atoms do not have spin; such, for example, are helium atoms with atomic weight 4. Both the electrons in the cloud of the helium atom and the protons and neutrons in the helium nucleus are in the 1s-state. They all go in pairs and by the Pauli principle the spins are antiparallel. Therefore, the resultant spin is zero.

From (39.30), the weight of the state of a spinless particle is

$$dg(\varepsilon) = \frac{V m^{3/2}}{2^{1/2}\pi^2} \frac{\sqrt{\varepsilon}\, d\varepsilon}{h^3}. \tag{43.1}$$

The normalization condition (39.23) looks like

$$\frac{V m^{3/2}}{2^{1/2}\pi^2 h^3} \int_0^\infty \frac{\sqrt{\varepsilon}\, d\varepsilon}{e^{\frac{\varepsilon-\mu}{\theta}} - 1} = N. \tag{43.2}$$

This condition can be satisfied only for negative μ. Indeed, if we suppose that μ is greater than zero, then the denominator of the integrand will be negative for $\varepsilon < \mu$ because then $e^{\frac{\varepsilon-\mu}{\theta}} < 1$. But this is impossible because the distribution function is, by its very meaning, a positive quantity.

Hence, $\mu < 0$. At high temperatures the Bose distribution passes into the Boltzmann distribution in accord with (40.6).

The sign of $\frac{d\mu}{d\theta}$. As the temperature diminishes, μ decreases in absolute value. This can be shown generally with the aid of (43.2). Differentiating this equation as an implicit function we have

$$\frac{\partial\mu}{\partial\theta} = -\frac{\frac{\partial}{\partial\theta}\int_0^\infty \frac{\sqrt{\varepsilon}\, d\varepsilon}{e^{\frac{\varepsilon-\mu}{\theta}} - 1}}{\frac{\partial}{\partial\mu}\int_0^\infty \frac{\sqrt{\varepsilon}\, d\varepsilon}{e^{\frac{\varepsilon-\mu}{\theta}} - 1}} = -\frac{\int_0^\infty \frac{\varepsilon-\mu}{\theta^2} \frac{e^{\frac{\varepsilon-\mu}{\theta}}\sqrt{\varepsilon}\, d\varepsilon}{\left(e^{\frac{\varepsilon-\mu}{\theta}} - 1\right)^2}}{\int_0^\infty \frac{1}{\theta} \frac{e^{\frac{\varepsilon-\mu}{\theta}}\sqrt{\varepsilon}\, d\varepsilon}{\left(e^{\frac{\varepsilon-\mu}{\theta}} - 1\right)^2}}. \tag{43.3}$$

The integrands in (43.3) are essentially positive quantities [$(\varepsilon - \mu) > 0$, because $\mu < 0$], and therefore $\frac{\partial\mu}{\partial\theta} < 0$. Hence, as θ decreases, the absolute value $|\mu|$ diminishes monotonically since μ must increase.

We shall now show that μ becomes zero at a temperature other than zero. To do this we put $\mu = 0$ in (43.2) and find the corresponding value $\theta = \theta_0$:

$$\frac{V m^{3/2}}{2^{1/2} \pi^2 h^3} \int_0^\infty \frac{\sqrt{\varepsilon}\, d\varepsilon}{e^{\frac{\varepsilon}{\theta_0}} - 1} = \frac{V m^{3/2} \theta_0^{3/2}}{2^{1/2} \pi^2 h^3} \int_0^\infty \frac{\sqrt{x}\, dx}{e^x - 1} = N. \tag{43.4}$$

The integral simply represents an abstract quantity: it is equal to ≈ 2.31 (see Appendix). Therefore equation (43.4) is satisfied by a value of θ_0 that is different from zero.

Bose condensation. What will happen when the temperature is reduced further? μ cannot go from negative to positive values since, as we have shown at the beginning of the section, this would lead to negative probability values. μ cannot become negative once again, because $\frac{\partial \mu}{\partial \theta}$ is always less than zero so that μ varies only monotonically, if it is at all capable of varying. Therefore, the only possibility is for μ to remain equal to zero after it has once attained its zero value. But then equation (43.2) is no longer satisfied if the temperature is less than θ_0, and N does not change. On the contrary, it can be seen from (43.4) that if we define the number of particles as

$$N' = \frac{V m^{3/2}}{2^{1/2} \pi^2 h^3} \int_0^\infty \frac{\sqrt{\varepsilon}\, d\varepsilon}{e^{\frac{\varepsilon}{\theta}} - 1} = \frac{2.31\, V m^{3/2} \theta^{3/2}}{2^{1/2} \pi^2 h^3} \tag{43.5}$$

for $\theta < \theta_0$, it decreases with the temperature in proportion to $\theta^{3/2}$. What happens to the remaining particles which number $N - N'$? As opposed to light quanta these particles cannot be absorbed. Therefore, they will pass into a state which is not taken into account in the normalizing integral (43.2). The only state of this kind possesses an energy equal to zero: due to factor $\sqrt{\varepsilon}$ it does not contribute anything to the integral (43.4). In normalization we can isolate the particles occurring in the zero state in a separate term. If a finite number of particles go to the zero-energy state, they will naturally fall out of the integral. N' particles remain continuously distributed, but with the value $\mu = 0$. Thus, at a temperature $\theta < \theta_0$, the whole distribution consists of an infinitely narrow "peak" at $\varepsilon = 0$ and of particles distributed according a $\left(e^{\frac{\varepsilon}{\theta}} - 1\right)^{-1}$ law. At absolute zero all the particles are in a zero state: this state of a Bose gas is obviously defined uniquely. It will be noted that a Boltzmann gas would behave in an entirely different way when the temperature tended to zero.

Liquid helium. Helium with atomic weight 4 obeys Bose statistics since the spin of its nuclei and of the electronic shells is equal to zero. It is therefore interesting to see whether anything like this "Bose condensation" is observed in helium.

It is difficult to give a unique answer because at low temperature helium is a liquid, and the Bose distribution, which relates to an ideal gas, does not apply. Nevertheless the qualitative aspect of the result obtained for a gas may still hold. Namely, it may be supposed that at a certain temperature part of the gas will pass into a zero energy state and, accordingly, will not contribute to the specific heat.

Liquid helium does, in fact, experience a peculiar change of state at a temperature of 2.19° K (at atmospheric pressure). Speaking of a monatomic liquid, which is what liquid helium is, it is difficult to imagine any change of state related to a rearrangement of the atoms in space. Therefore, it is interesting to compare the actual temperature of transition in liquid helium with the temperature at which Bose condensation would occur in gaseous helium of the same density.

The density of liquid helium is equal to 0.12 gm/cm³. Whence the ratio $\frac{N}{V} = \frac{0.12}{4} \times 6 \times 10^{23} = 0.18 \times 10^{23}$. Consequently, according to (43.4) the temperature θ_0 is

$$\theta_0 = \left(\frac{0.18 \cdot 10^{23} \cdot 9.86 \cdot 1.41 \cdot 1.18 \cdot 10^{-81}}{2.31 \cdot 17.1 \cdot 10^{-36}}\right)^{2/3} = 3.86 \cdot 10^{-16};$$

$$T_0 = \frac{3.86 \cdot 10^{-16}}{1.38 \cdot 10^{-16}} = 2.8° ,$$

which is close to the transition temperature. At the transition, the specific heat of helium experiences a discontinuity. In the case of a Bose gas, only the derivative of the specific heat with respect to temperature has a discontinuity.

Superfluidity. P. L. Kapitsa discovered that below the temperature of phase transition, liquid helium possesses a most remarkable property: it is capable of passing through the finest slit without exhibiting any signs of viscosity. This property was called superfluidity.

L. D. Landau developed a theory of superfluidity proceeding from the supposed quantum-level spectrum for a liquid. On the basis of this theory he built the hydrodynamics of a superfluid, which differs from conventional hydrodynamics in that each point possesses two velocities instead of one: a normal and a superfluid component. The occurrence of two velocities means that in a superfluid two types of sound vibrations may be propagated: ordinary sound, in which pressure and density oscillate, and "second sound," which is connected with the relative motion of the normal and superfluid components. The second sound was demonstrated in an experiment carried out by V. P. Peshkov using a method proposed by E. M. Lifshits. The experimentally found velocity of second sound (which is small compared with the velocity of conventional sound) is in excellent agreement with Landau's theory.

The question of the relationship between superfluidity and Bose condensation cannot be considered fully resolved. It may be suggested that the superfluid component corresponds to that part of the helium which has passed to the zero state. This hypothesis is strongly supported by the fact that the liquid isotope of helium with atomic weight 3 is not superfluid: the nuclear spin of helium 3 is equal to $\frac{1}{2}$, so that its atoms are subject to the statistics of Fermi and not Bose. Accordingly, they cannot all pass into the zero state together: the Pauli principle does not permit this.

N. N. Bogolyubov showed that a gas which is close to an ideal gas and consists of Bose particles possesses an energy spectrum which, according to Landau's theory, a superfluid liquid should have. However, no one has so far succeeded in proving theoretically that it is precisely liquid helium below the transition point that should possess such a spectrum.

Exercise

Calculate the energy and pressure of a Bose gas below the transition point. For the energy we have

$$\mathscr{E} = \frac{V m^{3/2}}{2^{1/2} \pi^2 h^3} \theta^{5/2} \int_0^\infty \frac{x^{3/2}\,dx}{e^x - 1} = \frac{1.78\, V m^{3/2} \theta^{5/2}}{2^{1/2} \pi^2 h^3}$$

(see Appendix). The pressure is determined from the general relationship (40.22):

$$p = \frac{2}{3} \frac{\mathscr{E}}{V} = \frac{1.18\, m^{3/2} \theta^{5/2}}{2^{1/2} \pi^2 h^3}.$$

Thus, the pressure of a Bose gas below the transition point is independent of volume and depends only upon the temperature. If we compress such a Bose gas its particles will go to the zero-energy state. Conversely, upon expansion the particles will come out of the zero-energy state until there are none left. If expansion continues the pressure will begin to decrease.

Sec. 44. Fermi Distribution

The form of the Fermi-distribution curve and its interpretation. The criterion for the transition from quantum statistics to classical statistics is that [see (40.7)]

$$\frac{N}{V} \ll \frac{g_{(0)}}{h^3} \left(\frac{m\theta}{2\pi}\right)^{3/2}.$$

If the inequality is reversed, then essentially quantum properties of the statistical distribution appear. In this section we shall consider the properties of the Fermi distribution when the inverse inequality

$$\frac{N}{V} \gg \frac{g_{(0)}}{h^3}\left(\frac{m\theta}{2\pi}\right)^{3/2} \tag{44.1}$$

or an equivalent inequality

$$\frac{\mu}{\theta} \gg 1 \tag{44.2}$$

is satisfied.

From (39.26) and (39.30), the Fermi-distribution curve is of the following form:

$$dn(\varepsilon) = \frac{V(2m^3)^{1/2}\varepsilon^{1/2}d\varepsilon}{\pi^2 h^3}\,\frac{1}{e^{\frac{\varepsilon-\mu}{\theta}}+1}\,. \tag{44.3}$$

Here, a weight factor 2 is introduced, since we have put $j=\frac{1}{2}$. The first factor in (44.3) represents the total number of states between ε and $\varepsilon+d\varepsilon$, while the second factor represents the probability that these states are occupied. We can interpret the function

$$f(\varepsilon) = \frac{1}{e^{\frac{\varepsilon-\mu}{\theta}}+1} \tag{44.4}$$

as a probability and as the mean number of particles, because $f(\varepsilon)$ is contained between zero and unity. A similar function in the Bose distribution could only denote the mean number of particles in one of the quantum states with a given energy, because the Bose-distribution function $\left(e^{\frac{\varepsilon-\mu}{\theta}}-1\right)^{-1}$ is sometimes even greater than unity and must not be interpreted as a probability.

Let us see how the curve $f(\varepsilon)$ behaves when $\frac{\mu}{\theta}\gg 1$. When $\varepsilon=0$ we obtain

$$f(0) = \frac{1}{e^{-\frac{\mu}{\theta}}+1} \simeq 1,$$

because $e^{-\frac{\mu}{\theta}}$ is a small number. The quantity $e^{\frac{\varepsilon-\mu}{\theta}}$ is also a small number as long as ε remains smaller than μ, while $f(\varepsilon)$ is close to unity, like $f(0)$. Only when $\varepsilon-\mu$ is comparable with θ, is $e^{\frac{\varepsilon-\mu}{\theta}}$ of the order of unity, so that $f(\varepsilon)$ begins to decrease noticeably with further increase of ε. For $\varepsilon=\mu$, $f(\mu)$ decreases to $\frac{1}{2}$:

$$f(\mu) = \frac{1}{e^0+1} = \frac{1}{2}\,.$$

For still greater values of ε, $f(\varepsilon)$ decreases exponentially because unity can then be neglected in the denominator, and, for $\varepsilon \gg \mu$, $f(\varepsilon)$ becomes the Boltzmann distribution

$$f(\varepsilon) \sim e^{\frac{\mu-\varepsilon}{\theta}}$$

The Bose distribution also has the same limiting form. The curve $f(\varepsilon)$ is roughly shown in Fig. 50. The region ε, where $f(\varepsilon)$ changes from unity to zero, has a width of the order θ, since $e^{\frac{\varepsilon-\mu}{\theta}}$ is comparable with unity only if $\varepsilon - \mu \sim \theta$: for smaller ε the exponential is considerably smaller than unity, while for larger ε the exponential is considerably greater than unity.

Fermi distribution at absolute zero. We shall call the region of transition of f from unity to zero the *spread* region of Fermi distribution. As the temperature decreases the spread region narrows and, at absolute zero, becomes a sharp discontinuity f, so that the distribution function takes the form of a right angle. Fig. 50 shows this step by a broken line. The value of μ at absolute zero is called μ_0. Hence, at $\theta = 0$, all states with energy less than μ_0 are occupied with unity probability (i.e., with certainty), while those with energy greater than μ_0 are empty, also with certainty.

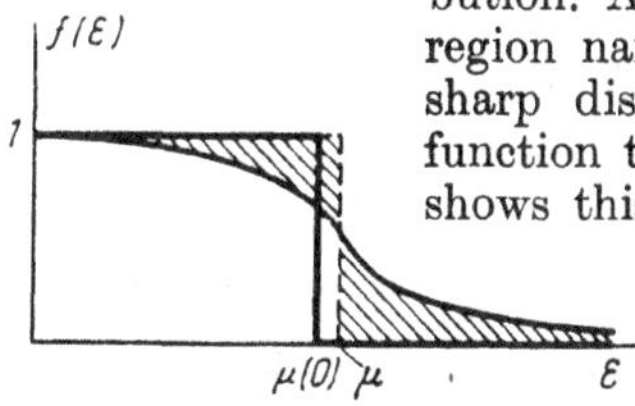

Fig. 50

This result can likewise be obtained directly from Pauli's principle without resorting to statistics. From (39.32), a definite interval of momentum-component values Δp_x, Δp_y, Δp_z corresponds to one state of particle motion. If the particle is contained in a box with sides a_1, a_2, a_3, then it follows from the uncertainty relation (23.4) that

$$\Delta p_x \sim \frac{2\pi h}{a_1}; \quad \Delta p_y \sim \frac{2\pi h}{a_2}; \quad \Delta p_z \sim \frac{2\pi h}{a_3},$$

since these quantities show by how much the momentum components of two particles must differ in order that the particles may be regarded as occurring in different states of motion.

This follows not only from the uncertainty relation, but can also be seen strictly when computing the states leading to formulae (25.23) and (39.32). Here, each state must be identified not with the volume of the parallelepiped, but with one of its vertices whose coordinates are given by the three integers n_1, n_2, n_3. The coefficient 2π in the uncertainty relations is taken so that both definitions for the number of states agree.

If we plot p_x, p_y, p_z on coordinate axes, then to each state of spatial motion of the electron there correspond three quantum numbers

n_1, n_2, n_3. These quantum numbers specify the number of the parallelepiped with sides Δp_x, Δp_y, Δp_z. It is shown in Fig. 51. All the space in which the axes p_x, p_y, p_z are drawn can be filled with such boxes. Since three quantum numbers correspond to a single box and, in addition, the state is also given by the spin, there may be two particles with spin $\frac{1}{2}$ having momentum projections in the same interval Δp_x, Δp_y, Δp_z. The spins of these two particles are antiparallel.

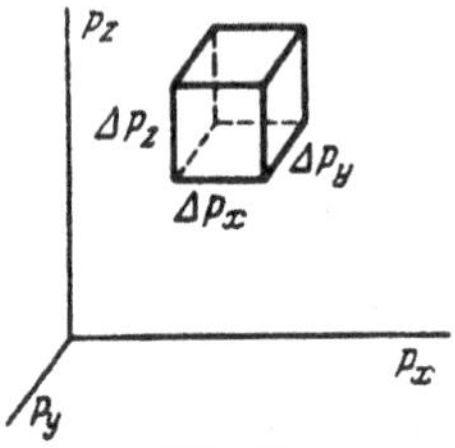

Fig. 51

Thus, the space p_x, p_y, p_z may be divided into boxes or cells with dimensions

$$\Delta p_x \Delta p_y \Delta p_z = \frac{(2\pi h)^3}{a_1 a_2 a_3} = \frac{(2\pi h)^3}{V}, \tag{44.5}$$

where there are no more than two particles in each cell.

The closer the cell to the coordinate origin, the less the energy it possesses, because the energy is equal to $\varepsilon = \frac{1}{2m}(p_x^2 + p_y^2 + p_z^2)$. In other words, it is proportional to the square of the distance of the cell from the origin.

Let us now consider the state of a gas at the absolute zero of temperature. If the gas consisted of only two particles, then at absolute zero the states of both particles would fill the cell closest to the origin. In accordance with the Pauli principle, the next two particles cannot enter the same cell: they are forced to take up positions further from the origin. As the number of particles increases, cells are filled which are situated further and further from the origin; but each time two particles are added they fall into a free cell closest to the origin, because, by definition, absolute zero corresponds to the least possible energy of the gas as a whole.

If there are very many particles, their cells will densely fill a sphere whose centre is the coordinate origin. All states inside the sphere are filled with unity probability, while those outside the sphere are free—also with certainty.

The limiting energy of Fermi distribution. If we denote the energy corresponding to the boundary of the sphere by ε_0, then it can be seen from Fig. 50 that $\varepsilon_0 = \mu_0$. μ_0 is the limiting energy of a particle at absolute zero. It is very easy to calculate ε_0 or μ_0. Since at absolute zero the function $f(\varepsilon)$ is equal to unity for all $\varepsilon < \mu_0$, the total number of particles N is, from (44.3),

$$N = \frac{V(2m^3)^{1/2}}{\pi^2 h^3} \int_0^{\varepsilon_0} \sqrt{\varepsilon}\, d\varepsilon = \frac{V(2m)^{3/2} \varepsilon_0^{3/2}}{3\pi^2 h^3}, \tag{44.6}$$

whence

$$\varepsilon_0 = 3^{2/3}\pi^{4/3}\frac{h^2}{2m}\left(\frac{N}{V}\right)^{2/3}. \tag{44.7}$$

The same can be seen without the aid of $f(\varepsilon)$. Indeed, the radius of the sphere of greatest energy is

$$p_0 = \sqrt{2m\varepsilon_0}\,.$$

Its volume is

$$\frac{4}{3}\pi p_0^3 = \frac{4\pi}{2}(2m\varepsilon_0)^{3/2}.$$

But this same quantity is equal to the number of filled elementary cells (with two particles per cell) multiplied by the volume of a single cell $\frac{(2\pi h)^3}{V}$. Consequently,

$$\frac{4}{3}\pi(2m\varepsilon_0)^{3/2} = \frac{N}{3}\frac{(2\pi h)^3}{V}, \tag{44.8}$$

whence equation (44.7) is again obtained.

At absolute zero the state of a Fermi gas as a whole is defined uniquely: in quantum statistics it is necessary to indicate which *states* are occupied by separate particles, but it is impossible to determine by which *particles* they are filled. In the given case all the states inside the sphere with limiting energy ε_0 are filled by particles.

The criterion for the closeness of the Fermi distribution to the distribution at absolute zero (based on the form of the distribution). At a temperature close to absolute zero thermal excitation can be imparted only to those particles whose energy is close to $\varepsilon_0 = \mu_0$. Indeed, as long as $\theta \ll \varepsilon_0$, a thermal excitation of the order θ cannot be imparted to a particle whose cell lies deep beneath the surface $\varepsilon = \varepsilon_0$, because the states between the surface and the given cell are occupied, and the energy θ is insufficient to remove the particle beyond the limits of the surface boundary. Therefore, only those particles whose energy differs from ε_0 by an amount of the order of θ can take up free places. Deeper states will remain densely filled as before. Thus, the filling probability will be almost equal to unity for all energies $\varepsilon < \varepsilon_0$, and will fall to zero in a region of the order of θ close to $\varepsilon \sim \varepsilon_0$, as shown in Fig. 50.

The criterion that the curve is close to the step is the inequality

$$\theta \ll \varepsilon_0\,, \tag{44.9}$$

and this agrees with (44.1) within the accuracy of the numerical factor. As we shall soon see, the concept of "closeness" of temperature to absolute zero, according to the criterion (44.9), differs greatly from conventional.

Electrons conducting electricity in metals are usually considered as an ideal gas. The main basis for this is the fact that we as yet have no better theoretical model. It has not been possible to consider electrostatic interactions between electrons sufficiently fully to obtain quantitative results that might compare with experiment. This is why the phrase "electron gas" in metals is used. In many cases the conclusions from such a model are in good agreement with experiment.

Without considering the electron theory of metals, we shall take the electron gas only as an example in which condition (44.9) is satisfied. Let us suppose that there is one conduction electron for each atom. This assumption appears to be satisfied for alkali metals, in which the outer electron is weakly bound and is separated from the atom in a lattice.

Let us find ε_0 for the electron gas in metallic sodium. The density of sodium is 0.97 and atomic weight 23. Hence, unit volume contains

$$\frac{0.97}{23} \cdot 6.02 \cdot 10^{23} = 0.25 \cdot 10^{23}$$

atoms and as many conduction electrons. Whence, from (44.7),

$$\varepsilon_0 = 2.1 \cdot 4.6 \frac{1.12 \cdot 10^{-54}}{1.8 \cdot 10^{-27}} \cdot 0.08 \cdot 10^{-16} = 4.8 \cdot 10^{-12}$$

[the sequence of the numbers is the same as in (44.7)]. In degrees ε_0 is 34,800. Hence, at all temperatures for which we can speak of sodium as a metal, the electron gas in it is close to a Fermi gas at absolute zero. Similar results are also obtained for non-alkali metals, though with a less reliable value of electron density.

The compressibility of alkali metals. Let us derive a formula for the compressibility of a Fermi gas at absolute zero. From (44.6), the energy at absolute zero is

$$\mathscr{E} = \int_0^{\varepsilon_0} \varepsilon dg(\varepsilon) = \frac{V(2m)^{3/2}}{5\pi^2 h^3} \varepsilon_0^{5/2}. \tag{44.10}$$

In accordance with the Bernoulli equation (40.22), the pressure is equal to two thirds the energy density, i.e.,

$$p = \frac{2}{15} \frac{(2m)^{3/2}}{\pi^2 h^3} \varepsilon_0^{5/2} = \frac{3^{2/3}\pi^{4/3}}{5} \frac{h^2}{m} \left(\frac{N}{V}\right)^{5/3}. \tag{44.11}$$

Whence

$$-\frac{\partial \ln V}{\partial p} = \frac{3}{5p} = \frac{3^{1/3}}{\pi^{4/3}} \frac{m}{h^2} \left(\frac{N}{V}\right)^{-5/3} = 0.273 \times 10^{27} \left(\frac{N}{V}\right)^{-5/3} \text{bar}^{-1}. \tag{44.12}$$

Ya. I. Frenkel noted that the compressibility of alkali metals is close to the compressibility of an electron gas.

Indeed, expressing N/V in terms of atomic weight and density, we obtain the following table:

	Li	Na	K	Rb	Cs
$-\frac{1}{V}\frac{\partial V}{\partial p} \times 10^6$ from equation (44.12)	4.7	13	37	52	79
$-\frac{1}{V}\frac{\partial V}{\partial p} \times 10^6$ from experimental data	8	15	32	40	61

In a crystal lattice there are, of course, not only forces of repulsion between particles, but also cohesive forces. The equilibrium of these forces with the forces of repulsion determines the characteristic volume which every condensed body, solid, or liquid has in the absence of external pressure. Ordinary atmospheric pressure gives a force which is negligibly small compared with these tremendous forces that keep bodies in their volumes. In order to change the volume of a body by only one per cent, pressures are required in the order of tens of thousands of atmospheres.

The coincidence of theoretical and experimental data indicates that when alkali metals are compressed the cohesive forces change insignificantly compared with the forces of repulsion. It is even conceivable that the state of the valence electrons in alkali metals is perturbed to a comparatively small degree by the atomic residues, and, to some extent, is close to an electron gas. Compression affects but little the electronic shells of the atomic residues, and therefore the compressibility of alkali metals is close to the compressibility of an ideal Fermi gas. That this should be so is, of course, not at all obvious beforehand.

Paramagnetism of alkali metals. According to Pauli, the paramagnetism of alkali metals can also be explained on the basis of the concept of a free electron gas.

If we place a Fermi gas (consisting of electrons) in a magnetic field, the energy of the electrons, whose spins are parallel to the field, will be equal to $-\beta H$, while the energy of electrons with opposite direction of spin will be equal to βH*. Therefore, if those electrons whose spin is antiparallel to the field reverse their spin directions, then the energy of the gas must decrease. But all the places inside the limiting-energy sphere are occupied; so for an electron to change its spin direction it must come out of the sphere into a free cell. But this increases its kinetic energy. Equilibrium is established between electrons with spins parallel and antiparallel to the field when their total energies become equal. Indeed, if there occurred a further transition of electrons into a state with spin parallel to the field, the increase in their

* Here the Bohr magneton is denoted by β instead of μ, so as to avoid confusion with the distribution parameter μ.

kinetic energy could not be compensated by a reduction in magnetic energy.

Let there be n electrons which have changed their spin directions. Then there remain $\frac{N}{2} - n$ electrons with spin antiparallel to the field, while $\frac{N}{2} + n$ have spins parallel to the field. The limiting energies are determined from formula (44.8), where we must put $\frac{N}{2} \pm n$ instead of $\frac{N}{2}$. Whence we obtain the following expression for the limiting kinetic energy of both types of electrons:

$$(\varepsilon_0)_\pm = \left[\frac{3}{4\pi V}\left(\frac{N}{2} \pm n\right)\right]^{2/3} \frac{(2\pi h)^2}{2m}, \tag{44.13}$$

and the equation for the total limiting energies is

$$\left(\frac{3}{4\pi V}\right)^{2/3} \frac{(2\pi h)^2}{2m}\left(\frac{N}{2} + n\right)^{2/3} - \beta H = $$
$$= \left(\frac{3}{4\pi V}\right)^{2/3} \frac{(2\pi h)^2}{2m}\left(\frac{N}{2} - n\right)^{2/3} + \beta H. \tag{44.14}$$

Since $n \ll \frac{N}{2}$, the binomials can be expanded in a series as follows:

$$\left(\frac{N}{2} \pm n\right)^{2/3} = \left(\frac{N}{2}\right)^{2/3} \pm \left(\frac{N}{2}\right)^{2/3} \frac{4n}{3N}.$$

Substituting this in (44.14), we find the number of electrons which change their spin directions in the magnetic field:

$$n = N\beta H \frac{3^{1/3}}{2\pi^{4/3}} \frac{m}{h^2} \left(\frac{V}{N}\right)^{2/3}. \tag{44.15}$$

Each of these electrons contributes a term 2β to the total magnetic moment of the whole gas, because its moment projection on the magnetic field has changed from $-\beta$ to β. The magnetic polarization (that is, the magnetic moment of unit volume) turns out equal to

$$M = 2\beta\frac{n}{V} = \frac{3^{1/3}}{\pi^{4/3}} \frac{m\beta^2}{h^2} \left(\frac{N}{V}\right)^{1/3} H, \tag{44.16}$$

while the magnetic polarizability α, defined as the coefficient of H on the right-hand side of this formula, depends only upon the density of the electron gas and not its temperature:

$$\alpha = \frac{3^{1/3}}{\pi^{4/3}} \frac{m\beta^2}{h^2} \left(\frac{N}{V}\right)^{1/3}. \tag{44.17}$$

Indeed, alkali metals have a paramagnetism independent of temperature. Let it be recalled that in accordance with the results of Sec. 40 [see (40.53)] atomic paramagnetism gives a magnetic polarizability which is inversely proportional to the temperature. Formula (44.17) agrees satisfactorily with experiment.

Diamagnetism of electrons. L. D. Landau has shown that the quantized motion of electrons in a magnetic field—this motion is similar to their classical motion in a spiral—leads, in a weak field, to the appearance of a magnetic moment equal to 1/3 of expression (44.16), and of opposite sign. The nature of this effect is purely quantum: if we regard the motion of electrons as classical then the additional magnetic moment becomes identically zero (see Sec. 46, exercise 13).

If βH is of the order of θ, then the polarizability does not depend monotonically upon the field and exhibits much oscillation as the field increases. The oscillatory variation of magnetic properties is, in fact, observed in very many metals.

The potential distribution in an atom. We shall now show how to find the general form for the electron-density distribution in atoms via the notion of a Fermi gas. To a certain approximation, the electrons in heavy atoms resemble a Fermi gas. However, it must be noted that each electron occurs in the inhomogeneous electric field formed by the nucleus and the entire configuration of the remaining electrons.

Let us first of all consider a Fermi gas at absolute zero in a potential field of the form shown in Fig. 52. $U=0$ for $0<x<a$, $U=U_0$ for

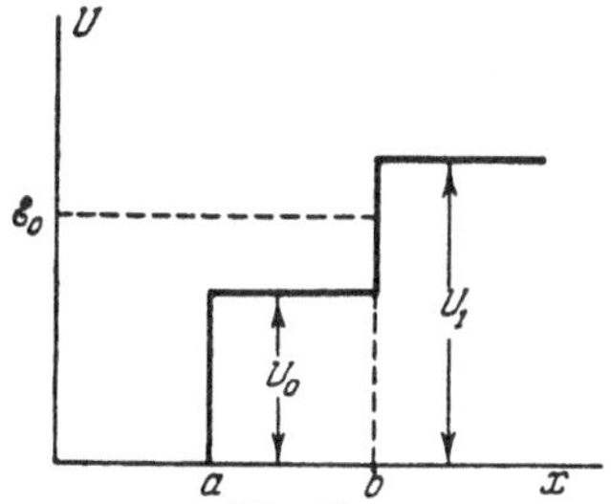

Fig. 52

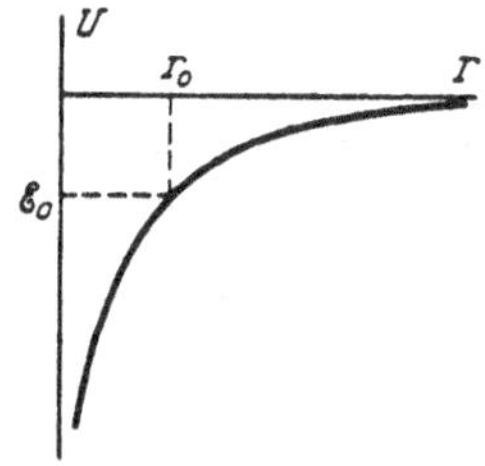

Fig. 53

$a<x<b$, $U=U_1$ for $x>b$. Then the limiting energy of the electrons ε_0 must be the same for $0<x<a$ and for $a<x<b$, because otherwise the electrons will pass into a region of lesser limiting energy according, as it were, to the law of communicating vessels. Of course, as a result of this the total energy will diminish. But the energy of a gas at absolute zero is the least possible, so that the limiting energy must be the same in any part of the gas.

The potential energy distribution in an atom is approximately as shown in Fig. 53. The potential energy is everywhere negative because we have taken it to be zero at infinity.

The limiting energy of electrons must not be positive anywhere, because electrons with a positive total energy could leave the atom for infinity. The limiting energy cannot anywhere be less than the potential energy. We shall show that it must be equal to zero. If, for example, it corresponded to the dashed line in Fig. 53, the electron density would become zero at the point $r=r_0$. But then the electric field would be zero for all values $r>r_0$, because in the case of spherical symmetrical charge distribution the action of all the electrons of a neutral atom balances the action of the nucleus. Accordingly, the potential would also be zero when $r=r_0$, because potential is the field integral: $\varphi=\int_\infty^r E\,dr$, i. e., the integral with integrand equal to zero at $r \geqslant r_0$.

Thus, the following three conditions would be satisfied at the point $r=r_0$: $n(r_0)=0$, $\varphi(r_0)=0$, $\left(\frac{d\varphi}{dr}\right)_0=0$ $\left(n=\frac{dN}{dV}\right.$, i.e., the electron density). The density is proportional to the 3/2 power of the kinetic energy [see (44.8)], $n \sim (\mathscr{E}_0+e\varphi)^{3/2}$. Here, $\mathscr{E}_0=\varepsilon_0-e\varphi$, i.e., the limiting total energy must be a constant quantity. But, applying this equation at the point r_0, we see that $\mathscr{E}_0$ must equal zero. It only remains to show that the point r_0 cannot occur at a finite distance away from the nucleus. As we shall see later, this follows from the equation for the distribution of potential.

Thus, putting $\mathscr{E}_0=0$ we obtain

$$(3\pi^2)^{2/3}\frac{h^2}{2m}n^{2/3}=\varepsilon_0=e\varphi. \tag{44.18}$$

Here, from formula (44.7), ε_0 is expressed in terms of density, i.e., the charge density is related to potential.

The equation for a self-consistent field. A second relationship between potential and density is given by the electrostatic equation (14.7). Since the electronic charge is negative, this equation should be written with a plus sign on the right-hand side:

$$\frac{1}{r^2}\frac{d}{dr}r^2\frac{d\varphi}{dr}=4\pi en. \tag{44.19}$$

The density $n=\frac{dN}{dV}$ must be eliminated from (44.18) and $e\varphi$ inserted in place of ε_0. Then we obtain

$$\frac{1}{r^2}\frac{d}{dr}r^2\frac{d\varphi}{dr}=\frac{2^{7/2}}{3\pi}\frac{m^{3/2}}{h^3}e^{5/2}\varphi^{3/2}. \tag{44.20}$$

We transform this equation like (19.6). For this, we substitute φ in the form

$$\varphi=\frac{Ze}{r}\psi. \tag{44.21}$$

The function ψ is nondimensional since $\frac{Ze}{r}$ has the dimensions of potential. In immediate proximity to the nucleus, φ is determined only by the nucleus because the potential of the nucleus tends to infinity like $\frac{Ze}{r}$, while the potential of the spatially distributed charge of the electrons remains finite.

Therefore, close to the nucleus (i.e., when $r=0$) $\psi=1$. At large distances from the nucleus, its charge is completely screened by the electron charge of opposite sign, so that the potential of an atom must tend to zero more rapidly than $\frac{1}{r}$. This shows that $\psi(\infty)=0$.

Substituting (44.21) in (44.20), we have an equation for ψ:

$$\frac{d^2\psi}{dr^2} = \frac{2^{7/2}}{3\pi} \cdot Z^{1/2} \frac{m^{3/2}}{h^3} e^3 \frac{\psi^{3/2}}{r^{1/2}}. \tag{44.22}$$

It is convenient to get rid of the dimensional factor on the right-hand side. To do this, we must introduce a new unit of length similar to the atomic unit [see (31.21)]:

$$r = \frac{(3\pi)^{2/3}}{2^{7/3}} \frac{1}{Z^{1/3}} \frac{h^2}{me^2} \cdot x. \tag{44.23}$$

This unit differs from the atomic unit by the factor $\frac{0.889}{Z^{1/3}}$. After the introduction of a nondimensional variable x, equation (44.22) reduces to standard form (the Thomas-Fermi equation):

$$\frac{d^2\psi}{dx^2} = \frac{\psi^{3/2}}{x^{1/2}}. \tag{44.24}$$

Now it does not involve atomic number. Both boundary conditions for ψ ($\psi(0)=1$ and $\psi(\infty)=0$) are also the same for all atoms. Therefore, it is sufficient to integrate equation (44.24) once with these boundary conditions.

If we return to the dimensional radius r, the function $\psi(x)$ gives the potential distribution for each Z:

$$\varphi = \frac{Ze}{r}\psi\left(1.125\, Z^{1/3} \frac{me^2}{h^2} \cdot r\right). \tag{44.25}$$

If the distance from the nucleus is expressed in terms of x, the electron density distribution is the same for all atoms to which the statistical method is applicable, i.e., for all elements of large and medium atomic weight. But the same x denotes a geometrical distance inversely proportional to $Z^{1/3}$, as can be seen from (44.23). Therefore, in heavy atoms, the main part of the electrons is concentrated closer to the nucleus than in the lighter atoms.

The accuracy of the Thomas-Fermi equation (44.24) is determined by the quantity $Z^{-2/3}$, as can be shown from a strictly quantum-mechanical derivation by using a quasi-classical approximation. Therefore, equation (44.24) cannot, of course, be applied to the very lightest atoms that contain few electrons.

Substantiation of the boundary conditions for equation (44.24). The integral curves of equation (44.24) begin at the point $\psi=1$ for $x=0$, and fall with increasing x, accounting thereby for the screening effect, i.e., weakening of the nuclear field by the atomic electrons. The diminishing function may either pass through a minimum, without attaining $\psi=0$, and then begin to increase, or it may intersect the x-axis at a certain point $x=x_0$, or it may tend to this axis asymptotically.

The first possibility must be rejected at once, because it results in an infinite total number of electrons proportional to $\int_0^\infty \psi^{3/2} x^{1/2}\, dx$, see (44.18) and (44.21) (if we take $\psi(\infty)>0$). It is impossible to cut off the integration at some x_0 when $\psi>0$, since this would correspond to a limiting total energy $\mathscr{E}_0$ not equal to zero.

If we take the second possibility, then the total number of electrons has a finite value and will be proportional to $\int_0^{x_0} \psi^{3/2} x^{1/2}\, dx$. The electron density and, hence, the electric field of a neutral atom also, must, by definition, become zero at the point $x=x_0$, since the nuclear charge in it is completely screened by electrons. In accordance with (44.21) the electric field will be determined by the expression

$$E=-\frac{d\varphi}{dr}=\frac{Ze\psi}{r^2}-\frac{Ze}{r}\frac{d\psi}{dr}.$$

But the condition $\frac{d\psi}{dr}=0$ is satisfied where $E=0$ and $\psi=0$. Therefore, the point $x=x_0$ must correspond to tangency of the integral curve with the x-axis, and not to intersection. In the general case, the integral curve close to the point of tangency has the following form:

$$\psi=a(x-x_0)^{2+k}+\ldots,$$

where k is a positive number. Terms with large values of k are denoted by the dots. Substituting this expansion in equation (44.20), we have

$$(2+k)(1+k)\,a(x-x_0)^k=\frac{a^{3/2}}{x_0^{1/2}}(x-x_0)^{3+\frac{3k}{2}},$$

whence it follows that $k=-6$, in spite of the assumption that $k\geqslant 0$. Hence, tangency of the integral curve with the x-axis is impossible at a finite distance from the origin, and asymptotic tangency must be assumed. And the condition $\frac{d\psi}{dx}=0$ is automatically satisfied at infinity.

The charge distribution in positive ions. In positive ions, the charge of all the electrons does not completely screen the nuclear charge, because, at the point where $\psi = 0$, the condition $\frac{d\psi}{dx} = 0$ should not be satisfied. The electron density distribution in an ion is given by the integral curves intersecting the x-axis. The point of intersection determines the radius of the ion x_0.

The order for filling the electron shells. From the potential distribution in an atom, we can determine the values of Z for which d- and f-electrons first appear in the atom.

We first of all note that the electron density distribution in an atom must be associated with the angular-momentum distribution of the electrons. As we have already indicated, the limiting momentum of electrons is proportional to the $^1/_3$ power of the electron density. Therefore, close to the nucleus, where the electron density is great, the limiting momentum is also great, while at large distances from the nucleus, the limiting momentum is small. But the angular momentum of an electron is determined by the product of the momentum by the distance to the nucleus, and close to the nucleus it is small despite the large limiting momentum. At large distances from the nucleus the angular momentum becomes small—this time as a result of the smallness of the limiting momentum. Hence, somewhere at medium distances, the angular momentum attains a maximum which is larger, the greater the electron density. Therefore, in heavy atoms with a large electron density, we find larger values of angular momentum. In order to find the greatest values of angular momentum that are possible for a given Z, we shall proceed from the classical expression for energy in a central field [see (5.7)]

$$\mathscr{E} = \frac{p_r^2}{2m} + \frac{M^2}{2mr^2} - \frac{Ze^2\psi}{r}. \qquad (44.26)$$

We must put $\mathscr{E} = 0$ for the boundary energy, in accordance with the basic assumption (44.18). Then, for the radial component of momentum we obtain the expression

$$p_r = \sqrt{\frac{2mZe^2\psi}{r} - \frac{M^2}{r^2}}. \qquad (44.27)$$

We can substitute $\hbar^2 l(l+1)$ in place of M^2. But since formula (44.26) is written to a quasi-classical approximation, a better result is obtained if we also take the quasi-classical approximation for M^2. It can be calculated using the same methods as those for determining the energy eigenvalues from formula (29.18). To this approximation $M^2 = \hbar^2\left(l + \frac{1}{2}\right)^2$. We notice that $\left(l + \frac{1}{2}\right)^2$ differs from $l\,(l+1)$ only by a quarter.

We write (44.27) in the following form:

$$p_r = \frac{h}{r}\sqrt{\frac{2me^2}{h^2}rZ\psi - \left(l+\frac{1}{2}\right)^2}. \tag{44.28}$$

Let us now express the factor r in the radicand in terms of the nondimensional quantity x according to formula (44.23). Then p_r will be

$$p_r = \frac{h}{r}\sqrt{1.778Z^{2/3}x\psi - \left(l+\frac{1}{2}\right)^2}. \tag{44.29}$$

For p_r to be a real quantity, the radicand must remain positive in a certain interval of values x. But since $x\psi = 0$ when $x = 0$ and $x = \infty$ this interval is finite and contains the maximum point of the function $x\psi$. The maximum is equal to 0.488. Thus, the whole interval in which p_r is a real quantity is contracted into a point for the value of Z at which

$$1.778 \cdot 0.488 \cdot Z^{2/3} = \left(l+\frac{1}{2}\right)^2, \tag{44.30}$$

the curve $y = 1.778\, Z^{2/3}\, x\psi$ being tangential to the constant straight line $y = \left(l+\frac{1}{2}\right)^2$.

It follows that a given value of l in an atom may occur when Z satisfies the condition

$$Z = 0.155\,(2l+1)^3. \tag{44.31}$$

According to this equation, electrons having $l = 2$ will occur for $Z = 19$, while f-electrons ($l = 3$) will occur when $Z = 53$. There will be better agreement if we take the coefficient 0.17 instead of 0.155.

Using the numerical form of the function $\psi(x)$, it can be shown that the d- and f-shells are formed mainly deep inside the atom, as was shown in Sec. 33.

The approximate integral formula for the Fermi distribution. In conclusion, let us consider a Fermi gas not at absolute zero, but at a temperature other than zero yet satisfying the inequality (44.9).

It is convenient first to derive a general formula for the integral of the Fermi distribution that holds for $\theta \ll \varepsilon_0$.

Let us take the integral

$$I = \int_0^\infty \frac{\frac{d\gamma(\varepsilon)}{d\varepsilon}\,d\varepsilon}{e^{\frac{\varepsilon-\mu}{\theta}}+1}, \tag{44.32}$$

where $\gamma(\varepsilon)$ is some power function, for example $\sqrt{\varepsilon}$, $\sqrt{\varepsilon^3}$, etc. We integrate (44.32) by parts:

$$I = \left.\frac{\gamma(\varepsilon)}{e^{\frac{\varepsilon-\mu}{\theta}}+1}\right|_0^\infty - \int_0^\infty \gamma(\varepsilon)\frac{d}{d\varepsilon}\left(\frac{1}{e^{\frac{\varepsilon-\mu}{\theta}}+1}\right)d\varepsilon =$$

$$= -\frac{\gamma(0)}{e^{-\frac{\mu}{\theta}}+1} + \frac{1}{\theta}\int_0^\infty \gamma(\varepsilon)\frac{e^{\frac{\varepsilon-\mu}{\theta}}}{\left(e^{\frac{\varepsilon-\mu}{\theta}}+1\right)^2}d\varepsilon. \qquad (44.33)$$

Let us write the second factor in the integrand thus:

$$\frac{e^{\frac{\varepsilon-\mu}{\theta}}}{\left(e^{\frac{\varepsilon-\mu}{\theta}}+1\right)^2} = \frac{1}{\left(e^{\frac{\varepsilon-\mu}{\theta}}+1\right)\left(e^{\frac{\mu-\varepsilon}{\theta}}+1\right)}. \qquad (44.34)$$

The denominator of this expression is large both for $\varepsilon<\mu$ and for $\varepsilon>\mu$. The exponential $e^{\frac{\mu-\varepsilon}{\theta}}$ is large in the first case, while $e^{\frac{\varepsilon-\mu}{\theta}}$ is large in the second case. Therefore, the whole expression differs noticeably from zero only in a narrow range of values ε, different from μ by an amount of the order of θ. Let us expand the function $\gamma(\varepsilon)$ within this range and let us terminate the expansion with the second term.

$$\gamma(\varepsilon) = \gamma(\mu) + (\varepsilon-\mu)\gamma'(\mu) + \frac{(\varepsilon-\mu)^2}{2}\gamma''(\mu). \qquad (44.35)$$

We substitute this expansion in (44.33). Taking into account that the second factor in the integrand is very small for $\varepsilon=0$, we can perform the integration to $\varepsilon=-\infty$ without making any perceivable error. In addition, we shall neglect the quantity $e^{-\frac{\mu}{\theta}}$ in the integrated term of (44.33). From this we obtain

$$I = -\gamma(0) - \gamma(\mu)\int_{-\infty}^{\infty} d\varepsilon\frac{d}{d\varepsilon}\left(\frac{1}{e^{\frac{\varepsilon-\mu}{\theta}}+1}\right) +$$

$$+\frac{\gamma'(\mu)}{\theta}\int_{-\infty}^{\infty} d\varepsilon\frac{\varepsilon-\mu}{\left(e^{\frac{\varepsilon-\mu}{\theta}}+1\right)\left(e^{\frac{\mu-\varepsilon}{\theta}}+1\right)} +$$

$$+\frac{\gamma''(\mu)}{2\theta}\int_{-\infty}^{\infty} d\varepsilon\frac{(\varepsilon-\mu)^2}{\left(e^{\frac{\varepsilon-\mu}{\theta}}+1\right)\left(e^{\frac{\mu-\varepsilon}{\theta}}+1\right)}. \qquad (44.36)$$

The first integral is calculated immediately; it is

$$\int_{-\infty}^{\infty} d\varepsilon \frac{d}{d\varepsilon}\left(\frac{1}{e^{\frac{\varepsilon-\mu}{\theta}}+1}\right) = \frac{1}{e^{\frac{\varepsilon-\mu}{\theta}}+1}\Bigg|_{-\infty}^{\infty} = -1. \tag{44.37}$$

We change the integration variable in the second and third integrals, assuming

$$\frac{\varepsilon-\mu}{\theta} = x.$$

Then the second integral reduces to the form:

$$\int_{-\infty}^{\infty} \frac{d\varepsilon\,(\varepsilon-\mu)}{\left(e^{\frac{\varepsilon-\mu}{\theta}}+1\right)\left(e^{\frac{\mu-\varepsilon}{\theta}}+1\right)} = \theta^2 \int_{-\infty}^{\infty} \frac{x\,dx}{(e^x+1)(e^{-x}+1)} = 0, \tag{44.38}$$

because the integrand is an odd function. Finally, the third integral (see Appendix) is

$$\int_{-\infty}^{\infty} \frac{(\varepsilon-\mu)^2\,d\varepsilon}{\left(e^{\frac{\varepsilon-\mu}{\theta}}+1\right)\left(e^{\frac{\mu-\varepsilon}{\theta}}+1\right)} = \theta^3 \int_{-\infty}^{\infty} \frac{x^2\,dx}{(e^x+1)(e^{-x}+1)} = \frac{\pi^2}{3}\theta^3. \tag{44.39}$$

Thus, the required integral appears in the form of the following expansion:

$$I = \gamma(\mu) - \gamma(0) + \frac{\pi^2}{6}\theta^2\gamma''(\mu) = \int_0^{\mu} \gamma'(\varepsilon)\,d\varepsilon + \frac{\pi^2}{6}\theta^2\gamma''(\mu). \tag{44.40}$$

The zero term in this expansion corresponds to the form that the Fermi distribution has at absolute zero; indeed, if $f=1$ for $0 \leqslant \varepsilon \leqslant \mu$, then I will just equal $\int_0^{\mu} \gamma'(\varepsilon)\,d\varepsilon$. The first term, which is linear in θ, drops out of the expansion. This is clearly evident from the following. The electrons which escape the limiting-energy sphere leave behind unoccupied levels, so-called "holes." To a first approximation, these holes are symmetrically distributed with respect to the occupied levels lying above the limiting energy. In Fig. 50 this can be seen from the fact that the shaded areas are approximately equal.

Finally, the quadratic term contributes the desired correction to the integral I.

The specific heat of a Fermi gas. We shall now apply the result (44.40) to calculation of specific heat. To do this we write down the expressions for the energy and the total number of particles:

$$\mathscr{E} = \frac{V(2m^3)^{1/2}}{\pi^2 h^3} \int_0^\infty \frac{\varepsilon^{3/2}\, d\varepsilon}{e^{\frac{\varepsilon-\mu}{\theta}} + 1}, \tag{44.41}$$

$$N = \frac{V(2m^3)^{1/2}}{\pi^2 h^3} \int_0^\infty \frac{\varepsilon^{1/2}\, d\varepsilon}{e^{\frac{\varepsilon-\mu}{\theta}} + 1}. \tag{44.42}$$

We apply formula (44.40) and obtain

$$\mathscr{E} = \frac{V(2m^3)^{1/2}}{\pi^2 h^3}\left(\frac{2}{5}\mu^{5/2} + \frac{3}{2}\frac{\pi^2}{6}\mu^{1/2}\theta^2\right), \tag{44.43}$$

$$N = \frac{V(2m^3)^{1/2}}{\pi^2 h^3}\left(\frac{2}{3}\mu^{3/2} + \frac{1}{2}\frac{\pi^2}{6}\mu^{-1/2}\theta^2\right), \tag{44.44}$$

because the function $\gamma'(\varepsilon)$ equalled $\varepsilon^{3/2}$ for the first integral and $\varepsilon^{1/2}$ for the second integral.

Using these formulae let us find the specific heat. From the definition of specific heat we have

$$C = \frac{\partial \mathscr{E}}{\partial \theta} = \frac{V(2m^3)^{1/2}}{\pi^2 h^3}\left(\mu^{3/2}\frac{\partial \mu}{\partial \theta} + \frac{\pi^2}{2}\theta\mu^{1/2}\right). \tag{44.45}$$

We calculate the derivative $\frac{\partial \mu}{\partial \theta}$ from the second equation, differentiating it as an implicit function:

$$\frac{\partial \mu}{\partial \theta} = -\frac{\frac{\partial N}{\partial \theta}}{\frac{\partial N}{\partial \mu}} = -\frac{\frac{\pi^2}{6}\mu^{-1/2}\theta}{\mu^{1/2}} = -\frac{\pi^2}{6}\frac{\theta}{\mu}. \tag{44.46}$$

We have both times omitted differentiating the coefficient of θ, because θ is regarded as small. Substituting (44.46) in (44.45), we write the specific heat as

$$C = \frac{\partial \mathscr{E}}{\partial \theta} = \frac{V(2m^3)^{1/2}}{3h^3}\mu^{1/2}\theta. \tag{44.47}$$

Finally, in place of μ we must substitute the expression for the limiting energy (44.7). Then the specific heat will be expressed in terms of the gas density and temperature:

$$C = N\left(\frac{\pi}{3}\right)^{2/3}\frac{m}{h^2}\left(\frac{N}{V}\right)^{-2/3}\theta = \frac{\pi^2}{2}\frac{\theta}{\varepsilon_0}N. \tag{44.48}$$

Thus, the specific heat per electron is approximately $5\frac{\theta}{\varepsilon_0}$, which, according to (44.9), is a very small quantity. For example, we estimated that for sodium $\varepsilon_0 = 34{,}800°$, so that $\frac{\theta}{\varepsilon_0} \sim 0.01$ at room temperature. The specific heat of a Fermi gas per electron at room temperature is 0.05. This must be compared with the specific heat for a Boltzmann gas, equal to 1.5 from Sec. 40 (if θ is expressed in ergs, the specific heat C is an abstract quantity).

It is easy to see why the specific heat of a Fermi gas is considerably less than the specific heat of a Boltzmann gas: not all the electrons in a Fermi distribution are capable of being thermally excited, but only those whose energy is close to the critical energy. This is why the specific heat of a Fermi gas turns out equal to a few per cent of N. A specific heat $\frac{3}{2}N$ is obtained only when all the electrons are capable of being thermally excited.

Difficulties in the classical electron theory of metals. Considerable difficulty was experienced in the prequantum theory of metals because the electron gas in a metal does not have an experimentally noticeable specific heat at room temperature. The specific heat of a metal does not exceed the value 3 per atom [see (42.32)]. Yet if the number of electrons present equalled the number of atoms, then, according to classical statistics, the metal would have a specific heat $3 + 3/2 = 9/2$ per atom, which is never observed.

If we apply Fermi statistics to electrons, then, as we have just seen, the difficulty with specific heat is removed.

At low temperature the specific heat of the crystal lattice of a metal is proportional to θ^3 [see (42.35)]. Therefore, if the temperature is sufficiently low, the electronic specific heat begins to predominate and can be measured. Measurements show that at very low temperatures the specific heat of metals is indeed proportional to θ. As can be seen from (44.48), if we know the specific heat we can also determine the number of electrons per atom. It is a curious fact that bismuth, which in many respects is not a typical metal, has a very small number of conduction electrons.

Exercises

1) Find the equilibrium concentration of electrons and positrons in some volume not containing charges at low temperature.

In place of the conservation of the number of particles we must take into account the conservation of charge in the formation and annihilation of electron-positron pairs. Denoting the number of electrons in a given quantum state by the letter f, and the number of positrons by the letter f', we have, in place of (39.23), the following supplementary condition:

$$\sum_k g_k (f_k - f'_k) = 0.$$

Determining f and f', which give the maximum of the function $S = \ln P$ with the supplementary condition indicated, we obtain the distribution functions for electrons and positrons:

$$f = \frac{1}{e^{\frac{\varepsilon - \mu}{\theta}} + 1}, \qquad f' = \frac{1}{e^{\frac{\varepsilon + \mu}{\theta}} + 1},$$

where the constant μ is the same. The total number of electrons must equal the total number of positrons, i.e.,

$$\int_0^\infty \frac{\sqrt{\varepsilon}\, d\varepsilon}{e^{\frac{\varepsilon - \mu}{\theta}} + 1} = \int_0^\infty \frac{\sqrt{\varepsilon}\, d\varepsilon}{e^{\frac{\varepsilon + \mu}{\theta}} + 1}.$$

This equation has a solution only for $\mu = 0$. Hence, the total number of electrons in unit volume is

$$\frac{2 \cdot 4\pi}{(2\pi h)^3} \int_0^\infty \frac{p^2\, dp}{e^{\frac{\varepsilon}{\theta}} + 1}.$$

Let us calculate this integral when $\theta \ll mc^2$. We can take the nonrelativistic approximation for the energy and represent the distribution function in the form $e^{-\frac{\varepsilon}{\theta}}$

Whence we have the equilibrium electron density

$$\frac{1}{\pi^2 h^3} e^{-\frac{mc^2}{\theta}} \int_0^\infty e^{-\frac{p^2}{2m\theta}} p^2\, dp = \frac{1}{2^{1/2} \pi^{3/2}} \left(\frac{mc}{h}\right)^3 \left(\frac{\theta}{mc^2}\right)^{3/2} e^{-\frac{mc^2}{\theta}}.$$

This quantity is equal to $1/\text{cm}^3$ for $\theta = \frac{mc^2}{64} \cong 8$ kev. The energy of the electromagnetic field per unit volume at the same temperatures is 0.6×10^{18} ergs, while only 1.6×10^{-6} erg is released in pair annihilation. The energy of electrons and positrons will be close to the electromagnetic field energy only when θ is of the order of mc^2.

2) Find the limiting energy of a superdense electron gas, for which the dependence of energy upon momentum is in the main extremely relativistic: $\varepsilon = cp$. Determine the density at which the gas may be regarded as ultrarelativistic.

In place of equation (44.8), we have

$$\frac{4\pi}{3} \frac{\varepsilon^3}{c^3} = \frac{N}{2} \frac{(2\pi h)^3}{V},$$

so that

$$\varepsilon_0 = \left(\frac{3}{8\pi}\right)^{1/3} \left(\frac{N}{V}\right)^{1/3} 2\pi hc.$$

The rest energy can be neglected if

$$\varepsilon_0 \gg mc^2,$$

so that the condition for the density is written in the form

$$\frac{N}{V} \gg \frac{1}{3\pi^2}\left(\frac{mc}{h}\right)^3 \approx 10^{30} \text{ electrons/cm}^3.$$

Since ε_0 involves $\left(\frac{N}{V}\right)^{1/3}$, the inequality must be great. The energy of such an ultrarelativistic gas is given by the expression

$$\frac{V}{\pi^2 h^2 c^3}\int_0^\infty \frac{\varepsilon^3\, d\varepsilon}{e^{\frac{\varepsilon-\mu}{\theta}}+1}.$$

3) Find the number of electrons passing through the surface of a metal in unit time if only those electrons can cross the surface for which the velocity component normal to the wall is greater than v_{0x}. This quantity satisfies the inequality

$$\frac{mv_{0x}^2}{2} - \mu(0) \gg \theta.$$

In other words, the energy of the emerging electrons differs from the limiting energy by an amount considerably greater than θ (thermionic emission).

The number of electrons with velocity v_x falling on a square centimetre of surface in one second is

$$v_x\, dn(v_x),$$

where $dn(v_x)$ is the density of electrons having a given value of velocity projection v_x. Like (44.3), we write $dn(v_x)$ in the form

$$dn(v_x) = \frac{2m^3\, dv_x\, dv_y\, dv_z}{(2\pi h)^3}\cdot\frac{1}{e^{\frac{\varepsilon-\mu}{\theta}}+1},$$

where $\varepsilon = \frac{m}{2}(v_x^2+v_y^2+v_z^2)$. The surface of a metal is crossed only by those electrons for which the difference $\varepsilon-\mu$ is considerably greater than θ, so that we are justified in passing from a Fermi distribution to a distribution of the Boltzmann type, but with the same value of μ as in the Fermi distribution. In other words, we take only the "tail" of the Fermi curve where $\varepsilon-\mu > \theta$. Whence, the required electron flux is

$$\frac{2m^3}{(2\pi h)^3} e^{\frac{\mu}{\theta}} \int_{v_{0x}}^\infty v_x\, dv_x\, e^{-\frac{mv_x^2}{2\theta}} \int_{-\infty}^\infty dv_y\, e^{-\frac{mv_y^2}{2\theta}} \int_{-\infty}^\infty dv_z\, e^{-\frac{mv_z^2}{2\theta}} =$$

$$= \frac{2m^3}{(2\pi h)^3} e^{\frac{\mu}{\theta}} \frac{\theta}{m} e^{-\frac{mv_{0x}^2}{2\theta}} \frac{2\pi\theta}{m} = \frac{m\theta^2}{2\pi^2 h^3} e^{\frac{1}{\theta}\left(\mu - \frac{mv_{0x}^2}{2}\right)}.$$

If we apply an electric field to the metal, the maximum current that can be extracted at a given temperature (saturation current) is determined by this formula. Since it relates to electrons in a metal, the quantity μ is close to μ_0, (i.e., to the limiting energy at absolute zero) and does not depend upon temperature.

It will be noticed that if we apply a very strong electric field to the metal, electrons will emerge from it overcoming the potential barrier which appears at the boundary under such conditions (cold emission). But this requires very large fields. Cold emission is analogous to the ionization of atoms in the Stark effect (see Sec. 35).

4) Calculate the total energy of the electrons in an atom in accordance with the Thomas-Fermi statistical model.

From (44.10), the kinetic energy of the electrons is

$$\mathscr{E}_{\text{kin}} = \frac{(2m)^{3/2}}{5\pi^2 \hbar^3} \int_0^\infty \varepsilon_0^{5/2} \cdot 4\pi r^2 \, dr = \frac{(2m)^{3/2}}{5\pi^2 \hbar^3} 4\pi \int_0^\infty (e\varphi)^{5/2} r^2 \, dr,$$

because the limiting kinetic energy of the electrons is $e\varphi$.

We substitute $\frac{Ze^2 \psi}{r}$ instead of $e\psi$ and go to non dimensional variables (44.23). Then for $\mathscr{E}_{\text{kin}}$ we obtain

$$\mathscr{E}_{\text{kin}} = \frac{12}{5} \left(\frac{2}{9\pi^2}\right)^{1/3} Z^{7/3} \frac{me^4}{\hbar^2} \int_0^\infty \psi^{5/2} \frac{dx}{\sqrt{x}} .$$

The potential energy is divided into two parts: the interaction energy of the electrons with the nucleus, equal to

$$\mathscr{E}_{\text{pot}}^{(1)} = -\int_0^\infty \frac{Ze^2}{r} n \cdot 4\pi r^2 \, dr ,$$

where the electron density is determined from (44.18), and the interaction energy between the electrons themselves,

$$\mathscr{E}_{\text{pot}}^{(2)} = \frac{1}{2} \int_0^\infty \frac{Ze^2}{r} (1 - \psi) \, n \cdot 4\pi r^2 \, dr.$$

The factor $\frac{1}{2}$ takes into account that each electron should be counted once. Combining both parts of the potential energy, we have

$$\mathscr{E}_{\text{pot}} = \mathscr{E}_{\text{pot}}^{(1)} + \mathscr{E}_{\text{pot}}^{(2)} = -\frac{1}{2} \int_0^\infty \frac{Ze^2}{r} (1 + \psi) \, n \cdot 4\pi r^2 \, dr.$$

Substituting the quantity n, we arrive at the following expression for the potential energy:

$$\mathscr{E}_{\text{pot}} = -2 \left(\frac{2}{9\pi^2}\right)^{1/3} Z^{7/3} \frac{me^4}{\hbar^2} \int_0^\infty (\psi^{3/2} + \psi^{5/2}) \frac{dx}{\sqrt{x}} .$$

The integrals appearing in the energy expression are easily calculated using equation (44.24). Namely,

$$\int_0^\infty \psi^{3/2} \frac{dx}{\sqrt{x}} = \int_0^\infty \psi'' \, dx = -\psi'(0) ,$$

because $\psi'(\infty) = 0$. The second integral is transformed by parts:

$$\int_0^\infty \psi^{5/2} \frac{dx}{\sqrt{x}} = 2\sqrt{x}\,\psi^{5/2}\Big|_0^\infty - 5\int_0^\infty \sqrt{x}\,\psi^{3/2}\psi'\,dx = -5\int_0^\infty x\psi''\psi'\,dx =$$

$$= -\frac{5}{2}\int_0^\infty x\frac{d}{dx}(\psi')^2\,dx = -\frac{5}{2}x(\psi')^2\Big|_0^\infty + \frac{5}{2}\int_0^\infty (\psi')^2\,dx,$$

since the integrated expressions are equal to zero. Further,

$$\int_0^\infty (\psi')^2\,dx = \psi\psi'\Big|_0^\infty - \int_0^\infty \psi\psi''\,dx = -\psi'(0) - \int_0^\infty \psi^{5/2}\frac{dx}{\sqrt{x}},$$

since $\psi(0) = 1$. Hence,

$$\int_0^\infty \psi^{5/2}\frac{dx}{\sqrt{x}} = -\frac{5}{7}\psi'(0).$$

Substituting these integral values in the expressions for $\mathscr{E}_{\text{kin}}$ and $\mathscr{E}_{\text{pot}}$, we notice that $\mathscr{E}_{\text{pot}} = -2\,\mathscr{E}_{\text{kin}}$, so that the total energy is $-\mathscr{E}_{\text{kin}}$ (this result is also obtained in the exact theory, and not only in the statistical model).

The quantity $\psi'(0)$ is equal to -1.589, whence we obtain the following formula for the total binding energy of all the electrons in an atom:

$$\mathscr{E} = -0.769\frac{me^4}{\hbar^2}\cdot Z^{7/3} = -20.94\,Z^{7/3}\,ev$$

For example, for uranium $\mathscr{E} = -8 \times 10^5$ ev, or $-1.6\,mc^2$.

A relationship of the form $Z^{7/3}$ is also easy to obtain, without calculation, in the following way. Coulomb forces fall off slowly with distance. Therefore, all the electrons interact with each other in pairs, so that there are about Z^2 pairs. From (44.23), the mean distance between the electrons decreases like $Z^{1/3}$. This is what yields $Z^{7/3}$. We notice that in the case of nuclei the total binding energy is proportional to the first power of the number of particles (within wide limits). This points to the short-range character of nuclear forces: each nucleon (i.e., proton or neutron) does not interact with all the other nucleons, but only with the "nearest."

Sec. 45. Gibbs Statistics

In this section we shall consider the general statistical method of Gibbs applied to any system consisting of a sufficiently large number of particles, irrespective of whether these systems are solid, liquid, or gaseous. It is very difficult to treat this method rigorously by proceeding only from the equations of quantum mechanics; it is probably still more difficult to do so classically, since the concept of probability does not exist in classical mechanics (to say nothing of the fact that the application of classical mechanics to the motion of microparticles is by no means always justified). The derivation of the basic principles of Gibbs statistics is somewhat intuitive in character and is justified by the fact that the statistics agree with a vast quantity of experimental facts.

Of course, in principle, it would be well to substantiate the statistical method in such a way as to be certain beforehand of its agreement with experiment, proceeding from the sole fact that quantum mechanics agrees with experiment; but, as yet, we have no such quantitative treatment at our disposal.

The quasi-closed system. Fundamental to statistics is the concept of a *quasi-closed* system, i.e., a system occurring in weak interaction with the surrounding medium. This interaction does not essentially destroy the structure of the system, but governs transitions between those of its states which correspond to close separate energy levels of the closed system. In a system consisting of a sufficiently large number of particles, an energy interval $d\mathscr{E}$ (due to the quasi-closed nature of the system) contains an exceptionally large number of separate energy levels or, more exactly, states corresponding to separate, exceptionally close, energy levels of an ideal closed system. It is this that makes the application of statistics possible.

Statistical equilibrium. As was shown in Sec. 39, all these separate states are equally probable; in other words, the system spends the same amount of time in each of them. If a study is being made of the behaviour of a macroscopic system, the essential thing to know is not its detailed state (characterized by a certain wave function), but a large group of states to which the state of the system belongs most of the time.

For example, let us consider N particles of an ideal monatomic gas that possess a total kinetic energy $\mathscr{E}$.

The separate states of the gas are equiprobable; in other words, the state, where a single particle has all the energy $\mathscr{E}$ and the remaining particles have zero kinetic energy, is as probable as that where all the particles have an equal energy $\frac{\mathscr{E}}{N}$ and a strictly identical direction of momentum along a chosen axis (for the time being, we ignore the possibility of Pauli exclusion). But for the greater part of the time the gas occurs in a state which is incomparably closer to the equilibrium distribution of energy than to the exceptional state where a single particle has obtained all the energy. Most probable is the state which is described sufficiently well by a Bose, or a Fermi, distribution, depending upon whether the gas particles have integral or half-integral spin. If the gas particles are close to the most probable distribution, for constant interaction conditions with the external medium, then its state will all the time be close to the most probable state. Any significant deviation from the most probable state is of vanishingly small probability.

The whole group of equally probable microscopic states in which a system exists for the greater part of the time is called the ***statistical*** *equilibrium* state of the system. It is defined in far less detail than in the case of the states in quantum mechanics, but fully enough for a description of the macroscopic system as a whole.

The concept of statistical equilibrium can be applied to any sufficiently large system of particles, irrespective of whether they interact as weakly as the particles of an ideal gas, or as strongly as the particles in a solid or liquid body. It will be recalled that in proving the equiprobability of microstates (Sec. 39) it was not assumed that the system consisted of noninteracting particles. The greater the number of microstates of a system, the more probable its state. Here, only those microstates are taken into account which are compatible with the law of conservation of energy, i.e., those belonging to the energy interval $d\mathscr{E}$ of a quasi-closed system.

Probability distribution in subsystems. Instead of considering a quasi-closed system in an external medium, it is more convenient to proceed from a large ideally closed system and divide it into individual quasi-closed *subsystems*. The quasi-closed nature of subsystems manifests itself when the surface layer of each subsystem, through which interaction with the surrounding subsystems is effected, produces but a small effect on the processes taking place inside the volume. Interaction between subsystems leads to the establishment of statistical equilibrium over the whole large system, while equilibrium in a subsystem is established by its internal interactions.

Let us suppose that equilibrium has been established inside a subsystem. What is the probability that its energy is between $\mathscr{E}$ and $\mathscr{E} + d\mathscr{E}$? To this interval there correspond $g(\mathscr{E})$ equally probable microstates. As we know, $g(\mathscr{E})$ is called the *weight of a state* with a given energy $\mathscr{E}$.

Since all the separate microstates are equally probable, the probability of $\mathbf{P}(\mathscr{E})$ states is directly proportional to $g(\mathscr{E})$:

$$P(\mathscr{E}) = \rho(\mathscr{E})\, g(\mathscr{E}), \tag{45.1}$$

where $\rho(\mathscr{E})$ is a function which we have got to determine in the present section.

Separate quasi-independent subsystems may be regarded as very large molecules of a Boltzmann gas. It is natural to consider that such a "gas" is subject to Boltzmann statistics since macroscopic subsystems differ from one another. It follows from this that the distribution function must be of Boltzmann form:

$$\rho(\mathscr{E}) \sim e^{-\frac{\varepsilon}{\theta}}.$$

This result is preliminary and intuitive. A more strict derivation is given below based on the properties of the function $\rho(\mathscr{E})$, which will now be established.

Liouville's theorem. We shall prove that the function $\rho(\mathscr{E})$ is constant during the interval of time within which a quasi-closed system may be regarded as closed, i.e., the remaining subsystems do not noticeably affect its state.

The weight of a state $g(\mathscr{E})$ is defined by the number of microstates whose energy is between $\mathscr{E}$ and $\mathscr{E}+d\mathscr{E}$. Each of these microstates is characterized by a definite set of integrals of motion (for example, for a monatomic ideal gas, the group of momenta of the separate systems). Therefore, $g(\mathscr{E})$ is a constant quantity.

The probability $P(\mathscr{E})$ is defined as $\lim \frac{t(\mathscr{E})}{t}$, when t tends to infinity (see Sec. 39). Here, t denotes the observation time for the whole closed system, which includes the given quasi-closed subsystem. Therefore, by its very meaning, $P(\mathscr{E})$ cannot depend upon time, because this is a resultant average quantity for large intervals of time. But if $P(\mathscr{E})$ is a constant quantity and $g(\mathscr{E})$, as a function of the integrals of motion, is also constant, then $\rho(\mathscr{E})$ is also independent of time and is an integral of motion. But since all the integrals of motion are, in principle, known from mechanics, ρ must be their function. In other words, ρ cannot depend upon quantities that vary with time, and, apart from $\mathscr{E}$, depends only upon the integrals of motion. More exactly, ρ remains constant over intervals of time for which the quasi-closed subsystem may be regarded as closed. The statement concerning the constancy of $\rho(\mathscr{E})$ is known as Liouville's theorem. At the end of this section, a classical formulation of Liouville's theorem will be given that is more vivid than a quantum formulation.

The theorem of multiplication of probabilities. Over a certain interval of time, quasi-closed subsystems may be regarded as independent. Then the well-known theorem of probability multiplication can be applied: the probability that one of the subsystems is in a state A and another in a state B is equal to the product of the probabilities corresponding to states A and B.

$$P_{AB} = P_A P_B. \tag{45.2}$$

The statistical weights of the states, g_A and g_B, are of course multiplied because they relate to different subsystems:

$$g_{AB} = g_A \cdot g_B. \tag{45.3}$$

Thus,

$$P_{AB} = P_A P_B = g_{AB} \cdot \rho_{AB} = g_A \rho_A \cdot g_B \rho_B.$$

It follows from formulae (45.2) and (45.3) that

$$\rho_{AB} = \rho_A \cdot \rho_B. \tag{45.4}$$

In other words, the probability density for two quasi-independent subsystems is a multiplicative function, i.e., it is obtained by multiplying the separate ρ functions.

Gibbs distribution. The logarithm of probability density is an additive quantity, i.e., it is equal to the sum of the logarithms of this quantity for each subsystem separately:

$$\ln \rho_{AB} = \ln \rho_A + \ln \rho_B. \tag{45.5}$$

We know from Liouville's theorem that $\ln \rho$ is, in addition, an integral of motion. Hence, $\ln \rho$ is an additive integral of motion.

In Sec. 4 of Part One we listed the additive integrals of motion: energy, linear momentum, and angular momentum. For $\ln \rho$ to be an additive integral of motion, it must depend *linearly* upon energy, linear momentum, and angular momentum. If we choose a reference system in which the subsystem as a whole does not move, then the linear momentum and angular momentum will be equal to zero and the logarithm of the probability density will turn out to be a linear function only of energy.

In other words, the following relationship results:

$$\ln \rho = a\mathscr{E} + b. \tag{45.6}$$

The coefficient a must be the same for all subsystems of the large system because, otherwise, $\ln \rho$ will not have the properties of an additive function. If a is the same for two subsystems, then these two subsystems yield

$$\begin{aligned} \ln \rho_{AB} &= \ln \rho_A + \ln \rho_B = a(\mathscr{E}_A + \mathscr{E}_B) + (b_A + b_B) = \\ &= a\mathscr{E}_{AB} + b_{AB}, \end{aligned} \tag{45.7}$$

whence the additivity of $\ln \rho$ can be seen.

The probability of an infinitely large energy must be infinitely small because $a < 0$.

We shall write

$$a \equiv -\frac{1}{\theta}. \tag{45.8}$$

The meaning of the quantity θ is the same as in the previous sections: it is the temperature multiplied by the Boltzmann constant. Indeed, for an ideal gas, a single molecule can be regarded as a separate subsystem, and then the Gibbs distribution of the form

$$e^{-\frac{\varepsilon}{\theta}} = e^{-\frac{\sum \varepsilon_i}{\theta}}$$

becomes the Boltzmann distribution

$$e^{-\frac{\varepsilon}{\theta}}.$$

In addition, we denote

$$b \equiv \frac{F}{\theta}. \tag{45.9}$$

Finally, the required distribution function is

$$\rho(\mathscr{E}) = e^{\frac{F-\mathscr{E}}{\theta}}. \tag{45.10}$$

The normalization condition. The following condition is imposed upon the function $\rho(\mathscr{E})$:

$$\sum_{\mathscr{E}} P(\mathscr{E}) = \sum_{\mathscr{E}} \lim \frac{t(\mathscr{E})}{t} = \lim \frac{\Sigma t(\mathscr{E})}{t} = 1 = \sum \rho(\mathscr{E})\, g(\mathscr{E}), \tag{45.11}$$

since $\sum t(\mathscr{E}) \equiv t$. This simply means that the probability of finding a subsystem in any of the possible states compatible with the conservation laws is equal to unity. With the aid of the normalization conditions (45.11) we can express the quantity F as a function of θ. It is sufficient for this to substitute the Gibbs distribution (45.10) into (45.11) and perform summation over all possible states. The factor $e^{\frac{F}{\theta}}$, as a constant quantity, is taken outside the summation. Hence, we have the following equation for finding F:

$$e^{-\frac{F}{\theta}} = \sum_{\mathscr{E}} e^{-\frac{\mathscr{E}}{\theta}} g(\mathscr{E}). \tag{45.12}$$

As we know, the expression on the right-hand side is called a *statistical sum.*

The mean energy of a subsystem. The mean values of quantities in statistics are determined in the following way. Let the quantity f assume a value f_A in any state A. Then, if the probability of this state is equal to P_A, the mean value will be defined as

$$\bar{f} = \sum_{A} f_A P_A. \tag{45.13}$$

For example,

$$\bar{\mathscr{E}} = \sum_{\mathscr{E}} \mathscr{E}\, \rho(\mathscr{E})\, g(\mathscr{E}), \tag{45.14}$$

because

$$P(\mathscr{E}) = \rho(\mathscr{E})\, g(\mathscr{E}).$$

Let us substitute the Gibbs distribution in (45.14). Then we obtain an expression for the mean energy:

$$\bar{\mathscr{E}} = \sum_{\mathscr{E}} \mathscr{E}\, e^{\frac{F-\mathscr{E}}{\theta}} g(\mathscr{E}). \tag{45.15}$$

Energy fluctuations. Generally, the state of a subsystem is, to a certain extent, characterized by the mean quantity. For this purpose, any statistics, and not only physical statistics, makes use of mean quantities: a constant mean quantity makes it possible to estimate the order of magnitude of a variable.

However, if a variable quantity exhibits a wide scatter, the mean does not describe it sufficiently well.

Therefore, in addition to the mean value of the energy in a subsystem, it is interesting to know its mean dispersion. These two mean quantities define the state of a subsystem considerably better than $\bar{\mathscr{E}}$ by itself. But if we average the quantity

$$\Delta \mathscr{E} \equiv \mathscr{E} - \bar{\mathscr{E}}, \tag{45.16}$$

the result is identical zero. Indeed, a second averaging of the constant quantity $\bar{\mathscr{E}}$ in no way changes it: the mean of $\bar{\mathscr{E}}$ is again equal to $\bar{\mathscr{E}}$. Whence

$$\overline{\Delta \mathscr{E}} = \bar{\mathscr{E}} - \bar{\mathscr{E}} = 0. \tag{45.17}$$

It is therefore expedient to average the quantity $(\Delta\mathscr{E})^2 = (\mathscr{E} - \bar{\mathscr{E}})^2$. Since it is essentially positive, a deviation of $\mathscr{E}$ from the mean value (to either side) makes a contribution. The desired mean quantity may be written somewhat differently:

$$\begin{aligned}\overline{(\Delta \mathscr{E})^2} &= \overline{(\mathscr{E} - \bar{\mathscr{E}})^2} = \overline{\mathscr{E}^2 - 2\mathscr{E}\bar{\mathscr{E}} + \bar{\mathscr{E}}^2} = \\ &= \overline{\mathscr{E}^2} - 2\bar{\mathscr{E}}\bar{\mathscr{E}} + \overline{\bar{\mathscr{E}}^2} = \overline{\mathscr{E}^2} - \bar{\mathscr{E}}^2,\end{aligned} \tag{45.18}$$

where we have taken advantage of the fact that the mean of a constant quantity $\bar{\mathscr{E}}^2$ is equal to itself, and also that the constant factor can be taken outside the averaging symbol:

$$\overline{\mathscr{E}\bar{\mathscr{E}}} = \bar{\mathscr{E}}\bar{\mathscr{E}} = \bar{\mathscr{E}}^2.$$

$\sqrt{\overline{(\Delta\mathscr{E})^2}}$ is called the *absolute fluctuation* of the energy. This quantity characterizes the average extent to which the energy deviates from its mean value. The ratio of the absolute fluctuation $\sqrt{\overline{(\Delta\mathscr{E})^2}}$ to the modulus $|\bar{\mathscr{E}}|$ is called the *relative fluctuation* of the energy. It is a measure of the relative fraction of the energy deviation from its mean value.

Naturally, the definitions of absolute and relative fluctuation retain their meaning also for other quantities that describe a subsystem, and not only for energy.

Calculating the energy fluctuation from the Gibbs distribution. We now apply the Gibbs distribution to the calculation of energy fluctuation in a subsystem. To do this, we differentiate, with respect to θ, the following two identities from which F and $\overline{\mathscr{E}}$ are determined:

$$\sum_{\mathscr{E}} e^{\frac{F-\mathscr{E}}{\theta}} g(\mathscr{E}) = 1 .$$

This expresses the normalization condition (45.11), and the definition of mean energy (45.15)

$$\sum_{\mathscr{E}} \mathscr{E} e^{\frac{F-\mathscr{E}}{\theta}} g(\mathscr{E}) = \overline{\mathscr{E}}.$$

$\mathscr{E}$ and $g(\mathscr{E})$ are purely mechanical quantities and do not depend upon the Gibbs distribution parameter θ. Therefore, only F, $\overline{\mathscr{E}}$, and, of course, θ itself need be differentiated with respect to θ. We thus obtain

$$\sum_{\mathscr{E}} \left(\frac{1}{\theta}\frac{\partial F}{\partial \theta} - \frac{F-\mathscr{E}}{\theta^2}\right) e^{\frac{F-\mathscr{E}}{\theta}} g(\mathscr{E}) = 0, \tag{45.19}$$

$$\sum_{\mathscr{E}} \left(\frac{1}{\theta}\frac{\partial F}{\partial \theta} - \frac{F-\mathscr{E}}{\theta}\right) \mathscr{E} e^{\frac{F-\mathscr{E}}{\theta}} g(\mathscr{E}) = \frac{\partial \overline{\mathscr{E}}}{\partial \theta}. \tag{45.20}$$

From (45.19) we have

$$\frac{1}{\theta}\frac{\partial F}{\partial \theta} = \frac{F}{\theta^2}\sum_{\mathscr{E}} e^{\frac{F-\mathscr{E}}{\theta}} g(\mathscr{E}) - \frac{1}{\theta^2}\sum_{\mathscr{E}} \mathscr{E} e^{\frac{F-\mathscr{E}}{\theta}} g(\mathscr{E}) = \frac{F-\overline{\mathscr{E}}}{\theta^2}.$$

Substituting this in (45.20) we find

$$\frac{\partial \overline{\mathscr{E}}}{\partial \theta} = \sum_{\mathscr{E}} \left(-\frac{F-\mathscr{E}}{\theta^2} + \frac{F}{\theta^2} - \frac{\overline{\mathscr{E}}}{\theta^2}\right) \mathscr{E} e^{\frac{F-\mathscr{E}}{\theta}} g(\mathscr{E}) =$$

$$= \frac{1}{\theta^2}\sum_{\mathscr{E}} (\mathscr{E}^2 - \overline{\mathscr{E}}\mathscr{E}) e^{\frac{F-\mathscr{E}}{\theta}} g(\mathscr{E}). \tag{45.21}$$

As a constant, the quantity $\overline{\mathscr{E}}$ may be taken outside the averaging sign. In accordance with (45.18), this gives

$$\theta^2 \frac{\partial \overline{\mathscr{E}}}{\partial \theta} = \overline{\mathscr{E}^2} - \overline{\mathscr{E}}^2 = \overline{(\Delta \mathscr{E})^2}. \tag{45.22}$$

Whence the relative fluctuation is

$$\frac{\sqrt{\overline{(\Delta \mathscr{E})^2}}}{|\overline{\mathscr{E}}|} = \frac{1}{|\overline{\mathscr{E}}|} \theta \sqrt{\frac{\partial \overline{\mathscr{E}}}{\partial \theta}}. \tag{45.23}$$

But this quantity is inversely proportional to the square root of the number of particles in the subsystem, because the energy, as an additive quantity, is proportional to the number of particles.

Let us illustrate this in the case of an ideal gas. From (40.17) $\overline{\mathscr{E}} = 3N\theta$. Hence, the relative fluctuation is $\sqrt{\frac{2}{3N}}$. For example, $N = 2.7 \times 10^{19}$ for 1 cm³ of a gas under normal conditions, so that the relative energy fluctuation is a few parts in 10^{10}.

For the greater part of its time, the energy of 1 cm³ of gas differs from its mean value by this small fraction. Nevertheless, in the later statistical development it is more convenient to regard the energy of a subsystem as only slightly fluctuating, and not strictly constant, as in an absolutely closed system.

Naturally, the relative fluctuation for a separate gas molecule is not a small quantity. Thus, from (40.14) and (40.15), the fluctuation in the velocity of a molecule is

$$\frac{\sqrt{\overline{v^2} - \overline{v}^2}}{\overline{v}} = \frac{\sqrt{\frac{3\theta}{m} - \frac{8\theta}{\pi m}}}{\sqrt{\frac{8\theta}{\pi m}}} = \sqrt{\frac{1.42}{8}} = 0.42.$$

Thus, we have shown that the probability of a given value of subsystem energy $\mathscr{E}$ has a very sharp maximum close to $\mathscr{E} = \overline{\mathscr{E}}$. This maximum becomes sharper, the larger the subsystem.

Entropy. Proceeding from the probability density for a subsystem, we can form an analogous function for a closed system. Utilizing the fact that probabilities are multiplicative quantities we obtain

$$P = \prod_i P_i = \prod \rho_i g_i = \prod_i \rho_i \cdot \prod_i g_i. \tag{45.24}$$

We now make use of the explicit form of the Gibbs distribution (45.10). Then for the product of all the functions ρ_i we obtain

$$\prod_i \rho_i = \prod_i e^{\frac{F_i - \mathscr{E}_i}{\theta}} = e^{\frac{\sum_i F_i - \sum_i \mathscr{E}_i}{\theta}} = e^{\frac{\sum_i F_i - \mathscr{E}}{\theta}}.$$

But if the large system is closed, $\sum \mathscr{E}_i = \mathscr{E} = \text{const}$ and, hence, $\prod_i \rho_i = \text{const}$ also. Thus, the probability of a state is proportional to the statistical weight of that state

$$P \sim \prod_i g_i = G(\mathscr{E}). \tag{45.25}$$

And all states of the system with the same energy are equally probable: the probability of $G(\mathscr{E})$ states is proportional to the number $G(\mathscr{E})$ (see Sec. 39).

As has already been repeatedly pointed out, ideally closed systems do not exist in nature. Speaking of a "closed" system, we mean that equilibrium for its subsystems is established more rapidly than the entire large system attains equilibrium with the surrounding medium. During the time that it takes for equilibrium to be established between the subsystems, the additive integrals of the large system do not have time to change noticeably. Hence, we can distinguish between the concept of statistical equilibrium in the whole system and in its subsystems.

It is obvious that statistical equilibrium in the large system is maintained for a longer period than the equilibrium established inside the subsystems alone. Therefore, the probability of the fuller equilibrium is, simply by definition, greater than the probability of the less full equilibrium. From (45.25), a measure of the probability for a large system is the statistical weight of its state. Therefore, the closer a system is to statistical equilibrium, the greater the statistical weight of the state of a closed system, so that $G(\mathscr{E})$ is a measure of the closeness of a large system to equilibrium. In this way we may also regard the quantity $g_i(\mathscr{E})$ of each ith subsystem as a measure of its closeness to equilibrium (internally) for those time intervals during which the subsystem may be regarded as quasi-closed.

For any, not too small, interval of time we can indicate systems which remain almost closed during this interval. For them the quantity G is a measure of the equilibrium of their states: the larger G is, the closer the subsystems of a given "closed" system are to equilibrium.

It is more convenient, as a measure of the closeness of a system to statistical equilibrium, to use $\ln G$ instead of the statistical weight G itself, since it possesses the property of additivity. $\ln G$ is called the *entropy* of a system and is denoted by the letter S:

$$S = \ln G. \tag{45.26}$$

It was shown in the previous section that the state of a Fermi gas at absolute zero is defined uniquely: $G = 1$. Hence, the entropy of a Fermi gas at $\theta = 0$ is $\ln 1 = 0$. At absolute zero, a Bose gas occurs completely in a zero energy state (see Sec. 43). Hence, its state is also uniquely defined, i. e., $S = 0$.

The entropy of a subsystem. By definition entropy is an additive quantity, so that

$$S = \ln G = \ln \prod_i g_i = \sum_i \ln g_i = \sum_i S_i. \tag{45.27}$$

It is seen from this equation that it is natural to call $\ln g_i = S_i$ the entropy of a subsystem. In order to calculate it, it is convenient to make use of the Gibbs distribution function for a subsystem. As was shown in this section, the energy of quasi-closed subsystem is very close to a constant value, namely, to its average value $\bar{\mathscr{E}}_i$, but is not strictly equal to it. Therefore, the formula for the entropy of a subsystem can be successfully applied also to a "closed" system, whose energy is strictly constant. The error here is determined by the relative fluctuation of quantities in the subsystem, i.e., it is negligibly small.

The entropy of a quasi-closed subsystem, equal to $\ln g_i(\mathscr{E}_i)$, should be represented as $\ln g_i(\bar{\mathscr{E}}_i)$, where $\bar{\mathscr{E}}_i$ is the mean value of the energy in a given subsystem in the case of "frozen" interaction with other subsystems. In other words, in finding this value $\bar{\mathscr{E}}_i$ it is taken that the subsystems do not arrive at a more complete equilibrium during the time considered.

We now take advantage of the fact that the energy fluctuations are small, and can replace the normalization condition (45.11) by the following simple relationship:

$$\sum_{\mathscr{E}_i} \rho(\mathscr{E}_i)\, g(\mathscr{E}_i) \approx \rho_i(\bar{\mathscr{E}}_i)\, g_i(\bar{\mathscr{E}}_i) = 1 . \tag{45.28}$$

Substituting $g_i(\bar{\mathscr{E}}_i)$ into the definition for the entropy of a subsystem, we find

$$S_i = \ln \frac{1}{\rho_i(\bar{\mathscr{E}}_i)} \cong \ln \frac{1}{\overline{\rho_i(\mathscr{E}_i)}} . \tag{45.29}$$

But since a logarithm is a slowly varying function, we can replace the logarithm of the mean value $\rho_i(\mathscr{E}_i)$ by the mean value of the logarithm $\ln \frac{1}{\rho_i}$:

$$S_i = \overline{\ln \frac{1}{\rho_i}} . \tag{45.30}$$

The resultant error is the smaller, the larger the subsystem, because the relative fluctuations tend to zero as the subsystem increases.

Substituting ρ_i from the Gibbs distribution (45.10) into (45.30), we find the following expression for entropy (omitting the index i):

$$S = \overline{\ln \frac{1}{\rho}} = \overline{\ln \left(e^{\frac{\mathscr{E} - F}{\theta}}\right)} = \frac{\bar{\mathscr{E}} - F}{\theta} . \tag{45.31}$$

Comparing this with (45.21), we obtain

$$S = -\frac{\partial F}{\partial \theta} . \tag{45.32}$$

Replacing F by $\bar{\mathscr{E}} - \theta S$ in this equation, we find, after differentiation

$$S = -\frac{\partial(\overline{\mathscr{E}} - \theta S)}{\partial\theta} = -\frac{\partial\overline{\mathscr{E}}}{\partial\theta} + S + \theta\frac{\partial S}{\partial\theta}\,;\quad \theta\frac{\partial S}{\partial\theta} = \frac{\partial\overline{\mathscr{E}}}{\partial\theta}\,.$$

Here, differentiation occurs with respect to θ under constant external conditions, of which $\overline{\mathscr{E}}$ and F can also be a function. Eliminating $\partial\theta$ we find

$$\theta = \frac{\partial\overline{\mathscr{E}}}{\partial S}\,. \tag{45.33}$$

Phase space. In conclusion we give a classical formulation of Liouville's theorem. In classical mechanics particle coordinates and momenta exist simultaneously. A system of N particles is characterized by a set of $3N$ coordinates q_i and $3N$ momenta p_i. As was shown in Sec. 10, these variables may be regarded as independent. In order to make the set of $6N$ variables more vivid, we shall plot them on the axes of a $6N$-dimensional coordinate system. This does not introduce anything fundamentally new but it somewhat simplifies the consideration due to associations with two- and three-dimensional space which are inherent in geometrical modes of expression. Such an imaginary $6N$-dimensional space is termed a *phase space* in mechanics. A single point in phase space specifies the state of a whole mechanical system, because the point is defined by all $3N$ coordinates and $3N$ momenta. As time passes the momenta and coordinates vary in accordance with Hamilton's equations (10.18):

$$\frac{dq_i}{dt} = \frac{\partial H}{\partial p_i}\,;\quad \frac{dp_i}{dt} = -\frac{\partial H}{\partial q_i}\,. \tag{45.34}$$

A point describing the state of a system moves in phase space describing a *path in phase space*. As an example, let us find the phase trajectory of a system with one degree of freedom—the linear harmonic oscillator. It is obtained from the energy conservation law

$$\frac{p^2}{2m} + \frac{m\omega^2 q^2}{2} = H = \mathscr{E} = \text{const.}$$

This is the equation of an ellipse with semiaxes $\sqrt{2m\mathscr{E}}$ and $\sqrt{\dfrac{2\mathscr{E}}{m\omega^2}}$.

Classical Liouville's theorem. We shall now consider a set of systems with the same Hamiltonian, but with somewhat different initial conditions. These systems will move along different phase paths. Let us suppose that the initial points densely fill some volume element in phase space. As each point moves, this volume element will be displaced similar to the volume of a flowing liquid. We shall prove that the magnitude of the volume remains unchanged when the points move, so that the motion resembles the flow of an incompressible liquid.

A volume element of phase space is expressed analogously to a three-dimensional volume element:

$$d\Gamma \equiv dp_1 \ldots dp_{3N}\, dq_1 \ldots dq_{3N}\,. \tag{45.35}$$

We assume that at the initial instant of time there are dn phase points in this volume element, i.e., the initial coordinates and momenta of the dn systems with the same Hamiltonian are contained between p_1 and $p_1 + dp_1$, p_2 and $p_2 + dp_2$, etc. The density of the systems in phase space is determined by the usual equation

$$dn = \rho\, d\Gamma. \tag{45.36}$$

We shall now prove that the density ρ does not change in the case of motion of the phase points. For simplicity in notation, we shall consider a two-dimensional space, though all the reasoning can be directly extended to a $6N$-dimensional space. We shall proceed from the fact that the total number of points dn is conserved in motion. Therefore, a decrease in the number of points inside some *fixed* volume $d\Gamma$ in unit time is equal to their flux across the surface bounding the volume. In the case of a two-dimensional region $d\Gamma = dp\, dq$, the flux across the side dp is equal to $\rho(q)\, \dot{q}(q)\, dp$ (see Fig. 54), and the flux across the opposite side is $\rho(q+dq)\, \dot{q}(q+dq)\, dp$. The total flux across all four sides is

$$\rho(q+dq)\, \dot{q}(q+dq)\, dp - \rho(q)\, \dot{q}(q)\, dp + \rho(p+dp)\, \dot{p}(p+dp)\, dq - \rho(p)\, \dot{p}(p)\, dq = \\ = \left[\frac{\partial}{\partial q}(\rho \dot{q}) + \frac{\partial}{\partial p}(\rho \dot{p})\right] dp\, dq\,,$$

where the same expansions have been used as in the derivation of the Gauss-Ostrogradsky theorem (Sec. 11). The resulting flux must be equated to the decrease of particles in the volume, i.e., to the quantity $-\frac{\partial \rho}{\partial t} d\Gamma$. As a result, the same equation is obtained as that of charge conservation in electrodynamics [see (12.18)]:

$$-\frac{\partial \rho}{\partial t} = \frac{\partial}{\partial q}(\rho \dot{q}) + \frac{\partial}{\partial p}(\rho \dot{p}) = \dot{q}\frac{\partial \rho}{\partial q} + \rho \frac{\partial \dot{q}}{\partial q} + \dot{p}\frac{\partial \rho}{\partial p} + \rho\frac{\partial \dot{p}}{\partial p}\,. \qquad (45.37)$$

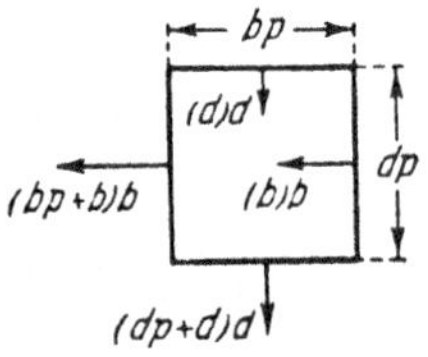

Fig. 54

But from Hamilton's equations (45.2)

$$\frac{\partial \dot{q}}{\partial q} = -\frac{\partial \dot{p}}{\partial p} = \frac{\partial^2 H}{\partial p\, \partial q}\,,$$

so that equation (45.5) may be rewritten as

$$\frac{\partial \rho}{\partial t} + \dot{q}\frac{\partial \rho}{\partial q} + \dot{p}\frac{\partial \rho}{\partial p} = \frac{d\rho}{dt} = 0\,. \qquad (45.38)$$

In other words, the *density of phase points is an integral of motion.* This is Liouville's theorem in classical formulation.

Strictly speaking, the classical density function ρ defined here has a meaning other than the quantum probability density in formula (45.1). We shall now show in what way these concepts are closely related.

If we observe the same quasi-closed subsystem very many times, taking different instants of time as origins, the result will be something in the nature of a set of identical systems with differing initial conditions. The definition of phase point density (45.36) is analogous to the definition of probability density (45.1), because, to the classical approximation, statistical weight is written in the following manner:

$$dG = \frac{d\Gamma}{(2\pi\hbar)^{3N}} \qquad (45.39)$$

[cf. (39.32)], putting $dV = dx\, dy\, dz$, i.e., for an infinitely small volume.

The difference in the two definitions of density is that the probability of a state in quantum mechanics can be determined directly as limit $\frac{t(\mathscr{E})}{t}$, while in classical mechanics we are to understand by probability the limit $\frac{n(\mathscr{E})}{n}$, where $n(\mathscr{E})$ denotes the number of times that a phase point appears in the region of the state under consideration with given energy $\mathscr{E}$, and n is the total number of observations. All the individual results of observations are regarded as equally probable. But the proof of this equiprobability is the basic difficulty in a nonquantum version of statistics.

Closed and quasi-closed systems. We have seen that two methods of statistical construction are possible.

1) Proceeding from an ideally closed system, in which the energy is strictly conserved, the problem is to determine the entropy as $\ln G(\mathscr{E})$. The state for which the entropy is maximum is the equilibrium state. The temperature is then determined as $\theta = \frac{\partial \overline{\mathscr{E}}}{\partial S}$.

2) Proceeding from a quasi-closed system, introduce temperature from the Gibbs distribution. Then entropy is calculated from the formula $\overline{\ln \frac{1}{\rho}}$

Both methods are, of course, equivalent, but it is far more convenient to use the Gibbs distribution in various applications.

Exercises

1) Verify Liouville's theorem in the case of four points in a gravitational field. At the initial instant of time, the points form a rectangle with sides Δz and Δp_z in a two-dimensional phase plane.

Two points possessing larger initial momenta will fall faster. Therefore, in the motion of the phase points, the rectangle becomes a parallelogram. Its height will equal Δp_z and its base Δz, so that its area will remain equal to the initial area of the rectangle. Conservation of phase volume is equivalent to conservation of phase density, i.e., to Liouville's theorem.

2) Verify Liouville's theorem for the three harmonic oscillators:

$$p_1 = \sqrt{2m\mathscr{E}}\sin\omega t; \quad p_2 = \sqrt{2m(\mathscr{E}+\Delta\mathscr{E})}\sin\omega t; \quad p_3 = \sqrt{2m\mathscr{E}}\sin(\omega t + \delta);$$

$$x_1 = \sqrt{\frac{2\mathscr{E}}{m\omega^2}}\cos\omega t; \quad x_2 = \sqrt{\frac{2(\mathscr{E}+\Delta\mathscr{E})}{m\omega^2}}\cos\omega t; \quad x_3 = \sqrt{\frac{2\mathscr{E}}{m\omega^2}}\cos(\omega t + \delta).$$

The area of the triangle is expressed in terms of the coordinates of its vertices in the following way:

$$F = \frac{1}{2}\left| (p_2 x_3 - p_3 x_2) - (p_1 x_3 - p_3 x_1) + (p_1 x_2 - p_2 x_1) \right|,$$

whence, substituting coordinates and momenta, the statement is proved.

3) Find how the entropy of an ideal Boltzmann gas, consisting of N separate atoms, depends upon energy and volume.

We proceed from the definitions (45.39) and (45.26):

$$S = \ln\Gamma = \ln \int \frac{d\tau_{p_1}\, d\tau_{p_2} \ldots d\tau_{p_N} \ldots dx_1 \ldots dz_{3N}}{(2\pi h)^{3N}} = \ln\left(V^N \frac{\int d\tau_{p_1} \ldots d\tau_{p_N}}{(2\pi h)^{3N}} \right),$$

where $d\tau_p = dp_x\, dp_y\, dp_z$ for a separate atom.

It is convenient, first of all, to evaluate the integral for all states in which the energy of the gas is less than the given value $\mathscr{E}$. The momenta of the atoms for all states with energy less than $\mathscr{E}$ satisfy the inequality

$$p_{x_1}^2 + p_{y_1}^2 + p_{z_1}^2 + p_{x_2}^2 + p_{y_2}^2 + p_{z_2}^2 + \ldots + p_{x_N}^2 + p_{y_N}^2 + p_{z_N}^2 < 2m\mathscr{E}$$

(the momenta of the atoms are numbered, corresponding to non quantum statistics). The $3N$-dimensional space region over which the integration is performed is analogous to a sphere in a three-dimensional space. The coordinates of the points inside the sphere satisfy the inequality

$$x^2 + y^2 + z^2 \leq R^2,$$

where R is the radius of the sphere. The radius of the $3N$-dimensional sphere is $\sqrt{2m\mathscr{E}}$. Therefore, it is clear from the dimensionality that the volume in a $3N$-dimensional space is proportional to $(\sqrt{2m\mathscr{E}})^{3N}$ just as in three dimensions it is proportional to R^3 (the N-dependent coefficient in this exercise will not be determined). The number of states between $\mathscr{E}$ and $\mathscr{E} + d\mathscr{E}$ is proportional to

$$\frac{d}{d\mathscr{E}}\int d\Gamma \sim (\sqrt{\mathscr{E}})^{3N-2}.$$

The entropy is equal to the logarithm of the statistical weight, so that it involves the term $\left(\frac{3N}{2} - 1\right)\ln \mathscr{E}$. Neglecting unity compared with $\frac{3N}{2}$, we obtain a formula expressing entropy in terms of the energy and volume of the gas:

$$S = \frac{3N}{2}\ln \mathscr{E} + N \ln V + \text{const.}$$

Sec. 46. Thermodynamic Quantities

Statistics and thermodynamics. The results of the previous section may appear somewhat abstract if they are not related to the real, measurable properties of macroscopic bodies, for example, specific heat, thermal expansion, compressibility under pressure, etc. In turn, these properties are determined as derivatives of energy and volume with respect to temperature, of volume with respect to pressure, and, in general, as derivatives of various mean values and parameters in the Gibbs distribution with respect to other quantities.

A quantity such as θ (called the distribution modulus) is essentially defined by the way that it appears in the Gibbs distribution, since, in calculating mean values, θ appears under the summation or integral sign as a parameter. Therefore θ is one of the quantities specifying the macroscopic state of a system, since the mean values characterizing this state are related to it.

The properties of the mean macroscopic quantities defining the state of a body form the subject of *thermodynamics*. These properties are expressed in the form of a series of relationships—differential and integral—which will be obtained and interpreted in the present section.

Historically, thermodynamics appeared before statistics. It was usual to base it upon two postulates or laws (see below) which were supported by a vast quantity of experimental facts. The "laws" are now no longer postulates, since they are based on statistical methods.

It must not be thought that with the advent of statistics thermodynamics lost its importance. Thermodynamics shows how real, experimentally observed macroscopic quantities defining the thermal, chemical, etc., properties of macroscopic bodies are interrelated. In those cases where it is impossible to calculate some quantity by statistical methods (due to a lack of knowledge of the elementary laws of force interaction, or because of great mathematical complexity), thermodynamics shows how this quantity may be found, directly or indirectly, from measurement.

But statistics is not only the basis of thermodynamics. Above all, statistics indicates the way that thermodynamic quantities can be calculated from the microscopic structure of bodies. In addition, statistics makes it possible to calculate in advance by how much the actual values differ from their mean values. As we have seen, this type of deviation is measured by the fluctuations of the quantities [see (45.22)]. Under certain conditions, fluctuations manifest themselves in such a way that they can be recorded experimentally (see Sec. 48).

Thermodynamics is one of the most important chapters of statistical physics. Therefore, we shall describe it on the basis of statistics, preferring a systematic to an historical account.

Quantity of heat. A quasi-closed macroscopic system spends the greater part of its time in a state of statistical equilibrium. In this state, the actual values of quantities are almost constant, and are close to their mean values.

The closeness to the equilibrium in a system is defined by the total entropy of all its subsystems, on the assumption that interaction of the system with the surrounding medium occurs considerably more slowly than interaction between its subsystems.

We shall now consider in what way interaction between subsystems affects the macroscopic quantities characterizing their states.

Let two bodies be brought into contact in such a way that the external conditions and the number of particles in each of them remain unchanged. Then, in accordance with the differential equation (45.33), the mean energy increment of each subsystem is proportional to its entropy increment:

$$d\bar{\mathscr{E}} = \theta\, dS, \tag{46.1}$$

where, after the stipulations concerning the constancy of external conditions and the dimensions of the bodies have been made, the partial differentials are replaced by total differentials.

The total energy increment for two bodies isolated from external action is equal to zero:

$$d\bar{\mathscr{E}}_1 + d\bar{\mathscr{E}}_2 = 0. \tag{46.2}$$

The total entropy increment is positive or equal to zero, because, as a result of interaction, the bodies will arrive at statistical equilibrium between themselves, and this equilibrium is, of course, more complete than the equilibrium inside each of them. Hence,

$$dS_1 + dS_2 \geqslant 0. \tag{46.3}$$

Using (46.1) and (46.2), we obtain

$$d\overline{\mathscr{E}}_1\left(\frac{1}{\theta_1} - \frac{1}{\theta_2}\right) \geqslant 0. \tag{46.4}$$

If $\theta_1 > \theta_2$, then $\overline{d\mathscr{E}_1} < 0$, i.e., the first system transmits energy to the second. The transmission of energy is entirely due to contact interaction, i.e., to the microscopic forces between molecules. The energy transferred in this manner is termed *heat*, so that heat is not "a form of energy" but a mode of energy *transmission* (we shall examine this question in more detail further on).

In formula (46.4), θ_1 and θ_2 are parameters in the Gibbs distribution for each of the subsystems separately. As long as these parameters differ, the subsystems cannot occur as a unit in a state of statistical equilibrium. Approximation to equilibrium occurs as a result of heat transfer, with the heat always going to the subsystem in which the parameter θ is least. Only then are θ_1 and θ_2 the same, the macroscopic quantities of heat are no longer transferred, and the energy of each subsystem exhibits only small fluctuations in the vicinity of its equilibrium value. If one of the systems is an ideal Boltzmann gas, then, as we know, θ is proportional to the absolute temperature, since the Gibbs distribution for a gas as a whole leads to a Boltzmann distribution for the individual molecules with the same parameter θ. The absolute temperature of a gas can be determined from independent (not thermal) measurements in the Clapeyron equation $pV = RT$. It is natural to consider that the quantity θ, for any system other than an ideal gas, is also nothing other than temperature. If a system is in equilibrium with an ideal gas, then its value θ is proportional to the absolute temperature of the gas. Thus, θ is the temperature measured in absolute units (ergs) if the ideal gas is taken as a *thermometric* substance. A little later in this section a definition of temperature will be given which does not depend upon the choice of the thermometric substance.

A Gibbs distribution occurs for any group of quasi-independent subsystems, including those that have not arrived at a state of mutual statistical equilibrium. Even though the quantity θ in this case, too, is, by *definition*, the same for all subsystems—which follows from the multiplicativity of the distribution function $\rho(\mathscr{E})$ [see (45.4)-(45.8)]—it must not be regarded as equal to the *temperature* of the large system, which, generally speaking, cannot be defined for a system not in equilibrium. If the subsystems in this case are in internal equilibrium, they are characterized by their *own* Gibbs distribution, which cannot be represented by a factor involved in the Gibbs distribution of the large system because the parameters θ of both distributions are different. Thus, the distribution modulus θ of an *equilibrium* system is a measure of its temperature.

Taking the example of temperature, it can be seen that quantities which are defined statistically can be identified with actually measured thermodynamic quantities. Any statistical quantity can be regarded as defined when, and only when, there is given a unique group of operations (of measurement and calculation) relating this quantity to real macroscopic quantities or to the microscopic parameters of a system which are found (or can be found) from experiment.

Work. The Hamiltonian function of a system usually depends not only upon generalized coordinates and momenta that vary according to dynamical laws, but also upon certain arbitrarily chosen parameters. The intensity of the external field, for example, may be such a parameter. The energy spectrum of the system, and hence the mean energy $\overline{\mathscr{E}}$ also, depends upon the parameters appearing in the Hamiltonian.

These parameters, transformed according to a given law, are termed the *external* parameters of the system. We denote them by the letter λ, where λ may mean any quantity of this type. As λ varies, the mean energy of the system also varies. It is obvious that it can only vary at the expense of some external source of energy. Since λ is a mechanical and not statistical quantity (λ is involved in the Hamiltonian!), the variation in λ is due to certain external mechanical work performed on the system, for example, a falling weight or a rotating motor.

The mechanical work performed with changing λ can be represented as

$$dA = \Lambda\, d\lambda, \tag{46.5}$$

where it is natural to call the quantity Λ the generalized force (since work is equal to the product of "force" Λ and "distance" $d\lambda$). If the entire energy change is due only to change in the external parameter λ, then

$$d\overline{\mathscr{E}} = \frac{\partial \overline{\mathscr{E}}}{\partial \lambda}\, d\lambda. \tag{46.6}$$

In formula (46.5), dA is the work performed on an external object due to a decrease in the energy of the system, so that $dA = -d\overline{\mathscr{E}}$. Comparing (46.6) and (46.5), we see that the mean quantity $\frac{\partial \overline{\mathscr{E}}}{\partial \lambda}$ is equal to the generalized force taken with opposite sign:

$$\Lambda = -\frac{\partial \overline{\mathscr{E}}}{\partial \lambda}. \tag{46.7}$$

The most frequent external parameter of a system is the volume that it occupies. In mechanical terms, this may be visualized by considering that the potential energy of any particle included in the given system is equal to infinity beyond the boundaries of the volume, i.e., an infinite amount of work is required even to remove a single particle from the volume. This is how the volume appears in the Hamiltonian of a system.

We shall consider that a system occupies the volume of some cylinder with a movable piston. The force acting on unit area of the piston is called the pressure and is denoted by the letter p. Then, if the whole area of the piston is f, the force acting on it is pf. When the piston is displaced through a distance dx, a quantity of work $dA = pf\,dx$ is performed on it. But the product $f dx$ is equal to the volume increment dV of the system. Hence, the change in energy for the system is

$$d\overline{\mathscr{E}} = -p\,dV \tag{46.8}$$

This type of energy change, produced by a change in the external parameters, is called work performed on a system.

It can be seen from formula (46.8) that pressure is a generalized force Λ related to a volume increment dV.

The first law of thermodynamics. It has already been pointed out that energy can be transmitted from one body to another by purely contact interaction, without any change in the macroscopic parameters or in particle interchange. This type of energy transfer is called heat transfer. It can be seen from this definition that the total energy increment of a body is made up of the work dA, performed by the body, and the quantity of heat dQ transmitted to the body:

$$d\overline{\mathscr{E}} = dQ - dA. \tag{46.9}$$

The sign in front of dA denotes that, if the body performs work, its energy diminishes.

The statement (46.9) looks like an identity from which the quantity of heat dQ is defined. Indeed, proceeding from a statistical interpretation of thermodynamics, we can be sure beforehand that the law of conservation of energy is applicable to thermal processes. Any energy imparted to a system without change of its external parameters must be transmitted by a contact. It is this energy that we have called the *quantity of heat*.

But thermodynamics appeared before statistics. Equation (46.9), interpreted thermodynamically, signifies that a quantity of heat can be measured in units of mechanical work, or that work can be measured in heat-quantity units. In other words, equation (46.9) extends the law of conservation of energy to thermal phenomena. Therefore, the establishment of the mechanical equivalent of heat by Mayer, Joule and Helmholtz signified a whole stage in the development of physical knowledge. Even though the earlier views on heat, as the disguised motion of molecules, were close to the modern statistical interpretation of thermal phenomena, they did not as yet contain any quantitative relationships. Therefore, the theory of heat could develop only after the relationship between thermal and mechanical quantities had been experimentally demonstrated. After the basic principles of thermodynamics had already been formulated, statistics began to develop as a physical, quantitative, theory.

Equation (46.9) can be reduced to another form. For this we must note that the energy of a body is a unique function of its state. Imagine a certain periodic process in which heat is delivered to a body and work is taken from it; this occurs in heat engines. We integrate equation (46.9) over one work cycle:

$$\int d\bar{\mathscr{E}} = \int dQ - \int dA\,. \tag{46.10}$$

The energy has the same value at the beginning and at the end of the cycle, thus producing the periodicity condition. Therefore, the total energy change in one cycle $\int d\bar{\mathscr{E}}$ is equal to zero. Hence,

$$\int dQ = \int dA\,. \tag{46.11}$$

The work performed by a heat engine in one cycle is equal to the heat delivered to the engine in that cycle. It is impossible to construct an engine which would work without an external supply of energy (heat). This statement is called "the first law of thermodynamics." An imaginary engine, performing work without an external source of energy, is called a perpetual motion engine of the first kind. The inevitable lack of success of all attempts to build such an engine finally led to the negative postulate called the first law of thermodynamics. Of course, if a statistical consideration lies at the basis of thermodynamics, then the first law emerges from the purely mechanical law of conservation of energy.

Neither work nor quantity of heat, taken separately, characterize the state of the body to which they are transmitted. In accordance with equation (46.11), a body may perform any number of work cycles, returning each time to the initial state. In doing so it obtains any amount of heat and performs any amount of work equal to this quantity of heat. Therefore, it is not correct to speak of the "heat reserve" that a body possesses. It only possesses an energy reserve which varies due to heat transfer and performance of work. Work and heat are not "forms of energy," but modes—macroscopic and microscopic—of energy transfer. This can be seen mathematically from the fact that dA and dQ are not total differentials of any quantities. For example, $dA = -p\,dV$. But pressure is a function not only of volume, but also of temperature. Thus, for an ideal gas $P = \frac{N\theta}{V}$, so that $dA = -\frac{N\theta}{V}dV$. This equation cannot be integrated until we know how the temperature θ varies with volume V in the given process. And so the quantity of heat and work characterize a process performed by a body and do not characterize the state of the body.

In certain cases the quantity of heat transferred in a process can be expressed very simply. If, for example, the volume of a body does not vary (an isochoric process) then $dA = 0$. In general, $dA = 0$ if the external parameters λ are constant, the quantity of heat being equal to the change in energy of the body:

$$dQ = d\bar{\mathscr{E}},\ Q = \Delta\bar{\mathscr{E}}\,(V = \text{const}). \tag{46.12}$$

If the pressure does not change (an isobaric process), $dA = -p\,dV = -d\,(pV)$. Then

$$dQ = d\bar{\mathscr{E}} + d\,(pV) = d\,(\bar{\mathscr{E}} + pV).$$

The quantity

$$\bar{\mathscr{E}} + pV \equiv I, \tag{46.13}$$

like energy, is a unique function of the state of the body. It is called the *heat function*, or the *enthalpy*, of a body and is denoted by the letter I. Thus, the quantity of heat in an isobaric process is equal to the change of enthalpy of the body

$$dQ = dI,\ Q = \Delta I\,(p = \text{const}). \tag{46.14}$$

Reversible processes. A definite state of statistical equilibrium corresponds to each value of the external parameters λ that describe a subsystem of any closed system. We can, for example, visualize a substance under a piston in a nonthermally isolated cylinder. In this case the substance and its external medium should be regarded as a single system. The external parameter defining the state of the system is, in this case, the volume V occupied by the substance.

For every value of the volume, a state of statistical equilibrium is established between the substance and its surrounding medium when the temperature of the substance and the medium is the same, and the total entropy has a maximum corresponding to a given value of the total energy of the whole system and to the volume V under the piston.

Let us now suppose that the external parameter of the subsystem λ varies so slowly that for every value of λ a total equilibrium inside the system has time to build up. In other words, the state of the system depends only upon the value of λ at the given instant. Maximum entropy corresponds to this value of λ, so that the system is all the time in a state of statistical equilibrium. But this means that in such a process the system never approaches a more complete statistical equilibrium, because it is not brought out of equilibrium. And since entropy is a measure of the fullness of the equilibrium, it does not vary with λ.

The constancy of entropy for slow variations in λ can be explained in the following way. Entropy is the logarithm of the number of equiprobable states of a system in a certain range of energy values close to $\mathscr{E}$. If λ varies infinitely slowly, the entire large system may be considered at each instant as *conservative* so that all its states remain equally probable (a rapid variation may stimulate transitions in some definite direction and, in this way, destroy the equiprobable nature of the states that follows from the principle of detailed equilibrium). The total number of states is conserved for slow variations of λ. The most probable *range* of states having an equal probability of occurrence is, in principle, determined (over all states, if they are known), purely by combinatorial analysis, and cannot therefore depend upon the particular value of λ for which the states are taken. For this reason, the number of states in the most probable region, and hence the entropy also, are conserved.

Thus, to each value of λ there corresponds a completely defined state of the system, quite independently of the way in which the value of λ varied prior to this, provided it varied slowly enough. Let λ change first from λ_1 to λ_2, and then from λ_2 to λ_1. Then, when λ varies in the reverse direction, the system goes through the same series of values that it passed through in the direct variation. This is why the process is termed *reversible*.

We can imagine the following two limiting cases.

1) The subsystem and the surrounding medium are all the time in statistical equilibrium, so that their temperature is the same. If the external medium is sufficiently large, its temperature does not change at all, and, hence, the temperature of the subsystem also remains unchanged in the process. Such a reversible process is termed *isothermal*. In an isothermal process, the total entropy of the whole system is conserved while the entropy of the subsystem, and, hence, of the medium also, varies.

2) The parameter λ varies so rapidly that an approach to statistical equilibrium between the medium and subsystem does not have time to occur, but at the same time the variation is so slow that the equilibrium inside the subsystem is not affected. Such a process would occur if the system were separated from the medium by an ideal thermally insulating barrier. Since the thermal transfer process is, in general, rather slow, we can easily imagine such rapid variations in

the parameter λ that there is not time enough for the heat to be transferred. In this process, the entropy of the subsystem and medium is conserved separately. For this reason, it is termed *isentropic* (or adiabatic).

Later we shall also consider some irreversible processes.

The second law of thermodynamics. Let us find an expression for the quantity of heat received by a system in a reversible process. As usual, we shall consider the given system to be a subsystem of some large closed system. The state of such a quasi-equilibrium subsystem is completely determined at each instant by its entropy and external parameters. According to (46.1) and (46.6), the energy increment, for a constant number of particles, is expressed in terms of the entropy increment in the following way:

$$d\bar{\mathscr{E}} = \theta\, dS + \overline{\frac{\partial \mathscr{E}}{\partial \lambda}}\, d\lambda, \tag{46.15}$$

which, from (46.5) and (46.7), can be also written as

$$d\bar{\mathscr{E}} = \theta\, dS - \Lambda\, d\lambda = \theta\, dS - dA\,, \tag{46.16}$$

whence it follows that

$$\theta\, dS = d\bar{\mathscr{E}} + dA\,. \tag{46.17}$$

But the right-hand side of the last equation is nothing other than the quantity of heat dQ received by the system. Hence, in a reversible process

$$dQ = \theta\, dS\,. \tag{46.18}$$

This is one of the most important equations in thermodynamics. It determines the entropy increment of a system in terms of the quantity of heat directly measured from experiment. It is significant that the quantity of heat obtained by a system in a process depends upon the development of the process, while the entropy increment is determined only by the initial and final states of the system. The quotient obtained from the division of an infinitely small quantity of heat (received by the subsystem in a reversible process) by the temperature is the total differential

$$\frac{dQ}{\theta} = dS\,. \tag{46.19}$$

If an irreversible process occurs inside the system, equation (46.19) may not hold. Indeed, let the system consist of two subsystems at different temperatures. In the process of temperature equalization, such a system approaches statistical equilibrium and its entropy increases. But no heat reaches the system from outside, so that dQ for the whole system is equal to zero and $dS > 0$.

Let us consider another example of an irreversible process. Let a gas expand into a vacuum. The phase volume $\Delta\Gamma$ [see (45.35), (45.39)] naturally increases, since the geometrical volume increases. But this means that the entropy also increases. When expanding into a vacuum, the gas does not perform work (since there are no opposition forces) and does not receive heat. In other words, it may be regarded as a closed system whose entropy increases as statistical equilibrium is approached (when a gas expands isothermally in a cylinder situated in an external medium, the entropy of the gas also increases, but the entropy of the medium decreases to the same extent). Consequently, the entropy increment in the case of an irreversible expansion of a gas into a vacuum is positive, and the quantity of heat transferred is equal to zero.

The two foregoing examples show that if an irreversible process occurs *inside* a system, then

$$\frac{dQ}{\theta} \leqslant dS, \tag{46.20}$$

since entropy increases without any heat transfer.

But if the given system irreversibly exchanges heat with other systems, and no irreversible processes occur inside it, then equation (46.19) is applicable.

Let us now determine, with the aid of equation (46.18), the amount of work that can be performed by a heat engine. By this term we understand a machine which periodically obtains heat from some thermal reservoir and, at the expense of this heat, performs work. According to the first law of thermodynamics, the total work performed by the engine in a single work cycle is equal to the quantity of heat received in that cycle (46.10). If the engine operates reversibly, then the quantity of heat is expressed in accordance with (46.18). Therefore

$$\int dA = \int \theta\, dS. \tag{46.21}$$

It follows from this that if the temperature of the working substance remains constant over one working cycle, the work is identically zero:

$$\int dA = \theta \int dS = 0, \tag{46.22}$$

since the initial state in a periodic process coincides with the final state, and the entropy is a single-valued function of the state, so that $\int dS = 0$. $dQ < \theta dS$ in irreversible processes, so that $\int dA \leqslant 0$ if the temperature is constant. By keeping the temperature constant, we can obtain a periodic process only at the expense of external work.

It follows from equation (46.22) that a heat engine cannot work using heat received from the surrounding medium, because by definition the medium is at a constant temperature. The statement formulated here is known as the *second law* of thermodynamics.

An imaginary engine operating from heat obtained from the surrounding medium is called a *perpetual motion engine of the second kind.* In an axiomatic account of thermodynamics, the impossibility of constructing such an engine is postulated (naturally on the basis of numerous failures to make a perpetual motion engine of the second kind), and the subsequent proofs are indirect: first it is assumed that the statement to be proved is false, and then the feasibility of such a perpetual motion engine is derived therefrom (see exercise 5).

A perpetual motion engine must not be confused with the so-called *free* engine, like the wind engine, which gets its energy from the sun heating up the earth.

Efficiency. For a heat engine to work, it is necessary to have the working cycle at two temperatures at least. The higher temperature θ_1 is usually called the *source* temperature, while the lower temperature θ_2 is the *sink* temperature. The work in one cycle is equal to

$$\int dA = \theta_1 \int_a^b dS + \theta_2 \int_b^a dS, \tag{46.23}$$

where the limit a refers to the initial and final state, and the limit b refers to the intermediate state. But $\int_a^b dS = -\int_b^a dS$ so that the work in one cycle is

$$\int dA = (\theta_1 - \theta_2) \int_a^b dS. \tag{46.24}$$

The total quantity of heat given up by the source is $\int dQ = \theta_1 \int_a^b dS$. The *efficiency of an engine* is the term used for the ratio of the work obtained from it to the quantity of heat taken from the source, since the principal losses are associated with obtaining this heat. From (46.24), the efficiency of a reversible engine (denoted by E) is

$$\mathrm{E} = \frac{\int dA}{\int dQ} = \frac{\theta_1 - \theta_2}{\theta_1} = 1 - \frac{\theta_2}{\theta_1}. \tag{46.25}$$

This equation shows that the efficiency of a reversible engine depends only upon the temperatures of the source and sink, and nothing else. Actually, θ_2 is either the temperature of the surrounding medium or a somewhat higher temperature. To increase E we must increase θ_1.

Equation (46.25) shows that the efficiency of a reversible engine can be taken to determine the absolute thermodynamic scale of temperature, independent of a thermometric substance. This scale coincides with the gas thermometer scale.

The efficiency of an irreversible engine is less than the efficiency of a reversible engine operating at the same temperature difference and at a given source temperature. Indeed, when (46.20) is taken into account, equation (46.24) is replaced by the inequality $\int dA \leqslant (\theta_1 - \theta_2)(S_a - S_b)$. This is why, given the same quantity of heat taken from the source, the work done by an irreversible engine is less than that of a reversible engine.

In axiomatic thermodynamics this statement is proved indirectly (see exercise 5). The efficiency of an irreversible engine is less because part of the heat obtained from the source is not spent in useful work, but in overcoming frictional forces or it is dissipated into the ambient medium in the engine itself, through the walls of the working cylinder, for example.

It should be noted that an ideally reversible engine would have to operate infinitely slowly, for otherwise finite deviations from statistical equilibrium would arise in it. Approximation to statistical equilibrium is always an irreversible process.

Thermodynamic identities for energy and enthalpy. Proceeding from the general equation (46.9), we can write down a general equation for the differential of the mean energy of a system in the case of a constant number of particles, if we take the volume as an external parameter

$$d\overline{\mathscr{E}} = \theta\, dS - p\, dV. \tag{46.26}$$

dS in this formula denotes the entropy increment, which is due to the reversible process in the system and the interaction with the surrounding medium. The state of a homogeneous system with a constant number of particles is defined by two quantities: volume and entropy. This can be seen from the number of independent parameters appearing in the Gibbs distribution: S and $\lambda = V$ can be taken instead of θ and $\lambda = V$; the energy of such a system is a single-valued function of entropy and volume. Let us take the total differential of the function $\overline{\mathscr{E}}(S, V)$:

$$d\overline{\mathscr{E}} = \left(\frac{\partial \overline{\mathscr{E}}}{\partial S}\right)_V dS + \left(\frac{\partial \overline{\mathscr{E}}}{\partial V}\right)_S dV, \tag{46.27}$$

where the index of the derivative denotes which quantity is kept constant during differentiation. Comparing (46.26) and (46.27), we have

$$\theta = \left(\frac{\partial \overline{\mathscr{E}}}{\partial S}\right)_V; \quad p = -\left(\frac{\partial \overline{\mathscr{E}}}{\partial V}\right)_S. \tag{46.28}$$

Differentiating θ with respect to V, and p with respect to S, we obtain an equation between the cross derivatives:

$$\left(\frac{\partial \theta}{\partial V}\right)_S = -\left(\frac{\partial p}{\partial S}\right)_V = \frac{\partial^2 \overline{\mathscr{E}}}{\partial V\, \partial S}. \tag{46.29}$$

The enthalpy or heat function I is connected with the energy by the relationship

$$\overline{\mathscr{E}} = I - pV$$

[see (46.13)]. Whence an expression follows for the total differential of enthalpy

$$dI = \theta\, dS + V dp, \tag{46.30}$$

which gives a series of differential relations

$$\theta = \left(\frac{\partial I}{\partial S}\right)_p; \quad V = \left(\frac{\partial I}{\partial p}\right)_S; \quad \left(\frac{\partial \theta}{\partial p}\right)_S = \left(\frac{\partial V}{\partial S}\right)_p = \frac{\partial^2 I}{\partial p\, \partial S}. \tag{46.31}$$

Equations (46.26) and (46.30) are known as thermodynamic identities. These identities permit expressing certain thermodynamic quantities in terms of others.

Free energy. If an irreversible process occurs in a system, then, from (46.20), $dQ \leqslant \theta\, dS$. Substituting this inequality into the equation of the first law (46.9), we have

$$d\overline{\mathscr{E}} \leqslant \theta\, dS - dA. \tag{46.32}$$

Thus, the work performed by the system will always satisfy the inequality

$$-dA \geqslant d\overline{\mathscr{E}} - \theta\, dS. \tag{46.33}$$

Let the process occur at a constant temperature. Then (46.33) can be written as a relationship between total differentials

$$-dA \geqslant d(\overline{\mathscr{E}} - \theta S). \tag{46.34 a}$$

If the system performs positive work, $dA > 0$. Reversing the sign of the inequality, we get

$$dA \leqslant -d(\overline{\mathscr{E}} - \theta S). \tag{46.34 b}$$

The quantity

$$\overline{\mathscr{E}} - \theta S \equiv F \tag{46.35}$$

[cf. (45.31)] appears in the Gibbs distribution (45.10); it is called the *free energy* of the system.

It follows from the inequality (46.34b) that the *greatest* amount of work that can be *performed by a system* at constant temperature is equal to the change in F, taken with opposite sign:

$$A_{\max} = F_1 - F_2. \tag{46.36}$$

Thus the work is equal to its maximum value in a reversible process.

The inequality (46.34a) has a somewhat different meaning: it determines the *least* amount of work which must be *performed on the system* in order to produce the given change of state in it:

$$A_{\min} = F_2 - F_1. \tag{46.37}$$

The entropy of the system and the surrounding medium (taken together) is conserved in these processes, and the inequality (46.32) becomes an equality.

Consider the following example. Let an ideal gas expand into a vacuum. No work is performed so that energy is conserved. But the energy of an ideal gas depends only upon its temperature (see Sec. 40), and not upon volume. Therefore, the temperature does not change during expansion into a vacuum. As we have seen, the entropy of the gas increases. Then the minimum work required to return the gas to its original volume at the same temperature is equal to the change in its free energy during expansion. The entropy of the gas will decrease in such a reversible compression, but on the other hand the entropy of the surrounding medium will increase to the same extent.

It is easy to obtain a thermodynamic identity for free energy. Differentiating the relationship between total and free energy, and substituting the identity (46.26), we obtain

$$dF = -S d\theta - p dV, \tag{46.38}$$

whence it is easy to find an expression for the entropy and pressure, and also an equation between the cross derivatives

$$p = -\left(\frac{\partial F}{\partial V}\right)_\theta, \quad S = -\left(\frac{\partial F}{\partial \theta}\right)_V, \quad \left(\frac{\partial p}{\partial \theta}\right)_V = \left(\frac{\partial S}{\partial V}\right)_\theta. \tag{46.39}$$

The relations (46.39) are convenient in that the independent variables are volume and temperature, which can be directly measured experimentally. Yet the thermodynamic identity for energy (46.26) involves entropy as an independent variable. But the entropy itself must be calculated, for example, by integration of (46.19).

From (45.12), the free energy F is expressed in terms of a statistical sum

$$F = -\theta \ln \sum e^{-\frac{\mathscr{E}}{\theta}}. \tag{46.40}$$

The right-hand side of this equation is expressed in terms of the temperature θ and the external parameters involved in the characteristic values of $\mathscr{E}$. But θ and λ are those very independent variables which are chosen in the identity (46.38). Therefore, for a determination of *all* the thermodynamic quantities it is sufficient to calculate the statistical sum $\sum e^{-\frac{\mathscr{E}}{\theta}}$. The actual calculation of this sum for an arbitrary system entails enormous mathematical difficulties. Actually, it is calculated only for ideal gases and crystals, and also for systems which deviate but little from ideal. It should be noted that even if it were possible to evaluate the statistical sum for some actual substance, say liquid water, the thermodynamic laws obtained with such very great difficulty would apply only to water and not to liquids generally. But the properties of ideal gases and crystals follow from statistics in a very general way.

Thermodynamic potential. Let us now determine the maximum work that can be performed by a system at constant temperature and pressure, equal to the temperature and pressure of the external medium. We note that in a homogeneous system with a constant number of particles, where there are no phase or chemical transitions, the state is completely defined by the temperature and pressure, since the thermodynamic identities for such systems involve two independent variables. In this case the specification of two quantities determines all the rest. But if a system consists of two phases of the same substance, for example, a liquid and its vapour, then the relationship between the fractions of liquid and gaseous substances may be quite arbitrary for a given temperature and pressure.

Work is performed in increasing the volume of a system. We can imagine, for example, a system in a cylinder under a piston, and the piston rod connected with some object capable of changing only its mechanical energy: by means of a flywheel or load. In addition, on expansion of the system work is done on the external medium. If we call the work on the object A, then the total work performed is equal to $-A-p\,\Delta\,V=-(A+\Delta\,pV)$. Here, p is the pressure in the external medium, which pressure in the process considered is equal to the pressure in the system. Since, by convention, the temperature of the system does not change, we have, from (46.33),

$$-(A+\Delta pV)\geqslant\Delta(\overline{\mathscr{E}}-\theta S)$$

or

$$-A\geqslant\Delta(\overline{\mathscr{E}}-\theta S+pV). \tag{46.41}$$

The quantity $\overline{\mathscr{E}}-\theta S+pV$ is, obviously, a function of the state of the system. It is called the *thermodynamic potential* and is denoted by the letter Φ:

$$\Phi\equiv\overline{\mathscr{E}}-\theta S+pV. \tag{46.42}$$

Thus, the maximum work which a system can perform at constant temperature and pressure is equal to the change in its thermodynamic potential (with reversed sign)

$$A_{\max}=\Phi_1-\Phi_2. \tag{46.43}$$

This work is performed in a reversible process.

When equilibrium is established in a system, work cannot be performed. Then Φ attains a minimum, because, according to (46.43), the work is performed at the expense of a decrease in Φ. When

$A_{\max}=0$, Φ cannot decrease further. It has already been pointed out that the process can occur at constant temperature and pressure with a phase transition or chemical transformation; hence, the equilibrium condition here is that Φ should be minimum.

Let us now find the thermodynamic relationships for Φ. From (46.42)

$$\Phi = F + pV. \tag{46.44}$$

Differentiating this equation and substituting dF from (46.38), we obtain

$$d\Phi = dF + pdV + Vdp = -Sd\theta - pdV + pdV + Vdp = \\ = -Sd\theta + Vdp. \tag{46.45}$$

Whence it follows, in familiar fashion, that

$$S = -\left(\frac{\partial \Phi}{\partial \theta}\right)_p, \quad V = \left(\frac{\partial \Phi}{\partial p}\right)_\theta, \quad \left(\frac{\partial S}{\partial p}\right)_\theta = -\left(\frac{\partial V}{\partial \theta}\right)_p = -\frac{\partial^2 \Phi}{\partial p \, \partial \theta}. \tag{46.46}$$

The thermodynamic potential depends only upon quantities that characterize the state of a body: its temperature and pressure. At the same time Φ is, of course, an additive quantity: if two equal volumes of the same substance are joined at the same temperature and pressure, the common thermodynamic potential will be twice as great as it was for each part separately. Therefore, we can write

$$\Phi = N\mu(p, \theta). \tag{46.47}$$

Here, μ is the thermodynamic potential related to a single molecule of the substance. μ is also called the chemical potential. We shall show later on that for ideal gases it is identical to the parameter μ in the distribution function (see Sec. 39). It is obvious that

$$\mu = \left(\frac{\partial \Phi}{\partial N}\right)_{p, \theta}. \tag{46.48}$$

If the system consists of several types of molecule, for example, a solution of one substance in another or a mixture of gases, then the state is determined not only by the temperature and pressure, but also by the concentrations of the substances. The concentration of the ith substance in a mixture is

$$c_i = \frac{N_i}{\sum\limits_k N_k}. \tag{46.49}$$

The chemical potential of the ith substance in a mixture is naturally expressed by analogy with (46.48):

$$\mu_i = \left(\frac{\partial \Phi}{\partial N_i}\right)_{p, \theta}, \tag{46.50}$$

where μ_i depends upon p, θ and all the concentrations: $c_1, c_2, \ldots, c_i, \ldots$.

Regarding N_i as variables, we can write the total differential of Φ in the following way:

$$d\Phi = -S d\theta + V dp + \sum_i \mu_i dN_i. \tag{46.51}$$

This equation generalizes (46.45) for the case of a variable number of particles.

Since the transition from $\bar{\mathscr{E}}$ to F and Φ does not involve the number of particles N_i, we can similarly generalize the differential relations (46.26) and (46.38):

$$d\bar{\mathscr{E}} = \theta dS - p dV + \sum_i \mu_i dN_i, \tag{46.52}$$

$$dF = -S d\theta - p dV + \sum_i \mu_i dN_i. \tag{46.53}$$

For a constant volume and for one type of molecule, (46.52) reduces to the form

$$d\bar{\mathscr{E}} = \theta dS + \mu dN. \tag{46.54}$$

But this equation coincides with (39.18), whence it can be seen that the quantity S introduced in Sec. 39 is the entropy of a gas and μ is its chemical potential.

Entropy in classical and quantum statistics. Let us compare the definition of entropy based on classical and on quantum laws of motion. In the latter case, entropy is defined as the logarithm of the number of states of a system for a certain energy value. When passing to a quasi-classical approximation, the number of states of the system is equal to the phase volume $\Delta\Gamma$ it occupies divided by $(2\pi\hbar)^n$, where n is the number of degrees of freedom [see (45.39)]. The logarithm of this ratio represents the entropy in the corresponding approximation. But statistics appeared before quantum mechanics. Therefore, entropy was originally defined in statistics as the logarithm of the denominate number $\Delta\Gamma$. In this definition, entropy depends upon the choice of units: if, for example, the unit of mass is changed by a factor two, then $n \ln 2$ must be added to the entropy. But since the units of measurement are arbitrary, it follows from this that in classical statistics entropy was defined only within the accuracy of an arbitrary additive constant. Only the *change* of entropy had strict meaning.

In quantum statistics, entropy is defined as the logarithm of an abstract number, and therefore does not depend upon the choice of units of measurement.

The temperature of a system is equal to absolute zero when the system is in the ground state, i.e., when it has the least possible energy. This state has a weight equal to unity, so that the entropy, or logarithm of the weight, becomes zero at the absolute zero of temperature. This statement is known as *Nernst's theorem*, which is sometimes called the third law of thermodynamics. Certain consequences of Nernst's theorem will be considered below (see exercise 6).

Exercises

1) Find the ratio between the specific heats at constant volume and at constant pressure.

With the aid of (46.18), we find, from the definition of specific heat,

$$C_V = \left(\frac{\partial Q}{\partial \theta}\right)_V = \theta\left(\frac{\partial S}{\partial \theta}\right)_V, \quad C_p = \left(\frac{\partial Q}{\partial \theta}\right)_p = \theta\left(\frac{\partial S}{\partial \theta}\right)_p.$$

The derivatives can be rewritten in the following way:

$$\left(\frac{\partial S}{\partial \theta}\right)_V = -\frac{\left(\frac{\partial V}{\partial \theta}\right)_S}{\left(\frac{\partial V}{\partial S}\right)_\theta}, \quad \left(\frac{\partial S}{\partial \theta}\right)_p = -\frac{\left(\frac{\partial p}{\partial \theta}\right)_S}{\left(\frac{\partial p}{\partial S}\right)_\theta}$$

from the formulae for the derivatives of implicit functions. The partial derivatives with the same subscript may be cancelled like fractions, since the differentials in them have the same sense. This gives

$$\frac{C_p}{C_V} = \frac{\left(\frac{\partial p}{\partial V}\right)_S}{\left(\frac{\partial p}{\partial V}\right)_\theta} = \frac{\left(\frac{\partial V}{\partial p}\right)_\theta}{\left(\frac{\partial V}{\partial p}\right)_S},$$

so that specific heats are related in the way that isothermal compressibility relates to isentropic compressibility. It is sufficient to measure only three of the four quantities C_p, C_V, $\left(\frac{\partial V}{\partial p}\right)_\theta$, $\left(\frac{\partial V}{\partial p}\right)_S$.

2) Calculate the derivative $\left(\frac{\partial \bar{\mathscr{E}}}{\partial V}\right)_\theta$.

We substitute $\bar{\mathscr{E}} = F + \theta S$; and, according to (46.39), transform the derivatives

$$\left(\frac{\partial \bar{\mathscr{E}}}{\partial V}\right)_\theta = \left(\frac{\partial F}{\partial V}\right)_\theta + \theta\left(\frac{\partial S}{\partial V}\right)_\theta = -p + \theta\left(\frac{\partial p}{\partial \theta}\right)_V.$$

If the pressure is known as a function of temperature and volume, the energy can be calculated only to the accuracy of an arbitrary temperature function

$$\bar{\mathscr{E}} = \int dV\left[-p + \theta\left(\frac{\partial p}{\partial \theta}\right)_V\right] + f(\theta).$$

Therefore, it must always be remembered that a determination of the relationship $p = p(V, \theta)$ does not yield complete information about the thermodynamic properties of a substance. In addition, any pressure term depending linearly upon temperature will not affect the energy, since it is eliminated from the equation obtained. For example, in all ideal gases $p = \frac{N\theta}{V}$, and the energy depends upon the temperature in a rather complicated way if discrete quantum levels must be taken in the statistical summations.

3) Find $\left(\frac{\partial I}{\partial p}\right)_\theta$. Answer: $V + \theta\left(\frac{\partial V}{\partial \theta}\right)_p$.

4) Find the difference between the specific heats at constant volume and at constant pressure.

The quantity of heat at constant pressure is equal to dI, and at constant volume, to $d\bar{\mathscr{E}}$ [see (46.14) and (46.12)]:

$$C_p = \left(\frac{\partial I}{\partial \theta}\right)_p, \quad C_V = \left(\frac{\partial \bar{\mathscr{E}}}{\partial \theta}\right)_V.$$

We transform C_p:

$$C_p = \left(\frac{\partial (\bar{\mathscr{E}} + pV)}{\partial \theta}\right)_p = \left(\frac{\partial \bar{\mathscr{E}}}{\partial \theta}\right)_p + p\left(\frac{\partial V}{\partial \theta}\right)_p.$$

Further, representing energy as $\bar{\mathscr{E}} = \bar{\mathscr{E}}\,[\theta, V(p, \theta)]$, we write the derivative $\left(\frac{\partial \bar{\mathscr{E}}}{\partial \theta}\right)_p$ in the form

$$\left(\frac{\partial \bar{\mathscr{E}}}{\partial \theta}\right)_p = \left(\frac{\partial \bar{\mathscr{E}}}{\partial \theta}\right)_V + \left(\frac{\partial \bar{\mathscr{E}}}{\partial V}\right)_\theta \left(\frac{\partial V}{\partial \theta}\right)_p = C_V + \left(\frac{\partial \bar{\mathscr{E}}}{\partial V}\right)_\theta \left(\frac{\partial V}{\partial \theta}\right)_p.$$

Whence

$$C_p - C_V = \left(\frac{\partial V}{\partial \theta}\right)_p \left[p + \left(\frac{\partial \bar{\mathscr{E}}}{\partial V}\right)_\theta\right] = \theta \left(\frac{\partial V}{\partial \theta}\right)_p \left(\frac{\partial p}{\partial \theta}\right)_V,$$

where we have used the result of exercise 2. The derivative $\left(\frac{\partial V}{\partial \theta}\right)_p$ is transformed thus:

$$\left(\frac{\partial V}{\partial \theta}\right)_p = -\frac{\left(\frac{\partial p}{\partial \theta}\right)_V}{\left(\frac{\partial p}{\partial V}\right)_\theta}.$$

Whence it follows that

$$C_p - C_V = -\theta \frac{\left(\frac{\partial p}{\partial \theta}\right)_V^2}{\left(\frac{\partial p}{\partial V}\right)_\theta}.$$

It will later be shown rigorously that $\left(\frac{\partial p}{\partial V}\right)_\theta < 0$, i.e., the pressure can only increase with decrease in volume (otherwise the state of the system is dynamically unstable, which is obvious as it is). Therefore $C_p > C_V$ always, and also

$$\left(\frac{\partial p}{\partial V}\right)_S > \left(\frac{\partial p}{\partial V}\right)_\theta.$$ For ideal gases $\left(\frac{\partial p}{\partial \theta}\right)_V = \frac{N}{V}$, $\left(\frac{\partial p}{\partial V}\right)_\theta = -\frac{N\theta}{V^2}$,

so that $C_p - C_V = N$.

5) Accepting the second law of thermodynamics as a postulate, prove that the efficiency of a reversible engine is always greater than the efficiency for an irreversible engine, working with the same temperature difference between source and sink.

The proof is indirect. Let a reversible engine and an irreversible engine obtain the same quantity of heat Q_1 from a source, but let the irreversible engine give a smaller quantity of heat Q_2' to the sink than the reversible engine. The reversible engine may be made to work as a refrigerator, i.e., to take heat from a cold reservoir and to deliver it to a hot reservoir at the expense of external work. In order to return a quantity of heat Q_1 to the hot reservoir, in accordance with our assumption, the reversible engine must take a larger quantity of heat from the cold reservoir than the irreversible engine delivered. But it will then turn out that the hot reservoir does not deliver heat at all, and the cold reservoir delivers a quantity of heat $Q_2 - Q_2'$, at the expense of which useful work is performed equal to the difference between the work of the irreversible engine and the work of the reversible engine operating as a refrigerator. The surrounding medium can be taken as the cold reservoir, so that useful work will be performed at the expense of heat obtained from the surrounding medium; this contradicts the second law of thermodynamics.

6) Prove that the specific heat of a system tends to zero when the temperature tends to absolute zero. Do the same for

$$\left(\frac{\partial V}{\partial \theta}\right)_p.$$

The entropy is related to the specific heat C by the relation

$$S = \int_0^\theta \frac{C}{\theta}\, d\theta,$$

where the lower limit of the integral is put equal to zero from Nernst's theorem. For the integral to have meaning, we must demand that $\lim_{\theta \to 0} C = 0$. In addition $\left(\frac{\partial V}{\partial \theta}\right)_p = -\left(\frac{\partial S}{\partial p}\right)_\theta$, and the limit of the last derivative is also equal to zero as θ tends to zero, because $\lim_{0 \to \theta} S = 0$ in the case of an arbitrary pressure.

7) Show that the sum of the enthalpy and kinetic energy is conserved in the motion of a substance without any internal heat exchange and without exchange of heat with the external medium.

Let a certain mass of substance be transferred from a volume V_1, pressure p_1, and energy $\bar{\mathscr{E}}_1$ to V_2, p_2, and $\bar{\mathscr{E}}_2$, respectively. In order to displace this mass from the volume V_1 at a pressure p_1, an amount of work $p_1 V_1$ must be done. Therefore, in going to p_2, V_2, a work $p_1 V_1 - p_2 V_2$ is done. The total change of energy of the given mass, in a reference system fixed in it, is equal to $\bar{\mathscr{E}}_1 - \bar{\mathscr{E}}_2 + p_1 V_1 - p_2 V_2 = I_1 - I_2$. Since there is no heat exchange, this quantity can be equal only to the change in kinetic energy

$$\frac{mv_2^2}{2} - \frac{mv_1^2}{2} = I_1 - I_2.$$

Hence,

$$\frac{mv_1^2}{2} + I_1 = \frac{mv_2^2}{2} + I_2.$$

In future we shall relate this equation to unit mass of the substance, and write it in the form:

$$I + \frac{v^2}{2} = \text{const},$$

where I is the enthalpy of unit mass.

8) Find the propagation velocity of small isentropic disturbances in an isotropic medium [in other words, neglecting heat transmission and considering that $p = p(\rho)$].

If the initial position of a particle is described by a single coordinate a, and the displaced position by the coordinate x, measured in the same direction, then the equation of conservation of mass is the following:

$$\rho_0\, da = \rho\, dx$$

(ρ is the density, ρ_0 is the initial density). Whence

$$\frac{\rho}{\rho_0} = \left(\frac{\partial a}{\partial x}\right)_t .$$

The force acting on an element of mass $\rho\, dx$ is

$$-p(x + dx) + p(x) = -\left(\frac{\partial p}{\partial x}\right)_t dx .$$

According to Newton's Second Law, this force is equal to the product of mass and acceleration, i.e.,

$$\rho\, dx \cdot \left(\frac{\partial^2 x}{\partial t^2}\right)_a = -\left(\frac{\partial p}{\partial x}\right)_t dx ,$$

whence we have

$$\left(\frac{\partial^2 x}{\partial t^2}\right)_a = -\frac{1}{\rho}\left(\frac{\partial p}{\partial x}\right)_t = -\frac{1}{\rho_0}\left(\frac{\partial p}{\partial a}\right)_t = -\frac{1}{\rho_0}\left(\frac{\partial p}{\partial \rho}\right)_S\left(\frac{\partial \rho}{\partial a}\right)_t =$$

$$= -\left(\frac{\partial p}{\partial \rho}\right)_S \frac{\partial}{\partial a}\frac{1}{\left(\frac{\partial x}{\partial a}\right)_t} = \frac{1}{\left(\frac{\partial x}{\partial a}\right)_t^2}\left(\frac{\partial p}{\partial \rho}\right)_S\left(\frac{\partial^2 x}{\partial a^2}\right)_t .$$

Considering the displacements small, we see that the derivative $\left(\frac{\partial x}{\partial a}\right)_t$ is close to unity, so that the second derivative is a small quantity. The result, therefore, is the approximate equation

$$\frac{\partial^2 x}{\partial a^2} - \frac{1}{\left(\frac{\partial p}{\partial \rho}\right)_S}\frac{\partial^2 x}{\partial t^2} = 0 .$$

It coincides with the wave equation of the form (17.4), which describes the propagation of disturbances with velocity c. In the given case, the propagation velocity of the process is nothing other than the velocity of sound. It is equal to $u = \sqrt{\left(\frac{\partial p}{\partial \rho}\right)_S}$.

9) A substance flows in a tube of constant cross-section without heat exchange, but with internal friction. Show that the maximum entropy is attained where the flow rate is equal to the velocity of sound.

The following conservation laws apply:

$$\rho v = j = \text{const},$$

$$I + \frac{v^2}{2} = \text{const}.$$

Substituting the velocity from the first equation, we obtain

$$I + \frac{j^2}{2\rho^2} = \text{const}.$$

Let us differentiate this equation, considering that the enthalphy is expressed in terms of the independent variables S and p:

$$\left(\frac{\partial I}{\partial S}\right)_p dS + \left(\frac{\partial I}{\partial p}\right)_S dp - \frac{j^2}{\rho^2} d\rho = 0.$$

Close to the entropy maximum $dS = 0$, the derivative $\left(\frac{\partial I}{\partial p}\right)_S = V = \frac{1}{\rho}$, so that

$$\left(\frac{\partial p}{\partial \rho}\right)_{dS=0} = \frac{j^2}{\rho^2} = v^2.$$

For constant entropy $\left(\frac{\partial p}{\partial \rho}\right)_S = u^2$, so that $v^2 = u^2$ at maximum entropy.

If there is a flux in the tube for which $v < u$ ("subsonic flow"), the value $v = u$ can only be attained at the tube outlet because, otherwise, the entropy, on reaching a maximum somewhere in the tube, would have to decrease in the subsequent flow, which is impossible.

10) A substance flows without heat exchange or friction, i.e., isentropically, in a tube of continuously variable cross-section f. Show that the velocity in subsonic flow increases with decrease in cross-section, but in supersonic flow, it increases with f. The flow is considered as one-dimensional because of the smooth variation of f.

From the law of conservation of mass

$$f\rho v = \text{const},$$

whence it follows that

$$\frac{df}{f} + \frac{d\rho}{\rho} + \frac{dv}{v} = 0.$$

Taking into account that entropy is constant, we can write:

$$\frac{d\rho}{\rho} = \frac{1}{\rho}\left(\frac{\partial \rho}{\partial p}\right)_S dp = \frac{1}{\rho}\frac{dp}{u^2}.$$

Differentiating the equation $I + \frac{v^2}{2} = $ const at constant entropy, we have

$$\left(\frac{\partial I}{\partial p}\right)_S dp + v\,dv = \frac{dp}{\rho} + v\,dv = 0.$$

Whence it follows that

$$\frac{d\rho}{\rho} = -\frac{v\,dv}{u^2}$$

and finally

$$\frac{dv}{v}\left(1 - \frac{v^2}{u^2}\right) = -\frac{df}{f}$$

which proves the statement. In order to obtain flow with supersonic velocity at the outlet of the tube, we must pass it through a *Laval nozzle*, i.e., along a tube whose aperture first decreases, so that $v = u$ at the narrowest place, after which v becomes greater than u and continues to increase.

11) A piston is in movement with constant velocity v, into a cylinder with cross-section f, filled with a substance with initial pressure p_0 and initial density ρ_0. The enthalpy I per unit mass is regarded as a known function of p and ρ. Formulate a system of equations, from which it is possible to determine the displacement velocity of the boundary between the compressed and non-compressed substance, and also the density and pressure of the compressed substance.

The compressed substance moves with velocity v equal to the piston velocity. The boundary between the compressed and noncompressed substance has a certain velocity D. The compressed substance has a velocity $v - D$ relative to this boundary, and the noncompressed substance has velocity D.

Let us pass to a reference system moving together with the interface. Then the mass conservation law is expressed as follows:

$$f \rho_0 D = f\rho (D - v). \qquad (*)$$

We shall consider a cylindrical volume of the substance passing through the boundary in unit time. The length of this cylinder in the compressed substance is equal to $D - v$, while its mass is $f \rho (D - v)$, so that its momentum is equal to $f \rho (D - v)^2$. The momentum in the noncompressed substance equalled $f \rho_0 D^2$. A resultant force $(p_0 - p) f$ acted on this cylinder, whence the conservation equation

$$f (p_0 + \rho_0 D^2) = f [p + \rho (D - v)^2]. \qquad (**)$$

The third equation expresses the absence of heat exchange (see exercise 7)

$$I_0 + \frac{D^2}{2} = I + \frac{(D - v)^2}{2}. \qquad (***)$$

At the interface, a discontinuous change occurs in the density, pressure, and velocity of the substance. This surface is called a ***shock wave***. We can determine D, p, and ρ from the three conservation laws, if the form of the function I (p, ρ) is known. These quantities will be determined specifically in exercise 7, Sec. 47, where it will also be shown that the compression process in a shock wave is irreversible.

12) Show that the classical expression for a statistical sum does not depend upon the constant magnetic field in which the system is situated.

The classical expression for the statistical sum (or more exactly, for the integral) is

$$Z = \int e^{-\frac{\mathscr{E}(\mathbf{p}_1, \mathbf{p}_2, \mathbf{p}_3, \ldots, \mathbf{p}_N; \mathbf{r}_1, \mathbf{r}_2, \ldots, \mathbf{r}_N)}{\theta}} d\tau_{\mathbf{p}_1} d\tau_{\mathbf{p}_2} d\tau_{\mathbf{p}_3} \ldots d\tau_{\mathbf{p}_N} \cdot dV_1 dV_2 \ldots dV_N.$$

When the system is placed in a magnetic field the momenta of the particles change according to the formula $\mathbf{p} \to \mathbf{p} - \frac{e}{c} \mathbf{A} = \mathbf{P}$. Passing (for the phase-volume element) from $d\tau_{\mathbf{p}}$ to $d\tau_{\mathbf{P}}$, we find that the statistical sum appears the same as in the absence of field because the new notation for the integration variable does not change anything:

$$Z = \int e^{-\frac{\mathscr{E}(\mathbf{P}_1, \mathbf{P}_2, \ldots, \mathbf{P}_N; \mathbf{r}_1, \mathbf{r}_2, \ldots, \mathbf{r}_N)}{\theta}} d\tau_{\mathbf{p}_1} d\tau_{\mathbf{p}_2} \ldots d\tau_{\mathbf{p}_N} \cdot dV_1 dV_2 \ldots dV_N.$$

Thus, classical mechanics cannot describe the magnetic properties of a substance.

13) Express the entropy of an ideal gas in terms of the occupation n_k for all three statistics ($g_k = 1$ everywhere).

Using the expressions for S in equations (39.14) and (39.25), we find:
Bose statistics:

$$S = \sum_k [(n_k + 1) \ln (n_k + 1) - n_k \ln n_k] ;$$

Fermi statistics:

$$S = -\sum_k [(1 - n_k) \ln (1 - n_k) + n_k \ln n_k] ;$$

Boltzmann statistics corresponding to $n_k \ll 1$:

$$S = -\sum_k n_k \ln \frac{n_k}{e} .$$

If the weight is not equal to unity, then, introducing $n_k \equiv g_k f_k$, we obtain for all three statistics:

$$S_{\text{Bose}} = \sum_k g_k [(f_k + 1) \ln (f_k + 1) - f_k \ln f_k],$$

$$S_{\text{Fermi}} = -\sum_k g_k [(1 - f_k) \ln (1 - f_k) + f_k \ln f_k],$$

$$S_{\text{Boltzmann}} = -\sum_k g_k f_k \ln \frac{f_k}{e} .$$

Sec. 47. The Thermodynamic Properties of Ideal Gases in Boltzmann Statistics

In this section we shall consider certain consequences that follow from the general principles of thermodynamics as applied to ideal gases. We shall suppose that the gas density is sufficiently small for Boltzmann statistics to be applied to its molecules. This does not mean that the motion of the molecules should be regarded as nonquantum; the quantization of rotational, vibrational (and all the more so, electronic) levels of a molecule must be taken into account in all cases when the spacing between neighbouring levels is comparable with θ (i.e., kT) or greater than θ. Even when the level spacing is sufficiently small compared with θ, as is the case of translational motion, the quantum of action should be left in the formula for the statistical weight of the states, since it would be impossible otherwise to obtain a unique expression for entropy.

Deviations from Boltzmann statistics that occur in gases at low temperatures or high densities are sometimes called "degeneracies." One should differentiate between deviations from the characteristic ideal gas state, due to the interaction between molecules, and quantum deviations from classical statistics. Of course, there also arise corrections which are due to the effect of both factors together.

Free energy of an ideal gas. As was indicated in the preceding section, it is convenient, when calculating thermodynamic quantities, to proceed from the expression for free energy.

We shall start with formula (46.40), reducing the statistical sum to the form that it takes for a Boltzmann gas. For this it is necessary to take into account that, by definition, a statistical sum is calculated over all the physically different states of a gas. But the state of the gas does not change if all possible molecular permutations are performed over the individual states; in nonquantum statistics such a permutation can be defined. The number of permutations of N molecules is equal to $N!$.

The total energy of an ideal gas separates into the sum of the energies of all of its molecules:

$$\mathscr{E} = \sum_{l=1}^{N} \varepsilon_l^{(k)},$$

where k is the number of the quantum state.

Substituting the expression for $\mathscr{E}$ into the statistical sum (46.40), and dividing this sum by the number of permutations of the molecules, we obtain

$$\sum e^{-\frac{\varepsilon}{\theta}} = \frac{\sum\limits_{(k)} e^{-\frac{1}{\theta}\sum\limits_{l=1}^{N} \varepsilon_l^{(k)}}}{N!} = \frac{\prod\limits_{l=1}^{N}\sum\limits_{k} e^{-\frac{\varepsilon^{(k)}}{\theta}}}{N!} = \frac{\left(\sum\limits_{k} e^{-\frac{\varepsilon^{(k)}}{\theta}}\right)^N}{N!}. \tag{47.1}$$

The second summation over (k) relates to all possible combinations of the energy of the separate molecules $\varepsilon_l^{(k)}$. Here, we have made use of the fact that the energy spectrum is the same for all molecules (if the gas consists of molecules of one type). The summation in (47.1) is performed over the spectrum of a single molecule. Replacing $N!$ by its expression in Stirling's formula, we arrive at a general formula for the free energy of an ideal gas under Boltzmann statistics.

$$F = -N\theta \ln \frac{e\sum\limits_{k} e^{-\frac{\varepsilon^{(k)}}{\theta}}}{N}. \tag{47.2}$$

Summation over translational degrees of freedom. It is expedient, in the statistical summation over the states of a separate molecule, to separate the translational degrees of freedom and represent the energy in the form

$$\varepsilon = \frac{p^2}{2m} + \varepsilon^{(i)}. \tag{47.3}$$

It is taken here that the energy does not depend upon the coordinates of the centre of mass of the molecule.

The statistical weight of a state with momentum p is equal to

$$g = g^{(i)} \frac{dp_x\, dp_y\, dp_z\, dx\, dy\, dz}{(2\pi h)^3}. \tag{47.4}$$

$g^{(i)}$ denotes the weight referring to an energy level $\varepsilon^{(i)}$. Integration over x, y, z contributes the factor $\int dx,\, dy,\, dz = V$ to the statistical sum. The integration over momenta is performed in a familiar manner:

$$\int\limits_{-\infty}^{\infty} e^{-\frac{p_x^2}{2m\theta}} dp_x = \sqrt{2\pi m\theta}. \tag{47.5}$$

Thus, the free energy for an ideal gas reduces to the following form:

$$F = -N\theta \ln\left[\frac{eV}{N}\left(\frac{m\theta}{2\pi h^2}\right)^{3/2}\left(\sum g^{(i)} e^{-\frac{\varepsilon^{(i)}}{\theta}}\right)\right] \equiv -N\theta \ln \frac{eV}{N} f(\theta). \tag{47.6}$$

A relationship is obtained here between free energy and volume. The function $f(\theta)$ depends upon the molecular structure.

Thermodynamic quantities of an ideal gas. It is easy to determine pressure from formula (47.6). From (46.39) we obtain

$$p = -\frac{\partial F}{\partial V} = \frac{N\theta}{V}, \tag{47.7}$$

i.e., the well-known Clapeyron equation. The thermodynamic potential is

$$\Phi = F + pV = F + N\theta = -N\theta \ln \frac{V}{N} f(\theta),$$

but here it is expressed in terms of volume. To be able to use identity (46.45) we must, in addition, replace $\frac{V}{N}$ by $\frac{\theta}{p}$, whence a final formula is obtained for the thermodynamic potential of an ideal gas:

$$\Phi = -N\theta \ln \frac{\theta f(\theta)}{p}. \tag{47.8}$$

We find the chemical potential with the aid of (46.47) or (46.48):

$$\mu = -\theta \ln \frac{\theta f(\theta)}{p}. \tag{47.9}$$

The entropy of an ideal gas is

$$S = -\frac{\partial F}{\partial \theta} = N \ln \frac{eV}{N} f(\theta) + N\theta \frac{f'(\theta)}{f(\theta)}. \tag{47.10}$$

This expression does not agree with Nernst's theorem. In actual fact, of course, we must apply to a gas at very low temperatures, not Boltzmann statistics but quantum statistics, even neglecting the fact that at low temperatures the gas actually condenses.

The energy is equal to $\overline{\mathscr{E}} = F + \theta S$, or

$$\overline{\mathscr{E}} = N\theta^2 \frac{f'(\theta)}{f(\theta)} = N\theta \frac{d \ln f}{d \ln \theta}. \tag{47.11}$$

Thus, the energy of an ideal Boltzmann gas, expressed in terms of temperature, does not depend upon volume at all. The mean energy of a single molecule $\overline{\varepsilon} = \frac{\overline{\mathscr{E}}}{N}$ depends only upon the temperature of the gas. This is not only because there is no force interaction between the gas molecules, but also because the properties of the gas are described by classical statistics. In the quantum statistics of ideal gases, the energy of a molecule depends both upon volume and temperature. It should be noted that the variables in formula (47.11) do not correspond to the identity (46.26). In order to make use of this identity, temperature must be eliminated from (47.10) and substituted in (47.11), but this is very difficult to do in the general form.

The enthalpy of an ideal gas is

$$I = \overline{\mathscr{E}} + pV = N \frac{\theta^2 f'(\theta)}{f(\theta)} + N\theta = N\theta \frac{d \ln \theta f(\theta)}{d \ln \theta}. \tag{47.12}$$

Like energy, it depends only on the temperature.

A mixture of ideal gases. Since the molecules of ideal gases do not interact, the free energy of a mixture is additive and reduces to the sum of the free energies of all the components:

$$F = -\sum_i N_i \theta \ln \frac{eV f_i(\theta)}{N}. \tag{47.13}$$

The pressure of the mixture is calculated in the usual way:

$$p = -\frac{\partial F}{\partial V} = \frac{\sum_i N_i \theta}{V}. \tag{47.14}$$

If we introduce the *partial* pressure of the ith component of the mixture, i.e., its contribution p_i to the total gas pressure p, then

$$p_i \equiv \frac{N_i \theta}{V} = \frac{N_i p}{\sum_k N_k}, \tag{47.15}$$

so that the total pressure appears as the sum of partial pressures.

This, of course, refers only to ideal gases. The thermodynamic potential of a mixture is

$$\Phi = -\sum_i N_i \theta \ln \frac{\theta f_i(\theta)}{p_i}. \tag{47.16}$$

The chemical potential of the ith component is determined from formula (46.50). It turns out to be equal to

$$\mu_i = -\theta \ln \frac{\theta f_i(\theta)}{p_i}. \tag{47.17}$$

These formulae are very important in the theory of chemical equilibria in gases.

Rotational energy of a gas. Let us now calculate the statistical sum over the rotational degrees of freedom of molecules. Since we wish to obtain simple thermodynamic formulae, in this section we shall confine ourselves to the case of nonquantum rotational motion (the quantized case was considered in Sec. 41). This means that the temperature satisfies the condition

$$\theta \gg \frac{h^2}{2J}. \tag{47.18}$$

Here, J is the molecular moment of inertia. At room temperatures, the condition (47.18) is satisfied for all gases, including hydrogen.

The expression for the mean energy of a molecule (47.11) involves only the logarithmic derivative of the statistical sum, so that the constant factors in this sum are not essential. But the value of the sum as such is important in many statistical applications (chemical equilibria). In order to calculate this value we must take into account that the summation is taken over the physically different rotational states. For example, diatomic molecules of H_2 or O_2 take on coincident positions when rotated through 180° about an axis perpendicular to the line joining the nuclei. The position of a diatomic molecule in space is given by two angles, the azimuthal and polar angles, and can be represented by a single point on the surface of a sphere of unit radius. But the physically different molecular orientations correspond only to half of this sphere.

The position of a nonlinear molecule in space is given with the aid of the Eulerian angles (see Sec. 9). If the molecule possesses any form of symmetry, then the statistical summation over all possible orientations should be divided by the number of rotations leaving the molecule invariant. For example, the ammonia molecule NH_3 has the form of a pyramid with a regular triangular base. Its rotational statistical sum, taken over all rotations in space, must be divided by 3. The benzene molecule C_6H_6 has a regular hexagonal form. A hexagon does not only coincide with itself for 60° rotation in its plane, but also for 180° rotation about an axis joining opposite vertices. Hence, its statistical sum is taken over $1/2\cdot 6 = 1/12$ part of all orientations.

It is now easy to write down a classical expression for the rotational statistical sum of a diatomic (or, in general, linear) molecule:

$$\sum_{\text{rot}} = \frac{4\pi}{\sigma}\iint e^{-\frac{M_1^2+M_2^2}{2J\theta}}\frac{dM_1\,dM_2}{(2\pi\hbar)^2} \tag{47.19}$$

the 4π factor takes into account all orientations in space. If the diatomic molecule consists of different atoms, $\sigma = 1$; if it consists of identical atoms, $\sigma = 2$. The quantum statistical sum for an oxygen molecule, whose nuclei do not have spin, is taken only over even rotational states (see Sec. 41). In the classical limit, this is taken into account by the factor 1/2. When the nuclei of a molecule possess spins we must multiply the statistical sum by the quantities $2s+1$, taken for all the nuclei. In the linear triatomic molecule CO_2 ($O=C=O$), we must also put $\sigma = 2$. Thus, for a linear molecule

$$\sum_{\text{rot}} = \frac{4\pi}{\sigma}\frac{2\pi J\theta}{(2\pi\hbar)^2} = \frac{2J\theta}{\sigma\hbar^2}. \tag{47.20}$$

The position of a nonlinear molecule in space is given by the orientation of its arbitrary axis and the rotation angle about this axis. Consequently, all rotations in space introduce the factor $4\pi\cdot 2\pi = 8\pi^2$ into the statistical sum for a nonlinear molecule, and the result is the following expression:

$$\sum_{\text{rot}} = \frac{8\pi^2}{\sigma}\iiint e^{-\frac{1}{2\theta}\left(\frac{M_1^2}{J_1}+\frac{M_2^2}{J_2}+\frac{M_3^2}{J_3}\right)}\cdot\frac{dM_1\,dM_2\,dM_3}{(2\pi\hbar)^3} =$$

$$= \frac{8\pi^2}{\sigma}\frac{(\sqrt{2\pi\theta})^3\sqrt{J_1J_2J_3}}{(2\pi\hbar)^3} = \frac{(2\theta)^{3/2}\sqrt{\pi J_1J_2J_3}}{\sigma\hbar^3}. \tag{47.21}$$

Thus, the contribution of the rotational energy to the total energy of the gas—equal to $N\theta\dfrac{d\ln\sum_{\text{rot}}}{d\ln\theta}$—amounts to $N\theta$ in the case of a linear molecule, and $\frac{3}{2}N\theta$ in the case of a nonlinear molecule.

The vibrational energy of a molecule. The energy of a molecule performing small oscillations can be represented, according to Sec. 8, in the form

$$\varepsilon_{\text{vib}} = \frac{1}{2}\sum_{\alpha=1}^{n}(P_\alpha^2 + \omega_\alpha^2 Q_\alpha^2) + U_0, \tag{47.22}$$

where Q_α are the normal oscillation coordinates, P_α are the corresponding momenta, U_0 is the potential energy in the equilibrium position (including the energy of the vibrational ground state). If $\hbar\omega_\alpha \ll \theta$, then the vibrational summation may be taken for nonquantized motion:

$$\sum_{\text{vib}} = \prod_{\alpha=1}^{n} \iint e^{-\frac{1}{2\theta}(P_\alpha^2 + \omega_\alpha^2 Q_\alpha^2) - \frac{U_0}{\theta}} \frac{dP_\alpha\, dQ_\alpha}{(2\pi\hbar)^n} =$$

$$= \frac{(\sqrt{2\pi\theta})^{2n}}{(2\pi\hbar)^n} \prod_{\alpha=1}^{n} \omega_\alpha^{-1} \cdot e^{-\frac{U_0}{\theta}} = \frac{\theta^n \prod_{\alpha=1}^{n} \omega_\alpha^{-1}}{\hbar^n} e^{-\frac{U_0}{\theta}}. \tag{47.23}$$

The vibrational motion makes a contribution $Nn\theta - NU_0$ to the mean energy of the molecule. Usually $\hbar\omega_\alpha \gg \frac{\hbar^2}{J}$ so that a temperature region exists that satisfies the two inequalities

$$\hbar\omega_\alpha \gg \theta \gg \frac{\hbar^2}{J}. \tag{47.24}$$

The vibrational quanta of the molecules are not yet excited at these temperatures, while the rotational specific heat is already constant. Thus, in the case of nitrogen and oxygen the total specific heat amounts to $\frac{3}{2}N + N = \frac{5}{2}N$ for temperatures from several tens to many hundreds of degrees. Under these conditions, gases (for instance, air) are subject to the equipartition law with a reduced number of degrees of freedom. The last expression means that each dynamical variable coordinate or momentum entering into the Hamiltonian quadratically gives, in the classical limit, an amount $\frac{N\theta}{2}$ in the mean energy of the gas.

The energy of the higher vibrational states is not governed by the simple formula (47.22). This can be seen from Fig. 47, where the lower curve is close to a parabola (harmonic oscillations) near the equilibrium position, and very different from a parabola (anharmonic oscillations) close to the dissociation limit. If the temperature is so high that oscillations (with large quantum numbers) close to the dissociation limit are excited, then the greater part of the molecules will have separated into atoms. At lower temperatures, the anharmonic nature of the oscillations affects but little the value of the statistical sum.

Thermodynamic quantities for a gas governed by the equipartition law. The specific heat of a gas governed by the principle of equipartition of energy is constant over a wide range of temperatures. Hence, the ratio of specific heats

$$\gamma = \frac{c_p}{c_V} = \frac{c_V + N}{c_V} \tag{47.25}$$

is also constant.

It will be convenient, in certain further applications, to express the thermodynamic quantities in terms of γ. The function $f(\theta)$ is proportional to $\theta^{\frac{C_V}{N}} e^{-\frac{U_0}{\theta}} = \theta^{\frac{1}{\gamma-1}} e^{-\frac{U_0}{\theta}}$.

Whence we obtain an expression for the energy of a gas, which energy we shall write here to the accuracy of the constant term $N U_0$:

$$\overline{\mathscr{E}} = C_V \theta = \frac{N\theta}{\gamma - 1} = \frac{pV}{\gamma - 1} \tag{47.26}$$

The enthalpy is equal to

$$I = \overline{\mathscr{E}} + pV = \frac{\gamma pV}{\gamma - 1}. \tag{47.27}$$

To the accuracy of the constant term, the entropy is

$$S = N \ln V + \frac{N}{\gamma - 1} \ln \theta \cong \frac{N}{\gamma - 1} \ln pV^{\gamma}. \tag{47.28}$$

Whence we get an equation for an adiabatic process in a gas governed by the equipartition principle:

$$pV^{\gamma} = \text{const.} \tag{47.29}$$

γ is often called the adiabatic index. From exercise 9 of the preceding section we find an expression for the velocity of sound

$$u = \sqrt{\left[\frac{\partial p}{\partial\left(\frac{1}{V}\right)}\right]_S} = \sqrt{\gamma pV}, \tag{47.30}$$

where V is the volume of *unit mass.*

Let us bring together the laws from which the specific heat of a gas subject to the equipartition principle is calculated. At a sufficiently high temperature there is a specific heat $\frac{N}{2}$ per rotational degree of freedom, and also per translational degree of freedom, since each such degree of freedom contributes one squared term to the energy expression of the form $\frac{p^2_x}{2m}$ or $\frac{M_1^2}{2J_1}$.

Hence, the integral of the distribution function acquires either a factor $\sqrt{2\pi m \theta}$ or $\sqrt{2\pi J_1 \theta}$; this yields the term $\frac{\theta}{2}$ when calculating the energy. If we can replace the summation by an integral, the vibrational degree of freedom contains two variables appearing quadratically in the energy [see (47.22)], thus yielding the mean energy θ.

To summarize, at a sufficiently high temperature, each vibrational degree of freedom, if it is strongly excited, makes a contribution N to the specific heat. If we apply the equipartition principle, then the specific heat of a molecule consisting of i atoms, which are not in one line ($i > 2$), is equal to $\left(3i - \frac{5}{2}\right) N$, and if the atoms form a line in the equilibrium position, it is $(3i - 3) N$.

Thus, for a triatomic molecule of triangular form (for example H_2O), a specific heat $6N$ is obtained for full excitation of all the degrees of freedom (besides electronic), and the ratio $C_p/C_V = 7/6$. If the vibrations are not yet excited, then $C_V = 3N$ and $C_p/C_V = 4/3$. At the lowest temperature, only the translational degrees of freedom remain, as in the case of a monatomic gas, which gives $C_V = \frac{3}{2} N$ and $C_p/C_V = 5/3$.

If the atoms of a triatomic molecule form a line (for example CO_2), then the maximum specific heat $C_V = \frac{13}{2} N$ and $C_p/C_V = 15/13$, i.e., C_V is greater for a linear molecule than for a triangular molecule. But if vibrations are not excited then $C_V = \frac{5}{2} N$, which is now less than for a triangular molecule. Such an intersection of the specific heat curves of CO_2 and H_2O with change of temperature is actually observed.

Adiabatic demagnetization. Of great interest is the process of isentropic (adiabatic) demagnetization. In Sec. 40 we considered the paramagnetism of the salts of rare-earth elements due to the free rotation of the magnetic moments of unfilled shells. Such moments may be interpreted as a "gas."

Let us suppose that a salt is magnetized to saturation at low temperature and is then suddenly demagnetized. Its entropy does not have time to change. But if all the moments are orientated in one direction, the entropy is small because this state is obtained in a small number of ways (in one way, in the limit). When the field is rapidly removed, the entropy will remain small only due to a big drop in the temperature. This method has been used to obtain temperatures of several thousandths of a degree above absolute zero.

Exercises

1) Find the work and the quantity of heat obtained by a gas in an isothermal process.

The work is equal to the change in free energy:

$$A = -N\theta \ln \frac{V_2}{V_1}.$$

The quantity of heat is expressed in terms of entropy change:

$$Q = \theta (S_2 - S_1) = N\theta \ln \frac{V_2}{V_1}.$$

A and Q are equal and opposite in sign, because energy remains unchanged at constant temperature.

2) Two portions of different gases occurring at the same temperature and pressure are mixed. Find the increase in entropy.

$$\Delta S = N_1 \ln \frac{\theta f_1(\theta)}{p_1} + N_2 \ln \frac{\theta f_2(\theta)}{p_2} - N_1 \ln \frac{\theta f_1(\theta)}{p} - N_2 \ln \frac{\theta f_2(\theta)}{p},$$

where p_1 and p_2 are the partial pressures of both gases after mixing. Whence

$$\Delta S = N_1 \ln \frac{p}{p_1} + N_2 \ln \frac{p}{p_2} = N_1 \ln \frac{N_1 + N_2}{N_1} + N_2 \ln \frac{N_1 + N_2}{N_2}.$$

If two portions of the same gas are mixed under the same conditions, the entropy will equal $(N_1 + N_2) \ln \frac{\theta f_1}{p}$, after mixing, so that $\Delta S = 0$ as it should be. This would not have occurred if the factor $N!$ in the statistical sum had not been introduced into the expression for free energy. Due to this factor, only the summation over physically different states of a gas appears in the free energy, and the entropy cannot change when two portions of the same gas are combined at the same temperature and at equal pressure.

3) Calculate the free energy of a gas in a centrifuge, of radius R and length l, rotating with angular velocity ω. Find the mean square distance of the particle from the axis.

The centrifugal force is equal to $m\omega^2 r$, which corresponds to an effective potential energy $U = -\int m\omega^2 r\, dr = -\frac{m\omega^2 r^2}{2}$.

Whence we obtain an expression for the free energy

$$F = -N\theta \ln\left(\frac{ef(\theta)}{N}\, l \int_0^R e^{\frac{m\omega^2 r^2}{2\theta}}\, r\, dr\right) = -N\theta \ln\left[\frac{ef(\theta)}{N}\frac{\theta l}{m\omega^2}\left(e^{\frac{m\omega^2 R^2}{2\theta}} - 1\right)\right].$$

The free energy satisfies the general relation $dF = -S\, d\theta - \Lambda\, d\lambda$ [cf. (46.38), where $\lambda = V$].

Regarding ω^2 as an external parameter λ, we determine the mean square distance of the particle from the axis:

$$\overline{r^2} = -\frac{2}{Nm}\frac{\partial F}{\partial(\omega^2)} = \left\{\frac{1}{1 - e^{-\frac{m\omega^2 R^2}{2\theta}}} - \frac{2\theta}{m\omega^2 R^2}\right\} R^2,$$

because, if $\omega^2 = \lambda$, then $\frac{m\overline{r^2}}{2} = -\frac{\overline{\partial \mathscr{E}}}{\partial \omega^2} = -\frac{\partial F}{\partial \lambda} = \Lambda$ [cf. (46.7)].

At very large angular velocities $\overline{r^2} \to R^2$, and at small velocities $\overline{r^2} = \frac{R^2}{2}$.

4) Find the velocity of stationary exit of a gas in vacuum, considering the adiabatic index as constant

$$v = \sqrt{2I} = \sqrt{\frac{2\gamma}{\gamma - 1} pV} = \sqrt{\frac{2}{\gamma - 1}} u.$$

5) Find the maximum flux density of a gas for an adiabatic transition from a pressure p_0 and specific volume V_0 to a pressure p.

From the adiabatic equation,

$$V = V_0 \left(\frac{p_0}{p}\right)^{\frac{1}{\gamma}}.$$

Whence we find the final enthalpy:

$$I = \frac{\gamma}{\gamma - 1} pV = \frac{\gamma}{\gamma - 1} p_0^{\frac{1}{\gamma}} V_0 p^{1 - \frac{1}{\gamma}}.$$

The flux density is

$$j = \frac{v}{V} = \frac{1}{V_0}\left(\frac{p}{p_0}\right)^{\frac{1}{\gamma}} \sqrt{\frac{2\gamma}{\gamma-1}\left(p_0 V_0 - p_0^{\frac{1}{\gamma}} p^{1-\frac{1}{\gamma}} V_0\right)}.$$

Denoting $P \equiv p/p_0$, we obtain

$$j = \sqrt{\frac{2\gamma}{\gamma-1}\frac{p_0}{V_0}}\, P^{\frac{1}{\gamma}} \sqrt{1 - P^{1-\frac{1}{\gamma}}}.$$

Hence, the maximum flux density is obtained at

$$P_{\max} = \left(\frac{2}{\gamma+1}\right)^{\frac{\gamma}{\gamma-1}}.$$

6) Show that the flux density is maximum when the flow velocity is equal to the velocity of sound at the given point.

Note. If the pressure at the output is made less than $p_0 P_{\max}$, the exit velocity will no longer vary and will remain equal to the local value of the velocity of sound. Any disturbance occurring in the flow is propagated with the velocity of sound. When the flow velocity is equal to the velocity of sound, the disturbance is carried away by the flow. Therefore, if the external pressure p_1 is less than $p_0 P_{\max}$, it will no longer affect the exit. In order to obtain supersonic exit, a Laval nozzle must be used (see exercise 10 of the previous section) for the flow velocity to equal the local velocity of sound at the narrowest part of the nozzle.

7) Derive a relationship between pressure and density in a shock wave propagated in an ideal gas of constant specific heat, and also find the sudden entropy change in the shock wave (see exercise 11, Sec. 46).

From equation (*) in the exercise mentioned, p. 534, we have a relationship between v and D:

$$\frac{v}{D} = \frac{\rho - \rho_0}{\rho}.$$

Whence, from equation (**), we obtain

$$v^2 = \frac{(p - p_0)(\rho - \rho_0)}{\rho\rho_0}, \qquad D^2 = \frac{\rho}{\rho_0}\frac{p - p_0}{\rho - \rho_0}.$$

We now substitute the expressions for v^2 and $\frac{v}{D}$, and also I from (47.27), into equation (***), and find a relation between density and pressure for shock compression

$$\frac{p}{p_0} = \frac{(\gamma+1)\rho - (\gamma-1)\rho_0}{(\gamma+1)\rho_0 - (\gamma-1)\rho}.$$

It must not be thought that for a shock compression from p_0 to p, the pressure actually follows this equation, passing through the entire intermediate series of values corresponding to the intermediate values of ρ. The relationship obtained shows what final values of ρ may be obtained when the gas is compressed with given initial p_0 and ρ_0 in the shock wave. The greatest density increase (for $p/p_0 \to \infty$) corresponds to $\frac{\rho}{\rho_0} = \frac{\gamma+1}{\gamma-1}$, i.e., the density in the shock wave cannot be increased by more than $\frac{\gamma+1}{\gamma-1}$ times. For example, when $\gamma = 7/5$, the quantity $\rho/\rho_0 < 6$. Using the formula (47.28) for entropy, we find

$$\Delta S = \frac{N\gamma}{\gamma - 1}\left(\frac{1}{\gamma}\ln\frac{(\gamma+1)\rho - (\gamma-1)\rho_0}{(\gamma+1)\rho_0 - (\gamma-1)\rho} - \ln\frac{\rho}{\rho_0}\right).$$

If the density does not change significantly, we can expand this expression in a series in powers of $\varepsilon = \frac{\rho}{\rho_0} - 1$. It then turns out that

$$\Delta S = \frac{N\gamma(\gamma+1)}{12}\varepsilon^3.$$

Hence, ΔS is third order with respect to ε or, what is just the same, with respect to the gas-compression velocity v. Therefore, we can neglect the change in entropy for a slow gas compression, and consider that the gas is compressed reversibly.

To the same approximation, the shock wave velocity assumes the following form:

$$D = \sqrt{\gamma\frac{p_0}{\rho_0}} = \sqrt{\gamma\frac{RT_0}{M}},$$

which coincides with the velocity of sound in an ideal gas. Thus, a weak shock wave degenerates to a sound wave. This holds for any substance.

A compression sound wave sent behind another sound wave in a substance will overtake it, because it will be moving in a substance having a certain velocity and, in addition, initially compressed by the first wave. A sufficiently large number of such waves, sent one after another, must finally merge to form a shock wave (of finite amplitude) propagated in the substance faster than sound.

Sec. 48. Fluctuations

The reversibility of mechanical equations with respect to time. The initial and final states of a system in classical mechanics are uniquely interrelated: one of them completely determines the other. This is expressed mathematically by the fact that if the signs of all the velocities are reversed, the entire motion of the system will be in the reverse direction. Changing the sign of the velocities is formally equivalent to changing the sign of the time. But if the sign of the time is changed, the form of Lagrange's equations will not change. It is still easier to see the invariance of the equations of Newton's Second Law with respect to a change of t to $-t$: these equations involve only second derivatives with respect to time, which preserve their sign in such a change.

In order to perform the same transition in electrodynamical equations, we must first of all change the signs of all the currents. In order that Maxwell's equations (12.24)-(12.27) should not change, we must also change the sign of the magnetic field together with the current, and leave the sign of the electric current unchanged. Since the magnetic field is an axial vector, or a pseudovector (see Sec. 16), the choice of sign is a matter of pure convenience.

Quantum mechanical equations also preserve their form when t is changed to $-t$. In the simplest case, when the Hamiltonian operator is real (i.e., when it does not involve i), the transition $t \to -t$ simply denotes the simultaneous transition $\psi \to \psi^*$, which can be directly seen from Schrödinger's equation (24.11). But the function ψ^* is completely equivalent to ψ (no matter which of them is regarded as conjugate). In more complicated cases, when the operator $\hat{\mathscr{H}}$ is complex, we can likewise always pass from the function ψ to another (physically fully equivalent to it) together with the transition from t to $-t$.

Statistics and reversibility in time. We shall now examine the way that the laws of statistical mechanics relate to time inversion. Statistical mechanics states that if at some initial instant of time a system is deviated from statistical equilibrium, then, in the overwhelming majority of cases, it will subsequently approach equilibrium. A system which is already in equilibrium will remain in equilibrium, no matter what imaginable changes in the sign of time are performed in the mechanical equations describing the detailed microscopic state of the system. Therefore, a situation arises that is rather paradoxical at first sight: statistical laws, which appear noninvariant with respect to time inversion, are derived from the equations of mechanics!

The problem stated in classical mechanics and statistics. Let us examine this paradox in the limits of the classical laws of motion. First take the following example. Let a gas occupy one half of a vessel divided by a partition. After this partition is removed the gas will occupy the whole vessel. Let us follow the motion of each molecule of the gas in this irreversible process (in classical mechanics this is, in principle, possible). The motion of all the molecules is represented in phase space by the displacement of a single point along a phase trajectory. If in the state of statistical equilibrium we mentally change the signs of all the velocities, the phase point in its imaginary movement will be displaced in the reverse direction, and all the gas will collect in one half of the vessel. Since any equilibrium state of the gas is attainable from a nonequilibrium state, and both velocity signs are, *a priori*, equiprobable, the gas must come out of the statistical equilibrium state as often as it enters it—which, it would appear, is never observed.

In actual fact, in statistics, equilibrium is not just any strictly defined state, but a whole range of states in which a closed system spends the greater part of its time. The phase point roams about in the equilibrium range for an extremely long time before spontaneously leaving it for any considerable distance. Through the vast majority of phase points in the statistical equilibrium region there pass trajectories which almost never enter regions that correspond noticeably to nonequilibrium states.

If we choose a certain section of the equilibrium range, we may say that the system emerges from it just as frequently as it returns to it, but that in the vast majority of cases it does not go "far."

Therefore, the apparent irreversibility of statistics is due to the way the problem is stated in it: the system does not remain for long in nonequilibrium states, and therefore rapidly enters equilibrium states; it remains for a very long time in equilibrium states, so that the probability of spontaneously leaving these states can, in the majority of cases, be neglected.

Quantum mechanics and the irreversibility of transitions. The principle of detailed balance (see Sec. 39) is fundamental to quantum statistics. In accordance with this principle, the probabilities of direct and inverse transitions are equal between two states having the same statistical weight. However, it by no means follows from this principle that the probability of transition from an equilibrium to a nonequilibrium state is the same as for a transition from a nonequilibrium to an equilibrium state. A statistically equilibrium state includes very many equiprobable microstates, while a nonequilibrium state contains a comparatively small number of microstates: the reason why the system spends the greater part of its time in equilibrium is that there are incomparably more equilibrium states than nonequilibrium states. Each given microstate, belonging to the set of statistical equilibrium states, passes to another state from the same range with overwhelming probability, while to a nonequilibrium state it passes with a negligible probability. A nonequilibrium state passes preferentially to an equilibrium state because the transition to a state of less equilibrium can occur in an incomparably smaller number of equally probable ways. It is for this reason that a system "tends," as it were, to equilibrium, despite the identical probability for direct and inverse transitions between any two initially equiprobable microstates.

Poisson's formula. The spontaneous transition of a system from an equilibrium state to a noticeably nonequilibrium state is of very small probability, but is not completely impossible. Deviations of actual values from their averages are more probable, the smaller the system in which they occur. If, for example, gas molecules are observed in a cube with a side 10^{-6} cm, then, under normal conditions (0° C, 760 mm Hg) the mean number of molecules is in all 27. The molecules may leave for neighbouring portions, so that their actual number in a certain volume will exhibit a very noticeable deviation from the number 27.

It is very easy to determine the probability that there will be N molecules in a given volume V, if there are N_0 molecules contained in the total volume V_0. The probability of finding a single molecule in the volume V is obviously equal to $\frac{V}{V_0}$. Therefore, the probability of finding N molecules in the volume V, and N_0—N molecules in the remaining portion of the volume, is equal to

$$\frac{N_0!}{(N_0-N)!\,N!}\left(\frac{V}{V_0}\right)^N\left(1-\frac{V}{V_0}\right)^{N_0-N}. \tag{48.1}$$

An analogous formula was derived in Sec. 39 for the probability of obtaining tails k times.

Let the total number of molecules N_0 be arbitrarily large and let N be any number, though considerably less than N_0. We replace the factorial ratio thus:

$$\frac{N_0!}{(N_0-N)!} = N_0(N_0-1)(N_0-2)\dots(N_0-N) =$$
$$= N_0^N\left(1-\frac{1}{N_0}\right)\left(1-\frac{2}{N_0}\right)\dots = N_0^N.$$

We represent the quantities $\left(\frac{V}{V_0}\right)^N$ and $\left(1-\frac{V}{V_0}\right)^{N_0-N}$ as

$$\left(1-\frac{V}{V_0}\right)^{N_0-N} = \left[\left(1-\frac{\overline{N}}{N_0}\right)^{N_0}\right]^{\left(1-\frac{N}{N^0}\right)} = (e^{-\overline{N}})^{\left(1-\frac{N}{N^0}\right)} \sim e^{-\overline{N}};$$
$$\left(\frac{V}{V_0}\right)^N = \frac{\overline{N}^N}{N_0^N},$$

where $\overline{N} = \frac{V}{V_0}$ by definition of $\overline{N}$. Substituting all the obtained expressions into the initial formula, we find the required probability

$$w_N = e^{-\overline{N}}\frac{(\overline{N})^N}{N!} \tag{48.2}$$

(Poisson's formula). It will be shown in exercise 1 that at large N the distribution (48.2) has a very sharp maximum at $N = \overline{N}$.

Fluctuation probability. Here, we shall obtain a general formula for fluctuation probability in a subsystem of a large system. The small volume of gas just considered may be taken as a special case of such a subsystem.

Let it be that a certain deviation from statistical equilibrium has occurred in the subsystem. The entire large system will thus have deviated somewhat from equilibrium. The ratio of the probabilities for the equilibrium and nonequilibrium states of the large system is equal to the ratio of the statistical weights of the states

$$\frac{w}{w_0} = \frac{G}{G_0}, \tag{48.3}$$

where w and G refer to the large system. The index 0 denotes the equilibrium value.

Expressing the statistical weight in terms of entropy ($S = \ln G$), we obtain

$$\frac{w}{w_0} = e^{S-S_0}. \tag{48.4}$$

Formula (48.4) can be given a somewhat different form. Since the large system is closed, its energy remains unchanged for fluctuation in the subsystem: $\overline{\mathscr{E}}=\overline{\mathscr{E}}_0$. But the total energy and the free energy F are related thus: $F=\overline{\mathscr{E}}-\theta S$, $F_0=\overline{\mathscr{E}}_0-\theta S_0$. It follows from these equations that the change in entropy of a system undergoing fluctuation is equal to the change in free energy, taken with opposite sign, divided by the temperature:

$$S-S_0=\frac{F_0-F}{\theta}=-\frac{A_{\min}}{\theta}. \tag{48.5}$$

The change in free energy is expressed, in accordance with (46.37), in terms of the minimum work.

Here, $A_{\min}$ is the minimum *external* work which must be performed on the system in order to produce this fluctuation reversibly, i.e., without change in entropy.

Thus, the fluctuation probability is defined by the following formula derived by Einstein:

$$w\sim e^{-\frac{A_{\min}}{\theta}}. \tag{48.6}$$

It must be borne in mind that fluctuation occurs spontaneously, without expending any external work. This was taken into account by the equation $\overline{\mathscr{E}}=\overline{\mathscr{E}}_0$. The same deviation of actual values from equilibrium values in the subsystem can be produced without expecting a fluctuation in it—by performing work $A_{\min}$ reversibly.

Fluctuations of thermodynamic quantities. The minimum work expression may be reduced to a more convenient form for actual calculations. We shall consider that the large system has been divided into two parts: a small part, in which a fluctuation occurs and statistical equilibrium is spontaneously disrupted, and the remaining part, in which the variation of quantities is reversible. In other words, the fluctuation has produced a deviation from equilibrium only in a small part of the system. The quantities referring to this part will be written without any indices, while those relating to the entire remaining part will be primed, and equilibrium quantities will be written with 0 index.

By definition, the minimum work is calculated in the case of constant entropy of the whole system; i.e., as if instead of a fluctuation occurring there is some change in the quantities at the expense of external action which does not destroy the statistical equilibrium. Given external action, the work is equal to the change in the energy of the system:

$$A_{\min}=\Delta\overline{\mathscr{E}}+\Delta\overline{\mathscr{E}}'. \tag{48.7}$$

The work here is equal to the energy change taken with positive sign (48.7) because, by definition, $A_{\min}$ is performed on the system.

The changes in the quantities in the large system are very small, being less the larger the system, so that $\Delta\overline{\mathscr{E}}'$ may be replaced from the thermodynamic identity (46.26):

$$\Delta\overline{\mathscr{E}}' = \theta_0 \Delta S' - p_0 \Delta V'. \tag{48.8}$$

As already pointed out, $A_{\min}$ is calculated in the case of reversible process. Therefore, $\Delta S' = -\Delta S$ and, in addition, $\Delta V' = -\Delta V$, of course. Hence,

$$A_{\min} = \Delta\overline{\mathscr{E}} - \theta_0 \Delta S + p_0 \Delta V. \tag{48.9}$$

Large fluctuations are highly improbable. Therefore, the quantities ΔS and ΔV should be regarded as small in the subsystem also; but it is now necessary, here, to make a series expansion up to second order quantities since, otherwise, $A_{\min}$ would be identically equal to zero (close to the maximum, the entropy expansion can begin only with quadratic terms):

$$\Delta\overline{\mathscr{E}} = \left(\frac{\partial\overline{\mathscr{E}}}{\partial S}\right)_0 \Delta S + \left(\frac{\partial\overline{\mathscr{E}}}{\partial V}\right)_0 \Delta V + \frac{1}{2}\left(\frac{\partial^2\overline{\mathscr{E}}}{\partial V^2}\right)_0 (\Delta V)^2 +$$
$$+ \left(\frac{\partial^2\overline{\mathscr{E}}}{\partial V \partial S}\right)_0 \Delta V \Delta S + \frac{1}{2}\left(\frac{\partial^2\overline{\mathscr{E}}}{\partial S^2}\right)_0 (\Delta S)^2.$$

But $\left(\frac{\partial\overline{\mathscr{E}}}{\partial S}\right)_0 = \theta_0$, $\left(\frac{\partial\overline{\mathscr{E}}}{\partial V}\right)_0 = -p_0$, so that only quadratic terms remain in the expression for $A_{\min} = (\Delta\overline{\mathscr{E}} + \Delta\overline{\mathscr{E}}')$. These terms may be represented in somewhat different form. Taking advantage of the fact that

$$\left(\frac{\partial^2\overline{\mathscr{E}}}{\partial V^2}\right)_S = -\left(\frac{\partial p}{\partial V}\right)_S; \quad \left(\frac{\partial^2\overline{\mathscr{E}}}{\partial S^2}\right)_V = \left(\frac{\partial \theta}{\partial S}\right)_V; \quad \frac{\partial^2\overline{\mathscr{E}}}{\partial V \partial S} = -\left(\frac{\partial p}{\partial S}\right)_V = \left(\frac{\partial \theta}{\partial V}\right)_S,$$

we write $A_{\min}$ as

$$A_{\min} = -\frac{1}{2}\Delta V\left[\left(\frac{\partial p}{\partial V}\right)_0 \Delta V + \left(\frac{\partial p}{\partial S}\right)_0 \Delta S\right] +$$
$$+ \frac{1}{2}\Delta S\left[\left(\frac{\partial \theta}{\partial V}\right)_0 \Delta V + \left(\frac{\partial \theta}{\partial S}\right)_0 \Delta S\right] = \frac{1}{2}(\Delta\theta\, \Delta S - \Delta p\, \Delta V). \tag{48.10}$$

And so Einstein's formula for fluctuation probability is transformed thus:

$$w = e^{\frac{1}{2\theta}(\Delta p \Delta V - \Delta\theta \Delta S)}, \tag{48.11}$$

where the 0 index is omitted from θ.

Let us find the probability for volume and temperature fluctuations. To do this we replace Δp and ΔS by their expressions in terms of volume and temperature:

$$\Delta p = \left(\frac{\partial p}{\partial V}\right)_\theta \Delta V + \left(\frac{\partial p}{\partial \theta}\right)_V \Delta\theta ,$$
$$\Delta S = \left(\frac{\partial S}{\partial V}\right)_\theta \Delta V + \left(\frac{\partial S}{\partial \theta}\right)_V \Delta\theta .$$

But according to (46.39) $\left(\frac{\partial p}{\partial \theta}\right)_V = \left(\frac{\partial S}{\partial V}\right)_\theta$, so that the right-hand side of equation (48.11) is represented as the product of two factors dependent only upon ΔV and $\Delta\theta$:

$$w \sim e^{\frac{1}{2\theta}\left(\frac{\partial p}{\partial V}\right)_0 (\Delta V)^2} \cdot e^{-\frac{1}{2\theta}\left(\frac{\partial S}{\partial \theta}\right)_V (\Delta\theta)^2} . \tag{48.12}$$

It is now easy to determine the mean square fluctuations $\overline{(\Delta V)^2}$ and $\overline{(\Delta\theta)^2}$. For the time being we write

$$\frac{1}{2\theta}\left(\frac{\partial p}{\partial V}\right)_\theta \equiv -\alpha. \tag{48.13}$$

Then the square of the volume fluctuation is easily written in the form

$$\overline{(\Delta V)^2} = -\frac{\partial}{\partial \alpha} \ln \int_{-\infty}^{+\infty} e^{-\alpha(\Delta V)^2} d(\Delta V) = -\frac{\partial}{\partial \alpha} \ln \sqrt{\frac{\pi}{\alpha}} = \frac{1}{2\alpha} . \tag{48.14}$$

The integration was justifiably extended from $-\infty$ to ∞, because the integrand is very small for large ΔV.

We finally arrive at the formula

$$\overline{(\Delta V)^2} = -\frac{\theta}{\left(\frac{\partial p}{\partial V}\right)_\theta} . \tag{48.15}$$

It must be remembered that this is not a volume fluctuation, generally, but only at constant temperature. At constant entropy, for example, the expression would have been different. The square of the temperature fluctuation, too, is found analogously:

$$\overline{(\Delta\theta)^2} = \theta \left(\frac{\partial \theta}{\partial S}\right)_V = \frac{\theta^2}{C_V} . \tag{48.16}$$

This fluctuation is calculated for constant volume.

We notice that the square of the fluctuation of volume is directly proportional to the first power of the additive quantity $\left(\frac{\partial V}{\partial p}\right)_\theta$ since volume is an additive quantity. Hence, the relative volume fluctuation $\sqrt{\overline{(\Delta V)^2}}/V$ is inversely proportional to the square root of the dimensions of the system. This statement, as applied to energy, was expressed in Sec. 45. The temperature fluctuation $\sqrt{\overline{(\Delta\theta)^2}}$ is inversely proportional to the square root of the specific heat and, for this reason, naturally, also decreases together with the dimensions of the subsystem.

The quantity θ is the modulus of the Gibbs distribution for the entire large system. When a fluctuation occurs in the subsystem, θ, naturally, does not coincide with its temperature, i.e., with its distribution modulus, which refers to the time interval during which the subsystem is quasi-independent of the large system. During such a time interval, θ is not the temperature of the large system either, since θ has the meaning of temperature only in equilibrium.

The temperature and energy of a system are not related quite unambiguously: at a given energy, temperature can experience slight fluctuations, and at a given temperature, the energy fluctuates.

Thermodynamic inequalities. From formulae (48.15) and (48.16) there follow the very important thermodynamic inequalities:

$$\left(\frac{\partial p}{\partial V}\right)_\theta < 0, C_V > 0. \tag{48.17}$$

The state of a substance can be stable only when these inequalities are satisfied. If the equation of state of a substance indicates that these inequalities break down at certain p, V, and θ, then the substance is unstable for such p, V, and θ and must break up into separate phases (liquid and vapour, for example), to which other values of V correspond.

The mean product of the fluctuations of two quantities. Let us now consider together the fluctuations of volume and entropy. In this case, the formula for fluctuation probability looks like this:

$$w = e^{-\frac{1}{2\theta}\left[-\left(\frac{\partial p}{\partial V}\right)_S (\Delta V)^2 + 2\left(\frac{\partial \theta}{\partial V}\right)_S \Delta V \Delta S + \left(\frac{\partial \theta}{\partial S}\right)_V (\Delta S)^2\right]}. \tag{48.18}$$

Here, the expression on the right-hand side no longer separates into the product of two factors that depend on each variable separately. Therefore, besides the volume fluctuation at constant entropy, and the entropy fluctuation at constant volume, the mean value of the product of their fluctuations also differs from zero: $\overline{\Delta V \Delta S} \neq 0$. Let us calculate this mean from formula (48.18). We write (48.18) in shortened notation:

$$w \sim e^{-\frac{1}{2}[\alpha_{11}(\Delta V)^2 + 2\alpha_{12}\Delta V \Delta S + \alpha_{22}(\Delta S)^2]}. \tag{48.19}$$

In this notation, the required quantity appears thus:

$$\overline{\Delta V \Delta S} = -\frac{\partial}{\partial \alpha_{12}} \ln \int_{-\infty}^{\infty}\int_{-\infty}^{\infty} e^{-\frac{1}{2}[\alpha_{11}(\Delta V)^2 + 2\alpha_{12}\Delta V \Delta S + \alpha_{22}(\Delta S)^2]}\, d(\Delta V)\, d(\Delta S).$$

In order to calculate the integral, we write the quadratic expression in the exponent in the form of a sum of quadratic terms

$$\left(\sqrt{\alpha_{11}}\,\Delta V+\frac{\alpha_{12}}{\sqrt{\alpha_{11}}}\,\Delta S\right)^2+\frac{\alpha_{11}\alpha_{22}-\alpha_{12}^2}{\alpha_{11}}(\Delta S)^2.$$

After this we change the variables in the integral, denoting

$$\Delta V+\frac{\alpha_{12}}{\alpha_{11}}\Delta S\equiv\xi.$$

The integration variable ξ varies within the same limits as ΔV and ΔS, i.e., from $-\infty$ to ∞. The integral in (48.19) is

$$\int_{-\infty}^{\infty}\int_{-\infty}^{\infty}e^{-\frac{1}{2}\left(\alpha_{11}\xi^2+\frac{\alpha_{11}\alpha_{22}-\alpha_{12}^2}{\alpha_{11}}(\Delta S)^2\right)}d(\Delta S)\,d\xi=$$

$$=2\sqrt{\frac{\pi}{\alpha_{11}}}\sqrt{\frac{\pi\alpha_{11}}{\alpha_{11}\alpha_{22}-\alpha_{12}^2}}=\frac{2\pi}{\sqrt{\alpha_{11}\alpha_{22}-\alpha_{12}^2}}. \qquad (48.20)$$

From this we obtain the required mean value:

$$\overline{\Delta V\Delta S}=-\frac{\partial}{\partial\alpha_{12}}\ln\frac{2\pi}{\sqrt{\alpha_{11}\alpha_{22}-\alpha_{12}^2}}=\frac{\alpha_{12}}{\alpha_{12}^2-\alpha_{11}\alpha_{22}}=$$

$$=\frac{\theta\left(\frac{\partial\theta}{\partial V}\right)_S}{\left(\frac{\partial\theta}{\partial V}\right)_S^2+\left(\frac{\partial p}{\partial V}\right)_S\left(\frac{\partial\theta}{\partial S}\right)_V}. \qquad (48.21)$$

We shall now show that this mean quantity is nothing other than $-\theta\left(\frac{\partial S}{\partial p}\right)_\theta=\theta\left(\frac{\partial V}{\partial\theta}\right)_p$. We consider the inverse quantity

$$\left(\frac{\partial\theta}{\partial V}\right)_S+\left(\frac{\partial p}{\partial V}\right)_S\frac{\left(\frac{\partial\theta}{\partial S}\right)_V}{\left(\frac{\partial\theta}{\partial V}\right)_S}=\left(\frac{\partial\theta}{\partial V}\right)_S-\left(\frac{\partial p}{\partial V}\right)_S\left(\frac{\partial V}{\partial S}\right)_\theta=$$

$$=-\left(\frac{\partial p}{\partial S}\right)_V-\left(\frac{\partial p}{\partial V}\right)_S\left(\frac{\partial V}{\partial S}\right)_\theta.$$

But if the pressure is represented as a function of entropy and volume in the form $p=p\,[S,\,V\,(S,\theta)]$, then the latter expression is $-\left(\frac{\partial p}{\partial S}\right)_\theta$, whence it follows that

$$\overline{\Delta V\Delta S}=\theta\left(\frac{\partial V}{\partial\theta}\right)_p. \qquad (48.22)$$

The volume and entropy fluctuations are said to be related, or correlated. This is understandable, since if the volume of a system increases, then the statistical weight of its state (i.e., its entropy) also increases.

Scattering of light by fluctuations. Because of fluctuations, no medium can be completely homogeneous. For this reason, electrodynamical equations, for which the constants of the medium ε and χ

are regarded as being fixed and everywhere the same, are, strictly speaking, nowhere valid. There always exist small imperfections in the homogeneity which must affect the propagation of light in the medium. Plane waves cannot be propagated in a nonhomogeneous medium; fluctuations cause scattering of the waves. Let us consider the quantity of scattered energy as a function of the frequency.

We shall consider that only the dielectric constant ε experiences fluctuations (since χ is always close to unity in transparent media). The wavelength of visible light λ is about half a micron, which is considerably greater than the mean dimensions of regions in which any noticeable fluctuations occur; this is because very many molecules are still contained in subsystems of volume 10^{-13} cm$^3 \sim \lambda^3$, and even in gases under normal conditions.

The period of oscillation in a light wave is of the order 10^{-15} sec, and is considerably less than the time during which fluctuation occurs. A time of at least $10^{-10} = 10^{-11}$ sec is required for the establishment of statistical equilibrium in the very smallest subsystem. For example, under normal conditions, the time interval between two collisions of a gas molecule is about 10^{-9} sec, and there is absolutely no reason for equilibrium to be established in the condensed phase a million times faster. The velocities of the molecules are very close in all phases at the same temperature, and interaction, in establishing equilibrium, must be transmitted over distances not less than $10^{-6} = 10^{-7}$ cm.

We can, therefore, consider that in the region where fluctuation has occurred the parameters of state of the medium, including polarizability, have changed somewhat. The polarization of the medium, produced in this region by a harmonic light wave, depends upon time in accordance with the same law that determines the electric field of an incident wave, i.e., like $e^{-i\omega t}$. Since the dimensions of the region are very much smaller than the wavelength of the incident light, the polarization has the same phase over the whole region. Consequently, the polarization may be integrated over the whole region, so that a resultant dipole moment is obtained proportional to $e^{-i\omega t}$. The light scattering problem must be considered here to a dipole approximation, in accord with the condition $r \ll \lambda$ being satisfied (see Sec. 19).

The total scattered energy is proportional to the square of the second dipole-moment derivative, i.e., ω^4 or $\frac{1}{\lambda^4}$ provided we neglect the way that polarizability depends upon frequency.

When light from the sun passes through the earth's atmosphere, the blue rays are scattered more than the red, because the wavelengths of the blue rays are shorter. Therefore, the blue portions of the solar spectrum predominate in the scattered light from the sky. This explains the colour of the sky.

Exercises

1) Write down Poisson's formula (48.2) for large N and $\overline{N}$. We represent (48.2) as

$$w_N = \frac{1}{\sqrt{2\pi\overline{N}}} e^{-\overline{N} + N\ln\overline{N} - N\ln N + N},$$

where $N!$ is written to the same accuracy as in exercise 1, Sec. 39.

Further, we must express $\ln\frac{\overline{N}}{N}$ as $-\ln\left(1 + \frac{N - \overline{N}}{\overline{N}}\right)$ and expand in a series up to the second term inclusively. This leads to the Gaussian distribution:

$$w_N = \frac{1}{\sqrt{2\pi\overline{N}}} e^{-\frac{1}{2}\frac{(N-\overline{N})^2}{\overline{N}}}, \overline{(\Delta N)^2} = \overline{N}.$$

The same value $\overline{(\Delta N)^2}$ is obtained from the exact Poisson formula if we write

$$\overline{N^2} = \sum N^2 w_N = e^{-\overline{N}}\,\overline{N}\frac{\partial}{\partial\overline{N}}\overline{N}\frac{\partial}{\partial\overline{N}}\sum\frac{\overline{N}^N}{N!} = e^{-\overline{N}}\,\overline{N}\frac{\partial}{\partial\overline{N}}\overline{N}\frac{\partial}{\partial\overline{N}}e^{\overline{N}} =$$
$$= \overline{N}^2 + \overline{N};\ \overline{N^2} - \overline{N}^2 = \overline{N}.$$

2) Find the pressure fluctuation for constant entropy, and the entropy fluctuation for constant pressure.
Answers:

$$\overline{(\Delta S)^2} = C_p;\ \overline{(\Delta p)^2} = -\theta\left(\frac{\partial p}{\partial V}\right)_S.$$

3) Find the mean value of $\overline{\Delta\theta\Delta p}$.
Answer:

$$\overline{\Delta\theta\Delta p} = \frac{\theta^2}{C_V}\left(\frac{\partial p}{\partial\theta}\right)_V.$$

4) Find the fluctuation of the energy and the number of quanta for an electromagnetic field of given frequency.
Proceeding from the expression

$$\overline{\mathscr{E}}_\omega = \frac{h\omega}{e^{\frac{h\omega}{\theta}} - 1},$$

we obtain, with the aid of the formula derived in Sec. 45 (45.22)

$$\overline{(\Delta\overline{\mathscr{E}}_\omega)^2} = \frac{(h\omega)^2 e^{\frac{h\omega}{\theta}}}{\left(e^{\frac{h\omega}{\theta}} - 1\right)^2}.$$

This formula can be represented thus:

$$\overline{(\Delta\overline{\mathscr{E}}_\omega)^2} = (h\omega)^2\left(\frac{1}{e^{\frac{h\omega}{\theta}} - 1} + \frac{1}{\left(e^{\frac{h\omega}{\theta}} - 1\right)^2}\right).$$

Introducing the number of quanta of given frequency, $N_\omega = \frac{\overline{\mathscr{E}}\omega}{h\omega}$, we have

$$\overline{(\Delta N_\omega)^2} = \overline{N}_\omega + \overline{N}_\omega^2.$$

The fluctuation of the number of quanta appears differently from the fluctuation of the number of particles of a Boltzmann gas (cf. exercise 1).

5) A vertically hanging mathematical pendulum performs fluctuation oscillations about the equilibrium position. Find the mean square deviation angle.

Let us denote the length of the pendulum by l and its mass by m. The potential energy at a deflection angle φ is equal to $\frac{1}{2} mlg\, \varphi^2$. In this case it is the minimum work appearing in the fluctuation probability. This gives

$$\overline{\varphi^2} = \frac{\theta}{mgl}.$$

6) Determine by how much the energy flux of a plane electromagnetic wave decreases in unit length in a gas as a result of density fluctuations.

The dielectric constant of the gas is

$$\varepsilon = 1 + 4\pi n \beta,$$

where n is the number of molecules in unit volume, β is the polarizability of a single molecule. The additional dipole moment introduced in some volume V by the density fluctuation is equal to

$$d = E_0 (N - \overline{N})\,\beta = E_0 \cdot \Delta N \frac{\overline{\varepsilon} - 1}{4\pi \overline{n}}.$$

The square of its time derivative is

$$\ddot{d}^2 = \omega^4 E_0^2 (\Delta N)^2 \left[\frac{\overline{\varepsilon} - 1}{4\pi \overline{n}}\right]^2.$$

After averaging over fluctuations, we obtain

$$\overline{\ddot{d}^2} = \omega^4 E_0^2 \overline{N} \left[\frac{\overline{\varepsilon} - 1}{4\pi \overline{n}}\right]^2.$$

The attenuation in the energy flux of a plane light wave over unit length is equal to

$$\frac{2}{3}\frac{\overline{\ddot{d}^2}}{c^3 V} \Big/ \frac{c}{4\pi} E_0^2 = \frac{\omega^4}{6\pi c^4}\frac{(\overline{\varepsilon} - 1)^2}{\overline{n}} = \frac{8\pi^3}{3}\frac{(\overline{\varepsilon} - 1)^2}{\lambda^4 \overline{n}}.$$

Sec. 49. Phase Equilibrium

Separation into phases. A substance consisting of molecules of a single type is characterized by four quantities: the number of particles, the temperature, the pressure, and the volume. Only three of the four quantities are independent, since the equation of state must always be satisfied. Thus, for an ideal gas the Clapeyron equation $pV = N\theta$ holds.

An ideal gas uniformly fills the whole of its permissible volume and in this sense is more an exception than the rule. Thus, for example, if we take one gram of water at a temperature of 20° C, then, no matter what the positive pressure, it is impossible to make it uniformly occupy a volume 10 cm³ (concerning negative pressures, see below in this section). One gram of water at 20° C placed in such a volume separates into two parts—liquid and gaseous; in other words, it does not remain homogeneous. And a certain, very definite, equilibrium pressure is established in the system.

In the state of statistical equilibrium, the mean number of molecules going from water to steam in unit time is equal to the mean number of molecules going from steam to water. It will be seen immediately that this condition cannot be satisfied for all pressures: the number of molecular impacts against the liquid surface is directly proportional to the pressure, whereas the number of evaporating molecules depends very weakly upon the pressure. Therefore, at a given temperature, only one pressure corresponds to equilibrium between liquid and vapour. Under other conditions, separation may occur into a liquid and a solid, into a gas and a solid, or into solids of various crystalline modifications or, in general, into phases.

The condition for phase equilibrium. Equilibrium pressure can be determined by the methods of statistical physics and does not require a detailed examination of the transition from one phase to another.

In the equilibrium state, the temperature and pressure in both phases are, of course, the same. This condition is necessary though not sufficient for equilibrium. In addition, a sufficient condition is that the thermodynamic potential Φ be a minimum (see Sec. 46). The termodynamic potential is additive: it is equal to the sum of the potentials of both phases, and the condition of it being a minimum is written as follows:

$$d\Phi = d\Phi_1 + d\Phi_2. \tag{49.1}$$

For a given temperature and pressure, the entire change of Φ_1 and Φ_2 can occur only due to a change in the number of particles:

$$d\Phi_1 = \mu_1 \, dN_1, \quad d\Phi_2 = \mu_2 \, dN_2. \tag{49.2}$$

But as many molecules leave one phase as enter another:

$$dN_1 = -\,dN_2.$$

Whence it follows that

$$(\mu_1 - \mu_2)\, dN_1 = 0. \tag{49.3}$$

Since dN_1 is any number, the phase-equilibrium condition consists in the equality of chemical potentials:

$$\mu_1 (p, \theta) = \mu_2 (p, \theta). \tag{49.4}$$

This equation may be represented in the form of a curve in the p, θ plane. In other words, to a certain temperature there corresponds very definite pressure.

And three phases of the same substance can occur in equilibrium. In this case the equilibrium condition is

$$\mu_1(p, \theta) = \mu_2(p, \theta) = \mu_3(p, \theta). \tag{49.5}$$

These two equations define a *single* point in the p, θ plane (the triple point). Out of it come the equilibrium curves between each two of the three phases (see Fig. 56).

Heat of transition. Usually, two phases of the same substance differ greatly from one another: their specific volume, entropy, energy, and other additive quantities experience a discontinuity at the transition point.

Let us find the quantity of heat released (or absorbed) at the transition point. Since the transition occurs at constant pressure, the quantity of heat is equal to the change in the heat function. We shall refer this heat to a single molecule, so that the heat function must also be referred to a single molecule. Such a heat function will be denoted by i (to distinguish it from I) while the entropy referred to a single molecule will be denoted by —s. The heat of transition to a single molecule is correspondingly equal to

$$q = i_2 - i_1. \tag{49.6}$$

The heat function is connected with the thermodynamic potential by the relation $I = \Phi + \theta S$. Going over to quantities which relate to a single molecule, and applying (46.48), we obtain

$$i = \mu + \theta s. \tag{49.7}$$

Whence

$$q = \mu_2 - \mu_1 + \theta(s_2 - s_1).$$

But in equilibrium $\mu_2 = \mu_1$ so that the heat of transition is equal to the temperature multiplied by the entropy change:

$$q = \theta(s_2 - s_1). \tag{49.8}$$

This result is quite understandable since phase transition is a reversible process.

The Clausius-Clapeyron equation. Let us consider two phases of the same substance occurring in mutual equilibrium. We suppose that the temperature in the equilibrium system is changed somewhat. It is required to determine how the pressure must be changed so as to keep the phase equilibrium intact. In other words, the derivative $\frac{dp}{d\theta}$ must be determined along the equilibrium curve.

The dependence of equilibrium pressure on temperature is given in the form of an implicit function (49.4). Hence, the derivative is found according to the usual rule:

$$\frac{dp}{d\theta} = -\frac{\left[\frac{\partial(\mu_1 - \mu_2)}{\partial\theta}\right]_p}{\left[\frac{\partial(\mu_1 - \mu_2)}{\partial p}\right]_\theta}. \tag{49.9}$$

From (46.48)

$$\left(\frac{\partial\mu}{\partial\theta}\right)_p = -s, \left(\frac{\partial\mu}{\partial p}\right)_\theta = v, \tag{49.10}$$

where v is the volume referred to a single molecule. Multiplying the numerator and denominator of the right-hand side of (49.9) by θ, and making use of (49.8), we obtain the required equation:

$$\frac{dp}{d\theta} = \frac{q}{\theta(v_2 - v_1)}, \tag{49.11}$$

which is known as the Clausius-Clapeyron equation.

Let us assume that a transition is considered for which q is positive, for example, fusion. Then the sign of the derivative $\frac{dp}{d\theta}$ is dependent upon which phase has the greatest specific volume: liquid or solid. For example, the specific volume of water at the melting point is less than the specific volume of ice, so that $\frac{dp}{d\theta}$ is a negative quantity. If the pressure above an equilibrium system of water and ice is raised, the melting temperature falls.

In the transition to the gaseous phase (vapourization, if the transition is by a liquid, or sublimation, if the transition is by a solid body) we have the inequality $v_2 \gg v_1$. Neglecting v_1 in equation (49.11) and replacing v_2 by $\frac{\theta}{p}$, we obtain

$$\frac{d \ln p}{d \ln \theta} = \frac{q}{\theta}. \tag{49.12}$$

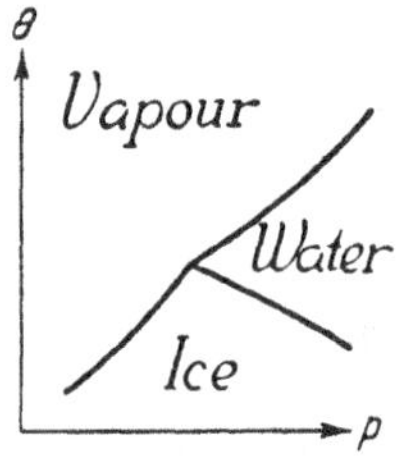

Fig. 55

This derivative is always positive. Therefore, the water equilibrium curves close to the triple point may be represented approximately as shown in Fig. 55. The equilibrium curve between water and ice has a negative derivative in accordance with what has been said.

Van der Waals' equation. We shall now show how, from the equation of state of a substance, it is possible to ascertain the necessity of a phase transition. It is convenient for this to make use of the well-known van der Waals equation of state for "real gases." This equation cannot strictly be derived from the fundamentals of statistical mechanics on any assumptions, and neither is it supported by accurate quantitative experiments. Nevertheless, it is the simplest of the equations suited for a qualitative description of a very wide range of states, from an ideal gas to its condensation into a liquid. Let us remind ourselves how the van der Waals equation is formed.

It is first of all assumed that a gas cannot be compressed indefinitely, but only to a certain volume b, which is related to the characteristic volume of all the molecules. This is taken into account in the Clapeyron equation $pV=N\theta$ by putting $V-b$ instead of V (in actual fact this has no strict foundation, even if the molecules are regarded as solid spheres). Over large distances between molecules the acting forces are those of attraction, which fall off rapidly; in the absence of such forces, condensation into a liquid would be impossible altogether. These forces reduce the pressure. The reduction in pressure is inversely proportional to the square of the volume occupied by the gas; this can be shown by the following reasoning. The gas pressure on a wall is proportional to the density of its kinetic energy. The kinetic energy of a molecule incident on a wall decreases due to the attraction of this molecule to the other molecules occurring in the volume. This attraction is due principally to the couplings of the molecules, because the contribution of triple interactions is slight in the case of small gas densities. The quantity of interacting pairs is proportional to the square of the gas density, or inversely proportional to the square of the volume it occupies. Since the energy stems from the attractive forces, its density is negative, and it results in a reduced pressure. The van der Waals equation is finally written thus:

$$p=\frac{N\theta}{V-b}-\frac{a}{V^2}, \tag{49.13}$$

where the second term takes into account the attractive forces. This term in the equation of state can also be rigorously substantiated by the methods of statistical mechanics, but, of course, only at densities which are still very much less than the density of the liquid phase, where each molecule is in constant interaction with many neighbouring molecules. Therefore, the exact equation of state for a real liquid must be immeasurably more complex than the van der Waals equation. It is doubtful whether it is possible to write a single exact equation applicable to a wide class of liquids.

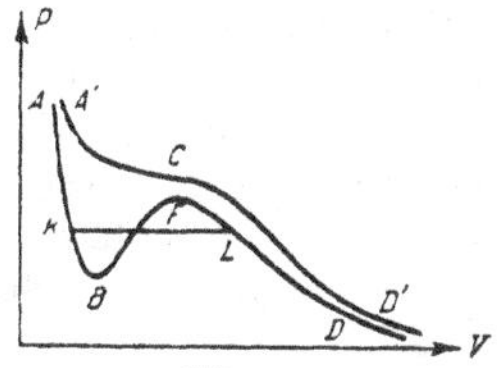

Fig. 56

Van der Waals' equation and phase transition. We shall now show how, from the van der Waals equation, the existence of a range of states, in which the substance separates into gaseous and liquid phases, can be demonstrated. Equation (49.13) is third degree in volume. It must have three real roots for certain values of θ and p. In other words, the pressure-versus-volume curve at constant temperature (isotherm) is of the form $ABFD$, as shown in Fig. 56. But the derivative $\left(\frac{\partial p}{\partial V}\right)_\theta$ is positive between the points B and F and, in accordance with the first inequality of (48.17), the state of the substance is unstable if $\left(\frac{\partial p}{\partial V}\right)_\theta > 0$. Hence, the necessity for the separation of the substance into two phases in this region.

The portion of the curve AB corresponds to the liquid state (small volume). As the pressure is reduced the liquid expands to the point K, after which change occurs along the straight line KL. The points K and L are uniquely defined from the equality condition of the chemical potentials (49.4), and the intermediate points along the line correspond to a mixture of liquid in a state corresponding to K, and vapour in a state L. We notice that the position of the point K, at a temperature corresponding to the given isotherm, is defined uniquely.

The portion KB is not absolutely unstable, since on it $\left(\frac{\partial p}{\partial V}\right)_\theta < 0$. The states of this portion can be attained without allowing the formation of vapour bubbles in the liquid (a superheated liquid). For this the liquid must be free from foreign agents, for example, bubbles of dissolved gases, which favour vapourization. Sometimes the portion KL lies partly below the abscissa axis, thus corresponding to a negative pressure, i.e., an extension of the liquid. A liquid can indeed be extended if it adheres everywhere to the walls of the vessel and does not have a free surface. The portion FL corresponds to a supercooled vapour, which can be obtained if condensation centres are prevented from forming. Such condensation nuclei or centres easily arise from ions, for example. This is the underlying principle of the Wilson cloud-chamber for the observation of the tracks of charged particles.

The critical point. At a sufficiently high temperature, the first term on the right in the van der Waals equation predominates over the second. The equation then becomes very similar to the Clapeyron equation for a volume $V-b$. But this equation has only one real root for each value of p. This corresponds to the well-known fact that at high temperatures a substance does not split into two phases at all.

Let us find the temperature at which separation into phases ceases. On the corresponding isotherm $A'CD'$ (Fig. 56) the points B and F, where the derivative $\left(\frac{\partial p}{\partial V}\right)_\theta$ becomes zero, merge into one point C, and the region of unstable states disappears. All three roots of equation (49.13) merge at the point C, so that C corresponds to the triple root of this equation. But the expansion of the function with respect to the difference $V - V_C$ must begin with a third-order term if V_C is a triple root. The linear and quadratic terms in the expansion become zero if the first and second pressure derivatives with respect to volume are equal to zero at the point C. It is easy from this to determine the position of the point C from the van der Waals equation.

Let us write down the condition that the first and second derivatives become zero:

$$\left(\frac{\partial p}{\partial V_C}\right)_{\theta_C} = -\frac{N\theta_C}{(V_C - b)^2} + \frac{2a}{V^3_C} = 0, \tag{49.14}$$

$$\left(\frac{\partial^2 p}{\partial V^2_C}\right)_{\theta_C} = \frac{2N\theta_C}{(V_C - b)^3} - \frac{6a}{V^4_C} = 0. \tag{49.15}$$

From this we obtain

$$\frac{V_C - b}{2} = \frac{V_C}{3},$$

or

$$V_C = 3b. \tag{49.16}$$

Then, from (49.14), we find

$$a = \frac{N\theta_C V^3_C}{2(V_C - b)^2} = \frac{27}{8} N\theta_C b,$$

so that

$$\theta_C = \frac{8}{27}\frac{a}{Nb}. \tag{49.17}$$

The pressure at the point C is determined from the van der Waals equation:

$$p_C = \frac{N\theta_C}{V_C - b} - \frac{a}{V_C^2} = \frac{1}{27}\frac{a}{b^2}. \tag{49.18}$$

If we represent the phase equilibrium curve in the p, θ plane, then this curve will end in the point $p = p_C$, $\theta = \theta_C$. C is called the critical point. Separation into phases does not occur at temperatures $\theta > \theta_C$.

The critical point can exist only on the equilibrium curve between two such phases, which have no feature that is incapable of varying continuously. An example of such a feature is the regularity of crystal structure: in principle, the position of an atom in an ideal crystal defines the position of the whole crystal (with the exception, naturally, of its orientation in space). Yet, the position of an atom in a liquid affects only the position of its closest neighbours. And so for certain substances a continuous transition between the solid crystalline phase and the liquid phase is impossible. The curve dividing the crystalline and liquid phases cannot end and cannot, therefore, have a critical point.

The law of corresponding states. Eliminating the constants a, b and N with the aid of (49.16), (49.17), and (49.18), we have

$$b = \frac{V_C}{3}, a = 3V_C^2 p_C, N = \frac{8}{3}\frac{p_C V_C}{\theta_C}. \qquad (49.19)$$

The last of these three equations shows by how much the equation of state of a substance differs, at the critical point, from that of an ideal gas: $p_C V_C = \frac{3}{8} N\theta_C$. But we should note that, as a general rule this relationship is not really satisfied. As has already been mentioned, the van der Waals equation is qualitative in character, and so there is nothing surprising in the fact that for real substances $p_C V_C \neq \frac{3}{8} N\theta_C$.

If we now substitute (49.19) in (49.13), we get

$$\frac{p}{p_C} = \frac{8\,\theta/\theta_C}{(3V/V_C)-1} - 3\left(\frac{V}{V_C}\right)^2. \qquad (49.20)$$

Formula (49.20) expresses a special form of the so-called law of corresponding states: for two different substances, the ratios $\frac{p}{p_C}$, $\frac{V}{V_C}$ and $\frac{\theta}{\theta_C}$ are related by a single universal equation. It should be noted that in general form the law of corresponding states, especially for substances of similar structure, is satisfied better in practice than the specific formula (49.20) based on the interpolation van der Waals equation, because this, more general, law does not impose a definite functional form on the equation of state. However, there are, of course, deviations from the law of corresponding states also: the ratios $\frac{V}{V_C}$ are not strictly the same for two substances having identical $\frac{p}{p_C}$ and $\frac{\theta}{\theta_C}$.

The properties of a substance close to the critical point. Let us now investigate the properties of a substance close to the critical point in general form, without assuming that the van der Waals equation (49.13) holds. We shall only make use of the fact that the derivative $\left(\frac{\partial p}{\partial V}\right)_\theta$, close to the critical point, must tend to zero like the square of the difference $V - V_C$, because the expansion of p on the critical isotherm begins with a term proportional to $(V - V_C)^3$. At temperatures sufficiently close to critical, $\left(\frac{\partial p}{\partial V}\right)_\theta$ differs from $\left(\frac{\partial p}{\partial V}\right)_{\theta=\theta_C}$ by first-order quantities in $(\theta - \theta_C)$, because the relationship between pressure and temperature close to the critical point does not exhibit any peculiarities. Thus, the expansion $\left(\frac{\partial p}{\partial V}\right)_\theta$ in the critical region is of the form

$$\left(\frac{\partial p}{\partial V}\right)_\theta = -\lambda(V - V_C)^2 - \nu(\theta - \theta_C). \qquad (49.21)$$

Here, $\nu > 0$, because the inequality $\left(\frac{\partial p}{\partial V}\right)_\theta < 0$ must always be satisfied at temperatures higher than critical. Therefore, $\lambda > 0$ also. At temperatures lower than critical, $\left(\frac{\partial p}{\partial V}\right)_\theta$ becomes zero at two points. They correspond to B and F in Fig. 56

$$V_B - V_C = -\sqrt{\frac{\nu}{\lambda}(\theta_C - \theta)}, \quad V_F - V_C = \sqrt{\frac{\nu}{\lambda}(\theta_C - \theta)}. \qquad (49.22)$$

Let us now find the points on the isotherm that correspond to K and L, i.e., to phase equilibrium.

For this we make use of the phase equilibrium condition $\mu_K = \mu_L$. It is conveniently written in the form of an integral taken along the isotherm on which the points K and L lie:

$$\int_K^L d\mu = 0. \qquad (49.23)$$

Multiplying by N and then replacing $d\Phi$ by $V\,dp$ for $\theta = \text{const}$, we obtain

$$\int_K^L d\mu = \int_K^L V\,dp = \int_K^L (V - V_C)\,dp,$$

because the integral $\int_K^L dp$ is equal to $p_L - p_K$ and becomes zero according to the condition $p_L = p_K$. Now substituting the initial expression (49.21) we see that the equality of chemical potentials reduces to the requirement

$$\int_{V_K}^{V_L} (V - V_C)\left[\lambda (V - V_C)^2 + \nu(\theta - \theta_C)\right] dV = 0. \qquad (49.24)$$

The integrand is odd in $V - V_C$. Therefore the integral becomes zero, if at the integration limits (i.e., $V_K - V_C$ and $V_L - V_C$) the values of $V - V_C$ are equal and opposite in sign or, in other words, the volumes V_L and V_K differ equally from critical.

We also represent the condition of pressure equality in integral form

$$\int_K^L dp = \int_{V_K}^{V_L} \frac{\partial p}{\partial V}\,dV = 0. \qquad (49.25)$$

Substituting (49.21) here and integrating, we obtain

$$\frac{\lambda}{3}(V_L-V_C)^3+\nu(V_L-V_C)(\theta-\theta_C)-\frac{\lambda}{3}(V_K-V_C)^3-$$
$$-\nu(V_K-V_C)(\theta-\theta_C)=0\,.$$

Making use of the fact that $V_K-V_C=-(V_L-V_C)$, we obtain the required equation

$$\frac{\lambda}{3}(V_L-V_C)^3+\nu(V_L-V_C)(\theta-\theta_C)=0,$$

from which it follows that

$$V_L-V_C=V_C-V_K=\sqrt{\frac{3\nu(\theta_C-\theta)}{\lambda}}\,. \tag{49.26}$$

Thus, close to the critical point, the region of absolutely unstable states is narrower, in the ratio $\frac{1}{\sqrt{3}}$, than the whole region where phase separation occurs.

Let us now find the heat of transition close to the critical point. By definition we have

$$Q\cong\theta_C(S_L-S_K)=\theta_C\left(\frac{\partial S}{\partial V}\right)_{\theta C}(V_L-V_K)\,. \tag{49.27}$$

At the critical point, the derivative $\frac{\partial S}{\partial V}$ maintains a finite value $\left[\text{it is equal to}\left(\frac{\partial p}{\partial\theta}\right)_V\right]$. Therefore, the heat of transition is proportional to $(\theta_C-\theta)$. Right at the critical point it becomes zero, as expected.

Close to the critical point, the density of the substance experiences large statistical fluctuations, because density fluctuations are inversely proportional to $\left(\frac{\partial p}{\partial V}\right)_\theta$. But, as we have seen in the previous section, this results in a strong scattering of light. As a result of this scattering, the substance acquires a certain turbidity, similar to the turbidity of opal (critical opalescence).

Phase transitions of the second kind. At the phase transition point the thermodynamic potentials of both phases are equal. The other additive quantities (such as entropy, energy, and volume) experience discontinuities. But there also exist phase transitions for which not the additive quantities themselves are discontinuous, but only their derivatives—specific heat, compressibility, etc. An example of such a transition was already given in Sec. 43: this is the transition of helium at a temperature of 2.2° K.

The specific heat at the transition point changes discontinuously. Another example is the transition of iron from the ferromagnetic to the nonferromagnetic state at 770° C (the Curie point).

Phase transitions of the second kind are very frequently observed in crystals. In this case they correspond to a certain change in the translational or vibrational symmetry of the lattice. Since the form of the symmetry cannot change continuously (the symmetry property either exists or it does not), symmetry always changes discontinuously. If an entropy discontinuity is then experienced, we have a phase transition of the *first* kind; if the entropy is continuous, and the derivatives experience discontinuities, the transition is of the *second* kind.

Let us interrelate the derivative discontinuities of various quantities on the lines of phase transitions of the second kind. Since entropy and volume are continuous, we write

$$\Delta S = S_2 - S_1 = 0; \;\; \Delta V = V_2 - V_1 = 0. \tag{49.28}$$

Let us differentiate these equations with respect to temperature along the transition line. We then obtain

$$\Delta\left(\frac{\partial S}{\partial \theta}\right)_p + \Delta\left(\frac{\partial S}{\partial p}\right)_\theta \frac{dp}{d\theta} = 0, \tag{49.29}$$

$$\Delta\left(\frac{\partial V}{\partial \theta}\right)_p + \Delta\left(\frac{\partial V}{\partial p}\right)_\theta \frac{dp}{d\theta} = 0. \tag{49.30}$$

Here $\frac{dp}{d\theta}$ denotes the derivative of pressure with respect to temperature along the transition curve. Further, $\left(\frac{\partial S}{\partial p}\right)_\theta = -\left(\frac{\partial V}{\partial \theta}\right)_p$ [see (46.46)] and $\left(\frac{\partial S}{\partial \theta}\right)_p = \frac{C_p}{\theta}$. Whence, after eliminating $\Delta\left(\frac{\partial S}{\partial p}\right)_\theta$, we obtain

$$\Delta C_p = \theta\left(\frac{dp}{d\theta}\right)^2 \Delta\left(\frac{\partial V}{\partial p}\right)_\theta. \tag{49.31}$$

Thus, along phase-transition lines of the second kind, the specific heat discontinuity at constant pressure is associated with a compressibility discontinuity. A similar expression can also easily be found for the discontinuity in specific heat at constant volume.

Sometimes, phase transition lines of the first kind become a phase transition line of the second kind at some point. If the transition is associated with a change in symmetry, then neither line can simply terminate.

The thermodynamic theory of phase transitions of the second kind has been developed by L. D. Landau (see L. D. Landau and E. M. Lifshits, *Statistical Physics*, Gostekhizdat, 1951.)

Exercises

1) Find the specific heat of one of the phases of a substance along the curve of phase transitions.

From the definition of specific heat

$$C = \theta\frac{\partial S}{\partial \theta} = \theta\left[\left(\frac{\partial S}{\partial \theta}\right)_p + \frac{dp}{d\theta}\left(\frac{\partial S}{\partial p}\right)_\theta\right] = C_p - \frac{q}{\theta(V_2 - V_1)}\left(\frac{\partial V}{\partial \theta}\right)_p.$$

2) Show that C_p becomes infinite at the critical point.
Use the result of exercise 4, Sec. 46, and the condition defining the critical point.

3) Find the discontinuity in specific heat (expressed in terms of compressibility discontinuity) at constant volume along a phase-transition line of the second.

$$\Delta C_V = \theta \left(\frac{dV}{d\theta}\right)^2 \Delta \left(\frac{\partial p}{\partial V}\right)_\theta .$$

Sec. 50. Weak Solutions

Weak solutions and ideal gases. Weak solutions exhibit many regularities which make them similar to ideal gases. The reason for this similarity can be easily seen in the fact that the molecules of a dissolved substance in a weak solution interact just as little as ideal gas molecules. But the molecules of a dissolved substance interact strongly with the surrounding molecules of the solvent, whence the differences between a solution and a gas.

The thermodynamic potential of a weak solution. We shall proceed from the general expression for free energy in classical statistics:

$$F = -\theta \ln \int e^{-\frac{\mathscr{E}}{\theta}} d\Gamma \tag{50.1}$$

[we have omitted the unessential factor $(2\pi h)^N$]. The integral is taken over all physically different states of the system. If one takes into account the identity of all the N molecules of the solvent and the n molecules of the dissolved substance, he can extend the statistical integral over the entire phase space and divide it by the total number of permutations of all identical particles. The number of such permutations is $N!\, n!$.

We now write down the thermodynamic potential of a weak solution

$$\Phi = F + pV = \left(-\theta \ln \int e^{-\frac{\mathscr{E}}{\theta}} d\Gamma + \theta \ln N! + pV\right) + \theta \ln n!\,, \tag{50.2}$$

where the integral extends over the whole phase space of the system. Let us expand the expression in brackets, on the right-hand side of (50.2), in powers of the small quantity $\frac{n}{N}$, taking into account that the zeroth term in the expansion is the thermodynamic potential of a pure solvent Φ_0. In addition, we replace $\ln n!$ by $n \ln \frac{n}{e}$ according to Stirling's formula:

$$\Phi = \Phi_0 + \frac{n}{N} B(p, \theta, N) + n\theta \ln \frac{n}{e}\,. \tag{50.3}$$

We can refine the dependence of $B(p, \theta, N)$ on the number of solvent particles N by noting that the thermodynamic potential must be an additive function of N and n. In other words, if N and n increase a certain number of times, for example, twice, Φ must also increase by a factor of two. But in (50.3), this requirement is satisfied directly only by the potential of the pure solvent Φ_0, equal to $N\mu_0$, where μ_0 is the chemical potential of the pure solvent. For the second and third terms to be additive, let us first write the third term in the form

$$\theta n \ln \frac{n}{e} = \theta n \ln \frac{n}{eN} + \theta n \ln N.$$

After this, the thermodynamic potential will look like

$$\Phi = N\mu_0(p, \theta) + n\theta \ln \frac{n}{eN} + n\left(\frac{B(p, \theta, N)}{N} + \theta \ln N\right).$$

In order to obtain an additive expression, we must demand that the function $\frac{B}{N} + \theta \ln N$ should not be dependent on N at all. The result is a general expression for the thermodynamic potential of a weak solution:

$$\Phi = N\mu_0(p, \theta) + n\theta \ln \frac{n}{eN} + n\mathrm{X}(p, \theta). \tag{50.4}$$

The chemical potential of a solvent in solution is equal to

$$\mu = \frac{\partial \Phi}{\partial N} = \mu_0 - \frac{n\theta}{N}, \tag{50.5}$$

while the chemical potential of the solute is

$$\mu' = \frac{\partial \Phi}{\partial n} = \theta \ln \frac{n}{N} + \mathrm{X}(p, \theta). \tag{50.6}$$

Osmotic pressure. Certain semipermeable membranes pass solvent molecules freely, but do not pass molecules of the dissolved substance. The solvent must be in statistical equilibrium on both sides of such a membrane. But this is possible only when the chemical potentials of the pure solvent and the solvent in the solution beyond the membrane are equal. The temperature of the substance on both sides of the membrane is, of course, the same, for otherwise equilibrium could not set in. Only the pressure can differ, provided the pressure difference is held in check by the membrane. Denoting the pressure difference by Δp, we obtain the equilibrium condition:

$$\mu_0(p, \theta) = \mu(p + \Delta p, \theta) = \mu_0(p + \Delta p, \theta) - \frac{n\theta}{N}. \tag{50.7}$$

Let us expand μ_0 in a series in powers of Δp, to the linear approximation. This expansion is justified, since the pressure difference for a liquid Δp is a small quantity. Therefore,

$$\mu_0(p + \Delta p, \theta) = \mu_0(p, \theta) + \frac{\partial \mu_0}{\partial p}\Delta p. \tag{50.8}$$

But the derivative $\frac{\partial \mu_0}{\partial p}$ is equal to the volume of a single molecule of pure solvent:

$$\frac{\partial \mu_0}{\partial p} = \frac{V}{N}.$$

Whence we obtain the equation

$$V \cdot \Delta p = n\theta. \tag{50.9}$$

The excess pressure Δp in the solution is called the osmotic pressure. Equation (50.9) bears a striking resemblance to the Clapeyron equation for ideal gases. It was originally found experimentally and served as the basis for the formulation of a thermodynamic theory of solutions. We obtained equation (50.9) by proceeding from the general principles of statistics.

Phase equilibrium of a solvent (Raoult's laws). We shall now consider another case, when equilibrium is also established between solvent molecules. Let the solution occur in equilibrium with another phase of the solvent, while the solute does not pass into this phase. We find the displacement of the phase equilibrium curve in the p, θ plane.

Let us call the chemical potential of the phase into which the solute does not pass, μ_1. Then the phase-equilibrium condition of the pure solvent is determined by the equation

$$\mu_1(p, \theta) = \mu_0(p, \theta), \tag{50.10}$$

while the equilibrium of the other phase of the solvent and solution is displaced and is given by the following condition:

$$\mu_1(p + \Delta p, \theta + \Delta\theta) = \mu_0(p + \Delta p, \theta + \Delta\theta) - \frac{n\theta}{N}. \tag{50.11}$$

Let us expand the chemical potentials in a series in Δp and $\Delta\theta$:

$$\begin{aligned} \mu_1(p + \Delta p, \theta + \Delta\theta) - \mu_0(p + \Delta p, \theta + \Delta\theta) &= \mu_1(p, \theta) - \mu_0(p, \theta) + \\ &+ \left[\frac{\partial}{\partial p}(\mu_1 - \mu_0)\right]\Delta p + \left[\frac{\partial}{\partial \theta}(\mu_1 - \mu_0)\right]\Delta\theta = \\ &= (v_1 - v_0)\Delta p - (s_1 - s_0)\Delta\theta. \end{aligned} \tag{50.12}$$

We shall now assume that the pressure in the system is the same as above a pure solvent, i.e., that $\Delta p = 0$. Then the equilibrium-temperature displacement $\Delta\theta$ will be defined:

$$\Delta\theta = \frac{n\theta^2}{Nq} = \frac{n\theta^2}{Q}, \tag{50.13}$$

where $Q = N\theta\,(s_1 - s_0)$ is the heat of the phase transition of the pure solvent. For vapourization $Q > 0$; therefore, $\Delta\theta > 0$ if the solute does not pass into vapour, so that the equilibrium temperature is raised. Indeed, the solution has a higher boiling point than the pure solvent.

Let us now suppose that the solute does not pass into the solid phase of the solvent. Then Q is the heat of solidification, $Q<0$. It is seen from this that the fusion temperature of the solution is lower than that of the pure solvent. The use of cooling mixtures is based on this property of solutions.

Let us now consider equilibrium at a given temperature, $\Delta\theta=0$. Then the reduction in equilibrium pressure over the solution is determined from (50.12):

$$\Delta p = -\frac{n\theta}{(v_1 - v_0)N}. \tag{50.14}$$

If a solution is in equilibrium with vapour, then $v_1 \gg v_0$. The product Nv_1 is the volume of the entire solvent in the vapour state. If it were possible to transform the solute to vapour together with the solvent, the partial pressure of the molecules of the substance would equal the reduction in the equilibrium pressure above the solution. The relative pressure reduction $-\Delta p/p$ is equal to the concentration of the solution n/N.

Solute equilibrium. A solution is termed saturated if it is in equilibrium with the dissolved substance. The equilibrium condition consists in that the chemical potential of a pure solute, μ'_0, is equal to its chemical potential in the dissolved state:

$$\mu'_0 = \mu' = \theta \ln\frac{n_0}{N} + \mathrm{X}(p, \theta). \tag{50.15}$$

We have supposed that the saturated solution is also still regarded as weak, i.e., $n_0 \ll N$.

If a pure substance occurs in a gaseous state, then its chemical potential depends upon pressure according to the law [see (47.17)]

$$\mu'_0 = \theta \ln p + f_1(\theta). \tag{50.16}$$

The function $\chi(p, \theta)$ is but slightly dependent on the external pressure: $\chi(p, \theta)$ is determined by the properties of the condensed phase, which do not change when the external pressure varies over several atmospheres. Comparing (50.15) and (50.16) and taking antilogarithms, we find that the equilibrium concentration of the dissolved gas is proportional to its pressure above the liquid (Henry's law)

$$\frac{n_0}{N} = a(\theta)\,p. \tag{50.17}$$

The coefficient of p depends very weakly on pressure.

Heat of solution. The heat of solution is equal to the difference in the heat functions of the substances comprising a solution before and after being dissolved. The heat function is related to the thermodynamic potential in the following way:

$$I = \Phi - \theta \left(\frac{\partial \Phi}{\partial \theta} \right)_p = -\theta^2 \frac{\partial}{\partial \theta} \left(\frac{\Phi}{\theta} \right). \tag{50.18}$$

Therefore, the heat of solution is equal to

$$Q = -\theta^2 \frac{\partial}{\partial \theta} \frac{1}{\theta} \left(N \mu_0 + n \theta \ln \frac{n}{eN} + n \mathrm{X} - n \mu'_0 - N \mu_0 \right), \tag{50.19}$$

where μ'_0 is the chemical potential of the dissolved substance. The quantities appearing here may be expressed in terms of the concentration of the saturated solution n_0/N with the aid of the saturation condition (50.15). This yields

$$Q = -n \theta^2 \frac{\partial}{\partial \theta} \ln \frac{n}{n_0}. \tag{50.20}$$

The heat of solution for one molecule is equal to $Q/n = q$, or

$$q = -\theta^2 \frac{\partial}{\partial \theta} \ln \frac{n}{n_0} = \frac{\theta^2}{n_0} \frac{\partial n_0}{\partial \theta}. \tag{50.21}$$

Thus, if the concentration of a saturated solution increases with temperature, then heat is absorbed in dissolution.

The Le Chatelier-Braun principle. Let us suppose that heat is supplied to a saturated solution in equilibrium with the solute. Then, if $\frac{\partial n_0}{\partial \theta} > 0$, part of the substance will further dissolve, and the heat is spent not only in raising the temperature but also in dissolving. But if $\frac{\partial n_0}{\partial \theta} < 0$, then some of the substance comes out of solution, on which, in accordance with (50.21), heat is also expended. In both cases, changes occur in the equilibrium system that counteract the external action (raising of the temperature). The foregoing example illustrates a general rule, known in thermodynamics as the Le Chatelier-Braun principle. The Clausius-Clapeyron equation can be examined on the basis of this principle.

The phase rule. Let us suppose that there are k substances (components) distributed in the form of solutions of arbitrary concentration over f phases. How many parameters define the equilibrium state of such a system?

The chemical potentials of the substances depend upon temperature, pressure, and relative concentrations. The concentrations of all the substances in any phase satisfy the equations

$$\sum_{l=1}^{k} c_l^f = 1, \tag{50.22}$$

since, by definition, the concentrations are equal to

$$c_l^f = \frac{n_l^f}{\sum_{l'=1}^{k} n_{l'}^f}.$$

The equilibrium condition consists in the equality of the chemical potentials of each of the k substances over all f phases:

$$\mu_1^1 (p, \theta, c_1^1 c_2^1, \ldots, c_k^1) = \mu_1^2 (p, \theta, c_1^2, c_2^2, \ldots, c_k^2) = \ldots \mu_1^f (p, \theta, c_1^f, c_2^f, \ldots, c_k^f)$$

$$\ldots\ldots\ldots\ldots\ldots\ldots\ldots\ldots\ldots\ldots\ldots\ldots\ldots\ldots\ldots\ldots\ldots\ldots\ldots\ldots$$

$$\mu_k^1 (p, \theta, c_1^1, c_2^1, \ldots, c_k^1) = \ldots = \mu_k^f (p, \theta, c_1^f, c_2^f, \ldots, c_k^f). \tag{50.23}$$

Here the superscript always denotes phase, while the subscript denotes the substance.

Equation (50.23) involves k concentrations in f phases, and two other variables (temperature and pressure), so that there are $kf+2$ variables in all.

There are $f-1$ equations (50.23) for each substance and, in addition, the concentrations satisfy f equations (50.22), so that in all there are $k\,(f-1)+f$ equations for determining $kf+2$ variables. The number of independent variables which may vary arbitrarily is equal to the difference between the number of variables and the number of equations, i.e.,

$$r = kf + 2 - k\,(f-1) - f = k - f + 2. \tag{50.24}$$

The quantity r is called the *number of thermodynamic degrees of freedom* of the system. (50.24) expresses the Gibbs *phase rule*: the number of degrees of freedom is equal to the number of components, minus the number of phases, plus two.

For example, if a single substance occurs in equilibrium in two phases, then $r=1$; in such a system, one may change arbitrarily a single variable: temperature or pressure. In a two-component, two-phase system, there are two degrees of freedom: the component concentration in one of the phases can be varied together with the temperature or pressure.

Strong electrolytes. The thermodynamic properties of solutions of strong electrolytes exhibit noticeable deviations from the laws obtained in this section for the solutions of neutral substances. It is natural to look for the cause of these deviations in the fact that ions interact electrostatically. This type of interaction was not taken into account at all in the theory of weak solutions.

Aqueous electrolyte solutions exhibit very strong dissociation, the cause of which can be qualitatively understood if we take into account that the static dielectric constant of water is equal to 81. The potential energy for the atomic interactions of a heteropolar molecule in a solution is less than in vacuum by roughly ε times, so that in water the atoms of such a molecule are more weakly bound by a factor of 81.

Thermal motion in the solution disrupts the bonds, and instead of molecules we have ions in the solution. The interaction forces between ions are Coulomb forces; they fall off with distance much more slowly than the interaction forces between neutral molecules. Let us determine the correction to the chemical potential for a weak solution, which correction is due to the interaction forces between ions.

The ionic atmosphere. For simplicity, we shall consider an electrolyte which contains only singly charged ions of both signs, for example, H^+ and Cl^- in a weak aqueous solution. Both positive and negative ions can occur close to a positive ion. The density of positive ions near an ion of the same sign is reduced, compared with the mean density, by the Boltzmann factor $e^{-\frac{e\varphi}{\theta}}$, where φ is the potential produced by the charge distribution at a given point in the vicinity of the positive ion. The negative-charge density at the same point is increased by the Boltzmann factor $e^{-\frac{e\varphi}{\theta}}$. If ρ_0 is the mean ion density, then the charge density at the point of the solution considered is

$$e\rho_0 e^{-\frac{e\varphi}{\theta}} - e\rho_0 e^{\frac{e\varphi}{\theta}} = -2\rho_0 e \operatorname{sh}\frac{e\varphi}{\theta}.$$

From equation (14.6), this quantity, multiplied by $-\frac{4\pi}{\varepsilon}$, is equal to the Laplacian of potential [we must introduce $\frac{1}{\varepsilon}$ into equation (14.6) in order to take into account the reduction of the electric field in the medium]. This equation of electrostatics, written down straightway in spherical coordinates, is of the foım

$$\frac{1}{r^2}\frac{d}{dr}r^2\frac{d\varphi}{dr} = \frac{8\pi}{\varepsilon}e\rho_0 \operatorname{sh}\frac{e\varphi}{\theta}. \qquad (50.25)$$

We shall consider the solution to this equation for points in space for which the inequality $e\varphi \ll \theta$ is satisfied. It is only this region which is of practical interest. If the indicated inequality is satisfied, then $\operatorname{sh}\frac{e\varphi}{\theta}$ is replaced by its argument, and equation (50.25) becomes linear:

$$\frac{1}{r^2}\frac{d}{dr}r^2\frac{d\varphi}{dr} = \frac{8\pi\rho_0 e^2}{\varepsilon\theta}\varphi. \qquad (50.26)$$

We shall look for a solution to (50.26) in the form of the usual Coulomb potential corrected by the screening factor $\xi(r)$:

$$\varphi = \frac{e}{\varepsilon r}\xi(r). \qquad (50.27)$$

Then, substituting into (50.26) leads to an equation for ξ:

$$\frac{d^2\xi}{dr^2} = \frac{8\pi\rho_0 e^2}{\varepsilon\theta}\xi. \qquad (50.28)$$

Putting

$$\varkappa = \sqrt{\frac{8\pi \rho_0 e^2}{\varepsilon \theta}}, \tag{50.29}$$

we obtain a solution of (50.28) in the form

$$\xi(r) = e^{-\varkappa r}. \tag{50.30}$$

Of the two solutions of (50.28), only the one with a minus sign in the exponent is retained. The constant of integration is put equal to unity, since screening should not have any effect at small distances from the ion; close to the ion, $\varphi(r)$ becomes the usual Coulomb potential.

From (50.27) and (50.30) we have

$$\varphi = \frac{e}{\varepsilon r} e^{-\varkappa r}. \tag{50.31}$$

The potential decreases e times for a distance $1/\chi$ from the ion. $1/\chi$ is termed the *radius of the ion cloud* surrounding the given ion. In actual fact, there is no "cloud," but simply an increased charge density of opposite sign and a reduced charge density of the same sign, which leads to the screening of the given ion field at a distance $1/\chi$.

Thermodynamic quantities for strong electrolytes. The potential produced by the ion cloud at a certain point is equal to the difference between the potential $\varphi(r)$ and the potential of the free charge not surrounded by ions. At the point where the original positive ion is situated, this additional potential is

$$\delta\varphi = \lim_{r\to 0} \frac{e}{\varepsilon r}(e^{-\varkappa r} - 1) = -\frac{e\varkappa}{\varepsilon}.$$

It is of finite magnitude. It was for this reason that the solution (50.31) could be extrapolated to such small distances away from the ion that the linear approximation (50.26) does not, strictly speaking, hold. The additional free energy of the ion can be calculated as the work produced in taking the charge e to a point of potential $\delta\varphi$. It is necessary, here, to take into account that the potential $\delta\varphi$ is produced by the charge e itself. Therefore, the work done in introducing the charge is not equal to $e\,\delta\varphi$, but to the integral $\int_0^e \delta\varphi\, d e$. Here, the charge is the external parameter λ, while the potential $\delta\varphi$ is the generalized force Λ (see Sec. 46):

$$\delta F = \delta A = \int_0^e \delta\varphi\, de = -\frac{e^2 \varkappa}{3\varepsilon}, \tag{50.32}$$

where the fact that $e\varkappa \sim e^2$ was taken into account in the integration. From this it is easy to obtain the addition to the thermodynamic potential for the ion, which is associated with the correction to the free energy:

$$\delta \mu_i = \delta A - V \frac{\partial \delta A}{\partial V} \doteq - V^2 \frac{\partial}{\partial V} \left(\frac{\delta A}{V} \right). \tag{50.33}$$

$\varkappa$ is inversely proportional to the square root of the volume, so that $\frac{\partial}{\partial V}\left(\frac{\delta F}{V}\right) = -\frac{3}{2}\frac{\delta F}{V^2}$. Since $\delta \mu$ refers to a single ion, the addition per ion pair (comprising a single molecule) is the required correction to the chemical potential of the solute

$$\delta \mu' = 2 \cdot \frac{3}{2} \delta A = -\sqrt{\frac{8\pi}{\varepsilon^3} \frac{n}{N} \frac{\rho e^6}{\theta}}, \tag{50.34}$$

where ρ is the solvent density $\left(\frac{\text{molecules}}{\text{cm}^3}\right)$.

To the thermodynamic potential of the whole solution we add the quantity

$$\delta \Phi = \int \delta \mu' \, dn = -\frac{2}{3}\sqrt{\frac{8\pi}{\varepsilon^3} \frac{n^3}{N} \frac{\rho e^6}{\theta}}, \tag{50.35}$$

while for the chemical potential of the solvent of a strong electrolyte we obtain the correction

$$\delta \mu = \frac{\partial \delta \Phi}{\partial N} = \frac{1}{3}\sqrt{\frac{8\pi}{\varepsilon^3} \frac{n^3}{N^3} \frac{\rho e^6}{\theta}}. \tag{50.36}$$

This theory of strong electrolytes was formulated by Debye and Hückel.

Implications of the Debye-Hückel theory. If we introduce $\delta \mu$ into equation (50.7) for the osmotic pressure, then instead of (50.9) we get

$$\Delta p \cdot V = 2n\theta\left(1 - \frac{1}{6}\sqrt{\frac{8\pi}{\varepsilon^3} \frac{n}{\theta^3} \frac{\rho e^6}{N}}\right) \tag{50.37}$$

(the factor 2 takes the complete dissociation into account). The correction factor inside the brackets tends to unity as the solution concentration n/N decreases, but its derivative with respect to concentration tends to infinity.

The additional term involved in equation (50.15), which determines the concentration of a saturated solution, is proportional to $\sqrt{n}$. It also contributes infinite additions to the derivatives as n tends to zero.

Sec. 51. Chemical Equilibria

Reversible and irreversible reactions. Like all processes whose velocities do not coincide with the rate of change of the external parameters of a statistical system, chemical reactions, which proceed with finite velocity, are irreversible. For example, when an explosive mixture (hydrogen and oxygen) burns, water vapour is produced irreversibly.

If a certain quantity of the oxyhydrogen mixture is prepared in a closed vessel, the state of the mixture will be thermodynamically unstable with respect to the reaction. True, the reaction by no means proceeds directly according to the "gross equation" $2\,H_2+O_2=$ $=2\,H^2O$. To do so, the molecules would have to overcome very high potential barriers. In actual fact, the reaction must proceed through stages involving the intermediate unstable substances OH, H, O with nonsaturated valences; these are the so-called active centres.

The initial formation of active centres is very difficult, so at room temperature an oxyhydrogen mixture may be preserved indefinitely. But if active centres are somehow produced (by a powerful electric spark, for example) then they are renewed and multiplied in the course of the reaction (a chain reaction).* When the multiplication of active centres is fast enough, the reaction proceeds explosively.

But chemical reactions never proceed to the end. If an explosion is produced in a sufficiently strong vessel (bomb), then the finite equilibrium state will contain hydrogen, oxygen, and water vapour in certain (strictly definite) concentrations that depend upon the temperature and pressure and the initial composition of the mixture. This finite equilibrium state is termed chemical equilibrium.

When the state in an equilibrium system is changed slowly, the equilibrium will shift in one direction or other, i.e., the quantity of initial or final products may increase. But these chemical reactions proceed with the same velocity as that with which the external conditions change. Hence, such reactions are reversible—as are all processes whose velocity is not established spontaneously but is all the time equal to the velocity of change of the quantities that the equilibrium state of the system is given by.

Chemical equilibrium. The state of chemical equilibrium can be found with the aid of the thermodynamic functions of the substances involved in the "gross equation" reaction, quite independently of the mechanism by which the reaction proceeds. This is why the theory of chemical equilibria had already been formulated in the nineteenth century, while the study of the velocity of chemical reactions is still developing vigorously at the present time. In this sense, the situation is similar to that in statistics in general, i.e., in the science of equilibria, and in kinetics—the science of the velocities of macroscopic processes.

---

* The majority of chain reactions are associated with an active centre. This was established by N. N. Semyonov, the discoverer of chain reactions, and his pupils (and independently by C. N. Hinshelwood).

For a given temperature and pressure, chemical equilibrium is attained only when the thermodynamic potential in the reacting mixture has a minimum

$$d\Phi = 0. \tag{51.1}$$

When p=const and θ=const, the minimum condition, $d\Phi$=0, appears thus:

$$d\Phi = \sum_i \mu_i \, dN_i. \tag{51.2}$$

Here, μ_i is the chemical potential of the ith substance appearing in the "gross equation" reaction. For an oxyhydrogen mixture, for example, the only substances of this kind are hydrogen, oxygen, and water vapour. But the numbers dN_i are not arbitrary: they change as the reaction proceeds, and are therefore interrelated by the reaction equation. In other words, N_i may vary only in equivalent (stoichiometric) quantities. For example, if we take the reaction

$$2\,CO + O_2 = 2\,CO_2,$$

then $dN_{CO}: dN_{O_2}: dN_{CO_2} = -2:-1:2$. In the reaction of the thermal dissociation of hydrogen

$$H_2 = 2\,H$$

$dN_{H_2}: dN_H = -1:2$. In general, the number dN_i is proportional to the equivalent of the given substance in the reaction ν_i; equation (51.2) can also be rewritten thus:

$$\sum_i \mu_i \nu_i = 0. \tag{51.3}$$

This equation expresses the condition of chemical equilibrium in a system.

Law of mass action. Equation (51.3) is especially useful when we have an explicit expression for the chemical potential of the reacting substances, as in a weak solution or in an ideal gas, for example. In the latter case, the equilibrium concentrations of the substances may be determined if there are sufficient data about the structure of all the molecules in equilibrium.

The chemical potential for a gas in a mixture of ideal gases is [see (47.17)]

$$\mu_i = -\theta \ln \frac{\theta f_i(\theta)}{p_i}, \tag{51.4}$$

where f_i is a statistical sum taken over all momentum values of the molecule as a whole, and also over all its rotational, vibrational, and electronic states. The latter are essential only when they occur close to the ground state of the molecule and are far from its dissociation limit. If they occur close to the dissociation limit, the molecule decomposes before such highly excited states can in any way affect the values of the statistical sums (see exercise 2).

Substituting the expression for chemical potential into the chemical equilibrium condition (51.3), and eliminating θ, we obtain

$$\sum_i \nu_i \ln p_i = \sum \nu_i \ln \theta f_i .$$

Taking antilogarithms of this equation, we obtain the equilibrium condition (expressed in terms of partial pressures) from the formula

$$\prod_i p_i^{\nu_i} = \prod_i (\theta f_i)^{\nu_i} \equiv K . \tag{51.5}$$

This equation can also be written in terms of the relative concentrations of the substances by replacing the partial pressures with the aid of (47.15):

$$\prod_i c_i^{\nu_i} = p^{-\sum_i \nu_i} \prod_i (\theta f_i)^{\nu_i} = p^{-\sum_i \nu_i} K . \tag{51.6}$$

Here, c_i denotes the concentration of the ith component of the mixture:

$$c_i = \frac{N_i}{N} . \tag{51.7}$$

The pressure on the right-hand side of (51.6) has still to be expressed in terms of the initial pressure or in terms of the initial density; this can always be done easily with the Clapeyron equation, if we take into account the change in the number of particles (relative to the initial number of particles), for a given equilibrium of the chemical reaction.

The component concentrations depend upon the initial quantities of the original substances involved in the reaction. Thus, also the equilibrium concentrations depend upon these quantities (or masses). Therefore, equation (51.6) expresses the so-called *law of mass action.*

The quantity appearing on the right-hand side of equation (51.5) is called the *equilibrium constant* of the given reaction, because it does not involve the concentrations of the mixture. Its dimensionality is $\left[p_i^{\Sigma \nu_i}\right]$.

Heat of reaction. The heat of a chemical reaction occurring at constant pressure is defined as the difference in the heat functions of the reacting substances before and after the reaction. It is convenient to write this heat as calculated for a single elementary act of the reaction

$$q = \delta i = -\theta^2 \frac{\partial}{\partial \theta} \frac{\delta \Phi}{\theta} \tag{51.8}$$

[cf. (50.18)]. But in an elementary act $\delta\Phi = \sum_i \mu_i \nu_i$, so that the heat of reaction is

$$q = -\theta^2 \frac{\partial}{\partial\theta}\left(\frac{\sum \mu_i \nu_i}{\theta}\right). \tag{51.9}$$

This expression denotes the heat *absorbed* during the reaction. The heat *evolved* would have to be defined with opposite sign.

When the law of mass action applies, the heat of reaction is expressed in terms of the equilibrium constant K:

$$q = \theta^2 \frac{\partial \ln K}{\partial \theta}. \tag{51.10}$$

Formula (51.10) agrees with the Le Chatelier-Braun principle, which can be easily seen from the following argument. If $\frac{\partial \ln K}{\partial\theta} > 0$, then, as the temperature increases, the equilibrium tends towards a predominance of those substances that enter into the reaction equation with positive coefficients ν_i. The concentrations of these substances appear in the numerator on the left-hand side of equation (51.6). But then, according to (51.10), the system absorbs heat, so that reactions occur in it which oppose the rise in temperature. The increase or decrease in the temperature of an equilibrium system can produce reversible reactions in it in any desired direction.

Exercises

1) Write down the equations of the law of mass action for the reaction $2CO + O_2 = 2CO_2$, if a moles of CO and b moles of O_2 initially take part in the reaction.

Let x moles of O_2 react; then $2x$ moles of CO enter into the reaction with them, and $2x$ moles of CO_2 are formed. In all, there are $a + b - 3x + 2x = a + b - x$ moles of different substances in the system. The concentrations are, respectively,

$$c_{CO} = \frac{a - 2x}{a + b - x},$$

$$c_{O_2} = \frac{b - x}{a + b - x},$$

$$c_{CO_2} = \frac{2x}{a + b - x},$$

so that the equilibrium equation appears thus:

$$\frac{(2x)^2 (a + b - x)}{(a - 2x)^2 (b - x)} = pK.$$

Here, p is the equilibrium pressure, which differs from the pressure of the original substances p_0 (for the same temperature) by the factor $\frac{a + b - x}{a + b}$. Whence the equation for the required quantity x is $\frac{x^2}{(a - 2x)^2 (b - x)} = \frac{p_0 K}{4(a + b)}$.

2) Calculate the equilibrium constant for the thermal dissociation of nitrogen, using the following data.

The ground state for a nitrogen atom is ^4S. The first excited state lies 2.4 ev higher (^2D), the next lies 3.5 ev above the ground state (^2P).

The formation energy of an N_2 molecule at absolute zero is equal to 9.76 ev (this value is now reliably established).

The moment of inertia for the ground state N_2 is $J = 13.84 \times 10^{-40}$ gm/cm^2. The vibrational quantum of a molecule is equal to 0.287 ev. In the ground state of the molecule, the orbital and spin angular momenta of the electrons do not have projections on the line joining the nuclei. The first excited state lies higher than 6 ev above the ground state of the molecule.

The statistical sum for the atoms is

$$f_N = \frac{(2\pi m_N \theta)^{3/2}}{(2\pi h)^3}\left(4 + 2\cdot 5\cdot e^{-\frac{2.4}{\theta}} + 2\cdot 3\cdot e^{-\frac{3.5}{\theta}}\right).$$

Here and in future, it is convenient to express θ in electron-volts, taking into account that 1 ev corresponds to 11,600°. We shall confine ourselves to temperatures for which the statistical sum for a molecule involves only the electronic ground state. Then we obtain [see (47.21)]:

$$f_{N^2} = \frac{(2\pi m_{N^2}\theta)^{3/2}}{(2\pi h)^3}\frac{2\pi J\theta}{(2\pi h)^2}\frac{4\pi}{2}\frac{e^{-\frac{9.76}{\theta}}}{1 - e^{-\frac{h\omega}{\theta}}}.$$

The equilibrium constant for the reaction $N_2 = 2N$ is, from (51.5), equal to

$$K = \frac{\theta f^2{}_N}{f_{N_2}} = \left(4 + 10\,e^{-\frac{2.4}{\theta}} + 6\,e^{-\frac{3.5}{\theta}}\right)^2 \cdot \frac{2}{Jh}\left(\frac{m_N\theta}{\pi}\right)^{3/2}\left(1 - e^{-\frac{h\omega}{\theta}}\right)e^{-\frac{9.76}{\theta}}.$$

To illustrate, let us find the fraction of dissociated molecules, if the temperature is equal to 1 ev and there are 2.7×10^{19} molecules in 1 cm^3. Then the equation of the law of mass action equation is

$$\frac{4x^2}{1-x} = 0.494 \cdot 10^6 \cdot 5.90 \cdot 10^{-5} = 24.2.$$

Here, the factor before the exponential is equal to $\sim 5 \times 10^5$, while the exponential is equal to 5.9×10^{-5}. The equilibrium degree of dissociation is $x = 0.88$. Thus, when the temperature is equal to only one tenth of the dissociation energy, 88% of all the molecules have already dissociated. The predominance of the pre-exponential factor over the exponential is explained, for such relatively low temperatures, by the fact that the statistical weight of the dissociated state is determined by the entire volume occupied by the gas, while the nondissociated state is determined only by the volume of the molecules; therefore, at atmospheric density of the gas (2.7×10^{19} mole/cm^3), dissociation is already highly probable.

3) Find the degree of thermal ionization for helium as a function of its temperature and pressure. Do not take second ionization into account.

The first ionization potential of helium is 24.47 ev, while the first excited state lies 20.5 ev above the ground state.

The ionization equilibrium satisfies the law of mass action

$$\frac{c_e\, c_{He^+}}{c_{He}} = p^{-1}K.$$

The statistical sums here are:

$$f_e = 2\,\frac{(2\pi\, m_e\, \theta)^{3/2}}{(2\pi\, h)^3},$$

$$f_{\mathrm{He}^+} = 2\,\frac{(2\pi\, m_{\mathrm{He}}\, \theta)^{3/2}}{(2\pi\, h)^3},$$

$$f_{\mathrm{He}} = \frac{(2\pi\, m_{\mathrm{He}}\, \theta)^{3/2}}{(2\pi\, h)^3}\, e^{\frac{24.47}{\theta}}$$

(the factor 2 takes into account the spin of the electron and He^+ ion).

From this we express the equilibrium constant as

$$\mathrm{K} = 4\,\frac{(2\pi\, m_e\, \theta)^{3/2}}{(2\pi\, h)^3}\,\theta \cdot e^{-\frac{24.47}{\theta}} = \frac{1}{h^3}\sqrt{\frac{2\, m^3\, \theta^5}{\pi^3}}\, e^{-\frac{24.47}{\theta}}.$$

If the initial pressure of helium is p_0, then the ionization-equilibrium equation assumes the following form:

$$\frac{x^2}{1-x} = \frac{\mathrm{K}}{p_0}.$$

To illustrate, we shall take a temperature of 4 ev and a molecular density 2.7×10^{19} mole/cm^3. Then the equation of ionization equilibrium is written numerically thus:

$$\frac{x^2}{1-x} = 3.47 \cdot 10^3 \cdot 2.19 \cdot 10^{-3} = 7.6;\quad x = 0.90.$$

Here, as in the previous example, the pre-exponential factor predominates over the exponential, equal to 2.19×10^{-3}, as a result of the large statistical weight of the ionized state. Excited states of the helium atom make a negligible contribution to the statistical sum. At higher temperatures, the first ionization is practically complete, so that there are simply no neutral atoms which could be excited.

4) Relate the e.m.f. of a primary cell to the heat of the chemical reaction occurring in it.

By definition, the e.m.f. is the work done in carrying unit charge around a conducting circuit. If a primary cell is connected in the circuit, then, in traversing it, reversible chemical reactions occur that neutralize the ions at the electrodes. The work done in a reversible reaction is equal to the change in thermodynamic potential, while the heat is equal to the change in the heat function. Whence, from equation (51.9), we obtain

$$q = -\,\theta^2 \frac{\partial}{\partial \theta}\left(\frac{\mathrm{e.m.f.}}{\theta}\right).$$

Sec. 52. Surface Phenomena

The thermodynamic potential of a surface layer. We have so far considered only the volume properties of a substance, so that all the results relating to phase and chemical equilibria, and to equilibria in solutions, refer, strictly speaking, to very large systems.

The surface layers separating different substances or different phases of the same substance exhibit special properties, which, however, depend both on the nature and the states of the volumes in contact.

The thermodynamic potential for unit surface of contact of two media depends upon the temperature θ and pressure p in the surrounding media. In equilibrium, θ and p are constant over the whole surface. Interaction between the contacting portions occurs across their boundaries. The dimensions of the boundaries are proportional to the first power of the linear dimensions of these portions, while their areas are proportional to the squares of the linear dimensions. Therefore, for sufficiently large dimensions, the portions of the surface layer may be regarded as quasi-independent subsystems, like the way in which we considered volume subsystems. Hence, the thermodynamic potential of a surface is additive for the same reason as the volume potential is. If the thermodynamic potential for unit interface surface of two media is denoted by α and the magnitude of the whole contact surface is ζ, then, by virtue of additivity, the potential for the whole surface will be equal to

$$\Phi = \alpha \zeta. \tag{52.1}$$

Surface tension. The work done at constant pressure and temperature is equal to the change in thermodynamic potential (Sec. 46). Therefore, in changing surface area by unity, work is performed equal to α. This work is called the surface tension of two given media.

It is easy to demonstrate the relation between this definition of surface tension and its elementary definition. Let a film of liquid be stretched on a rigid Π-shaped wire frame, closed by a movable member, completing a rectangular liquid surface. If the length of the movable member is equal to unity, then a force acts on it (from the direction of the film) equal to twice the surface tension of the film (because the film has two sides). When the cross-member is displaced by unit length, the force of surface tension performs work numerically equal to double its magnitude. But the total surface of the film increases in this case by two units, so that the work done in increasing the surface by unity is indeed equal to the "force" of surface tension.

When the surface area is increased, part of the atoms leave the volume for the surface layer; to do so it is necessary in part to overcome the attractive forces due to other atoms. This is what explains the origin of the work that is lost (or gained, depending upon the nature of the volumes in contact) in increasing the area. The surface tension of a condensed phase at the boundary with a vacuum is, of course, always positive.

The thermodynamic potential is at a minimum when in equilibrium. In this case, the minimum is attained simply for the least area ζ. Therefore, a liquid film, stretched on a certain (in the general case a nonplane) frame assumes the least possible surface area for the given frame. A liquid drop in ideal equilibrium assumes a spherical form that has the least surface area for a given volume.

The heat of surface increase. When surface area increases, heat is evolved in addition to the performance of work. Since the process of surface increase is reversible, the heat is determined from the general formula (46.18) $Q=\theta\Delta S$. The entropy of the surface is calculated from formula (46.46). Substituting the thermodynamic potential for the surface (52.1) into this formula, we find an expression for the heat:

$$Q=-\theta(\zeta_2-\zeta_1)\frac{\partial\alpha}{\partial\theta}. \tag{52.2}$$

Thus, heat may be evolved or absorbed depending upon the sign of $\frac{\partial\alpha}{\partial\theta}$.

The equilibrium vapour pressure above a drop. The phase equilibrium condition changes if the surface thermodynamic potential is taken into account in addition to that for volume. Of course, the general condition $d\Phi=0$ still holds, but it no longer reduces to the form (49.4) $\mu_1=\mu_2$. Instead, it is necessary to write, generally,

$$\frac{\partial\Phi_1}{\partial N}=\frac{\partial\Phi_2}{\partial N}. \tag{52.3}$$

Let the subscript 1 refer to the vapour phase contained in a large volume, and the subscript 2 relate to a small liquid drop of radius R. Then, for the first phase,

$$\frac{\partial\Phi_1}{\partial N}=\mu_1, \tag{52.4}$$

while for the second phase,

$$\frac{\partial\Phi_2}{\partial N}=\frac{\partial}{\partial N}(N\mu_2+\alpha\zeta). \tag{52.5}$$

The derivative of the second term is calculated in the following way:

$$\frac{\partial}{\partial N}\alpha\zeta=\alpha\frac{\partial\zeta}{\partial N}=8\pi\alpha R\frac{\partial R}{\partial N}. \tag{52.6}$$

If the density of the liquid is ρ mole/cm³, then $R=\left(\frac{N}{\frac{4}{3}\pi\rho}\right)^{1/3}$, so that

$$\frac{\partial R}{\partial N}=\frac{1}{3}\frac{R}{N}. \tag{52.7}$$

Substituting this in (52.6) and now expressing N in terms of R, we obtain

$$\alpha\frac{\partial\zeta}{\partial N}=\frac{8\pi\alpha R^2}{3\cdot\frac{4}{3}\pi R^3\rho}=\frac{2\alpha}{\rho R}. \tag{52.8}$$

Thus, the equilibrium condition between the vapour and liquid drop is

$$\mu_1(p,\theta)=\mu_2(p,\theta)+\frac{2\alpha}{\rho R}. \tag{52.9}$$

We represent the pressure p as $p_0 + \Delta p$, where p_0 is the equilibrium pressure on a plane surface. The expansion of chemical potential in powers of Δp gives [cf. (50.12)]

$$(v_1 - v_2)\,\Delta p = \frac{2\alpha}{\rho R}. \tag{52.10}$$

Neglecting the specific volume of the liquid compared with the specific volume of the vapour, we find a final expression for the excess pressure:

$$\Delta p = p_0 \frac{2\alpha}{\rho R \theta}. \tag{52.11}$$

The analogous expression for the pressure in a vapour bubble within a liquid leads to the same formula, but with opposite sign.

The stability of supersaturated phases. We have thus seen that the equilibrium vapour pressure over the convex surface of a drop is greater than that over a plane surface, while that over the concave surface of a bubble is less. This explains the relative stability of supersaturated phases, which was mentioned in Sec. 49. If a liquid drop appears in a supersaturated vapour with pressure $p' > p_0$, whose radius is less than

$$\frac{2\alpha p_0}{\rho\,\theta\,(p' - p_0)}, \tag{52.12}$$

then this drop evaporates once again, and further condensation on it is highly improbable, since it is a fluctuation phenomenon. Only if inequality (52.12) is reversed can the drop begin to grow. But the spontaneous formation of a large drop, like any large fluctuation, is highly improbable. Therefore, condensation usually begins on small nuclei that are already in the vapour, for example, ions.

In exactly the same way we can explain why a highly purified superheated liquid does not boil. Boiling of a liquid consists in the formation of vapour bubbles in its volume. In order that such a bubble should not collapse due to external pressure, the equilibrium pressure of the vapour must be at least that of the external atmospheric pressure above the liquid. But if the equilibrium vapour pressure above a plane surface is only equal to the external pressure, then there is not sufficient pressure in the bubble for equilibrium. Therefore, a bubble of insufficient size cannot grow.

APPENDIX

An integral of the form

$$\int_0^\infty \frac{x^n\,dx}{e^x \pm 1}$$

is calculated in the following way. The function $(e^x \pm 1)^{-1}$ is expanded in a series

$$\frac{1}{(e^x \pm 1)} = \sum_{k=1}^{\infty} (\mp)^{k+1} e^{-kx}.$$

The series is integrated term by term, and the integral for each separate term is calculated thus:

$$\int_0^\infty e^{-kx} x^n\,dx = \frac{1}{k^{n+1}} \int_0^\infty e^{-z} z^n\,dz.$$

The latter integral is equal to $n!$ for integral n, as may easily be shown from integration by parts. For n half-integral, it may be taken according to the formulae derived in exercise 3, Sec. 39. Indeed, putting $\sqrt{z} = u$, we obtain

$$\int_0^\infty e^{-z} \sqrt{z}\,dz = 2\int_0^\infty e^{-u^2} u^2\,du = \frac{\sqrt{\pi}}{2}$$

and, in general

$$\int_0^\infty e^{-z} z^{m-1/2}\,dz = 2\int_0^\infty e^{-u^2} u^{2m}\,du = \frac{1\cdot 3\cdot 5\ldots(2m-1)\sqrt{\pi}}{2^m}.$$

We shall call this quantity $\left(m - \frac{1}{2}\right)!$, so that in general

$$\int_0^\infty e^{-z} z^n\,dz = n!$$

Hence,

$$\int_0^\infty \frac{z^n\,dz}{e^z \pm 1} = n! \sum (\mp)^{k+1} \frac{1}{k^{n+1}}.$$

The summation with the upper sign may be reduced to the summation with the lower sign. Indeed,

$$1 - \frac{1}{2^{n+1}} + \frac{1}{3^{n+1}} - \frac{1}{4^{n+1}} + \ldots =$$
$$= 1 + \frac{1}{2^{n+1}} + \frac{1}{3^{n+1}} + \frac{1}{4^{n+1}} + \ldots - 2 \cdot \frac{1}{2^{n+1}} \left(1 + \frac{1}{2^{n+1}} + \frac{1}{3^{n+1}} + \ldots\right) =$$
$$= \left(1 - \frac{1}{2^n}\right)\left(1 + \frac{1}{2^{n+1}} + \frac{1}{3^{n+1}} + \frac{1}{4^{n+1}} + \ldots\right).$$

Finally, the summation involving positive signs has the following values:

$n = \frac{1}{2}$	1	$\frac{3}{2}$	2	$\frac{5}{2}$	3
$\sum\limits_{k=1}^{\infty} \frac{1}{k^{n+1}} = 2.612$	1.645	1.341	1.202	1.127	1.0823

For odd n, the following formulae obtain:

$$\sum_{k=1}^{\infty} \frac{1}{k^2} = \frac{\pi^2}{6}, \quad \sum_{k=1}^{\infty} \frac{1}{k^4} = \frac{\pi^4}{90}.$$

Therefore,

$$\int_0^{\infty} \frac{x^3\,dx}{e^x - 1} = 3! \sum_{k=1}^{\infty} \frac{1}{k^4} = \frac{\pi^4}{15}.$$

We also note that

$$\int_0^{\infty} \frac{x^{1/2}\,dx}{e^x - 1} = \frac{\sqrt{\pi}}{2} \sum_{k=1}^{\infty} \frac{1}{k^{3/2}} = \frac{\sqrt{\pi}}{2} \cdot 2.612 = 2.31.$$

We have met with the integral (44.39):

$$\int_{-\infty}^{\infty} \frac{x^2\,dx}{(e^x + 1)(1 + e^{-x})} = -2\int_0^{\infty} x^2 \frac{d}{dx}\left(\frac{1}{e^x + 1}\right) dx =$$
$$= 4\int_0^{\infty} \frac{x\,dx}{e^x + 1} = 4\left(1 - \frac{1}{2}\right)\frac{\pi^2}{6} = \frac{\pi^2}{3}.$$

BIBLIOGRAPHY

Part I

1. *L. D. Landau and E. M. Lifshits, *Mechanics*, Gostekhizdat, Moscow-Leningrad (1958).
2. *G. K. Suslov, *Theoretical Mechanics*, Gostekhizdat, Moscow-Leningrad (1944).
3. *T. Levi-Civita and I. Amaldi, *Lezioni di meccanica razionale*, Bologna (1930).
4. E. T. Whittaker, *Analytical Dynamics* (1937). *A Treatise on the Analytical Dynamics of Particles and Rigid Bodies, Cambridge* (1927).
5. A. Sommerfeld, *Mechanik* (1944).
6. G. Goldstein, *Classical Mechanics*, Cambridge (1950).

Part II

7. *N. E. Kochin, *Vector Calculus*, GITTL (1934).
8. L. D. Landau and E. M. Lifshits, *Field Theory*, Gostekhizdat, Moscow-Leningrad (1948).
9. *I. E. Tamm, *Fundamentals of the Theory of Electricity*, Gostekhizdat, Moscow-Leningrad (1954).
10. *Abraham and Becker, *Theorie der Elektrizität*, Bd. I, Leipzig (1932).
11. *P. Becker, *Elektronentheorie*.
12. A. Einstein, *The Meaning of Relativity*, Princeton (1953).
13. P. G. Bergman, *Introduction to the Theory of Relativity* (1942).
14. **Einstein and Modern Physics*, Gostekhizdat (1956).

Part III

15. L. D. Landau and E. M. Lifshits, *Quantum Mechanics*, Part 1 (1948).
16. *D. I. Blokhintsev, *Fundamentals of Quantum Mechanics* (1944), 2nd ed. (1949).
17. *V. A. Fok, *The Principles of Quantum Mechanics* (1932).
18. H. A. Bethe and E. E. Salpeter, *Quantum Mechanics of One- and Two-Electron Systems*, Springer-Verlag (1957).
19. P. A. M. Dirac, *The Principles of Quantum Mechanics*, Oxford (1957).
20. E. V. Shpolskii, *Atomic Physics*, Parts 1 and 2 (1950).

Part IV

21. L. D. Landau and E. M. Lifshits, *Statistical Physics* (1950).
22. *V. G. Levich, *Introduction to Statistical Physics* (1954).
23. *M. A. Leontovich, *Statistical Physics* (1944).
24. *M. A. Leontovich, *Introduction to Thermodynamics* (1950).
25. *E. Fermi, *Molecules and Crystals* (1947); (Moleküle und Kristalle, 1938).
26. *A. G. Samoilovich, *Statistical Physics* (1955).

* As far as the author knows, these have not been translated into English.

SUBJECT INDEX